Essential environment

THE SCIENCE BEHIND THE STORIES 6th Edition

Jay Withgott

Matthew Laposata

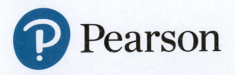
330 Hudson Street, NY NY 10013

Courseware Portfolio Management, Director: Beth Wilbur
Executive Editor: Alison Rodal
Sponsoring Editor: Cady Owens
Editorial Assistant: Alison Candlin
Courseware Director, Content Development: Ginnie Simione Jutson
Courseware Sr. Analyst, Content Development: Mary Ann Murray
Managing Producer, Science: Michael Early
Content Producer, Science: Margaret Young
Rich Media Content Producers: Nicole Constantine, Libby Reiser, Kimberly Twardochleb
Production Management: Norine Strang, Cenveo® Publisher Services

Composition: Cenveo® Publisher Services
Design Manager: Maria Guglielmo Walsh
Interior and Cover Design: Lisa Buckley
Illustrators: ImagineeringArt.com Inc.
Rights & Permissions Manager: Ben Ferrini
Rights & Permissions Project Manager: Eric Schrader
Manufacturing Buyer: Stacey Weinberger, LSC Communications
Photo Researcher: Kristin Piljay
Director of Product Marketing: Allison Rona
Product Marketing Manager: Christa Pesek Pelaez
Field Marketing Manager: Kelly Galli

Cover Photo Credit: Tim Laman/naturepl.com

Library of Congress Cataloging-in-Publication Data

Essential Environment
Library of Congress Cataloging in Publication Control Number: 2017035269

1   18

ISBN 10: 0-13-471488-1; ISBN 13: 978-0-13-471488-2 (Student Edition)
ISBN 10: 0-13-481873-3; ISBN 13: 978-0-13-481873-3 (Books a la Carte)

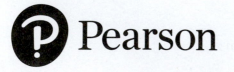

www.pearson.com

About the Authors

Jay Withgott has authored *Essential Environment* as well as its parent volume, *Environment: The Science behind the Stories*, since their inception. In dedicating himself to these books, he works to keep abreast of a diverse and rapidly changing field and continually seeks to develop new and better ways to help today's students learn environmental science.

As a researcher, Jay has published scientific papers in ecology, evolution, animal behavior, and conservation biology in journals ranging from *Evolution* to *Proceedings of the National Academy of Sciences*. As an instructor, he has taught university lab courses in ecology and other disciplines. As a science writer, he has authored articles for numerous journals and magazines including *Science, New Scientist, BioScience, Smithsonian,* and *Natural History*. By combining his scientific training with prior experience as a newspaper reporter and editor, he strives to make science accessible and engaging for general audiences. Jay holds degrees from Yale University, the University of Arkansas, and the University of Arizona.

Jay lives with his wife, biologist Susan Masta, in Portland, Oregon.

Matthew Laposata is a professor of environmental science at Kennesaw State University (KSU). He holds a bachelor's degree in biology education from Indiana University of Pennsylvania, a master's degree in biology from Bowling Green State University, and a doctorate in ecology from The Pennsylvania State University.

Matt is the coordinator of KSU's two-semester general education science sequence titled Science, Society, and the Environment, which enrolls over 5000 students per year. He focuses exclusively on introductory environmental science courses and has enjoyed teaching and interacting with thousands of nonscience majors during his career. He is an active scholar in environmental science education and has received grants from state, federal, and private sources to develop and evaluate innovative curricular materials. His scholarly work has received numerous awards, including the Georgia Board of Regents' highest award for the Scholarship of Teaching and Learning.

Matt resides in suburban Atlanta with his wife, Lisa, and children, Lauren, Cameron, and Saffron.

about our SUSTAINABILITY INITIATIVES

Pearson recognizes the environmental challenges facing this planet, and acknowledges our responsibility in making a difference. This book is carefully crafted to minimize environmental impact. The paper is Forest Stewardship Council® (FSC®) certified. The binding, cover, and paper come from facilities that minimize waste, energy consumption, and the use of harmful chemicals. We collect and map data on the country of origin of the paper we purchase in order to ensure we do not contribute to deforestation and illegal logging. Pearson closes the loop by recycling every out-of-date textbook returned to our warehouse. Along with developing and exploring digital solutions to our market's needs, Pearson has a strong commitment to achieving carbon neutrality, and achieved a cumulative reduction of 40% in our global climate emissions between 2009 and 2016. The future holds great promise for reducing our impact on Earth's environment, and Pearson is proud to be leading the way. We strive to publish the best books with the most up-to-date and accurate content, and to do so in ways that minimize our impact on Earth.

To learn more about our initiatives, please visit https://www.pearson.com/sustainability.html.

FSC
www.fsc.org
MIX
Paper from
responsible sources
FSC® C132124

Contents

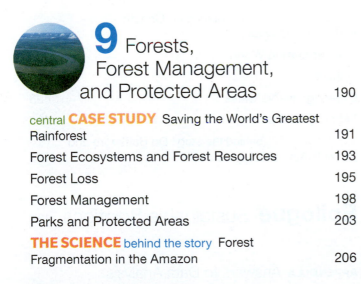

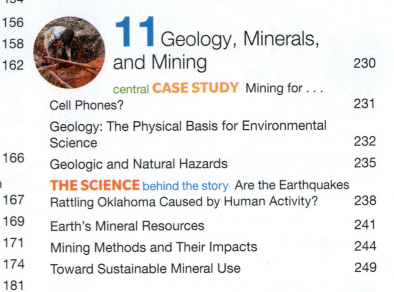
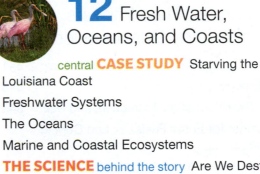
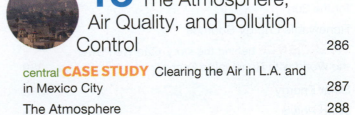

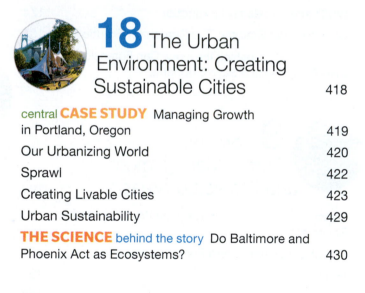

Preface

Dear Student,

You are coming of age at a unique and momentous time in history. Within your lifetime, our global society must chart a promising course for a sustainable future. The stakes could not be higher.

Today we live long lives enriched with astonishing technologies, in societies more free, just, and equal than ever before. We enjoy wealth on a scale our ancestors could hardly have dreamed of. However, we have purchased these wonderful things at a steep price. By exploiting Earth's resources and ecological services, we are depleting our planet's ecological bank account. We are altering our planet's land, air, water, nutrient cycles, biodiversity, and climate at dizzying speeds. More than ever before, the future of our society rests with how we treat the world around us.

Your future is being shaped by the phenomena you will learn about in your environmental science course. Environmental science gives us a big-picture understanding of the world and our place within it. Environmental science also offers hope and solutions, revealing ways to address the problems we confront. Environmental science is more than just a subject you study in college. It provides you basic literacy in the foremost issues of the 21st century, and it relates to everything around you throughout your lifetime.

We have written this book because today's students will shape tomorrow's world. At this unique moment in history, the decisions and actions of your generation are key to achieving a sustainable future for our civilization. The many environmental challenges we face can seem overwhelming, but you should feel encouraged and motivated. Remember that each dilemma is also an opportunity. For every problem that human carelessness has created, human ingenuity can devise a solution. Now is the time for innovation, creativity, and the fresh perspectives that a new generation can offer. Your own ideas and energy can, and *will*, make a difference.

—*Jay Withgott and Matthew Laposata*

Dear Instructor,

You perform one of our society's most vital functions by educating today's students—the citizens and leaders of tomorrow—on the processes that shape the world around them, the nature of scientific inquiry, and the pressing environmental challenges we face. We have written this book to assist you in this endeavor because we feel that the crucial role of environmental science in today's world makes it imperative to engage, educate, and inspire a broad audience of students.

In *Essential Environment: The Science behind the Stories*, we strive to show students how science informs our efforts to bring about a sustainable society. We aim to encourage critical thinking and to maintain a balanced approach as we flesh out the vibrant social debate that accompanies environmental issues. As we assess the challenges facing our civilization and our planet, we focus on providing realistic, forward-looking solutions, for we truly feel there are many reasons for optimism.

As environmental science has grown, so has the length of textbooks that cover it. With this volume, we aim to meet the needs of instructors who favor a more succinct and affordable book. We have distilled the most essential content from our full-length book, *Environment: The Science behind the Stories*, now in its sixth edition. We have streamlined our material, updated our coverage, and carefully crafted our writing to make *Essential Environment* every bit as readable, informative, and engaging as its parent volume.

New to This Edition

This sixth edition includes an array of revisions that enhance our content and presentation while strengthening our commitment to teach science in an engaging and accessible manner.

- **SUCCESS** STORY   This brand-new feature highlights one discrete story per chapter of successful efforts to address environmental problems, ranging from local examples (such as prairie restoration in Chicago) to national and global successes (such as halting ozone depletion by treaty, or removing lead from gasoline). Our book has always focused on positive solutions, but the new emphasis the *Success Story* feature brings should help encourage students by showing them that sustainable solutions are within reach. Students can explore the data behind these solutions with new *Success Story Coaching Activities* in *Mastering Environmental Science*.

- central **CASE STUDY**   Three *Central Case Studies* are completely new to this edition, complementing the seven new *Central Case Studies* added in the fifth edition. All other *Central Case Studies* have been updated as needed to reflect recent developments. These updates provide fresh stories and new ways to frame emerging issues in environmental science. In our new *Central Case Studies,* students will learn of the changes that Asian carp and other invasive species are having on North American waterways, wrestle with the challenges of conserving the

Amazon rainforest, and examine how Miami-area residents are coping with sea level rise.

- **Chapter 4:** Leaping Fish, Backwards River: Asian Carp Threaten the Great Lakes
- **Chapter 9:** Saving the World's Greatest Rainforest
- **Chapter 14:** Rising Seas Threaten South Florida

- closing **THE LOOP** Also new to this edition, each chapter now concludes with a brief section that "closes the loop" by revisiting the *Central Case Study* while reviewing key principles from the chapter. This new *Closing the Loop* section enhances our long-standing and well-received approach of integrating each *Central Case Study* throughout its chapter. A further step in this direction is the new **CASE STUDY CONNECTION** question feature. These questions, in the *Seeking Solutions* section at the end of each chapter, place students in a scenario and empower them to craft solutions to issues raised in the *Central Case Study*.

- **THE SCIENCE** behind the story Nine of our 18 *Science behind the Story* features are new to this edition, giving you a current and exciting selection of scientific studies to highlight. Students will follow along as researchers discover how Hawaiian birds evolved, trace ecological recovery at Mt. St. Helens, sleuth out the mystery of honeybee declines, use DNA fingerprinting to combat poaching, reveal synthetic chemicals in fast food, determine whether fracking is causing earthquakes, predict the future of drought in the American West, ask whether renewable energy alone can power civilization, and seek to enhance recycling efforts on campus.

  - **Chapter 3:** How Do Species Form in Hawaii's "Natural Laboratory" of Evolution?
  - **Chapter 4:** How Do Communities Recover after Disturbance?
  - **Chapter 7:** What Role Do Pesticides Play in the Collapse of Bee Colonies?
  - **Chapter 8:** Can Forensic DNA Analysis Help Save Elephants?
  - **Chapter 10:** Are Endocrine Disruptors Lurking in Your Fast Food?
  - **Chapter 11:** Are the Earthquakes Rattling Oklahoma Caused by Human Activity?
  - **Chapter 12:** Are We Destined for a Future of "Mega-droughts" in the United States?
  - **Chapter 16:** Can We Power the World with Renewable Energy?
  - **Chapter 17:** Can Campus Research Help Reduce Waste?

- **New and revised DATA Q, FAQ, and Weighing the Issues features** Incorporating feedback from instructors across North America, we have examined each example of these three features that boost student engagement, and have revised them and added new examples as appropriate.

- **Currency and coverage of topical issues** To live up to our book's hard-won reputation for currency, we have incorporated the most recent data possible and have enhanced coverage of emerging issues. As climate change and energy concerns play ever-larger roles in today's world, our coverage has evolved to keep pace. This edition highlights the tremendous growth and potential of renewable energy, yet also makes clear how we continue reaching further for fossil fuels, using ever more powerful technologies. The text tackles the complex issue of climate change in depth, while connections to this issue proliferate among topics in every chapter. And in a world newly shaken by dynamic political forces amid concerns relating to globalization, trade, immigration, health care, jobs, national security, and wealth inequality, our introduction of ethics, economics, and policy early in the book serves as a framework to help students relate the scientific findings they learn about to the complex cultural aspects of the society around them.

- **Enhanced style and design** We have significantly refreshed and improved the look and clarity of our presentation throughout the text. A more appealing layout, striking visuals, additional depth in the *Central Case Studies*, and an inviting new style all make the book more engaging for students. More than 40% of the photographs, graphs, and illustrations in this edition are new or have been revised to reflect current data or to enhance clarity or pedagogy.

Existing Features

We have also retained the major features that made the first five editions of our book unique and that are proving so successful in classrooms across North America:

- **A focus on science and data analysis** We have maintained and strengthened our commitment to a rigorous presentation of modern scientific research while simultaneously making science clear, accessible, and engaging to students. Explaining and illustrating the *process* of science remains a foundational goal of this endeavor. We also continue to provide an abundance of clearly cited data-rich graphs, with accompanying tools for data analysis. In our text, our figures, and our online features, we aim to challenge students and to assist them with the vital skills of data analysis and interpretation.

- **An emphasis on solutions** For many students, today's deluge of environmental dilemmas can lead them to feel that there is little hope or that they cannot personally make a difference. We have consistently aimed to counter this impression by highlighting innovative solutions being developed on campuses and around the world—a long-standing approach now enhanced by our new *Success Story* feature. While taking care not to paint too rosy a picture of the challenges that lie ahead, we demonstrate that there is ample reason for optimism, and we encourage action and engagement.

- **central CASE STUDY integrated throughout the chapter** We integrate each chapter's *Central Case Study* into the main text, weaving information and elaboration throughout the chapter. In this way, compelling stories about real people and real places help to teach foundational concepts by giving students a tangible framework with which to incorporate novel ideas. Students can explore the locations featured in each Central Case Study with new Case Study Video Tours in *Mastering Environmental Science.*

- **THE SCIENCE behind the story** Because we strive to engage students in the scientific process of testing and discovery, we feature *The Science behind the Story* in each chapter. By guiding students through key research efforts, this feature shows not merely *what* scientists discovered, but *how* they discovered it.

- **DATA** These data analysis questions help students to actively engage with graphs and other data-driven figures, and challenge them to practice quantitative skills of interpretation and analysis. To encourage students to test their understanding as they progress through the material, answers are provided in Appendix A. Students can practice data analysis skills further with *Interpreting Graphs and Data: DataQs* in *Mastering Environmental Science.*

- **FAQ** The *FAQ* feature highlights questions frequently posed by students, thereby helping to address widely held misconceptions and to fill in common conceptual gaps in knowledge. By also including questions students sometimes hesitate to ask, the *FAQs* show students that they are not alone in having these questions, thereby fostering a spirit of open inquiry in the classroom.

- **weighing the ISSUES** These questions aim to help develop the critical-thinking skills students need to navigate multifaceted issues at the juncture of science, policy, and ethics. They serve as stopping points for students to reflect on what they have read, wrestle with complex dilemmas, and engage in spirited classroom discussion.

- **Diverse end-of-chapter features** *Testing Your Comprehension* provides concise study questions on main topics, while *Seeking Solutions* encourages broader creative thinking aimed at finding solutions. "Think It Through" questions place students in a scenario and empower them to make decisions to resolve problems. *Calculating Ecological Footprints* enables students to quantify the impacts of their choices and measure how individual impacts scale up to the societal level.

Mastering Environmental Science

With this edition we continue to offer expanded opportunities through *Mastering Environmental Science*, our powerful yet easy-to-use online learning and assessment platform. We have developed new content and activities specifically to support features in the textbook, thus strengthening the connection between online and print resources. This approach encourages students to practice their science literacy skills in an interactive environment with a diverse set of automatically graded exercises. Students benefit from self-paced activities that feature immediate wrong-answer feedback, while instructors can gauge student performance with informative diagnostics. By enabling assessment of student learning outside the classroom, *Mastering Environmental Science* helps the instructor to maximize the impact of classroom time. As a result, both educators and learners benefit from an integrated text and online solution.

New to this edition *Mastering Environmental Science* for this edition of *Essential Environment: The Science behind the Stories* offers new resources that are designed to grab student interest and help them develop quantitative reasoning skills.

- NEW *GraphIt!* activities help students put data analysis and science reasoning skills into practice in a highly interactive and engaging format. Each of the 10 *GraphIt!* activities prompts students to manipulate a variety of graphs and charts to develop an understanding of how data can be used in decision making about environmental issues. Topics range from agriculture to fresh water to air pollution. These mobile-friendly activities are accompanied by assessment in *Mastering Environmental Science.*

- NEW *Case Study Video Tours* use Google Earth to take students on a virtual tour of the locations featured in each *Central Case Study.*

- NEW *Success Story Coaching Activities* pair with the new in-text *Success Story* features and give students the opportunity to explore the data behind each solution.

- NEW *Everyday Environmental Science* videos highlight current environmental issues in short (5 minutes or less) video clips and are produced in partnership with BBC News. These videos will pique student interest, and can be used in class or assigned as a high-interest out-of-class activity.

Existing features *Mastering Environmental Science* also retains its popular existing features:

- *Process of Science* activities help students navigate the scientific method, guiding them through explorations of experimental design using *Science behind the Story* features from the current and former editions. These activities encourage students to think like a scientist and to practice basic skills in experimental design.

- *Interpreting Graphs and Data: Data Q* activities pair with the in-text *Data Q* questions, coaching students to further develop skills related to presenting, interpreting, and thinking critically about environmental science data.

- *First Impressions Pre-Quizzes* help instructors determine their students' existing knowledge of core content areas in environmental science at the outset of the academic term, providing class-specific data that can then be employed for powerful teachable moments throughout the term. Assessment items in the Test Bank connect to each quiz item, so instructors can formally assess student understanding.

- *Video Field Trips* enable students to visit real-life sites that bring environmental issues to life. Students can tour a power plant, a wind farm, a wastewater treatment facility, a site combating invasive species, and more—all without leaving campus.

Essential Environment has grown from our experiences in teaching, research, and writing. We have been guided in our efforts by input from hundreds of instructors across North America who have served as reviewers and advisors. The participation of so many learned, thoughtful, and committed experts and educators has improved this volume in countless ways.

We sincerely hope that our efforts are worthy of the immense importance of our subject matter. We invite you to let us know how well we have achieved our goals and where you feel we have fallen short. Please write to us in care of our editor, Cady Owens (**cady.owens@pearson.com**), at Pearson Education. We value your feedback and are eager to know how we can serve you better.

—*Jay Withgott and Matthew Laposata*

Instructor Supplements

A robust set of instructor resources and multimedia accompanies the text and can be accessed through *Mastering Environmental Science*. Organized chapter-by-chapter, everything you need to prepare for your course is offered in one convenient set of files. Resources include Video Field Trips, Everyday Environmental Science Videos, PowerPoint Lecture presentations, Instructor's Guide, Active Lecture questions to facilitate class discussions (for use with or without clickers), and an image library that includes all art and tables from the text.

The Test Bank files, offered in both MS Word and TestGen formats, include hundreds of multiple-choice questions plus unique graphing and scenario-based questions to test students' critical-thinking abilities.

The *Mastering Environmental Science* platform is the most effective and widely used online tutorial, homework, and assessment system for the sciences.

NEW to this edition, Ready-to-Go Teaching Modules on key environmental issues provide instructors with assignments to use before and after class, as well as in-class activities that use clickers or Learning Catalytics for assessment.

Acknowledgments

This textbook results from the collective labor and dedication of innumerable people. The two of us are fortunate to be supported by a tremendous publishing team.

Sponsoring editor Cady Owens coordinated our team's efforts for this sixth edition of *Essential Environment*. She has been a pleasure to work with, and we are grateful for her guidance, deft touch, and sound judgment. We were also thrilled to welcome back Courseware Sr. analyst Mary Ann Murray, whose past work for our books has stood the test of time. Mary Ann again brought an intense work ethic and a mix of creativity, big-picture smarts, and focus on detail that we truly appreciate. Content producer Margaret Young once more ably steered us through the complex logistical tangles of the textbook process. Executive editor Alison Rodal oversaw the project and lent her steady hand, and we thank director Beth Wilbur for her support of this book through its six editions and for helping to invest the resources that our books continue to enjoy.

Editorial assistant Ali Candlin managed the review process and provided timely assistance, while Courseware director Ginnie Simione-Jutson oversaw our development needs. Bonnie Boehme provided meticulous copy editing, and photo researcher Kristin Piljay helped to acquire quality photos. Eric Schrader managed permissions for our figures. Alicia Elliot of Imagineering Art did a wonderful job executing our art program, and Lisa Buckley designed our engaging new text and cover style. We offer a big thank-you to Norine Strang for her extensive work with our compositor to help guide our book through production.

As always, a select number of top instructors from around North America teamed with us to produce the supplementary materials, and we are grateful for their work. Our thanks go to Danielle DuCharme for updating our Instructor's Guide, Todd Tracy for his help with the Test Bank, James Dauray for revising the PowerPoint lectures, Jenny Biederman for updating the clicker questions, Donna Bivans for revising the reading questions, Julie Stoughton for correlating the shared media, and Karyn Alme for accuracy reviewing the Dynamic Study Modules, reading questions, and practice tests.

As we expand our online offerings with *Mastering Environmental Science*, we thank Sarah Jensen, Nicole Constantine, Libby Reiser, Kimberly Twardochleb, and Todd Brown for their work on *Mastering Environmental Science* and our media supplements.

We give thanks to marketing managers Christa Pesek Pelaez and Mary Salzman. And we admire and appreciate the work and commitment of the many sales representatives who help communicate our vision, deliver our product, and work with instructors to ensure their satisfaction.

Finally, we each owe debts to the people nearest and dearest to us. Jay thanks his parents and his many teachers and mentors over the years for making his own life and education so enriching. He gives loving thanks to his wife, Susan, who has patiently provided caring support throughout this book's writing and revision over the years. Matt thanks his family, friends, and colleagues, and is grateful for his children, who give him three reasons to care passionately about the future. Most important, he thanks his wife, Lisa, for being a wonderful constant within a whirlwind of change and for lending him her keen insight and unwavering support. The talents, input, and advice of Susan and of Lisa have been vital to this project, and without their support our own contributions would not have been possible.

We dedicate this book to today's students, who will shape tomorrow's world.

—*Jay Withgott and Matthew Laposata*

Reviewers

We wish to express special thanks to the dedicated reviewers who shared their time and expertise to help make this sixth edition the best it could be. Their efforts built on those of the nearly 700 instructors and outside experts who have reviewed material for the previous five editions of this book and the six editions of this book's parent volume, where they are acknowledged in full. Here we thank those who contributed in particular to this sixth edition of *Essential Environment*—in most cases with multiple in-depth chapter reviews despite busy teaching schedules. Our sincere gratitude goes out to all of them. If the thoughtfulness and thoroughness of these reviewers are any indication, we feel confident that the teaching of environmental science is in excellent hands!

Donna Bivans, *Pitt Community College*
Martha Bollinger, *Winthrop University*
Lynn Corliss, *South Puget Sound Community College*
James Daniels, *Huntingdon College*
Eden Effert-Fanta, *Eastern Illinois University*
Jeff Fennell, *Everett Community College*
Paul Gier, *Huntingdon College*
Kelley Hodges, *Gulf Coast State College*
Ned Knight, *Linfield College*
Kurt Leuschner, *College of the Desert*
Heidi Marcum, *Baylor University*
Eric Myers, *South Suburban College*
Craig Phelps, *Rutgers University*
Julie Stoughton, *University of Nevada at Reno*
Jamey Thompson, *Hudson Valley Community College*
Pat Trawinski, *Erie Community College*
Daniel Wagner, *Eastern Florida State College*

Engage students in science through current environmental issues

Essential Environment: The Science Behind the Stories, **6th Edition,** by Jay Withgott and Matt Laposata, is the #1 book in the introductory environmental science market, known for its student-friendly narrative style, its integration of real stories and case studies, and its presentation of the latest science and research.

CHAPTER 14

Global Climate Change

Upon completing this chapter, you will be able to:

- Describe Earth's climate system and explain the factors that influence global climate
- Identify greenhouse gases, and characterize human influences on the atmosphere and on climate
- Summarize how researchers study climate
- Outline current and expected future trends and impacts of climate change in the United States and across the world
- Suggest and assess ways we may respond to climate change

Bulldozing beach sand off a Fort Lauderdale boulevard after a storm surge

central **CASE STUDY**

Rising Seas Threaten South Florida

"Miami, as we know it today, is doomed. It's not a question of if. It's a question of when."
—Dr. Harold Wanless, University of Miami geologist

"Miami Beach is not going to sit back and go underwater."
—Philip Levine, mayor of Miami Beach

It happens now in Miami at least six times a year. Salty water bubbles up from drains, seeps up from the ground, fills the streets, and spills across lawns and sidewalks. Under the dazzling sun of a South Florida sky, floodwaters stall car traffic, creep into doorways, force businesses to close, and keep people from crossing the street. Employees struggle to get to work while tourists stand around, baffled.

The flooding is most severe in Miami Beach, the celebrated strip of glamorous hotels, clubs, shops, and restaurants that rises from a 7-mile barrier island just offshore from Miami. The carefree affluent image of Miami Beach, with its sun and fun, is increasingly jeopardized by the grimy reality of these unwelcome saltwater intrusions. By 2030, flooding is predicted to strike Miami and Miami Beach about 45 times per year—becoming no longer a curious inconvenience, but an existential threat.

These mysterious floods that seem to come out of nowhere are a recent phenomenon, so Miami-area residents are just now realizing that their coastal metropolis is slowly being swallowed by the ocean. The cause? Rising sea levels driven by global climate change.

The world's oceans rose 20 cm (8 in.) in the 20th century as warming temperatures expanded the volume of seawater and caused glaciers and ice sheets to melt, discharging water into the oceans. These processes are accelerating today, and scientists predict that sea level will rise another 26–98 cm (10–39 in.) or more in this century as climate change intensifies.

As sea levels rise, coastal cities across the globe—from Venice to Amsterdam to New York to San Francisco—are facing challenges. In the United States, scientists find that the Atlantic Seaboard and the Gulf Coast are especially vulnerable. The hurricane-prone shores of Florida, Louisiana, Texas, and the Carolinas are at risk, as are coastal cities such as Houston and New Orleans. From Cape Cod to Corpus Christi, millions of Americans who live in shoreline communities are beginning to suffer significant expense, disruption to daily life, and property damage as beaches erode, neighborhoods flood, aquifers are fouled, and storms strike with more force.

Perhaps nowhere in America is more vulnerable to sea level rise than Miami and its surrounding communities in South Florida. Six million people live in this region, and three-quarters of them inhabit low-lying coastal areas that also hold most of the region's wealth and property. Experts calculate that Miami alone has more than $400 billion in assets at risk from sea level rise—more than any other city in the

Flooding in Miami after Hurricane Irma in 2017 ▲

311

Integrated Central Case Studies begin and are woven throughout each chapter, highlighting the real people, real places, and real data behind environmental issues. Revised throughout and updated with current stories, the Central Case Studies draw students in, providing a contextual framework to make science memorable and engaging.

New Topics Include:

- **Chapter 4:** Leaping Fish, Backwards River: Asian Carp Threaten the Great Lakes
- **Chapter 9:** Saving the World's Greatest Rainforest
- **Chapter 14:** Rising Seas Threaten South Florida

Help students see the big picture by making connections

closing THE LOOP

Many factors influence Earth's climate, and human activities have come to play a major role. Climate change is well underway, and additional greenhouse gas emissions will intensify global warming and cause progressively severe and diverse impacts. Sea level rise and other consequences of global climate change are affecting locations worldwide—from Miami to the Maldives, Alaska to Bangladesh, and New York to the Netherlands. As scientists and political leaders come to better understand anthropogenic climate disruption and its consequences, more and more of them are urging immediate action.

Policymakers at the international and national levels have struggled to take meaningful steps to slow greenhouse gas emissions, so increasingly, people at local and regional levels are the ones making a difference. In South Florida, citizens and local leaders are investing time, thought, money, and creativity into finding solutions to rising sea levels. They are seeking to mitigate climate change by reducing greenhouse gas emissions and to adapt to climate change by building pumping systems, raising streets and foundations, and tailoring financial and insurance incentives to guide development toward upland areas. Like people anywhere who love their homes, residents of South Florida are girding themselves for a long battle to protect their land, communities, and quality of life while our global society inches its way toward emissions reductions. For all of us across the globe, taking steps to mitigate and adapt to climate change represents the foremost challenge for our future.

CASE STUDY CONNECTION You are the city manager for a coastal U.S. city that scientists predict will be hit hard by sea level rise, with risks and impacts trailing those in Miami by just a few years. You have recently returned from a professional conference in Florida, where you toured Miami Beach and learned of the efforts being made there to adapt to climate change. What steps would you take to help your own city prepare for rising sea level? How would you explain the risks and impacts of climate disruption to your fellow city leaders to gain their support? Of the measures being taken in Florida communities, which would you choose to study closely, which would you want to begin right away, and which would be highest priority in the long run? Explain your choices.

Encourage students with a focus on sustainable solutions

Using Wetlands to Aid Wastewater Treatment

Long before people built the first wastewater treatment plants, natural wetlands filtered and purified water. Recognizing this, engineers have begun manipulating wetlands—and have even constructed new wetlands—to employ as tools to cleanse wastewater. In this approach, wastewater that has gone through primary or secondary treatment at a conventional facility is pumped into the wetland, where microbes living amid the algae and aquatic plants decompose the remaining pollutants. Water cleansed in the wetland can then be released into waterways or allowed to percolate underground.

The Arcata Marsh and Wildlife Sanctuary

One of the first constructed wetlands was established in Arcata, a town on northern California's scenic Redwood Coast. This 35-hectare (86-acre) engineered wetland system was built in the 1980s after residents objected to a $50 million plan to build a large treatment plant that would have pumped treated wastewater into the ocean. The wetland, which cost just $7 million to build, not only treats wastewater but also serves as a haven for wildlife and human recreation. In fact, the Arcata Marsh and Wildlife Sanctuary has brought the town's waterfront back to life, with more than 200,000 people visiting each year. Additionally, wildlife has flourished in the area, with 100 species of plants, 300 species of birds, 6 species of fish, and a diversity of mammals and invertebrates populating the wetlands. The practice of treating wastewater with artificial wetlands is growing fast; today more than 500 artificially constructed or restored wetlands in the United States are performing this valuable service.

EXPLORE THE DATA at **Mastering** Environmental Science

NEW! Success Story feature, included in every chapter, highlights successful efforts to address local, national, and global environmental issues. These Success Stories encourage students by showing them that sustainable solutions are within reach.

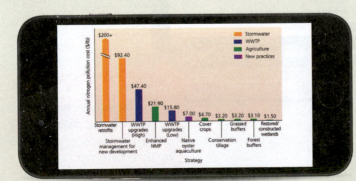

NEW! Success Story Coaching Activities pair with the new in-text Success Story feature, giving students the opportunity to explore the data behind each sustainable solution.

Assign activities that use real data to help students develop problem-solving skills

NEW! 10 GraphIt! Coaching Activities help students develop the skills they need to read, interpret, and plot graphs. Each activity uses real data focusing on a current environmental issue—such as the carbon footprint of food, fresh water availability, and ocean acidification—in an entirely new mobile experience.

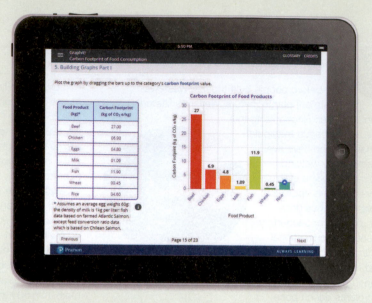

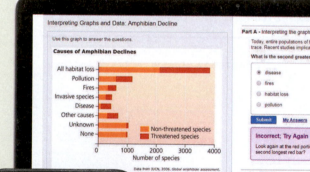

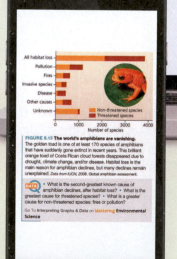

Interpreting Graphs and Data activities help students develop basic data analysis skills and practice applying those skills by interpreting real data related to environmental issues. Each activity is paired with one of the Data Q questions that appear with a number of figures throughout the text.

Use Ready-to-Go Teaching Modules to get the most out of your course

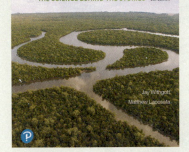

Withgott/Laposata
*Essential Environment:
The Science Behind the Stories, Sixth Edition*
**Ready-To-Go
Teaching Modules**

ESSENTIAL
environment
THE SCIENCE BEHIND THE STORIES   6th Edition

Jay Withgott
Matthew Laposata

Ready-To-Go Teaching Modules provide instructors with easy-to-use teaching tools for the toughest topics in Environmental Science.

Assign ready-made activities and assignments for before, during, and after class.

Incorporate active learning with class-tested resources from environmental science instructors.

Take full advantage of Mastering™ Environmental Science and Learning Catalytics™, the powerful "bring your own device" student assessment system.

Ready-to-Go Teaching Modules on key topics provide instructors with assignments for before and after class, as well as ideas for in-class activities. These modules incorporate the best that the text, Mastering Environmental Science, and Learning Catalytics have to offer, and are accessible through the Instructor Resources area of Mastering Environmental Science.

Instructors can easily incorporate **active learning** into their courses using suggested activity ideas and questions.

Learning Catalytics™ helps instructors to customize lectures, generate class discussions, and promote peer-to-peer learning with real-time analytics using students' smartphones, tablets, or laptops. Learning Catalytics allows instructors to:

- Engage students in more interactive tasks, helping them to develop critical thinking skills.
- Monitor student responses to determine where they are struggling.
- Use real-time data to adjust a teaching strategy.
- **Misconception Questions** can be used during class to spark discussion and reveal common misconceptions about environmental issues. Test Bank questions connect to Learning Catalytics, allowing instructors to reinforce concepts and test student understanding.

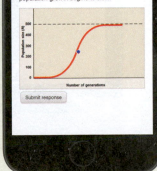

Give students access to their textbook, anytime, anywhere

NEW! Pearson eText integrates an electronic version of the text with rich media assets, such as Case Study Video Tours, in a customizable format that students can use on smartphones, tablets, and computers.

The Pearson eText mobile app offers offline access and can be downloaded for most iOS and Android phones/tablets from the Apple App Store or Google Play, providing:

- Seamlessly integrated videos and other rich media assets
- ADA accessibility (screen-reader ready)
- Configurable reading settings, including resizable type and night reading mode
- Instructor and student note-taking, highlighting, bookmarking, and search capabilities

ESSENTIAL
environment

THE SCIENCE BEHIND THE STORIES
6th Edition

CHAPTER 1

Science and Sustainability

An Introduction to Environmental Science

Our Island, Earth

Viewed from space, our home planet resembles a small blue marble suspended in a vast inky-black void. Earth may seem enormous to us as we go about our lives on its surface, but the astronaut's view reveals that our planet is finite and limited. With this perspective, it becomes clear that as our population, technological power, and resource consumption all increase, so does our capacity to alter our surroundings and damage the very systems that keep us alive. Learning how to live peacefully, healthfully, and sustainably on our diverse and complex planet is our society's prime challenge today. The field of environmental science is crucial in this endeavor.

Our environment surrounds us

A photograph of Earth from space offers a revealing perspective, but it cannot convey the complexity of our environment. Our **environment** consists of all the living and nonliving things around us. It includes the continents, oceans, clouds, and ice caps you can see in a photo from space, as well as the animals, plants, forests, and farms of the landscapes in which we live. In a more inclusive sense, it also encompasses the structures, urban centers, and living spaces that people have created. In its broadest sense, our environment includes the complex webs of social relationships and institutions that shape our daily lives.

People commonly use the term *environment* in the narrowest sense—to mean a nonhuman or "natural" world apart from human society. This is unfortunate, because it masks the vital fact that people exist within the environment and are part of nature. As one of many species on Earth, we share dependence on a healthy, functioning planet. The limitations of language make it all too easy to speak of "people and nature," or "humans and the environment," as though they were separate and did not interact. However, the fundamental insight of environmental science is that we are part of the "natural" world and that our interactions with the rest of it matter a great deal.

Environmental science explores our interactions with the world

Understanding our relationship with the world around us is vital because we depend on our environment for air, water, food, shelter, and everything else essential for living. Throughout human history, we have modified our environment. By doing so, we have enriched our lives; improved our health; lengthened our life spans; and secured greater material wealth, mobility, and leisure time. Yet many of the changes we have made to our surroundings have degraded the natural systems that sustain us. Air and water pollution, soil erosion, species extinction, and other impacts compromise our well-being and jeopardize our ability to survive and thrive in the long term.

Environmental science is the scientific study of how the natural world works, how our environment affects us, and how we affect our environment. Understanding these interactions helps us devise solutions to society's many pressing challenges. It can be daunting to reflect on the sheer magnitude of dilemmas that confront us, but these problems also bring countless opportunities for creative solutions.

Environmental scientists study the issues most centrally important to our world and its future. Right now, global conditions are changing more quickly than ever. Right now, we are gaining scientific knowledge more rapidly than ever. And right now there is still time to tackle society's biggest challenges. With such bountiful opportunities, this moment in history is an exciting time to be alive—and to be studying environmental science.

| (a) Inexhaustible renewable natural resources | (b) Exhaustible renewable natural resources | (c) Nonrenewable natural resources |

FIGURE 1.1 Natural resources may be renewable or nonrenewable. Perpetually renewable, or inexhaustible, resources such as sunlight and wind energy **(a)** will always be there for us. Renewable resources such as timber, soils, and fresh water **(b)** are replenished on intermediate timescales, if we are careful not to deplete them. Nonrenewable resources such as minerals and fossil fuels **(c)** exist in limited amounts that could one day be gone.

We rely on natural resources

Islands are finite and bounded, and their inhabitants must cope with limitations in the materials they need. On our island—planet Earth—there are limits to many of our **natural resources,** the substances and energy sources we take from our environment and that we rely on to survive (**FIGURE 1.1**).

Natural resources that are replenished over short periods are known as **renewable natural resources.** Some renewable natural resources, such as sunlight, wind, and wave energy, are perpetually renewed and essentially inexhaustible. Others, such as timber, water, animal populations, and fertile soil, renew themselves over months, years, or decades. These types of renewable resources may be used at sustainable rates, but they may become depleted if we consume them faster than they are replenished. **Nonrenewable natural resources,** such as minerals and fossil fuels, are in finite supply and are formed far more slowly than we use them. Once we deplete a nonrenewable resource, it is no longer available.

We rely on ecosystem services

If we think of natural resources as "goods" produced by nature, then we soon realize that Earth's natural systems also provide "services" on which we depend. Our planet's ecological processes purify air and water, cycle nutrients, regulate climate, pollinate plants, and recycle our waste. Such essential services are commonly called **ecosystem services** (**FIGURE 1.2**). Ecosystem services arise from the normal functioning of natural systems and are not meant for our benefit, yet we could not survive without them. The ways that ecosystem services support our lives and civilization are countless and profound (pp. 39, 101–102, 172).

Just as we may deplete natural resources, we may degrade ecosystem services when, for example, we destroy habitat or generate pollution. For years, our depletion of nature's goods and our disruption of nature's services have intensified, driven by rising resource consumption and a human population that grows larger every day.

Population growth amplifies our impact

For nearly all of human history, fewer than a million people populated Earth at any one time. Today our population has grown beyond 7.5 *billion* people. For every one person who used to exist more than 10,000 years ago, several thousand people exist today! **FIGURE 1.3** shows just how recently and suddenly this monumental change has taken place.

FIGURE 1.2 We rely on the ecosystem services that natural systems provide. For example, forested hillsides help people living below by purifying water and air, cycling nutrients, regulating water flow, preventing flooding, and reducing erosion, as well as by providing game, wildlife, timber, recreation, and aesthetic beauty.

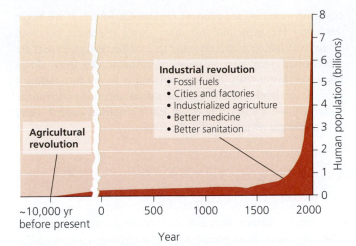

FIGURE 1.3 The global human population increased after the agricultural revolution and then skyrocketed following the industrial revolution. Note that the tear in the graph represents the passage of time and a change in *x*-axis values. *Data from U.S. Census Bureau, U.N. Population Division, and other sources.*

DATA Q For every person alive in the year 1800, about how many are alive today?

NOTE: Each **DATA Q** in this book asks you to examine the figure carefully so that you understand what it is showing. Once you take the time to understand what it shows, the rest is a breeze!

Because this is the first **DATA Q** of our book, let's walk through it together. You would first note that in the graph, time is shown on the *x*-axis and population size on the *y*-axis. You would find the year 1800 (three-fifths of the way between 1500 and 2000 on the *x*-axis) and trace straight upward to determine the approximate value of the data in that year. You'd then do the same for today's date at the far right end of the graph. To calculate roughly how many people are alive today for every one person alive in 1800, you would simply divide today's number by the number for 1800.

For each **DATA Q**, you can check your answers in **APPENDIX A** in the back of the book.

Go to **Interpreting Graphs & Data** on **Mastering** Environmental Science

Two phenomena triggered our remarkable increase in population size. The first was our transition from a hunter-gatherer lifestyle to an agricultural way of life. This change began about 10,000 years ago and is known as the **agricultural revolution.** As people began to grow crops, domesticate animals, and live sedentary lives on farms and in villages, they produced more food to meet their nutritional needs and began having more children.

The second phenomenon, known as the **industrial revolution,** began in the mid-1700s. It entailed a shift from rural life, animal-powered agriculture, and handcrafted goods toward an urban society provisioned by the mass production of factory-made goods and powered by **fossil fuels** (nonrenewable energy sources such as coal, oil, and natural gas; pp. 343, 346). Industrialization brought dramatic advances in technology, sanitation, and medicine. It also enhanced food production through the use of fossil-fuel-powered equipment and synthetic pesticides and fertilizers (pp. 142–143).

The factors driving population growth have brought us better lives in many ways. Yet as our world fills with people, population growth has begun to threaten our well-being. We must ask how well the planet can accommodate the nearly 10 billion people forecast by 2050. Already our sheer numbers are putting unprecedented stress on natural systems and the availability of resources.

Resource consumption exerts social and environmental pressures

Besides stimulating population growth, industrialization increased the amount of resources each of us consumes. By mining energy sources and manufacturing more goods, we have enhanced our material affluence—but have also consumed more and more of the planet's limited resources.

One way to quantify resource consumption is to use the concept of the ecological footprint, developed in the 1990s by environmental scientists Mathis Wackernagel and William Rees. An **ecological footprint** expresses the cumulative area of biologically productive land and water required to provide the resources a person or population consumes and to dispose of or recycle the waste the person or population produces (**FIGURE 1.4**). It measures the total area of Earth's biologically productive surface that a given person or population "uses" once all direct and indirect impacts are summed up.

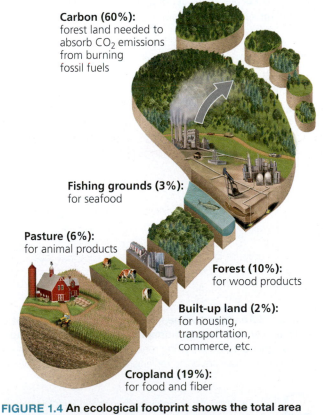

Carbon (60%): forest land needed to absorb CO_2 emissions from burning fossil fuels

Fishing grounds (3%): for seafood

Pasture (6%): for animal products

Forest (10%): for wood products

Built-up land (2%): for housing, transportation, commerce, etc.

Cropland (19%): for food and fiber

FIGURE 1.4 An ecological footprint shows the total area of biologically productive land and water used by a given person or population. Shown here is a breakdown of major components of the average person's footprint. *Data from Global Footprint Network, 2017.*

For humanity as a whole, Wackernagel and his colleagues at the Global Footprint Network calculate that we are now using 68% more of the planet's resources than are available on a sustainable basis. That is, we are depleting renewable resources by using them 68% faster than they are being replenished. To look at this another way, it would take 1.68 years for the planet to regenerate the renewable resources that people use in just 1 year. The practice of consuming more resources than are being replenished is termed **overshoot** because we are overshooting, or surpassing, Earth's capacity to sustainably support us (**FIGURE 1.5**).

Scientists debate how best to calculate footprints and measure overshoot. Indeed, any attempt to boil down complicated issues to a single number is perilous, even if the general concept is sound and useful. Yet some things are clear; for instance, people from wealthy nations such as the United States have much larger ecological footprints than do people from poorer nations. Using the Global Footprint Network's calculations, if all the world's people consumed resources at the rate of Americans, we would need the equivalent of almost five planet Earths!

Conserving natural capital is like maintaining a bank account

We can think of our planet's vast store of resources and ecosystem services—Earth's **natural capital**—as a bank account. To keep a bank account full, we need to leave the principal intact and spend only the interest, so that we can continue living off the account far into the future. If we begin depleting the principal, we draw down the bank account. To live off nature's interest—the renewable resources that are replenished year after year—is sustainable. To draw down resources faster than they are replaced is to eat into nature's capital, the bank account for our planet and our civilization. Currently we are drawing down Earth's natural capital—and we cannot get away with this for long.

Environmental science can help us learn from the past

Historical evidence suggests that civilizations can crumble when pressures from population and consumption overwhelm resource availability. Historians have inferred that environmental degradation contributed to the fall of the Greek and Roman empires; the Angkor civilization of Southeast Asia; and the Maya, Anasazi, and other civilizations of the Americas. In Syria, Iraq, and elsewhere in the Middle East, areas that today are barren desert had earlier been lush enough to support the origin of agriculture and thriving ancient societies. Easter Island has long been held up as a society that self-destructed after depleting its resources, although new research paints a more complex picture (see **THE SCIENCE BEHIND THE STORY**, pp. 8–9).

In today's globalized society, the stakes are higher than ever because our environmental impacts are global. If we cannot forge sustainable solutions to our problems, then the resulting societal collapse will be global. Fortunately, environmental science holds keys to building a better world. By studying environmental science, you will learn to evaluate the whirlwind of changes taking place around us and to think critically and creatively about ways to respond.

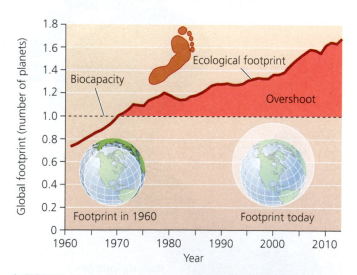

FIGURE 1.5 **Analyses by one research group indicate that we have overshot Earth's biocapacity—its capacity to support us—by 68%.** We are using renewable natural resources 68% faster than they are being replenished. *Data from Global Footprint Network, 2017.*

DATA Q How much larger is the global ecological footprint today than it was half a century ago?

Go to **Interpreting Graphs & Data** on **Mastering** Environmental Science

The Nature of Environmental Science

Environmental scientists examine how Earth's natural systems function, how these systems affect people, and how we influence these systems. Many environmental scientists are motivated by a desire to develop solutions to environmental problems. These solutions (such as new technologies, policies, or resource management strategies) are *applications* of environmental science. The study of such applications and their consequences is, in turn, also part of environmental science.

Environmental science is interdisciplinary

Studying our interactions with our environment is a complex endeavor that requires expertise from many academic disciplines, including ecology, earth science, chemistry,

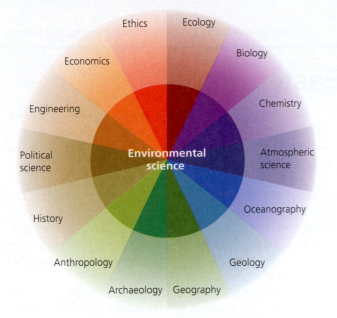

FIGURE 1.6 Environmental science is an interdisciplinary pursuit. It draws from many different established fields of study across the natural sciences and social sciences.

biology, geography, economics, political science, demography, ethics, and others. Environmental science is **interdisciplinary,** bringing techniques, perspectives, and research results from multiple disciplines together into a broad synthesis (**FIGURE 1.6**).

Traditional established disciplines are valuable because their scholars delve deeply into topics, developing expertise in particular areas and uncovering new knowledge. In contrast, interdisciplinary fields are valuable because their practitioners consolidate and synthesize the specialized knowledge from many disciplines and make sense of it in a broad context to better serve the multifaceted interests of society.

Environmental science is especially broad because it encompasses not only the **natural sciences** (disciplines that examine the natural world) but also the **social sciences** (disciplines that address human interactions and institutions). Most environmental science programs focus more on the natural sciences, whereas programs that emphasize the social sciences often use the term **environmental studies.** Whichever approach one takes, these fields bring together many diverse perspectives and sources of knowledge.

Environmental science is not the same as environmentalism

Although many environmental scientists are interested in solving problems, it would be incorrect to confuse environmental science with environmentalism or environmental activism. They are very different. Environmental science involves the scientific study of the environment and our interactions with it. In contrast, **environmentalism** is a social

movement dedicated to protecting the natural world—and, by extension, people—from undesirable changes brought about by human actions.

The Nature of Science

Science is a systematic process for learning about the world and testing our understanding of it. The term *science* is also used to refer to the accumulated body of knowledge that arises from this dynamic process of observing, questioning, testing, and discovery.

Knowledge gained from science can be applied to address society's needs—for instance, to develop technology or to inform policy and management decisions. From the food we eat to the clothing we wear to the health care we rely on, virtually everything in our lives has been improved by the application of science. Many scientists are motivated by the potential for developing useful applications. Others are motivated simply by a desire to understand how the world works.

Scientists test ideas by critically examining evidence

Science is all about asking and answering questions. Scientists examine how the world works by making observations, taking measurements, and testing whether their ideas are supported by evidence. The effective scientist thinks critically and does not simply accept conventional wisdom from others. The scientist becomes excited by novel ideas but is skeptical and judges ideas by the strength of evidence that supports them.

A great deal of scientific work is **descriptive science,** research in which scientists gather basic information about organisms, materials, systems, or processes that are not yet well known. In this approach, researchers explore new frontiers of knowledge by observing and measuring phenomena to gain a better understanding of them.

Once enough basic information is known about a subject, scientists can begin posing questions that seek deeper explanations about how and why things are the way they are. At this point scientists may pursue **hypothesis-driven science,** research that proceeds in a more targeted and structured manner, using experiments to test hypotheses within a framework traditionally known as the scientific method.

FAQ

Aren't environmental scientists also environmentalists?

Not necessarily. Although environmental scientists search for solutions to environmental problems, they strive to keep their research rigorously objective and free from advocacy. Of course, like all human beings, scientists are motivated by personal values and interests—and like any human endeavor, science can never be entirely free of social influence. However, whereas personal values and social concerns may help shape the questions scientists ask, scientists do their utmost to carry out their work impartially and to interpret their results with wide-open minds. Remaining open to whatever conclusions the data demand is a hallmark of the effective scientist.

What Are the Lessons of Easter Island?

Terry Hunt and Carl Lipo on Easter Island

A mere speck of land in the vast Pacific Ocean, Easter Island is one of the most remote spots on the globe. Yet this far-flung island—called Rapa Nui by its inhabitants—is the focus of an intense debate among scientists seeking to solve its mysteries. The debate shows how, in science, new information can challenge existing ideas—and also how interdisciplinary research helps us to tackle complex questions.

Ever since European explorers stumbled upon Rapa Nui on Easter Sunday, 1722, outsiders have been struck by the island's barren landscape. Early European accounts suggested that the 2000–3000 people living on the island seemed impoverished, subsisting on a few meager crops and possessing only stone tools. Yet the forlorn island also featured hundreds of gigantic statues of carved rock. How could people without wheels or ropes, on an island without trees, have moved 90-ton statues as far as 10 km (6.2 mi) from the quarry where they were chiseled to the coastal sites where they were erected? Apparently, some calamity must have befallen a once-mighty civilization on the island.

Researchers who set out to solve Rapa Nui's mysteries soon discovered that the island had once been lushly forested. Scientist John Flenley and his colleagues drilled cores deep into lake sediments and examined ancient pollen grains preserved there, seeking to reconstruct, layer by layer, the history of vegetation in the region. Finding a great deal of palm pollen, they inferred that when Polynesian people colonized the island (A.D. 300–900, they estimated), it was covered with palm trees.

By studying pollen and the remains of wood from charcoal, archaeologist Catherine Orliac found that at least 21 other plant species—now gone—had also been common. Clearly the island had once supported a diverse forest. Forest plants would have provided fuelwood, building material for houses and canoes, fruit to eat, fiber for clothing—and, researchers guessed, logs and fibrous rope to help move statues.

But pollen analysis showed that trees began declining after human arrival and were replaced by ferns and grasses. Then between 1400 and 1600, pollen levels plummeted. Charcoal in the soil proved the forest had been burned, likely for slash-and-burn farming. Researchers concluded that the islanders, desperate for forest resources and cropland, had deforested their own island.

With the forest gone, soil eroded away (data from lake bottoms showed a great deal of accumulated sediment). Erosion would have lowered yields of bananas, sugarcane, and sweet potatoes, perhaps leading to starvation and population decline.

Further evidence indicated that wild animals disappeared. Archaeologist David Steadman analyzed 6500 bones and found that at least 31 bird species provided food for the islanders. Today, only one native bird species is left. Remains from charcoal fires show that early islanders feasted on fish, sharks, porpoises, turtles, octopus, and shellfish—but in later years they consumed little seafood.

As resources declined, researchers concluded, people fell into clan warfare, revealed by unearthed weapons and skulls with head wounds. Rapa Nui appeared to be a tragic case of ecological suicide: A once-flourishing civilization depleted its resources and destroyed itself. In this interpretation—popularized by scientist Jared Diamond in his best-selling 2005 book *Collapse*—Rapa Nui seemed to offer a clear lesson: We on our global island, planet Earth, had better learn to use our limited resources sustainably.

When Terry Hunt and Carl Lipo began research on Rapa Nui in 2001, they expected simply to help fill gaps in a well-understood history. But science is a process of discovery, and sometimes evidence leads researchers far from where they anticipated. For Hunt, an anthropologist at the University of Hawai'i at Manoa, and Lipo, an archaeologist at California State University, Long Beach, their work led them to conclude that the traditional "ecocide" interpretation didn't tell the whole story.

First, their radiocarbon dating (dating of items using radio-isotopes of carbon; p. 31) indicated that people had not colonized the island until about A.D. 1200, suggesting that deforestation occurred rapidly after their arrival. How could so few people have destroyed so much forest so fast? Hunt and Lipo's answer: rats. When Polynesians settled new islands, they brought crop plants and chickens and other domestic animals. They also brought rats—intentionally as a food source or unintentionally as stowaways. In either case, rats can multiply quickly, and they soon overran Rapa Nui.

Researchers found rat tooth marks on old nut casings, and Hunt and Lipo suggested that rats ate so many palm nuts and shoots that the trees could not regenerate. With no young trees growing, the palm went extinct once mature trees died.

Diamond and others counter that plenty of palm nuts on Easter Island escaped rat damage, that most plants on other islands survived rats introduced by Polynesians, and that more than 20 additional plant species went extinct on Rapa Nui. Moreover, people brought the rats, so even if rats destroyed the forest, human colonization was still to blame.

Despite the forest loss, Hunt and Lipo argue that islanders were able to persist and thrive. Archaeology shows how islanders adapted to Rapa Nui's poor soil and windy weather by developing rock gardens to protect crop plants and nourish the soil. Hunt and Lipo contended that tools viewed by previous researchers as weapons were actually farm implements; lethal injuries were rare; and no evidence of battle or defensive fortresses was uncovered.

Hunt, Lipo, and others also unearthed old roads and inferred that the statues could have been moved by tilting and rocking them upright, much as we might move a refrigerator. Islanders had adapted to their resource-poor environment by becoming a peaceful and cooperative society, they maintained, with the statues providing a harmless outlet for competition among family clans over status and prestige.

Altogether, the evidence led Hunt and Lipo to propose that far from destroying their environment, the islanders had acted as responsible stewards. The collapse of this sustainable civilization, they argue, came with the arrival of Europeans, who unwittingly brought contagious diseases to which the islanders had never been exposed. Indeed, historical journals of sequential European voyages depict a society falling progressively into disarray as if reeling from epidemics.

Peruvian ships then began raiding Rapa Nui and taking islanders away into slavery. Foreigners acquired the land, forced the remaining people into labor, and introduced thousands of sheep, which destroyed the few native plants left on the island.

Thus, the new hypothesis holds that the collapse of Rapa Nui's civilization resulted from a barrage of disease, violence, and slave raids following foreign contact. Before that, Hunt and Lipo say, Rapa Nui's people boasted 500 years of a peaceful and resilient society.

Hunt and Lipo's interpretation, put forth in a 2011 book, *The Statues That Walked*, would represent a paradigm shift (p. 14) in how we view Easter Island. Debate between the two camps remains heated, however, and research continues as scientists look for new ways to test the differing hypotheses. In 2015, a six-person research team set out to estimate when human land use began to decline for each of three sites on the island. They did this by measuring how long ago pieces of obsidian rock at each site were unearthed from the soil and exposed to the air (obsidian absorbs water molecules very slowly, chemically changing over many years). The researchers found that land use had declined prior to European contact at a dry site and at a site with naturally poor soil, but that land use had continued at a moist site with fertile soil for farming. They proposed that the true picture was complex: Perhaps islanders had indeed degraded their environment in areas where conditions were sensitive, but had sustained themselves in areas where conditions were more forgiving.

Like the people of Rapa Nui, we are all stranded together on an island with limited resources. What, then, is the lesson of Easter Island for our global island, Earth? Perhaps there are two: Any island population must learn to live within its means—but with care and ingenuity, there is hope that we can.

Were the haunting statues of Rapa Nui erected by a civilization that collapsed after devastating its environment or by a sustainable civilization that fell because of outside influence?

The scientific method is a traditional approach to research

The **scientific method** is a technique for testing ideas with observations. There is nothing mysterious about the scientific method; it is merely a formalized version of the way any of us might use logic to resolve a question. Because science is an active, creative process, innovative researchers may depart from the traditional scientific method when particular situations demand it. Moreover, scientists in different fields approach their work differently because they deal with dissimilar types of information. Nonetheless, scientists of all persuasions broadly agree on fundamental elements of the process of scientific inquiry. As practiced by individual researchers or research teams, the scientific method (**FIGURE 1.7**) typically follows the steps outlined below.

Make observations Advances in science generally begin with the observation of some phenomenon that the scientist wishes to explain. Observations set the scientific method in motion and play a role throughout the process.

Ask questions Curiosity is in our human nature. Just observe young children exploring a new environment—they want to touch, taste, watch, and listen to everything, and as soon as they can speak, they begin asking questions. Scientists, in this respect, are kids at heart. Why is the ocean salty? Why are storms becoming more severe? What is causing algae to cover local ponds? When pesticides poison fish or frogs, are people also affected? How can we help restore

populations of plants and animals? All of these are questions environmental scientists ask.

Develop a hypothesis Scientists address their questions by devising explanations that they can test. A **hypothesis** is a statement that attempts to explain a phenomenon or answer a scientific question. For example, a scientist investigating why algae are growing excessively in local ponds might observe that chemical fertilizers are being applied on farm fields nearby. The scientist might then propose a hypothesis as follows: "Agricultural fertilizers running into ponds cause the amount of algae in the ponds to increase."

Make predictions The scientist next uses the hypothesis to generate **predictions,** specific statements that can be directly and unequivocally tested. In our algae example, a researcher might predict: "If agricultural fertilizers are added to a pond, the quantity of algae in the pond will increase."

Test the predictions Scientists test predictions by gathering evidence that could potentially refute the predictions and thus disprove the hypothesis. The strongest form of evidence comes from experiments. An **experiment** is an activity designed to test the validity of a prediction or a hypothesis. It involves manipulating **variables,** or conditions that can change.

For example, a scientist could test the prediction linking algal growth to fertilizer by selecting two identical ponds and adding fertilizer to one of them. In this example, fertilizer input is an **independent variable,** a variable the scientist manipulates, whereas the quantity of algae that results is the **dependent variable,** a variable that depends on the fertilizer input. If the two ponds are identical except for a single independent variable (fertilizer input), then any differences that arise between the ponds can be attributed to changes in the independent variable. Such an experiment is known as a **controlled experiment** because the scientist controls for the effects of all variables except the one he or she is testing. In our example, the pond left unfertilized serves as a **control,** an unmanipulated point of comparison for the manipulated **treatment** pond.

Whenever possible, it is best to replicate one's experiment; that is, to stage multiple tests of the same comparison. Our scientist could perform a replicated experiment on, say, 10 pairs of ponds, adding fertilizer to one of each pair.

Analyze and interpret results Scientists record **data,** or information, from their studies (**FIGURE 1.8**). Researchers particularly value quantitative data (information expressed using numbers), because numbers provide precision and are easy to compare. The scientist conducting the fertilization experiment, for instance, might quantify the area of water surface covered by algae in each pond or might measure the dry weight of algae in a certain volume of water taken from each. It is vital, however, to collect data that are representative. Because it is impractical to measure a pond's total algal

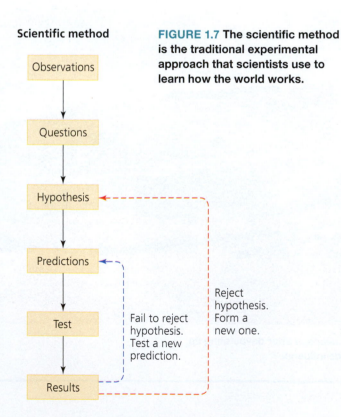

Scientific method

FIGURE 1.7 The scientific method is the traditional experimental approach that scientists use to learn how the world works.

growth, our researcher might instead sample from multiple areas of each pond. These areas must be selected in a random manner. Choosing areas with the most growth or the least growth, or areas most convenient to sample, would not provide a representative sample.

Even with the precision that numerical data provide, experimental results may not be clear-cut. Data from treatments and controls may vary only slightly, or replicates may yield different results. Researchers must therefore analyze their data using statistical tests. With these mathematical methods, scientists can determine objectively and precisely the strength and reliability of patterns they find.

If experiments disprove a hypothesis, the scientist will reject it and may formulate a new hypothesis to replace it. If experiments fail to disprove a hypothesis, this lends support to the hypothesis but does not *prove* it is correct. The scientist may choose to generate new predictions to test the hypothesis in different ways and further assess its likelihood of being true. In this way, the scientific method loops back on itself, giving rise to repeated rounds of hypothesis revision and experimentation (see Figure 1.7).

If repeated tests fail to reject a hypothesis, evidence in favor of it accumulates, and the researcher may eventually conclude that the hypothesis is well supported. Ideally, the scientist would want to test all possible explanations. For instance, our researcher might formulate an additional hypothesis, proposing that algae increase in fertilized ponds because chemical fertilizers diminish the numbers of fish or invertebrate animals that eat algae. It is possible, of course, that both hypotheses could be correct and that each may explain some portion of the initial observation that local ponds were experiencing algal blooms.

We test hypotheses in different ways

An experiment in which the researcher actively chooses and manipulates the independent variable is known as a *manipulative experiment.* A manipulative experiment provides strong evidence because it can reveal causal relationships, showing that changes in an independent variable cause changes in a dependent variable. In practice, however, we cannot run manipulative experiments for all questions, especially for processes involving large spatial scales or long timescales. For example, to study global climate change (Chapter 14), we cannot run a manipulative experiment adding carbon dioxide to 10 treatment planets and 10 control planets and then compare the results! Thus, it is common for researchers to run *natural experiments,* which compare how dependent variables are expressed in naturally occurring, but different, contexts. In such experiments, the independent variable varies naturally, and researchers test their hypotheses by searching for **correlation,** or statistical association among variables.

For instance, let's suppose our scientist studying algae surveys 50 ponds, 25 of which happen to be fed by fertilizer runoff from nearby farm fields and 25 of which are not. Let's

FIGURE 1.8 Researchers gather data to test predictions in experiments. Here, a scientist samples algae from a pond.

say he or she finds seven times more algal growth in the fertilized ponds. The scientist may conclude that algal growth is correlated with fertilizer input, that is, that one tends to increase along with the other.

This type of evidence is not as strong as the causal demonstration that manipulative experiments can provide, but often a natural experiment is the only feasible approach. Because many questions in environmental science are complex and exist at large scales, they must be addressed with correlative data. As such, environmental scientists cannot always provide clear-cut answers to questions from policymakers and the public. Nonetheless, good correlative studies can make for very strong science, and they preserve the real-world complexity that manipulative experiments often sacrifice. Whenever possible, scientists try to integrate natural experiments and manipulative experiments to gain the advantages of each.

Scientists use graphs to represent data visually

To summarize and present the data they obtain, scientists often use graphs. Graphs help to make patterns and trends in the data visually apparent and easy to understand. **FIGURE 1.9** shows a few examples of how different types of graphs can be used to present data. Each of these types is illustrated clearly and explained further in **APPENDIX B** at the back of this book. The ability to interpret graphs is a skill you will find useful throughout your life. We encourage you to consult Appendix B closely as you begin your environmental science course.

You will also note that many of the graphs in this book are accompanied by **DATA Q** questions. These questions are designed to help you interpret scientific data and build your graph-reading skills. You can check your answers to these questions by referring to **APPENDIX A** at the back of this book.

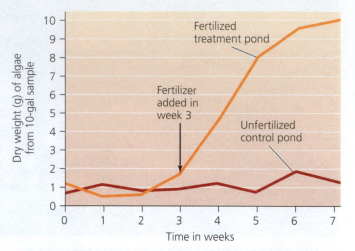

(a) Line graph of algal density through time in a fertilized treatment pond and an unfertilized control pond

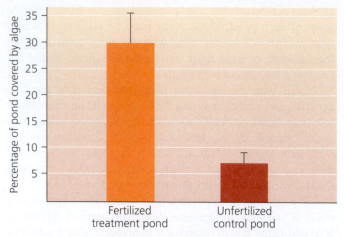

(b) Bar chart of mean algal density in several fertilized treatment ponds and unfertilized control ponds

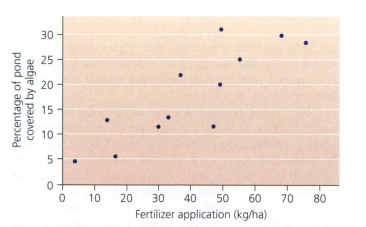

(c) Scatter plot of algal density correlated with fertilizer use on surrounding farmland

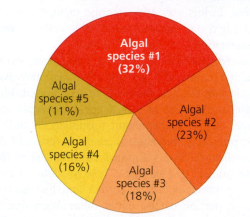

(d) Pie chart of species of algae in a sample of pond water

FIGURE 1.9 Scientists use graphs to present and visualize their data. For example, in **(a)**, a line graph shows how the amount of algae increased when fertilizer was added to a treatment pond in an experiment yet stayed the same in an unfertilized control pond. In **(b)**, a bar chart shows how fertilized ponds, on average, have several times more algae than unfertilized ponds. In **(c)**, a scatter plot shows how ponds with more fertilizer tend to contain more algae. In **(d)**, a pie chart shows the relative abundance of five species of algae in a sample of pond water. See **APPENDIX B** to learn more about how to interpret these types of graphs.

DATA Q • In part **(a)**, is time in weeks shown on the *x*-axis or the *y*-axis? • In part **(b)**, what is the dependent variable? • In part **(c)**, do the data show a positive correlation or a negative correlation? • In part **(d)**, which species is most numerous? Which is least numerous? • What are the thin black lines atop the colored bars in part **(b)** called? Explain what these lines indicate.

Go to **Interpreting Graphs & Data** on **Mastering** Environmental Science

The scientific process continues beyond the scientific method

Scientific research takes place within the context of a community of peers. To have impact, a researcher's work must be published and made accessible to this community (**FIGURE 1.10**).

Peer review When a researcher's work is complete and the results are analyzed, he or she writes up the findings and submits them to a journal (a scholarly publication in which scientists share their work). The journal's editor asks several other scientists who specialize in the subject area to examine the manuscript, provide comments and criticism (generally anonymously), and judge whether the work merits publication in the journal. This procedure, known as **peer review,** is an essential part of the scientific process.

Peer review is a valuable guard against faulty research contaminating the literature (the body of published studies) on which all scientists rely. However, because scientists are human, personal biases and politics can sometimes creep into the review process. Fortunately, just as individual scientists strive to remain objective in conducting their research, the scientific community does its best to ensure fair review of all work.

Grants and funding To fund their research, most scientists need to spend a great deal of time requesting money from private foundations or from government agencies such as the National Science Foundation. Grant applications undergo peer review just as scientific papers do, and competition for funding is generally intense.

Scientists' reliance on funding sources can occasionally lead to conflicts of interest. A researcher who obtains data showing his or her funding source in an unfavorable light may be reluctant to publish the results for fear of losing funding. This situation can arise, for instance, when an industry funds research to test its products for health or safety. Most scientists resist these pressures, but whenever you are assessing a scientific study, it is always a good idea to note where the researchers obtained their funding.

Conference presentations Scientists frequently present their work at professional conferences, where they interact with colleagues and receive comments on their research. Such interactions can help improve a researcher's work and foster collaboration among researchers.

Repeatability The careful scientist may test a hypothesis repeatedly in various ways. Following publication, other scientists may attempt to reproduce the results in their own experiments. Scientists are inherently cautious about accepting a novel hypothesis, so the more a result can be reproduced by different research teams, the more confidence scientists will have that it provides a correct explanation.

Theories If a hypothesis survives repeated testing by numerous research teams and continues to predict experimental outcomes and observations accurately, it may be incorporated into a theory. A **theory** is a widely accepted, well-tested explanation of one or more cause-and-effect relationships that have been extensively validated by a great amount of research. Whereas a hypothesis is a simple explanatory statement that may be disproved by a single experiment, a theory consolidates many related hypotheses that have been supported by a large body of data.

Note that scientific use of the word *theory* differs from popular usage of the word. In everyday language when we say something is "just a theory," we are suggesting it is a speculative idea without much substance. However, scientists mean just the opposite when they use the term. In a scientific context, a theory is a conceptual framework that explains a phenomenon and has undergone extensive and rigorous testing, such that confidence in it is extremely strong.

For example, Darwin's theory of evolution by natural selection (pp. 50–51) has been supported and elaborated by many thousands of studies over 160 years of intensive research. Observations and experiments have shown repeatedly and in great detail how plants and animals change over generations, or evolve, expressing characteristics that best promote survival and reproduction. Because of its strong support and explanatory power, evolutionary theory is the central unifying principle of modern biology. Other prominent scientific theories include atomic theory, cell theory, big bang theory, plate tectonics, and general relativity.

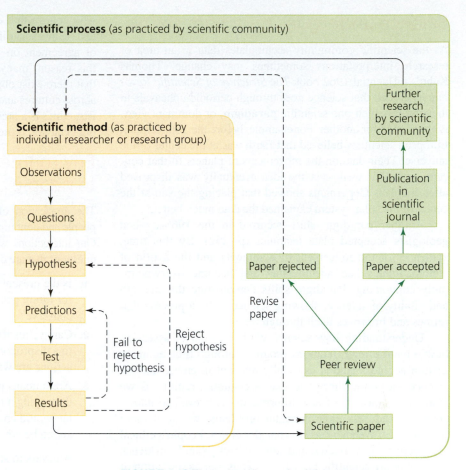

FIGURE 1.10 The scientific method that research teams follow is part of a larger framework—the overall process of science carried out by the scientific community. This process includes peer review and publication of research, acquisition of funding, and the elaboration of theory through the cumulative work of many researchers.

weighing the **ISSUES**

Follow the Money

Let us say you are a research scientist wanting to study the impacts of chemicals released into lakes by pulp-and-paper mills. Obtaining research funding has been difficult. Then a large pulp-and-paper company contacts you and offers to fund your research examining how its chemical effluents affect water bodies. What are the benefits and drawbacks of this offer? Would you accept the offer? Why or why not?

Science undergoes paradigm shifts

As the scientific community accumulates data in an area of research, interpretations sometimes may change. Thomas Kuhn's influential 1962 book *The Structure of Scientific Revolutions* argued that science goes through periodic upheavals in thought, in which one scientific **paradigm,** or dominant view, is abandoned for another. For example, before the 16th century, European scientists believed that Earth was at the center of the universe. Their data on the movements of planets fit that concept somewhat well—yet the idea eventually was disproved after Nicolaus Copernicus showed that placing the sun at the center of the solar system explained the data much better.

Another paradigm shift occurred in the 1960s, when geologists accepted plate tectonics (p. 232). By this time, evidence for the movement of continents and the action of tectonic plates had accumulated and become overwhelmingly convincing. Paradigm shifts demonstrate the strength and vitality of science, showing science to be a process that refines and improves itself through time.

Understanding how science works is vital to assessing how scientific interpretations progress through time as information accrues. This is especially relevant in environmental science—a young discipline that is changing rapidly as we attain vast amounts of new information. However, to understand and address environmental problems, we need more than science. We also need to consider ethics. People's ethical perspectives, worldviews, and cultural backgrounds influence how we apply scientific knowledge. Thus, our examination of ethics (and of economics and policy in Chapter 5) will help us learn how values shape human behavior and how information from science is interpreted and put to use in our society.

Environmental Ethics

Ethics is a branch of philosophy that involves the study of good and bad, of right and wrong. The term *ethics* can also refer to the set of moral principles or values held by a person or a society. Ethicists examine how people judge right from wrong by clarifying the criteria that people use in making these judgments. Such criteria are grounded in values—for instance, promoting human welfare, maximizing individual freedom, or minimizing pain and suffering.

Ethical standards are the criteria that help differentiate right from wrong. One classic ethical standard is the *categorical imperative* proposed by German philosopher Immanuel Kant, which advises us to treat others as we would prefer to be treated ourselves. In Christianity this standard is called the "Golden Rule," and most of the world's religions teach this same lesson. Another ethical standard is the *principle of utility,* elaborated by British philosophers Jeremy Bentham and John Stuart Mill. The utilitarian principle holds that something is right when it produces the greatest practical benefits for the most people. We all employ ethical standards as we make countless decisions in our everyday lives.

People of different cultures or worldviews may differ in their values and thus may disagree about actions they consider to be right or wrong. This is why some ethicists are **relativists,** who believe that ethics do and should vary with social context. However, different societies show a remarkable extent of agreement on what moral standards are appropriate. For this reason, many ethicists are **universalists,** who maintain that there exist objective notions of right and wrong that hold across cultures and contexts. For both relativists and universalists, ethics is a prescriptive pursuit; rather than simply describing behavior, it prescribes how we *ought* to behave.

Environmental ethics pertains to people and the environment

The application of ethical standards to relationships between people and nonhuman entities is known as **environmental ethics.** Our interactions with our environment can give rise to ethical questions that are difficult to resolve. Consider some examples:

1. Is the present generation obligated to conserve resources for future generations? If so, how much should we conserve?

2. Can we justify exposing some communities of people to a disproportionate share of pollution? If not, what actions are warranted to prevent this?

3. Are humans justified in driving species to extinction? If destroying a forest would drive extinct a species of bird but would create jobs for hundreds of people, would that action be ethically admissible?

Answers to such questions depend partly on what ethical standard(s) a person adopts. They also depend on the breadth of the person's domain of ethical concern. A person who values the welfare of animals or ecosystems would answer the third pair of questions very differently from a person whose domain of ethical concern ends with human beings. We can think about how peoples' domains of ethical concern can vary by dividing the continuum of attitudes toward the natural world into three ethical perspectives, or worldviews: anthropocentrism, biocentrism, and ecocentrism (**FIGURE 1.11**).

FIGURE 1.11 We can categorize people's ethical perspectives as anthropocentric, biocentric, or ecocentric.

Anthropocentrism describes a human-centered view of our relations with the environment. An anthropocentrist denies, overlooks, or devalues the notion that nonhuman entities have rights and inherent value. An anthropocentrist evaluates the costs and benefits of actions solely according to their impact on people. For example, if cutting down a forest for farming or ranching would provide significant economic benefits while doing little harm to aesthetics or human health, the anthropocentrist would conclude this was worthwhile, even if it would destroy many plants and animals. Conversely, if protecting the forest would provide greater economic, spiritual, or other benefits to people, an anthropocentrist would favor its protection. In the anthropocentric perspective, anything not providing benefit to people is considered to be of negligible value.

In contrast, **biocentrism** ascribes inherent value to certain living things or to the biotic realm in general. In this perspective, human life and nonhuman life both have ethical standing. A biocentrist might oppose clearing a forest if this would destroy a great number of plants and animals, even if it would increase food production and generate economic growth for people.

Ecocentrism judges actions in terms of their effects on whole ecological systems, which consist of living and nonliving elements and the relationships among them. An ecocentrist values the well-being of entire species, communities, or ecosystems (we study these in Chapters 2–4) over the welfare of a given individual. Implicit in this view is that preserving systems generally protects their components, whereas protecting components may not safeguard the entire system. An ecocentrist would respond to a proposal to clear forest by broadly assessing the potential impacts on water quality, air quality, wildlife populations, soil structure, nutrient cycling, and ecosystem services. Ecocentrism is a more holistic perspective than biocentrism or anthropocentrism. It encompasses a wider variety of entities at a larger scale and seeks to preserve the connections that tie them together into functional systems.

Conservation and preservation arose with the 20th century

As industrialization proceeded, raising standards of living but amplifying human impacts on the environment, more people began adopting biocentric and ecocentric worldviews. In the 19th and 20th centuries, the worldviews of people in the United States evolved as the nation pushed west, urbanized, and exploited the continent's resources, boosting affluence and dramatically altering the landscape in the process.

A key voice for restraint during this period of rapid growth and change was **John Muir** (1838–1914), a Scottish immigrant who made California's Yosemite Valley his wilderness home. Muir lived in his beloved Sierra Nevada for long stretches of time, but also became politically active and won fame as a tireless advocate for the preservation of wilderness (**FIGURE 1.12**).

Muir was motivated by the rapid environmental change he witnessed during his life and by his belief that the natural world should be treated with the same respect we give to cathedrals. Today he is associated with the **preservation ethic,**

FIGURE 1.12 A pioneering advocate of the preservation ethic, John Muir helped establish the Sierra Club, a major environmental organization. Here Muir **(right)** is shown with President Theodore Roosevelt in Yosemite National Park in 1903. After this wilderness camping trip with Muir, the president expanded protection of areas in the Sierra Nevada.

which holds that we should protect the natural environment in a pristine, unaltered state. Muir argued that nature deserved protection for its own sake (an ecocentrist argument), but he also maintained that nature promoted human happiness (an anthropocentrist argument based on the principle of utility). "Everybody needs beauty as well as bread," he wrote, "Places to play in and pray in, where nature may heal and give strength to body and soul alike."

Some of the factors that motivated Muir also inspired **Gifford Pinchot** (1865–1946; **FIGURE 1.13**). Pinchot founded what would become the U.S. Forest Service and served as its chief in President Theodore Roosevelt's administration. Like Muir, Pinchot opposed the deforestation and unregulated development of American lands. However, Pinchot took a more anthropocentric view of how and why we should value nature. He espoused the **conservation ethic,** which holds that people should put natural resources to use but that we have a

FIGURE 1.13 Gifford Pinchot was a leading proponent of the conservation ethic. This ethic holds that we should use natural resources in ways that ensure the greatest good for the greatest number of people for the longest time.

responsibility to manage them wisely. The conservation ethic employs a utilitarian standard and contends that we should allocate resources to provide the greatest good to the greatest number of people for the longest time. Whereas preservation aims to preserve nature for its own sake and for our aesthetic and spiritual benefit, conservation promotes the prudent, efficient, and sustainable extraction and use of natural resources for the good of present and future generations.

The contrasting ethical approaches of Pinchot and Muir often pitted them against one another on policy issues of the day. However, both men opposed a prevailing tendency to promote economic development without regard to its negative consequences—and both men left legacies that reverberate today.

Aldo Leopold's land ethic inspires many people

As a young forester and wildlife manager, **Aldo Leopold** (1887–1949; FIGURE 1.14) began his career in the conservationist camp after graduating from Yale Forestry School, which Pinchot had helped to establish. As a forest manager in Arizona and New Mexico, Leopold embraced the government policy of shooting predators, such as wolves, to increase populations of deer and other game animals.

At the same time, Leopold followed the advance of ecological science. He eventually ceased to view certain species as "good" or "bad" and instead came to see that healthy ecological systems depend on protecting all their interacting parts. Drawing an analogy to mechanical maintenance, he wrote, "to keep every cog and wheel is the first precaution of intelligent tinkering."

FIGURE 1.14 Aldo Leopold, a wildlife manager, author, and philosopher, articulated a new relationship between people and the environment. In his essay "The Land Ethic" he called on people to embrace the land in their ethical outlook.

It was not just science that pulled Leopold from an anthropocentric perspective toward a more holistic one. One day he shot a wolf, and when he reached the animal, Leopold was transfixed by "a fierce green fire dying in her eyes." The experience remained with him for the rest of his life and helped lead him to an ecocentric ethical outlook. Years later, as a University of Wisconsin professor, Leopold argued that people should view themselves and "the land" as members of the same community and that we are obligated to treat the land in an ethical manner.

Leopold intended the land ethic to help guide decision making. "A thing is right," he wrote, "when it tends to preserve the integrity, stability, and beauty of the biotic community. It is wrong when it tends otherwise." Leopold died before seeing his seminal 1949 essay "The Land Ethic" and his best-known book, *A Sand County Almanac*, in print, but today many view him as the most eloquent philosopher of environmental ethics.

Environmental justice seeks fair treatment for all people

Our society's domain of ethical concern has been expanding from rich to poor, and from majority races and ethnic groups to minority ones. This ethical expansion involves applying a standard of fairness and equality, and it has given rise to the environmental justice movement. **Environmental justice** involves the fair and equitable treatment of all people with respect to environmental policy and practice, regardless of their income, race, or ethnicity.

The struggle for environmental justice has been fueled by the recognition that poor people tend to be exposed to more pollution, hazards, and environmental degradation than are richer people (FIGURE 1.15). Advocates of environmental justice also note that racial and ethnic minorities tend to suffer more than their share of exposure to most hazards. Indeed, studies repeatedly document that poor and nonwhite communities each tend to bear heavier burdens of air pollution, lead poisoning, pesticide exposure, toxic waste exposure, and workplace hazards. This is thought to occur because lower-income and minority communities often have less access to information on environmental health risks, less political power with which to protect their interests, and less money to spend on avoiding or alleviating risks.

A protest in the 1980s by residents of Warren County, North Carolina, against a proposed toxic waste dump in their community helped to ignite the environmental justice movement. The state had chosen to site the dump in the county with the highest percentage of African Americans. Warren County residents lost their battle and the dump was established—but the protest inspired countless efforts elsewhere.

Like African Americans, Native Americans have encountered many environmental justice issues. Uranium mining on lands of the Navajo nation employed many Navajo in the 1950s and 1960s. Although uranium mining had been linked to health problems and premature death, neither the mining industry nor the U.S. government provided the miners information or safeguards against radiation and its risks. As cancer began to appear among Navajo miners, a later generation of Americans

(a) Migrant farm workers in Colorado

(b) Homes near a coal-fired power plant

(c) Children in New Orleans after Hurricane Katrina

FIGURE 1.15 Environmental justice efforts are inspired by the fact that the poor are often exposed to more hazards than are the rich. For example, Latino farm workers **(a)** may experience health risks from pesticides, fertilizers, and dust. Low-income white Americans in Appalachia **(b)** may suffer air pollution from nearby coal-fired power plants. African American communities in New Orleans **(c)** were most susceptible to flooding and were devastated by Hurricane Katrina in 2005.

perceived negligence and discrimination. They sought relief through the Radiation Exposure Compensation Act of 1990, a federal law compensating Navajo miners who suffered health impacts from unprotected work in the mines.

Likewise, low-income white residents of the Appalachian region have long been the focus of environmental justice concerns. Mountaintop coal-mining practices (pp. 246, 352) in this economically neglected region provide jobs to local residents but also pollute water, bury streams, destroy forests, and cause flooding. Low-income residents of affected Appalachian communities continue to have little political power to voice complaints over the impacts of these mining practices.

Today the world's economies have grown, but the gaps between rich and poor have widened. And despite much progress toward racial equality, significant inequities remain. Environmental laws have proliferated, but minorities and the poor still suffer substandard environmental conditions. Yet today, more people are fighting environmental hazards in their communities and winning. One ongoing story involves Latino farm workers in California's San Joaquin Valley. These workers harvest much of the U.S. food supply of fruits and vegetables yet suffer some of the nation's worst air pollution. Industrial agriculture generates pesticide emissions, dairy feedlot emissions, and windblown dust from eroding farmland, yet this pollution was not being regulated. Valley residents enlisted the help of organizations including the Center on Race, Poverty, and the Environment, a San Francisco–based environmental justice law firm. Together they persuaded California regulators to enforce Clean Air Act provisions and convinced California legislators to pass new laws regulating agricultural emissions.

As we explore environmental issues from a scientific standpoint throughout this book, we will also encounter the social, political, ethical, and economic aspects of these issues, and the concept of environmental justice will arise again and again. Environmental justice is a key component in pursuing the environmental, economic, and social goals of the modern drive for sustainability and sustainable development.

Sustainability and Our Future

Recall the ethical question posed earlier (p. 14): "Is the present generation obligated to conserve resources for future generations?" This question cuts to the core of **sustainability,** a guiding principle of modern environmental science and a concept you will encounter throughout this book.

Sustainability means living within our planet's means, such that Earth can sustain us—and all life—for the future. It means leaving our children and grandchildren a world as rich and full as the world we live in now. Sustainability means conserving Earth's resources so that our descendants may enjoy them as we have. It means developing solutions that work in the long term. Sustainability requires maintaining fully functioning ecological systems, because we cannot sustain human civilization without sustaining the natural systems that nourish it.

weighing the ISSUES

Environmental Justice?

Consider the area where you grew up. Where were the factories, waste dumps, and polluting facilities located, and who lived closest to them? Who lives nearest them in the town or city that hosts your campus? Do you think the concerns of environmental justice advocates are justified? If so, what could be done to ensure that poor communities do not suffer more hazards than wealthy ones?

Population and consumption drive environmental impact

Each day, we add over 200,000 people to the planet. This is like adding a city the size of Augusta, Georgia, on Monday; Akron, Ohio, on Tuesday; Richmond, Virginia, on Wednesday; Rochester, New York, on Thursday; Amarillo, Texas, on Friday; and on and on, day after day. This ongoing rise in human population (Chapter 6) amplifies nearly all of our environmental impacts.

Our consumption of resources has risen even faster than our population. The modern rise in affluence has been a positive development for humanity, and our conversion of the planet's natural capital has made life better for most of us so far. However, like rising population, rising per capita consumption magnifies the demands we make on our environment.

The world's people have not benefited equally from society's overall rise in affluence. Today the 20 wealthiest nations boast over 55 times the per capita income of the 20 poorest nations—three times the gap that existed just two generations ago. Within the United States, the richest 10% of people now claim half of the total income and more than 70% of the total wealth. The ecological footprint of the average citizen of a developed nation such as the United States is considerably larger than that of the average resident of a developing country (**FIGURE 1.16**).

Our growing population and consumption are intensifying the many environmental impacts we examine in this book, including erosion and other impacts from agriculture (Chapter 7), deforestation (Chapter 9), toxic substances (Chapter 10), mineral extraction and mining impacts (Chapter 11), fresh water depletion (Chapter 12), fisheries declines (Chapter 12), air and water pollution (Chapters 12 and 13), waste generation (Chapter 17), and, of course, global climate change (Chapter 14). These impacts degrade our health and quality of life, and they alter the landscapes in which we live. They also are driving the loss of Earth's biodiversity (Chapter 8)—perhaps our greatest problem, because extinction is irreversible. Once a species becomes extinct, it is lost forever.

weighing the **ISSUES**

Leaving a Large Footprint

What do you think accounts for the variation in per capita ecological footprints among societies? Do you feel that people with larger footprints have an ethical obligation to reduce their environmental impact, so as to leave more resources available for people with smaller footprints? Why or why not?

Energy choices will shape our future

Our reliance on fossil fuels intensifies virtually every impact we exert on our environment. Yet fossil fuels have also helped to bring us the material affluence we enjoy. By exploiting the concentrated energy in coal, oil, and natural gas, we've been able to power the machinery of the industrial revolution, produce chemicals that boost crop yields, run vehicles and transportation networks, and manufacture and distribute countless consumer goods (Chapter 15).

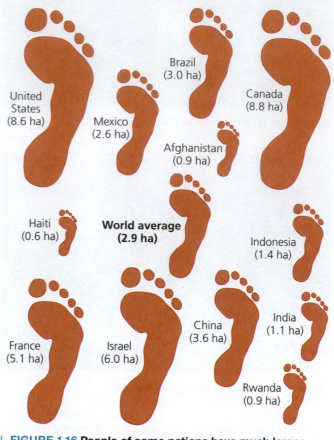

FIGURE 1.16 People of some nations have much larger ecological footprints than people of others. Shown are ecological footprints for average citizens of several nations, along with the world's average per capita footprint of 2.9 hectares. One hectare (ha) = 2.47 acres. *Data from Global Footprint Network, 2017.*

DATA
- Which nation shown here has the largest footprint?
- How many times larger is it than that of the nation shown here with the smallest footprint?

Go to **Interpreting Graphs & Data** on **Mastering** Environmental Science

However, in extracting coal, oil, and natural gas, we are splurging on a one-time bonanza, because these fuels are nonrenewable and in finite supply. Attempts to reach further for new fossil fuel sources threaten more impacts for relatively less fuel. The energy choices we make now will greatly influence the nature of our lives for the foreseeable future.

Sustainable solutions abound

Humanity's challenge is to develop solutions that enhance our quality of life while protecting and restoring the environment that supports us. Many workable solutions are at hand:

- Renewable energy sources (Chapter 16) are beginning to replace fossil fuels.
- Scientists and farmers are pursuing soil conservation, high-efficiency irrigation, and organic agriculture (Chapter 7).
- Energy efficiency efforts continue to gain ground (Chapter 15).

Removing Lead from Gasoline

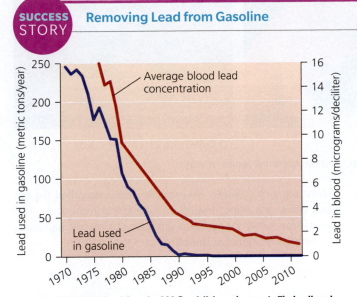

Levels of lead in the blood of U.S. children (ages 1–5) declined as lead use in U.S. gasoline was reduced. *Data from National Health and Nutrition Examination Survey (CDC) and other sources.*

Did you ever wonder why unleaded gas is called "unleaded?" It's because we used to add lead to gasoline to make cars run more smoothly—even though scientific research showed that emissions of this toxic heavy metal from tailpipes caused severe health problems, including brain damage and premature death. Back in 1970, air pollution was severe in many American cities, and motor vehicles accounted for 78% of U.S. lead emissions. In response, environmental scientists, medical researchers, auto engineers, and policymakers all merged their knowledge and skills into a process that brought about the removal of lead from gasoline. The ban on lead was phased in gradually because some older vehicles required leaded gas, but by 1996, all gasoline sold in the United States was unleaded, and the nation's largest source of atmospheric lead pollution was eliminated. As a result, levels of lead in people's blood fell dramatically, producing one of America's greatest public health successes.

EXPLORE THE DATA at **Mastering** Environmental Science

- Laws and new technologies have reduced air and water pollution in wealthier societies (Chapters 5, 12, and 13).

- Conservation biologists are protecting habitat and endangered species (Chapter 8).

- Governments, businesses, and individuals are taking steps to reduce emissions of the greenhouse gases that drive climate change (Chapter 14).

- Better waste management is helping us to conserve resources (Chapter 17).

These are a few of the many efforts we will examine in the course of this book while exploring sustainable solutions to our challenges. In each chapter, a **SUCCESS STORY** will feature one specific example. Additionally, you will encounter many further solutions to problems, addressed in many ways, throughout the text. Then in the **EPILOGUE** at the end of the book, we will review 10 major strategies or approaches that can help us generate sustainable solutions (p. 436).

Students are promoting solutions on campus

As a college student, you can help to design and implement sustainable solutions on your own campus. Proponents of **campus sustainability** seek ways to help colleges and universities reduce their ecological footprints. Although we tend to think of colleges and universities as enlightened and progressive institutions that benefit society, they are also centers of lavish resource consumption. Classrooms, offices, research labs, dormitories, dining halls, sports arenas, vehicle fleets, and road networks all consume resources and generate waste. Together the 4500 campuses in the United States generate

about 2% of U.S. carbon emissions. Reducing the ecological footprint of a campus can be challenging, yet students, faculty, staff, and administrators on thousands of campuses are working together to make the operations of educational institutions more sustainable (**FIGURE 1.17**).

Students are running recycling programs, promoting efficient transportation options, restoring native plants, growing organic gardens, and fostering sustainable dining halls. They are finding ways to improve energy efficiency and water conservation and are pressing for green buildings. To address climate change, students are urging their institutions to reduce greenhouse gas emissions, divest from fossil fuel corporations, and use and invest in renewable energy.

In our **Epilogue** and throughout this book you will encounter examples of campus sustainability efforts (for example, pp. 139, 399, and 435). Should you wish to pursue such efforts on your own campus, information and links in the **Selected Sources and References** online at **MasteringEnvironmentalScience** point you toward organizations and resources that can help.

Environmental science prepares you for the future

By taking a course in environmental science, you are preparing yourself for a lifetime in a world increasingly dominated by concerns over sustainability. The course for which you are using this book right now likely did not exist a generation ago. But as society's concerns have evolved, colleges and universities have adapted their curricula, helping students learn how to tackle the challenges of creating a sustainable future.

Still, at most schools, fewer than half of students take even a single course on the basic functions of Earth's natural systems, and still fewer take courses on the links between human activity and sustainability. As a result, many educators

(a) Urging divestment from fossil fuels   (b) Recycling   (c) Collecting electronic waste

FIGURE 1.17 Students are helping to make their campuses more sustainable in all kinds of ways.

worry that most students graduate lacking **environmental literacy,** a basic understanding of Earth's physical and living systems and how we interact with them. By taking an environmental science course, you will gain a better understanding of how the world works. You will be better qualified for the green-collar job opportunities of today and tomorrow. And you will be better prepared to navigate the many challenges of creating a sustainable future.

closing THE LOOP

Finding effective ways of living peacefully, healthfully, and sustainably on our diverse and complex planet requires a solid ethical grounding as well as a sound scientific understanding of natural and social systems. Environmental science helps us comprehend our intricate relationship with our environment and informs our attempts to solve and prevent environmental problems. Although many of today's trends may cause concern, a multitude of inspiring success stories give us reason for optimism. Identifying a problem is the first step toward devising a solution, and addressing environmental problems can move us toward health, longevity, peace, and prosperity. Science in general, and environmental science in particular, can help us develop balanced, workable, sustainable solutions and create a better world now and for the future.

TESTING Your Comprehension

1. How and why did the agricultural revolution affect human population size? How and why did the industrial revolution affect human population size? Explain what benefits and what environmental impacts have resulted.

2. What is an *ecological footprint*? Explain what is meant by the term *overshoot*.

3. What is *environmental science*? Name several disciplines that environmental science draws upon.

4. Compare and contrast the two meanings of the term *science*. Name three applications of science.

5. Describe the scientific method. What is its typical sequence of steps? What needs to occur before a researcher's results are published? Why is this process important?

6. Compare and contrast anthropocentrism, biocentrism, and ecocentrism. Explain how individuals with each perspective might evaluate the development of a shopping mall atop a wetland in your town or city.

7. Differentiate the preservation ethic from the conservation ethic. Explain the contributions of John Muir and Gifford Pinchot in the history of environmental ethics.

8. Describe Aldo Leopold's land ethic. How did Leopold define the "community" to which ethical standards should be applied?

9. Explain the concept of environmental justice. Give an example of an inequity relevant to environmental justice that you believe exists in your city, state, or country.

10. Describe in your own words what you think is meant by the term *sustainability*. Name three ways that students, faculty, or administrators are seeking to make their campuses more sustainable.

SEEKING Solutions

1. Resources such as soils, timber, fresh water, and biodiversity are renewable if we use them in moderation, but can become nonrenewable if we overexploit them (see Figure 1.1). For each of these four resources, describe one way we sometimes overexploit the resource and name one thing we could do to conserve the resource. For each, what might constitute sustainable use? (Feel free to look ahead and peruse coverage of these issues throughout this book.)

2. What do you think is the lesson of Easter Island? What more would you like to learn or understand about this island and its people? What similarities do you perceive between Easter Island and our own modern society? What differences do you see between their predicament and ours?

3. Describe your ethical perspective, or worldview, as it pertains to your relationship with your environment. Do you feel that you fit into any particular category discussed in this chapter? How do you think your culture has influenced your worldview? How has your personal experience shaped it? What environmental problem do *you* feel most acutely yourself?

4. Find out what sustainability efforts are being made on your campus. What results have these efforts produced so far? What further efforts would you like to see pursued on your campus? Do you foresee any obstacles to these efforts? How could these obstacles be overcome? How could you become involved?

5. **THINK IT THROUGH** You have become head of a major funding agency that grants money to scientists pursuing research in environmental science. You must give your staff several priorities to determine what types of scientific research to fund. What environmental problems would you most like to see addressed with research? Describe the research you think would need to be completed in order to develop workable solutions. What else, beyond scientific research, might be needed to develop sustainable solutions?

CALCULATING Ecological Footprints

Researchers at the Global Footprint Network continue to refine their method of calculating ecological footprints—the amount of biologically productive land and water required to produce the energy and natural resources we consume and to absorb the wastes we generate. According to their most recent data, there are 1.71 hectares (4.23 acres) available per person in the world, yet we use on average 2.87 ha (7.09 acres) per person, creating a global ecological deficit, or overshoot (p. 6), of 68%.

Compare the ecological footprints of each nation listed in the table. Calculate their proportional relationships to the world population's average ecological footprint and to the area available globally to meet our ecological demands.

NATION	ECOLOGICAL FOOTPRINT (HECTARES PER PERSON)	PROPORTION RELATIVE TO WORLD AVERAGE FOOTPRINT	PROPORTION RELATIVE TO WORLD AREA AVAILABLE
Bangladesh	0.7	0.3 (0.7 ÷ 2.9)	0.4 (0.7 ÷ 1.7)
Tanzania	1.3		
Colombia	1.9		
Thailand	2.6		
Mexico	2.6		
Sweden	6.5		
United States	8.6		
World average	2.9	1.0 (2.9 ÷ 2.9)	1.7 (2.9 ÷ 1.7)
Your personal footprint (see Question 4)			

Data from Global Footprint Network, 2017.

1. Why do you think the ecological footprint for people in Bangladesh is so small?

2. Why do you think the ecological footprint is so large for people in the United States?

3. Based on the data in the table, how do you think average per capita income is related to ecological footprints? Name some ways in which you believe a wealthy society can decrease its ecological footprint.

4. Go to an online footprint calculator such as the one at http://www.footprintnetwork.org/en/index.php/GFN/page/personal_footprint and take the test to determine your own personal ecological footprint. Enter the value you obtain in the table, and calculate the other values as you did for each nation. How does your footprint compare to that of the average person in the United States? How does it compare to that of people from other nations? Name three actions you could take to reduce your footprint.

Mastering Environmental Science

CHAPTER 2

Environmental Systems

Matter, Energy, and Ecosystems

The Vanishing Oysters of the Chesapeake Bay

> **I'm 60. Danny's 58. We're the young ones.**
> —Grant Corbin, oysterman in Deal Island, Maryland

> **The Bay continues to be in serious trouble. And it's really no question why this is occurring. We simply haven't managed the Chesapeake Bay as a system the way science tells us we must.**
> —Will Baker, President, Chesapeake Bay Foundation

Upon completing this chapter, you will be able to:

- Describe the nature of environmental systems

- Explain the fundamentals of matter and chemistry, and apply them to real-world situations

- Differentiate among forms of energy and explain the first and second laws of thermodynamics

- Distinguish photosynthesis, cellular respiration, and chemosynthesis, and summarize their importance to living things

- Define ecosystems and discuss how living and nonliving entities interact in ecosystem-level ecology

- Outline the fundamentals of landscape ecology and ecological modeling

- Explain ecosystem services and describe how they benefit our lives

- Compare and contrast how water, carbon, nitrogen, and phosphorus cycle through the environment, and explain how human activities affect these cycles

Chesapeake Bay oystermen haul in their catch on the shores of the Chesapeake Bay.

A visit to Deal Island, Maryland, on the Chesapeake Bay reveals a situation that is all too common in modern America: The island, which was once bustling with productive industries and growing populations, is falling into decline. Economic opportunities in the community are few, and its populace is shrinking and "graying" as more and more young people leave to find work elsewhere. In 1930, Deal Island had a population of 1237 residents. In 2010, it was a mere 471 people.

Unlike other parts of the country with similar stories of economic decline, the demise of Deal Island and other bayside towns was not caused by the closing of a local factory, steel mill, or corporate headquarters; it was caused by the collapse of the Chesapeake Bay oyster fishery.

The Chesapeake Bay was once a thriving system of interacting microbes, plants, and animals, including economically important blue crabs, scallops, and fish. Nutrients carried to the bay by streams in its roughly 168,000 km^2 (64,000 mi^2) **drainage basin,** or **watershed**—the land area that funnels water to a given body of water—nourished fields of underwater grasses that provided food and refuge to juvenile fish, shellfish, and crabs. Hundreds of millions of oysters kept the bay's water clear by filtering nutrients and phytoplankton (microscopic photosynthetic algae, protists, and cyanobacteria that drift near the surface) from the water column.

Oysters had been eaten locally since the region was populated, but the intensive harvest of bay oysters for export didn't begin until the 1830s. By the 1880s the bay boasted the world's largest oyster fishery. People flocked to the Chesapeake to work on oystering ships or in canneries, dockyards, and shipyards. Bayside towns like Deal Island prospered along with the oyster industry and developed a unique maritime culture that defined the region.

But by 2010 the bay's oyster populations had been reduced to a mere 1% of their abundance prior to the start of commercial harvesting, and the oyster industry in the area was all but wiped out. Perpetual overharvesting, habitat destruction, virulent oyster diseases, and water pollution had nearly eradicated this economically and ecologically important organism from bay waters. The monetary losses associated with the oyster fishery collapse in the Chesapeake Bay have been staggering, costing the economies of Maryland and Virginia an estimated $4 billion from 1980 to 2010.

In addition to overharvesting, one of the biggest impacts on oysters in recent decades is the pollution of the

Sorting oysters from the Chesapeake Bay ▲

bay with high levels of the nutrients **nitrogen** and **phosphorus** from agricultural fertilizers, animal manure, stormwater runoff, and atmospheric compounds produced by fossil fuel combustion. Oysters naturally filter nutrients from water, but with so few oysters in the bay, elevated nutrient levels have caused phytoplankton populations to increase. When phytoplankton die, settle to the bay bottom, and are decomposed by bacteria, oxygen in the water is depleted—a condition called **hypoxia**—which creates "dead zones" in the bay. Grasses, oysters, crabs, fish, and other organisms perish in these dead zones or are forced to flee to less optimal areas of the bay. Hypoxia and other human impacts ultimately landed the Chesapeake Bay on the Environmental Protection Agency's (EPA's) list of dangerously polluted waters in the United States.

However, recent events in the Chesapeake have offered hope for the recovery of the Chesapeake Bay system. After 25 years of failed pollution control agreements and nearly $6 billion spent on cleanup efforts, the Chesapeake Bay Foundation (CBF), a non-profit organization dedicated to conserving the bay, sued the EPA in 2009 for failing to use its available powers under the Clean Water Act to clean up the bay. This spurred federal action, and in 2010 a comprehensive "pollution budget" and restoration plan was developed and implemented by the EPA—with the assistance of the District of Columbia, Delaware, Maryland, New York, Pennsylvania, Virginia, and West Virginia—aiming to substantially reduce inputs of nitrogen and phosphorus into the bay by 2025. Further, oyster restoration efforts are finally showing promise in the Chesapeake (see THE SCIENCE BEHIND THE STORY, pp. 26–27). If these initiatives can begin to restore the bay to health, Deal Island and other oyster-fishing communities may again enjoy the prosperity they once did on the scenic shores of the Chesapeake.

Earth's Environmental Systems

Understanding the rise and fall of the oyster industry in the Chesapeake Bay, as with many other human impacts on the environment, involves comprehending the complex, interlinked systems that make up Earth's environment. A **system** is a network of relationships among parts, elements, or components that interact with and influence one another through the exchange of energy, matter, or information. Earth's natural systems include processes that shape the landscape, affect planetary climate, govern interactions between species and the nonliving entities around them, and cycle chemical elements vital to life. Because we depend on these systems and processes for our very survival, understanding how they function and how human activities affect them is an important aspect of environmental science.

There are many ways to delineate natural systems. For instance, scientists sometimes divide Earth's components into structural spheres to help make our planet's dazzling complexity comprehensible. The **lithosphere** (p. 232) is the rock and sediment beneath our feet, the planet's uppermost mantle and crust. The **atmosphere** (p. 288) is composed of the air surrounding our planet. The **hydrosphere** (p. 257) encompasses all water—salt or fresh; liquid, ice, or vapor—in surface bodies, underground, and in the atmosphere. The **biosphere** (p. 59) consists of all the planet's organisms and the abiotic (nonliving) portions of the environment with which they interact.

Natural systems seldom have well-defined boundaries, so deciding where one system ends and another begins sometimes can be difficult. As an analogy, consider a smartphone. It is certainly a system—a network of circuits and parts that interact and exchange energy and information—but where are its boundaries? Is the system merely the phone itself, or does it include the other phones you call and text, the websites you access on it, and the cellular networks that keep it connected? What about the energy grid that recharges the phone's battery with its many miles of transmission lines and distant power plants?

No matter how we attempt to isolate or define a system, we soon see that it has connections to systems larger and smaller than itself. Systems may exchange energy, matter, and information with other systems, and they may contain or be contained within other systems. Thus, where we draw boundaries may depend on the spatial (space) or temporal (time) scale on which we choose to focus.

Assessing questions holistically by taking a "systems approach" is helpful in environmental science, where so many issues are multifaceted and complex. Taking a broad and integrative approach poses challenges, because systems often show behavior that is difficult to predict. However, environmental scientists are rising to the challenge of studying systems holistically, helping us to develop comprehensive solutions to complicated problems such as those faced in the Chesapeake Bay.

Systems involve feedback loops

Earth's environmental systems receive inputs of energy, matter, or information; process these inputs; and produce outputs. The Chesapeake Bay system receives inputs of fresh water, sediments, nutrients, and pollutants from the rivers that empty into it. Oystermen, crabbers, and fishermen harvest some of the bay system's output: matter and energy in the form of seafood. This output subsequently becomes input to the nation's economic system and to the body systems of people who consume the seafood.

Sometimes a system's output can serve as input to that same system, a circular process described as a **feedback loop,** which can be either negative or positive. In a **negative feedback loop,** output that results from a system moving in one direction acts as input that moves the system in the other direction. Input and output essentially neutralize one another's effects, stabilizing the system. As an example, negative feedback regulates body temperature in humans

(**FIGURE 2.1a**): If we get too hot, our sweat glands pump out water that then evaporates, which cools us down. Or we just move into the shade. If we get too cold, we shiver, creating heat, or we just move into the sun or put on a sweater. Most systems in nature involve such negative feedback loops. Negative feedback enhances stability, and over time only those systems that are stable will persist.

In a system stabilized by negative feedback, when processes move in opposing directions at equivalent rates so that their effects balance out, they are said to be in **dynamic equilibrium.** Processes in dynamic equilibrium can contribute to **homeostasis,** the tendency of a system to maintain relatively constant or stable internal conditions. A system (such as an organism) in homeostasis keeps its internal conditions within a range that allows it to function. However, the steady state of a homeostatic system may itself change slowly over time. For instance, Earth has experienced gradual changes in atmospheric composition and ocean chemistry over its long history, yet life persists and our planet remains, by most definitions, a homeostatic system.

Rather than stabilizing a system, **positive feedback loops** drive the system further toward an extreme. In positive feedback, increased output from a system leads to increased input, leading to further increased output, and so on. Exponential growth in a population (p. 63) is one such example—the more individuals there are, the more offspring that can be produced.

Another positive feedback cycle that is of great concern to environmental scientists today involves the melting of glaciers and sea ice in the Arctic as a result of global warming (p. 322). Because ice and snow are white, they reflect sunlight and keep surfaces cool. But if the climate warms enough to melt the ice and snow, darker surfaces of land and water are exposed, and these darker surfaces absorb sunlight. This absorption warms the surface, causing further melting, which in turn exposes more dark surface area, leading to further warming (**FIGURE 2.1b**). Runaway cycles of positive feedback are rare in untouched nature, but they are common in natural systems altered by human activities, and such feedback loops can destabilize those systems.

FAQ

But isn't positive feedback "good" and negative feedback "bad"?

In daily life, positive feedback, such as a compliment, can act as a stabilizing force ("Keep up the good work, and you'll succeed") whereas negative feedback, such as a criticism, can destabilize ("You need to change your approach to succeed"). But in environmental systems, it's the opposite! Negative feedback resists change in systems, enhancing stability and typically keeping conditions within ranges beneficial to life. Positive feedback, in contrast, exerts destabilizing effects that push conditions to extremes, threatening organisms adapted to the system's normal conditions. Thus, negative feedback in environmental systems typically aids living things, whereas positive feedback often harms them.

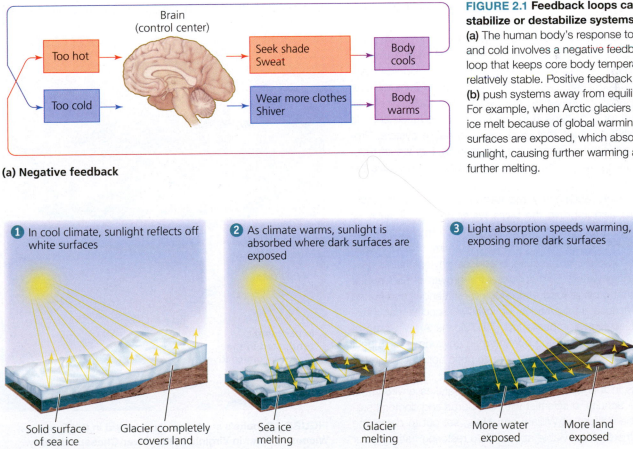

Brain (control center)

Too hot → Seek shade / Sweat → Body cools

Too cold → Wear more clothes / Shiver → Body warms

(a) Negative feedback

FIGURE 2.1 Feedback loops can stabilize or destabilize systems.
(a) The human body's response to heat and cold involves a negative feedback loop that keeps core body temperatures relatively stable. Positive feedback loops **(b)** push systems away from equilibrium. For example, when Arctic glaciers and sea ice melt because of global warming, darker surfaces are exposed, which absorb more sunlight, causing further warming and further melting.

❶ In cool climate, sunlight reflects off white surfaces

❷ As climate warms, sunlight is absorbed where dark surfaces are exposed

❸ Light absorption speeds warming, exposing more dark surfaces

Solid surface of sea ice | Glacier completely covers land

Sea ice melting | Glacier melting

More water exposed | More land exposed

(b) Positive feedback

Are We "Turning the Tide" for Native Oysters in Chesapeake Bay?

In 2001, the Eastern oyster (*Crassostrea virginica*) was in dire trouble in the Chesapeake Bay. Populations had dropped by 99%, and the Chesapeake's oyster industry, once the largest in the world, had collapsed. Poor water quality, reef destruction, virulent diseases spread by transplanted oysters, and 200 years of overharvesting all contributed to the collapse.

Restoration efforts had largely failed. Moreover, when scientists or resource managers proposed rebuilding oyster populations by significantly restricting oyster harvests or establishing oyster reef "sanctuaries," these initiatives were typically defeated by the politically powerful oyster industry. All this had occurred in a place whose very name (derived from the Algonquin word *Chesepiook*) means "great shellfish bay."

With the collapse of the native oyster fishery and with political obstacles blocking restoration projects for native oysters, support grew among the oyster industry, state resource managers, and some scientists for the introduction of Suminoe oysters (*Crassostrea ariakensis*) from Asia. This species seemed well suited for conditions in the bay and showed resistance to the parasitic diseases that were ravaging native oysters. Proponents argued that introducing Suminoe oysters would reestablish thriving oyster populations in the bay and revitalize the oyster fishery.

Proponents additionally maintained that introducing oysters would also improve the bay's water quality, because as oysters feed, they filter phytoplankton and sediments from the water column. Filter-feeding by oysters is an important ecological service in the bay because it reduces phytoplankton densities, clarifies waters, and supports the growth of underwater grasses that provide food and refuge for waterfowl and young crabs. Because introductions of invasive species can have profound ecological impacts (pp. 78–79), the Army Corps of Engineers was directed to coordinate an environmental impact statement (EIS, p. 108) on oyster restoration approaches in the Chesapeake.

It was in this politically charged, high-stakes environment that Dave Schulte, a scientist with the Corps and doctoral student at the College of William and Mary, set out to determine whether there was a viable approach to restoring native oyster

populations. The work he and his team began would help turn the tide in favor of native oysters in the bay's restoration efforts.

One of the biggest impacts on native oysters was the destruction of oyster reefs by a century of intensive oyster harvesting. Oysters settle and grow best on the shells of other oysters, and over long periods this process forms reefs (underwater outcrops of living oysters and oyster shells) that solidify and become as hard as stone. Throughout the bay, massive reefs that at one time had jutted out of the water at low tide had been reduced to rubble on the bottom from a century of repeated scouring by metal dredgers used by oyster-harvesting ships. The key, Schulte realized, was to construct artificial reefs like those that once existed, to get oysters off the bottom—away from smothering sediments and hypoxic waters—and up into the plankton-rich upper waters.

In 2004, armed with the resources available to the Corps, Schulte opted to take a landscape ecology approach to restore patches of reef habitat on nine complexes of reefs, creating a total of 35.3 hectares (87 acres) of oyster sanctuary near the mouth of the Great Wicomico River in the lower Chesapeake Bay (**FIGURE 1**)—a much larger restoration effort than any

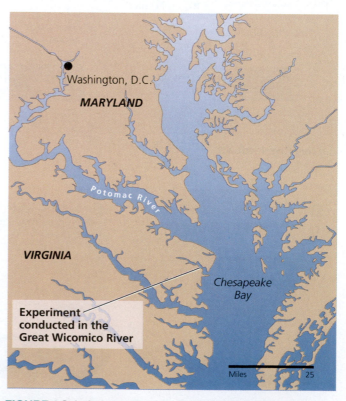

FIGURE 1 Schulte's study was conducted in the Great Wicomico River in Virginia in the lower Chesapeake Bay.

David Schulte, U.S. Army Corps of Engineers.

previously attempted. Schulte and his team constructed artificial reefs by spraying oyster shells off barges (**FIGURE 2**), which then drifted to the river bottom, forming high-relief reefs (in which shells were piled to create a reef that was 25–45 cm above the river bottom) and low-relief reefs (with shells piled to 8–12 cm above the river bottom) that the oysters could colonize, safe from harvesting. Other areas of the river bottom were "unrestored" and left in their natural state.

Oyster populations on the constructed reefs were sampled in 2007, and the results were stunning. The reef complex supported an estimated 185 million oysters, a number nearly as large as the wild population of 200 million oysters estimated to live on the remaining degraded habitat in all of Maryland's waters. Higher constructed reefs supported an average of more than 1000 oysters per square meter—four times more than the lower constructed reefs and 170 times more than unrestored bottom (**FIGURE 3**). Like natural reefs, the constructed reefs began to solidify, providing a firm foundation for the settlement of spat—young, newly settled oysters. In 2009, Schulte's research made a splash when his team published its findings in the journal *Science*, bringing international attention to their study.

After reviewing eight alternative approaches to oyster restoration that involved one or more oyster species, the Corps advocated an approach that avoided the introduction of non-native oysters. Instead it proposed a combination of native oyster restoration, a temporary moratorium on oyster harvests (accompanied by a compensation program for the oyster industry), and enhanced support for oyster aquaculture in the bay region.

Schulte's restoration project cost roughly $3 million and will require substantial investments if it is to be repeated elsewhere in the bay. This is particularly true in upper portions of the bay, where water conditions are poorer, the oysters are less resistant to disease, and oyster reproduction levels are lower, requiring restored reefs to be "seeded" with oysters. Many scientists contend that expanded reef restoration efforts are worth the cost because they enhance oyster populations and provide a vital service to the bay through water filtering. Some scientists also see value in promoting oyster farming, in which restoration efforts would be supported by businesses instead of taxpayers.

These efforts are encouraged by the continued success of the project. By summer 2016 the majority of high-relief reef acreage was thriving, despite pressures from poachers and several years of hypoxic conditions. Moreover, many of the low-relief reefs that were originally constructed eventually accumulated enough new shell to be as tall as the high-relief reefs in the initial experiment—the reef treatment that showed the highest oyster densities in the original experiment. Further, oyster reproduction rates in 2012 were among the highest Schulte had seen during the project, and a follow-up study in 2013 found that spat from the sanctuary reefs were seeding other parts of the Great Wicomico River and increasing oyster populations outside protected areas.

Protected sites for oyster restoration efforts are now being established elsewhere in the bay. Maryland recently designated

FIGURE 2 **A water cannon blows oyster shells off a barge and onto the river bottom to create an artificial oyster reef for the experiment.**

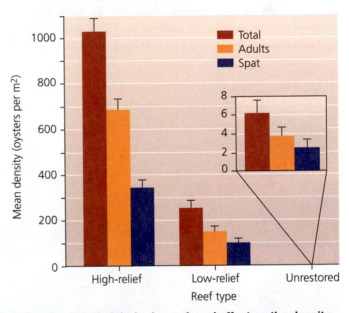

FIGURE 3 **Reef height had a profound effect on the density of adult oysters and spat.** Schulte's work suggested that native oyster populations could rebound in portions of Chesapeake Bay if they were provided elevated reefs and were protected from harvest. *Data from Schulte, D.M., R.P. Burke, and R.N. Lipicus, 2009. Unprecedented restoration of a native oyster metapopulation. Science 325: 1124–1128.*

3640 hectares (9000 acres) of new oyster sanctuaries—25% of existing oyster reefs in state waters—and seeded these reefs with more than a billion hatchery-raised spat. This movement toward increased protection for oyster populations, coupled with findings of increased disease resistance in bay oysters, has given new hope that native oysters may once again thrive in the waters of the "great shellfish bay."

Environmental systems interact

The Chesapeake Bay and the rivers that empty into it provide an example of how systems interact. On a map, the rivers that feed into the bay are a branched and braided network of water channels surrounded by farms, cities, and forests (**FIGURE 2.2**). But where are the boundaries of this system? For a scientist interested in **runoff**—the precipitation that flows over land and enters waterways—and the flow of water, sediment, or pollutants, it may make the most sense to define the bay's watershed as a system. However, for a scientist interested in hypoxia and the bay's dead zones, it may be best to define the watershed together with the bay as the system of interest, because their interaction is central to the problem being investigated. Thus, in environmental science, identifying the boundaries of systems depends on the questions being asked.

If the question we are asking about the Chesapeake Bay relates to the dead zones in the bay, which are due to the extremely high levels of nitrogen and phosphorus delivered to its waters from the 6 states in its watershed and the 15 states in its **airshed**—the geographic area that produces air pollutants likely to end up in a waterway—then we'll want to define the boundaries of the system to include both the watershed and the airshed of the bay. In 2015, the bay received an estimated 121 million kg (267 million lb) of nitrogen and 7.2 million kg (15.8 million lb) of phosphorus.

Runoff from agriculture was a major source of these nutrients, contributing 43% of the nitrogen (**FIGURE 2.3a**) and 55% of the phosphorus (**FIGURE 2.3b**) entering the bay. In some parts of the bay's watershed, roughly one-third of nitrogen inputs come from atmospheric sources within this system.

Elevated nitrogen and phosphorus inputs cause phytoplankton in the bay's waters to flourish. High phytoplankton density leads to elevated mortality in the population due to increased competition for sunlight and nutrients. Dead phytoplankton drift to the bottom of the bay, where they are joined by the waste products of zooplankton, tiny creatures that feed on phytoplankton. The increase in organic material causes an explosion in populations of microbial decomposers, which deplete the oxygen in bottom waters as they consume the organic matter. Deprived of oxygen, bottom-dwelling organisms will either flee the area or suffocate. This process of nutrient overenrichment, increased production of organic matter, and subsequent ecosystem degradation is known as **eutrophication** (**FIGURE 2.4**).

Once oxygen levels at the bottom of the bay are depleted, they are slow to recover. Oxygenated fresh water entering the bay from rivers remains stratified in a layer at the surface and is slow to mix with the denser, saltier bay water, limiting the amount of oxygenated surface water that reaches the bottom-dwelling life that needs it. As a result, sedentary creatures living on the bay bottom, such as oysters, suffocate and die.

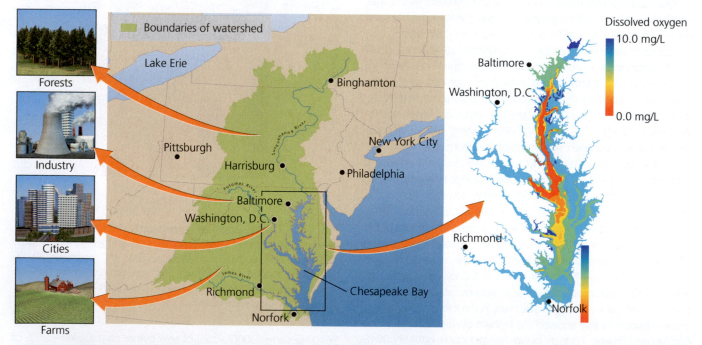

FIGURE 2.2 The Chesapeake Bay watershed encompasses 168,000 km² (64,000 mi²) of land area in six states and the District of Columbia. Tens of thousands of streams carry water, sediment, and pollutants from a variety of sources downriver to the Chesapeake, where nutrient pollution has given rise to large areas of hypoxic waters. The zoomed-in map **(at right)** shows dissolved oxygen concentrations in the Chesapeake Bay in 2016. Oysters, crabs, and fish typically require a minimum of 3 mg/L of oxygen and are therefore excluded from large portions of the bay where oxygen levels are too low. *Source: Figure at right adapted from* National Oceanic and Atmospheric Administration and U.S. Geological Survey, *https://coastalscience.noaa.gov/news/?p=15670*

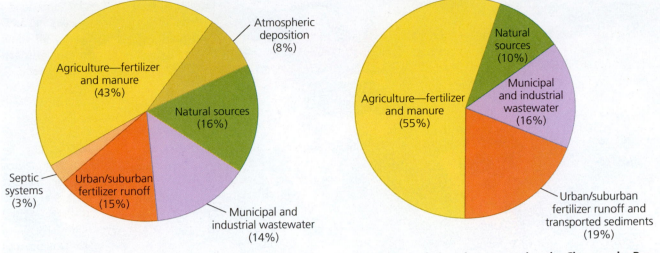

(a) Sources of nitrogen entering the Chesapeake Bay

(b) Sources of phosphorus entering the Chesapeake Bay

FIGURE 2.3 The Chesapeake Bay receives inputs of (a) nitrogen and (b) phosphorus from many sources in its watershed. *Data from Chesapeake Bay Program Office, 2015*, Watershed Model Phase 5.3.2 *(Chesapeake Bay Program Office, 2016)*. Totals for nitrogen do not equal 100% due to rounding.

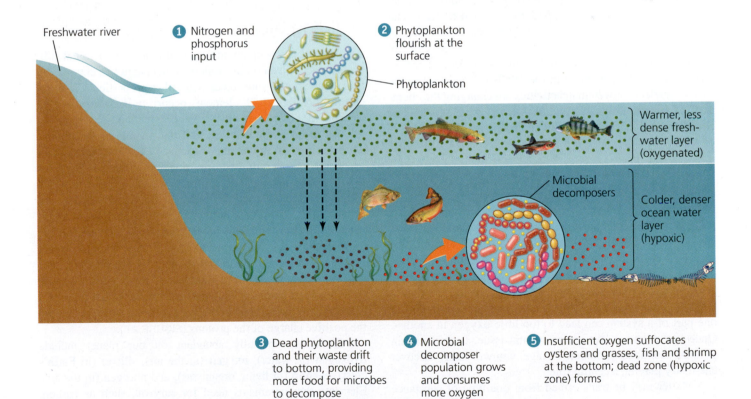

FIGURE 2.4 Excess nitrogen and phosphorus cause eutrophication in aquatic systems such as the Chesapeake Bay. Coupled with stratification (layering) of water, eutrophication can severely deplete dissolved oxygen. ❶ Nutrients from river water ❷ boost growth of phytoplankton, ❸ which die and are decomposed at the bottom by bacteria. Stability of the surface layer prevents deeper water from absorbing oxygen to replace ❹ oxygen consumed by decomposers, and ❺ the oxygen depletion suffocates or drives away bottom-dwelling marine life. This process gives rise to hypoxic zones like those in the bay. The process of eutrophication occurs in both fresh water and marine environments, and in water bodies of all sizes—from small ponds to large expanses of coastal ocean waters.

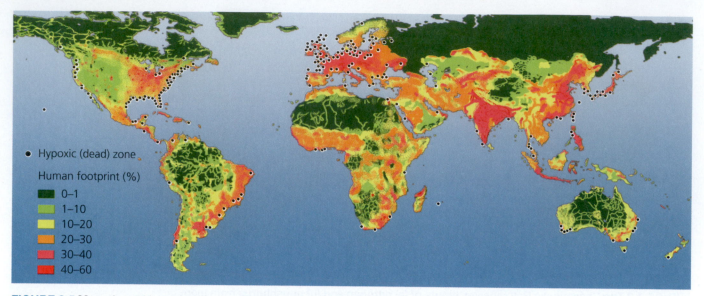

FIGURE 2.5 More than 500 marine dead zones have been recorded across the world. Dead zones (shown by dots on the map) occur mostly offshore from areas of land with the greatest human ecological footprints (here, expressed on a scale of 0 to 100, with higher numbers indicating bigger human footprints). *Data from World Resources Institute, 2016, http://www.wri.org/our-work/project/eutrophication-and-hypoxia; and Diaz, R., and R. Rosenberg, 2008. Spreading dead zones and consequences for marine ecosystems. Science 321: 926–929. Reprinted with permission from AAAS.*

The Chesapeake Bay is not the only water body suffering from eutrophication. Nutrient pollution has led to more than 500 documented hypoxic dead zones (**FIGURE 2.5**), including one that forms each year near the mouth of the Mississippi River (p. 277). The increase in the number of dead zones—there were 162 documented in the 1980s and only 49 in the 1960s—reflects how human activities are changing the chemistry of waters around the world. Let's now take a look at chemistry and its applications in environmental science.

Matter, Chemistry, and the Environment

All material in the universe that has mass and occupies space—solid, liquid, and gas alike—is called **matter.** The study of types of matter and their interactions is called **chemistry.** Chemistry plays a central role in addressing the environmental challenges facing the Chesapeake Bay, as it helps us understand how too much nitrogen or phosphorus in one part of a system can lead to too little oxygen in another. Once you examine any environmental issue, from acid rain to toxic chemicals to climate change, you will likely discover chemistry playing a central role.

Matter may be transformed from one type of substance into others, but it cannot be created or destroyed. This principle is referred to as the **law of conservation of matter.** In environmental science, this principle helps us understand that the amount of matter stays constant as it is recycled in ecosystems and nutrient cycles (pp. 39–44). It also makes it clear that we cannot simply wish away "undesirable" matter, such as nuclear waste or toxic pollutants. Because harmful substances can't be destroyed, we must take steps to minimize their impacts on the environment.

Atoms and elements are chemical building blocks

An **element** is a fundamental type of matter, a chemical substance with a given set of properties that cannot be chemically broken down into substances with other properties. Chemists currently recognize 98 elements occurring in nature, as well as about 20 others they have created in the lab.

An **atom** is the smallest unit that maintains the chemical properties of an element. Atoms of each element contain a specific number of **protons,** positively charged particles in the atom's nucleus (its dense center), and this number is called the element's atomic number. (Elemental carbon, for instance, has six protons in its nucleus; thus, its atomic number is 6.) Most atoms also contain **neutrons,** particles in the nucleus that lack an electrical charge. An element's mass number denotes the combined number of protons and neutrons in the atom. An atom's nucleus is surrounded by negatively charged particles known as **electrons,** which are equal in number to the protons in the nucleus of an atom, balancing the positive charge of the protons (**FIGURE 2.6**).

Elements especially abundant on our planet include **hydrogen** (in water), **oxygen** (in the air), **silicon** (in Earth's crust), **carbon** (in living organisms), and nitrogen (in the air). Elements that organisms need for survival, such as carbon, nitrogen, calcium, and phosphorus, are called **nutrients.** Each element is assigned an abbreviation, or chemical symbol; for instance, "H" stands for hydrogen and "O" for oxygen. The periodic table of the elements organizes the elements according to their chemical properties and behavior (see **APPENDIX D**).

Isotopes Although all atoms of a given element contain the same number of protons, they do not necessarily contain the same number of neutrons. Atoms with differing numbers

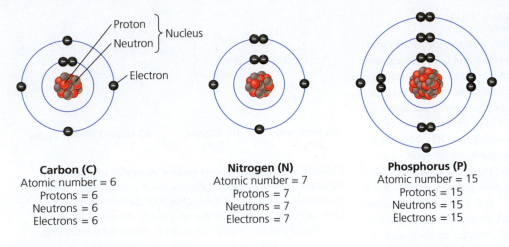

Carbon (C)
Atomic number = 6
Protons = 6
Neutrons = 6
Electrons = 6

Nitrogen (N)
Atomic number = 7
Protons = 7
Neutrons = 7
Electrons = 7

Phosphorus (P)
Atomic number = 15
Protons = 15
Neutrons = 15
Electrons = 15

FIGURE 2.6 In an atom, protons and neutrons stay in the nucleus, and electrons move about the nucleus. Each chemical element has its own particular number of protons. Carbon possesses 6 protons; nitrogen, 7; and phosphorus, 15. These schematic diagrams are meant to clearly show and compare numbers of electrons for these three elements. In reality, however, electrons do not orbit the nucleus in rings as shown; they move through space in more complex ways, forming a negatively-charged "cloud" around the nucleus.

of neutrons are **isotopes** (**FIGURE 2.7a**). Isotopes are denoted by their elemental symbol preceded by the mass number, or combined number of protons and neutrons in the atom. For example, ^{12}C (carbon-12), the most abundant carbon isotope, has six protons and six neutrons in the nucleus, whereas ^{14}C (carbon-14) has eight neutrons (and six protons). Because they differ slightly in mass, isotopes of an element differ slightly in their behavior.

Some isotopes, called **radioisotopes,** are **radioactive** and "decay" by changing their chemical identity as they shed subatomic particles and emit high-energy radiation. The radiation released by radioisotopes harms organisms because it focuses a great deal of energy in a very small area, which can be damaging to living cells. Radioisotopes decay into lighter and lighter radioisotopes until they become stable isotopes (isotopes that are not radioactive). Each radioisotope decays at a rate determined by that isotope's **half-life,** the amount of time it takes for one-half the atoms to give off radiation and decay. Different radioisotopes have very different half-lives,

ranging from fractions of a second to billions of years. The radioisotope uranium-235 (^{235}U) is the primary source of energy for commercial nuclear power (pp. 366–371). It decays into a series of daughter isotopes, eventually forming lead-207 (^{207}Pb), and has a half-life of about 700 million years.

Ions Atoms may also gain or lose electrons, thereby becoming **ions,** electrically charged atoms or combinations of atoms (**FIGURE 2.7b**). Ions are denoted by their elemental symbol followed by their ionic charge. For instance, a common ion used by mussels and clams to form shells is Ca^{2+}, a calcium atom that has lost two electrons and thus has a charge of positive 2.

Atoms bond to form molecules and compounds

Atoms bond together because of an attraction for one another's electrons. They can bond together to form **molecules,** combinations of two or more atoms. Some common molecules contain only a single element, such as hydrogen and oxygen, which can be written as "H_2" and "O_2," respectively, using their chemical formulas as a shorthand way to indicate the type and number of atoms in the molecule. A molecule composed of atoms of two or more different elements is called a **compound.** One compound is **water,** which is composed of two hydrogen atoms bonded to one oxygen atom and is denoted by the chemical formula H_2O. Another compound is **carbon dioxide,** consisting of one carbon atom bonded to two oxygen atoms; its chemical formula is CO_2.

Some compounds are made up of ions of differing charge that bind with one another to form **ionic bonds.** A crystal of table salt, sodium chloride (NaCl), is held together by ionic bonds between the positively charged sodium ions (Na^+) and the negatively charged chloride ions (Cl^-). Atoms that lack an electrical charge combine by "sharing" electrons. For example, two atoms of hydrogen bind together to form hydrogen gas (H_2) by sharing their electrons. This type of bond is called a **covalent bond.**

Elements, molecules, and compounds can also come together without chemically bonding, in mixtures called solutions. Air in the atmosphere is a solution formed of constituents such as nitrogen, oxygen, water vapor, and carbon dioxide. Other solutions include ocean water, petroleum, and metal alloys such as brass.

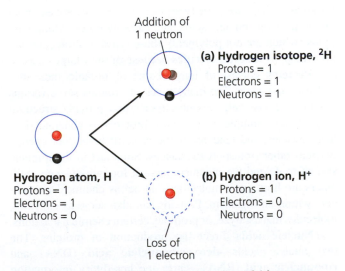

Addition of 1 neutron

(a) Hydrogen isotope, ^2H
Protons = 1
Electrons = 1
Neutrons = 1

Hydrogen atom, H
Protons = 1
Electrons = 1
Neutrons = 0

(b) Hydrogen ion, H^+
Protons = 1
Electrons = 0
Neutrons = 0

Loss of 1 electron

FIGURE 2.7 Hydrogen has a mass number of 1 because a typical atom of this element contains one proton and no neutrons. Deuterium (hydrogen-2 or ^2H), an isotope of hydrogen **(a)**, contains a neutron as well as a proton and thus has greater mass than a typical hydrogen atom; its mass number is 2. The hydrogen ion, H^+ **(b)**, occurs when an electron is lost; it therefore has a positive charge.

Hydrogen ions determine acidity

In any aqueous solution, a small number of water molecules split apart, each forming a hydrogen ion (H^+) and a hydroxide ion (OH^-). The product of hydrogen and hydroxide ion concentrations is always the same; as one increases, the other decreases. Pure water contains equal numbers of these ions. Solutions in which the H^+ concentration is greater than the OH^- concentration are **acidic,** whereas solutions in which the OH^- concentration exceeds the H^+ concentration are **basic,** or alkaline.

The **pH** scale (**FIGURE 2.8**) quantifies the acidity or alkalinity of solutions. It runs from 0 to 14; pure water is neutral, with a hydrogen ion concentration of 10^{-7} and a pH of 7. Solutions with a pH less than 7 are acidic, and those with a pH greater than 7 are basic. The pH scale is logarithmic, so each step on the scale represents a 10-fold difference in hydrogen ion concentration. Thus, a substance with a pH of 6 contains 10 times as many hydrogen ions as a substance with a pH of 7 and 100 times as many hydrogen ions as a substance with a pH of 8.

Most biological systems have a pH between 6 and 8, and substances that are strongly acidic (battery acid) or strongly basic (sodium hydroxide) are harmful to living things. Human activities can change the pH of water or soils and make conditions less amenable to life. Examples include the acidification of soils and water from acid rain (pp. 303–306) and from acidic mine drainage (p. 245).

Matter is composed of organic and inorganic compounds

Beyond their need for water, living things also depend on organic compounds. **Organic compounds** consist of carbon atoms (and generally hydrogen atoms) joined by covalent

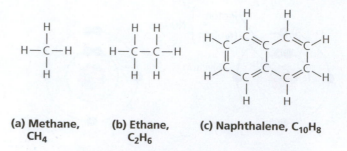

(a) Methane, CH₄   **(b) Ethane, C₂H₆**   **(c) Naphthalene, C₁₀H₈**

FIGURE 2.9 Hydrocarbons have a diversity of chemical structures. The simplest hydrocarbon is methane **(a)**. Many hydrocarbons consist of linear chains of carbon atoms with hydrogen atoms attached; the shortest of these is ethane **(b)**. The air pollutant naphthalene **(c)** is a ringed hydrocarbon.

bonds, and they may also include other elements, such as nitrogen, oxygen, sulfur, and phosphorus. Inorganic compounds, in contrast, lack carbon–carbon bonds.

Carbon's unusual ability to bond together in chains, rings, and other structures to build elaborate molecules has resulted in millions of different organic compounds. One class of such compounds that is important in environmental science is **hydrocarbons,** which consist solely of bonded atoms of carbon and hydrogen (although other elements may enter these compounds as impurities) (**FIGURE 2.9**). Fossil fuels and the many petroleum products we make from them (Chapter 15) consist largely of hydrocarbons.

Macromolecules are building blocks of life

Just as carbon atoms in hydrocarbons may be strung together in chains, organic compounds sometimes combine to form long chains of repeated molecules. These chains are called **polymers.** There are three types of polymers that are essential to life: proteins, nucleic acids, and carbohydrates. Along with lipids (which are not polymers), these types of molecules are referred to as **macromolecules** because of their large sizes.

Proteins consist of long chains of organic molecules called amino acids. The many types of proteins serve various functions. Some help produce tissues and provide structural support; for example, animals use proteins to generate skin, hair, muscles, and tendons. Some proteins help store energy, whereas others transport substances. Some act in the immune system to defend the organism against foreign attackers. Still others are hormones, molecules that act as chemical messengers within an organism. Proteins can also serve as enzymes, molecules that catalyze, or promote, certain chemical reactions.

Nucleic acids direct the production of proteins. The two nucleic acids—**deoxyribonucleic acid (DNA)** and **ribonucleic acid (RNA)**—carry the hereditary information for organisms and are responsible for passing traits from parents to offspring. Nucleic acids are composed of a series of nucleotides, each of which contains a sugar molecule, a phosphate group, and a nitrogenous base. DNA contains four types of nucleotides and can be pictured as a ladder twisted into a spiral, giving the molecule a shape called a double helix

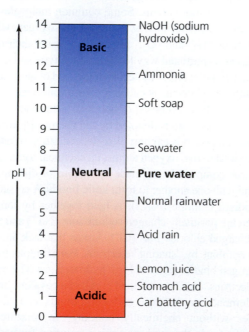

FIGURE 2.8 The pH scale measures how acidic or basic (alkaline) a solution is. The pH of pure water is 7, the midpoint of the scale. Acidic solutions have a pH less than 7, whereas basic solutions have a pH greater than 7.

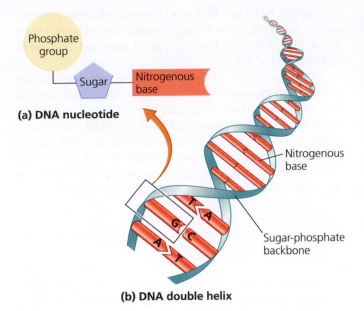

Phosphate group

Sugar

Nitrogenous base

(a) DNA nucleotide

Nitrogenous base

Sugar-phosphate backbone

(b) DNA double helix

FIGURE 2.10 Nucleic acids encode genetic information in the sequence of nucleotides, small molecules that pair together like rungs of a ladder. DNA includes four types of nucleotides **(a)**, each with a different nitrogenous base: adenine (A), guanine (G), cytosine (C), and thymine (T). Adenine (A) pairs with thymine (T), and cytosine (C) pairs with guanine (G). In RNA, thymine is replaced by uracil (U). DNA **(b)** twists into the shape of a double helix.

(**FIGURE 2.10**). Regions of DNA coding for particular proteins that perform particular functions are called **genes.**

Carbohydrates include simple sugars that are three to seven carbon atoms long. Glucose ($C_6H_{12}O_6$) fuels living cells and serves as a building block for complex carbohydrates, such as starch. Plants use starch to store energy, and animals eat plants to acquire starch. Plants and animals also use complex carbohydrates to build structure. Insects and crustaceans form hard shells from the carbohydrate chitin. Cellulose, the most abundant organic compound on Earth, is a complex carbohydrate found in the cell walls of leaves, bark, stems, and roots.

Lipids include fats and oils (for energy storage), phospholipids (for cell membranes), waxes (for structure), and steroids (for hormone production). Although chemically diverse, these compounds are grouped together because they do not dissolve in water.

Energy: An Introduction

Creating and maintaining organized complexity—of a cell, an organism, or an ecological system—requires energy. Energy is needed to organize matter into complex forms, to build and maintain cellular structure, to govern species' interactions, and to drive the geologic forces that shape our planet. Energy is involved in nearly every chemical, biological, and physical phenomenon.

But what is energy? **Energy** is the capacity to change the position, physical composition, or temperature of matter. Scientists differentiate two types of energy: **potential energy,** or the energy of position; and **kinetic energy,** the energy of motion. Consider river water held behind a dam. By preventing water from moving downstream, the dam causes the water to accumulate potential energy. When the dam gates are opened, the potential energy is converted to kinetic energy as the water rushes downstream.

Energy conversions take place at the atomic level every time a chemical bond is broken or formed. **Chemical energy** is essentially potential energy stored in the bonds among atoms. Bonds differ in their amounts of chemical energy, depending on the atoms they hold together. Converting molecules with high-energy bonds (such as the carbon–carbon bonds of fossil fuels) into molecules with lower-energy bonds (such as the bonds in water or carbon dioxide) releases energy and produces motion, action, or heat. Just as automobile engines split the hydrocarbons of gasoline to release chemical energy and generate movement, our bodies split glucose molecules in our food for the same purpose (**FIGURE 2.11**).

Potential energy

Food molecules

Kinetic energy

| $C_6H_{12}O_6$ | + | O_2 | → | CO_2 | + | H_2O | + | Heat |
| Glucose | | Oxygen | | Carbon dioxide | | Water | | |

FIGURE 2.11 Energy is released when potential energy is converted to kinetic energy. Potential energy **(a)** stored in sugars (such as glucose) in the food we eat, combined with oxygen, becomes kinetic energy **(b)** when we exercise, releasing carbon dioxide, water, and heat as by-products.

Besides occurring as chemical energy, potential energy can occur as nuclear energy, the energy that holds atomic nuclei together. Nuclear power plants use this energy when they break apart the nuclei of large atoms within their reactors. Mechanical energy, such as the energy stored in a compressed spring, is yet another type of potential energy. Kinetic energy can also express itself in different forms, including thermal energy, light energy, sound energy, and electrical energy—all of which involve the movement of atoms, subatomic particles, molecules, or objects.

Energy is always conserved, but it changes in quality

Energy can change from one form to another, but it cannot be created or destroyed. Just as matter is conserved, the total energy in the universe remains constant and thus is said to be conserved. Scientists refer to this principle as the **first law of thermodynamics.** The potential energy of the water behind a dam will equal the kinetic energy of its eventual movement downstream. Likewise, we obtain energy from the food we eat and then expend it in exercise, apply it toward maintaining our body and all of its functions, or store it in fat. We do not somehow create additional energy or end up with less energy than the food gives us. Any particular system in nature can temporarily increase or decrease in energy, but the total amount in the universe always remains constant.

Although the overall amount of energy is conserved in any conversion of energy, the **second law of thermodynamics** states that the nature of energy will change from a more-ordered state to a less-ordered state, as long as no force counteracts this tendency. That is, systems tend to move toward increasing disorder, or entropy. For instance, a log of firewood—the highly organized and structurally complex product of many years of tree growth—transforms in a campfire to a residue of carbon ash, smoke, and gases such as carbon dioxide and water vapor, as well as the light and the heat of the flame (**FIGURE 2.12**). With the help of oxygen, the complex biological polymers that make up the wood are converted into a disorganized assortment of rudimentary molecules and heat and light energy. When energy transforms from a more-ordered to less-ordered state, it cannot

accomplish tasks as efficiently. For example, the potential energy available in ash (a less-ordered state of wood) is far lower than that available in a log of firewood (the more-ordered state of wood).

The second law of thermodynamics specifies that systems tend to move toward disorder. How, then, does any system maintain its order? The order of an object or system can be increased by the input of energy from outside the system. Living organisms, for example, maintain their highly ordered structure by consuming energy. When they die and these inputs of energy cease, the organisms undergo decomposition and attain a less-ordered state.

Light energy from the sun powers most living systems

The energy that powers Earth's biological systems comes primarily from the sun. The sun releases radiation across large portions of the electromagnetic spectrum, although our atmosphere filters much of this out and we see only some of this radiation as visible light (**FIGURE 2.13**).

Some organisms use the sun's radiation directly to produce their own food. Such organisms, called **autotrophs,** or primary producers, include green plants, algae, and cyanobacteria. Through the process of **photosynthesis** (**FIGURE 2.14**), autotrophs use sunlight to power a series of chemical reactions that transform molecules with lower-energy bonds—water and carbon dioxide—into sugar molecules with many high-energy bonds. Photosynthesis is an example of a process that moves toward a state of lower entropy, and so requires a substantial input of outside energy, in this case from sunlight.

Photosynthesis occurs within cellular organelles called chloroplasts, where the light-absorbing pigment chlorophyll (the substance that makes plants green) uses solar energy to initiate a series of chemical reactions called ❶ the light reactions. During these reactions, water molecules split, releasing electrons whose energy is used to produce the high-energy molecules ATP (from the addition of a phosphate group to ADP) and NADPH (where a pair of electrons and a hydrogen ion are added to $NADP^+$); ATP and NADPH are then used to fuel reactions in the Calvin cycle. During ❷ the Calvin cycle,

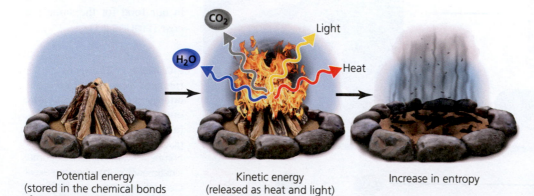

Potential energy
(stored in the chemical bonds
of organic molecules in wood)

Kinetic energy
(released as heat and light)

Increase in entropy

FIGURE 2.12 The burning of firewood demonstrates energy conversion from a more-ordered to a less-ordered state. This increase in entropy reflects the second law of thermodynamics.

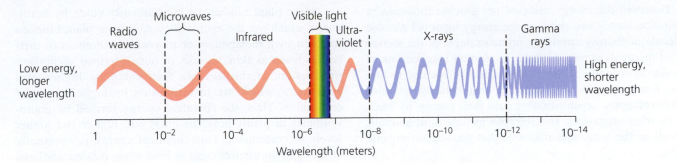

FIGURE 2.13 The sun emits radiation from many portions of the electromagnetic spectrum. Visible light makes up only a small proportion of this energy.

carbon atoms—from the carbon dioxide in air that enters the plant through its leaves—are linked together to produce sugars. Thus in photosynthesis, plants draw up water from the ground through their roots, absorb carbon dioxide from the air through their leaves, and harness the power of sunlight with the light-absorbing pigment chlorophyll. With these ingredients, green plants create sugars for their growth and maintenance, and in turn provide chemical energy to any organism that eats them. Plants also release oxygen as a by-product of photosynthesis, forming the oxygen gas in the air we breathe.

Photosynthesis is a complex process, but the overall reaction can be summarized in the following equation:

$$6\,CO_2 + 6\,H_2O + \text{the sun's energy} \longrightarrow C_6H_{12}O_6 + 6\,O_2 \text{(sugar)}$$

Not all primary production requires sunlight, however. On the deep ocean floor, jets of water heated by magma in the crust gush into the icy-cold depths. These hydrothermal vents can host entire communities of specialized organisms that thrive in the extreme high-temperature, high-pressure conditions.

Hydrothermal vents are so deep underwater that they completely lack sunlight, so the energy flow of these communities cannot be fueled through photosynthesis. Instead, bacteria in deep-sea vents use the chemical-bond energy of hydrogen sulfide (H_2S) to transform inorganic carbon into organic carbon compounds in a process called **chemosynthesis.** Chemosynthesis occurs in various ways, and one way is defined by the following equation:

$$6\,CO_2 + 6\,H_2O + 3\,H_2S \longrightarrow C_6H_{12}O_6 + 3\,H_2SO_4 \text{(sugar)}$$

Energy from chemosynthesis passes through the deep-sea-vent animal community as consumers such as gigantic clams, tubeworms, mussels, fish, and shrimp gain nutrition from chemoautotrophic bacteria and one another.

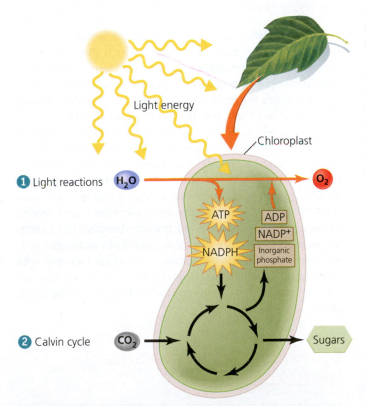

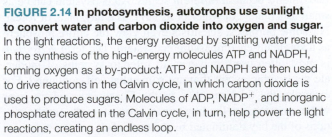

FIGURE 2.14 In photosynthesis, autotrophs use sunlight to convert water and carbon dioxide into oxygen and sugar. In the light reactions, the energy released by splitting water results in the synthesis of the high-energy molecules ATP and NADPH, forming oxygen as a by-product. ATP and NADPH are then used to drive reactions in the Calvin cycle, in which carbon dioxide is used to produce sugars. Molecules of ADP, NADP$^+$, and inorganic phosphate created in the Calvin cycle, in turn, help power the light reactions, creating an endless loop.

Cellular respiration releases chemical energy

Organisms make use of the chemical energy created by photosynthesis in a process called **cellular respiration.** In this process, cells use oxygen to release the chemical energy of glucose, converting it back into its original starting materials: water and carbon dioxide. The energy released during this process is used to power all of the biochemical reactions that sustain life. The net equation for cellular respiration is the exact opposite of that for photosynthesis:

$$C_6H_{12}O_6 + 6\,O_2 \longrightarrow 6\,CO_2 + 6\,H_2O + \text{energy} \text{(sugar)}$$

However, the energy released per glucose molecule in respiration is only two-thirds of the energy input per glucose molecule in photosynthesis—a prime example of the second law of thermodynamics. Cellular respiration is a continuous process occurring in all living things and is essential to life. Thus, it occurs in the autotrophs that create glucose and also in **heterotrophs,** organisms that gain their energy by feeding on other organisms. Heterotrophs include most animals, as well as the fungi and microbes that decompose organic matter.

Ecosystems

Let's now apply our knowledge of chemistry and energy to see how energy, matter, and nutrients move through the living and nonliving environment. An **ecosystem** consists of all organisms and nonliving entities that occur and interact in a particular area at the same time. Animals, plants, water, soil, nutrients—all these and more help compose ecosystems.

The ecosystem concept originated with scientists who recognized that biological entities are tightly intertwined with the chemical and physical aspects of their environment. For instance, in the Chesapeake Bay **estuary**—a water body where rivers flow into the ocean, mixing fresh water with saltwater—aquatic organisms are affected by the flow of water, sediment, and nutrients from the rivers that feed the bay and from the land that feeds those rivers. In turn, the photosynthesis, respiration, and decomposition that these organisms undergo influence the chemical and physical conditions of the Chesapeake's waters.

Ecologists soon began analyzing ecosystems as an engineer might analyze the operation of a machine. In this view, ecosystems are systems that receive inputs of energy, process and transform that energy while cycling chemical nutrients internally, and produce outputs (such as heat, water flow, and animal waste products) that enter other ecosystems.

Energy flows and matter cycles through ecosystems

Energy flows in one direction through ecosystems. As autotrophs, such as green plants and phytoplankton, convert solar energy to the energy of chemical bonds in sugar through the process of photosynthesis, they perform **primary production.** The total amount of chemical energy produced by autotrophs is termed **gross primary production.** Autotrophs use most of this production to power their own metabolism by cellular respiration, releasing heat energy to the environment as a by-product. The energy that remains after respiration and that is used to generate biomass (such as leaves, stems, and roots) is called **net primary production.** Thus, net primary production equals gross primary production minus the energy used in respiration.

Some plant biomass is subsequently eaten by herbivores, which use the energy they gain from plant biomass for their own metabolism or to generate biomass in their bodies (such as skin, muscle, or bone), termed **secondary production.** Herbivores are then eaten by higher-level consumers, which are in turn eaten by higher levels of consumers. Thus, the chemical energy formed by photosynthesis in plants provides energy to higher and higher levels of consumers. This chemical energy is eventually released to the environment as heat when it is metabolized by producers, consumers, or decomposers (**FIGURE 2.15**). Then, when producers and consumers die, their biomass is consumed and metabolized by detritivores and decomposers.

In contrast to chemical energy, nutrients are generally recycled within ecosystems. Energy and nutrients pass among organisms through trophic interactions in food-web relationships (p. 75). Chemical nutrients are recycled because when organisms die and decay, their nutrients remain in the system—unlike the chemical energy that eventually leaves the ecosystem once it is metabolized.

Ecosystems vary in their productivity

Ecosystems differ in the rate at which autotrophs convert energy to biomass. The rate at which this conversion occurs is termed **productivity,** and the energy or biomass that remains in an ecosystem after autotrophs have metabolized enough for their own maintenance through cellular respiration is called **net primary productivity.** Ecosystems whose plants convert solar energy to biomass rapidly are said to have high net primary productivity. Freshwater wetlands, tropical forests, coral reefs, and algal beds tend to have the highest net primary productivities, whereas deserts, tundra, and open ocean tend to have the lowest (**FIGURE 2.16**). Variation among ecosystems and among biomes (see Chapter 4) in net primary productivity results in geographic patterns across the globe. In terrestrial ecosystems, net primary productivity tends to increase with temperature and precipitation. In aquatic ecosystems, net primary productivity tends to rise with light and the availability of nutrients.

Ecosystems interact across landscapes

Ecosystems occur at different scales. An ecosystem can be as small as a puddle of water or as large as a bay, lake, or forest. For some purposes, scientists even view the entire biosphere as a single all-encompassing ecosystem. The term *ecosystem* is most often used, however, to refer to systems of moderate geographic extent that are somewhat self-contained. For example, the tidal marshes in the Chesapeake where river water empties into the bay are an ecosystem, as are the sections of the bay dominated by oyster reefs.

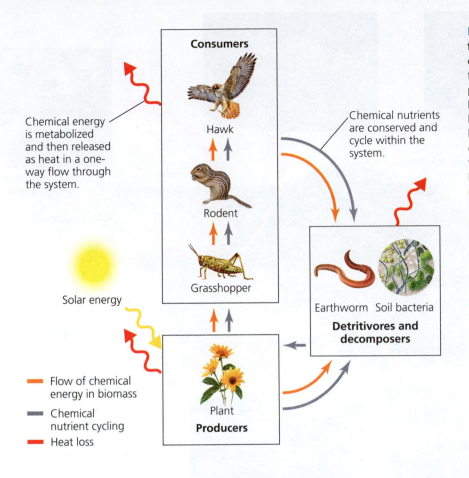

FIGURE 2.15 In systems, energy flows in one direction, whereas chemical nutrients cycle. Light energy from the sun **(yellow arrow)** drives photosynthesis in producers, which begins the transfer of chemical energy in biomass **(orange arrows)** among trophic levels (pp. 73–75) and detritivores and decomposers. Energy exits the system through cellular respiration in the form of heat **(red arrows)**. Chemical nutrients **(gray arrows)** cycle within the system. For simplicity, various abiotic components (such as water, air, and inorganic soil content) of ecosystems have been omitted.

Adjacent ecosystems may share components and interact extensively. Rivers, tidal marshes, and open waters in estuaries all may interact, as do forests and prairie where they converge. Areas where ecosystems meet may consist of transitional zones called **ecotones,** in which elements of each ecosystem mix.

Because components of different ecosystems may intermix, ecologists often find it useful to view these systems on larger geographic scales that encompass multiple ecosystems. In such a broad-scale approach, called **landscape ecology,** scientists study how landscape structure affects the abundance, distribution, and interaction of organisms. Taking a view across the landscape is important in studying birds that migrate long distances, mammals that move seasonally between mountains and valleys, and fish such as salmon that swim upriver from the ocean to reproduce.

For a landscape ecologist, a landscape is made up of **patches** (of ecosystems, communities, or habitat) arrayed

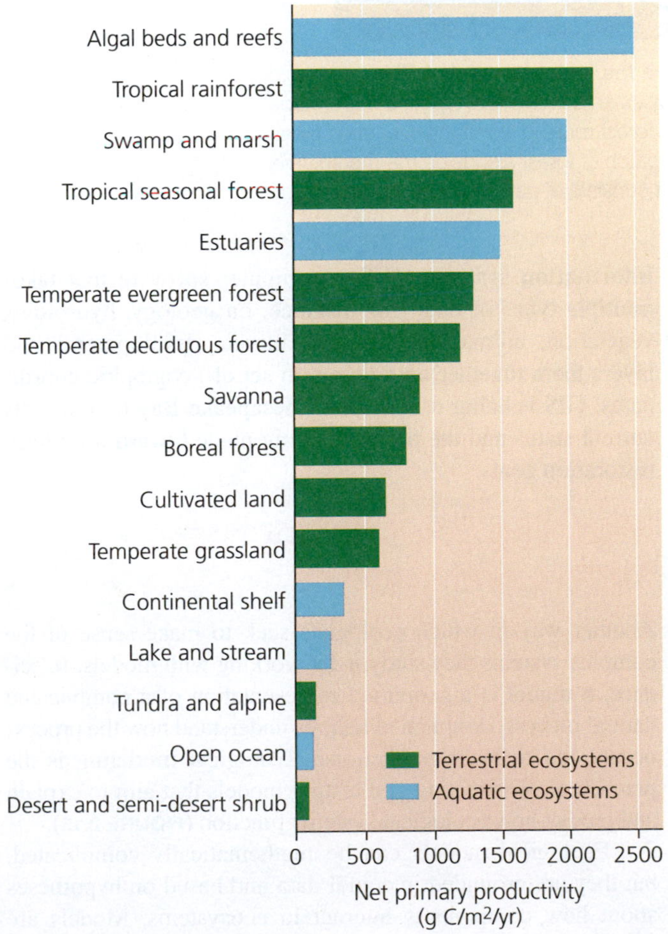

FIGURE 2.16 Net primary productivity varies greatly between ecosystem types. Freshwater wetlands, tropical forests, coral reefs, and algal beds show high values on average, whereas deserts, tundra, and the open ocean show low values. *Data from Whittaker, R.H., 1975.* Communities and ecosystems, 2nd ed. New York, NY: Macmillan.

DATA Q If a farmer in the Amazon basin converts a hectare of tropical rainforest to cultivated land, how many times less productive will that hectare of land be?

Go to **Interpreting Graphs & Data** on **Mastering** Environmental Science

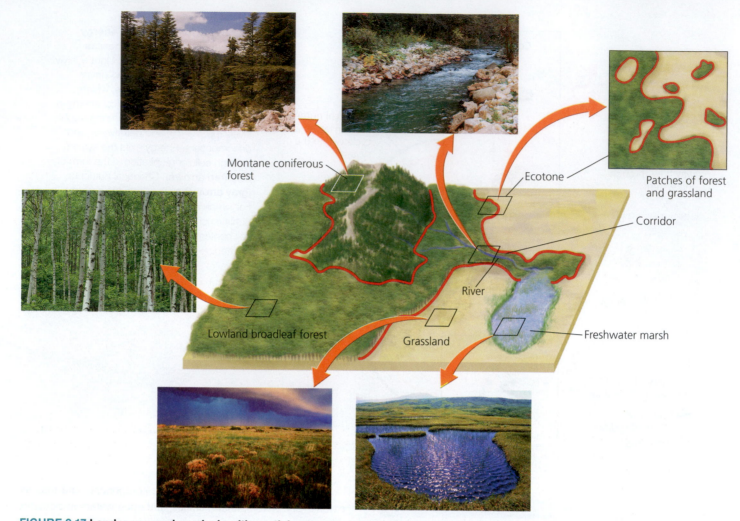

FIGURE 2.17 Landscape ecology deals with spatial patterns above the ecosystem level. This generalized diagram of a landscape shows a mosaic of patches of five ecosystem types (three terrestrial types, a marsh, and a river). Thick red lines indicate ecotones. A stretch of lowland broadleaf forest running along the river serves as a corridor connecting the large region of forest on the left to the smaller patch of forest alongside the marsh. The inset shows a magnified view of the forest-grassland ecotone and how it consists of patches on a smaller scale.

spatially over a landscape in a **mosaic** (**FIGURE 2.17**). Landscape ecology is of great interest to **conservation biologists** (p. 181), scientists who study the loss, protection, and restoration of biodiversity. Populations of organisms have specific habitat requirements and so occupy suitable patches across the landscape. Of particular concern is the fragmentation of habitat into small and isolated patches (p. 205)—something that often results from human development pressures. Establishing corridors of habitat (see Figure 2.17) to link patches and allow animals to move among them is one approach that conservation biologists pursue as they attempt to maintain biodiversity in the face of human impact.

Landscape-level analyses have been greatly aided by satellite imaging and **geographic information systems (GIS)**—computer software that takes multiple types of data (for instance, on geology, hydrology, vegetation, animal species, and human development) and layers them together on a common set of geographic coordinates. GIS is being used in the Chesapeake Bay to assess its current status and the progress being made toward long-term restoration goals.

Modeling helps ecologists understand systems

Another way in which ecologists seek to make sense of the complex systems they study is by working with models. In science, a **model** is a simplified representation of a complicated natural process, designed to help us understand how the process occurs and to make predictions. **Ecological modeling** is the practice of constructing and testing models that aim to explain and predict how ecological systems function (**FIGURE 2.18**).

Ecological models can be mathematically complicated, but they are grounded in actual data and based on hypotheses about how components interact in ecosystems. Models are

weighing the
ISSUES

Ecosystems Where You Live

Think about the area where you live, and briefly describe its ecosystems. How do these systems interact? If one ecosystem were greatly modified (say, if a large apartment complex were built atop a wetland or amid a forest), what impacts on nearby ecosystems might result? (Note: If you live in a city, realize that urban areas can be thought of as ecosystems, too.)

used to make predictions about how large, complicated systems will behave under different conditions. Modeling is a vital pursuit in the scientific study of Earth's changing climate (pp. 317–328), and ecological models are used to predict the responses of fish, crabs, oysters, and underwater grasses to changing water conditions in the Chesapeake Bay.

Ecosystem services sustain our world

Human society depends on healthy, functioning ecosystems. When Earth's ecosystems function normally and undisturbed, they provide goods and services that we could not survive without. As we've seen, we rely not just on natural resources (which can be thought of as goods from nature) but also on the ecosystem services (p. 4) that our planet's systems provide (TABLE 2.1).

Ecological services are the natural processes that humans benefit from, such as the way soil nourishes our crops, estuaries purify the water we drink, insects pollinate the food plants we eat, and bacteria break down some of the waste and pollution we generate. The negative feedback cycles that are typical of ecosystems regulate and stabilize the climate and help to dampen the impacts of disturbances humans create in natural systems.

One of the most important ecosystem services is the cycling of nutrients. Through the processes that take place within and among ecosystems, the chemical elements and compounds that we need—water, carbon, nitrogen, phosphorus, and many more—cycle through our environment in intricate ways.

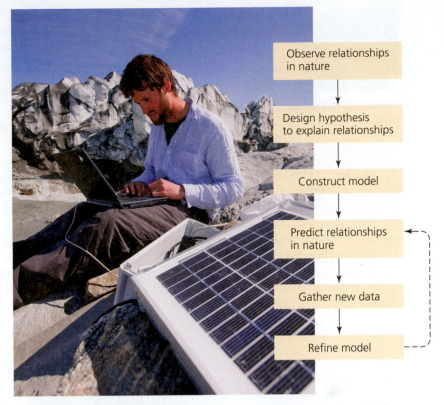

FIGURE 2.18 **Ecological modelers observe relationships among variables in nature and then construct models to explain those relationships and make predictions.** They test and refine the models by gathering new data from nature and seeing how well the models predict those data.

Biogeochemical Cycles

Just as nitrogen and phosphorus from fertilizer on Pennsylvania corn fields end up in Chesapeake Bay oysters, all nutrients move through the environment in intricate ways. As we have discussed, whereas energy enters an ecosystem from the sun, flows from organism to organism, and dissipates to the atmosphere as heat, the physical matter of an ecosystem is circulated over and over again.

Nutrients circulate through ecosystems in biogeochemical cycles

Nutrients move through ecosystems in **nutrient cycles** (or **biogeochemical cycles**) that circulate elements or molecules through the lithosphere, atmosphere, hydrosphere, and biosphere. A carbon atom in your fingernail today might have been in the muscle of a cow a year ago, may have resided in a blade of grass a month before that, and may have been part of a dinosaur's tooth 100 million years ago. After we die, the nutrients in our bodies will disperse into the environment, and could be incorporated into other organisms far into the future.

TABLE 2.1 Ecosystem Services

ECOLOGICAL PROCESSES DO MANY THINGS THAT BENEFIT US:
• Cycle carbon, nitrogen, phosphorus, and other nutrients
• Regulate oxygen, carbon dioxide, stratospheric ozone, and other atmospheric gases
• Regulate temperature and precipitation by means of ocean currents, cloud formation, and so on
• Store and regulate water supplies in watersheds and aquifers
• Form soil by weathering rock, and prevent soil erosion
• Protect against storms, floods, and droughts, mainly by the moderating effects of vegetation
• Filter waste, remove toxic substances, recover nutrients, and control pollution
• Pollinate plants and control crop pests
• Produce fish, game, crops, nuts, and fruits that people eat
• Supply lumber, fuel, metals, fodder, and fiber
• Provide recreation such as ecotourism, fishing, hiking, birding, hunting, and kayaking
• Provide aesthetic, artistic, educational, spiritual, and scientific amenities

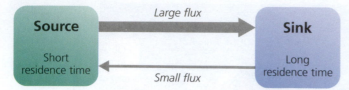

FIGURE 2.19 The main components of a biogeochemical cycle are reservoirs and fluxes. A source releases more materials than it accepts, and a sink accepts more materials than it releases.

Nutrients and other materials move from one **reservoir**, or pool, to another, remaining in each reservoir for varying amounts of time (the **residence time**). The dinosaur, the grass, the cow, and your body are each reservoirs for carbon atoms. The rate at which materials move between reservoirs is termed a **flux**. When a reservoir releases more materials than it accepts, it is called a **source**, and when a reservoir accepts more materials than it releases, it is called a **sink**. **FIGURE 2.19** illustrates these concepts in a simple manner.

As we will see in the following sections, human activities affect the cycling of nutrients by altering fluxes, residence times, and the relative amounts of nutrients in reservoirs.

The water cycle affects all other cycles

Water is so integral to life and to Earth's fundamental processes that we frequently take it for granted. Water is the essential medium for all manner of biochemical reactions, and it plays key roles in nearly every environmental system, including each of the nutrient cycles we are about to discuss. Water carries nutrients, sediments, and pollutants from the continents to the oceans via surface runoff, streams, and rivers. These materials can then be carried thousands of miles on ocean currents. Water also carries atmospheric pollutants to Earth's surface when they dissolve in falling rain or snow. The **hydrologic cycle**, or *water cycle* (**FIGURE 2.20**), summarizes how water—in liquid, gaseous, and solid forms—flows through our environment.

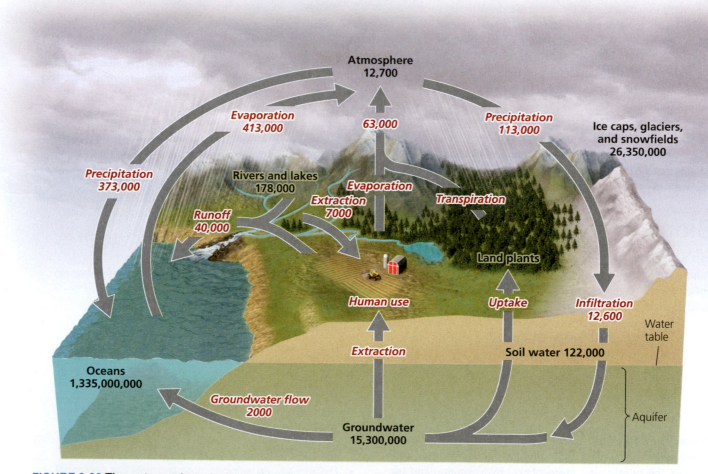

FIGURE 2.20 The water cycle, or hydrologic cycle, summarizes the many routes that water molecules take as they move through the environment. Gray arrows represent fluxes among reservoirs for water. Oceans hold 97% of our planet's water, whereas most fresh water resides in groundwater and ice caps. Water vapor in the atmosphere condenses and falls to the surface as precipitation, then evaporates from land and transpires from plants to return to the atmosphere. Water flows downhill into rivers, eventually reaching the oceans. In the figure, reservoir names are printed in black type, and numbers in black type represent reservoir sizes expressed in units of cubic kilometers (km³). Processes give rise to fluxes, both of which are printed in italic red type and expressed in km³ per year. *Data from Schlesinger, W.H., 2013.* Biogeochemistry: An analysis of global change, *3rd ed. London, England: Academic Press.*

The oceans are the largest reservoir in the hydrologic cycle, holding more than 97% of all water on Earth. The fresh water we depend on for our survival accounts for the remaining water, and two-thirds of this small amount is tied up in glaciers, snowfields, and ice caps (p. 257). Thus, considerably less than 1% of the planet's water is in a form that we can readily use—groundwater, surface fresh water, and rain from atmospheric water vapor.

Water moves from oceans, lakes, ponds, rivers, and moist soil into the atmosphere by **evaporation,** the conversion of a liquid to gaseous form. Water also enters the atmosphere by **transpiration,** the release of water vapor by plants through their leaves, or by evaporation from the surfaces of organisms (such as sweating in humans). Water returns from the atmosphere to Earth's surface as **precipitation** when water vapor condenses and falls as rain or snow. Precipitation may be taken up by plants and used by animals, but much of it flows as runoff (p. 259) into streams, rivers, lakes, ponds, and oceans.

Some water soaks down through soil and rock through a process called infiltration, recharging underground reservoirs known as **aquifers.** Aquifers are porous regions of rock and soil that hold **groundwater,** water found within the soil. The upper limit of groundwater held in an aquifer is referred to as the **water table.** Groundwater becomes surface water when it emerges from springs or flows into streams, rivers, lakes, or the ocean from the soil (p. 258).

Human activity affects every aspect of the water cycle. By damming rivers, we slow the movement of water from the land to the sea, and we increase evaporation by holding water in reservoirs. We remove natural vegetation by clear-cutting and developing land, which increases surface runoff, decreases infiltration and transpiration, and promotes evaporation. Our withdrawals of surface water and groundwater for agriculture, industry, and domestic uses deplete rivers, lakes, and streams and lower water tables. This can lead to water shortages and conflicts over water supplies (p. 275). (We will revisit the water cycle, water resources, and human impacts in more detail in Chapter 12.)

The carbon cycle circulates a vital nutrient

As the definitive component of organic molecules, carbon is an ingredient in carbohydrates, fats, and proteins, and occurs in the bones, cartilage, and shells of all living things. The **carbon cycle** describes the routes that carbon atoms take through the environment (**FIGURE 2.21**). Autotrophs pull carbon dioxide out of the atmosphere and out of surface water to use in photosynthesis. They use some of the carbohydrates from photosynthesis to fuel cellular respiration, thereby releasing some of the carbon back into the atmosphere and oceans as CO_2. When producers are eaten by consumers, which in turn are eaten by other animals, more carbohydrates are broken down in cellular respiration and released as carbon dioxide. The same process occurs as decomposers consume waste and dead organic matter.

The largest reservoir of carbon, sedimentary rock (p. 235), is formed in oceans and freshwater wetlands. When organisms in these habitats die, their remains can settle in sediments, and as layers of sediment accumulate, the older layers are buried more deeply and experience high pressure for long periods. These conditions can convert soft tissues into fossil fuels—coal, oil, and natural gas—and can turn shells and skeletons into sedimentary rock, such as limestone. Although any given carbon atom spends a relatively short time in the atmosphere, carbon trapped in sedimentary rock may reside there for hundreds of millions of years. Carbon trapped in sedimentary rocks and fossil fuel deposits may eventually be released into the oceans or atmosphere by geologic processes such as uplift, erosion, and volcanic eruptions. It also reenters the atmosphere when we extract and burn fossil fuels.

Ocean waters are the second-largest reservoir of carbon on Earth. Oceans absorb carbon-containing compounds from the atmosphere, terrestrial runoff, undersea volcanoes, and the detritus of marine organisms. The rates at which the oceans absorb and release carbon depend on many factors, including temperature and the numbers of marine organisms converting CO_2 into carbohydrates and shells and skeletons.

Human activity affects the carbon cycles through our uses of coal, oil, and natural gas. By combusting fossil fuels, we release carbon dioxide and greatly increase the flux of carbon from the ground to the air. In addition, cutting down forests removes carbon from the pool of vegetation and releases it to the air. And if less vegetation is left on the surface, there are fewer plants to draw CO_2 back out of the atmosphere.

As a result, scientists estimate that today's atmospheric carbon dioxide reservoir is the largest that Earth has experienced in the past 1 million years, and likely in the past 20 million years. The ongoing flux of carbon into the atmosphere is one driving force behind today's anthropogenic global climate change (Chapter 14).

Some of the excess CO_2 in the atmosphere is now being absorbed by ocean water. This is causing ocean water to become more acidic, leading to problems that threaten many marine organisms (pp. 283–284).

Our understanding of the carbon cycle is not yet complete. Scientists remain baffled by the so-called missing carbon sink. Of the carbon dioxide we emit by fossil fuel combustion and deforestation, researchers have measured how much goes into the atmosphere and oceans, but there remain roughly 2.3–2.6 billion metric tons unaccounted for. Many scientists think this CO_2 is probably taken up by plants or soils of the temperate and boreal forests (pp. 85–89). They'd like to know for sure, though, because if certain forests are acting as a major sink for carbon, conserving these ecosystems is particularly vital.

The nitrogen cycle involves specialized bacteria

Nitrogen makes up 78% of our atmosphere by mass and is the sixth most abundant element on Earth. It is an essential ingredient in proteins, DNA, and RNA and, like phosphorus, is an

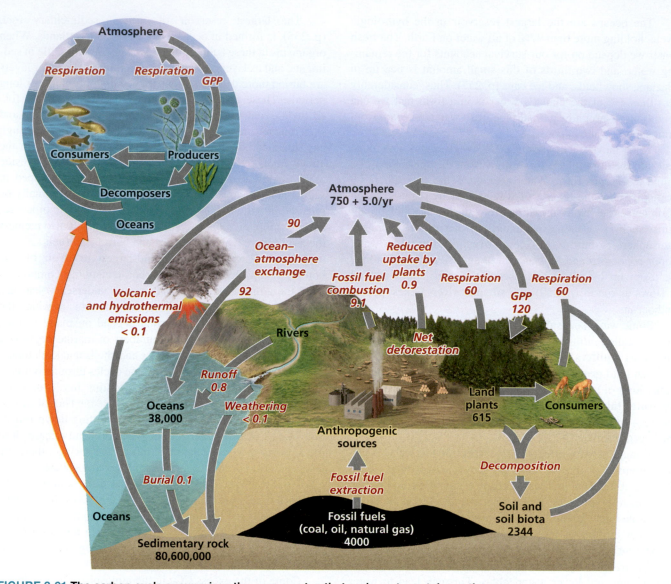

FIGURE 2.21 The carbon cycle summarizes the many routes that carbon atoms take as they move through the environment. Gray arrows represent fluxes among reservoirs for carbon. In the carbon cycle, plants use carbon dioxide from the atmosphere for photosynthesis (gross primary production, or "GPP" in the figure). Carbon dioxide is returned to the atmosphere through cellular respiration by plants, their consumers, and decomposers. The oceans sequester carbon in their water and in deep sediments. The vast majority of the planet's carbon is stored in sedimentary rock. In the figure, reservoir names are printed in black type, and numbers in black type represent reservoir sizes expressed in petagrams (units of 10^{15} g) of carbon. Processes give rise to fluxes, both of which are printed in italic red type and expressed in petagrams of carbon per year.
Data from Schlesinger, W.H., 2013. Biogeochemistry: An analysis of global change, *3rd ed. London, England: Academic Press.*

essential nutrient for plant growth. Thus the **nitrogen cycle** (**FIGURE 2.22**) is of vital importance to all organisms. Despite its abundance in the air, nitrogen gas (N_2) is chemically inert and cannot cycle out of the atmosphere and into living organisms without assistance from lightning, highly specialized bacteria, or human intervention. However, once nitrogen undergoes the right kind of chemical change, it becomes biologically active and available to the organisms that need it, and can act as a potent fertilizer.

To become biologically available, inert nitrogen gas (N_2) must be "fixed," or combined with hydrogen in nature to form ammonia (NH_3), whose water-soluble ions of ammonium (NH_4^+) can be taken up by plants. **Nitrogen fixation** can be

accomplished in two ways: by the intense energy of lightning strikes or by particular types of **nitrogen-fixing bacteria** that inhabit the top layer of soil. These bacteria live in a mutualistic relationship (p. 73) with many types of plants, including soybeans and other legumes, providing them nutrients by converting nitrogen to a usable form. Other types of bacteria then perform a process known as **nitrification,** converting ammonium ions first into nitrite ions (NO_2^-), then into nitrate ions (NO_3^-). Plants can take up these ions, which also become available after atmospheric deposition on soils or in water or after application of nitrate-based fertilizer.

Animals obtain the nitrogen they need by consuming plants or other animals. Decomposers obtain nitrogen from

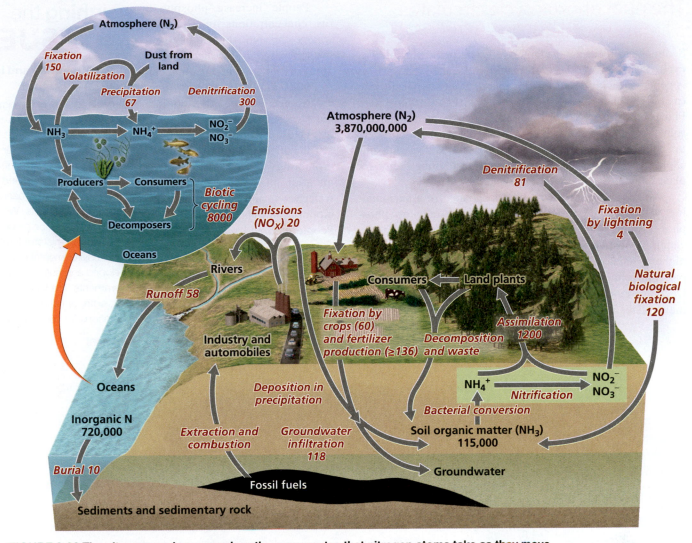

FIGURE 2.22 The nitrogen cycle summarizes the many routes that nitrogen atoms take as they move through the environment. Gray arrows represent fluxes among reservoirs for nitrogen. In the nitrogen cycle, specialized bacteria play key roles in "fixing" atmospheric nitrogen and converting it to chemical forms that plants can use. Other types of bacteria convert nitrogen compounds back to the atmospheric gas, N_2. In the oceans, inorganic nitrogen is buried in sediments, whereas nitrogen compounds are cycled through food webs as they are on land. In the figure, reservoir names are printed in black type, and numbers in black type represent reservoir sizes expressed in teragrams (units of 10^{12} g) of nitrogen. Processes give rise to fluxes, both of which are printed in italic red type and expressed in teragrams of nitrogen per year. *Data from Schlesinger, W.H., 2013. Biogeochemistry: An analysis of global change, 3rd ed. London, England: Academic Press.*

dead and decaying plant and animal matter, and from the urine and feces of animals. Once decomposers process the nitrogen-rich compounds, they release ammonium ions, making these available to nitrifying bacteria to convert again to nitrates and nitrites. The next step in the nitrogen cycle occurs when **denitrifying bacteria** convert nitrates in soil or water to gaseous nitrogen. Denitrification thereby completes the cycle by releasing nitrogen back into the atmosphere as a gas.

Historically, nitrogen fixation was a **bottleneck,** a step that limited the flux of nitrogen out of the atmosphere and into water-soluble forms. Once people discovered how to fix nitrogen on massive scales, a process called industrial fixation, we accelerated its flux into other reservoirs. Today,

our species is fixing at least as much nitrogen artificially as is being fixed naturally, and we are overwhelming nature's denitrification abilities.

While the impacts of nitrogen runoff have become painfully evident to oystermen and scientists in the Chesapeake Bay, hypoxia in waters is by no means the only human impact on the nitrogen cycle. Oddly enough, the overapplication of nitrogen-based fertilizers can strip the soil of other vital nutrients, such as calcium and potassium, thereby reducing soil fertility. Additionally, burning fossil fuels, forests, or fields generates nitrogenous compounds in the atmosphere that act as greenhouse gases (p. 314), cause acid deposition (p. 303), promote eutrophication, and contribute to photochemical smog (p. 296).

The phosphorus cycle circulates a limited nutrient

Sedimentary rocks are the largest reservoir in the **phosphorus cycle** (FIGURE 2.23). The vast majority of Earth's phosphorus is contained within rocks and is released only by weathering (p. 144), which releases phosphate ions (PO_4^{3-}) into water. Phosphates dissolved in lakes or in the oceans precipitate into solid form, settle to the bottom, and reenter the lithosphere in sediments. The scarcity of phosphorus in waters and soils explains why phosphorus is frequently a limiting factor for plant growth.

Aquatic producers take up phosphates from surrounding waters, whereas terrestrial producers take up phosphorus from soil water through their roots. Herbivores acquire phosphorus from plant tissues and pass it on to higher predators when they are consumed. Animals also pass phosphorus to the soil through the excretion of waste. Decomposers break down phosphorus-rich organisms and their wastes and, in so doing, return phosphorus to the soil.

People increase phosphorus concentrations in surface waters through runoff of the phosphorus-rich fertilizers we apply to lawns and farmlands. A 2008 study determined that an average hectare of land in the Chesapeake Bay region received a net input of 4.52 kg (10 lb) of phosphorus per year, promoting phosphorus accumulation in soils, runoff into waterways, and phytoplankton blooms and hypoxia in the bay. People also add phosphorus to waterways through releases of treated wastewater rich in phosphates from the detergents we use to wash our clothes and dishes.

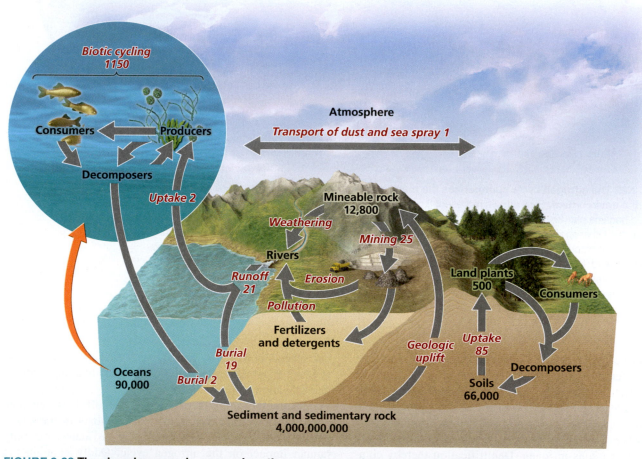

FIGURE 2.23 The phosphorus cycle summarizes the many routes that phosphorus atoms take as they move through the environment. Gray arrows represent fluxes among reservoirs for phosphorus. Most phosphorus resides underground in rock and sediment. Rocks containing phosphorus are uplifted geologically and slowly weathered away. Small amounts of phosphorus cycle through food webs, where this nutrient is often a limiting factor for plant growth. In the figure, reservoir names are printed in black type, and numbers in black type represent reservoir sizes expressed in teragrams (units of 10^{12} g) of phosphorus. Processes give rise to fluxes, both of which are printed in italic red type and expressed in teragrams of phosphorus per year. *Data from Schlesinger, W.H., 2013. Biogeochemistry: An analysis of global change, 3rd ed. London, England: Academic Press.*

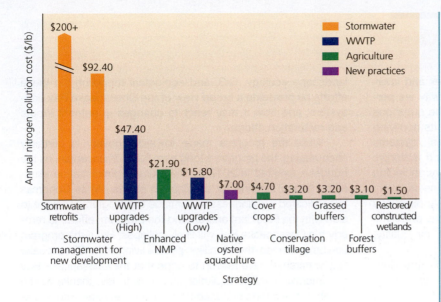

FIGURE 2.24 Costs for reducing nitrogen inputs into the Chesapeake Bay vary widely. Approaches that slow runoff to waterways avoid nitrogen inputs for a few dollars per pound, whereas upgrades to wastewater treatment plants (WWTP), enhanced nutrient management plans (NMP—careful regulation of nutrient applications), and stormwater upgrades can be considerably more expensive. *Data from Jones, C., et al., 2010. How nutrient trading could help restore the Chesapeake Bay.* WRI Working Paper. *Washington D.C.: World Resources Institute.*

DATA For what it costs to remove 1 pound of nitrogen by using enhanced nutrient management programs (NMP), how many pounds of nitrogen could be kept out of waterways by planting forested buffers around streams instead?

Go to **Interpreting Graphs & Data** on **Mastering** Environmental Science

Tackling nutrient enrichment requires diverse approaches

Given our reliance on synthetic fertilizers for food production and fossil fuels for energy, nutrient enrichment of ecosystems will pose a challenge for many years to come. Fortunately, a number of approaches are available to control nutrient pollution in the Chesapeake Bay and other waterways affected by eutrophication, including:

- Reducing fertilizer use on farms and lawns and timing its application to reduce water runoff
- Planting and maintaining vegetation "buffers" around streams to trap nutrient and sediment runoff

- Using natural and constructed wetlands (p. 280) to filter stormwater and farmland runoff
- Improving technologies in sewage treatment plants to enhance nitrogen and phosphorus capture
- Upgrading stormwater systems to capture runoff from roads and parking lots
- Reducing fossil fuel combustion to minimize atmospheric inputs of nitrogen to waterways

Some of these methods cost more than others for similar results. For example, planting vegetation buffers and restoring wetlands can reduce nutrient inputs into waterways at a fraction of the cost of some other approaches, such as upgrading wastewater treatment plants (**FIGURE 2.24**).

SUCCESS STORY | **Considering Cost When Saving the Bay**

The federal government and the states around the Chesapeake Bay are now managing the bay as a holistic system—an approach that, at last, offers prospects for recovery. This systemic approach is showing early signs of success. The Chesapeake Bay Foundation's most recent "State of the Bay" report (published in 2016) concluded that the bay's health rating in 2016 was the highest it had been since CBF's founding in 1964, with meaningful improvements in pollution reduction, fisheries recovery, and the restoration of natural habitats in and around the bay.

A forested buffer lining a waterway in Maryland.

One reason for the recent success is due to farmers, residents, resource managers, and local, state, and federal government agencies embracing a variety of approaches to reduce nutrient inputs into the bay. Some of these methods cost more than others for similar results. For example, planting vegetation buffers and restoring wetlands can reduce nutrient inputs into waterways at a fraction of the cost of some other approaches, such as upgrading wastewater treatment plants. By educating people about the many inexpensive yet effective steps that can be taken in yards, farms, businesses, and local communities to reduce nutrient inputs into the Chesapeake Bay, saving the bay became something for which everyone can do his or her part.

EXPLORE THE DATA at Mastering Environmental Science

closing THE LOOP

Earth hosts many complex and interacting systems, and the way we perceive them depends on the questions we ask. Life interacts with its nonliving environment in ecosystems, systems through which energy flows and materials are recycled. Energy and chemistry, including the cycles that circulate vital nutrients, are tied to nearly every environmental issue examined in this textbook. Applications of chemistry can provide solutions to environmental problems involving agricultural practices, water resources, air quality, energy policy, and environmental health.

The Chesapeake Bay provides a case study that illustrates the importance of understanding systems, chemistry, and the need for taking a systems-level approach to restore ecosystems degraded by human activities. Tools such as landscape ecology, GIS, and ecological modeling aid these efforts by providing a broad view of the Chesapeake Bay ecosystem, and how it may react to changes in nutrient inputs and restoration efforts.

While the progress made toward recovery is certainly encouraging, the program's long term future is uncertain, as the budget submitted in 2017 by the Trump administration eliminated federal funding for its cleanup efforts. While Congress has indicated a willingness to continue the program, if it were to side with the President and defund the program, efforts to remedy the Chesapeake Bay will likely collapse. But if the program can be continued, the 18 million people living in the Chesapeake Bay watershed have reason to hope that the Chesapeake Bay of tomorrow may be healthier than it is today, thanks to the collaborative efforts of concerned citizens, advocacy organizations, and the federal and bay-state governments.

TESTING Your Comprehension

1. Which type of feedback loop is more common in nature, and which more commonly results from human action? For either type of feedback loop, provide an example that was not mentioned in the text.

2. Describe how hypoxic conditions can develop in aquatic ecosystems such as the Chesapeake Bay.

3. Differentiate an ion from an isotope.

4. Describe the two major forms of energy and give examples of each. Compare and contrast the first law of thermodynamics with the second law of thermodynamics.

5. What substances are produced by the process of photosynthesis? By cellular respiration?

6. Compare and contrast the typical movements of energy and matter through an ecosystem.

7. List five ecosystem services provided by functioning ecosystems, and rank them according to your perceived value of each.

8. What role does each of the following play in the carbon cycle?

   - Photosynthesis
   - Automobiles
   - The oceans
   - Earth's crust

9. Distinguish the function performed by nitrogen-fixing bacteria from that performed by denitrifying bacteria.

10. How has human activity altered the hydrologic cycle? The carbon cycle? The phosphorus cycle? The nitrogen cycle? What environmental problems have been produced by these alterations?

SEEKING Solutions

1. Can you think of an example of an environmental problem *not* mentioned in this chapter that a good knowledge of chemistry could help us solve? Explain your answer.

2. Consider the ecosystem(s) that surround(s) your campus. How is each affected by human activities?

3. For a conservation biologist interested in sustaining populations of the organisms that follow, why would it be helpful to take a landscape ecology perspective? Explain your answer in each case.

   - A forest-breeding warbler that suffers poor nesting success in small, fragmented forest patches
   - A bighorn sheep that must move seasonally between mountains and lowlands
   - A toad that lives in upland areas but travels cross-country to breed in localized pools each spring

4. **CASE STUDY CONNECTION** Suppose you are a Pennsylvania farmer who has learned that the government is offering incentives to farmers to help reduce fertilizer runoff into the Chesapeake Bay. What

types of approaches described in this chapter might you be willing to try, and why?

5. **THINK IT THROUGH** You are an oysterman in the Chesapeake Bay, and your income is decreasing because the dead zone is making it harder to harvest oysters. One day your senator comes to town, and you have a one-minute audience with her. What steps would you urge her to take in Washington, D.C., to try to help alleviate the dead zone and bring back the oyster fishery?

CALCULATING Ecological Footprints

The second law of thermodynamics has profound implications for human impacts on the environment, as it affects the efficiency with which we produce our food. In ecological systems, a rough rule of thumb is that when energy is transferred from plants to plant-eaters or from prey to predator, the efficiency is only about 10% (p. 74). Much of this inefficiency is a consequence of the second law of thermodynamics. Another way to think of this is that eating 1 Calorie of meat from an animal is the ecological equivalent of eating 10 Calories of plant material. So, when we raise animals for meat using grain, it is less energetically efficient than if we ate the grain directly.

Humans are considered omnivores because we can eat both plants and animals. The choices we make about what to eat have significant ecological consequences. With this in mind, calculate the ecological energy requirements for four different diets, each of which provides a total of 2000 dietary Calories per day.

DIET	SOURCE OF CALORIES	NUMBER OF CALORIES CONSUMED	ECOLOGICALLY EQUIVALENT CALORIES	TOTAL ECOLOGICALLY EQUIVALENT CALORIES
100% plant	Plant			
0% animal	Animal			
90% plant	Plant	1800	1800	3800
10% animal	Animal	200	2000	
50% plant	Plant			
50% animal	Animal			
0% plant	Plant			
100% animal	Animal			

1. How many ecologically equivalent Calories would it take to support you for a year for each of the four diets listed?

2. How does the ecological impact from a diet consisting strictly of animal products (e.g., dairy products, eggs, and meat) compare with that of a strictly vegetarian diet? How many additional ecologically equivalent Calories do you consume each day by including as little as 10% of your Calories from animal sources?

3. What percentages of the Calories in your own diet do you think come from plant versus animal sources? Estimate the ecological impact of your diet, relative to a strictly vegetarian one.

4. List the major factors influencing your current diet (e.g., financial considerations, convenience, access to groceries, taste preferences). Do you envision your diet's distribution of plant and animal Calories changing in the near future? Why or why not?

Mastering Environmental Science

Students Go to **Mastering Environmental Science** for assignments, the etext, and the Study Area with practice tests, videos, current events, and activities.

Instructors Go to **Mastering Environmental Science** for automatically graded activities, current events, videos, and reading questions that you can assign to your students, plus Instructor Resources.

CHAPTER 3

Evolution, Biodiversity, and Population Ecology

Saving Hawaii's Native Forest Birds

HAWAI`I

Pacific Ocean

> **When an entire island avifauna . . . is devastated almost overnight because of human meddling, it is, quite simply, a tragedy.**
> —H. Douglas Pratt, ornithologist and expert on Hawaiian birds

> **To keep every cog and wheel is the first precaution of intelligent tinkering.**
> —Aldo Leopold

Jack Jeffrey stopped in his tracks. "I hear one!" he said. "Over there in those trees!"

Jeffrey led his group of ecotourists through a lush and misty woodland of ferns, shrubs, and vines toward an emphatic chirping sound. They ducked under twisting gnarled limbs covered with moss and lichens, beneath stately ancient 'ōhi'a-lehua trees offering bright red flowers loaded with nectar and pollen. At last, in the branches of a koa tree, they spotted the bird—an 'akiapōlā'au, one of fewer than 1500 of its kind left alive in the world.

The 'akiapōlā'au (or "aki" for short) is a sparrow-sized wonder of nature—one of many exquisite birds that evolved on the Hawaiian Islands and exists only here (see photo below). For millions of years, this chain of islands in the middle of the Pacific Ocean has acted as a cradle of evolution, generating new and unique species. Yet today many of these species are going from the cradle to the grave. Half of Hawaii's native bird species (70 of 140) have gone extinct in recent times, and many of those that remain—like the aki—teeter on the brink of extinction.

The aki is a type of Hawaiian honeycreeper. The Hawaiian honeycreepers include 18 living species (and at least 38 species recently extinct), all of which originated from an ancestral species that reached Hawai'i several million years ago. As new volcanic islands emerged from the ocean and then eroded away, and as forests expanded and contracted over millennia, populations were split and new honeycreeper species evolved.

As honeycreeper species diverged from their common ancestor and from one another, they evolved different colors, sizes, body shapes, feeding behaviors, mating preferences, diets, and bill shapes. Bills in some species became short and straight, allowing birds to glean insects from leaves. In other species, bills became long and curved, enabling birds to probe into flowers to sip nectar. The bills of still other species became thick and strong for cracking seeds. Some birds evolved highly specialized bills: The aki uses the short, straight lower half of its bill to peck into dead branches to find beetle grubs, then uses the long, curved upper half to reach in and extract them.

Hawaii's honeycreepers thrived for several million years in the islands' forests, amid a unique community of plants found nowhere else in the world. Yet today these native Hawaiian forests are under siege. The crisis began 750 or more years ago as Polynesian settlers colonized the islands, cutting down trees and introducing non-native animals. Europeans arrived more than 200 years ago and did more of the same. Pigs, goats, and cattle ate their way through the

The endangered 'akiapōlā'au ▲

Upon completing this chapter, you will be able to:

- **Explain natural selection and cite evidence for this process**

- **Describe how evolution generates and shapes biodiversity**

- **Discuss the factors behind species extinction and identify Earth's known mass extinction events**

- **List the levels of ecological organization**

- **Outline the characteristics of populations that help predict population growth**

- **Explain how logistic growth, limiting factors, carrying capacity, and other fundamental concepts affect population ecology**

- **Identify and discuss challenges and current efforts in conserving biodiversity**

Native Hawaiian forest at Hakalau Forest National Wildlife Refuge

native plants, transforming luxuriant forests into ragged grasslands. Rats, cats, dogs, and mongooses destroyed the eggs and young of native birds. Foreign plants from Asia, Europe, and America, whose seeds accompanied the people and animals, spread across the altered landscape.

Foreign diseases also arrived, including strains of pox and malaria that target birds. The native fauna were not adapted to resist pathogens they had never encountered. Avian pox and avian malaria, carried by introduced mosquitoes, killed off native birds everywhere except on high mountain slopes, where it was too cold for mosquitoes. Today few native forest birds exist anywhere on the Hawaiian Islands below 1500 m (4500 ft) in elevation.

The aki being watched by Jeffrey's group inhabits the Hakalau Forest National Wildlife Refuge, high atop the slopes of Mauna Kea, a volcano on the island of Hawai'i, the largest island in the chain (**FIGURE 3.1**). At Hakalau, native birds find a rare remaining patch of disease-free native forest.

Jeffrey was a biologist at Hakalau for 20 years before his retirement, and led innovative projects to save native plants and birds from extinction. Staff and volunteers at Hakalau fenced out pigs and planted thousands of native plants in areas deforested by cattle grazing. Young restored native forest is now regrowing on thousands of acres. More birds are using this restored forest every year.

However, today global climate change is presenting new challenges. As temperatures climb, mosquitoes move upslope, and malaria and pox spread deeper into the remaining forests,

FIGURE 3.1 The Hakalau Forest National Wildlife Refuge lies on the slopes of Mauna Kea on the island of Hawai'i.

so that even protected areas such as Hakalau are not immune. The next generation of managers will need to innovate further to fend off extinction for the island's native species.

Plenty of challenges remain, but the successes at Hakalau Forest so far provide hope that through responsible management we can restore Hawaii's native flora and fauna, prevent further impacts, and preserve the priceless bounty of millions of years of evolution on this extraordinary chain of islands.

Evolution: The Source of Earth's Biodiversity

The animals and plants native to the Hawaiian Islands help reveal how our world became populated with the remarkable diversity of life we see today—a lush cornucopia of millions of species (**FIGURE 3.2**).

A **species** is a particular type of organism. More precisely, it is a population or group of populations whose members share characteristics and can freely breed with one another and produce fertile offspring. A **population** is a group of individuals of a given species that live in a particular region at a particular time. Over vast spans of time, the process of evolution has shaped populations and species, giving us the vibrant abundance of life that enriches Earth today.

In its broad sense, the term *evolution* means change over time. In its biological sense, **evolution** consists of change in populations of organisms from generation to generation. Changes in genes (p. 33) often lead to modifications in appearance or behavior.

Evolution is one of the best-supported and most illuminating concepts in all of science, and it is the very foundation of modern biology. Perceiving how species adapt to their environments and change over time is crucial for comprehending ecology and for learning the history of life. Evolutionary processes influence many aspects of environmental science, including agriculture, pesticide resistance, medicine, and environmental health.

Natural selection shapes organisms

Natural selection is a primary mechanism of evolution. In the process of **natural selection,** inherited characteristics that enhance survival and reproduction are passed on more frequently to future generations than characteristics that do not, thereby altering the genetic makeup of populations through time.

Natural selection is a simple concept that offers a powerful explanation for patterns evident in nature. The idea of natural selection follows logically from a few straightforward facts that are readily apparent to anyone who observes the life around us:

- Organisms face a constant struggle to survive and reproduce.
- Organisms tend to produce more offspring than can survive to maturity.
- Individuals of a species vary in their attributes.

Variation is due to differences in genes, the environments in which genes are expressed, and the interactions between genes and environment. As a result of this variation, some individuals of a species will be better suited to their environment than others and will be better able to reproduce.

Attributes are passed from parent to offspring through the genes, and a parent that produces many offspring will pass on more genes to the next generation than a parent that produces few or no offspring. In the next generation, therefore, the genes of better-adapted individuals will outnumber those of individuals that are less well adapted. From one generation to another through time, characteristics, or traits, that lead

to better and better reproductive success in a given environment will evolve in the population. This process is termed **adaptation,** and a trait that promotes reproductive success is also called an *adaptation* or an *adaptive trait.*

The concept of natural selection was first proposed in the 1850s by **Charles Darwin** and, independently, by **Alfred Russel Wallace,** two exceptionally keen British naturalists. By this time, scientists and amateur naturalists were widely discussing the idea that populations evolve, yet no one could say how or why. After spending years studying and cataloguing an immense variety of natural phenomena—in his English garden and across the world to the Galápagos Islands—Darwin finally concluded that natural selection helped explain the world's great variety of living things. Once he came to this conclusion, however, Darwin put off publishing his findings, fearing the social disruption that might ensue if people felt their religious convictions were threatened. Darwin was at last driven to go public when Wallace wrote to him from the Asian tropics, independently describing the idea of natural selection. The two men's shared ideas were presented together at a scientific meeting in 1858, and the next year Darwin published his groundbreaking book, *On the Origin of Species.*

With natural selection, humanity at last uncovered a precise and viable mechanism to explain how and why organisms evolve through time. Once geneticists worked out how traits are inherited, this understanding launched evolutionary biology. In the century and a half since Darwin and Wallace, legions of researchers have refined our understanding of evolution, powering dazzling progress in biology that has helped shape our society.

Understanding evolution is vital for modern society

Evolutionary processes play key roles in today's society and in our everyday lives. As we will see shortly, we depend on a working knowledge of evolution for the food we eat and the clothes we wear, each and every day, as these have been made possible by the selective breeding of crops and livestock. Applying an understanding of evolution to agriculture can also help us avoid antibiotic resistance in feedlots and pesticide resistance in crop-eating insects (p. 152).

Medical advances result from our knowledge of evolution, as well. Understanding evolution helps us determine how infectious diseases spread and how they gain or lose potency. It allows scientists to track the constantly evolving strains of influenza (flu), HIV (human immunodeficiency virus, which causes AIDS), and other pathogens. Armed with such information, biomedical experts can predict which flu strains will most likely spread in a given year and then design effective vaccines targeting them. Comprehending evolution also enables us to detect and respond to the evolution of antibiotic resistance by dangerous bacteria.

Additionally, applying our knowledge of evolution informs our technology. From studying how organisms adapt to challenges and evolve new abilities, we develop ideas on how to design technologies and engineer solutions.

Selection acts on genetic variation

For an organism to pass a trait along to future generations, genes in its DNA (p. 32) must code for the trait. In an organism's lifetime, its DNA will be copied millions of times by millions of cells. Amid all this copying, sometimes a mistake

FAQ

Isn't evolution based on just one man's beliefs?

Because Charles Darwin contributed so much to our early understanding of evolution, many people assume the concept itself hinges on his ideas. But scientists and laypeople had been observing nature, puzzling over fossils, and discussing the notion of evolution long before Darwin. Once he and Alfred Russel Wallace independently proposed the concept of natural selection, scientists finally gained a way of explaining how and why organisms change across generations. Later, geneticists discovered Gregor Mendel's research on inheritance and worked out how traits are passed on—and modern evolutionary biology was born. Twentieth-century scientists Ronald Fisher, Sewall Wright, Theodosius Dobzhansky, George Gaylord Simpson, Ernst Mayr, and others ran experiments and developed sophisticated mathematical models, documenting phenomena with extensive evidence and building evolutionary biology into one of science's strongest fields. Since then, evolutionary research by thousands of scientists has driven our understanding of biology and has facilitated spectacular advances in agriculture, medicine, and technology.

FIGURE 3.2 Hawai'i hosts a treasure trove of biodiversity, including the ① happyface spider, ② 'i'iwi, ③ nēnē, and ④ Haleakala silversword.

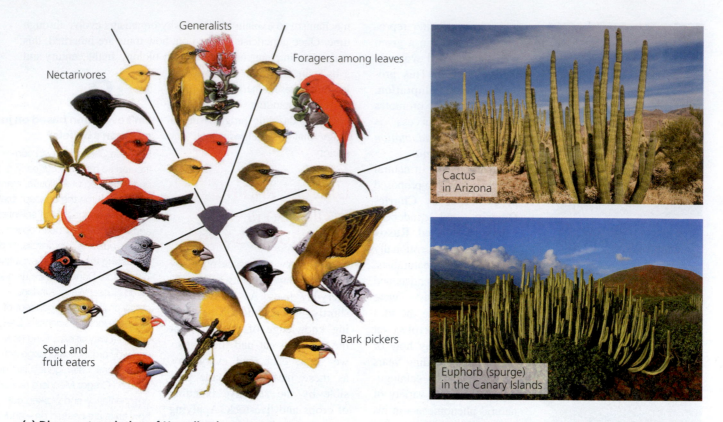

Generalists

Foragers among leaves

Nectarivores

Seed and
fruit eaters

Bark pickers

(a) Divergent evolution of Hawaiian honeycreepers

Cactus
in Arizona

Euphorb (spurge)
in the Canary Islands

(b) Convergent evolution of cactus and spurge

FIGURE 3.3 Natural selection can cause closely related species to diverge or distantly related species to converge. Hawaiian honeycreepers **(a)** diversified as they adapted to different food resources and habitats, as indicated by the diversity of their plumage colors and bill shapes. In contrast, cacti of the Americas and euphorbs of Africa **(b)** became similar to one another as they independently adapted to arid environments. These plants each evolved succulent tissues to hold water, thorns to keep thirsty animals away, and photosynthetic stems without leaves to reduce surface area and water loss.

is made. Accidental changes in DNA, called **mutations,** give rise to genetic variation among individuals. If a mutation occurs in a sperm or egg cell, it may be passed on to the next generation. Most mutations have little effect, but some can be deadly, and others can be beneficial. Those that are not lethal provide the genetic variation on which natural selection acts.

Genetic variation also results as organisms mix their genetic material through sexual reproduction. When organisms reproduce sexually, each parent contributes to the genes of the offspring, producing novel combinations of genes and generating variation among individuals.

Genetic variation can help populations adapt to changing environmental conditions. For example, one of the honeycreeper species of the Hakalau Forest, the 'amakihi, has recently been discovered in 'ōhi'a trees at low elevations where avian malaria has killed off other honeycreepers. Researchers determined that some of the 'amakihis living there when malaria arrived had genes that by chance gave them a natural resistance to the disease. These resistant birds survived malaria's onslaught, and their descendants that carried the malaria-resistant genes reestablished a population that is growing today.

Environmental conditions determine the pressures that natural selection will exert, and these selective pressures in turn affect which members of a population will survive and reproduce. Over many generations, this results

in the evolution of traits that enable success within a given environment. Closely related species that live in different environments tend to diverge in their traits as differing selective pressures drive the evolution of different adaptations (**FIGURE 3.3a**). Conversely, sometimes very unrelated species living in similar environments in separate locations may independently acquire similar traits as they adapt to selective pressures; this is called **convergent evolution** (**FIGURE 3.3b**).

Evidence of selection is all around us

The results of natural selection are all around us, visible in every adaptation of every organism. Moreover, scientists have demonstrated the rapid evolution of traits by selection in countless lab experiments with fast-reproducing organisms such as bacteria, yeast, and fruit flies.

The evidence for selection that may be most familiar to us is that which Darwin himself cited prominently in his work 160 years ago: our breeding of domesticated animals. In dogs, cats, and livestock, we have conducted selection under our own direction; that is, **artificial selection.** We have chosen animals possessing traits we like and bred them together, while culling out individuals with traits we do not like. Through such *selective breeding*, we have been able to augment particular traits we prefer.

(a) Ancestral wolf (*Canis lupus*) and derived dog breeds

Great Dane

Saint Bernard

Collie

Chihuahua

Cabbage

Broccoli

Brussels sprouts

Cauliflower

(b) Ancestral *Brassica oleracea* and derived crops

FIGURE 3.4 Artificial selection through selective breeding has given us many breeds of dogs and varieties of crops. The ancestral wild species for dogs **(a)** is the gray wolf (*Canis lupus*). By breeding like with like and selecting for traits we preferred, we produced breeds as different as Great Danes and Chihuahuas. By this same process we created an immense variety of crop plants **(b)**. Cabbage, brussels sprouts, broccoli, and cauliflower were all generated from a single ancestral species, *Brassica oleracea*.

Consider the great diversity of dog breeds (**FIGURE 3.4a**). People generated every type of dog alive today by starting with a single ancestral species and selecting for particular desired traits as individuals were bred together. From Great Dane to Chihuahua, all dogs are able to interbreed and produce viable offspring, yet breeders maintain differences among them by allowing only like individuals to breed.

Artificial selection through selective breeding has also given us the many crop plants and livestock we depend on for food, all of which people domesticated from wild species and carefully bred over years, centuries, or millennia (**FIGURE 3.4b**). Through selective breeding, we have created corn with bigger, sweeter kernels; wheat and rice with larger and more numerous grains; and apples, pears, and oranges with better taste. We have diversified single types into many—for instance, breeding variants of wild cabbage (*Brassica oleracea*) to create broccoli, cauliflower, cabbage, and brussels sprouts. Our entire agricultural system is based on artificial selection.

Evolution generates biodiversity

Just as selective breeding helps us create new types of pets, farm animals, and crop plants, natural selection can elaborate and diversify traits in wild organisms, helping to form new species and whole new types of organisms. Life's complexity can be expressed as **biological diversity,** or **biodiversity.** These terms refer to the variety of life across all levels, including the diversity of species, genes, populations, and communities.

Scientists have described about 1.8 million species, but many more remain undiscovered or unnamed. Estimates vary for the actual number of species in the world, but they range from 3 million up to 100 million. Hawaii's insect fauna provides one example of how much we have yet to learn. Scientists studying fruit flies in the Hawaiian Islands have described more than 500 species of them, but they have also identified about 500 others that have not yet been formally named and described. Still more fruit fly species probably exist but have not yet been found.

Subtropical islands such as Hawai'i are by no means the only places rich in biodiversity. Step outside just about anywhere, and you will find many species within close reach. Plants poke up from cracks in asphalt in every city in the world, and even Antarctic ice harbors microbes. A handful of backyard soil may contain an entire miniature world of life, including insects, mites, millipedes, nematode worms, plant seeds, fungi, and millions of bacteria. (We will examine Earth's biodiversity in detail in Chapter 8.)

Speciation produces new types of organisms

How did Earth come to have so many species? The process by which new species are generated is termed **speciation.** Speciation can occur in a number of ways, but the main mode is generally thought to be *allopatric speciation*, whereby species form from populations that become physically separated over some geographic distance. To understand allopatric speciation, begin by picturing a population of organisms. Individuals within the population possess many similarities that unify them as a species because they are able to breed with one another, sharing genetic information. However, if the population is split into two or more isolated areas, individuals from one area cannot reproduce with individuals from the others.

When a mutation arises in the DNA of an organism in one of these newly isolated populations, it cannot spread to the other populations. Over time, each population will independently accumulate its own set of mutations. Eventually, the populations may diverge, growing so different that their members can no longer mate with one another and produce viable offspring. (This can occur because of changes in reproductive organs, hormones, courtship behavior, breeding timing, or other factors.) Populations that no longer exchange genetic information will embark on their own independent paths as separate species.

For speciation to occur, populations must remain isolated for a very long time, generally thousands of generations.

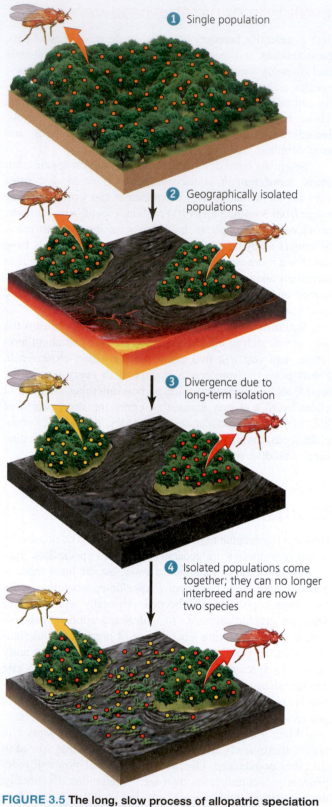

① Single population

② Geographically isolated populations

③ Divergence due to long-term isolation

④ Isolated populations come together; they can no longer interbreed and are now two species

FIGURE 3.5 The long, slow process of allopatric speciation begins when a geographic barrier splits a population—as when forest ① is destroyed by lava flowing from a volcano, but isolated patches of forest ② are left. Hawaiian fruit flies are weak fliers and become isolated in such forested patches, called *kipukas*. Over centuries, each population accumulates its own set of genetic changes ③ until individuals become unable to breed with individuals from the other population. The two populations now represent separate species and will remain so even if the geographic barrier disappears ④ and the new species intermix.

Populations can undergo long-term geographic isolation in various ways. Lava flows can destroy forest, leaving small isolated patches intact (**FIGURE 3.5**). Glacial ice sheets may move across continents during ice ages and split populations in two. Major rivers may change course or mountain ranges may be uplifted, dividing regions and their organisms. Sea level may rise, flooding low-lying regions and isolating areas of higher ground as islands. Drying climate may partially evaporate lakes, subdividing them into smaller bodies of water. Warming or cooling climate may cause plant communities to shift, creating new patterns of plant and animal distribution.

Alternatively, sometimes organisms colonize newly created areas, establishing isolated populations. Hawai'i provides an example. As shown in Figure 11.7 (Chapter 11, p. 237), the Pacific tectonic plate moves over a volcanic "hotspot" that extrudes magma into the ocean, building volcanoes that form islands once they break the water's surface. The plate inches northwest, dragging each island with it, while new islands are formed at the hotspot. The result, over millions of years, is a long string of islands, called an *archipelago*. As each new island is formed, plants and animals that colonize it may undergo allopatric speciation if they are isolated enough from their source population (see **THE SCIENCE BEHIND THE STORY**, pp. 56–57).

We can infer the history of life's diversification

Innumerable speciation events have generated complex patterns of diversity beyond the species level. Scientists represent this history of divergence by using branching diagrams called **phylogenetic trees.** Similar to family genealogies, these diagrams illustrate hypotheses proposing how divergence took place (**FIGURE 3.6**). Phylogenetic trees can show relationships among species, groups of species, populations, or genes. Scientists construct these trees by analyzing patterns of similarity among the genes or external traits of present-day organisms and by inferring which groups share similarities because they are related. Once a phylogenetic tree is created, traits can be mapped onto the tree according to which organisms possess them, and we can thereby trace how the traits have evolved.

Knowing how organisms are related to one another also helps scientists classify them and name them. *Taxonomists* use an organism's genetic makeup and physical appearance to determine its species identity. These scientists then group species by their similarity into a hierarchy of categories meant to reflect evolutionary relationships. Related species are grouped together into *genera* (singular: *genus*), related genera are grouped into families, and so on (**FIGURE 3.7**). Each species is given a two-part Latin or Latinized scientific name denoting its genus and species.

For example, the 'akiapōlā'au, *Hemignathus munroi*, is similar to other honeycreepers in the genus *Hemignathus*. These species are closely related in evolutionary terms, as indicated by the genus name they share. They are more distantly related to honeycreepers in other genera, but all honeycreepers are classified together in the family Fringillidae.

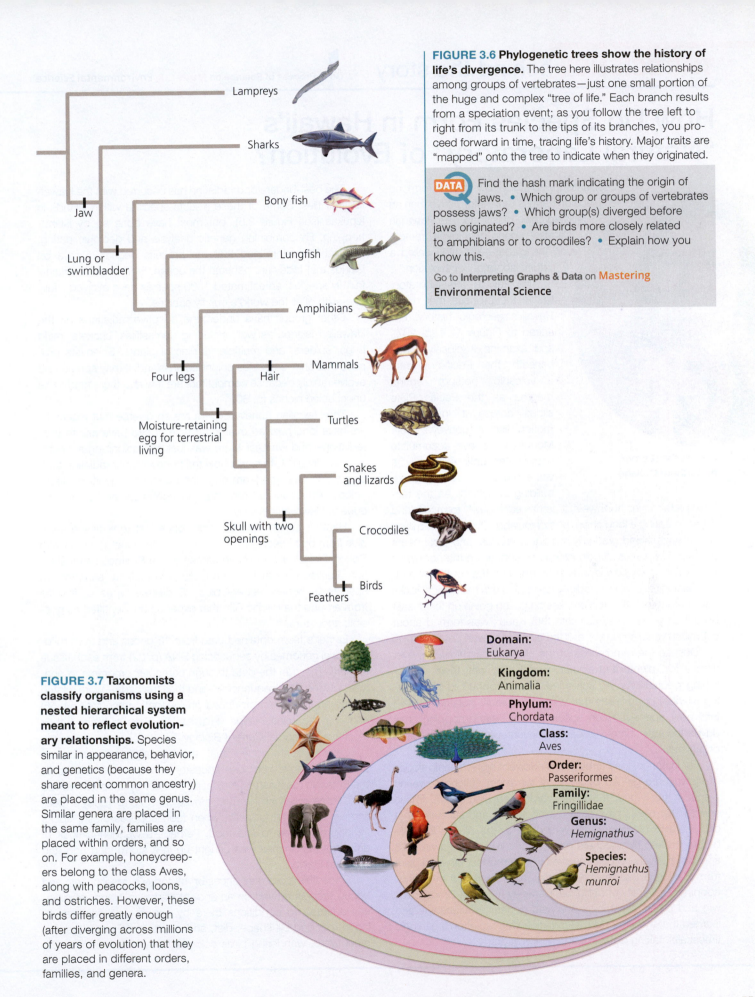

FIGURE 3.6 Phylogenetic trees show the history of life's divergence. The tree here illustrates relationships among groups of vertebrates—just one small portion of the huge and complex "tree of life." Each branch results from a speciation event; as you follow the tree left to right from its trunk to the tips of its branches, you proceed forward in time, tracing life's history. Major traits are "mapped" onto the tree to indicate when they originated.

DATA Find the hash mark indicating the origin of jaws. • Which group or groups of vertebrates possess jaws? • Which group(s) diverged before jaws originated? • Are birds more closely related to amphibians or to crocodiles? • Explain how you know this.

Go to **Interpreting Graphs & Data** on **Mastering Environmental Science**

Lampreys

Sharks

Jaw

Bony fish

Lung or swimbladder

Lungfish

Amphibians

Mammals

Four legs   Hair

Moisture-retaining egg for terrestrial living

Turtles

Snakes and lizards

Skull with two openings

Crocodiles

Birds

Feathers

FIGURE 3.7 Taxonomists classify organisms using a nested hierarchical system meant to reflect evolutionary relationships. Species similar in appearance, behavior, and genetics (because they share recent common ancestry) are placed in the same genus. Similar genera are placed in the same family, families are placed within orders, and so on. For example, honeycreepers belong to the class Aves, along with peacocks, loons, and ostriches. However, these birds differ greatly enough (after diverging across millions of years of evolution) that they are placed in different orders, families, and genera.

Domain: Eukarya

Kingdom: Animalia

Phylum: Chordata

Class: Aves

Order: Passeriformes

Family: Fringillidae

Genus: *Hemignathus*

Species: *Hemignathus munroi*

How Do Species Form in Hawaii's "Natural Laboratory" of Evolution?

Dr. Heather Lerner, of Earlham College

For scientists who study how species form, no place on Earth is more informative than an isolated chain of islands. The Hawaiian Islands—among the most remote on the planet—are often called a "natural laboratory of evolution."

The key to this laboratory lies in the process that drives Hawaii's geologic history. Flip ahead to Figure 11.7 (p. 237), and examine it closely. Deep beneath the Pacific Ocean, a volcanic "hotspot" spurts magma as the Pacific Plate slides across it in tectonic motion like a conveyor belt. Mountains of lava accumulate underwater until eventually a volcano rises above the waves, building an island. As the tectonic plate moves northwest, it carries each newly formed island with it, creating a long chain, or *archipelago*. Over several million years, each island gradually subsides, erodes, and disappears beneath the waves. As old islands disappear on the northwest end of the chain, new islands are formed on the southeast end.

Geologists analyzing radioisotopes (p. 31) in the islands' rocks have determined that this process has been going on for at least 85 million years. They estimate that Kaua'i was formed about 5.1 million years ago (mya), and the island of Hawai'i just 0.43 mya.

Despite the remoteness of the Hawaiian archipelago, over time a few plants and animals found their way there, establishing populations. As some individuals hopped to neighboring islands, populations that were adequately isolated evolved into new species. Such speciation by "island-hopping" has driven the radiation of Hawaiian honeycreepers and many other organisms.

For instance, the barren and windswept high volcanic slopes of Hawai'i are graced by some of the most striking flowering plants in the world, the silverswords (see Figure 3.2). These spectacular plants have spiky, silvery leaves and tall stalks that explode into bloom with flowers once in the plant's long life before it dies. Researchers have discovered that Hawaii's 28 species of silverswords all evolved from a modest tarweed plant from California that reached Hawai'i and diversified by island-hopping. University of California, Berkeley, botanist Bruce Baldwin and other researchers analyzed genetic relationships and learned that the silverswords' radiation was rapid (on a geologic timescale), taking place in just 5 million years.

The best-understood radiation has occurred with the Hawaiian fruit flies. Some of these insects speciate within islands in kipukas (see Figure 3.5), but most have done so by island-hopping. By combining genetic analysis and geologic dating, researchers determined that the process began 25 mya on islands that today are beneath the ocean. From a single original fruit fly species, an estimated 1000 species have evolved—fully one-sixth of all the world's fruit fly species.

Other groups have undergone adaptive radiations on the Hawaiian Islands as well, including damselflies, crickets, mirid bugs, spiders, and multiple families of plants. Scientists propose that once a species colonizes an island, it may spread and evolve rapidly because competitors are few and there tend to be unoccupied niches (p. 60).

The Hawaiian honeycreepers are so diverse that researchers have long puzzled over what type of bird gave rise to their radiation—and whether there was just one colonizing ancestor or many. In 2011, to clarify how the honeycreeper radiation took place, one research team combined genetic sequencing technology with resources from museum collections and our knowledge of Hawaiian geology.

Heather Lerner and five colleagues first took tissue samples from bird specimens in museum collections. Working with Robert Fleischer and Helen James at the Smithsonian Institution, Lerner, now at Earlham College in Indiana, sampled 19 species of honeycreepers plus 28 diverse types of finches from around the Pacific Rim that experts had identified as possible ancestors.

Lerner's team obtained data from 13 genes and from mitochondrial genomes by sequencing DNA (p. 32) from each tissue sample. They ran the data through computer programs to analyze how the DNA sequences—and thus the birds—were related to one another, then produced phylogenetic trees (pp. 54–55) showing the relationships (**FIGURE 1**). They published their results in the journal *Current Biology*.

Lerner's team found that the Hawaiian honeycreepers apparently derive from one ancestor and are most related to the Eurasian rosefinches, indicating that honeycreepers evolved after some rosefinch-like bird arrived from Asia. Today's rosefinches are partly nomadic; when food supplies crash, flocks fly long distances to find food. Perhaps a wandering flock of ancestral rosefinches was caught up in a storm long ago and blown to Hawai'i.

Once this common ancestor of today's rosefinches and honeycreepers arrived on an ancient Hawaiian island, its progeny adapted to conditions there by natural selection, resulting in modified bill shape, diet, and coloration. Every once in a great while, wandering birds colonized other islands, founding

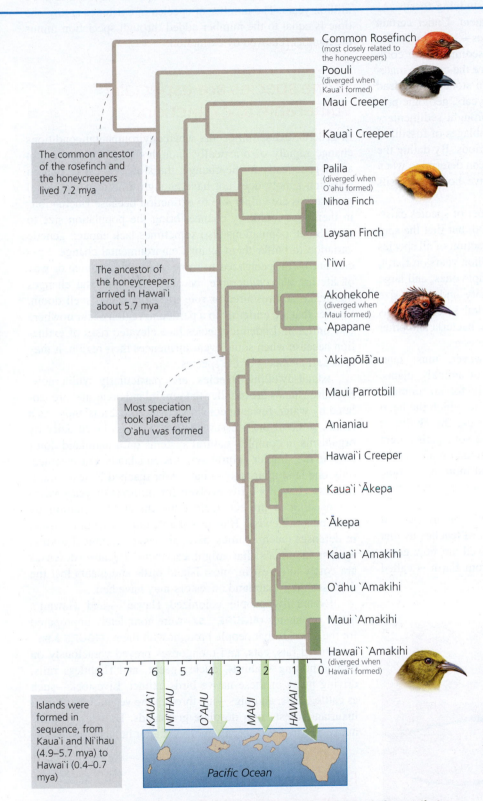

The common ancestor of the rosefinch and the honeycreepers lived 7.2 mya

The ancestor of the honeycreepers arrived in Hawai'i about 5.7 mya

Most speciation took place after O'ahu was formed

Common Rosefinch (most closely related to the honeycreepers)
Poouli (diverged when Kaua'i formed)
Maui Creeper
Kaua'i Creeper
Palila (diverged when O'ahu formed)
Nihoa Finch
Laysan Finch
'I'iwi
Akohekohe (diverged when Maui formed)
'Apapane
'Akiapōlā'au
Maui Parrotbill
Anianiau
Hawai'i Creeper
Kaua'i 'Ākepa
'Ākepa
Kaua'i 'Amakihi
O'ahu 'Amakihi
Maui 'Amakihi
Hawai'i 'Amakihi (diverged when Hawai'i formed)

8 7 6 5 4 3 2 1 0

KAUA'I   NI'IHAU   O'AHU   MAUI   HAWAI'I

Pacific Ocean

Islands were formed in sequence, from Kaua'i and Ni'ihau (4.9–5.7 mya) to Hawai'i (0.4–0.7 mya)

FIGURE 1 Using gene sequences, researchers generated this phylogenetic tree showing relationships among the Hawaiian honeycreepers. They then matched the history of the birds' diversification with the known geologic history of the islands' formation.

Adapted from Lerner, H.R.L., et al., 2011. Multilocus resolution of phylogeny and timescale in the extant adaptive radiation of Hawaiian honeycreepers. Curr. Biol. 21: 1838–1844.

populations that each adapted to local conditions and might eventually evolve into separate species.

Because the age of each island is known, Lerner's team could calibrate rates of evolutionary change in the birds' DNA sequences, and thus measure the age of each divergence. That is, they could tell how "old" each bird species is. They found that the rosefinch-like ancestor arrived by 5.7 mya, about the time that the oldest of today's main islands (Kaua'i and Ni'ihau) were forming. After O'ahu emerged 4.0–3.7 mya, the speciation process went into overdrive, giving rise to many new species with distinctively different colors, bill shapes, and habits. By the time Maui arose 2.4–1.9 mya, most of the major differences in body form and appearance had evolved (see bottom portion of Figure 1).

Thus, most major innovations arose midway through the island-formation process, when O'ahu and Kaua'i were the main islands in the chain. After this burst of innovation, major changes were fewer, perhaps because most evolutionary possibilities had been explored, or perhaps because the newer islands of Maui and Hawai'i were too close together to isolate populations adequately.

The team's data show that the age of each honeycreeper species does not neatly match the age of the island or islands it inhabits today. Instead, the island-hopping process was complex, with some birds hopping "backward" from newer islands to older ones. Moreover, within each island there is a great deal of variation in climate, topography, and vegetation, because windward slopes catch moisture from trade winds over the ocean and become lush and green, whereas leeward slopes in the rainshadow are arid. The varied habitats and rugged topography create barriers that can lead to speciation within islands.

For all these reasons, the "natural laboratory" of Hawai'i still has much to teach us about how the honeycreepers and other groups have evolved, and how new species are formed.

Fossils reveal life's history

Scientists also decipher life's history by studying fossils. As organisms die, some are buried by sediment. Under certain conditions, the hard parts of their bodies—such as bones, shells, and teeth—may be preserved as sediments are compressed into rock (p. 235). Minerals replace the organic material, leaving behind a **fossil,** an imprint in stone of the dead organism (**FIGURE 3.8**). Over millions of years, geologic processes have buried sediments and later brought sedimentary rock layers to the surface, revealing assemblages of fossilized plants and animals from different time periods. By dating the rock layers that contain fossils, scientists can determine when particular organisms lived. The cumulative body of fossils worldwide is known as the **fossil record.**

The fossil record shows that the number of species existing at any one time has generally increased, but that the species alive on Earth today are just a small fraction of all species that have ever existed. During life's 3.5 billion years on Earth, complex structures have evolved from simple ones, and large sizes from small ones. However, simplicity and small size have also evolved when favored by natural selection; it is easy to argue that Earth still belongs to the bacteria and other microbes, some of them little changed over eons.

Even enthusiasts of microbes, however, must marvel at some of the exquisite adaptations of animals, plants, and fungi: the heart that beats so reliably for an animal's entire lifetime; the complex organ system to which the heart belongs; the stunning plumage of a peacock; the ability of plants to lift water and nutrients from the soil, gather light from the sun, and produce food; and the human brain and its ability to reason. All these adaptations and more have come about as evolution has generated new species and whole new branches on the tree of life.

Although speciation generates Earth's biodiversity, it is only part of the equation; the fossil record teaches us that the vast majority of creatures that once lived are now gone. The disappearance of an entire species from Earth is called **extinction.** From studying fossils, paleontologists calculate that the average time a species spends on Earth is 1–10 million years. The number of species in existence at any one time is equal to the number added through speciation minus the number removed by extinction.

Some species are especially vulnerable to extinction

In general, extinction occurs when environmental conditions change rapidly or drastically enough that a species cannot adapt genetically to the change; the slow process of natural selection simply does not have enough time to work. Small populations are vulnerable to extinction because fluctuations in their size could, by chance, bring the population size to zero. Small populations also sometimes lack enough genetic variation to buffer them against environmental change. Species narrowly specialized to some particular resource or way of life are also vulnerable, because environmental changes that make that resource or role unavailable can spell doom. Species that are **endemic** to a particular region occur nowhere else on Earth. Endemic species face elevated risks of extinction because when some event influences their region, it may affect all members of the species.

Island-dwelling species are particularly vulnerable. Because islands are smaller than mainland areas and are isolated by water, fewer species reach and inhabit islands. As a result, some of the pressures and challenges faced daily by organisms in complex natural systems on a mainland don't exist in the simpler natural systems on islands. For instance, only one land mammal—a bat—ever reached Hawai'i naturally, so Hawaii's birds evolved for millions of years without needing to protect against the threat of predation by mammals. Likewise, Hawaii's plants did not need to invest in defenses (such as thick bark, spines, or chemical toxins) against mammals that might eat them. Because defenses are costly to invest in, most island birds and plants lost the defenses their mainland ancestors may have had.

Eventually, people colonized Hawai'i—and Hawaii's native organisms (**FIGURE 3.9a**) were completely unprepared for the animals that people brought with them (**FIGURE 3.9b**). Introduced rats, cats, and mongooses preyed voraciously on ground-nesting seabirds, ducks, geese, and flightless rails, driving many of these native birds extinct. Livestock—such as cattle, goats, and pigs—ate through the vegetation, turning luxuriant forests into desolate grasslands. Half of Hawaii's native birds were driven extinct soon after human arrival.

Earth has seen episodes of mass extinction

Most extinction occurs gradually, one species at a time, at a rate referred to as the **background extinction rate.** However, the fossil record reveals that Earth has seen at least five events of staggering proportions that killed off massive numbers of species at once. These episodes, called **mass extinction events,**

FIGURE 3.8 The fossil record helps reveal the history of life on Earth. Trilobites were once abundant, but today we know these extinct animals only from their fossils.

(a) Hawaiian petrel, a native species at risk

(b) Mongoose, an introduced species that preys on natives

FIGURE 3.9 **Island-dwelling species that have lost defenses are vulnerable to extinction when enemies are introduced.** The Hawaiian petrel **(a)**, a seabird that nests in the ground, is endangered as a result of predation by mammals introduced to Hawai'i, such as **(b)** the Indian mongoose.

have occurred at widely spaced intervals in our planet's history and have wiped out 50–95% of Earth's species each time (see Figure 8.7, p. 175).

The best-known mass extinction occurred 66 million years ago and brought an abrupt end to the dinosaurs (although today's birds are descendants of a type of dinosaur that survived). Evidence suggests that the collision of a gigantic asteroid with Earth caused this event. Still more catastrophic was the mass extinction 250 million years ago at the end of the Permian period (see **APPENDIX E** for Earth's geologic periods). Scientists estimate that 75–95% of all species perished during this event, described by one researcher as the "mother of all mass extinctions." Hypotheses as to what caused the end-Permian extinction event include massive volcanism, an asteroid impact, methane releases and global warming, or some combination of these factors.

The sixth mass extinction is upon us

Many biologists have concluded that Earth is currently entering its sixth mass extinction event—and that we are the cause. Changes to our planet's natural systems set in motion by human population growth, development, and resource depletion have driven many species to extinction and are threatening countless more. As we alter and destroy natural habitats; overhunt and overharvest populations; pollute air, water, and soil; introduce invasive non-native species; and alter climate, we set in motion processes that combine to threaten Earth's biodiversity (pp. 176–181). Because we depend on organisms for life's necessities—food, fiber, medicine, and vital ecosystem services (pp. 4, 39, 172)—biodiversity loss and extinction ultimately threaten our own survival.

Ecology and the Organism

Extinction, speciation, and other evolutionary forces play key roles in ecology. **Ecology** is the scientific study of the interactions among organisms and of the relationships between organisms and their environments. Ecology allows us to explain and predict the distribution and abundance of organisms in nature. It is often said that ecology provides the stage on which the play of evolution unfolds. The two are intertwined in many ways.

We study ecology at several levels

Life exists in a hierarchy of levels, from atoms, molecules, and cells (pp. 30–33) up through the **biosphere,** the cumulative total of living things on Earth and the areas they inhabit. **Ecologists** are scientists who study relationships at the higher levels of this hierarchy (**FIGURE 3.10**), namely at the levels of the organism, population, community, ecosystem, landscape, and biosphere.

At the level of the organism, ecology describes relationships between an organism and its physical environment. Organismal ecology helps us understand, for example, what aspects of a Hawaiian honeycreeper's environment are important to it, and why. In contrast, **population ecology** examines the dynamics of population change and the factors that affect the distribution and abundance of members of a population. It helps us understand why populations of some species decline while populations of others increase.

In ecology, a **community** consists of an assemblage of populations of interacting species that inhabit the same area. A population of 'akiapōlā'au, a population of koa trees, a population of wood-boring grubs, and a population of ferns, together with all the other interacting plant, animal, fungal, and microbial populations in the Hakalau Forest, would be considered a community. **Community ecology** (Chapter 4) focuses on patterns of species diversity and on interactions among species, ranging from one-to-one interactions up to complex interrelationships involving the entire community.

Ecosystems (p. 36) encompass communities and the abiotic (nonliving) materials and forces with which community members interact. Hakalau's cloud-forest ecosystem consists of its community plus the air, water, soil, nutrients, and energy used by the community's organisms. **Ecosystem ecology** (Chapter 2) addresses the flow of energy and nutrients by studying living and nonliving components of systems in conjunction. Today's warming climate (Chapter 14) is having ecosystem-level consequences as it affects Hakalau and other ecosystems across the world.

Biosphere

The sum total of living things on Earth and the areas they inhabit

Landscape

A geographic region including an array of ecosystems

Ecosystem

A functional system consisting of a community, its nonliving environment, and the interactions between them

Community

A set of populations of different species living together in a particular area

Population

A group of individuals of a species that live in a particular area

Organism

An individual living thing

FIGURE 3.10 Green sea turtles are part of a coral reef community that inhabits reef ecosystems along Hawaii's coasts.

Concerns such as climate change and habitat loss, together with technologies such as satellite imagery, are invigorating the study of how ecosystems are arrayed across the landscape. **Landscape ecology** (p. 37) helps us understand how and why ecosystems, communities, and populations are distributed across geographic regions. Indeed, as new technologies help scientists study the complex dynamics of natural systems at a global scale, ecologists are expanding their horizons to the biosphere as a whole.

Each organism has habitat needs

At the level of the organism, each individual relates to its environment in ways that tend to maximize its survival and reproduction. One key relationship involves the specific environment in which an organism lives, its **habitat.** A species' habitat consists of the living and nonliving elements around it, including rock, soil, leaf litter, humidity, plant life, and more. The 'akiapōlā'au lives in a habitat of cool, moist, montane forest of native koa and 'ōhi'a trees, where it is high enough in elevation to be safe from avian pox and malaria.

Each organism thrives in certain habitats and not in others, leading to nonrandom patterns of **habitat use.** Mobile organisms actively select habitats from among the range of options they encounter, a process called **habitat selection.** Each species assesses habitats differently because each species has different needs. A species' needs may vary with the time and place; many migratory birds use distinct breeding, wintering, and migratory habitats. In the case of plants and of stationary animals (such as sea anemones in the ocean), whose young disperse and settle passively, patterns of habitat use result from success in some habitats and failure in others.

Habitat is a vital concept in environmental science. Because habitats provide everything an organism needs, including nutrition, shelter, breeding sites, and mates, the organism's survival depends on the availability of suitable habitats. Often this need results in conflict with people who want to alter a habitat for their own purposes.

Organisms have roles in communities

Another way in which an organism relates to its environment is through its **niche,** its functional role in a community. A species' niche reflects its use of habitat and resources, its consumption of certain foods, its role in the flow of energy and matter, and its interactions with other organisms. The niche is a multidimensional concept, a kind of summary of everything an organism does. The pioneering ecologist Eugene Odum once wrote that "habitat is the organism's address, and the niche is its profession."

Organisms vary in the breadth of their niches. A species with narrow breadth (and thus very specific requirements) is said to be a **specialist.** One with broad tolerances, a "jack-of-all-trades" able to use a wide array of resources, is a **generalist.** A native Hawaiian honeycreeper like the 'akiapōlā'au is a specialist, because its unique bill is exquisitely adapted for feeding on grubs that tunnel through the wood of certain native trees. In contrast, the common myna (a bird introduced to Hawai'i from Asia) is a generalist; its unremarkable bill allows it to eat many foods in many habitats. As a result, the common myna has spread throughout the Hawaiian Islands wherever people have altered the landscape.

Generalists like the myna succeed by being able to live in many different places and withstand variable conditions, yet they do not thrive in any single situation as well as a specialist adapted for those specific conditions. (A jack-of-all-trades, as the saying goes, is a master of none.) Specialists succeed over evolutionary time by being extremely good at the things they do, yet they are vulnerable when conditions change and threaten the habitat or resource on which they have specialized. An organism's habitat preferences, niche, and degree of specialization each reflect adaptations of the species and are products of natural selection.

Population Ecology

A population, as we have seen, consists of individuals of a species that inhabit a particular area at a particular time. Population ecologists try to understand and predict how populations change over time. The ability to predict a population's growth or decline is useful in monitoring and managing wildlife, fisheries, and threatened and endangered species. For instance, biologists at Hawaii's Hakalau Forest survey populations of native birds to assess the success of forest restoration efforts there. Understanding population ecology is also crucial for predicting the dynamics of our human population (Chapter 6)—a central element of environmental science and one of the prime challenges for our society today.

Populations show features that help predict their dynamics

All populations—from humans to honeycreepers—exhibit attributes that help population ecologists predict their dynamics.

Population size Expressed as the number of individual organisms present at a given time, **population size** may increase, decrease, undergo cyclical change, or remain stable over time. Populations generally grow when resources are abundant and natural enemies are few. Populations can decline in response to loss of resources, natural disasters, or impacts from other species.

The passenger pigeon, now extinct, illustrates extremes in population size (**FIGURE 3.11**). Not long ago it was the most abundant bird in North America; flocks of passenger pigeons literally darkened the skies. In the early 1800s, ornithologist Alexander Wilson watched a flock of 2 billion birds 390 km (240 mi) long that took 5 hours to fly over and sounded like a tornado. Passenger pigeons nested in gigantic colonies in the forests of the upper Midwest and southern Canada. Once settlers began cutting the forests, the birds were easy targets for market hunters, who gunned down thousands at a time. The birds were shipped to market by the wagonload and sold for food. By 1890, the population had declined to such a low number that the birds could not form the large colonies they evidently needed to breed. In 1914, the last passenger pigeon on Earth died in the Cincinnati Zoo, bringing the continent's most numerous bird species to extinction in just a few decades.

Hawai'i offers a story with a happier ending. Hawaii's state bird is the nēnē (pronounced "nay-nay"), also called the Hawaiian goose (see Figure 3.2). Before people reached the Hawaiian Islands, nēnēs were common, and researchers estimate that the nēnē population numbered at least 25,000 birds. After human arrival, the nēnē was nearly driven to extinction by human hunting; livestock and plants that people introduced (which destroyed and displaced the vegetation it fed on); and rats, cats, dogs, pigs, and mongooses that preyed on its eggs and young. By the 1950s, these impacts had eliminated nēnēs from all islands except the island of Hawai'i, where the population size was down to just 30 individuals. Fortunately, dedicated conservation efforts have turned this decline around. Biologists and wildlife managers have labored to breed nēnēs in captivity and have reintroduced them into protected areas.

FIGURE 3.11 Flocks of passenger pigeons literally darkened the skies as billions of birds passed overhead. Still, hunting and deforestation drove North America's most numerous bird to extinction within decades.

(a) Passenger pigeon   **(b) 19th-century lithograph of pigeon hunting in Iowa**

These efforts are succeeding, and today nēnēs live in at least seven regions on four of the Hawaiian Islands, with a population size of more than 2000 birds.

Population density The flocks and breeding colonies of passenger pigeons showed high population density, another attribute that ecologists assess. **Population density** describes the number of individuals per unit area in a population. High population density makes it easier for organisms to group together and find mates, but it can also lead to competition and conflict if space, food, or mates are in limited supply. Overcrowding among individuals can also increase the transmission of infectious disease. In contrast, at low population densities, individuals benefit from more space and resources but may find it harder to locate mates and companions.

Population distribution **Population distribution** describes the spatial arrangement of organisms in an area. Ecologists define three distribution types: random, uniform, and clumped (**FIGURE 3.12**). In a *random distribution*, individuals are located haphazardly in no particular pattern. This type of distribution can occur when resources are plentiful throughout an area and other organisms do not strongly influence where members of a population settle.

A *uniform distribution*, in which individuals are evenly spaced, can occur when individuals compete for space. Animals may hold and defend territories. Plants need space for their roots to gather moisture, and they may exude chemicals that poison one another's roots as a means of competing for space. As a result, competing individuals may end up distributed at equal distances from one another.

A *clumped distribution* often results when organisms seek habitats or resources that are unevenly spaced. In arid areas, many plants grow in patches near water. Hawaiian honeycreepers tend to cluster near flowering trees that offer nectar. People frequently aggregate in villages, towns, and cities.

Sex ratio A population's **sex ratio** is its proportion of males to females. In monogamous species (in which each sex takes a single mate), a 1:1 sex ratio maximizes population growth, whereas an unbalanced ratio leaves many individuals without mates. Most species are not monogamous, however, so sex ratios vary from one species to another.

Age structure **Age structure** describes the relative numbers of individuals of different ages within a population. By combining this information with data on the reproductive potential of individuals in each age class, a population ecologist can predict how the population may grow or shrink.

Many plants and animals continue growing in size as they age, and in these species, older individuals often reproduce more. Older, larger trees in a population produce more seeds than smaller, younger trees. Larger, older fish produce more eggs than smaller, younger fish of the same species. Birds use the experience they gain with age to become more successful breeders at older ages.

Human beings are unusual because we often survive past our reproductive years. A human population made up largely of older (post-reproductive) individuals will tend to decline over time, whereas one with many young people (of reproductive or pre-reproductive age) will tend to increase. (We will use diagrams to explore these ideas further in Chapter 6, pp. 124–125, as we study human population growth.)

Populations may grow, shrink, or remain stable

Let's now take a more quantitative approach by examining some simple mathematical concepts used by population ecologists and by **demographers** (scientists who study human populations). Population change is determined by four factors:

- *Natality* (births within the population)
- *Mortality* (deaths within the population)
- *Immigration* (arrival of individuals from outside the population)
- *Emigration* (departure of individuals from the population)

(a) Random: Distribution of organisms displays no pattern.

(b) Uniform: Individuals are spaced evenly.

(c) Clumped: Individuals concentrate in certain areas.

FIGURE 3.12 Individuals in a population can spatially distribute themselves in three fundamental ways.

We can measure a population's **rate of natural increase** simply by subtracting the death rate from the birth rate:

(birth rate) − (death rate) = rate of natural increase

To obtain the **population growth rate,** the total rate of change in a population's size per unit time, we must also include the effects of immigration and emigration:

(birth rate − death rate) + (immigration rate − emigration rate) = population growth rate

The rates in these formulas are often expressed in numbers per 1000 individuals per year. For example, a population with a birth rate of 18 per 1000/yr, a death rate of 10 per 1000/yr, an immigration rate of 5 per 1000/yr, and an emigration rate of 7 per 1000/yr would have a population growth rate of 6 per 1000/yr:

(18/1000 − 10/1000) + (5/1000 − 7/1000) = 6/1000

Given these rates, a population of 1000 in one year will grow to 1006 in the next. If the population is 1,000,000, it will reach 1,006,000 the next year. Such population increases are often expressed in percentages, as follows:

population growth rate × 100%

Thus, a growth rate of 6/1000 would be expressed as:

6/1000 × 100% = 0.6%

By measuring population growth in percentages, we can compare changes among populations of far different sizes. We can also project changes into the future. Understanding and predicting such changes helps wildlife and fisheries managers regulate hunting and fishing to ensure sustainable harvests, informs conservation biologists trying to protect rare and declining species, and assists policymakers planning for human population growth in cities, regions, and nations.

Unregulated populations increase by exponential growth

When a population increases by a fixed percentage each year, it is said to undergo **exponential growth.** Imagine you put money in a savings account at a fixed interest rate and leave it untouched for years. As the principal accrues interest and grows larger, you earn still more interest, and the sum grows by escalating amounts each year. The reason is that a fixed percentage of a small number makes for a small increase, but the same percentage of a larger number produces a larger increase. Thus, as savings accounts (or populations) grow larger, each incremental increase likewise becomes larger in absolute terms. Such acceleration characterizes exponential growth.

We can visualize changes in population size by using population growth curves. The J-shaped curve in **FIGURE 3.13**

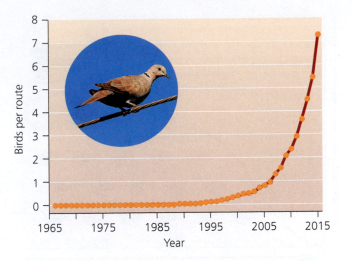

FIGURE 3.13 A population may grow exponentially when colonizing an unoccupied environment or exploiting an unused resource. The Eurasian collared dove is spreading across the United States, propelled by exponential growth. *Data from Sauer, J.R., et al., 2017.* The North American Breeding Bird Survey, results and analysis 1966–2015. *Version 2.07.2017. Laurel, MD: USGS Patuxent Wildlife Research Center.*

shows exponential growth. Populations increase exponentially unless they meet constraints. Each organism reproduces by a certain amount, and as populations grow, there are more individuals reproducing by that amount. If there are adequate resources and no external limits, ecologists expect exponential growth.

Normally, exponential growth occurs in nature only when a population is small, competition is minimal, and environmental conditions are ideal for the organism in question. Most often, these conditions occur when the organism arrives in a new environment that contains abundant resources. Mold growing on a piece of fruit, or bacteria decomposing a dead animal, are cases in point. Plants colonizing regions during primary succession (p. 77) after glaciers recede or volcanoes erupt also show exponential growth. In Hawai'i, many species that colonized the islands underwent exponential growth for a time after their arrival. One current example of exponential growth in mainland North America is the Eurasian collared dove (see Figure 3.13). Unlike its extinct relative the passenger pigeon, this species arrived here from Europe, thrives in areas disturbed by people, and has spread across the continent in a matter of years.

Limiting factors restrain growth

Exponential growth rarely lasts long. If even a single species were to increase exponentially for very many generations, it would blanket the planet's surface! Instead, every population eventually is constrained by **limiting factors**—physical, chemical, and biological attributes of the environment that restrain population growth. Together, these limiting factors determine the **carrying capacity,** the maximum population size of a species that a given environment can sustain.

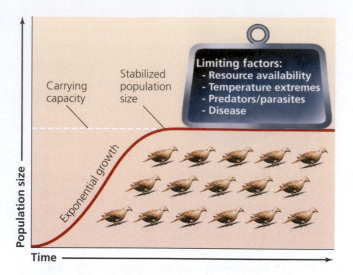

FIGURE 3.14 The logistic growth curve shows how population size may increase rapidly at first, then slow down, and finally stabilize at a carrying capacity.

Ecologists use the S-shaped curve in **FIGURE 3.14** to show how an initial exponential increase is slowed and eventually brought to a standstill by limiting factors. This phenomenon is called **logistic growth.** A logistic growth curve rises sharply at first but then begins to level off as the effects of limiting factors become stronger. Eventually the collective force of these factors stabilizes the population size at its carrying capacity.

We can witness this process by taking a closer look at data for the Eurasian collared dove, gathered by thousands of volunteer birders and analyzed by government biologists in the Breeding Bird Survey, a long-running citizen science project. The dove first appeared in Florida a few decades ago and then spread north and west. Today its numbers are growing fastest in western areas it has recently reached, and more slowly in southeastern areas where it has been present for longer. In Florida, it has apparently reached carrying capacity (**FIGURE 3.15**).

Many factors influence a population's growth rate and carrying capacity. For animals in terrestrial environments, limiting factors include temperature extremes; disease prevalence; predator abundance; and the availability of food, water, mates, shelter, and breeding sites. Plants are often limited by sunlight, moisture, soil chemistry, disease, and plant-eating animals. In aquatic systems, limiting factors include salinity, sunlight, temperature, dissolved oxygen, fertilizers, and pollutants.

A population's density can enhance or diminish the impact of certain limiting factors. Recall that high population density can help organisms find mates but may also increase competition, predation, and disease. Such limiting factors are said to be **density-dependent** because their influence rises and falls with population density. The logistic growth curve in Figure 3.14 represents the effects of density dependence. The larger the population size, the stronger the effects of the limiting factors.

Density-independent factors are those whose influence is independent of population density. Temperature extremes and catastrophic events such as floods, fires, and landslides are examples of density-independent factors, because they can eliminate large numbers of individuals without regard to their density.

The logistic curve is a simple model, and real populations in nature can behave differently. Some may cycle above and below the carrying capacity. Others may overshoot the carrying capacity and then crash, destined for either extinction or recovery.

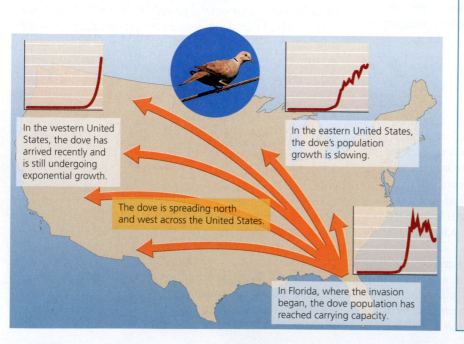

FIGURE 3.15 Exponential growth slows over time and gives way to logistic growth. By breaking down the continent-wide data for the Eurasian collared dove from Figure 3.13, we can track its spread west and north from Florida, where it first arrived. Today its population growth is fastest in the west, slower in the east (where the species has been present longer), and stable in Florida (where it has apparently reached carrying capacity). *Data from Sauer, J.R., et al., 2017. The North American Breeding Bird Survey, results and analysis 1966–2015. Version 2.07.2017. Laurel, MD: USGS Patuxent Wildlife Research Center.*

DATAQ Looking several decades into the future, what do you predict the population growth graph for the Eurasian collared dove in the western United States will look like? Explain why.

Go to **Interpreting Graphs & Data** on **Mastering Environmental Science**

Carrying capacities can change

Because environments are complex and ever-changing, carrying capacities can vary. If a fire destroys a forest, the carrying capacities for most forest animals will decline, whereas carrying capacities for species that benefit from fire will increase. Our own species has proved capable of intentionally altering our environment to raise our carrying capacity. When our ancestors began to build shelters and use fire for heating and cooking, they eased the limiting factors of cold climates and were able to expand into new territory. As human civilization developed, we overcame limiting factors time and again by creating new technologies and cultural institutions. People have managed so far to increase the planet's carrying capacity for our species, but we have done so by appropriating immense proportions of the planet's natural resources. In the process, we have reduced carrying capacities for countless other organisms that rely on those same resources.

weighing the ISSUES

Carrying Capacity and Human Population Growth

The global human population has surpassed 7 billion, far exceeding our population's size throughout our history on Earth. Name some ways in which we have raised Earth's carrying capacity for our species. Do you think we can continue to raise our carrying capacity? How might we do so? What limiting factors exist for the human population today? Might Earth's carrying capacity for us decrease? Why or why not?

FIGURE 3.16 Hawai'i protects some of its diverse natural areas, helping to stimulate its economy with ecotourism. Here, a scuba diver observes raccoon butterflyfish at a coral reef.

Conserving Biodiversity

Populations have always been affected by environmental change, but today human development, resource extraction, and population pressure are speeding the rate of change and bringing new types of impacts. Fortunately, committed people are taking action to safeguard biodiversity and to preserve and restore Earth's ecological and evolutionary processes (as we shall see more fully in our coverage of conservation biology in Chapter 8).

Innovative solutions are working

Amid all the challenges of Hawaii's extinction crisis, hard work is resulting in some inspirational success stories, and several species have been saved from imminent extinction. At Hakalau Forest, ranchland is being restored to forest, invasive plants are being removed, native ones are being planted, and nēnē are being protected while new populations of them are being established.

Early conservation work at Hawai'i Volcanoes National Park inspired the work at Hakalau, as well as efforts by managers and volunteers from the Hawai'i Division of Forestry and Wildlife, The Nature Conservancy of Hawai'i, Kamehameha Schools, and local watershed protection groups. Across Hawai'i, people are protecting land, removing alien mammals and weeds, and restoring native habitats. Offshore, Hawaiians are striving to protect their fabulous coral reefs, sea grass beds, and beaches from pollution and overfishing. The northwesternmost Hawaiian Islands are now part of the largest federally declared marine reserve (p. 283) in the world.

Hawaii's citizens are reaping economic benefits from their conservation efforts. The islands' wildlife and natural areas draw visitors from around the world, a phenomenon called **ecotourism** (**FIGURE 3.16**). A large proportion of Hawaii's tourism is ecotourism, and altogether tourism draws over 7 million visitors to Hawai'i each year, creates thousands of jobs, and pumps $12 billion annually into the state's economy.

Climate change poses a challenge

Traditionally, people sought to conserve populations of threatened species by preserving and managing tracts of land (or areas of ocean) designated as protected areas. However, global climate change (Chapter 14) now threatens this strategy. As temperatures climb and rainfall patterns shift, conditions within protected areas may become unsuitable for the species they were meant to protect.

Hawaii's systems are especially vulnerable. At Hakalau Forest on the slopes of Mauna Kea, mosquitoes and malaria are moving upslope toward the refuge as temperatures rise, exposing more and more birds to disease. Some research suggests that climate change will also reduce rainfall here, pushing the upper limit of the forest downward. If so, Hakalau's honeycreepers may become trapped within a shrinking band of forest by disease from below and drought from above.

The challenges posed by climate change mean that scientists and managers need to come up with new ways to save declining populations. In Hawai'i, management and ecotourism can help conserve natural systems, but resources to preserve habitat and protect endangered species will likely need to be stepped up. Restoring communities—as at Hakalau Forest—will also be necessary. Ecological restoration is one phenomenon we will examine in our next chapter, as we shift from populations to communities.

Restoring Hakalau's Forest

By the time the Hakalau Forest National Wildlife Refuge was established in 1985, much of Hawaii's native forest had been cleared for cattle ranching, while free-roaming pigs and invasive plants had degraded what remained. So Hakalau's managers swung into action to restore the native forest. They built fences to keep pigs and cattle out. They planted half a million native plants. They labored to remove invasive weeds. They located and protected the last remaining individuals of several endangered plant

The Hawai'i 'ākepa is one native species benefiting from forest restoration.

species. Gradually the forest recovered, and stands of young trees took root in what had been barren pasture. Biologists monitored nine species of native forest birds, including four federally endangered ones. After 21 years of surveys, they concluded that populations of most native birds were either stable or slowly increasing. Although data suggested that populations may have declined in the most recent 9 years, native birds at Hakalau were faring better than elsewhere on the island of Hawai'i, and the reforestation of Hakalau's upper slopes was creating new habitat into which birds were moving. The success at Hakalau is a hopeful sign that with research and careful management we can undo past ecological damage and preserve endangered island species, habitats, and communities.

EXPLORE THE DATA at Mastering Environmental Science

closing THE LOOP

The honeycreepers of Hakalau Forest National Wildlife Refuge, along with many other Hawaiian species, help to illuminate the fundamentals of evolution and population ecology that are integral to environmental science. Island chains like Hawai'i can be viewed as "laboratories of evolution," showing us how populations evolve and how new species arise. But just as islands are crucibles of speciation, today they are also hotspots of extinction. In our age of global

human travel, non-native species brought to Hawai'i have overrun the islands' landscapes and disrupted their ecological systems. As biologists monitor dwindling populations of native species, conservationists race to restore habitats, fight invasive species, and save native plants and animals. If we can study, understand, and act to address the challenges facing populations and biodiversity in places like Hawai'i, we can do so anywhere in the world—and do so we must if we are to protect Earth's biodiversity.

TESTING Your Comprehension

1. Define the concept of natural selection in your own words, and explain how it follows logically from a few common observations of nature.

2. Describe an example of evidence for natural selection and an example of evidence for artificial selection.

3. Describe the steps involved in allopatric speciation.

4. Name two organisms that have become extinct or are threatened with extinction. For each, give a probable reason for its decline.

5. Define the terms *species, population,* and *community*. How does a species differ from a population? How does a population differ from a community?

6. Define and contrast the concepts of habitat and niche.

7. List and describe each of the five major attributes of populations that help ecologists predict population growth or decline. Briefly explain how each attribute shapes population dynamics.

8. Can a species undergo exponential growth indefinitely? Explain your answer.

9. Describe how limiting factors relate to carrying capacity.

10. What are some advantages of ecotourism for a state like Hawai'i? Can you think of any potential disadvantages?

SEEKING Solutions

1. In what ways have artificial selection and selective breeding changed people's quality of life? Give examples. How might artificial selection and selective

breeding be used to improve our quality of life further? Can you envision a way they could be used to reduce our environmental impact?

2. In your region, what species are threatened with extinction? Why are they vulnerable? Suggest steps that could be taken to increase their populations.

3. Do you think the human species can continue raising its global carrying capacity? How so, or why not? Do you think we *should* try to keep raising our carrying capacity? Why or why not?

4. **CASE STUDY CONNECTION** Describe two of the threats facing native species at Hakalau Forest National Wildlife Refuge or elsewhere in Hawai'i, and two actions people have taken to address these threats. What new trend is now jeopardizing some of these efforts? What steps do you think might be required in the future to safeguard native species, populations, and communities in mountainous habitats in places like Hawai'i?

5. **THINK IT THROUGH** You are a population ecologist studying animals in a national park, and park managers are asking for advice on how to focus their limited conservation funds. How would you rate the following three species, from most vulnerable (and thus most in need of conservation attention) to least vulnerable? Give reasons for your choices.

- A bird that is a generalist in its use of habitats and resources
- A salamander endemic to the park that lives in high-elevation forest
- A fish that specializes on a few types of invertebrate prey and has a large population size

CALCULATING Ecological Footprints

Professional demographers delve deeply into the latest statistics from cities, states, and nations to bring us updated estimates on human populations. The table shows population estimates for two consecutive years that take into account births, deaths, immigration, and emigration. To calculate the population growth rate (p. 63) for each region, simply divide the 2016 data by the 2015 data, subtract 1.00, and multiply by 100:

pop. growth rate = [(2016 pop. / 2015 pop.) − 1] × 100

Florida was one of the fastest-growing U.S. states during this particular 1-year period, and Illinois was one of the slowest-growing states (in fact, Illinois decreased in population). You can find data for your own state, city, or metropolitan area by exploring the Web pages of the U.S. Census Bureau.

REGION	2015 POPULATION	2016 POPULATION	POPULATION GROWTH RATE
Hawai'i	1,425,157	1,428,557	
Florida	20,244,914	20,612,439	
Illinois	12,839,047	12,801,539	
Your state or city			
United States	320,896,618	323,127,513	0.70%
World	7,336,435,000	7,418,152,000	

Data from U.S. Census Bureau and Population Reference Bureau.

1. Assuming its population growth rate remained at the calculated rate, what would the population of the United States have been in 2017?

2. The birth rate of the United States in recent years has been 12.5 births per 1000 people, and the death rate has been 8.2 deaths per 1000 people. Using these data and the formula on p. 63, what was the rate of natural increase for the United States between 2015 and 2016? Now subtract this rate from the population growth rate to obtain the net migration rate. Was the United States experiencing more immigration or more emigration during this year?

3. At the growth rate you calculated for Florida, the population of that state will double in just 38 years. What impacts would you expect this to have on (1) food supplies, (2) drinking water supplies, (3) forests and other natural areas, and (4) wildlife populations?

4. How does your own state, city, or metropolitan area compare in its growth rate with other regions in the table? What steps could your region take to lessen potential impacts of population growth on (1) food supplies, (2) drinking water supplies, (3) forests and other natural areas, and (4) wildlife populations?

Mastering Environmental Science

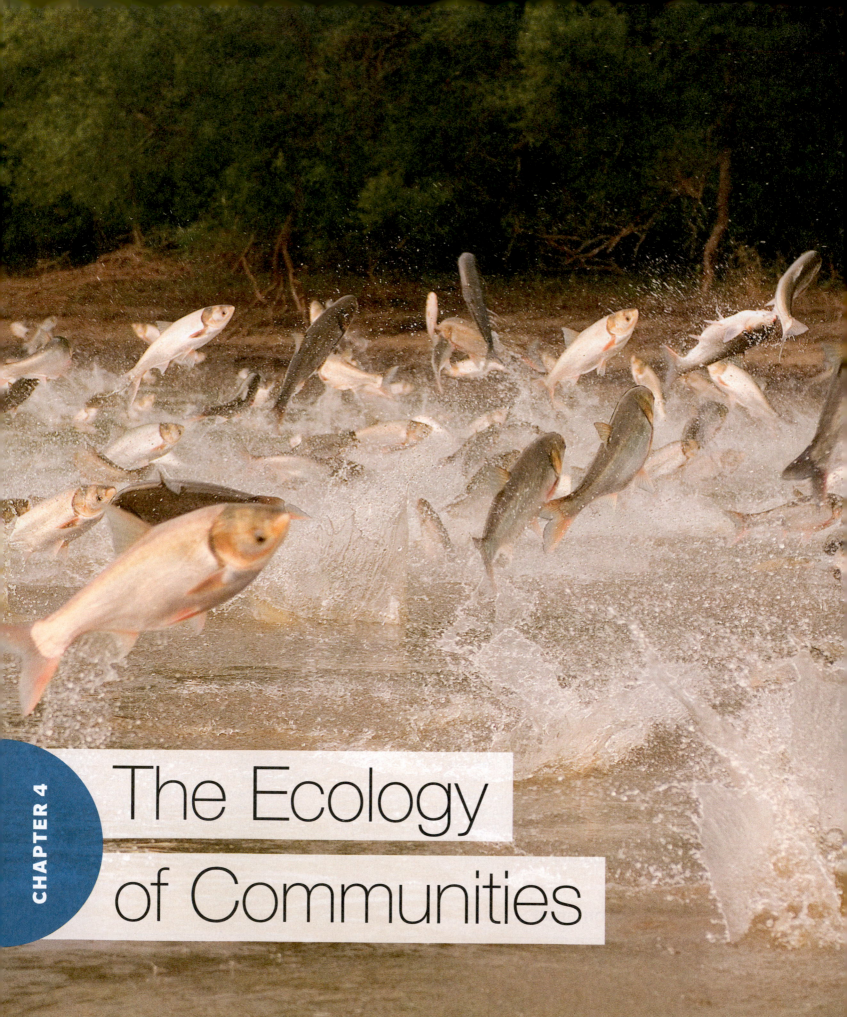

CHAPTER 4

The Ecology
of Communities

Leaping Fish, Backwards River:
Asian Carp Threaten the Great Lakes

CANADA
Great Lakes
Chicago   U.S.

> **Asian carp, if they come into the Great Lakes . . . could be very abundant. They could become one-third of the total fish weight. That is quite a lot.**
> —Hongyan Zhang, University of Michigan researcher

> **This is not about Asian carp. This is about two artificially connected watersheds that many people argue never should have been connected.**
> —Peter Annin, University of Notre Dame

Call it electroshock therapy for fish. The U.S. Army Corps of Engineers sends electricity into the waters of the Chicago Ship and Sanitary Canal to stun fish swimming toward Lake Michigan, causing them to drift back in the other direction, toward the Mississippi River. These electric barriers were an engineering solution to a biological challenge—to keep invasive Asian carp from reaching the Great Lakes.

But let's rewind the story a bit. It begins far away in China, where for centuries people have raised carp in aquaculture ponds for food. Starting in the 1960s and 1970s, managers of catfish farms and wastewater treatment plants in the southern and central United States began importing several species of these carp to help clean up infestations of algae and parasitic snails. Black carp eat snails and other mollusks. Grass carp scrape leafy aquatic plants off the bottom. Silver carp and bighead carp consume plankton (microscopic aquatic plants and animals) that they filter from the water.

In time, individuals of these four species, collectively nicknamed "Asian carp," escaped from these facilities and moved into streams, rivers, ponds, and lakes. Finding plenty of food, these alien fish grew rapidly, becoming too large for predators to capture. Reproducing quickly, their populations grew as they spread to waterways across the southern United States, especially in the lower Mississippi River Valley. As millions of these non-native fish invaded these ecosystems, they competed with native fish for food, preyed on native plants and animals, filtered out plankton, and stirred up sediment. These impacts reduced native populations, changed water quality, and altered aquatic communities. The silver carp even injures people: It leaps high out of the water when boats approach—and a hunk of scaly fish 1 m (3.3 ft) long weighing 27 kg (60 lb) can do real damage in a head-on collision!

Before long, Asian carp swam far up the Mississippi and Ohio rivers and their tributaries, invading waterways across the Midwest (**FIGURE 4.1**). As carp from the Mississippi River advanced up the Illinois River, people around the Great Lakes became alarmed. If the carp moved into Lakes Michigan, Huron, Superior, Ontario, and Erie, they

An invasive Asian carp ▲

Upon completing this chapter, you will be able to:

- **Summarize and compare the major types of species interactions**

- **Describe feeding relationships and energy flow, and use them to identify trophic levels and navigate food webs**

- **Discuss characteristics of a keystone species**

- **Characterize disturbance, succession, and notions of community change**

- **Describe the potential impacts of invasive species in communities, and offer solutions to biological invasions**

- **Explain the goals and methods of restoration ecology**

- **Identify and describe the terrestrial biomes of the world**

Silver carp leap from the waters of the Illinois River as a boat approaches.

FIGURE 4.1 Asian carp have spread from the southern United States up the Mississippi River and many of its tributaries. Red color indicates stretches of major rivers invaded by carp. The gray-shaded area indicates the region in which carp have invaded smaller rivers, lakes, and ponds. When Chicago **(inset)** built canals and reversed the flow of the Chicago River to flow into the Illinois River, this connected the watersheds of the Great Lakes and the Mississippi River, enabling species to move between them, affecting aquatic communities in each watershed. Today an electric barrier (see inset) is in place to try to stop Asian carp from reaching the Great Lakes.

might alter the ecology of the lakes and destroy multimillion-dollar sport and commercial fisheries. Millions of people could be affected in Wisconsin, Minnesota, Ohio, Michigan, Indiana, Illinois, Pennsylvania, New York, and Ontario.

An Asian carp invasion of the Great Lakes is possible only because of an amazing feat the city of Chicago pulled off a century ago. Chicago had always drawn its drinking water from Lake Michigan, but it also was polluting this drinking water source by dumping its sewage into the Chicago River, which flowed into the lake. So, in 1900 Chicago reversed the flow of the river! It engineered a canal system that connected the Chicago River to the Illinois River, whose waters flowed naturally to the Mississippi River (see inset in Figure 4.1). As a result, water was now pulled out of Lake Michigan and through the canals, sending Chicago's waste flowing southwest down the Illinois River and into the Mississippi, where it became other people's problem.

In this way, two of North America's biggest natural watersheds—those of the Mississippi River and the Great Lakes—were artificially joined. This was a boon for ship navigation and commerce, but it also allowed species that had evolved in the Great Lakes to enter the Mississippi Basin, and vice versa. And when species from one place are introduced to another, ecological havoc can result.

Meanwhile, hundreds of miles away at their eastern end, the Great Lakes were connected to the Atlantic Ocean when engineers completed the Saint Lawrence Seaway in 1959. This 600-km (370-mi) system of locks and canals along the Saint Lawrence River allowed ocean-going ships to travel into the lakes. It also allowed access for non-native species, including the

sea lamprey, an eel-like creature that began attacking and parasitizing the lakes' freshwater fish. All told, the Great Lakes today host 180 non-native species that are considered to be invasive.

Ships have been part of the problem. To maintain stability, ships take water into their hulls as they begin their voyage and then discharge it at their destination. Along with this *ballast water* come all the plants and animals that were taken up in the water. In this way, the zebra mussel was brought to America in the 1980s from seas in central Asia. Within a few years, this tiny mussel spread by the trillions through the Great Lakes and into waterways across half the nation, clogging pipes, covering boat propellers, outcompeting native mussels, and transforming fisheries. The zebra mussel invasion has been blamed for economic losses of hundreds of millions of dollars each year.

As zebra mussels sped down the Illinois River plastered to boat hulls, they passed Asian carp making their way up the river toward the Great Lakes. Fishermen and scientists alarmed about the carp bent the ear of policymakers, who now fund biologists to monitor lakes, rivers, and canals for signs of the alien fish. They are also funding numerous efforts to prevent and control their spread, including catching the carp; poisoning them; deterring their movements; introducing larger fish to eat them; and persuading restaurants to put them on the menu. Against this backdrop, Michigan and other Great Lakes states launched lawsuits against Chicago in an unsuccessful attempt to force it to shut down its canal system.

Time will tell whether Asian carp successfully invade the Great Lakes and, if so, what impacts they will have. What is clear is that we live in a dynamic world, and that as species are moved from place to place they can transform the communities around us.

Species Interactions

By interacting with many species in a variety of ways, Asian carp have set in motion an array of changes in the communities they have invaded. Interactions among species are the threads in the fabric of ecological communities. Ecologists organize species interactions into several main categories (**TABLE 4.1**).

TABLE 4.1 Species Interactions: Effects on Each Participant

TYPE OF INTERACTION	EFFECT ON SPECIES 1	EFFECT ON SPECIES 2
Competition	–	–
Predation, parasitism, herbivory	+	–
Mutualism	+	+

"+" denotes a positive effect; "–" denotes a negative effect.

Competition can occur when resources are limited

When multiple organisms seek the same limited resource, their relationship is said to be one of **competition**. Competing organisms do not usually fight with one another directly and physically. Instead, competition is generally more subtle and indirect, taking place as organisms vie with one another to procure resources. Such resources include food, water, space, shelter, mates, sunlight, and more. Competitive interactions can occur between members of the same species (intraspecific competition) or between members of different species (interspecific competition). As an example of interspecific competition, researchers have found that silver and bighead carp compete with native fish for plankton. In the Illinois River, these two species of Asian carp eat so much plankton that native fish have become smaller and have had to shift their diets.

If one species is a very effective competitor, it may exclude other species from resource use entirely. Alternatively, species may be able to coexist, adapting over evolutionary time as natural selection (p. 50) favors individuals that use slightly different resources or that use shared resources in different ways. For example, if two bird species eat the same type of seeds, natural selection might drive one species to specialize on larger seeds and the other to specialize on

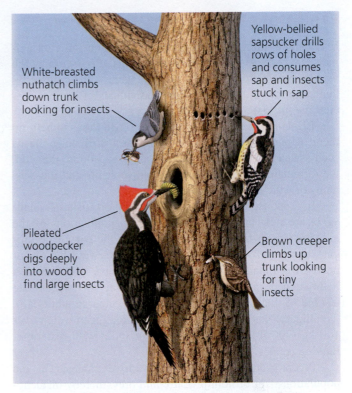

White-breasted nuthatch climbs down trunk looking for insects

Yellow-bellied sapsucker drills rows of holes and consumes sap and insects stuck in sap

Pileated woodpecker digs deeply into wood to find large insects

Brown creeper climbs up trunk looking for tiny insects

FIGURE 4.2 When species compete, they may partition resources. Birds that forage for insects on tree trunks use different portions of the trunk and seek different foods in different ways.

smaller seeds. This process is called **resource partitioning** because the species partition, or divide, the resources they use in common by specializing in different ways (**FIGURE 4.2**).

In competitive interactions, each participant exerts a negative effect on other participants by taking resources the others could have used. This is reflected in the two minus signs shown for competition in Table 4.1. In other types of interactions, some participants benefit while others are harmed; that is, one species exploits the other (note the +/– interactions in Table 4.1). Such exploitative interactions include predation, parasitism, and herbivory (**FIGURE 4.3**).

FIGURE 4.3 Predation, parasitism, and herbivory are exploitative interactions in which one participant benefits at the expense of another. In predation **(a)**, a predator kills and eats prey. In parasitism **(b)**, a parasite gains nourishment while harming its host. In herbivory **(c)**, an animal feeds on plants.

A snake devours a frog.

Sea lampreys suck blood from Great Lakes fish.

A caterpillar feeds on leaves.

(a) Predation

(b) Parasitism

(c) Herbivory

Predators kill and consume prey

Every living thing needs to procure food, and for most animals, this means eating other organisms. **Predation** is the process by which individuals of one species—the *predator*—hunt, capture, kill, and consume individuals of another species, the *prey* (see Figure 4.3a).

Black carp were introduced to U.S. fish farms to help rid them of parasitic snails because the carp prey on mollusks such as snails. However, once the carp escaped into the wild, these predators began preying on native snails and mussels, many of which became threatened.

In response to the spread of Asian carp, fisheries managers are now proposing to reestablish populations of the alligator gar, a fish that would prey on the carp. The alligator gar is native to North American waterways but disappeared from many rivers because in the past, fishermen felt it threatened sport fish, and managers set out to eradicate it. One of the few fish large enough to catch an adult carp, the alligator gar grows up to 3 m (9 ft) long, can weigh over 140 kg (300 lb), and features a long snout with fearsome-looking teeth.

Predator-prey interactions can drive changes in the population sizes of both predator and prey, thereby influencing the composition of ecological communities and affecting the structure of the food webs we will examine shortly. Predation also has evolutionary consequences. Individual predators that are more adept at capturing prey may live longer lives, reproduce more, and be better providers for their offspring. Natural selection (p. 50) will thereby lead to the evolution of adaptations (p. 51) that enhance hunting skills. Prey, however, face a stronger selective pressure—the risk of immediate death. As a result, predation pressure has driven the evolution of an elaborate array of defenses against being eaten (**FIGURE 4.4**).

Parasites exploit living hosts

Organisms can exploit other organisms without killing them. **Parasitism** is a relationship in which one organism, the *parasite*, depends on another, the *host*, for nourishment or some other benefit while doing the host harm (see Figure 4.3b). Unlike predation, parasitism usually does not result in an organism's immediate death.

Many types of parasites live inside their hosts. For example, tapeworms live in their hosts' digestive tracts, robbing them of nutrition. The larvae of parasitoid wasps burrow into the tissues of caterpillars and consume them from the inside. Other parasites are free-living. Cuckoos of Eurasia and cowbirds of the Americas lay their eggs in other birds' nests and let the host raise their young. Still other parasites live on the exterior of their hosts, such as fleas or ticks that suck blood through the skin. The sea lamprey is a tube-shaped vertebrate that grasps fish with its suction-cup mouth and rasping tongue, sucking their blood for days or weeks (see Figure 4.3b). Sea lampreys invaded the Great Lakes from the Atlantic Ocean after people dug canals to connect the lakes for shipping. The lampreys soon devastated economically important fisheries of chubs, lake herring, whitefish, and lake trout.

Parasites that cause disease in their hosts are called *pathogens*. Common human pathogens include the protists that cause malaria and amoebic dysentery, the bacteria that cause pneumonia and tuberculosis, and the viruses that cause hepatitis and AIDS.

Just as predators and prey evolve in response to one another, so do parasites and hosts, in a reciprocal process of adaptation and counter-adaptation called **coevolution.** Hosts and parasites may become locked in a duel of escalating adaptations, known as an *evolutionary arms race*. Like rival nations racing to stay ahead of one another in military

A gecko's camouflage hides it from predators.

(a) Cryptic coloration (camouflage)

A yellowjacket's coloring signals that it is dangerous.

(b) Warning coloration

This caterpillar's false eyespots startle predators by mimicking a snake's head.

(c) Mimicry

FIGURE 4.4 Natural selection to avoid predation has resulted in fabulous adaptations. Some prey species use cryptic coloration **(a)** to blend into their background. Others are brightly colored **(b)** to warn predators they are toxic, distasteful, or dangerous. Others use mimicry **(c)** to fool predators.

technology, host and parasite repeatedly evolve new responses to the other's latest advance. In the long run, though, it may not be in a parasite's best interest to do its host too much harm. In many cases, a parasite may leave more offspring by allowing its host to live longer.

Herbivores exploit plants

In **herbivory,** animals feed on the tissues of plants. Insects that feed on plants are the most common type of herbivore; nearly every plant in the world is attacked by insects (see Figure 4.3c). Herbivory generally does not kill a plant outright but may affect its growth and reproduction.

Like animal prey, plants have evolved an impressive arsenal of defenses against the animals that feed on them. Many plants produce chemicals that are toxic or distasteful to herbivores. Others arm themselves with thorns, spines, or irritating hairs. Still others attract insects like ants or wasps that attack herbivorous insects. In response, herbivores evolve ways to overcome these defenses, and the plant and the animal may embark on an evolutionary arms race.

Mutualists help one another

Unlike exploitative interactions, **mutualism** is a relationship in which two or more species benefit from interacting with one another. Generally, each partner provides some resource or service that the other needs (**FIGURE 4.5**).

Many mutualistic relationships—like many parasitic ones—occur between organisms that live in close physical contact. Physically close association is called **symbiosis,** and symbiosis can be either mutualistic or parasitic. Most terrestrial plant species depend on mutualisms with fungi; plant roots and some fungi together form symbiotic associations called *mycorrhizae*. In these relationships, the plant provides energy and protection to the fungus while the fungus helps the plant absorb nutrients from the soil. In the ocean, coral polyps, the tiny animals that build coral reefs (p. 267), provide

FIGURE 4.5 In mutualism, organisms of different species benefit one another. Hummingbirds visit flowers to gather nectar, and in the process they transfer pollen between flowers, helping the plant to reproduce.

housing and nutrients for specialized algae in exchange for food the algae produce through photosynthesis (pp. 34–35).

You, too, are part of a symbiotic mutualism. Your digestive tract is filled with microbes that help you digest food and carry out other bodily functions. In return, you provide these microbes a place to live. Without these mutualistic microbes, none of us would survive for long.

Not all mutualists live in close proximity. **Pollination** (p. 153) involves free-living organisms that may encounter each other only once. Bees, birds, bats, and other creatures transfer pollen (containing male sex cells) from flower to flower, fertilizing ovaries (containing female sex cells) that grow into fruits with seeds. These pollinators consume pollen or nectar (a reward the plant uses to entice them), and the plants are pollinated and reproduce. Bees pollinate three-fourths of our crops—from soybeans to potatoes to tomatoes to beans to cabbage to oranges.

Ecological Communities

A **community** is an assemblage of populations of organisms living in the same area at the same time (as we saw in Figure 3.10, p. 60). Members of a community interact in the ways discussed above, and these species interactions have indirect effects that ripple outward to affect other community members. Species interactions help determine the structure, function, and species composition of communities. Community ecology (p. 59) is the scientific study of species interactions and the dynamics of communities. Community ecologists study which species coexist, how they interact, how communities change through time, and why these patterns occur.

Food chains consist of trophic levels

Some of the most important interactions among community members involve who eats whom. As organisms feed on one another, matter and energy move through the community from one **trophic level,** or rank in the feeding hierarchy, to another (**FIGURE 4.6**; and compare Figure 2.15, p. 37). As matter and energy are transferred from lower trophic levels to higher ones, they are said to pass up a **food chain,** a linear series of feeding relationships.

Producers *Producers*, or *autotrophs* ("self-feeders," p. 34), make up the first (lowest) trophic level. Terrestrial green plants, cyanobacteria, and algae capture solar energy and use photosynthesis to produce sugars (p. 35). The chemosynthetic bacteria of hydrothermal vents use geothermal energy in a similar way to produce food (p. 35).

Consumers Organisms that consume producers are known as *primary consumers* and make up the second trophic level. Herbivorous grazing animals, such as deer and grasshoppers, are primary consumers. The third trophic level consists of *secondary consumers*, which prey on primary consumers. Wolves that prey on deer are secondary consumers, as are rodents and birds that prey on grasshoppers. Predators that feed at still higher trophic levels are known as *tertiary consumers*. Examples of tertiary

An aquatic food chain   **An terrestrial food chain**

Tertiary consumers

Cormorant   Hawk

Secondary consumers

Fish   Rodent

Primary consumers

Zooplankton   Grasshopper

Producers

Phytoplankton   Plant

FIGURE 4.6 A food chain organizes species hierarchically by trophic level. The diagram shows aquatic **(left)** and terrestrial **(right)** examples at each level. Arrows indicate the direction of energy flow. Producers synthesize food by photosynthesis, primary consumers (herbivores) feed on producers, secondary consumers eat primary consumers, and tertiary consumers eat secondary consumers. Detritivores and decomposers (not shown) feed on nonliving organic matter and "close the loop" by returning nutrients to the soil or the water column for use by producers.

consumers include hawks and owls that eat rodents that have eaten grasshoppers.

Detritivores and decomposers Detritivores and decomposers consume nonliving organic matter. Detritivores, such as millipedes and soil insects, scavenge the waste products or dead bodies of other community members. Decomposers, such as fungi, bacteria, and earthworms, break down leaf litter and other nonliving matter into simpler constituents that can be taken up and used by plants. These organisms enhance the topmost soil layers (pp. 144–145) and play essential roles as the community's recyclers, making nutrients from organic matter available for reuse by living members of the community.

In Great Lakes communities, the main producers are phytoplankton, consisting of microscopic algae, protists, and cyanobacteria that drift in open water and conduct

photosynthesis with sunlight that penetrates the water. The most abundant primary consumers are zooplankton, the tiny aquatic animals that eat phytoplankton. Some fish eat phytoplankton and thus are primary consumers, whereas fish that eat zooplankton are secondary consumers. Tertiary consumers include birds and larger fish that feed on zooplankton-eating fish. (The left side of Figure 4.6 shows these relationships in very generalized form.) Asian carp that eat both phytoplankton and zooplankton function on multiple trophic levels. When an organism dies and sinks to the bottom, detritivores scavenge its tissues and decomposers recycle its nutrients.

Energy decreases at higher trophic levels

At each trophic level in a food chain, organisms use energy in cellular respiration (p. 35), and most of the energy ends up being given off as heat. Only a small portion of the energy is transferred to the next trophic level through predation, herbivory, or parasitism. A general rule of thumb is that each trophic level contains about 10% of the energy of the trophic level below it (although the actual proportion varies greatly). This pattern can be visualized as a pyramid (**FIGURE 4.7**).

This pattern also tends to hold for numbers of organisms; in general, fewer organisms exist at high trophic levels than at low ones. A grasshopper eats many plants in its lifetime, a rodent eats many grasshoppers, and a hawk eats many rodents. Thus, for every hawk there must be many rodents, still more grasshoppers, and an immense number of plants. Because the difference in numbers of organisms among trophic levels tends to be large, the same pyramid-like relationship often holds true for **biomass,** the collective mass of living matter.

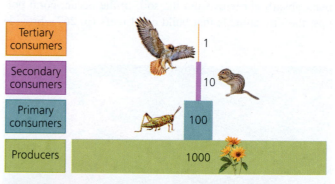

Tertiary consumers   1
Secondary consumers   10
Primary consumers   100
Producers   1000

FIGURE 4.7 Lower trophic levels contain more energy— and generally more organisms and biomass—than higher trophic levels. The 10:1 ratio shown here for energy is typical, but varies greatly.

DATA Q Using the ratios shown in this example, let's suppose that primary consumers have a total energy content of 3000 kcal/m²/yr. • How much energy would you expect among secondary consumers? • How much would you expect among tertiary consumers?

Go to **Interpreting Graphs & Data** on **Mastering** Environmental Science

This pyramid pattern illustrates why eating at lower trophic levels—being vegan or vegetarian, for instance—decreases a person's ecological footprint. Each amount of meat or other animal product we eat requires the input of a considerably greater amount of plant material (see Figure 7.14, p. 156). Thus, when we eat animal products, we use up far more energy per calorie that we gain than when we eat plant products.

Food webs show feeding relationships and energy flow

Thinking in terms of food chains is conceptually useful, but ecological systems are far more complex than simple linear chains. A fuller representation of the feeding relationships in a community is a **food web**—a visual map that shows the many paths along which energy and matter flow as organisms consume one another.

FIGURE 4.8 portrays a food web from a temperate deciduous forest of eastern North America. It is greatly simplified and leaves out most species and interactions. Note, however, that even within this simplified diagram we can pick out many

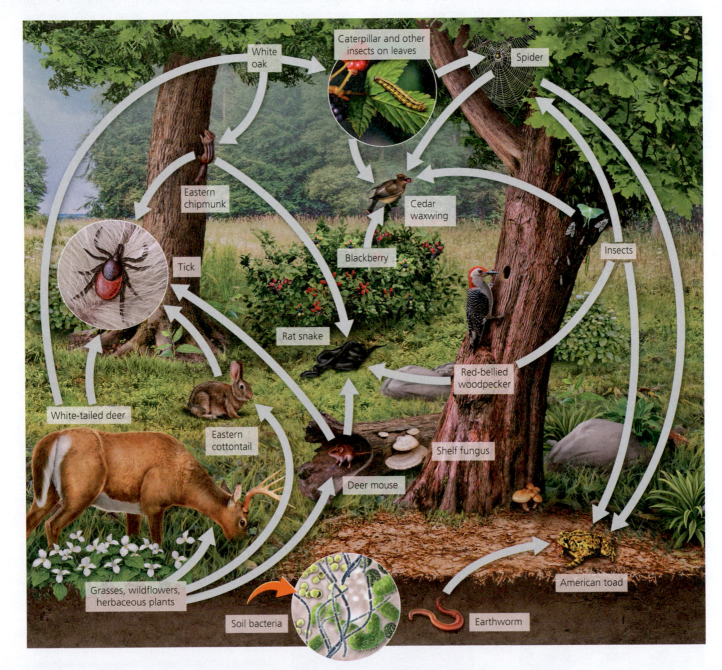

FIGURE 4.8 This food web shows feeding relationships among organisms in eastern North America's temperate deciduous forest. Arrows indicate the direction of energy flow as a result of predation, parasitism, or herbivory. The actual community contains many more species and interactions than can be shown.

food chains involving different sets of species. For instance, grasses are eaten by deer mice that may be consumed by rat snakes—while in another food chain, blackberry leaves are consumed by caterpillars that provide food for spiders, which in turn may be eaten by American toads.

A food web for the aquatic community of the Illinois River would include phytoplankton that photosynthesize; zooplankton that eat them; fish that eat phytoplankton and zooplankton; and herons, ducks, and raptors that eat fish. It would include many underwater plants and algae that provide food and shelter for animals. And it would include a diversity of insect larvae, mussels, snails, crayfish, and other freshwater invertebrates living amid the plants and on the muddy bottom. Today, an Illinois River food web would also include invasive Asian carp that are modifying many aspects of the community as they carve out their own roles within it (jump ahead to Figure 4.12 for a peek at some of these interactions).

In 2016, researchers assessed the diets of various fish native to Lake Erie to predict how the food web of the lake might change if Asian carp were to invade it. Their ecological model projected that competition for food from silver and bighead carp would reduce the biomass of plankton-eating fish such as rainbow smelt, gizzard shad, and emerald shiner by 13–37%, and that the biomass of predatory adult walleye would decrease in turn once they had fewer of those fish to eat. However, smallmouth bass were predicted to increase by 13–16% because they would benefit from consuming the newly abundant young of the invasive carp. Overall, the researchers predicted that in 20 years the carp would comprise 30% of the total fish biomass in Lake Erie, after causing populations of most native fish species to decline.

Some organisms play outsized roles

"Some animals are more equal than others," George Orwell wrote in his classic novel *Animal Farm*. Orwell was making wry sociopolitical commentary, but his remark hints at an ecological truth. In communities, some species exert greater influence than do others. A species that has strong or wide-reaching impact far out of proportion to its abundance is often called a **keystone species.** A keystone is the wedge-shaped stone at the top of an arch that holds the structure together. Remove the keystone, and the arch will collapse (**FIGURE 4.9a**). In an ecological community, removal of a keystone species will likewise have major consequences.

Often, secondary or tertiary consumers near the tops of food chains are considered keystone species. Top predators control populations of herbivores, which otherwise could multiply and greatly modify the plant community (**FIGURE 4.9b**). Thus, predators at high trophic levels can indirectly promote populations of organisms at low trophic levels by keeping species at intermediate trophic levels in check, a phenomenon referred to as a **trophic cascade.** For example, the U.S. government long paid bounties to promote the hunting of wolves and mountain lions, which were largely exterminated by the mid-20th century. In the absence of these predators, deer populations grew, overgrazing forest-floor vegetation and eliminating tree seedlings, leading to major changes in forest structure.

This removal of top predators was an uncontrolled large-scale experiment with unintended consequences, but ecologists have verified the keystone species concept in controlled scientific experiments. Classic research by biologist Robert Paine established that the predatory sea star *Pisaster ochraceus* shapes the community composition of intertidal organisms (p. 263) on North America's

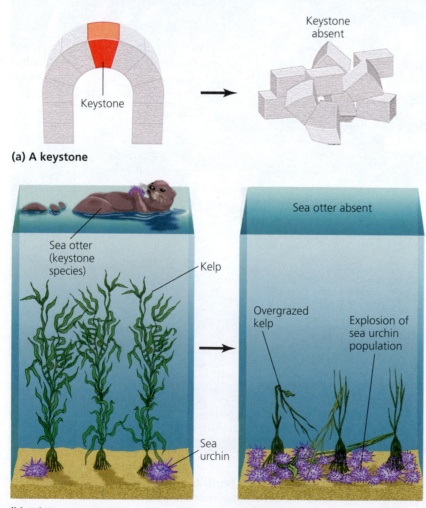

(a) A keystone

Keystone

Keystone absent

(b) A keystone species

Sea otter (keystone species)

Kelp

Sea urchin

Sea otter absent

Overgrazed kelp

Explosion of sea urchin population

FIGURE 4.9 Sea otters are a keystone species. A keystone **(a)** sits atop an arch, holding the structure together. A keystone species **(b)** is a species that exerts great influence on a community's composition and structure. Sea otters consume sea urchins that eat kelp in coastal waters of the Pacific. Otters keep urchin numbers down, allowing lush underwater forests of kelp to grow, providing habitat for many species. When otters are absent, urchins increase and devour the kelp, destroying habitat and depressing species diversity.

Pacific coast. When *Pisaster* is present in this community, species diversity is high, with various types of barnacles, mussels, and algae. When *Pisaster* is removed, the mussels it preys on become numerous and displace other species, suppressing species diversity.

Animals at high trophic levels—such as wolves, sea stars, sharks, and sea otters—are often viewed as keystone species, but other types of organisms also exert strong community-wide effects. "Ecosystem engineers" physically modify environments. Beavers build dams across streams, creating ponds and swamps by flooding land. Prairie dogs dig burrows that aerate the soil and serve as homes for other animals. Ants disperse seeds, redistribute nutrients, and selectively protect or destroy insects and plants near their colonies. Asian carp alter the communities they invade in many ways.

Less conspicuous organisms at low trophic levels can exert still more impact. Remove the fungi that decompose dead matter, or the insects that control plant growth, or the phytoplankton that support the marine food chain, and a community may change rapidly indeed. However, because there are usually more species at lower trophic levels, it is less likely that any single one of them alone has wide influence. Often if one species is removed, other species that remain may be able to perform many of its functions.

Communities respond to disturbance in various ways

The removal of a keystone species is just one type of disturbance that can modify a community. In ecological terms, a **disturbance** is an event that has drastic impacts on environmental conditions, resulting in changes to the community and ecosystem. A disturbance can be localized, such as when a tree falls in a forest, creating a gap in the canopy that lets in sunlight. Or it can be as large and severe as a hurricane, tornado, or volcanic eruption. Some disturbances are sudden, such as landslides or floods, whereas others are more gradual. Some recur regularly and are considered normal aspects of a system (such as periodic fire, seasonal storms, or cyclical insect outbreaks). Today, human impacts are major sources of disturbance for ecological communities worldwide.

Communities are dynamic and may respond to disturbance in several ways. A community that resists change and remains stable despite disturbance is said to show **resistance** to the disturbance. Alternatively, a community may show **resilience**, meaning that it changes in response to disturbance but later returns to its original state. Or, a community may be modified by disturbance permanently and never return to its original state.

Succession follows severe disturbance

If a disturbance is severe enough to eliminate all or most of the species in a community, the affected site may then undergo a predictable series of changes that ecologists have traditionally called **succession.** In the conventional view of this process, there are two types of succession (**FIGURE 4.10**). **Primary succession** follows a disturbance so severe that no vegetation or soil life remains from the community that had occupied the site. In primary succession, a community is built essentially from scratch. In contrast, **secondary succession** begins when a disturbance dramatically alters an existing

FIGURE 4.10 In succession, an area's plant community passes through a series of typical stages. Primary succession begins as organisms colonize a lifeless new surface (two panels at top left). Secondary succession occurs after some disturbance removes most vegetation from an area (panel at bottom left).

community but does not destroy all life and organic matter. In secondary succession, vestiges of the previous community remain, and these building blocks help shape the process.

At terrestrial sites, primary succession takes place after a bare expanse of rock, sand, or sediment becomes exposed to the atmosphere. This can occur when glaciers retreat, lakes dry up, or volcanic lava or ash covers a landscape. Species that arrive first and colonize the new substrate are referred to as **pioneer species.** Pioneer species are adapted for colonization; for instance, they generally have spores or seeds that can travel long distances.

The pioneers best suited to colonizing bare rock are the mutualistic aggregates of fungi and algae known as lichens. In lichens, the algal component provides food and energy via photosynthesis while the fungal component grips the rock and captures moisture. As lichens grow, they secrete acids that break down the rock surface, beginning the process that forms soil. Small plants and insects arrive, providing more nutrients and habitat. As time passes, larger plants and animals establish, vegetation increases, and species diversity rises.

Secondary succession begins when a fire, a storm, logging, or farming removes much of the biotic community. Consider a farmed field in eastern North America that has been abandoned. The site becomes colonized by pioneer species of grasses, herbs, and forbs that disperse well or were already in the vicinity. Soon, shrubs and fast-growing trees such as aspens and poplars begin to grow. As time passes, pine trees rise above the pioneer trees and shrubs, forming a pine forest. This forest attains an understory of hardwood trees, because pine seedlings do not grow well under a canopy but some hardwood seedlings do. Eventually the hardwoods outgrow the pines, creating a hardwood forest.

Succession occurs in many ecological systems. For instance, a pond may undergo succession as algae, microbes, plants, and zooplankton grow, reproduce, and die, gradually filling the water body with organic matter. The pond acquires further organic matter and sediments from streams and runoff, and eventually it may fill in, becoming a bog (p. 260) or a terrestrial system.

In the traditional view of succession described here, the process leads to a *climax community,* which remains in place until some

FAQ

Once we disturb a community, won't it return to its original state if we just leave the area alone?

Probably not, if the disturbance has been substantial. For example, if soil has become compacted or if water sources have dried up, then the plant species that grew at the site originally may no longer be able to grow. Different plant species may take their place—and among them, a different suite of animal species may find habitat. Sometimes a whole new community may arise. For instance, as deforestation has caused climate to become drier in parts of the Amazon, unprecedented fires have burned tropical rainforests, converting some of them to scrub-grassland. In the past, people didn't realize how permanent such changes could be, because we tended to view natural systems as static, predictable, and liable to return to equilibrium. Today ecologists recognize that systems are highly dynamic and can sometimes undergo rapid, extreme, and long-lasting change.

disturbance restarts succession. Early ecologists felt that each region had its own characteristic climax community, determined by climate.

Communities may undergo shifts

Today, ecologists recognize that community change is far more variable and less predictable than early models of succession suggested. Conditions at one stage may promote progression to another stage, or organisms may, through competition, inhibit a community's progression to another stage. The trajectory of change can vary greatly according to chance factors, such as which particular species happen to gain an early foothold. And climax communities are not determined solely by climate, but vary with soil conditions and other factors from one time or place to another. Ecologists came to modify their views about how communities respond to disturbance after observing changes during long-term field studies at locations such as Mount Saint Helens following its eruption (see THE SCIENCE BEHIND THE STORY, pp. 80–81).

Once a community is disturbed and changes are set in motion, there is no guarantee that it will return to its original state. Instead, sometimes communities may undergo a **regime shift,** or *phase shift,* in which the character of the community fundamentally changes. This can occur if some crucial climatic threshold is passed, a keystone species is lost, or a non-native species invades. For instance, many coral reef communities have become dominated by algae after people overharvested fish or turtles that consume algae. In some grasslands, livestock grazing and fire suppression have led shrubs and trees to invade, forming shrublands. And removing sea otters can lead to the loss of kelp forests (p. 76). Regime shifts show that we cannot count on being able to reverse damage caused by human disturbance, because some changes we set in motion may become permanent.

Today human disturbance and the introduction of non-native species are creating wholly new communities that have not previously occurred on Earth. These **novel communities,** or *no-analog communities,* are composed of novel mixtures of plants and animals and have no analog or precedent. Given today's fast-changing climate, habitat alteration, species extinctions, and species invasions, scientists predict that we will see more and more novel communities.

Introduced species may alter communities

Traditional concepts of communities involve species native to an area. But what if a species not native to the area arrives from elsewhere? In our age of global mobility and trade, people have moved countless organisms from place to place. As a result, today most non-native arrivals in a community are **introduced species,** species introduced by people.

In many cases, introductions have been intentional. People intentionally imported Asian carp to North America, believing these fish could offer valuable services to aquaculture and wastewater treatment facilities. Many other introductions have occurred by accident. Zebra mussels arrived in the Great Lakes as a result of global trade, inadvertently transported in the ballast water of cargo ships.

Most introduced species fail to establish populations, but a minority succeed. Over time, successful alien species may accumulate, and today many regions support novel communities rich in non-native species. Over 180 non-native species inhabit the Great Lakes. Across much of San Francisco Bay the majority of species are non-native, while 30% of California's plant species are non-native.

Of those species that successfully persist in their new home, some may do exceptionally well, spreading widely, pushing aside native species, and coming to dominate communities. Ecologists term such species **invasive species** because they "invade" native communities (**FIGURE 4.11**). The U.S. government defines invasive species as non-native species "whose introduction causes or is likely to cause economic or environmental harm or harm to human health."

Introduced species can become invasive when limiting factors (p. 63) that regulate their population growth are absent. Plants and animals brought to a new area may leave behind the predators, parasites, herbivores, and competitors that had exploited them in their native land. If few organisms in a new environment eat, parasitize, or compete with an introduced species, then it may thrive and spread. As the species proliferates, it may exert diverse influences on other community members (**FIGURE 4.12**). By altering communities, invasive species are one of the central ecological forces in today's world.

Examples abound of invasive species that have had major ecological impacts (pp. 178–180). The chestnut blight, an Asian fungus, killed nearly every mature American chestnut, a dominant tree species of eastern North American forests, between 1900 and 1930. Asian trees had evolved defenses against the fungus over millennia of coevolution, but the American chestnut had not. A different

FIGURE 4.11 When non-native species are introduced, they may sometimes become invasive. Garlic mustard has displaced native ground cover in many forests of eastern North America. Dedicated volunteers are doing their best to remove it.

fungus caused Dutch elm disease, which destroyed most of the American elms that once gracefully lined the streets of many U.S. cities. Alien grasses introduced in the American West for ranching have overrun entire regions, crowding out native vegetation and encouraging fires that damage native plants like sagebrush, furthering the dominance of the grasses. Hundreds of island-dwelling animal and plant species worldwide have been driven extinct by goats, pigs, rats, and other mammals introduced by human colonists and visitors (Chapter 3).

Because of such impacts, most ecologists view invasive species in a negative light, and this often carries over to their perceptions of non-native species in general. Yet many of us enjoy the beauty of introduced ornamental plants in our gardens. Some introduced species provide economic benefits, such as the European honeybee, which pollinates many of our crops (p. 153). And some organisms are introduced intentionally to control pests through biocontrol (p. 153).

In recent years, some scientists have begun pushing back against the conventional wisdom. They point out that the locations and habitats in which exotic species generally establish themselves tend to be those that are already severely disturbed by human impact. In such degraded areas, non-native species can actually be beneficial; they may provide resources to native organisms and, arguably, may help severely degraded ecosystems

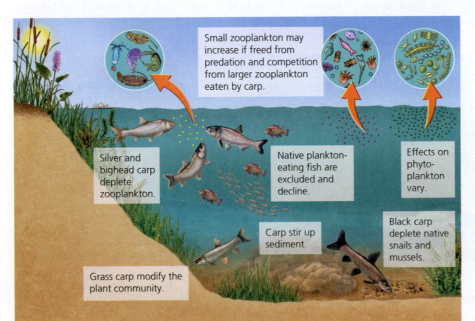

Small zooplankton may increase if freed from predation and competition from larger zooplankton eaten by carp.

Silver and bighead carp deplete zooplankton.

Native plankton-eating fish are excluded and decline.

Effects on phytoplankton vary.

Carp stir up sediment.

Black carp deplete native snails and mussels.

Grass carp modify the plant community.

FIGURE 4.12 Asian carp modify the aquatic communities they invade. Silver and bighead carp outcompete native fish for zooplankton, which can in turn affect phytoplankton in various ways. Grass and black carp alter water quality by stirring up sediment as they consume native plants and mollusks. *Based on data from Sass, G., et al., 2014. Invasive bighead and silver carp effects on zooplankton communities in the Illinois River, Illinois, USA. J. Great Lakes Research 40: 911–921, and various other studies.*

How Do Communities Recover after Disturbance?

The eruption of Mount Saint Helens offered ecologists a rare opportunity to study how communities recover from catastrophic disturbance. On May 18, 1980, this volcano in the state of Washington erupted in sudden and spectacular violence, with 500 times the force of the atomic blast at Hiroshima.

Dr. Virginia Dale at Mount Saint Helens

The massive explosion obliterated an entire landscape of forest as a scalding mix of gas, steam, ash, and rock was hurled outward. A pyroclastic flow (p. 237) sped downslope, along with the largest landslide in recorded history. Rock and ash rained down and mudslides and lahars (p. 240) raced down river valleys, inundating everything in their paths. Altogether, 4.1 km³ (1.0 mi³) of material was ejected from the mountain, affecting 1650 km² (637 mi²), an area twice the size of New York City.

In the aftermath of the blast, ecologists moved in to take advantage of the natural experiment of a lifetime. For them, the eruption provided an extraordinary chance to study how primary succession unfolds on a fresh volcanic surface. Which organisms would arrive first? What kind of community would emerge? How long it would take? These researchers set up study plots to examine how populations, communities, and ecosystems would respond.

Today, almost 40 years later, the barren gray moonscape that resulted from the blast is a vibrant green in many places (FIGURE 1), carpeted with shrubs, young trees, and colorful flowers. And what ecologists have learned has modified our view of primary succession and informed the entire study of disturbance ecology.

Given the ferocity and scale of the eruption, most scientists initially presumed that life had been wiped out completely over a large area. Based on traditional views of succession, they expected that pioneer species would colonize the area gradually, spreading slowly from the outside margins inward, and that over many years a community would be rebuilt in a systematic and predictable way.

Instead, researchers discovered that some plants and animals had survived the blast. Some were protected by deep snowbanks. Others were sheltered on steep slopes facing away from the blast. Still others were dormant underground when the eruption occurred. These survivors, it turned out, would play key roles in rebuilding the community.

Many of the ecologists drawn to Mount Saint Helens studied plants. Virginia Dale of Oak Ridge National Laboratory in Tennessee and her colleagues examined the debris avalanche, a landslide of rock and ash as deep as a 15-story building. This region appeared barren, yet small numbers of plants of 20 species had survived, growing from bits of root or stem carried down in the avalanche. However, most plant regrowth occurred from seeds blown in from afar. Dale's team used sticky traps to sample these seeds as they chronicled the area's recovery. After 30 years, the number of plant species had grown to 155, covering 50% of the ground surface.

One important pioneer species was the red alder. This tree grows quickly, deals well with browsing by animals, and produces many seeds at a young age. As a result, it has become the dominant tree species on the debris avalanche. Because it

(a) 1980   (b) 2011

FIGURE 1 Mount Saint Helens (a) after its eruption in 1980 and (b) three decades later.

fixes nitrogen (p. 42), the red alder enhances soil fertility and thereby helps other plants grow. Researchers predict that red alder will remain dominant for years or decades and that conifers such as Douglas fir will eventually outgrow the alders and establish a conifer-dominated forest.

Patterns of plant growth have varied in different areas. Roger del Moral of the University of Washington and his colleagues compared ecological responses on a variety of surfaces, including barren pumice, mixed ash and rock, mudflows, and the "blowdown zone" where trees were toppled like matchsticks. Numbers of species and percentage of plant cover increased in different ways on each surface, affected by a diversity of factors.

Parts of the blast zone were replanted by people. After the eruption, concerns about erosion led federal officials to disperse seeds of eight plant species (seven of them non-native) by helicopter over large areas in hopes of quickly stabilizing the surface with vegetation. Dale and her colleagues took the opportunity to compare plant regrowth in areas that were manually reseeded against regrowth in areas that recovered naturally. They found that manual seeding established plant cover on the ground more quickly, and that this effect was long-lasting: Even after 30 years, manually seeded areas had more plant cover than unseeded areas (**FIGURE 2a**). The same was true, to a lesser extent, with the number of plant species (**FIGURE 2b**). However, the manually reseeded areas contained a larger proportion of plants of non-native species than did the naturally seeded areas, even after 30 years. Researchers predict that eventually both types of areas will become more similar as conifer forest replaces the pioneer species—and Dale's team is finding evidence that this is already beginning to occur.

Among plants, windblown seeds accounted for most regrowth, but plants that happened to survive in sheltered "refugia" within the impact zone helped to repopulate areas nearby. Indeed, chance played a large role in determining which organisms survived and how vegetation recovered, researchers have found. Had the eruption occurred in late summer instead of spring, there would have been no snow, and many of the plants that survived would have died. Had the eruption occurred at night, nocturnal animals would have been hit harder.

Animals played major roles in the recovery. In fact, researchers such as Patrick Sugg and John Edwards of the University of Washington showed that insects and spiders arrived in great numbers before plants did. Insects fly, while spiders disperse by "ballooning" on silken threads, so in summer the atmosphere is filled with an "aerial plankton" of windblown arthropods. Trapping and monitoring at Mount Saint Helens in the months following the eruption showed that insects and spiders landed in the impact zone by the billions. Researchers estimated that more than 1500 species arrived in the first few years, surviving by scavenging or by preying on other arthropods. Most individuals soon died, but the nutrients from their bodies enriched the soil, helping the community to develop.

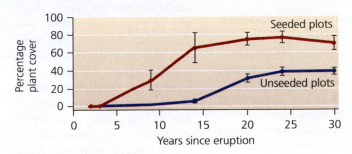

(a) Percentage plant cover

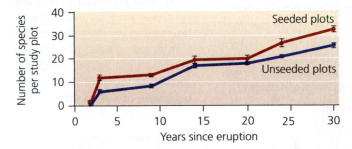

(b) Species richness

FIGURE 2 Plants recovered differently at manually seeded and unseeded sites at Mount Saint Helens. In the 30 years after the eruption, **(a)** percentage of ground covered by plants and **(b)** species richness of plants both increased, with cover and species richness being greater in manually reseeded areas than in unseeded areas recovering naturally. *Data from Dale, V.H. and E.M. Denton, in press. Plant succession on the Mount St. Helens debris avalanche deposit and the role of non-native species. In C.M. Crisafulli and V.H. Dale (Eds.),* Ecological responses at Mt. St. Helens: Revisited 35 years after the 1980 eruption. *New York, NY: Springer-Verlag.*

Once plants took hold, animals began exerting influence through herbivory. Caterpillars fed on plants, occasionally extinguishing small populations as they began to establish. Researchers led by Charles Crisafulli of the U.S. Forest Service studied mammals and found that their species diversity increased in a linear fashion. Fully 34 of the 45 mammal species present in the larger region colonized the blast area within 35 years—from tiny mice to elk and mountain goats.

All told, research at Mount Saint Helens has shown that succession is not a simple and predictable process. Instead, communities recover from disturbance in ways that are dynamic, complex, and highly dependent on chance factors. The results also show life's resilience. Even when the vast majority of organisms perish in a natural disaster, a few may survive, and their descendants may eventually build a new community.

Ecological change at Mount Saint Helens will continue for many decades more. All along, ecologists will continue to study and learn from this tremendous natural experiment.

Are All Introduced Species Harmful?

If we introduce a non-native species to a community and it greatly modifies the community, do you think that, in itself, is a bad thing? What if it causes a native species to disappear from the community? What if the non-native species arrived on its own, rather than through human intervention? What if it provides economic services, as the European honeybee does? What ethical standard(s) (p. 14) would you apply to assess whether we should reject, tolerate, or welcome an introduced species?

We can respond to invasive species with control, eradication, or prevention

Ever since U.S. policymakers began to recognize the impacts of invasive species in the 1990s, federal funding has become available for their control and eradication. Eradication (total elimination of a population) is extremely difficult, so managers usually aim merely to control populations—that is, to limit their growth, spread, and impact. Managers have tried to control Asian carp in heavily infested waterways like the Illinois River by netting juvenile fish; hiring contract fishermen to catch adults; bubbling carbon dioxide into the water; broadcasting noise into the water; and testing the effects of hot water and ozone. Researchers are also trying to develop a chemical microparticle that will poison carp while not affecting other fish, and they are aiming to reintroduce the alligator gar as a predator.

However, most of these are localized or short-term fixes. With one invasive species after another, managers find that control and eradication are so difficult and expensive that trying to prevent invasions in the first place represents a better investment. This explains why, with Asian carp, managers are working hard to prevent them from entering the Great Lakes. The Army Corps' electric barriers in the Chicago Ship and Sanitary Canal are one main prevention technique. Engineers have also built various dikes and berms to block floodwaters when heavy rains cause water to overflow from riverbanks and canals.

To meet the challenge of species transport across oceans in ballast water, the U.S. government and the international community each require ships to dump their freshwater ballast at sea (where freshwater organisms are killed by salt water) and exchange it with salt water before entering port areas. They also are working with industry researchers to test and approve methods of treating ballast water with

recover from human impact. Defenders of non-native species also stress that non-natives generally increase the species diversity of a community rather than decreasing it. Some even maintain that in a world shaped by human impact, the species that have proven themselves successful invaders may give biodiversity its best chance of persisting in the face of our dominance.

Whatever view one takes, the changes that introduced species—and particularly invasive species—bring to native populations and communities can be significant. These impacts are growing year by year with our increasing mobility and the globalization of our society.

chemicals, ultraviolet light, heat, electricity, and physical filtration to kill organisms onboard without exchanging water.

Communities can be restored

Invasive species are adding to the transformations that people have forced on natural systems through habitat alteration, pollution, overhunting, and climate change. Ecological systems support our civilization and all of life, so when systems cease to function, our health and well-being are threatened.

This realization has given rise to the science of **restoration ecology.** Restoration ecologists research the historical conditions of ecological communities as they existed before our industrialized civilization altered them. They then devise ways to restore altered areas to an earlier condition. In some cases, the intent is to restore the functionality of a system—to reestablish a wetland's ability to filter pollutants and recharge groundwater, for example, or a forest's ability to cleanse the air, build soil, and provide wildlife habitat. In other cases, the aim is to return a community to its natural "presettlement" condition. Either way, the science of restoration ecology informs the practice of **ecological restoration,** the on-the-ground efforts to carry out these visions and restore communities.

SUCCESS STORY · Restoring Native Prairie Near Chicago

Nearly all the tallgrass prairie in North America was converted to agriculture in the 1800s and early 1900s. Immense regions of land in states like Iowa, Kansas, and Illinois that today grow corn, wheat, and soybeans were once carpeted in a lush and diverse mix of grasses and wildflowers and inhabited by countless birds, bees, and butterflies. Today, people are trying to re-create these rich native landscapes through ecological restoration. Scientists and volunteers are restoring patches of prairie by planting native vegetation, weeding out

The restored prairie at Fermilab, outside Chicago, Illinois.

invaders and competitors, and introducing prescribed fire (p. 201) to mimic the fires that historically maintained prairie communities. The region outside Chicago, Illinois, boasts a number of prairie restoration projects, the largest of which is inside the gigantic ring at the Fermilab National Accelerator Laboratory. The project, begun in 1971 by ecologists Robert Betz and Raymond Schulenberg, today involves hundreds of community volunteers and has restored more than 400 ha (1000 acres) of prairie.

***EXPLORE THE DATA* at Mastering Environmental Science**

Ecological restoration involves trying to undo impacts of human disturbance and to reestablish species, populations, communities, and natural ecological processes. Restoration often involves removing invasive species and planting native vegetation. Sometimes it means reintroducing natural processes such as fire or flooding. It may also mean modifying the landscape to reduce erosion or influence patterns of water flow.

The world's largest restoration project is the ongoing effort to restore the Everglades, a vast ecosystem of marshes and seasonally flooded grasslands stretching across southern Florida. This wetland system has been drying out for decades because the water that feeds it has been manipulated for flood control and overdrawn for irrigation and development. Economically important fisheries have suffered greatly as a result, and the region's famed populations of wading birds have dropped by 90–95%. The 30-year, $7.8-billion restoration project intends to restore natural water flow by removing hundreds of miles of canals and levees. Because the Everglades provides drinking water for millions of Florida citizens, as well as considerable tourism revenue, restoring its ecosystem services (pp. 4, 39, 101) should prove economically beneficial as well as ecologically valuable.

Efforts in Florida have met with some success so far: Canals have been filled in and stretches of the Kissimmee River now flow freely (**FIGURE 4.13**), bringing water to Lake Okeechobee and then to the Everglades. But many challenges remain. Invasive species such as the Burmese python are spreading faster than they can be controlled, eating their way through the native fauna. Moreover, the ambitious project has struggled against budget shortfalls and political interference, and most of its goals have not yet been met. (We will explore ecological restoration projects again in Chapter 8; p. 185.)

FIGURE 4.13 The largest restoration project today is in Florida. Here, water from a canal has been returned to the Kissimmee River, which now winds freely toward the Everglades.

As our population grows and as development spreads, ecological restoration is becoming an increasingly vital conservation strategy. However, restoration is difficult, time-consuming, and expensive. It is therefore best, whenever possible, to protect natural systems from degradation in the first place.

Earth's Biomes

Across the world, each location is home to different sets of species, leading to endless variety in community composition. However, communities in far-flung places often share strong similarities, so we can classify communities into broad types. A **biome** is a major regional complex of similar communities—a large-scale ecological unit recognized primarily by its dominant plant type and vegetation structure. The world contains a number of terrestrial biomes, each covering large geographic areas (**FIGURE 4.14**).

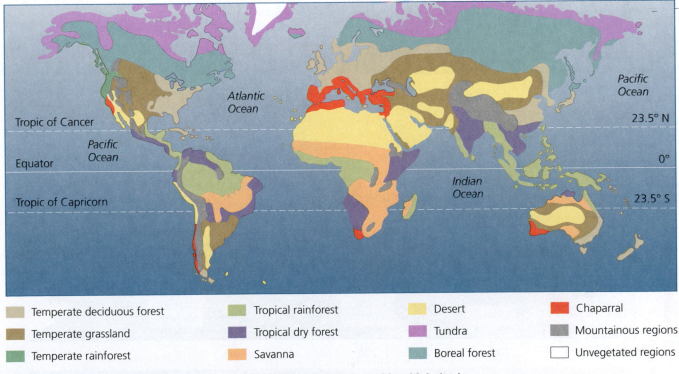

▨ Temperate deciduous forest	▨ Tropical rainforest	▨ Desert	▨ Chaparral
▨ Temperate grassland	▨ Tropical dry forest	▨ Tundra	▨ Mountainous regions
▨ Temperate rainforest	▨ Savanna	▨ Boreal forest	☐ Unvegetated regions

FIGURE 4.14 Biomes are distributed around the world, correlated roughly with latitude.

Climate helps determine biomes

Which biome covers each portion of the planet depends on a variety of abiotic factors, including temperature, precipitation, soil conditions, and the circulation patterns of wind in the atmosphere and water in the oceans. Among these factors, temperature and precipitation exert the greatest influence (**FIGURE 4.15**). Global climate patterns cause biomes to occur in large patches in different parts of the world. For instance, temperate deciduous forest occurs in Europe, China, and eastern North America. Note in Figure 4.14 that patches of any given biome tend to occur at similar latitudes. This is due to Earth's north–south gradients in temperature and to atmospheric circulation patterns (p. 290).

Scientists use **climate diagrams**, or *climatographs,* to depict information on temperature and precipitation. As we tour the world's terrestrial biomes, you will see climate diagrams from specific localities. The data in each graph are typical of the climate for the biome the locality lies within.

Aquatic systems resemble biomes

Our discussion of biomes focuses exclusively on terrestrial systems because the biome concept, as traditionally developed and applied, has done so. However, areas equivalent to biomes also exist in the oceans, along coasts, and in freshwater systems (Chapter 12). One might consider the shallows along the world's coastlines to represent one aquatic system, the continental shelves another, and the open ocean, the deep sea, coral reefs, and kelp forests as still others. Freshwater wetlands and many coastal systems—such as salt marshes, rocky intertidal communities, mangrove forests, and estuaries—share terrestrial and aquatic components.

Aquatic systems are shaped not by air temperature and precipitation, but by water temperature, salinity, dissolved nutrients, wave action, currents, depth, light levels, and type of substrate (e.g., sandy, muddy, or rocky bottom). Marine communities are also more clearly delineated by their animal life than by their plant life.

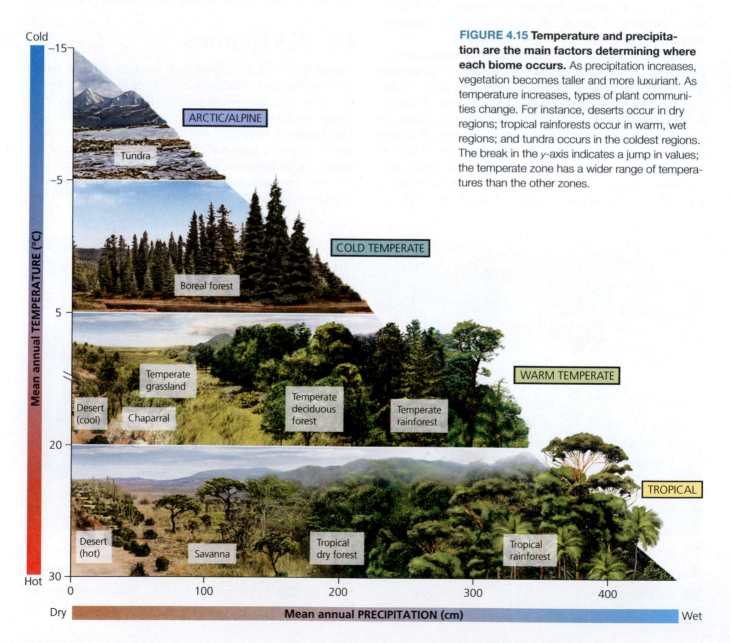

FIGURE 4.15 Temperature and precipitation are the main factors determining where each biome occurs. As precipitation increases, vegetation becomes taller and more luxuriant. As temperature increases, types of plant communities change. For instance, deserts occur in dry regions; tropical rainforests occur in warm, wet regions; and tundra occurs in the coldest regions. The break in the *y*-axis indicates a jump in values; the temperate zone has a wider range of temperatures than the other zones.

(a) Temperate deciduous forest

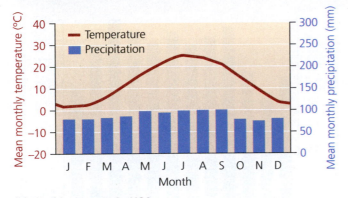

(b) Washington, D.C., USA

FIGURE 4.16 Temperate deciduous forests (a) experience fairly stable precipitation but temperatures that vary with the seasons. Scientists use climate diagrams **(b)** to illustrate average monthly precipitation and temperature. In these diagrams, the blue bars indicate precipitation and the red data lines indicate temperature, from month to month. Summer months are in the center of the *x*-axis for both Northern-Hemisphere and Southern-Hemisphere locations. *Climate diagram here and in the following figures adapted from Breckle, S.-W., 2002.* Walter's vegetation of the Earth: The ecological systems of the geo-biosphere, *4th ed. Berlin, Heidelberg: Springer-Verlag.*

(a) Temperate grassland

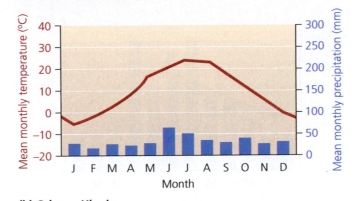

(b) Odessa, Ukraine

FIGURE 4.17 Temperate grasslands experience seasonal temperature variation and too little precipitation for trees to grow. *Climatograph adapted from Breckle, S.-W., 2002.*

DATA Compare this climate diagram to the one for temperate deciduous forest in Figure 4.16. • How do average temperatures for temperate grassland differ from those for temperate deciduous forest? • What differences do you see in precipitation between temperate grassland and temperate deciduous forest?

Go to **Interpreting Graphs & Data** on Mastering **Environmental Science**

We can divide the world into 10 terrestrial biomes

Temperate deciduous forest The **temperate deciduous forest** (**FIGURE 4.16**) that dominates the landscape around much of the Great Lakes and Mississippi River Valley is characterized by broad-leafed trees that are *deciduous*, meaning that they lose their leaves each fall and remain dormant during winter, when hard freezes would endanger leaves. These midlatitude forests occur in much of Europe, eastern China, and eastern North America—all areas where precipitation is spread relatively evenly throughout the year.

Soils of the temperate deciduous forest are fertile, but this biome contains far fewer tree species than do tropical

rainforests. Oaks, beeches, and maples are a few of the most common types of trees in these forests. Some typical animals of the temperate deciduous forest of eastern North America are shown in Figure 4.8 (p. 75).

Temperate grassland Traveling westward from the Great Lakes and the Mississippi River, temperature differences between winter and summer become more extreme, rainfall diminishes, and we find **temperate grassland** (**FIGURE 4.17**). This is because the limited precipitation in the Great Plains region supports grasses more easily than trees. Also known as *steppe* or *prairie*, temperate grasslands were once widespread in much of North and South America and central Asia.

(a) Temperate rainforest

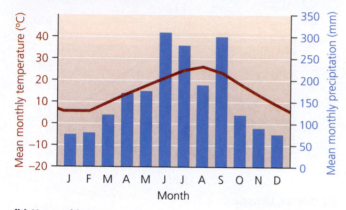

(b) Nagasaki, Japan

FIGURE 4.18 Temperate rainforests receive a great deal of precipitation and have moist, mossy interiors. *Climatograph adapted from Breckle, S.-W., 2002.*

(a) Tropical rainforest

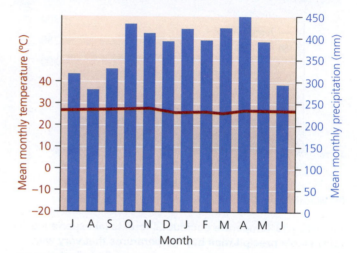

(b) Bogor, Java, Indonesia

FIGURE 4.19 Tropical rainforests, famed for their biodiversity, grow under constant, warm temperatures and a great deal of rain. *Climatograph adapted from Breckle, S.-W., 2002.*

Vertebrate animals of North America's native grasslands include American bison, prairie dogs, pronghorn antelope, foxes, coyotes, and ground-nesting birds such as meadowlarks and prairie chickens. People have converted most of the world's grasslands for farming and ranching, however, so most of these animals exist today at a tiny fraction of their historic population sizes.

Temperate rainforest Farther west in North America, the topography becomes varied, and biome types intermix. The coastal Pacific Northwest region, with its heavy rainfall and mild temperatures, features **temperate rainforest** (**FIGURE 4.18**). Coniferous trees such as cedars, spruces, hemlocks, and Douglas fir grow very tall in the temperate rainforest, and the forest interior is shaded and damp. Moisture-loving animals such as the bright yellow banana slug are common.

The soils of temperate rainforests are fertile but are susceptible to landslides and erosion when forests are cleared. We have long extracted commercially valuable products from temperate rainforests, but timber harvesting has eliminated most old-growth trees, driving species that rely on

old-growth forests, such as the spotted owl and marbled murrelet, toward extinction.

Tropical rainforest In tropical regions we see the same pattern found in temperate regions: Areas of high rainfall grow rainforests, areas of intermediate rainfall support dry or deciduous forests, and areas of low rainfall are dominated by grasses. However, tropical biomes differ from their temperate counterparts in other ways because they are closer to the equator and therefore warmer on average year-round. For one thing, they hold far greater biodiversity.

Tropical rainforest (**FIGURE 4.19**)—found in Central America, South America, Southeast Asia, West Africa, and other tropical regions—is characterized by year-round rain and uniformly warm temperatures. Tropical rainforests have dark, damp interiors; lush vegetation; and highly diverse communities, with more species of insects, birds, amphibians, and other animals than any other biome. These forests

(a) Tropical dry forest

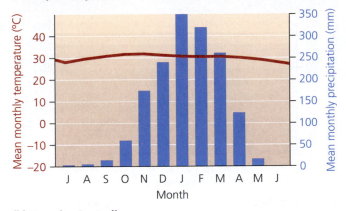

(b) Darwin, Australia

FIGURE 4.20 Tropical dry forests experience significant seasonal variation in precipitation but relatively stable, warm temperatures. *Climatograph adapted from Breckle, S.-W., 2002.*

(a) Savanna

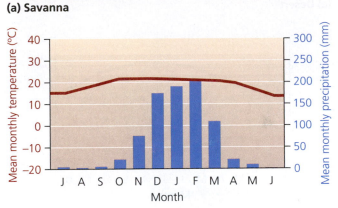

(b) Harare, Zimbabwe

FIGURE 4.21 Savannas are grasslands with clusters of trees. They experience slight seasonal variation in temperature but significant variation in rainfall. *Climatograph adapted from Breckle, S.-W., 2002.*

consist of high numbers of tree species intermixed, each at a low density. A tree may be draped with vines and loaded with orchids. Indeed, trees sometimes collapse under the weight of all the life they support!

Despite this profusion of life, tropical rainforests have poor, acidic soils that are low in organic matter. Nearly all nutrients in this biome are contained in the plants, not in the soil. An unfortunate consequence is that once people clear tropical rainforests, the soil can support agriculture for only a short time (p. 145). As a result, farmed areas are abandoned quickly, and farmers move on and clear more forest.

Tropical dry forest Tropical areas that are warm year-round but where rainfall is lower overall and highly seasonal give rise to **tropical dry forest**, or *tropical deciduous forest* (**FIGURE 4.20**), a biome widespread in India, Africa, South America, and northern Australia. Wet and dry seasons each span about half a year in tropical dry forest. Organisms that inhabit tropical dry forest have adapted to seasonal fluctuations in precipitation and temperature. For instance, plants leaf out and grow profusely with the rains, then drop their leaves during dry times of year.

Rains during the wet season can be heavy and can lead to severe soil loss where people have cleared forest. Across the globe, we have converted a great deal of tropical dry forest for agriculture. Clearing for farming or ranching is straightforward because vegetation is lower and canopies less dense than in tropical rainforest.

Savanna Drier tropical regions give rise to **savanna** (**FIGURE 4.21**), tropical grassland interspersed with clusters of acacias or other trees. The savanna biome is found across stretches of Africa, South America, Australia, India, and other dry tropical regions. Precipitation usually arrives during distinct rainy seasons, whereas in the dry season grazing animals concentrate near widely spaced water holes. Common herbivores on the African savanna include zebras, gazelles, and giraffes. Predators of these grazers include lions, hyenas, and other highly mobile carnivores. Science indicates that the African savanna was the ancestral home of the human species. The open spaces of this biome likely favored the evolution of our upright stance, running ability, and keen vision.

(a) Desert

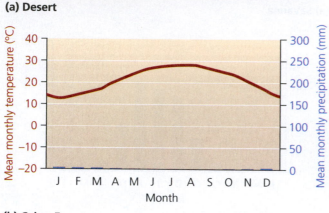

(b) Cairo, Egypt

FIGURE 4.22 **Deserts are dry year-round, but are not always hot.** *Climatograph adapted from Breckle, S.-W., 2002.*

(a) Tundra

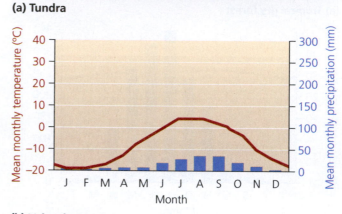

(b) Vaigach, Russia

FIGURE 4.23 **Tundra is a cold, dry biome found near the poles.** Alpine tundra occurs atop high mountains at lower latitudes. *Climatograph adapted from Breckle, S.-W., 2002.*

Desert Where rainfall is very sparse, **desert** (**FIGURE 4.22**) forms. The driest biome on Earth, most deserts receive well under 25 cm (10 in.) of precipitation per year, much of it during isolated storms months or years apart. Some deserts, such as Africa's Sahara, are mostly bare sand dunes. Others, such as the Sonoran Desert of Arizona and northwest Mexico, receive more rain and are more heavily vegetated.

Deserts are not always hot; the high desert of the western United States is positively cold in winter. Because deserts have low humidity and little vegetation to insulate them from temperature extremes, sunlight readily heats them in the daytime, but heat is quickly lost at night. As a result, temperatures vary greatly from day to night and from season to season. Desert soils can be saline and are sometimes known as lithosols, or stone soils, for their high mineral and low organic-matter content.

Desert animals and plants show many adaptations to deal with a harsh climate. Most reptiles and mammals, such as rattlesnakes and kangaroo mice, are active in the cool of night. Many Australian desert birds are nomadic, wandering long distances to find areas of recent rainfall and plant growth. Desert plants tend to have thick, leathery leaves to

reduce water loss, or green trunks so that the plant can conduct photosynthesis without leaves. The spines of cacti and other desert plants guard them from being eaten by herbivores desperate for the precious water they hold. Such traits have evolved by convergent evolution in deserts across the world (see Figure 3.3b, p. 52).

Tundra Nearly as dry as desert, **tundra** (**FIGURE 4.23**) occurs at very high latitudes in northern Russia, Canada, and Scandinavia. Extremely cold winters with little daylight and summers with lengthy days characterize this landscape of lichens and low, scrubby vegetation without trees. The great seasonal variation in temperature and day length results from this biome's high-latitude location, angled toward the sun in summer and away from the sun in winter.

Because of the cold climate, underground soil remains permanently frozen and is called *permafrost*. During winter, surface soil freezes as well. When the weather warms, the soil melts and produces pools of surface water, forming ideal habitat for the larvae of mosquitoes and other insects. The swarms of insects benefit bird species that migrate long distances to breed during the brief but productive summer.

(a) Boreal forest

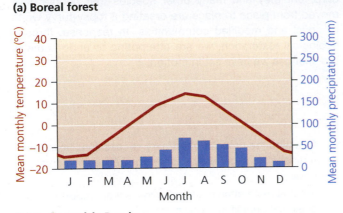

(b) Archangelsk, Russia

FIGURE 4.24 Boreal forest experiences long, cold winters, cool summers, and moderate precipitation. *Climatograph adapted from Breckle, S.-W., 2002.*

(a) Chaparral

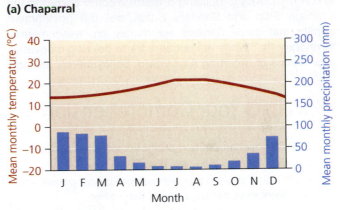

(b) Los Angeles, California, USA

FIGURE 4.25 Chaparral is a seasonally variable biome dominated by shrubs, influenced by marine weather, and dependent on fire. *Climatograph adapted from Breckle, S.-W., 2002.*

Caribou also migrate to the tundra to breed, then leave for the winter. Only a few animals, such as polar bears and musk oxen, can survive year-round in tundra. Today global climate change is warming high-latitude regions the most. This is melting large areas of permafrost, causing the greenhouse gas methane to seep out of the ground, driving climate change further.

Tundra also occurs as alpine tundra at the tops of mountains in temperate and tropical regions. Here, high elevation creates conditions similar to those of high latitude.

Boreal forest The northern coniferous forest, or **boreal forest,** often called *taiga* (**FIGURE 4.24**), extends across much of Canada, Alaska, Russia, and Scandinavia. A few species of evergreen trees, such as black spruce, dominate large stretches of forest, interspersed with many bogs and lakes. Boreal forests occur in cooler, drier regions than do temperate forests, and they experience long, cold winters and short, cool summers.

Soils are typically nutrient-poor and somewhat acidic. As a result of strong seasonal variation in day length, temperature, and precipitation, many organisms compress a year's

worth of feeding and breeding into a few warm, wet months. Year-round residents of boreal forest include mammals such as moose, wolves, bears, lynx, and rodents. Many insect-eating birds migrate here from the tropics to breed during the brief, intensely productive, summers. The boreal forests are vast, but are being lost and modified today as a result of logging, fossil fuel extraction, and fires and pest outbreaks driven by climate change (p. 202).

Chaparral In contrast to the boreal forest's broad, continuous distribution, **chaparral** (**FIGURE 4.25**) is limited to small patches widely flung around the globe. Chaparral consists mostly of evergreen shrubs and is densely thicketed. This biome is highly seasonal, with mild, wet winters and warm, dry summers—a climate influenced by ocean waters and often termed "Mediterranean." Besides ringing the Mediterranean Sea, chaparral occurs along the coasts of California, Chile, and southern Australia. Chaparral communities naturally experience frequent fire, and their plants are adapted to resist fire or even to depend on it for germination of their seeds. As a result, people living in regions of chaparral need to pay special attention to managing risks from wildfire.

closing THE LOOP

Ecological communities are shaped by many forces. Within communities, species interact through competition, predation, parasitism, herbivory, and mutualism. Communities are stable only until disturbed—and in today's world, invasive species such as Asian carp are one major and growing form of disturbance.

Scientists, policymakers, and managers are trying to limit the spread of Asian carp by controlling their numbers in the Mississippi and Ohio river basins, and are trying to prevent their establishment in the Great Lakes. The Army Corps is building a fourth electric barrier in the Chicago Ship and Sanitary Canal, but the government admits that these barriers are merely an "experimental, temporary fix." While researchers try to develop innovative new means of deterring Asian carp, policymakers are now considering whether to spend billions of dollars to reconfigure the canal system to somehow shut off access to alien species for good.

Meanwhile, along the waterways where these invasive fish are established, people bear the costs and adapt as best as they can. Fortunately, all biological invasions eventually slow and populations stop growing. Often, some native species discover the aliens and become their predators, parasites, or competitors. Some long-established invasive species in North America have begun to decline, and a few have even disappeared.

No one knows what the future holds in the case of Asian carp, but they and many other species that people have moved from place to place are creating a topsy-turvy world of novel and modified communities. In response, through ecological restoration, we are attempting to undo some of the changes we have set in motion.

TESTING Your Comprehension

1. Explain how competition promotes resource partitioning.

2. Compare and contrast the three main types of exploitative species interactions (predation, parasitism, and herbivory), explaining how they differ.

3. Give examples of symbiotic and nonsymbiotic mutualisms. Describe at least one way in which a mutualism affects your daily life.

4. Using the concepts of trophic levels and energy flow, explain why the ecological footprint of a vegetarian person is smaller than that of a meat-eater.

5. Differentiate a food chain from a food web. Which best represents the reality of communities, and why?

6. What is meant by the term *keystone species,* and what types of organisms are most often considered keystone species?

7. Describe the process of primary succession. How does it differ from secondary succession? Give an example of each.

8. What is restoration ecology? Why is it an important scientific pursuit in today's world?

9. What factors most strongly influence the type of biome that forms in a particular place on land? What factors determine the type of aquatic system that may form in a given location?

10. Draw a typical climate diagram for a tropical rainforest. Label all parts of the diagram, and describe all the types of information an ecologist could glean from such a diagram. Now draw a climate diagram for a desert. How does it differ from your rainforest diagram, and what does this tell you about how the two biomes differ?

SEEKING Solutions

1. Suppose you spot two species of birds feeding side by side, eating seeds from the same plant. You begin to wonder whether competition is at work. Describe how you might design scientific research to address this question. What observations would you try to make at the outset? Would you try to manipulate the system to test your hypothesis that the two birds are competing? If so, how?

2. Spend some time outside on your campus, in your yard, or in the nearest park or natural area. Find at least 10 species of organisms (plants, animals, or others), and observe each one long enough to watch it feed or to make an educated guess about how it derives its nutrition. Now, using Figure 4.8 as a model, draw a simple food web involving all the organisms you observed.

3. Can you think of one organism not mentioned in this chapter as a keystone species that you believe may be a keystone species? For what reasons do you suspect this? How could an ecologist experimentally test whether an organism is a keystone species?

4. **CASE STUDY CONNECTION** Describe three ecological changes to freshwater communities that have occurred since the invasion of Asian carp in North America. Describe one economic impact of the invasion. What is one way to prevent these fish from spreading to new areas?

5. **THINK IT THROUGH** A federal agency has put you in charge of devising responses to the invasion of Asian carp. Based on what you know from this chapter, how would you seek to control the spread of these species and reduce their impacts? What strategies would you consider pursuing immediately, and for which strategies would you commission further scientific research? For each of your ideas, name one benefit or advantage. For each idea, identify one obstacle it might face in being implemented.

CALCULATING Ecological Footprints

Environmental scientists David Pimentel, Rodolfo Zuniga, and Doug Morrison of Cornell University reviewed scientific estimates for the economic and ecological costs imposed by introduced and invasive species in the United States. They found that approximately 50,000 species had been introduced in the United States and that these accounted for over $120 billion in economic costs each year. These costs include direct losses and damage, as well as costs required to control the species. (The researchers did not quantify monetary estimates for losses of biodiversity, ecosystem services, and aesthetics, which they said would drive total costs several times higher.) Calculate values missing from the table to determine the number of introduced species of each type of organism and the annual cost that each imposes on our economy.

GROUP OF ORGANISM	PERCENTAGE OF TOTAL INTRODUCED	NUMBER OF SPECIES INTRODUCED	PERCENTAGE OF TOTAL ANNUAL COSTS	ANNUAL ECONOMIC COSTS
Plants	50.0	25,000	27.2	
Microbes	40.0		20.2	
Arthropods	9.0		15.7	
Fish	0.28		4.2	
Birds	0.19		1.5	$1.9 billion
Mollusks	0.18		1.7	
Reptiles and amphibians	0.11		0.009	
Mammals	0.04	20	29.4	$37.5 billion
TOTAL	100	50,000	100	$127.4 billion

Data from Pimentel, D., R. Zuniga, and D. Morrison, 2005. Update on the environmental and economic costs associated with alien-invasive species in the United States. Ecological Economics 52: 273–288.

1. Of the 50,000 species introduced into the United States, half are plants. Describe two ways in which non-native plants might be brought to a new location. How might we help prevent non-native plants from establishing in new areas and altering native communities?

2. Organisms that damage crop plants are the most costly of introduced species. Weeds, pathogenic microbes, and arthropods that attack crops together account for half of the costs documented by Pimentel's team. What steps can we—farmers, policymakers, and all of us as a society—take to minimize the impacts of invasive species on crops?

3. How might your own behavior influence the influx and ecological impacts of non-native species such as those listed above? Name three things you could personally do to help reduce the impacts of invasive species.

Mastering Environmental Science

Students Go to Mastering Environmental Science for assignments, the etext, and the Study Area with practice tests, videos, current events, and activities.

Instructors Go to Mastering Environmental Science for automatically graded activities, current events, videos, and reading questions that you can assign to your students, plus Instructor Resources.

CHAPTER 5

Economics, Policy, and Sustainable Development

Costa Rica Values Its Ecosystem Services

> **Costa Rica's PSA program has been one of the conservation success stories of the last decade.**
> —Stefano Pagiola, The World Bank

> **In the last 25 years, my home country has tripled its GDP while doubling the size of its forests.**
> —Carlos Manuel Rodriguez, former Minister of Energy and the Environment, Costa Rica

Few nations have transformed their path of development in just a few decades—but Costa Rica has. In the 1980s, this small Central American country was losing its forests as fast as any place on Earth. Yet today this nation of 4.9 million people has regained much of its forest cover, boasts a world-class park system, and stands as a global model for sustainable resource management.

Costa Rica took many steps on this impressive road to success. One key step was to begin paying landholders to conserve forest on private land, in a novel government program called *Pago por Servicios Ambientales* (PSA)—Payment for Environmental Services.

Nature provides ecosystem services (pp. 4, 39, 99), such as air and water purification, climate regulation, and nutrient cycling. For example, forests in Costa Rica's mountains capture rainfall and provide clean drinking water for towns and cities below. Ecosystem services are vital for our lives, but historically we have taken them for granted, and rarely do we acknowledge their value by paying for them in the marketplace. As a result, these services have diminished as we degrade the natural systems that provide them. For these reasons, many economists feel it is important to create financial incentives for conserving ecosystem services.

In Costa Rica, which had lost over three-quarters of its forest, political leaders adopted this approach in Forest Law 7575, passed in 1996. Since then, the Costa Rican government has been paying farmers and ranchers to preserve forest on their land, replant cleared areas, allow forest to regenerate naturally, and establish sustainable forestry systems. Payments are designed to be competitive with potential profits from farming or cattle ranching, and in recent years these payments have averaged $78/hectare (ha)/yr ($32/acre/yr).

The PSA program recognized four ecosystem services that forests provide:

1. *Watershed protection*: Forests cleanse water by filtering pollutants, and they conserve water and reduce soil erosion by slowing runoff.

2. *Biodiversity*: Tropical forests such as Costa Rica's are especially rich in life.

3. *Scenic beauty*: This encourages recreation and ecotourism, which bring money to the economy.

4. *Carbon sequestration*: By pulling carbon dioxide from the atmosphere, forests slow global warming.

To fund the PSA program, Costa Rica's government sought money from people and companies that benefit from these services. With watershed protection,

Costa Rica's forests are home to wildlife like this toucan ▲

(a) 1940　　　　**(b) 1987**　　　　**(c) 2005**

Forested area

FIGURE 5.1 **Forest cover in Costa Rica decreased between 1940 and 1987, but it increased thereafter.** *Data from FONAFIFO.*

for example, irrigators, bottlers, municipal water suppliers, and utilities that generate hydropower all made voluntary payments into the program, and a tariff on water users was added in 2005. For biodiversity and scenery, the country targeted ecotourism, while international lending agencies provided loans and donations. Because carbon dioxide is emitted when fossil fuels are burned, the nation used a 3.5% tax on fossil fuels to help fund the program. It also sought to sell carbon offsets in global carbon trading markets (p. 337).

Costa Rican landholders rushed to sign up for the PSA program. The agency administering it, *Fondo Nacional de Financiamiento Forestal* (FONAFIFO), signed landowners to contracts and sent agents to advise them on forest conservation and to monitor compliance. Through 2016, FONAFIFO had paid 186 billion colónes ($336 million in today's U.S. dollars) to more than 15,000 landholders and had registered 1.2 million ha (3.0 million acres)—23% of the nation's land area.

Deforestation slowed in Costa Rica in the wake of the program. Forest cover rose by 10% in the decade after 1996, and policymakers, economists, and environmental advocates cheered the PSA program's apparent success. However, some observers argued that forest loss had been slowing for other reasons and that the program itself was having little effect. They contended that payments were being wasted on people who had no plans to cut down their trees. Critics also lamented that large wealthy landowners utilized the program more than low-income small farmers. All these concerns were borne out by researchers (see **THE SCIENCE BEHIND THE STORY**, pp. 102–103).

In response, the government modified its policies, making the program more accessible to small farmers and targeting the payments to locations where forest is most at risk and environmental assets are greatest.

In recent years, forest cover in Costa Rica has been rising (**FIGURE 5.1**), from a low of 17% in 1983 to more than 53% today. The nation has thrived economically while protecting its environment. Since the PSA program began, Costa Ricans have enjoyed an increase in real, inflation-adjusted per capita income of more than 60%—a rise in wealth surpassing the vast majority of nations.

Many factors have contributed to Costa Rica's success in building a wealthier society while protecting its ecological assets. Back in 1948, Costa Rica abolished its armed forces and shifted funds from the military budget into health and education. (The only mainland nation in the world without a standing army, Costa Rica enjoys security from alliances with the United States and other nations, and has experienced seven decades of peace.) With a stable democracy and a healthy and educated citizenry, the stage was set for well-managed development, including innovative advances in conservation. The nation created one of the world's finest systems of national parks, which today covers fully one-quarter of its territory. Ecotourism at the parks brings wealth to the country: Each year more than 2 million foreign tourists inject over $2 billion into Costa Rica's economy.

As a result, Costa Ricans understand the economic value of protecting their natural capital. They see how innovative policies can help conserve resources while boosting the economy and enhancing the quality of people's everyday lives. By placing economic value on nature, Costa Rica is pointing the way toward truly sustainable development.

Economics and the Environment

An **economy** is a social system that converts resources into *goods* (material commodities made and bought by individuals and businesses) and *services* (work done for others as a form of business). **Economics** is the study of how people decide to use potentially scarce resources to provide goods and services that are in demand. The word *economics* and the word *ecology* come from the same Greek root, *oikos,* meaning "household." Economists traditionally have studied the household of human society, whereas ecologists study the broader household of all life. Just as the environment influences our economy, the economic decisions we all make from day to day have implications for the environment. For these reasons, understanding economics helps us appreciate the complex interface between environmental science and society.

Economies rely on goods and services from the environment

Our economies and our societies exist within the natural environment and depend on it in vital ways. Economies receive inputs (such as natural resources and ecosystem services) from the environment, process them, and discharge outputs (such as waste) into the environment (**FIGURE 5.2**).

These interactions are readily apparent, yet many mainstream economists still adhere to a worldview that overlooks their importance and largely ignores the environment (and instead considers only the tan box in the middle of Figure 5.2). This traditional view, which continues to drive many policy decisions, implies that natural resources are free and limitless and that wastes can be endlessly absorbed at no cost.

Newer schools of thought, however, recognize that economies exist within the environment and rely on its natural resources and ecosystem services. This explains why Costa Rica and other nations have taken steps to better protect their natural assets.

Natural resources (p. 4) are the substances and forces that sustain us: the fresh water we drink; the trees that supply our lumber; the rocks that provide our metals; and the energy from sun, wind, water, and fossil fuels. We can think of natural resources as "goods" produced by nature. Environmental systems also naturally function in a manner that supports our economies. Earth's ecological systems purify air and water, form soil, cycle nutrients, regulate climate, pollinate plants, and recycle waste. Such essential ecosystem services (pp. 4, 39, 99) support the very life that makes our economic activity possible. Together, nature's resources and services make up the natural capital (p. 6) on which we depend.

When we deplete natural resources and generate pollution, we degrade the capacity of ecological systems to function. Scientists with the Millennium Ecosystem Assessment, a worldwide review undertaken in 2005, concluded that 15 of 24 ecosystem services they surveyed globally were being degraded or used unsustainably. The degradation of ecosystem services can weaken economies. In Costa Rica, rapid forest loss up through the 1980s was causing soil erosion, water pollution, and biodiversity loss that increasingly threatened the nation's economic potential. Low-income small farmers were the first to feel these impacts. Indeed, across the world, ecological degradation tends to harm poor and marginalized people before wealthy ones, the Millennium Ecosystem Assessment found. As a result, restoring ecosystem services stands as a prime avenue for alleviating poverty.

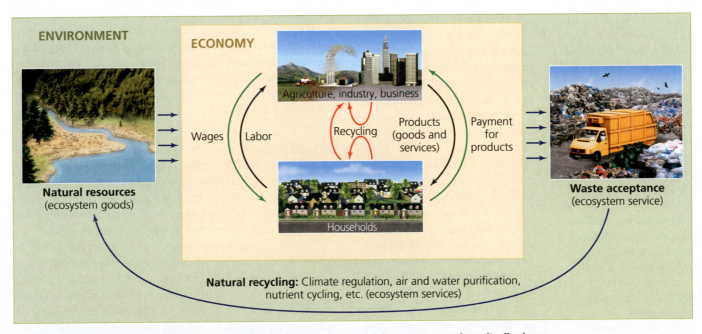

FIGURE 5.2 Economies exist within the natural environment, receiving resources from it, discharging waste into it, and benefiting from ecosystem services. Conventional neoclassical economics has focused only on processes of production and consumption between households and businesses **(tan box in middle)**, viewing the environment merely as an external factor. In contrast, environmental and ecological economists emphasize that economies exist within the natural environment and depend on all that it offers.

DATA Q Follow each arrow in this figure to make sure you understand what it indicates. Then answer these questions: • When you work at a job, what do you give and what do you receive? • When you buy a product, what do you give and what do you receive? • Describe three things that the environment provides to the economy, basing your answer on this diagram.

Go to **Interpreting Graphs & Data** on **Mastering** Environmental Science

Economic theory moved from "invisible hand" to supply and demand

As the field of economics developed in the 18th century, many people felt that individuals acting in their own self-interest harm society. However, Scottish philosopher Adam Smith (1723–1790) argued that self-interested economic behavior can benefit society, as long as the behavior is constrained by the rule of law and private property rights within a competitive marketplace. A founder of **classical economics,** Smith felt that when people pursue economic self-interest under these conditions, the marketplace will behave as if guided by "an invisible hand" to benefit society as a whole.

Today, Smith's philosophy remains a pillar of free-market thought, which many credit for the tremendous gains in material wealth that capitalist market economies have achieved. Others contend that free-market policies tend to worsen inequalities between rich and poor.

Economists subsequently adopted more quantitative approaches. **Neoclassical economics** examines consumer choices and explains market prices in terms of preferences for units of particular commodities. In neoclassical economics, buyers desire a low price whereas sellers desire a high price, so a compromise price is reached. Marketplace dynamics are expressed in terms of *supply*, the amount of a product offered for sale at a given price, and *demand*, the amount of a product people will buy at a given price if free to do so.

To evaluate an action or decision, economists use **cost-benefit analysis,** which compares the estimated costs of a proposed action with the estimated benefits. If benefits exceed costs, the action should be pursued; if costs exceed benefits, it should not. Given a choice of actions, the one with the greatest excess of benefits over costs should be chosen.

This reasoning seems eminently logical, but problems arise when not all costs and benefits can be easily identified, defined, or quantified. It may be simple to quantify the dollar value of bananas grown or cattle raised on a tract of Costa Rican land cleared for agriculture, yet difficult to assign monetary value to the complex ecological costs of clearing the forest. Because monetary benefits are usually more easily quantified than environmental costs, benefits tend to be over-represented in traditional cost-benefit analyses. As a result, environmental advocates often feel such analyses are biased toward economic development and against environmental protection.

Neoclassical economics has environmental consequences

Today's market systems operate largely in accord with the principles of neoclassical economics. These systems have generated unprecedented material wealth for our societies, yet four basic assumptions of neoclassical economics often contribute to environmental degradation.

Replacing resources One assumption is that natural resources and human resources (such as workers or technologies) are largely substitutable and interchangeable. This implies that once we have depleted a resource, we will always be able to find some replacement for it. As a result, the market imposes no penalties for depleting resources.

It is true that many resources can be replaced. However, Earth's material resources are ultimately limited. Nonrenewable resources (such as fossil fuels) can be depleted, and many renewable resources (such as soils, fish stocks, timber, and clean water) can be used up if we exploit them faster than they are replenished.

External costs A second assumption of neoclassical economics is that all costs and benefits associated with an exchange of goods or services are borne by individuals engaging directly in the transaction. In other words, it is assumed that the costs and benefits are "internal" to the transaction, experienced by the buyer and seller alone.

However, many transactions affect other members of society. When a landowner fells a forest, people nearby suffer poorer water quality, dirtier air, and less wildlife. When a factory, power plant, or mining operation pollutes the air or water, it harms the health of those who live nearby. In such cases, people who are not involved in degrading the environment end up paying the costs. A cost of a transaction that affects someone other than the buyer or seller is known as an **external cost** (TABLE 5.1). Often, whole communities suffer external costs while certain individuals enjoy private gain.

If market prices do not take the social, ecological, or economic costs of environmental degradation into account, then taxpayers bear the burden of paying them. When economists ignore external costs, this creates a false impression of the true and full consequences of our choices. External costs are one reason that governments develop environmental policy (p. 104).

FAQ

Doesn't environmental protection hurt the economy?

We often hear it said that protecting environmental quality costs too much money, interferes with progress, or leads to job loss. However, growing numbers of economists dispute this. Instead, they assert that environmental protection tends to *enhance* our economic bottom line and improve our quality of life. The view one takes depends in part on the timeframe. We often make economic judgments on short timescales, and in the short term many activities that cause environmental damage may be economically profitable. In the longer term, however, environmental degradation imposes economic costs. Moreover, when resource extraction or development degrades environmental conditions, often a few private parties benefit financially while the broader public is harmed.

Today, concerns over fossil fuels, pollution, and climate change have led many people to see immense opportunities in revamping our economies with clean and renewable energy technologies. The jobs, investment, and economic activity that come with building a green energy economy demonstrate how economic progress and environmental protection can go hand in hand.

TABLE 5.1 Common Examples of External Costs

Health impacts
Physical health problems, stress, anxiety, or other medical conditions caused by air pollution, water pollution, toxic chemicals, or other environmental impacts.

Depletion of resources
Declines in abundance of—or loss of access to—natural resources that provide wealth or sustenance, such as fish, game, and wildlife; timber and forest products; grazing land; or fertile soils.

Aesthetic damage
Degradation of scenery and harm to the enjoyment of one's physical surroundings, as from strip mining; clear-cutting; urban sprawl; erosion; or air, water, noise, and light pollution.

Financial loss
Loss of monetary value caused by pollution or by resource depletion, via declining real estate values, lost tourism revenue, costlier medical expenses, damage from climate change and sea level rise, or other means.

Discounting Third, neoclassical economics grants an event in the future less value than one in the present. In economic terminology, future effects are "discounted." **Discounting** is meant to reflect how people tend to grant more importance to present conditions than to future conditions. Just as you might prefer to have an ice cream cone today than to be promised one next month, market demand is greater for goods and services that are received sooner.

Unfortunately, giving more weight to current costs and benefits than to future costs and benefits encourages policymakers to play down the long-term consequences of decisions. Many environmental problems unfold gradually, yet discounting discourages us from addressing resource depletion, pollution buildup, and other cumulative impacts. Instead, discounting shunts the costs of dealing with such problems onto future generations. Discounting has emerged as a flashpoint in the debate over how to respond to climate change. Economists agree that climate change will impose major costs on society, but they differ on how much to discount future impacts—and so they differ on how much we should invest today to battle climate change.

Growth **Economic growth** can be defined as an increase in an economy's production and consumption of goods and services. Neoclassical economics assumes that economic growth is essential for maintaining social order, because a growing economy can alleviate the discontent of poorer people by creating opportunities for them to become wealthier. A rising tide raises all boats, as the saying goes; if we make the overall economic pie larger, then each person's slice can become larger (even if some people still have much smaller slices than others). However, critics of the growth paradigm maintain that growth cannot be sustained forever, because resources to support growth are ultimately limited.

How sustainable is economic growth?

Our global economy is eight times the size it was just half a century ago. All measures of economic activity are greater than ever before. Economic expansion has brought many people much greater wealth (although not equally, and gaps between rich and poor are wide and growing).

Economic growth can occur in two ways: (1) by an increase in inputs to the economy (such as more labor or natural resources) or (2) by improvements in the efficiency of production due to better methods or technologies (ideas or equipment that enable us to produce more goods with fewer inputs).

As our population and consumption rise, it is becoming clearer that we cannot sustain growth forever by using the first approach. Nonrenewable resources are finite in quantity, and renewable resources can also be exhausted if we overexploit them. As for the second approach to growth, we have used technological innovation to push back the limits on growth time and again. More efficient technologies for extracting minerals, fossil fuels, and groundwater allow us to mine these resources more fully with less waste. Better machinery and robotics in our factories speed manufacturing. We continue to make computer chips more powerful while also making them smaller. In such ways, we are producing more goods and services with relatively fewer resources.

Can we conclude, then, that technology and human ingenuity will allow us to overcome all environmental constraints and continue economic growth forever? We can certainly continue to innovate and achieve further efficiency. Yet ultimately, if our population and consumption continue to grow and we do not enhance the reuse and recycling of materials, we will continue to diminish our natural capital, putting ever-greater demands on our capacity to innovate.

Economists in the field of **environmental economics** feel we can modify neoclassical economic principles to make resource use more efficient and thereby attain sustainability within our current economic system. Environmental economists were the first to develop methods to tackle the problems of external costs and discounting.

Economists in the field of **ecological economics** feel that sustainability requires more far-reaching changes. They stress that in nature, every population has a carrying capacity (p. 63) and systems operate in self-renewing cycles. Ecological economists maintain that societies, like natural populations, cannot surpass environmental limitations. Many of these economists advocate economies that neither grow nor shrink, but are stable. Such a **steady-state economy** is intended to mirror natural systems. Critics of steady-state economies assert that to halt growth would dampen our quality of life. Proponents respond that technological advances would continue, behavioral changes (such as greater use of recycling) would accrue, and wealth and happiness would rise.

Attaining sustainability will certainly require the reforms pioneered by environmental economists and may require the fundamental shifts advocated by ecological economists. One approach they each take is to assign monetary values to ecosystem goods and services, so as to better integrate them into cost-benefit analyses.

We can assign monetary value to ecosystem goods and services

Ecosystems provide us essential resources and life-support services, including fertile soil, waste treatment, clean water, and clean air. Yet we often abuse the very ecological systems that sustain us. Why? From the economist's perspective, people overexploit natural resources and processes largely because the market assigns these entities no quantitative monetary value—or assigns values that underestimate their true worth.

Ecosystem services are said to have **nonmarket values,** values not usually included in the price of a good or service (**FIGURE 5.3**). For example, the aesthetic and recreational pleasure we obtain from natural landscapes is something of

FIGURE 5.3 Accounting for nonmarket values such as those shown here may help us make better environmental and economic decisions.

(a) Use value: The worth of something we use directly

(b) Existence value: The worth of knowing that something exists, even if we never experience it ourselves

(c) Option value: The worth of something we might use later

(d) Aesthetic value: The worth of something's beauty or emotional appeal

(e) Scientific value: The worth of something for research

(f) Educational value: The worth of something for teaching and learning

(g) Cultural value: The worth of something that sustains or helps define a culture

real value. Yet because we do not generally pay money for this, its value is hard to quantify and appears in no traditional measures of economic worth. Or consider Earth's water cycle (p. 40): Rain fills our reservoirs with drinking water; rivers give us hydropower and flush away our waste; and water evaporates, purifying itself of contaminants and later falling as rain. This natural cycle is vital to our very existence, yet because we do not pay money for it, markets impose no financial penalties when we disturb it.

For these reasons, economists have sought ways to assign market values to ecosystem services. They use surveys to determine how much people are willing to pay to protect or restore a resource. They measure the money, time, or effort people expend to travel to parks. They compare housing prices for similar homes in different settings to infer the dollar value of landscapes, views, or peace and quiet. They calculate how much it costs to restore natural systems

that are degraded, to replace their functions with technology, or to clean up pollution.

For example, in Costa Rica, a team led by Taylor Ricketts of Stanford University studied pollination (pp. 73, 153) by native bees at a coffee plantation. By carefully measuring how bees pollinated the coffee plants and comparing the resulting coffee production in areas near forest and far from forest, the researchers calculated that forests were providing the farm with pollination services worth $60,000 per year.

Researchers have even sought to calculate the total economic value of all the services that oceans, forests, wetlands, and other systems provide across the world. Teams headed by ecological economist Robert Costanza have combed the scientific literature and evaluated hundreds of studies that estimated dollar values for 17 major ecosystem services (**FIGURE 5.4**). The researchers reanalyzed the data using multiple valuation techniques to improve accuracy, then

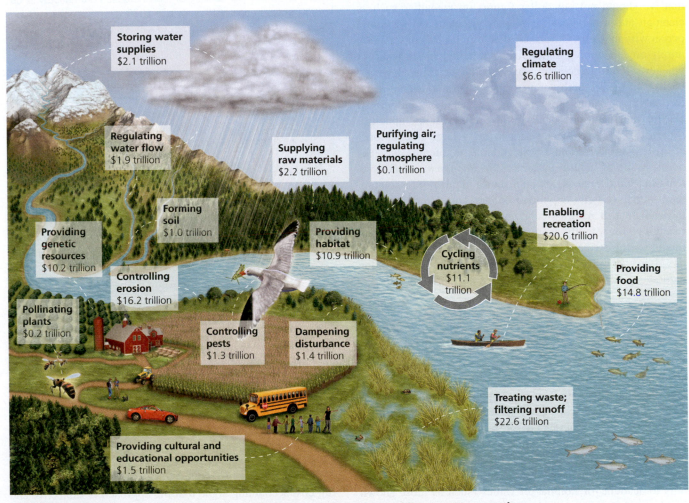

FIGURE 5.4 Ecological economists have estimated the value of the world's ecosystem services at more than $148 trillion (in 2017 dollars). This amount is an underestimate because it does not include ecosystems and services for which adequate data were unavailable. Shown are subtotals for each ecosystem service, in 2007 dollars. *Data from Costanza, R., et al., 2014. Changes in the global value of ecosystem services. Global Env. Change 26: 152–158.*

 Which three ecosystem services provide the greatest benefit to us, in dollar value, according to the data in this illustration?

Go to **Interpreting Graphs & Data** on **Mastering** Environmental Science

multiplied average estimates for each ecosystem by the global area it occupied. Their initial analysis in 1997 was groundbreaking, and in 2014 they updated their research. The 2014 study calculated that Earth's biosphere in total provides more than $125 trillion worth of ecosystem services each year, in 2007 dollars. This is equal to $148 trillion in 2017 dollars, an amount that exceeds the global annual monetary value of goods and services created by people!

Costanza also joined Andrew Balmford and 17 other colleagues to compare the benefits and costs of preserving natural systems intact versus converting wild lands for agriculture, logging, or fish farming. After reviewing many studies, they reported that a global network of nature reserves covering 15% of Earth's land surface and 30% of the ocean would be worth $4.4 to $5.2 trillion. This amount is 100 times greater than the value of those areas were they to be converted and exploited for direct use.

Such research has sparked debate. Some ethicists argue that we should not put dollar figures on amenities such as clean air and water, because they are priceless and we would perish without them. Others say that arguing for conservation purely on economic grounds risks not being able to justify it whenever it fails to deliver clear economic benefits. However, backers of the research counter that valuation does not argue for making decisions on monetary grounds alone, but instead clarifies and quantifies values that we already hold implicitly.

In 2010, researchers wrapped up a large international effort to summarize and assess attempts to quantify the economic value of natural systems. *The Economics of Ecosystems and Biodiversity* study published a number of fascinating reports that you can download online. This effort describes the valuation of nature's economic worth as "a tool to help recalibrate [our] faulty economic compass." It concludes that this is useful because "the invisibility of biodiversity values has often encouraged inefficient use or even destruction of the natural capital that is the foundation of our economies."

We can measure progress with full cost accounting

If assigning market values to ecosystem services gives us a fuller and truer picture of costs and benefits, then we can take a similar approach in measuring our economic progress as a society. For decades, we have assessed each nation's economy by calculating its **Gross Domestic Product (GDP),** the total monetary value of final goods and services a nation produces each year. Governments regularly use GDP to make policy decisions that affect billions of people. However, GDP fails to account for nonmarket values (such as those shown in Figure 5.3). It also lumps together all economic activity, desirable and undesirable. GDP can rise in response to crime, war, pollution, and natural disasters, because we spend money to protect ourselves from these things and to recover from them.

Environmental economists have developed indicators meant to distinguish desirable from undesirable economic activity and to better reflect our well-being. One such alternative to the GDP is the **Genuine Progress Indicator (GPI).**

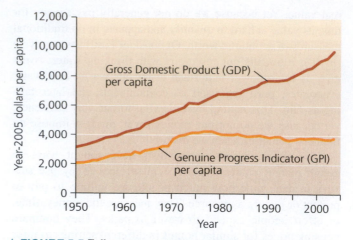

FIGURE 5.5 Full cost accounting indicators such as the GPI aim to measure progress and well-being more effectively than the GDP. Data from 17 major nations, combined here, show that per capita GDP has increased dramatically since 1950, yet per capita GPI peaked in 1978. *Data from Kubiszewski, I., et al., 2013. Beyond GDP: Measuring and achieving global genuine progress. Ecological Economics 93: 57–68. All data are adjusted for inflation using year-2005 dollars.*

DATA • What was the ratio of GDP to GPI in 1950? (Divide GDP by GPI using the values shown on the graph for that year.) • What was this ratio in the year you were born? • What was this ratio in 2004? • How has the ratio of GDP to GPI changed through time, and what does this indicate to you?

Go to **Interpreting Graphs & Data** on **Mastering** Environmental Science

To calculate GPI, we begin with conventional economic activity and add to it positive contributions not paid for with money, such as volunteer work and parenting. We then subtract negative impacts, such as crime and pollution.

GPI can differ strikingly from GDP: **FIGURE 5.5** compares these indices internationally across half a century. On a per-person basis, GDP rose greatly, but GPI has declined slightly since 1978. Data for the United States show a very similar pattern, with per capita GDP more than tripling over 50 years but GPI remaining flat for the latter 30 years. These discrepancies suggest that people in most nations—including the United States—have been spending more and more money but that their quality of life is not improving.

The GPI is an example of **full cost accounting** (also called *true cost accounting*) because it aims to account fully for all costs and benefits. Several U.S. states have begun using the GPI to measure progress and help guide policy. Critics of full cost accounting argue that the approach is subjective and too easily driven by ideology. Proponents respond that making a subjective attempt to measure progress is better than misapplying an indicator such as the GDP to quantify well-being—something it was never meant to do.

Today, attempts are gaining ground to measure happiness (rather than economic output) as the prime goal of national policy. The small Asian nation of Bhutan pioneered this approach with its measure of Gross National Happiness. Another indicator is the Happy Planet Index, which measures how much happiness we gain per amount of resources we consume. By this measure, Costa Rica was recently calculated to be the top-performing nation in the world.

FAQ

Does having more money make a person happier?

This age-old question has long been debated by philosophers, but in recent years researchers have analyzed the issue scientifically. Studies have found a surprising degree of consensus: In general, we become happier as we get wealthier, but once we gain a moderate level of wealth (roughly $50,000–$90,000 in yearly income), attaining further money no longer increases our happiness. Apparently, reaching a basic level of financial security alleviates day-to-day economic worries, but once those worries are taken care of, our happiness revolves around other aspects of our lives (such as family, friends, and the joy of helping others). Research on happiness can help us guide our personal life decisions. It also suggests that enhancing a society's happiness might best be achieved by raising many people's incomes by a little, rather than by raising some people's incomes by a lot.

Environmental Policy: An Overview

When a society recognizes a problem, its leaders may try to resolve the problem using **policy,** a formal set of general plans and principles intended to guide decision making. **Public policy** is policy made by people in government. **Environmental policy** pertains to our interactions with our environment. Environmental policy generally aims to regulate resource use or reduce pollution to promote human welfare or protect natural systems.

Forging effective policy requires input from science, ethics, and economics. Science provides information and

Costa Rica is also one of five nations working with the World Bank (p. 110) in a program to implement full cost accounting methods. Together they are addressing questions such as how much economic benefit the nation's forests, parks, and other natural amenities generate through tourism and watershed protection. Data so far show Costa Rica's forests to be contributing 10–20 times more to the economy than had been estimated from timber sales alone.

Markets can fail

When markets do not take into account the positive outside effects on economies (such as ecosystem services) or the negative side effects of economic activity (external costs), economists call this **market failure.** Traditionally, we have tried to counteract market failure with government intervention. Government can help by restraining improper individual and corporate behavior through laws and regulations. It can tax harmful activities. It can also design economic incentives that use market mechanisms to promote fairness, resource conservation, and economic sustainability. Paying for the conservation of ecosystem services, as Costa Rica does, is one way of deploying economic incentives toward policy goals. We will now examine these approaches in our discussion of environmental policy.

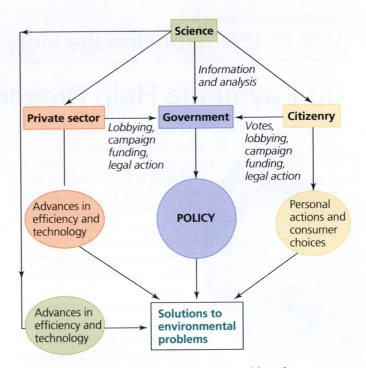

FIGURE 5.6 Policy plays a central role in addressing environmental problems.

analyses needed to identify and understand problems and devise solutions. Ethics and economics each offer criteria by which to assess problems and help clarify how we might address them. Government interacts with citizens, organizations, and the private sector to formulate policy (**FIGURE 5.6**).

Environmental policy addresses issues of fairness and resource use

Because market capitalism is driven by incentives for short-term economic gain, it provides businesses and individuals with little motivation to minimize environmental impacts, to seek long-term social benefits, or to equalize costs and benefits among parties. As we noted, such market failure has traditionally been viewed as justification for government involvement. Governments typically intervene in the marketplace for several reasons:

- To provide social services, such as national defense, health care, and education
- To provide "safety nets" (for the elderly, the poor, victims of natural disasters, and so on)
- To eliminate unfair advantages held by single buyers or sellers
- To manage publicly held resources
- To minimize pollution and other threats to health and quality of life.

Environmental policy aims to protect people's health and well-being, to safeguard environmental quality and conserve natural resources, and to promote equity or fairness in people's use of resources.

Do Payments Help Preserve Forest?

Costa Rica's program to pay for ecosystem services has garnered international praise and inspired other nations to implement similar policies. But have Costa Rica's payments actually been effective in preventing forest loss? A number of research teams have sought to answer this surprisingly difficult question by analyzing data from the PSA program.

Some early studies were quick to credit the PSA program for saving forests. A 2006 study conducted for FONAFIFO, the agency administering the program, concluded that PSA payments in the central region of the country had prevented 108,000 ha (267,000 acres) of deforestation—38% of the area under contract. Indeed, deforestation rates fell as the program proceeded; rates of forest clearance in 1997–2000 were half what they were in the preceding decade.

However, some researchers hypothesized that PSA payments were not responsible for this decline and that forest loss would have slowed anyway because of other factors. To test this hypothesis, a team led by G. Arturo Sanchez-Azofeifa of the University of Alberta and Alexander Pfaff of Duke University worked with FONAFIFO's payment data, as well as data on land use and forest cover from satellite surveys. They layered these data onto maps using a geographic information system (GIS) (p. 38), and then explored the patterns revealed.

In 2007, in the journal *Conservation Biology,* they reported that only 7.7% of PSA contracts were located within 1 km of regions where forest was at greatest risk of clearance. PSA contracts were only slightly more likely to be near such a region than far from it. This meant, they argued, that PSA contracts were not being targeted to regions where they could have the most impact.

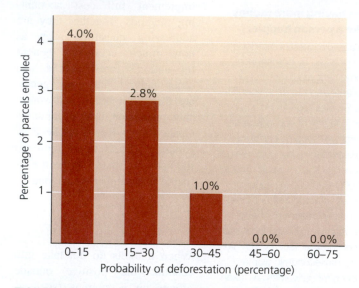

Costa Rican farmers take PSA payments into account when judging whether to clear forest.

FIGURE 1 In areas at greater risk of deforestation, lower percentages of land parcels were enrolled in the PSA program. This is because land more profitable for agriculture was less often enrolled. *Data from Pfaff, A., et al., 2008.* Payments for environmental services: Empirical analysis for Costa Rica. *Working Papers Series SAN08-05, Terry Sanford Institute of Public Policy, Duke University.*

Moreover, since enrollment was voluntary, most landowners applying for payments likely had land unprofitable for agriculture and were not actually planning to clear forest (**FIGURE 1**). In a 2008 paper, these researchers compared lands under PSA contracts with similar lands not under contracts. PSA lands experienced no forest loss, whereas the deforestation rate on non-PSA lands was 0.21%/yr. However, their analyses indicated that PSA lands stood only a 0.08%/yr likelihood of being cleared in the first place, suggesting that the program prevented only 0.08%/yr of forest loss, not 0.21%/yr. Other research was bearing this out; at least two studies found that many PSA participants, when interviewed, said they would have retained their forest even without the PSA program.

These researchers argued that Costa Rica's success in halting forest loss was likely due to other factors. In particular,

The tragedy of the commons When publicly accessible resources are open to unregulated exploitation, they tend to become overused, damaged, or depleted. So argued environmental scientist Garrett Hardin in his 1968 essay "The Tragedy of the Commons." Basing his argument on an age-old scenario, Hardin explained how in a public pasture (or "common") open to unregulated grazing, each person who grazes animals will be motivated by self-interest to increase the number of his or her animals in the pasture. Because

no single person owns the pasture, no one has incentive to expend effort taking care of it. Instead, each person takes what he or she can until overgrazing causes grass growth to collapse, hurting everyone. This scenario, known as the **tragedy of the commons,** pertains to many types of resources held and used in common by the public: forests, fisheries, clean air, clean water—even global climate.

When shared resources are being depleted or degraded, it is in society's interest to develop guidelines for their use.

Forest Law 7575, which had established the PSA system, had also banned forest clearing nationwide. This top-down government mandate, assuming it was enforceable, in theory made the PSA payments unnecessary. However, the PSA program made the mandate far more palatable to legislators, and Forest Law 7575 might never have passed had it not included the PSA payments.

Despite the PSA program's questionable impact in preserving existing forest, scientific studies show that it has been effective in regenerating new forest. In Costa Rica's Osa Peninsula, Rodrigo Sierra and Eric Russman of the University of Texas at Austin found in 2006 that PSA farms had five times more regrowing forest than did non-PSA farms. Interviews with farmers indicated that the program encouraged them to let land grow back into forest if they did not soon need it for production.

In the nation's northern Caribbean plain, a team led by Wayde Morse of the University of Idaho combined satellite data with on-the-ground interviews, finding that PSA payments plus the clearance ban reduced deforestation rates from 1.43%/year to 0.10%/year and that the program encouraged forest regrowth still more. Meanwhile, research by Rodrigo Arriagada indicated that the regeneration of new forest seemed to be the PSA program's major effect at the national level as well.

Most researchers today hold that Costa Rica's forest recovery results from a long history of conservation policies and economic developments. Indeed, deforestation rates had been dropping before the PSA program was initiated (**FIGURE 2**). There are several major reasons:

- Earlier policies (tax rebates and tax credits for timber production) encouraged forest cover.
- The creation of national parks fed a boom in ecotourism, so Costa Ricans saw how conserving natural areas could bring economic benefits.
- Falling market prices for meat discouraged ranching.
- After an economic crisis roiled Latin America in the 1980s, Costa Rica ended subsidies that had encouraged ranchers and farmers to expand into forested areas.

To help the PSA program make better use of its money, most researchers today feel that PSA payments should be targeted. Instead of paying equal amounts to anyone who applies,

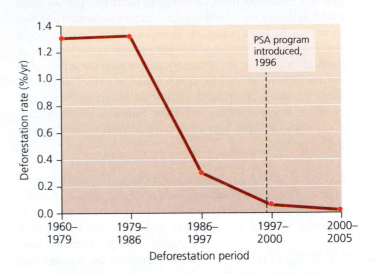

FIGURE 2 Forest recovery was underway in Costa Rica before the PSA program began. Deforestation rates had already dropped steeply, indicating that other factors were responsible. *Data from Sanchez-Azofeifa, G.A., et al., 2007. Costa Rica's payment for environmental services program: intention, implementation, and impact. Conservation Biology 21: 1165–1173.*

FONAFIFO should prioritize applicants, or pay more money, in regions that are ecologically most valuable or that are at greatest risk of deforestation. In a 2008 study, Tobias Wünscher of Bonn, Germany, and colleagues modeled and tested seven possible ways to target the payments, using data from Costa Rica's Nicoya Peninsula. Wünscher's team suggested using auctions, in which applicants for PSA funds put in bids stating how much they were requesting. Because applicants have outnumbered available contracts 3 to 1, FONAFIFO could favor the lower bids to keep costs down, while the auction system could make differential payments politically acceptable.

Costa Rica's government is responding to suggestions from researchers by aiming payments toward regions of greater environmental value and by making the program more accessible to low-income farmers in undeveloped regions. The government has also raised the payment amounts considerably. Researchers—and other nations—are watching closely to see how the program develops.

In Hardin's example, guidelines might limit the number of animals each person can graze or might require pasture users to help restore and manage the resource. These two concepts—management and restriction of use—are central to environmental policy today.

Public oversight through government is a standard way to alleviate the tragedy of the commons, but we can also address it in other ways. Resource users can voluntarily cooperate to prevent overexploitation. This may be effective

if the resource is localized and enforcement is simple, but these conditions are rare. Alternatively, the resource can be subdivided and allotments sold into private ownership, so that each owner gains incentive to manage his or her portion. Privatization may be effective if property rights can be clearly assigned (as with land), but it tends not to work with resources such as air or water. Privatization also opens the door to short-term profit taking at the long-term expense of the resource.

Free riders A second reason we develop policy for publicly held resources is the predicament caused by **free riders.** Let's say a community on a river suffers from water pollution that emanates from 10 different factories. The problem could in theory be solved if every factory voluntarily agrees to reduce its own pollution. However, once they all begin reducing their pollution, it becomes tempting for any one of them to stop doing so. A factory that avoids the efforts others are making would in essence get a "free ride." If enough factories take a free ride, the whole collective endeavor will collapse. Because of the free-rider problem, private voluntary efforts are often less effective than efforts mandated by public policy.

External costs Environmental policy also aims to promote fairness by dealing with external costs (p. 96). For example, a factory that discharges waste into a river imposes external costs (water pollution, health impacts, reduced fish populations, aesthetic impacts) on downstream users of the river. If the government forces the company running the factory to clean up its pollution, pay fees, or reimburse affected residents for damage, this helps to "internalize" costs. The costs would then be paid by the company, which would likely pass them on to consumers by raising the prices of its products. Higher market prices may reduce demand for the products, and consumers may instead favor less expensive products that impose fewer costs on society.

weighing the ISSUES

Internalizing External Costs

Imagine that we were to use policy to internalize all the external costs of gasoline (pollution, health risks, climate change, impacts from oil drilling and transport, etc.) and that as a result, gas prices rise to $13 per gallon. What effects do you think this would have on the choices we make as consumers (such as driving behavior and types of vehicles purchased)? What influence might it have on the types of vehicles produced and the types of energy sources developed? What effects might it have on our taxes and our health insurance premiums? In the long run, do you think that internalizing external costs in this way would end up costing society more money or saving society money? What factors might be important in determining the outcome?

Various factors can obstruct environmental policy

If environmental policy brings clear benefits, then why are environmental laws and regulations often challenged? One reason is the perception that environmental protection requires economic sacrifice (see FAQ, p. 96). Businesses often view regulations as restrictive and costly. Landowners may fear that zoning (p. 424) or protections for endangered species (p. 182) will restrict how they can use their land. Developers complain of time and money lost in obtaining permits; reviews by government agencies; and required environmental controls, monitoring, and mitigation.

Another hurdle for environmental policy stems from the nature of environmental problems, which often develop gradually over long time periods. Human behavior is geared toward addressing short-term needs, and this is reflected in our social institutions. Businesses usually opt for short-term financial gain. The news media focus coverage on new and sudden events. Politicians often act in their short-term interest because they depend on reelection. For all these reasons, environmental policy may be obstructed.

Policy in general can be held up for many reasons, even if a majority of people favor it. Checks and balances in a constitutional democracy seek to ensure that new policy is implemented only after extensive review and debate. However, less desirable factors can also hinder policy. In democracies such as the United States, each person has a political voice and can make a difference—yet money wields influence. People, organizations, industries, or corporations with enough wealth to buy access to power exert disproportionate influence over policymakers.

Science informs policy but is sometimes disregarded

Policy that is effective is generally informed by scientific research. For instance, Costa Rica's PSA program was inspired by research into the value of ecosystem services. Later, once scientists diagnosed shortcomings in how the program was being run, policymakers responded with remedies. When deciding whether to regulate a substance that may pose a public health risk, government agencies may comb the scientific literature for information or commission new studies to resolve outstanding questions. When crafting a bill to reduce pollution, a legislator may consult scientific data that quantify impacts of the pollution or that predict benefits from its reduction. In today's world, a nation's strength depends on its commitment to science. This explains why governments devote a portion of their tax revenue to fund scientific research.

Unfortunately, sometimes policymakers allow factors other than science to determine their decision making on scientific matters. Politicians may ignore scientific consensus on well-established matters such as evolution, vaccination, or climate change if it suits their political needs or if they are motivated chiefly by political or religious ideology. Some may reject or distort scientific advice if this helps to please campaign contributors or powerful constituencies.

Whenever taxpayer-funded science is suppressed or distorted for political ends, society is harmed. We cannot take for granted that science will play a role in policy. As educated citizens of a democracy, we need to stay vigilant and help ensure that our representatives in government are making proper use of the tremendous scientific assets we have at our disposal. Each of us can contribute by reelecting those politicians who consult and appreciate science and by voting out of office those who do not.

U.S. Environmental Law and Policy

The United States provides a good focus for understanding environmental policy in constitutional democracies worldwide, for several reasons. First, the United States has pioneered innovative environmental policy. Second, U.S. policies serve as models—of both success and failure—for other nations and for international bodies. Third, the United States exerts a great deal of influence on the affairs of other nations. Finally, understanding U.S. policy at the federal level helps us to understand policy at local, state, and international levels.

Federal policy arises from the three branches of government

Federal policy in the United States results from actions of the legislative, executive, and judicial branches of government. Congress creates laws, or **legislation,** by crafting bills that can become law with the signature of the head of the executive branch, the president. Once a law is enacted, its implementation and enforcement are assigned to an administrative agency in the executive branch. Administrative agencies create **regulations,** specific rules intended to achieve objectives of a law. These agencies also monitor compliance with laws and regulations. Several dozen administrative agencies influence U.S. environmental policy, ranging from the Environmental Protection Agency to the Forest Service to the Food and Drug Administration to the Bureau of Land Management.

The judiciary, consisting of the Supreme Court and various lower courts, is charged with interpreting law and is an important arena for environmental policy. Grass-roots environmental advocates and organizations use lawsuits to help level the playing field with large corporations and agencies. Conversely, the courts hear complaints from businesses and individuals challenging the constitutional validity of environmental laws they feel to be infringing on their rights. Individuals and organizations also lodge suits against government agencies when they feel the agencies are failing to enforce their own regulations.

The structure of the federal government is mirrored at the state level with governors, legislatures, judiciaries, and agencies. States, counties, and municipalities all generate policy of their own. They can act as laboratories experimenting with novel ideas, so that policies that succeed may be adopted elsewhere. In the "cooperative federalism" approach, a federal agency sets national standards and then works with state agencies to achieve them in each state.

Early U.S. environmental policy promoted development

Environmental policy in the United States was created in three periods. Laws enacted during the first period, from the 1780s to the late 1800s, accompanied the westward expansion of the nation and were intended mainly to promote settlement and the extraction and use of the continent's abundant natural resources (**FIGURE 5.7**).

Among these early laws were the General Land Ordinances of 1785 and 1787, by which the new federal government gave itself the right to manage the lands it was expropriating from Native American nations. These laws created a grid system for surveying these lands and readying them for private ownership. Subsequently, the government promoted settlement in the Midwest and West, and doled out millions

(a) Settlers in Nebraska, circa 1860

(b) Loggers felling an old-growth tree, Washington

FIGURE 5.7 Early U.S. environmental policy promoted settlement and natural resource extraction. The Homestead Act of 1862 allowed settlers **(a)** to claim 160 acres of public land by paying $16, living there for 5 years, and farming or building a home. The timber industry was allowed to cut the nation's ancient forests **(b)** with little policy to encourage conservation.

of acres to its citizens and to railroad companies, encouraging settlers, entrepreneurs, and land speculators to move west.

Western settlement was meant to provide U.S. citizens with means to achieve prosperity while relieving crowding in Eastern cities. It expanded the geographic reach of the United States at a time when the young nation was still jostling with European powers for control of the continent. It also wholly displaced the millions of Native Americans whose ancestors had inhabited these lands for millennia. U.S. environmental policy of this era reflected a perception that the vast western lands were inexhaustible in natural resources.

The second wave of U.S. environmental policy encouraged conservation

In the late 1800s, as the continent became more populated and its resources were increasingly exploited, public policy toward natural resources began to shift. Reflecting the emerging conservation and preservation ethics (p. 15) in American society, laws of this period aimed to alleviate some of the environmental impacts of westward expansion.

In 1872, Congress designated Yellowstone the world's first national park. In 1891, Congress authorized the president to create forest reserves to prevent overharvesting and protect forested watersheds. In 1903, President Theodore Roosevelt created the first national wildlife refuge. These acts launched the creation of a national park system, national forest system, and national wildlife refuge system that still stand as global models (pp. 203, 199, 204). These developments reflected a new understanding that the continent's resources were exhaustible and required legal protection.

Land management policies continued through the 20th century, targeting soil conservation in the wake of the Dust Bowl (p. 149) and wilderness preservation with the Wilderness Act of 1964 (p. 204).

The third wave responded to pollution

Further social changes in the 20th century gave rise to the third major period of U.S. environmental policy. In a more densely populated nation driven by technology, industry, and intensive resource consumption, Americans found themselves better off economically but living amid dirtier air, dirtier water, and more waste and toxic chemicals. Events in the 1960s and 1970s triggered greater awareness of environmental problems, bringing about a profound shift in public policy.

A landmark event was the 1962 publication of *Silent Spring,* a best-selling book by American scientist and writer Rachel Carson (**FIGURE 5.8**). *Silent Spring* awakened the public to the ecological and health impacts of pesticides and industrial chemicals (p. 216). The book's title refers to Carson's warning that pesticides might kill so many birds that few would be left to sing in springtime.

Ohio's Cuyahoga River also drew attention to pollution hazards. The Cuyahoga was so polluted with oil and industrial

FIGURE 5.8 Scientist and writer Rachel Carson revealed the effects of DDT and other pesticides in her 1962 book, *Silent Spring*.

waste that the river actually caught fire near Cleveland a number of times in the 1950s and 1960s (**FIGURE 5.9**). Such spectacles, coupled with an oil spill offshore from Santa Barbara, California, in 1969, moved the public to prompt Congress and the president to better safeguard water quality and public health.

Leaders in government and academia responded to the growing public desire for environmental protection. Stewart Udall, secretary of the interior from 1961 to 1969, helped shape key laws and oversaw the creation of more than 100 federal parks and refuges for conservation and public use. Legal scholar Joseph Sax in 1970 published a seminal paper developing the

FIGURE 5.9 Ohio's Cuyahoga River was so polluted with oil and waste that the river caught fire multiple times in the 1950s and 1960s and would burn for days at a time.

public trust doctrine, which holds that natural resources such as air, water, soil, and wildlife should be held in trust for the public and that government should protect them from exploitation by private parties. Wisconsin Senator Gaylord Nelson founded Earth Day in 1970, a now-annual event, which galvanized public support for action to address pollution problems.

With the aid of such leaders, public demand for a cleaner environment during this period inspired a number of major laws that underpin modern U.S. environmental policy (**TABLE 5.2**). You will encounter most of these laws again later in this book, and they have already helped to shape the quality of your life.

TABLE 5.2 Major U.S. Environmental Protection Laws, 1963–1980

Clean Air Act
1963; amended 1970, 1990

Sets standards for air quality, restricts emissions from new sources, enables citizens to sue violators, funds research on pollution control, and established an emissions trading program for sulfur dioxide. As a result, the air we breathe today is far cleaner (pp. 292–295).

Resource Conservation and Recovery Act
1976

Sets standards and permitting procedures for the disposal of solid waste and hazardous waste (p. 405). Requires that the generation, transport, and disposal of hazardous waste be tracked "from cradle to grave."

Endangered Species Act
1973

Seeks to protect species threatened with extinction. Forbids destruction of individuals of listed species or their critical habitat on public and private land, provides funding for recovery efforts, and allows negotiation with private landholders (pp. 182–183).

Clean Water Act
1977

Regulates the discharge of wastes, especially from industry, into rivers and streams (p. 278). Aims to protect wildlife and human health, and has helped to clean up U.S. waterways.

Safe Drinking Water Act
1974

Authorizes the EPA to set quality standards for tap water provided by public water systems, and to work with states to protect drinking water sources from contamination.

Soil and Water Conservation Act
1977

Directs the U.S. Department of Agriculture to survey and assess soil and water conditions across the nation and prepare conservation plans. Responded to worsening soil erosion and water pollution on farms and rangeland as production intensified.

Toxic Substances Control Act
1976; amended, 2016

Directs the EPA to monitor thousands of industrial chemicals and gives it power to ban those found to pose too much health risk (p. 228). However, the number of chemicals continues to increase far too quickly for adequate testing.

CERCLA ("Superfund")
1980

Funds the Superfund program to clean up hazardous waste at the nation's most polluted sites (p. 414). Costs were initially charged to polluters but most are now borne by taxpayers. The EPA continues to progress through many sites that remain. Full name is the Comprehensive Environmental Response Compensation and Liability Act.

Historians suggest that major advances in environmental policy occurred in the 1960s and 1970s because (1) environmental problems became readily apparent and were directly affecting people's lives, (2) people could visualize policies to deal with the problems, and (3) citizens were politically active and leaders were willing to act. In addition, photographs from NASA's space program allowed humanity to see, for the first time ever, images of Earth from space (see photos on pp. 2 and 437). It is hard for us to comprehend the power those images had at the time, but they revolutionized many people's worldviews by making us aware of the finite nature of our planet.

Today, largely because of policies enacted since the 1960s, our health is better protected and the nation's air and water are considerably cleaner. Thanks to the many Americans who worked tirelessly in grass-roots efforts, and to policymakers who listened and chose to make a difference in people's lives, we now enjoy a cleaner environment where industrial chemicals, waste disposal, and resource extraction are more carefully regulated. Much remains to be done—and we will always need to stand ready to defend our advances— but all of us alive today owe a great deal to the dedicated people who inspired policy to tackle pollution during this period.

Passage of NEPA and creation of the EPA were milestones

One of the foremost U.S. environmental laws is the **National Environmental Policy Act (NEPA),** drafted by Indiana University political scientist Lynton Caldwell and signed into law by Republican President Richard Nixon in 1970. NEPA created an agency called the Council on Environmental Quality and required that an **environmental impact statement (EIS)** be prepared for any major federal action that might significantly affect environmental quality. An EIS summarizes results from studies that assess environmental impacts that could result from development projects undertaken or funded by the federal government.

The EIS process forces government agencies and the businesses that contract with them to evaluate impacts using a cost-benefit approach (p. 96) before proceeding with a new dam, highway, or building project. The EIS process rarely halts development projects, but it serves as an incentive to lessen environmental damage. NEPA grants ordinary citizens input into the policy process by requiring that EISs be made publicly available and that policymakers solicit and consider public comment on them.

In 1970 policymakers also created the **Environmental Protection Agency (EPA).** The EPA was charged with conducting and evaluating research, monitoring environmental quality, setting and enforcing standards for pollution levels, assisting the states in meeting the standards, and educating the public. Since then, the EPA has played a central role in environmental policy. Industries whose practices are regulated by the EPA complain constantly of expense and bureaucracy, but studies that examine how regulations can reduce external costs (pp. 96–97, 104) show that EPA actions have brought net benefits worth hundreds of billions of dollars to the American public each and every year (p. 117).

The social context for policy evolves

In the 1980s Congress strengthened, broadened, and elaborated upon the laws of the 1970s. But the political climate in the United States soon changed. Although public support for the goals of environmental protection remained high, many people began to feel that the regulatory means used to achieve these goals too often imposed economic burdens on businesses or individuals. Attempts were made to roll back environmental policy, beginning with the presidential administration of Ronald Reagan and continuing with the George W. Bush administration and most congressional sessions since 1994. The administration of Barack Obama strengthened environmental policy, particularly on issues pertaining to climate change, but was repeatedly obstructed by Congress in these efforts. The rise of Donald Trump to the presidency, along with Republican control of Congress, brought an aggressive backlash against environmental policy on multiple fronts.

Today in the United States, legal protections for public health and environmental quality remain strong in some areas but have been weakened in others. Past policies restricting toxic substances such as lead and DDT have improved public health, but scientists and regulators cannot keep up with the flood of new chemicals being introduced by industry. And as demand for energy rises while concern over climate change intensifies, people continue to suffer impacts from fossil fuel use and extraction while striving to find a path toward clean and renewable energy.

Amid the heightened partisanship of U.S. politics today, environmental policy has gotten caught in the political crosshairs. Despite the fact that some of the greatest early conservationists were Republicans, and even though the words *conservative* and *conservation* share the same root meaning, environmental issues have today become identified as a predominantly Democratic concern. As a result, significant bipartisan advances rarely occur, and most environmental policy is now being crafted at the state and local level.

Environmental policy advances today on the international stage

Although the United States has ceded much of its leadership internationally on environmental policy, other nations are forging ahead with innovative policy. Germany has used policy to make impressive strides with solar energy (pp. 375–376). Sweden maintains a thriving society while promoting progressive environmental policies. Small developing nations such as Costa Rica are bettering their citizens' lives while protecting and restoring their natural capital. Even China, despite becoming the world's biggest polluter, is taking the world's biggest steps toward renewable energy, reforestation, and pollution control.

Worldwide, we have now embarked on a fourth wave of environmental policy, defined by two main goals. One is to develop solutions to climate change (Chapter 14), an issue of unprecedented breadth and global reach (**FIGURE 5.10**). The second goal is to achieve sustainability through sustainable development (pp. 114–115), finding ways to safeguard natural systems while raising living standards for the world's people. International conferences (pp. 335, 115) have brought together the world's nations to grapple with each issue, and these efforts seem bound to continue for the foreseeable future.

International Environmental Policy

Environmental systems pay no heed to political boundaries, and neither do environmental problems. Climate change is a global issue because carbon pollution from any one nation spreads through the atmosphere and oceans, affecting all nations. Because one nation's laws have no authority in other nations, international policy is vital to solving "transboundary" problems in our globalizing world.

Globalization makes international institutions vital

We live in an era of rapid and profound change. **Globalization** describes the process by which the world's societies have become more interconnected, linked by trade, diplomacy, and communication technologies in countless ways. Globalization has brought us many benefits by facilitating the spread of ideas and technologies that empower individuals and enhance our lives. Billions of people enjoy a degree of access to news, education, arts, and science that we could barely have imagined in the past, and billions also now live under governments that are more democratic. Over recent decades, we have gained a richer awareness of other cultures, and warfare has declined.

Yet as globalization proceeds and social conditions change, many people are feeling anxious or threatened by the loss of traditional cultural norms. As people, goods, and ideas flow freely across national borders, debates over trade, jobs, immigration, and identity are fueling populist reactions against globalization, forcing difficult conversations, and undercutting social and political stability in North America and Europe. At the same time, ecological systems are undergoing change at unprecedented rates and scales. People are moving organisms from one continent to another, allowing invasive species to affect ecosystems everywhere. Multinational corporations operate outside the reach of national laws and rarely have incentive to conserve resources or limit pollution while moving from nation to nation. Today our biggest environmental challenges are global in scale (such as climate change, ozone depletion, overfishing, and biodiversity loss). For all these reasons, in our globalizing world the institutions that shape international law and policy play increasingly vital roles.

International law includes customary law and conventional law

International law known as **customary law** arises from long-standing practices, or customs, held in common by most cultures. International law known as **conventional law** arises from conventions, or treaties (written contracts), into which nations enter. One example of a treaty is the United

TABLE 5.3 Major International Environmental Treaties

CONVENTION OR PROTOCOL	YEAR IT CAME INTO FORCE	NATIONS THAT HAVE RATIFIED IT	STATUS IN UNITED STATES
CITES: Convention on International Trade in Endangered Species of Wild Fauna and Flora (p. 183)	1975	175	Ratified
Ramsar Convention on Wetlands of International Importance	1975	159	Ratified
Montreal Protocol, of the Vienna Convention for the Protection of the Ozone Layer (p. 303)	1989	196	Ratified
Basel Convention on the Control of Transboundary Movements of Hazardous Wastes and Their Disposal (p. 414)	1992	172	Signed but not ratified
Convention on Biological Diversity (p. 183)	1993	168	Signed but not ratified
Stockholm Convention on Persistent Organic Pollutants (p. 228)	2004	152	Signed but not ratified
Paris Agreement, of the UN Framework Convention on Climate Change (p. 335)	2016	151, as of 2017	Signed but not ratified; U.S. may withdraw

Nations Framework Convention on Climate Change, which in 1994 established a framework for agreements to reduce greenhouse gas emissions that contribute to climate change. The Kyoto Protocol (a *protocol* is an amendment or addition to a convention) and the Paris Agreement each later specified the agreed-upon details of the emissions limits (p. 335). **TABLE 5.3** shows a selection of major environmental treaties ratified (legally approved by a government) by most of the world's nations.

Treaties are also signed among pairs or groups of nations. The United States, Mexico, and Canada entered into the North American Free Trade Agreement (NAFTA) in 1994. NAFTA eliminated trade barriers such as tariffs on imports and exports, making goods cheaper to buy. Yet NAFTA also threatened to undermine protections for workers and the environment by steering economic activity to regions where regulations were most lax. Side agreements were negotiated to try to address these concerns, and NAFTA's impacts on jobs and on environmental quality in the three nations have been complex. Many U.S. jobs moved to Mexico, but fears that pollution would soar and regulations would be gutted largely did not come to pass—indeed, some sustainable products and practices spread from nation to nation. Debates recur with each proposed free trade agreement, as negotiators try to find ways to gain the benefits of free trade while avoiding environmental damage and economic harm to working people.

Several organizations shape international environmental policy

In our age of globalization, a number of international institutions act to influence the policy and behavior of nations by providing funding, applying political or economic pressure, and directing media attention.

The United Nations Founded in 1945 and including representatives from virtually all nations of the world, the **United Nations (UN)** seeks to maintain peace, security, and friendly relations among nations; to promote respect for human rights and freedoms; and to help nations cooperate to resolve global challenges. Headquartered in New York City, the United Nations plays an active role in environmental policy by sponsoring conferences, coordinating treaties, and publishing research.

The World Bank Established in 1944 and based in Washington, D.C., the **World Bank** is one of the largest sources of funding for economic development and major infrastructure projects. In fiscal year 2016, the World Bank provided $61 billion in loans and support for projects designed to benefit low-income people in developing countries. Despite its admirable mission, the World Bank is often criticized for funding unsustainable projects that cause environmental impacts, such as dams that generate electricity but also flood valuable forests and farmland. Providing for the needs of growing human populations in poor nations while minimizing damage to the ecological systems on which people rely can be a tough balancing act. Environmental scientists agree that the concept of sustainable development must be the guiding principle for such efforts.

The World Trade Organization Founded in 1995 and based in Geneva, Switzerland, the **World Trade Organization (WTO)** represents multinational corporations. It promotes free trade by reducing obstacles to international commerce and enforcing fairness among nations in trading practices. The WTO has authority to impose financial penalties on nations that do not comply with its directives.

The WTO has interpreted some national environmental laws as unfair barriers to trade. For instance, in 1995, the U.S. EPA issued regulations requiring cleaner-burning gasoline in U.S. cities. Brazil and Venezuela filed a complaint with the WTO, saying the new rules discriminated against the dirtier-burning petroleum they exported to the United States. The WTO agreed, ruling that even though the dirty gasoline posed a threat to human health in the United States, the EPA rules

SUCCESS STORY   Using a Treaty to Halt Ozone Depletion

In the 1980s, the world was shocked to learn of a global threat that seemed to come from out of nowhere. Scientific monitoring revealed that ozone in Earth's

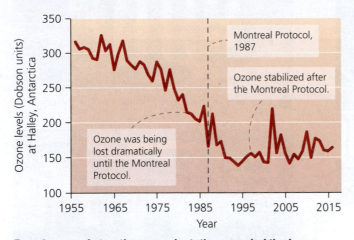

Data from an Antarctic research station revealed the loss of ozone in the stratosphere—and shows its stabilization following treaties to address the problem.

upper atmosphere was being lost rapidly, stripping our planet of a compound that protects life from dangerous ultraviolet radiation from the sun. Researchers discovered that artificially manufactured chemicals such as chlorofluorocarbons (CFCs)—widely used in aerosol spray cans, as refrigerants, and for other purposes—were destroying ozone once they were released into the atmosphere. Policymakers and industry responded quickly to the science, and in 1987 the world's nations signed the Montreal Protocol, a treaty to cut CFC production in half. Follow-up treaties deepened the cuts and restricted other ozone-depleting substances. Industry successfully switched to alternative chemicals for products. As a result, we stopped atmospheric ozone loss in the 1990s, and Earth's ozone layer is on track to fully recover later this century. Thus, by harnessing international cooperation on policy, humanity succeeded in resolving a major global problem. (Read the full story in Chapter 13, pp. 301–303.) Now, we need the world's nations to act together to tackle today's major global problem: climate change. The Montreal Protocol provides hope that such an effort can succeed.

EXPLORE THE DATA at Mastering Environmental Science

---

weighing the
ISSUES

Trade Barriers and Environmental Protection

If Canada has stricter laws for environmental protection than Mexico, and if these laws limit Mexico's ability to export its goods to Canada, then by WTO policy Canada's laws could be overruled in the name of free trade. Do you think this is fair? Now consider that Canada is wealthier than Mexico and that Mexico could use an economic boost. Does this affect your response?

Approaches to Environmental Policy

When most of us think of environmental policy, what comes to mind are major laws or regulations. However, environmental policy is diverse.

were an illegal trade barrier. The ruling forced the United States to weaken its regulations.

Nongovernmental organizations Many **nongovernmental organizations (NGOs)**—nonprofit, mission-driven organizations not overseen by any government—have become international in scope and exert influence over policy. Environmental NGOs such as the Nature Conservancy focus on conservation objectives on the ground (such as purchasing and managing land and habitat for rare species) without becoming politically involved. Other groups, such as Greenpeace, Conservation International, and Population Connection, attempt to shape policy through research, education, lobbying, or protest.

Policy can follow three approaches

Environmental policy can use a variety of strategies within three major approaches (**FIGURE 5.11**).

Lawsuits in the courts Prior to the legislative push of recent decades, most environmental policy questions were addressed with lawsuits in the courts. Individuals suffering external costs from pollution would sue polluters, one case at a time. The courts sometimes punished polluters by ordering them to stop their operations or pay damages to the affected parties. However, as industrialization proceeded and population grew, pollution became harder to avoid, and judges became reluctant to hinder industry. People began to view legislation and regulation as more effective means of protecting public health and safety.

Command-and-control policy Most environmental laws and regulations use a **command-and-control** approach, in which a regulating agency such as the EPA sets rules, standards, or limits on certain actions and threatens punishment for violations. This simple and direct approach has brought residents of the United States and other nations cleaner air, cleaner water, safer workplaces, and many other advances. The relatively safe, healthy, comfortable lives most of us enjoy today owe much to the command-and-control environmental policy of recent decades.

Even in plain financial terms, command-and-control policy has been effective. Each year the White House Office of Management and Budget analyzes U.S. policy to calculate the economic costs and benefits of regulations. These analyses have consistently revealed that benefits far outweigh

PROBLEM
Pollution from factory harms people's health

FIGURE 5.11 Three major policy approaches exist to resolve environmental problems. To address pollution from a factory, we might ➊ seek damages through lawsuits, ➋ limit pollution through legislation and regulation, or ➌ reduce pollution using market-based strategies.

SOLUTIONS
Three policy approaches

➊ **Lawsuits in the courts:** People can sue factories

➋ **Command-and-control policy:** Governments can regulate emissions

➌ **Economic policy tools:** Policy can create incentives; a factory that pollutes less (right) will outcompete one that pollutes more (left) through permit trading, avoiding green taxes, or selling ecolabeled products

costs and that environmental regulations have been most beneficial of all. You can explore some of these data in *Calculating Ecological Footprints* (p. 117).

Economic policy tools Despite the successes of command-and-control policy, many people dislike the top-down nature of government mandates that dictate particular solutions to problems. As an alternative approach, we can aim to channel the innovation and economic efficiency of market capitalism in ways that benefit the public. Economic policy tools use financial incentives to promote desired outcomes by encouraging private entities competing in a marketplace to innovate and generate new or better solutions at lower cost.

Each of these three approaches has strengths and weaknesses, and each is best suited to different conditions. The approaches may also be used together. For instance, Costa Rica's Forest Law 7575 was a command-and-control law that banned forest clearing, but it also established the PSA program as an economic policy tool to help the policy succeed. Government regulation is often needed to frame market-based efforts, and citizens can use the courts to ensure that regulations are enforced. Let's now explore several types of economic policy tools: taxes, subsidies, emissions trading, and ecolabeling.

Green taxes discourage undesirable activities

In taxation, money passes from private parties to the government, which uses it to pay for services to benefit the public. Taxing undesirable activities helps to internalize external costs by making these costs part of the normal expense of doing business. A tax on an environmentally harmful activity or product is called a **green tax.**

Under green taxation, a firm owning a polluting factory might pay taxes on the pollution it discharges—the more pollution, the higher the tax payment. This gives factory owners a financial incentive to reduce pollution while allowing them the freedom to decide how to do so. One polluter might choose to invest in pollution control technology if this is more affordable than paying the tax. Another polluter might choose to pay the tax—funds the government could then use to reduce pollution in some other way.

Costa Rica uses a green tax to help fund its PSA program. It applies a tax of 3.5% to sales of fossil fuels and then uses the revenue to pay for conserving forests, which soak up carbon emissions from fossil fuel combustion. In the United States, similar "sin taxes" on cigarettes and alcohol are long-accepted tools of social policy. Taxes on pollution are more common in Europe, where many nations have adopted the **polluter-pays principle,** which specifies that the party creating pollution be held responsible for covering the costs of its impacts. Today many nations and states are experimenting with carbon taxes—taxes on gasoline, coal-based electricity, and fossil-fuel-intensive products—to fight climate change (p. 332).

Subsidies promote certain activities

Another economic policy tool is the **subsidy,** a government giveaway of money or resources that is intended to support or promote an industry or activity. Subsidies take many forms, and one is the *tax break*, which relieves the tax burden on an industry, firm, or individual. Costa Rica's PSA program subsidizes the conservation and restoration of forests by transferring public money to landowners who conserve or restore forests. Ironically, much of the nation's deforestation had resulted from ranching and farming that the government had previously been subsidizing.

Subsidies like Costa Rica's payments for ecological services promote environmentally sustainable activities—but all too often subsidies are used to prop up unsustainable ones. In the United States, subsidies for grazing, timber extraction (p. 200), and mineral extraction (p. 249) on public lands all benefit private parties that may profit from, and deplete, publicly held resources.

Fossil fuels have been a major recipient of subsidies over the years. From 1950 to 2010, the U.S. government gave $594 billion of its citizens' money to oil, gas, and coal corporations (most of this in tax breaks), according to one recent compilation (**FIGURE 5.12**). In comparison, just $171 billion was granted to renewable energy, and most of these subsidies went to hydropower and to corn ethanol, which is not widely viewed as a sustainable fuel (p. 393). In recent years, global fossil fuel subsidies have outpaced renewable energy subsidies by four to eight times, according to the International Energy Agency.

In 2009, President Obama and other leaders of the Group of 20 (G-20) nations resolved to gradually phase out their collective $300 billion of annual fossil fuel subsidies. Doing so would hasten a shift to cleaner renewable energy sources and accomplish half the greenhouse gas emissions cuts needed to hold global warming to 2°C. However, since that time, fossil fuel subsidies have *grown*, not shrunk. One reason is lobbying by fossil fuel corporations. Another is that consumers accustomed to artificially low prices for gasoline and electricity might punish policymakers who lift subsidies and let these prices rise.

Emissions trading uses markets

Subsidies and green taxes each create financial incentives in direct and selective ways. However, we may also pursue policy goals by establishing financial incentives and then letting marketplace dynamics run their course, as in **emissions trading.** In an emissions trading system, a government

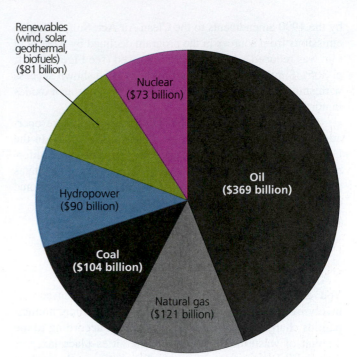

FIGURE 5.12 **The well-established fossil fuel industries (slices in shades of gray) receive the majority of U.S. energy subsidies.** Cumulative data for the United States from 1950 to 2010 are shown. Apportionments remain similar today. *Data from Management Information Services, Inc., 2011.* 60 years of energy incentives: Analysis of federal expenditures for energy development. *Washington, D.C.: Management Information Services.*

DATA • How many dollars in subsidies have gone to fossil fuels (oil, coal, and natural gas) for every dollar that has gone to other energy sources? • How many dollars in subsidies have gone to fossil fuels for every dollar that has gone to energy from wind, solar, geothermal, and biofuels?

Go to **Interpreting Graphs & Data** on **Mastering** Environmental Science

creates a market in permits for the emission of pollutants, and companies, utilities, or industries then buy and sell the permits among themselves. In a **cap-and-trade** emissions trading system, the government first caps the overall amount of pollution it will allow, then grants or auctions off permits to polluters that allow them each to emit a certain fraction of that amount. As polluters trade these permits, the government progressively lowers the cap of overall emissions allowed (see Figure 14.28, p. 333).

Suppose you own an industrial plant with permits to release 10 units of pollution, but you find that you can make your plant more efficient and release only 5 units instead. You now have a surplus of permits, which you can sell to some other plant owner who needs them. Doing so generates income for you and meets the needs of the other plant, while the total amount of pollution does not rise. By providing firms an economic incentive to reduce pollution, emissions trading can lower expenses for industry relative to a conventional regulatory system.

The United States pioneered the cap-and-trade approach with its program to reduce sulfur dioxide emissions, established by the 1990 amendments to the Clean Air Act. Sulfur dioxide emissions from sources in the program declined by 67%, acid rain was reduced, and air quality improved (see Figure 13.21, p. 305). Similar cap-and-trade programs have shown success with smog in the Los Angeles basin and with nitrogen oxides in northeastern states.

To address climate change, European nations are operating a market in greenhouse gas emissions (p. 333). In the United States, carbon trading markets are running in California and among northeastern states (p. 333), while other states and nations are establishing programs. (We will assess some of these efforts in Chapter 14.)

Market incentives are diverse at the local level

You are most likely already taking part in transactions involving financial incentives as policy tools. Many municipalities charge residents for waste disposal according to the amount of waste they generate. Some cities place taxes or disposal fees on items whose safe disposal is costly, such as tires and motor oil. Others give rebates to residents who buy water-efficient toilets and appliances, because rebates can cost a city less than upgrading its wastewater treatment system. Likewise, power utilities may offer discounts to customers who buy high-efficiency appliances, because doing so is less costly than expanding the generating capacity of their plants.

Ecolabeling empowers consumers

In the approach known as **ecolabeling,** sellers who use sustainable practices in growing, harvesting, or manufacturing products advertise this fact on their labels, hoping to win approval from buyers (**FIGURE 5.13**). Examples include labeling recycled paper (p. 404), organic foods (p. 162), dolphin-safe tuna, shade-grown and fair-trade coffee, and sustainably harvested lumber (p. 202).

In many cases, ecolabeling grew from initial steps taken by governments to require the disclosure of information to consumers. Once established, however, ecolabeling can spread in a free market as businesses seek to win consumer confidence and to outcompete less sustainable brands. Of course, some businesses may try to mislead us into thinking their products are more sustainable than they actually are—a phenomenon called **greenwashing.** Independent certification by outside parties can help ensure that consumers get accurate information. When labeling is accurate, each of us as consumers can provide businesses and industries a powerful incentive to switch to more sustainable processes when we buy ecolabeled products.

The creative use of economic policy tools is growing and diversifying, while command-and-control regulation and legal action in the courts continue to play vital roles in environmental policy. As a result, we have a variety of strategies available as we seek sustainable solutions to our society's challenges.

Sustainable Development

Today's search for sustainable solutions centers on **sustainable development,** economic progress that maintains resources for the future. The United Nations defines sustainable development as development that "meets the needs of the present without sacrificing the ability of future generations to meet their own needs." Sustainable development is an economic pursuit shaped by policy and informed by science. It is also an ethical pursuit because it asks us to manage our resource use so that future generations can enjoy similar access to resources.

Sustainable development involves environmental protection, economic well-being, and social equity

Economists use the term **development** to describe the use of natural resources for economic advancement (as opposed to simple subsistence, or survival). Development involves

(a) Organic foods

(b) Energy-efficient appliances

(c) Fair-trade products

FIGURE 5.13 Ecolabeling enables each of us to promote sustainable business practices through our purchasing decisions. Among the many ecolabeled products now widely available are **(a)** organic foods, **(b)** energy-efficient appliances, and **(c)** fair-trade coffee.

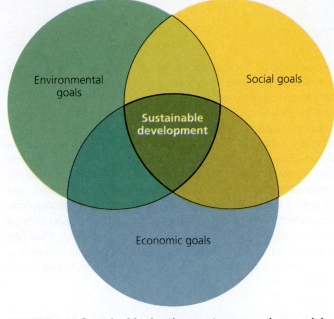

FIGURE 5.14 Sustainable development occurs when social, economic, and environmental goals overlap.

TABLE 5.4 U.N. Sustainable Development Goals

- End poverty in all its forms everywhere
- End hunger, achieve food security, and promote sustainable agriculture
- Promote health and well-being for all
- Ensure quality education for all
- Achieve gender equality and empower women and girls
- Ensure water and sanitation for all
- Ensure access to affordable, reliable, and sustainable energy sources
- Promote sustainable economic growth and employment
- Build resilient infrastructure, promote sustainable industry, and foster innovation
- Reduce inequality within and among nations
- Make cities safe, resilient, and sustainable
- Ensure sustainable consumption and production
- Take urgent action to combat climate change and its many impacts
- Conserve marine resources
- Protect and restore terrestrial ecosystems and halt biodiversity loss
- Promote peaceful, inclusive, and just institutions
- Renew partnerships for sustainable development

Adapted from the United Nations Division for Sustainable Development (www.un.org/sustainabledevelopment).

making purposeful changes intended to improve our quality of life. Construction of homes, schools, hospitals, power plants, factories, and transportation networks are all examples of development. In the past, many advocates of development felt that protecting the environment threatened people's economic needs, while many advocates of environmental protection felt that development degraded the environment, jeopardizing the improvements in quality of life that were intended. Today, however, people increasingly perceive how we all depend on a healthy and functional natural environment.

We also now recognize that society's poorer people tend to suffer the most from environmental degradation. As a result, advocates of environmental protection, economic development, and social justice began working together toward common goals. This cooperation gave rise to the modern drive for sustainable development, which seeks ways to promote social justice, economic well-being, and environmental quality at the same time (**FIGURE 5.14**). Governments, businesses, industries, and organizations pursuing sustainable development aim to satisfy a **triple bottom line,** a trio of goals including economic advancement, environmental protection, and social equity.

Programs that pay for ecosystem services are one example of a sustainable development approach that seeks to satisfy a triple bottom line. Costa Rica's PSA program aims to enhance its citizens' well-being by conserving the country's natural assets while compensating affected landholders for any economic losses. The intention is to achieve a win-win-win result that pays off in economic, social, and environmental dimensions.

Sustainable development is global

Sustainable development has blossomed as an international movement. The United Nations, the World Bank, and other organizations sponsor conferences, fund projects, publish research, and facilitate collaboration across borders among governments, businesses, and nonprofit organizations.

The Earth Summit at Río de Janeiro, Brazil, in 1992 was the world's first major gathering focused on sustainable development. With representatives from over 200 nations, this conference gave rise to notable achievements, including the Convention on Biological Diversity (p. 183) and the Framework Convention on Climate Change (p. 334). Ten years later, nations met in Johannesburg, South Africa, at the 2002 World Summit on Sustainable Development. Then in 2012, the world returned to Río de Janeiro for the Rio+20 conference.

In 2015, world leaders met at the United Nations and adopted 17 **Sustainable Development Goals** for humanity (**TABLE 5.4**). Each broad goal for sustainable development has a number of specific underlying targets—169 in all—that may be met by implementing concrete strategies. Many of the Sustainable Development Goals were given a 2030 target date, and we are making better progress on some than on others. We still have a long way to go to resolve the many challenges facing humanity. Pursuing solutions that meet a triple bottom line of environmental, economic, and social goals can help pave the way for a truly sustainable global society.

closing THE LOOP

Environmental policy is a problem-solving tool that makes use of science, ethics, and economics. Command-and-control legislation and regulation remain our most common policy approaches, but innovative market-based policy tools are also being deployed. Environmental and ecological economists are quantifying the value of ecosystem services and devising alternative means of measuring progress, thereby helping to show how economic progress is tied to environmental protection and resource conservation. In pursuing sustainable development, we recognize that economic, social, and environmental well-being depend on one another and can be mutually reinforcing.

The nation of Costa Rica provides one useful model of a pathway toward sustainable development. A series of decisions by its political leaders has enabled the nation to make impressive progress in social, economic, and environmental dimensions. By paying farmers and ranchers to preserve and restore forest on private land, for example, the country's citizenry reaps the rewards of a cleaner and healthier environment, which in turn has enhanced economic progress. As Costa Rica's leaders use research-based feedback to refine and improve the PSA program, the program should be able to better accomplish its goals. The government, businesses, and people of Costa Rica recognize how economic health depends on environmental protection, and seem poised to build on their success so far. In Costa Rica and across the world, if we can enhance our economic and social well-being while conserving natural resources, then truly sustainable solutions will be within reach.

TESTING Your Comprehension

1. Name and describe two key contributions that the natural environment makes to our economies.

2. Describe four ways in which neoclassical economic approaches can contribute to environmental problems.

3. Compare and contrast the views of neoclassical economists, environmental economists, and ecological economists, particularly regarding the issue of economic growth.

4. What are ecosystem services? Give several examples. Describe ways in which some economists have assigned monetary values to ecosystem services.

5. Describe two of the major justifications for environmental policy. Now articulate three problems that environmental policy commonly seeks to address.

6. Summarize how the first, second, and third waves of environmental policy in U.S. history differed from one another. Describe two current priorities in international environmental policy.

7. What did the National Environmental Policy Act accomplish? Briefly describe the origin and mission of the U.S. Environmental Protection Agency.

8. Compare and contrast the three major approaches to environmental policy: lawsuits, command-and-control, and economic policy tools. Describe an advantage and disadvantage of each.

9. Explain how each of the following works: a green tax, a subsidy, and an emissions trading system.

10. Define *sustainable development*. What is meant by the *triple bottom line*? Why is it important to pursue sustainable development?

SEEKING Solutions

1. Do you think that a steady-state economy is a practical alternative to our current approach that prioritizes economic growth? Why or why not?

2. Do you think we should attempt to quantify and assign market values to ecosystem services? Why or why not? What consequences might this have?

3. Reflect on causes for the transitions in U.S. history from one type of environmental policy to another. Now peer into the future, and consider how life and society might be different in 25, 50, or 100 years. What would you predict about the environmental policy of the future, and why? What issues might future policy address? Do you predict we will have more or less environmental policy?

4. **CASE STUDY CONNECTION** Suppose you are a Costa Rican farmer who needs to decide whether to clear a stand of forest or apply to receive payments to preserve it through the PSA program. Describe all the types of information you would want to consider before making your decision. Now, describe what you think each of the following people would recommend to you if you were to ask him or her for advice: (a) a neoclassical economist and (b) an ecological economist.

5. **THINK IT THROUGH** You have just returned from serving in the U.S. Peace Corps in Costa Rica, where you worked closely with farmers, foresters, ecologists, and policymakers on issues related to Costa Rica's PSA program. You have now been hired as an adviser

on natural resource issues to the governor of your state. Think about the condition of the forests, land and soil, water supplies, and other natural resources in your state. Given what you learned in Costa Rica, would you advise your governor to institute some kind of program to pay residents to conserve ecosystem services? Why or why not? Describe what policies you would advocate to best conserve your state's resources and ecosystem services while advancing the economic and social condition of its people.

CALCULATING Ecological Footprints

Critics of command-and-control policy often argue that regulations are costly to business and industry, yet cost-benefit analyses (p. 96) have repeatedly shown that regulations bring citizens more benefits than costs, overall. Each year the U.S. Office of Management and Budget assesses costs and benefits of major federal regulations of administrative agencies.

Results from the most recent report, covering the decade from 2005 to 2015, are shown below. This decade includes periods of both Republican and Democratic control of the presidency and of Congress. Subtract costs from benefits, and enter these values for each agency in the third column. Divide benefits by costs, and enter these values in the fourth column.

Costs and Benefits of Major U.S. Federal Regulations, 2005–2015
(average values from ranges of estimates, in billions of dollars)

AGENCY	BENEFITS	COSTS	BENEFITS MINUS COSTS	BENEFIT : COST RATIO
Department of Energy	25.9	9.1	16.8	2.8
Department of Health and Human Services	13.9	3.7		
Department of Transportation	28.6	10.9		
Environmental Protection Agency (EPA)	426.8	47.1		
Other departments	76.8	22.3		
Total	572.0	93.1		

Data from U.S. Office of Management and Budget, 2016. 2016 Draft report to Congress on the benefits and costs of federal regulations and agency compliance with the Unfunded Mandates Reform Act. Washington, D.C.: OMB.

1. For how many of the agencies shown do regulations exert more costs than benefits? For how many do regulations provide more benefits than costs?

2. Which agency's regulations have the greatest excess of benefits over costs? Which agency's regulations have the greatest ratio of benefits to costs?

3. What percentage of total benefits from regulations comes from EPA regulations? Most of the benefits and costs from EPA regulations are from air pollution rules resulting from the Clean Air Act and its amendments. Judging solely by these data, would you say that Clean Air Act legislation has been a success or a failure for U.S. citizens? Why?

Mastering Environmental Science

Students Go to **Mastering** Environmental Science for assignments, the etext, and the Study Area with practice tests, videos, current events, and activities.

Instructors Go to **Mastering** Environmental Science for automatically graded activities, current events, videos, and reading questions that you can assign to your students, plus Instructor Resources.

CHAPTER 6

Human Population

Will China's New "Two-Child Policy" Defuse Its Population "Time Bomb"?

> **We don't need adjustments to the family-planning policy. What we need is a phaseout of the whole system.**
> —Gu Baochang, Chinese demographer at People's University, Beijing, referring to the nation's "one-child" policy in 2013

> **As you improve health in a society, population growth goes down.... Before I learned about it, I thought it was paradoxical.**
> —Bill Gates, Founder, Microsoft Corporation

The People's Republic of China is the world's most populous nation, home to one-fifth of the more than 7 billion people living on Earth. It is also the site of one of the most controversial social experiments in history.

When Mao Zedong founded the country's current regime in 1949, roughly 540 million people lived in a mostly rural, war-torn, impoverished nation. Mao's policies encouraged population growth, and by 1970 improvements in food production, food distribution, and public health allowed China's population to swell to 790 million people. At that time, Chinese women gave birth to an average of 5.8 children in their lifetimes and China's population grew by 2.8% annually.

However, the country's burgeoning population and its industrial and agricultural development were eroding the nation's soils, depleting its water, and polluting its air. Realizing that the nation might not be able to continue to feed its people, Chinese leaders decided in 1970 to institute a population control program that prohibited most Chinese couples from having more than one child. The "one-child" program applied mostly to families in urban areas. Many farmers and ethnic minorities in rural areas were allowed more than one child, because success on the farm often depends on having multiple children.

The program encouraged people to marry later and have fewer children, and increased accessibility to contraceptives and abortion. Families with only one child were rewarded with government jobs and better housing, medical care, and access to schools. Families with more than one child, meanwhile, were subjected to costly monetary fines, employment discrimination, and social scorn. The experiment was a success in slowing population growth: The nation's growth rate is now down to 0.5%, and Chinese women now have only an average of 1.6 children in their lifetimes.

However, the one-child policy also produced a population with a shrinking labor force, increasing numbers of older people, and too few women. These unintended consequences led some demographers to question whether China's one-child policy simply traded one population problem—overpopulation—for other population problems.

Upon completing this chapter, you will be able to:

- **Describe the scope of human population growth**

- **Explain how human population, affluence, and technology affect the environment**

- **Explain the fundamentals of demography**

- **Describe the concept of demographic transition**

- **Explain how family planning, the status of women, and affluence affect population growth**

Crowded street in Shanghai, one of China's largest cities

Will the two-child policy balance China's skewed sex ratio? ▲

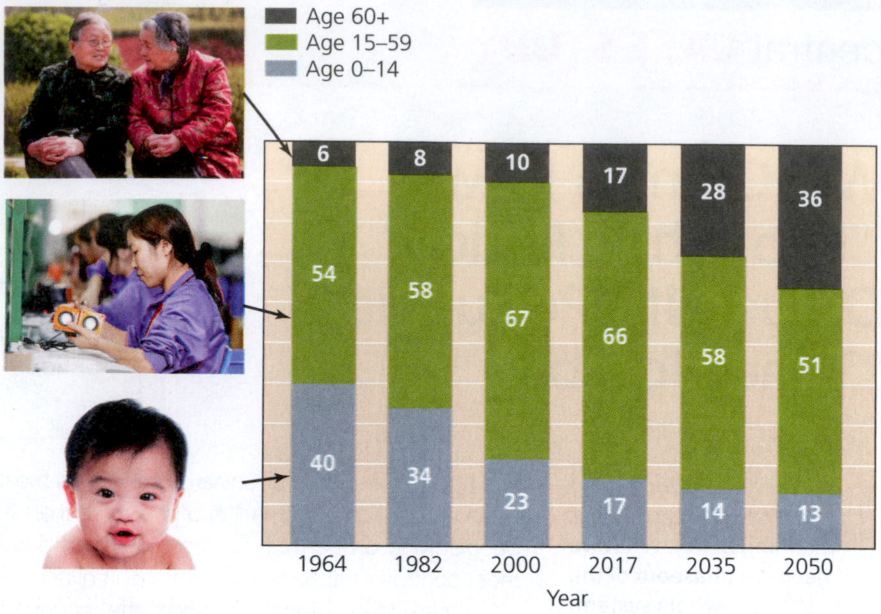

FIGURE 6.1 China's one-child policy is leading to a shrinking workforce and rising numbers of older citizens. Values in the figure represent the percentage of the Chinese population in each age group. *Source: Figure from Population Reference Bureau, 2004. China's Population: New Trends and Challenges. Data for 2017–2050 from U.S. Census Bureau International Database, www.census.gov/population/international/data/idb/.*

The rapid reduction in fertility that resulted from this policy drastically changed China's age structure (**FIGURE 6.1**). Once consisting predominantly of young people, China's population has shifted, such that the numbers of children and older people are now more even. This means there will be relatively fewer workers for China's growing economy, which is driving up wages and encouraging companies with factories in China to seek out more inexpensive labor in other nations. The growing number of older Chinese individuals poses problems because the Chinese government lacks the resources to fully support them, putting a heavy economic burden on the millions of only children produced under the one-child policy as they help provide for their retired parents.

Modern China also has too few women. Chinese culture has traditionally valued sons because they carry on the family name, assist with farm labor in rural areas, and care for aging parents. Daughters, in contrast, will most likely marry and leave their parents, as the traditional culture dictates. Thus, when faced with being limited to just one child, many Chinese couples preferred a son to a daughter. Tragically, this led in some instances to selective abortion and the killing of female infants. This has caused a highly unbalanced ratio of young men and women in China, leading to the social instability that arises when large numbers of young men are unable to find brides and remain longtime bachelors.

Until recently, Chinese authorities attempted to address this looming population "time bomb" of an aging population with skewed ratios of men and women by occasionally loosening the one-child policy. For example, the government announced in 2013 that if either member of a married couple is an only child, the couple would be allowed to have a second child—but only 1.5 million of the 11 million citizens eligible for this exemption applied for it. Faced with the prospect of continued population issues, the Chinese government announced in October 2015 that the former one-child policy would immediately become a two-child policy, and couples would be permitted to have two children without penalty.

It is unclear, however, if Chinese couples, used to the material wealth and urban lifestyle many enjoy, will embrace the opportunity to grow their families—and accept the costs of raising a second child—now that it is allowed. A survey from 2008 by China's family-planning commission, for example, reported that only 19% of the people they surveyed wished to have a second child if the one-child policy was relaxed. But in 2016, the first full year since the relaxing of the one-child policy, birth rates were 7.9% higher than in 2015, suggesting that many couples are choosing to have a second child. It therefore remains to be seen if China's relaxing of the one-child policy came too late to defuse the pending time bomb the nation may experience as China's population grays in the midst of its rapid industrialization.

China's reproductive policies have long elicited intense criticism worldwide from people who oppose government intrusion into personal reproductive choices, and such intrusion continues today, albeit with a higher allowable family size. The policy, however, has proven highly effective in slowing population growth rates and aiding the economic rise of modern China. As other nations become more crowded and seek to emulate China's economic growth, might their governments also feel forced to turn to drastic policies that restrict individual freedoms? In this chapter, we examine human population dynamics in China and worldwide, consider their causes, and assess their consequences for the environment and human society.

Our World at Seven Billion

China receives a great deal of attention with regard to population issues because of its unique reproductive policies and its status as the world's most populous nation. But China is not alone in struggling with population matters. India soon will surpass China in possessing the world's largest population (**FIGURE 6.2**). India was the first nation to implement comprehensive population control policies, but when India's policymakers introduced forced sterilization in the 1970s, the resulting outcry forced the government to change its policies. Since then, India's efforts have been more modest and far less coercive, focusing on family planning and reproductive health care.

Like India, many of the world's poorer nations continue to experience substantial population growth, which leads to stresses on society, the environment, and people's well-being. In our world of now more than 7.4 *billion* people, one of our greatest challenges is finding ways to slow the growth of the human population without coercive measures such as those used in China, but rather by establishing conditions that lead people to desire to have fewer children.

The human population continues to grow

Our global population grows by over 80 million people each year, which means that we add more than two people to the planet *every second*. Take a look at **FIGURE 6.3** and note just how recent and sudden our rapid increase has been. It took until after 1800, virtually all of human history, for our population to reach 1 billion. Yet by 1930 we had reached 2 billion, and 3 billion in just 30 more years. Our population added its next billion in just 15 years, and it has taken only 12 years to add each of the next three installments of a billion people.

What accounts for our unprecedented growth? Exponential growth—the increase in a quantity by a fixed percentage per unit time—accelerates the increase in population size over time, just as compound interest accrues in a savings account (p. 63). The reason, you will recall, is that a fixed percentage of a small number makes for a small increase, but that same percentage of a large number

FAQ

How big is a billion?

It can be difficult to conceptualize huge numbers. As a result, we often fail to recognize the true magnitude of a number such as 7 billion. Although we know that a billion is bigger than a million, we tend to view both numbers as impossibly large and therefore similar in size. For example, guess (without calculating) how long it would take a banker to count out $1 million if she did so at a rate of a dollar a second for 8 hours a day, 7 days a week. Now guess how long it would take to count $1 billion at the same rate. It may surprise you to learn that counting $1 million would take a mere 35 days, whereas counting $1 billion would take 95 years! Living 1 million seconds takes only 12 days, while living for 1 billion seconds requires more than 31 years. Examples like these can help us appreciate the *b* in *billion*.

(a) Reproductive counseling in India

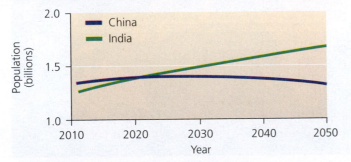

(b) Population projections for China and India

FIGURE 6.2 India will likely soon surpass China as the most populous nation. Both nations have embraced population control initiatives **(a)**, but China's more stringent reproductive policies will likely lead to its population stabilizing in coming decades, while India's population will continue to grow **(b)**. China's rate of growth is now lower than India's as a result of China's aggressive population policies. *Data from U.S. Census Bureau International Database, www.census.gov/population/international/data/idb/.*

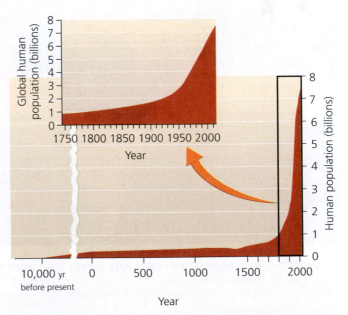

FIGURE 6.3 We have risen from fewer than 1 billion in 1800 to more than 7.4 billion today. Viewing global human population size over a long timescale **(bottom graph)** and growth since the industrial revolution **(inset top graph)** shows that nearly all growth has occurred in just the past 200 years. *Data from U.S. Census Bureau.*

produces a large increase. Thus, even if the growth *rate* remains steady, population *size* will increase by greater increments with each successive generation.

For much of the 20th century, the growth rate of the human population rose from year to year. This rate peaked at 2.1% during the 1960s and has declined to 1.2% since then. Although 1.2% may sound small, a hypothetical population starting with one man and one woman that grows at 1.2% gives rise to a population of 112,695 after only 60 generations.

At a 1.2% annual growth rate, a population doubles in size in only 58 years. We can roughly estimate doubling times with a handy rule of thumb. Just take the number 70 (which is 100 times 0.7, the natural logarithm of 2) and divide it by the annual percentage growth rate: $70/1.2 = 58.3$. China's current growth rate of 0.5% means it would take roughly 140 years ($70/0.5 = 140$) for its current population to double, but India's current growth rate of 1.5% predicts a doubling in only about 47 years ($70/1.5 = 46.7$). Had China not instituted its one-child policy and its growth rate remained at 2.8%, it would have taken only 25 years ($70/2.8 = 25$) to double in size. Although the global growth rate is 1.2%, rates vary widely from region to region and are highest in nations with developing economies (**FIGURE 6.4**).

Is there a limit to human population growth?

Our spectacular growth in numbers has resulted largely from technological innovations, improved sanitation, better medical care, increased agricultural output, and other factors that have brought down death rates. Birth rates have not declined as much, so births have outpaced deaths for many years now, leading to population growth. But can the human population continue to grow indefinitely?

Environmental factors set limits on the growth of populations (p. 64), but environmental scientists who have tried to pin a number to the human carrying capacity (p. 63) have come up with wildly differing estimates. The most rigorous estimates range from 1–2 billion people living prosperously in a healthy environment to 33 billion people living in extreme poverty in a degraded world of intensive cultivation without natural areas.

The difficulty in estimating how many humans our planet can support is that we have repeatedly increased our carrying capacity by developing technology to overcome the natural limits on our population growth. For example, British economist Thomas Malthus (1766–1834) argued in his influential work, *An Essay on the Principle of Population* (1798), that if society did not reduce its birth rate, then rising death rates would reduce the population through war, disease, and starvation. Although his contention was reasonable at the time, agricultural improvements in the 19th century increased food supplies, and his prediction did not come to pass. Similarly, biologist Paul Ehrlich predicted in his 1968 book, *The Population Bomb,* that human population growth would soon outpace food production and unleash massive famine and conflict in the latter 20th century. However, thanks to the way the "Green Revolution" (p. 142) increased food production in developing regions in the decades after his book, Ehrlich's dire forecasts did not fully materialize.

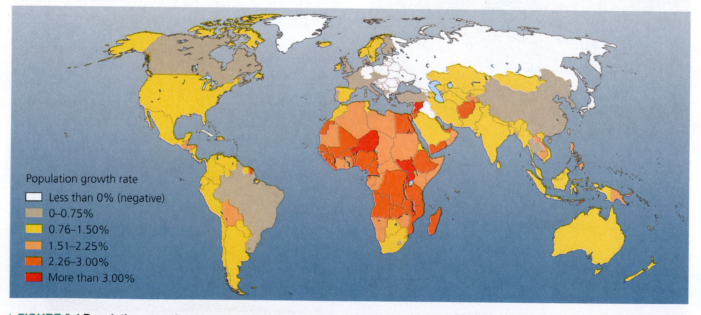

Population growth rate
- ☐ Less than 0% (negative)
- 0–0.75%
- 0.76–1.50%
- 1.51–2.25%
- 2.26–3.00%
- More than 3.00%

FIGURE 6.4 Population growth rates vary greatly from place to place. Populations are growing fastest in poorer nations, while populations are beginning to decrease in some highly industrialized nations. Shown are rates of natural increase as of 2016. *Data from U.S. Census Bureau International Database, www.census.gov/population/international/data/idb/.*

DATA Q • Which world region has the highest population growth rates? • Which world region has the lowest population growth rates?

Go to **Interpreting Graphs & Data** on **Mastering** Environmental Science

Does this mean we can disregard the concerns of Malthus and Ehrlich? Some economists say yes. Under the Cornucopian view that many economists hold, population growth poses no problem if new resources can be found or created to replace depleted ones. In contrast, environmental scientists recognize that not all resources can be replaced—such as the ecosystem services provided by a species that was driven to extinction. Thus, the environmental scientists argue, population growth is indeed a problem if it depletes resources, stresses social systems, and degrades the natural environment, such that our quality of life declines.

Population is one of several factors that affect the environment

One widely used formula gives us a handy way to think about population and other factors that affect environmental quality. Nicknamed the **IPAT model,** it is a variation of a formula proposed in 1974 by Paul Ehrlich and John Holdren. The IPAT model represents how our total impact (I) on the environment results from the interaction among population (P), affluence (A), and technology (T):

$$I = P \times A \times T$$

We can interpret impact in various ways, but can generally boil it down either to unsustainable resource consumption or to the degradation of ecosystems by pollution. Increased population intensifies impact on the environment as more individuals take up space, use natural resources, and generate waste. Increased affluence magnifies environmental impact through greater per capita resource consumption, which generally has accompanied enhanced wealth. Technology that enhances our abilities to exploit minerals, fossil fuels, old-growth forests, or fisheries generally increases impact, but technology to reduce smokestack emissions, harness renewable energy, or improve manufacturing efficiency can decrease impact. One reason our population has kept growing, despite limited resources, is that we have developed technology—the T in the IPAT equation—time and again to increase efficiency, alleviate our strain on resources, and allow us to expand further.

We might also add a fourth factor, sensitivity (S), to the equation to denote how sensitive a given environment is to human pressures. For instance, the arid lands of western China are more sensitive to human disturbance than the moist regions of southeastern China. Plants grow more slowly in the arid west, making the land more vulnerable to deforestation and soil degradation. Thus, adding an additional person to western China has more environmental impact than adding one to southeastern China. We could refine the IPAT equation further by adding terms for the influence of social factors such as education, laws, ethical standards, and social stability and cohesion. Such factors all affect how population, affluence, and technology translate into environmental impact.

Modern-day China shows how all elements of the IPAT formula can combine to cause tremendous environmental impact in little time. Although China boasts one of the world's fastest-growing economies, the country is battling unprecedented environmental challenges brought about by this rapid economic development. Intensive agriculture has expanded westward out of the nation's moist rice-growing regions, causing farmland to erode and blow away, much like the Dust Bowl tragedy that befell the U.S. heartland in the 1930s (p. 149). China has overpumped aquifers and has drawn so much water for irrigation from the Yellow River that this once-mighty waterway now dries up in many stretches. Although China is reducing its air pollution from industry and charcoal-burning homes, the country faces new threats to air quality from rapidly rising numbers of automobiles. The air in Beijing is so polluted (p. 300), for example, that simply breathing it on a daily basis damages the lungs to the same extent as smoking 40 cigarettes. As the world's developing countries try to attain the material prosperity that industrialized nations enjoy, China is a window into what much of the rest of the world could soon become.

Demography

People do not exist outside nature. We exist within our environment as one species of many. As such, all the principles of population ecology that drive biological change in the natural world (Chapter 3) apply to humans as well. The application of principles from population ecology to the study of statistical change in human populations is the focus of **demography.**

Demography is the study of human populations

Demographers study population size, density, distribution, age structure, sex ratio, and rates of birth, death, immigration, and emigration of people, just as population ecologists study these characteristics in other organisms. Each is useful for predicting population dynamics and environmental impacts.

Population size Our global human population of more than 7.4 billion is spread among 200 nations with populations ranging up to China's 1.38 billion, India's 1.33 billion, and the 324 million citizens of the United States (**FIGURE 6.5**).

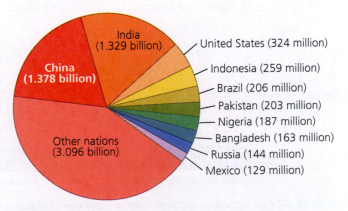

FIGURE 6.5 Almost one in five people in the world lives in China, and more than one of every six live in India. Three of every five people live in one of the 10 most populous nations. *Data from Population Reference Bureau, 2016. 2016 World population data sheet.*

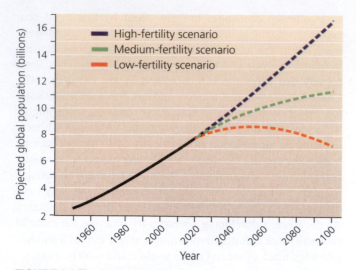
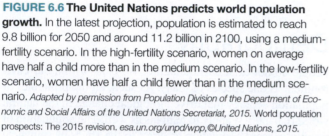

FIGURE 6.6 The United Nations predicts world population growth. In the latest projection, population is estimated to reach 9.8 billion for 2050 and around 11.2 billion in 2100, using a medium-fertility scenario. In the high-fertility scenario, women on average have half a child more than in the medium scenario. In the low-fertility scenario, women have half a child fewer than in the medium scenario. *Adapted by permission from Population Division of the Department of Economic and Social Affairs of the United Nations Secretariat, 2015. World population prospects: The 2015 revision. esa.un.org/unpd/wpp,©United Nations, 2015.*

The United Nations Population Division estimates that by the year 2050, the global population will surpass 9.8 billion (**FIGURE 6.6**). However, population size alone—the absolute number of individuals—doesn't tell the whole story. Rather, a population's environmental impact depends on its density, distribution, and composition (as well as on affluence, technology, and other factors outlined earlier).

Population density and distribution People are distributed unevenly across our planet. In ecological terms, our distribution is clumped (p. 62) at all spatial scales. At the global scale, population density is highest in regions with temperate, subtropical, and tropical climates and lowest in regions with extreme-climate biomes, such as desert, rainforest, and tundra. Human population is dense along seacoasts and rivers, and less dense farther away from water. At more local scales, we cluster together in cities and towns.

This uneven distribution means that certain areas bear more environmental impact than others. Just as the Yellow River experiences pressure from Chinese farmers, the world's other major rivers all receive more than their share of human impact. At the same time, some areas with low population density are sensitive (a high *S* value in our revised IPAT model) and thus vulnerable to impact. Deserts and arid grasslands, for instance, are easily degraded by agriculture and ranching that commandeer too much water.

Age structure Age structure describes the relative numbers of individuals of each age class within a population (p. 62). Data on age structure are especially valuable to demographers trying to predict future dynamics of human populations. A population made up mostly of individuals past reproductive age will tend to decline over time. In contrast, a population with many individuals of reproductive age or pre-reproductive age is likely to increase. A population with an even age distribution will likely remain stable as births keep pace with deaths.

Age structure diagrams, often called population pyramids, are visual tools scientists use to illustrate age structure (**FIGURE 6.7**). The width of each horizontal bar represents the number of males or females in each age class. A pyramid with a wide base denotes a large proportion of people who have not yet reached reproductive age—and this indicates a population soon capable of rapid growth. As an example, compare age structures for Canada and Nigeria (**FIGURE 6.8**). Nigeria's large concentration of individuals in younger age classes predicts a great deal of future reproduction. Not surprisingly, Nigeria has a higher population growth rate than Canada.

Today, populations are aging in many nations, and the global population is "grayer" than in the past. The global median age today is 28, but it is predicted to be 38 by the year 2050. By causing dramatic reductions in the number of children born since 1970, China's former one-child policy virtually guaranteed that the nation's population age structure

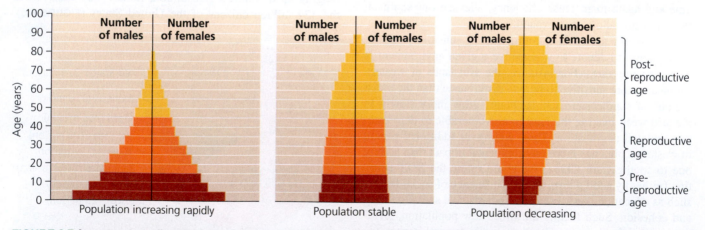

FIGURE 6.7 Age structure diagrams show numbers of males and females of different age classes in a population. A diagram like that on the left is weighted toward young age classes, indicating a population that will grow quickly. A diagram like that on the right is weighted toward old age classes, indicating a population that will decline. Populations with balanced age structures, like the one shown in the middle diagram, will remain relatively stable in size.

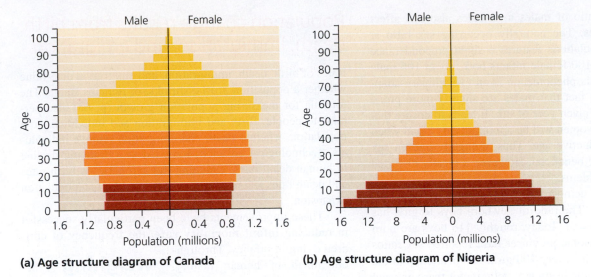

(a) Age structure diagram of Canada

(b) Age structure diagram of Nigeria

FIGURE 6.8 Canada (a) shows a fairly balanced age structure, whereas Nigeria (b) shows an age distribution heavily weighted toward young people. Nigeria's population growth rate (2.6%) is over eight times greater than Canada's (0.3%). *Data from U.S. Census Bureau International Database, www.census.gov/population/ international/data/idb/.*

(a) Billboard promoting China's "one child" policy

would change (**FIGURE 6.9**). Indeed, in 1970 the median age in China was 20; by 2050 it is predicted to be 45.

Changing age distributions have caused concerns in some nations that have declining numbers of workers and strong social welfare programs for retirees (which are supported by current workers), such as the Social Security program in the United States. Despite the long-term benefits associated with smaller populations, many policymakers find it difficult to let go of the notion that population growth increases a nation's economic, political, and military strength. So, while China and India struggle to get their population growth under control, some national governments—such as Canada—offer financial and social incentives that encourage their own citizens to have more children. These incentives include free health care, extended maternity and paternity leave, subsidized child care, and tax breaks for larger families.

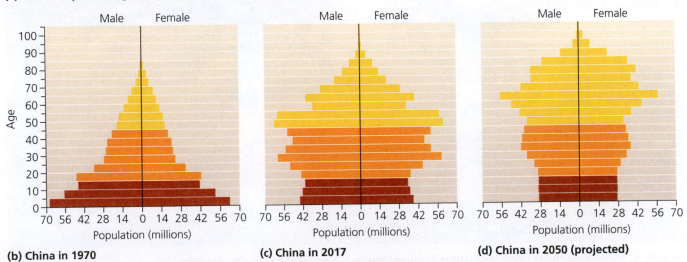

(b) China in 1970

(c) China in 2017

(d) China in 2050 (projected)

FIGURE 6.9 As China's population ages, older people will outnumber the young. China's one-child policy **(a)** was highly successful in reducing birth rates but also in significantly changing China's age structure. Population pyramids show the predicted graying of the Chinese population from **(b)** 1970 to **(c)** 2017 to **(d)** what is predicted for 2050. *Data from U.S. Census Bureau International Database, www.census.gov/population/international/data/idb./data/idb.*

Sex ratios The ratio of males to females also can affect population dynamics. The naturally occurring sex ratio at birth in human populations features a slight preponderance of males; for every 100 female infants born, about 106 male infants are born. This phenomenon is an evolutionary adaptation (p. 51) to the fact that males are slightly more prone to death during any given year of life. It tends to ensure that the ratio of men to women will be approximately equal when people reach reproductive age. Thus, a slightly uneven sex ratio at birth may be beneficial. However, a greatly distorted ratio can lead to problems.

In recent years, demographers have witnessed an unsettling trend in China: The ratio of newborn boys to girls has become strongly skewed. Today, roughly 116 boys are born for every 100 girls. Some provinces have reported sex ratios as high as 138 boys for every 100 girls. As mentioned in the opening Case Study, the leading hypothesis for these unusual sex ratios is that some parents learn the gender of their fetus by ultrasound and selectively abort female fetuses. A 2016 report by American researchers has suggested that some of these "missing girls" in rural areas may actually have been born, but they just weren't reported. Local authorities simply "looked the other way" when the girls were born, and then added them to official registries when they became school-aged.

China's skewed sex ratio may further lower population growth rates. However, it has the undesirable social consequence of leaving large numbers of Chinese men single. Without the anchoring effect a wife and family provide, many of these men leave their native towns and find work elsewhere as migrant workers. Living as bachelors far from home, these men often engage in more risky sexual activity than their married counterparts. Researchers speculate that this could lead to higher incidence of HIV infection in China in coming decades, as tens of millions of bachelors find work as migrant workers.

Population change results from birth, death, immigration, and emigration

Rates of birth, death, immigration, and emigration determine whether a population grows, shrinks, or remains stable. The formula for measuring population growth (p. 63) also pertains to people: Birth and immigration add individuals to a population, whereas death and emigration remove individuals. Technological advances have led to a dramatic decline in human death rates, widening the gap between birth rates and death rates and resulting in the global human population expansion.

These improvements have been particularly successful in reducing **infant mortality rate,** the frequency of children dying in infancy. Throughout much of human history, parents needed to have larger families as insurance against the likelihood that one or more of their children would die during infancy. Poor nutrition, disease, exposure to hostile elements, and limited medical care claimed the lives of many infants in their first year of life. As societies have industrialized and become more affluent, infant mortality rates have plummeted as a result of better nutrition, prenatal care, and the presence of medically trained practitioners during birth.

As shown in **FIGURE 6.10**, infant mortality rates vary widely around the world and are closely tied to a nation's level of industrialization. China, for example, saw its infant

weighing the
ISSUES

China's Reproductive Policy

Consider the benefits as well as the problems associated with a reproductive policy such as China's. Do you think a government should be able to enforce strict penalties for citizens who fail to abide by such a policy? If you disagree with China's policy, what alternatives can you suggest for dealing with the resource demands of a rapidly growing population?

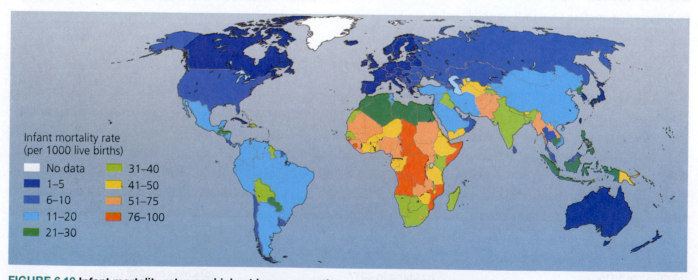

Infant mortality rate
(per 1000 live births)

No data | 31–40
1–5 | 41–50
6–10 | 51–75
11–20 | 76–100
21–30

FIGURE 6.10 Infant mortality rates are highest in poorer nations, such as those in sub-Saharan Africa, and lowest in wealthier nations. Industrialization brings better nutrition and medical care, which greatly reduce the number of children dying in their first year of life. *Data from Population Reference Bureau, 2016.* 2016 World population data sheet.

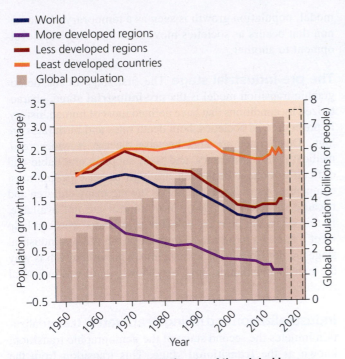

Legend:
- World
- More developed regions
- Less developed regions
- Least developed countries
- Global population

FIGURE 6.11 The annual growth rate of the global human population peaked in the late 1960s and has declined since then. Growth rates of developed nations have fallen since 1950, whereas those of developing nations have fallen since the global peak in the late 1960s. For the world's least developed nations, growth rates began to fall in the 1990s. Although growth rates are declining, global population size is still growing about the same amount each year, because smaller percentage increases of ever-larger numbers produce roughly equivalent additional amounts. *Data from Population Division of the Department of Economic and Social Affairs of the United Nations Secretariat, 2011. World population prospects: The 2010 revision. esa.un.org/unpd/wpp. © United Nations, 2011. Data updates for 2011–2016 from Population Reference Bureau, 2011–2016 World population data sheets. Global population in 2020 is projected.*

mortality rate drop from 47 children per 1000 live births in 1980 to 11 children per 1000 live births in 2016, as the nation industrialized and prospered. Many other industrializing nations enjoyed similar success in reducing infant mortality during this time period.

In recent decades, falling growth rates in many countries have led to an overall decline in the global growth rate (**FIGURE 6.11**). This decline has come about, in part, from a steep drop in birth rates. Note, however, that although the rate of growth is slowing, the absolute size of the population continues to increase, as growth rates remain positive.

Total fertility rate influences population growth

One key statistic demographers calculate to examine a population's potential for growth is the **total fertility rate (TFR)**, the average number of children born per woman during her lifetime. **Replacement fertility** is the TFR that keeps the size of a population stable. For humans, replacement fertility roughly equals a TFR of 2.1. (Two children replace the mother and father, and the extra 0.1 accounts for the risk of a child dying before reaching reproductive age.) If the TFR drops below 2.1, population size in a given country (in the absence of immigration) will shrink.

Factors such as industrialization, improved women's rights (pp. 133–134), access to family planning, and quality health care have driven the TFR downward in many nations in recent years. All these factors have come together in Europe, where the TFR has dropped from 2.6 to 1.6 in the past half-century. Nearly every European nation now has a fertility rate below the replacement level, and populations are declining in 15 of 45 European nations. In 2016, Europe's overall annual **rate of natural increase** (also called the *natural rate of population change*)—change due to birth and death rates alone, excluding migration—was between 0.0% and 0.1%. Worldwide by 2016, 84 countries had fallen below the replacement fertility of 2.1. These low-fertility countries make up a sizeable portion of the world's population and include China (with a TFR of 1.6). **TABLE 6.1** shows total fertility rates of major continental regions.

weighing the ISSUES

What Are the Consequences of Low Fertility?

In the United States, Canada, and almost every European nation, the total fertility rate is now at or below the replacement fertility rate (although some of these nations are still growing because of immigration). What economic or social consequences do you think might result from below-replacement fertility rates? Would you rather live in a society with a growing population, a shrinking population, or a stable population? Why?

Many nations have experienced the demographic transition

Many nations with lowered birth rates and TFRs are experiencing a common set of interrelated changes. In countries with reliable food supplies, good public sanitation, and effective health care, more people than ever before are living long lives. As a result, over the past 50 years the life expectancy for the average person worldwide has

TABLE 6.1 Total Fertility Rates for Major Regions

REGION	TOTAL FERTILITY RATE (TFR)
Africa	4.7
Australia and the South Pacific	2.3
Latin America and the Caribbean	2.1
Asia	2.1
North America	1.8
Europe	1.6

Data from Population Reference Bureau, 2016. 2016 World population data sheet.

increased from 46 to 71 years, as the worldwide death rate has dropped from 20 deaths per 1000 people to 8 deaths per 1000 people. **Life expectancy** is the average number of years that an individual in a particular age group is likely to continue to live, but often people use this term to refer to the average number of years a person can expect to live from birth. Much of the increase in life expectancy is due to reduced rates of infant mortality. Societies going through these changes are generally those that have undergone urbanization and industrialization and have generated personal wealth for their citizens.

To make sense of these trends, demographers developed a concept called the **demographic transition.** This is a model of economic and cultural change—first proposed in the 1940s and 1950s by demographer Frank Notestein—to explain the declining death rates and birth rates that have occurred in Western nations as they industrialized. Notestein argued that nations move from a stable pre-industrial state of high birth and death rates to a stable post-industrial state of low birth and death rates (**FIGURE 6.12**). Industrialization, he proposed, causes these rates to fall by first decreasing mortality and then lessening the need for large families. Parents thereafter choose to invest in quality of life rather than quantity of children. Because death rates fall before birth rates fall, a period of net population growth results. Thus, under the demographic transition model, population growth is seen as a temporary phenomenon that occurs as societies move from one stage of development to another.

The pre-industrial stage

The first stage of the demographic transition model is the **pre-industrial stage,** characterized by conditions that have defined most of human history. In pre-industrial societies, both death rates and birth rates are high. Death rates are high because disease is widespread, medical care rudimentary, and food supplies unreliable and difficult to obtain. Birth rates are high because people must compensate for infant mortality by having many children and because reliable methods of birth control are not available. In this stage, children are valuable as workers who can help meet a family's basic needs. Populations in the pre-industrial stage are not likely to experience much growth, which is why the human population grew very slowly until the industrial revolution.

Industrialization and falling death rates

Industrialization initiates the second stage of the demographic transition, known as the **transitional stage.** This transition from the pre-industrial stage to the industrial stage is generally characterized by declining death rates due to increased food production better sanitation, and improved medical care. Birth rates in the transitional stage remain high, however, because

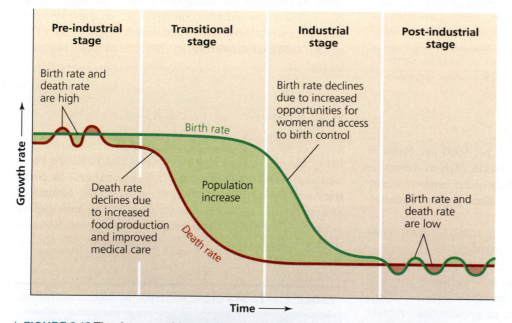

FIGURE 6.12 The demographic transition models a process that has taken some populations from a pre-industrial stage of high birth rates and high death rates to a post-industrial stage of low birth rates and low death rates. In this diagram, the wide green area between the two curves illustrates the gap between birth and death rates that causes rapid population growth during the middle portion of this process. *Adapted from Kent, M., and K. Crews, 1990. World population: Fundamentals of growth. By permission of the Population Reference Bureau.*

DATA • In which stage of the demographic transition does population increase the most?
• Is growth greatest toward the beginning or end of this stage?

Go to **Interpreting Graphs & Data** on **Mastering** Environmental Science

people have not yet grown used to the new economic and social conditions. As a result, population growth surges.

The industrial stage and falling birth rates

The third stage in the demographic transition is the **industrial stage.** Industrialization increases opportunities for employment outside the home, particularly for women. Children become less valuable, in economic terms, because they do not help meet family food needs as they did in the pre-industrial stage. If couples are aware of this, and if they have access to birth control, they may choose to have fewer children. Birth rates fall, closing the gap with death rates and reducing population growth.

The post-industrial stage

In the final stage, called the **post-industrial stage,** both birth and death rates have fallen to low and stable levels. Population sizes stabilize or decline slightly. The society enjoys the fruits of industrialization without the threat of runaway population growth.

Is the demographic transition inevitable?

The demographic transition has occurred in European countries, the United States, Canada, Japan, and a number of other developed nations over the past 200–300 years. It is a model that may or may not apply to all developing nations as they industrialize now and in the future. On the one hand, note that, as shown in Figure 6.11, growth rates fell first for more developed nations, then for less developed nations, and finally for least developed nations. This pattern suggests that it may merely be a matter of time before all nations experience the transition. On the other hand, some developing nations may already be suffering too greatly from the impacts of large populations to replicate the developed world's transition, a phenomenon called **demographic fatigue.** Demographically fatigued governments face overwhelming challenges related to population growth, including educating and employing swelling ranks of young people. When these stresses are coupled with large-scale environmental degradation or disease epidemics, the society may never complete the demographic transition.

Moreover, natural scientists estimate that for people of all nations to attain the material standard of living that North Americans now enjoy, we would need the natural resources of four and one-half more planet Earths. Whether developing nations (which include the vast majority of the planet's people) pass through the demographic transition is one of the most important questions for the future of our civilization and Earth's environment.

Population and Society

Demographic transition theory links the quantitative study of how populations change with the societal factors that influence (and are influenced by) population dynamics. Many factors affect fertility in a given society. They include public health issues, such as people's access to contraceptives and the rate of infant mortality. They also include cultural factors—such as the level of women's rights, the relative acceptance of contraceptive use, and even cultural influences like television programs (see **THE SCIENCE BEHIND THE STORY**, pp. 130–131). There are also effects from economic factors, such as the society's level of affluence, the importance of child labor, and the availability of governmental support for retirees. Let's now examine a few of these societal influences on fertility more closely.

Family planning is a key approach for controlling population growth

Perhaps the greatest single factor enabling a society to slow its population growth is the ability of women and couples to engage in **family planning,** the effort to plan the number and spacing of one's children. Family-planning programs and clinics offer information and counseling to potential parents on reproductive issues.

An important component of family planning is **birth control,** the effort to control the number of children one bears, particularly by reducing the frequency of pregnancy. Birth control relies on **contraception,** the deliberate attempt to prevent pregnancy despite sexual intercourse. Common methods of modern contraception include condoms, spermicide, hormonal treatments (birth control pill/hormone injection), intrauterine devices (IUDs), and permanent sterilization through tubal ligation or vasectomy. Many family-planning organizations aid clients by offering free or discounted contraceptives.

Worldwide in 2016, 56% of women aged 15–49 reported using modern contraceptives, with rates of use varying widely among nations. China and the United Kingdom, at 84%, had the highest rate of contraceptive use of any nations. Eight European nations showed rates of contraceptive use

weighing the ISSUES

Should the United States Abstain from International Family Planning?

Over the years, the United States has joined 180 other nations in providing millions of dollars to the United Nations Population Fund (UNFPA), which advises governments on family planning, sustainable development, poverty reduction, reproductive health, and AIDS prevention in many nations, including China. Starting with the Reagan administration in 1984, Republican presidential administrations have withheld funds from UNFPA, saying that U.S. law prohibits funding any organization that "supports or participates in the management of a program of coercive abortion or involuntary sterilization," maintaining that the Chinese government has been implicated in both. Democratic presidential administrations, however, provided funds to the program, noting that money from the United States is not used to perform abortions and citing the value of the family-planning services that UNFPA provides. What do you think U.S. policy should be? Should the United States fund family-planning efforts in other nations? What conditions, if any, should it place on the use of such funds?

Did Soap Operas Help Reduce Fertility in Brazil?

**Eliana La Ferrara,
Bocconi University**

Over the past 50 years, the South American nation of Brazil experienced the second-largest drop in fertility among developing nations with large populations—second only to China. In the 1960s, the average woman in Brazil had six children. Today, Brazil's total fertility rate is 1.8 children per woman, which is lower than that of the United States. Brazil's drastic decrease in fertility is interesting because, unlike in China, it occurred without intrusive governmental policies to control its citizens' reproduction.

Brazil accomplished this, in part, by providing women equal access to education and opportunities to pursue careers outside the home. Women now make up 40% of the workforce in Brazil and graduate from college in greater numbers than men. In 2010, Brazilians elected a woman, Dilma Rousseff, as their nation's president.

The Brazilian government also provides family planning and contraception to its citizens free of charge. Eighty percent of married women of childbearing age in Brazil currently use contraception, a rate higher than that in the United States or Canada. Universal access to family planning has given women control over their desired family size and has helped reduce fertility across all economic groups, from the very rich to the very poor. It is interesting to note that induced abortion is not used in Brazil as it is in China; the procedure is illegal except in rare circumstances.

As Brazil's economy grew with industrialization, people's nutrition and access to health care improved, greatly reducing infant mortality rates. Increasing personal wealth promoted materialism and greater emphasis on career and possessions over family and children. The nation also urbanized as people flocked to growing cities such as Río de Janeiro and São Paulo. This brought about the fertility reductions that typically occur when people leave the farm for the city.

It turns out, however, that Brazil may also have had a rather unique influence on its fertility rates over the past several decades—soap operas (**FIGURE 1**). Brazilian soap operas, called *telenovelas* or *novelas*, are a cultural phenomenon and are watched religiously by people of all ages, races, and incomes. Each *novela* follows the activities of several fictional families, and these TV shows are wildly popular because they have characters, settings, and plot lines with which everyday Brazilians can identify.

Telenovelas do not overtly address fertility issues, but they do promote a vision of the "ideal" Brazilian family. This family is typically middle- or upper-class, materialistic, individualistic, and full of empowered women. By challenging existing cultural and religious values through their characters, *novelas* had, and continue to have, a profound impact on Brazilian society. In essence, these programs provided a model family for Brazilians to emulate—with small family sizes being a key characteristic.

FIGURE 1 *Telenovelas* **are a surprising force for promoting lower fertility rates.** Here, residents gather outside a café in Río de Janeiro to watch the popular program *Avenida Brasil*.

In a 2012 paper in the *American Economic Journal: Applied Economics*, a team of researchers from Bocconi University in Italy, George Washington University, and the Inter-American Development Bank (based in Washington, D.C.) analyzed various parameters to investigate statistical relationships between *telenovelas* and fertility patterns in Brazil from 1965 to 2000. Rede Globo, the network that has a virtual monopoly on the most popular *novelas*, increased the number of areas that received its signal in Brazil over those 35 years (**FIGURE 2**), and it now reaches 98% of Brazilian households. By combining data on Rede Globo broadcast range with demographic data, the researchers were able to compare changes in fertility patterns over time in areas of Brazil that received access to *novelas* with areas of Brazil that did not.

The team, led by Dr. Eliana La Ferrara, found that women in areas that received the Globo signal had significantly lower fertility than those in areas not served by Rede Globo. They also found that fertility declines were age-related, with substantial reductions in fertility occurring in women aged 25–44, but not in younger women (**FIGURE 3**). The authors hypothesized that this effect was likely because women between 25 and 44 were closer in age to the main female characters in *novelas*, who typically had no children or only a single child. The depressive effect on fertility among women in areas served by Globo was therefore attributed to wider spacing of births and earlier ending of reproduction by women over 25, rather than to younger women delaying the birth of their first child.

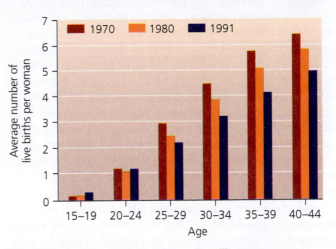

FIGURE 3 **Fertility declines among Brazilian women between 1970 and 1991 were most pronounced in later age classes.** The authors attribute some of this decline to women in those age classes emulating the low fertility of lead female characters in *novelas*. Source: La Ferrara, E., et al., 2012. Soap operas and fertility: Evidence from Brazil. Am. Econ. J. Appl. Econ. *4: 1–31.*

The researchers determined that access to television alone did not depress fertility. For example, comparisons of fertility rate in areas with access to a different television network, Sistema Brasileiro de Televisão, found no relationship. The study authors concluded that this was likely due to the reliance of Sistema Brasileiro de Televisão on programming imported from other nations, with which everyday Brazilians did not connect as they did with *novelas* from Rede Globo.

Television's ability to influence fertility is not limited to Brazil. A 2014 study found that in the United States, tweets and Google searches for terms such as "birth control" increased significantly the day following the airing of new episodes of MTV's *16 and Pregnant*. By correlating geographic patterns in viewership with fertility data, the study authors concluded that MTV's *Teen Mom* series may have been responsible for reducing teenage births in the United States by up to 20,000 per year.

The factors that affect human fertility can be complex and vary greatly from one society to another. As this is a correlative study (p. 11), it does not prove causation between watching *telenovelas* and reduced fertility. It does show, however, that effects on fertility may come from intentional factors, such as a government increasing the availability of birth control, and at other times may come from unexpected and unintentional factors—such as popular television shows.

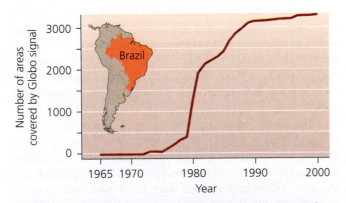

FIGURE 2 **The Globo television network expanded over time and now reaches nearly all households in Brazil.** Fertility declines were correlated with the availability of Globo, and its *novelas,* over the time periods in the study. Source: La Ferrara, E., et al., 2012. Soap operas and fertility: Evidence from Brazil. Am. Econ. J. Appl. Econ. *4: 1–31.*

Family Planning without Coercion: Thailand's Population Program

No nation has pursued a sustained population control program as intrusive as China's, but other rapidly growing nations have implemented successful family-planning programs. The government of Thailand, for example, facing many of the same challenges as its populous neighbor to the north, instituted a comprehensive population program in 1971. At the time, Thailand's growth rate was 2.3%, and the nation's TFR was 5.4. Unlike the Chinese, Thais were given control over their own reproductive choices, but were provided with family-planning counseling and modern contraceptives supported by an engaging public education campaign. Aided by a relatively high level of women's rights in Thai society, this program—and the fertility reductions that accompanied the nation's economic development over the past 45 years—reduced the growth rate to 0.4%, with a TFR of 1.6 children per woman in 2016. The success of this program, and similar initiatives in nations such as Brazil, Cuba, Iran, and Mexico, show that population programs need not be as intrusive as China's to produce similar declines in population growth.

EXPLORE THE DATA at **Mastering** Environmental Science

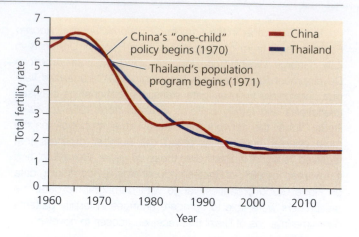

China and Thailand instituted population control programs at roughly the same time and showed similar patterns in fertility declines over the subsequent 45 years, despite utilizing very different approaches. *Data from World Bank, 2016, data.worldbank.org.*

of 70% or more, as did Brazil, Canada, Colombia, Costa Rica, Cuba, Micronesia, New Zealand, Nicaragua, Paraguay, Puerto Rico, South Korea, Thailand, and Uruguay. At the other end of the spectrum, 12 African nations had rates at or below 10%.

Low usage rates for contraceptives in some societies are caused by limited availability, especially in rural areas. As the need for contraceptives can be continuous, women in isolated villages can therefore experience "gaps" in birth control. This occurs when couples use up their supply of contraceptives before reproductive counselors once again visit their village. In others, low usage may be due to religious doctrine or cultural influences that hinder family planning, denying counseling and contraceptives to people who might otherwise use them. This can result in family sizes that are larger than the parents desire and lead to elevated rates of population growth.

In a physiological sense, access to family planning (and the civil rights to demand its use) gives women control over their **reproductive window,** the period of their life—beginning with sexual maturity and ending with menopause—in which they may become pregnant. A healthy woman can potentially bear up to 25 children within this window (**FIGURE 6.13**), but she may choose to delay the birth of her first child to pursue education and employment. She may also use contraception to delay her first child, space births within the window, and "close" her reproductive window after achieving her desired family size.

Family-planning programs are working around the world

Data show that funding and policies that encourage family planning can lower population growth rates in all types of nations, even those that are least industrialized. No other nation has pursued a sustained population control program as intrusive as China's, but some rapidly growing nations have implemented programs that are less restrictive but have nonetheless been very effective in lowering rates of population growth.

The effects of an effective approach to reproductive initiatives is best seen when comparing nations that have similar cultures and levels of economic development but very different approaches to family planning—such as Bangladesh and Pakistan. When both nations were faced with rapid population growth due to high fertility in the 1970s (with TFR in both nations hovering around 7), Bangladesh instituted a government-supported program to improve access to contraception and reproductive counseling to its citizens in an effort to reduce its rate of population growth. Pakistan took a far less aggressive and coordinated approach, which made access to family planning by Pakistani women far less reliable than that for Bangladeshi women. After 40 years of differing approaches to reproductive issues, the results are striking. While Bangladesh's TFR in 2016 had fallen to 2.3, Pakistan's TFR was 3.7 children per woman—one of the highest in southern Asia.

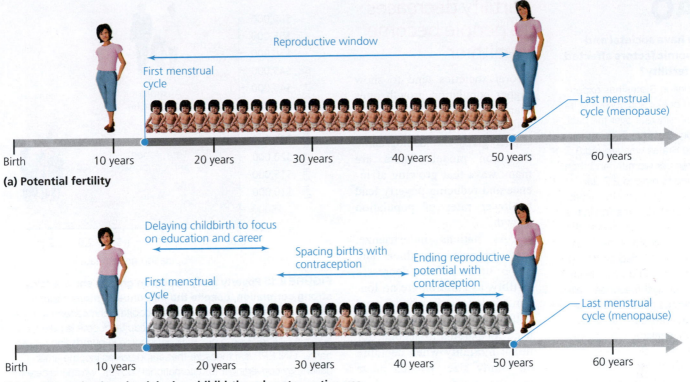

Reproductive window

First menstrual cycle

Last menstrual cycle (menopause)

Birth | 10 years | 20 years | 30 years | 40 years | 50 years | 60 years

(a) Potential fertility

Delaying childbirth to focus on education and career

Spacing births with contraception

Ending reproductive potential with contraception

First menstrual cycle

Last menstrual cycle (menopause)

Birth | 10 years | 20 years | 30 years | 40 years | 50 years | 60 years

(b) Fertility reductions by delaying childbirth and contraceptive use

FIGURE 6.13 Women can potentially have very high fertility within their "reproductive window" but can choose to reduce the number of children they bear. They may do this by delaying the birth of their first child, or by using contraception to space pregnancies or to end their reproductive window.

Empowering women reduces fertility rates

Today, many social scientists and policymakers recognize that for population growth to slow and stabilize, women in societies worldwide should be granted equality in both decision-making and access to education and job opportunities. In addition to providing a basic human right, empowerment of women would have many benefits with respect to fertility rates: Studies show that where women are freer to decide whether and when to have children, fertility rates fall, and children are better cared for, healthier, and better educated.

For women, one benefit of equal rights is the ability to make reproductive decisions. In some societies, men restrict women's decision-making abilities, including decisions as to how many children they will bear. Birth rates have dropped the most in nations where women have gained reliable access to contraceptives and to family planning. This trend indicates that giving women the right to control their reproduction reduces fertility rates.

Expanding educational opportunities for women is an important component of equal rights. In many nations, girls are discouraged from pursuing an education or are kept out of school altogether. Worldwide, more than two-thirds of people who cannot read are women. And data clearly show that as women become educated, fertility rates decline (**FIGURE 6.14**). Education encourages women to delay childbirth as they pursue careers, and gives them more knowledge of reproductive options and greater say in reproductive decisions.

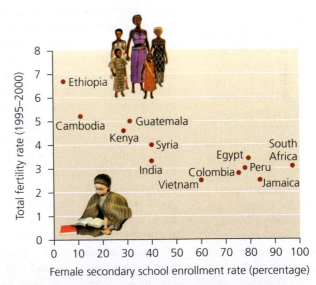

FIGURE 6.14 Increasing female literacy is strongly associated with reduced birth rates in many nations. *Data from McDonald, M., and D. Nierenberg, 2003.* Linking population, women, and biodiversity. State of the world 2003. *Washington, D.C.: Worldwatch Institute.*

DATA Is the relationship between total fertility rate and the rate of enrollment of girls in secondary school positive (as TFR increases, so does school enrollment), negative (as TFR increases, school enrollment decreases), or is there no obvious relationship (increases in TFR are not correlated with changes in school enrollment)?

Go to **Interpreting Graphs & Data** on **Mastering** Environmental Science

FAQ

How have societal and economic factors affected U.S. fertility?

The United States has experienced peaks and valleys in its total fertility rate over the past century. In 1913, TFR in the United States was around 3.5 children per woman, and then plunged to around 2.2 during the trying economic times of the Great Depression in the 1930s. Fertility climbed as the nation pulled out of the Great Depression, leveled off during World War II, and then climbed again during the economically prosperous period following the war (the post-war "baby boom"), peaking at 3.7 in 1957. Fertility then fell sharply in the mid-1960s, as modern contraception became widely available and women enjoyed increased opportunities to pursue higher education and employment outside the home.

Fertility rates have been largely stable around the replacement level (2.1) since 1970, but may soon fall. A 2015 report by the Urban Institute found that U.S. women born after the early 1980s (the "millennial" generation) are having children at the lowest rate in American history. Surveys find that millennial women value children as much as previous generations did at their age, but many are delaying marriage and childbirth due to financial insecurity. That is, many young people state that they'd like to marry and have children but simply can't afford to do so when faced with uncertain employment prospects and relatively high levels of debt from college.

Fertility decreases as people become wealthier

Poorer societies tend to show higher population growth rates than do wealthier societies (**FIGURE 6.15**), as one would expect given the demographic transition model. There are many ways that growing affluence and reducing poverty lead to lower rates of population growth.

As nations industrialize, they become wealthier and more urban. This depresses fertility, as children are no longer needed as farmhands and better healthcare reduces the need for parents to account for infant mortality when deciding on family size. Women move into the workforce and modern contraception becomes available and affordable. Moreover, if a government provides some form of social security to retirees, parents need fewer children to support them in their old age.

Economic factors are tied closely to population growth. Poverty exacerbates population growth, and rapid population growth worsens poverty. This connection is important because a vast majority of the next billion people to be added to the global population will be born into nations in Africa, Asia, and Latin America that have emerging, industrializing economies (**FIGURE 6.16**). Some of these nations have rather high rates of poverty, and adding more people can lead to increased environmental degradation. People who depend on agriculture and live in areas of poor farmland, for instance, may need to farm even if doing so degrades the soil and is not sustainable. Poverty also drives people to cut forests and to deplete biodiversity as they seek to support their families. For example, impoverished settlers and miners hunt large mammals for "bush meat" in Africa's forests, including the great apes that are now heading toward extinction.

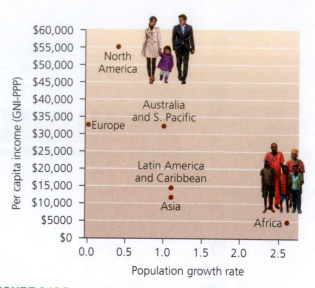

FIGURE 6.15 Poverty and population growth show a fairly strong correlation, despite the influence of many other factors. Regions with the lowest per capita incomes tend to have the most rapid population growth. Per capita income is here measured in GNI PPP, or "gross national income in purchasing power parity." GNI PPP is a measure that standardizes income among nations by converting it to "international" dollars, which indicate the amount of goods and services one could buy in the United States with a given amount of money. *Data from Population Reference Bureau, 2016.* 2016 World population data sheet.

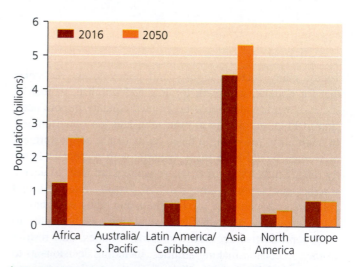

FIGURE 6.16 The vast majority of future population growth will occur in developing regions. Africa will experience the greatest population growth of any region in coming decades. The highly industrialized regions of Europe and North America are predicted to experience only minor population change. *Data from Population Reference Bureau, 2016.* 2016 World population data sheet.

DATA Q • Which region will add the most people between 2016 and 2050? • Which will increase by the greatest percentage during this time period? • Propose one explanation for why the fastest-growing region will increase faster than other regions.

Go to **Interpreting Graphs & Data** on **Mastering** Environmental Science

Average per U.S. resident:
• Economic activity: $56,430
• Footprint: 8.2 hectares
• CO_2: 17.0 metric tons

(a) A family living in the United States

Average per India resident:
• Economic activity: $6020
• Footprint: 1.2 hectares
• CO_2: 1.7 metric tons

(b) A family living in India

FIGURE 6.17 Material wealth varies widely from nation to nation. A typical U.S. family **(a)** may own a large house with a wealth of material possessions. A typical family in a developing nation such as India **(b)** may live in a much smaller home with far fewer material possessions. Compared with the average resident of India, the average U.S. resident shares in 9 times more economic activity, has an ecological footprint 7 times higher, and emits 10 times more carbon dioxide. Economic activity is calculated by dividing the gross national income (GNI) of each country by its population. *Data for ecological footprints are for 2012 and are from Global Footprint Network, www.footprintnetwork.org; data for economic activity and carbon dioxide emissions are from World Bank, 2016, data.worldbank.org.*

Expanding wealth can escalate a society's environmental impacts

Poverty can lead people into environmentally destructive behavior, but wealth can produce even more severe and far-reaching environmental impacts. The affluence of a society such as the United States, Japan, or Germany is built on levels of resource consumption unprecedented in human history. Consider that the richest one-fifth of the world's people possess over 80 times the income of the poorest one-fifth and use 85% of the world's resources (**FIGURE 6.17**). This is meaningful, as the environmental impact of human activities depends not only on the number of people involved but also on the way those people live (recall the *A* for *affluence* in the IPAT equation).

An *ecological footprint* represents the cumulative amount of Earth's surface area required to provide the raw materials a person or population consumes and to dispose of or recycle the waste produced (p. 400). Individuals from affluent societies leave considerably larger per capita ecological footprints (see Figure 1.16, p. 18). In this sense, the addition of one American to the world has as much environmental impact as the addition of 3.4 Chinese, 8 Indians, or 14 Afghans. This fact reminds us that the "population problem" does not lie solely with the developing world.

Indeed, just as population is rising, so is consumption. Researchers have found that humanity's global ecological footprint surpassed Earth's capacity to support us in 1971 (p. 6), and that our species is now living 50% beyond its means (**FIGURE 6.18**). We are running a global ecological deficit, gradually draining our planet of its natural capital and its long-term ability to support our civilization. The rising consumption that is accompanying the rapid industrialization of China, India, and other populous nations makes it all the more urgent for us to reverse this trend and find a path to global sustainability.

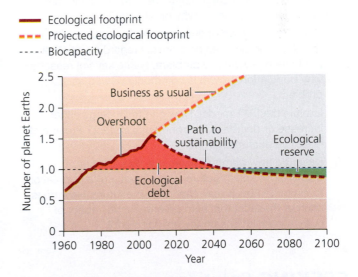

—— Ecological footprint
- - - Projected ecological footprint
- - - - Biocapacity

FIGURE 6.18 The global ecological footprint of the human population is estimated to be 50% greater than what Earth can bear. If population and consumption continue to rise **(orange dashed line)**, we will increase our ecological deficit, or degree of overshoot, until systems give out and populations crash. If, instead, we pursue a path to sustainability **(red dashed line)**, we can eventually repay our ecological debt and sustain our civilization. *Adapted from WWF, 2008. Living planet report 2008. Gland, Switzerland: WWF International.*

closing THE LOOP

China has demonstrated that it is possible to rapidly and drastically slow a nation's population growth, but its one-child policy created numerous demographic problems and raised important human rights issues. China's populous neighbor India started employing population control initiatives at roughly the same time as China, but its initial draconian policies caused public outcry. India's subsequent relaxed policies, along with the societal changes brought about by industrialization, have acted to reduce fertility rates, but not to the extent of its neighbor to the north—and India will soon surpass China as the world's most populous nation.

Today's human population is larger than at any time in the past. Our growing population and our growing consumption of resources affect the environment and our ability to meet the needs of all the world's people. However, there are at least two reasons to be encouraged. First, although global population is still rising, the rate of growth has decreased nearly everywhere, and some countries are even seeing population declines. Most developed nations have passed through the demographic transition, showing that it is possible to lower death rates while stabilizing population and creating more prosperous societies. Second, progress has been made in expanding rights for women worldwide. Although there is still a long way to go, women are obtaining better education, more economic independence, and more ability to control their reproductive decisions than they had in the past.

Human population cannot continue to rise forever. The question is how it will stop rising: Will it be through the gentle and benign process of the demographic transition as is happening in India, through restrictive governmental intervention such as China's policies, or through the miserable Malthusian checks of disease and social conflict caused by overcrowding and competition for scarce resources? How we answer this question today will determine not only the quality of the world in which we live but also the quality of the world we leave to our children and grandchildren.

TESTING Your Comprehension

1. What is the approximate current human global population? How many people are being added to the population each day?

2. Why has the human population continued to grow despite environmental limitations? Do you think this growth is sustainable? Why or why not?

3. Contrast the views of environmental scientists with those of Cornucopian economists regarding whether population growth is a problem. Name several reasons why population growth is commonly viewed as a problem.

4. Explain the IPAT model. How can technology either increase or decrease environmental impact? Provide at least two examples.

5. Describe how demographers use size, density, distribution, age structure, and sex ratio of a population to estimate how it may change in the future. How does each of these factors help determine the impact of human populations on the environment?

6. What is the total fertility rate (TFR)? Why is the replacement fertility for humans approximately 2.1? How is Europe's TFR affecting its rate of natural increase?

7. How does the demographic transition model explain the increase in population growth rates in recent centuries? How does it explain the recent decrease in population growth rates in many countries?

8. Why have fertility rates fallen in many countries?

9. Why is the empowerment of women and the pursuit of gender equality viewed as important to controlling population growth? Describe the aim of family-planning programs.

10. Why do poorer societies tend to have higher population growth rates than wealthier societies? How does poverty affect the environment? How does affluence affect the environment?

SEEKING Solutions

1. The World Bank estimates that 10% of the world's people live in extreme poverty on less than $2 per day. How do you think this situation affects the political stability of the world? Explain your answer.

2. Apply the IPAT model to the example of China provided in the chapter. How do population, affluence, technology, and ecological sensitivity each affect China's environment? Now consider your own country or your own state. How do the same factors each affect your environment? How can we minimize the environmental impacts of growth in the human population?

3. Do you think that all of today's developing nations will complete the demographic transition and reach a permanent state of low birth and death rates? Why or why not? What steps might we as a global society take to help ensure they do? And will developed nations such as the United States and Canada continue to lower and stabilize their birth and death rates in a state of prosperity? What factors might affect whether they do so?

4. **CASE STUDY CONNECTION** China's reduction in birth rates is leading to significant change in the nation's age structure. Review Figure 6.9, which shows that the population is growing older, as shown by the top-heavy population pyramid for the year 2050. If you were tasked with maximizing the contributions of the growing number of retirees to Chinese society, what sorts of programs would you devise?

5. **THINK IT THROUGH** India's prime minister puts you in charge of that nation's population policy. India has a population growth rate of 1.5% per year, a TFR of 2.3, a 47% rate of contraceptive use, and a population that is 67% rural. What policy steps would you recommend to reduce growth rates, and why?

CALCULATING Ecological Footprints

A nation's population size and the affluence of its citizens each influence its resource consumption and environmental impact. As of 2016, the world's population passed 7.4 billion, average per capita income was $15,415 per year, and the latest estimate for the world's average ecological footprint was 2.8 hectares (ha) per person. The sampling of data in the table will allow you to explore patterns in how population, affluence, and environmental impact are related.

NATION	POPULATION (MILLIONS OF PEOPLE)	AFFLUENCE (PER CAPITA INCOME, IN GNI PPP)[1]	PERSONAL IMPACT (PER CAPITA FOOTPRINT, IN HA/PERSON)	TOTAL IMPACT (NATIONAL FOOTPRINT, IN MILLIONS OF HA)
Belgium	11.3	$44,100	7.4	83.6
Brazil	206.1	$15,020	3.1	
China	1378.0	$14,160	3.4	
Ethiopia	101.7	$1,620	1.0	
India	1328.9	$6,020	1.2	
Japan	125.3	$38,870	5.0	
Mexico	128.6	$17,150	2.9	
Russia	144.3	$23,790	5.7	
United States	323.9	$56,430	8.2	2656.0

[1]GNI PPP (gross national income in purchasing power parity) is a measure that standardizes income among nations by converting it to "international" dollars, the amount of goods and services one could buy in the United States with a given amount of money.

Sources: Population and affluence data are from Population Reference Bureau, 2016 World population data sheet. Footprint data are from Global Footprint Network, www.footprintnetwork.org/ecological_footprint_nations/. All data are for 2012.

1. Calculate the total impact (national ecological footprint) for each country.

2. Draw a graph illustrating per capita impact (on the *y* axis) versus affluence (on the *x* axis). What do the results show? Explain why the data look the way they do.

3. Draw a graph illustrating total impact (on the *y* axis) in relation to population (on the *x* axis). What do the results suggest to you?

4. Draw a graph illustrating total impact (on the *y* axis) in relation to affluence (on the *x* axis). What do the results suggest to you?

5. You have just used three of the four variables in the IPAT equation. Now give one example of how the T (technology) variable could potentially increase the total impact of the United States, and one example of how it could potentially decrease the U.S. impact.

Mastering Environmental Science

CHAPTER 7

Soil, Agriculture, and the Future of Food

Farm to Table—And Back Again: The Commons at Kennesaw State University

Atlanta • Kennesaw State University

GEORGIA

> **What we eat has changed more in the last 40 years than in the previous 40,000.**
> —Eric Schlosser, *Fast Food Nation* (2005)

> **There are two spiritual dangers in not owning a farm. One is the danger of supposing that breakfast comes from the grocery, and the other that heat comes from the furnace.**
> —Aldo Leopold, conservationist and philosopher

Kennesaw State University's Hickory Grove Farm.

It's not surprising to see phrases such as "think globally, eat locally," and "farm to table" when you're dining at a trendy restaurant, but would you expect to see them at your campus dining hall?

Believe it or not, campus dining services around the country are now among the industry leaders in culinary sustainability, a pursuit that embraces the use of fresh, healthy, locally produced foods to provide diners delicious, nutritious meals. One leader in sustainable dining is Kennesaw State University (KSU) in suburban Atlanta, Georgia. In 2009, the university opened The Commons, a dining facility that serves more than 5000 students each day and was granted gold-level certification as a sustainable building by the Leadership in Energy and Environmental Design (LEED) program (p. 429). The next year, KSU launched a Farm-to-Campus program when it acquired a plot of farmland just off campus.

Today, the university runs three farms near the campus, on 27 hectares (67 acres) of land that grow thousands of pounds of produce each year, supplying many of the fruits and vegetables served to students in The Commons. But KSU does even more, and is working to create a fully closed-loop system. Uneaten food waste from The Commons is fed into an aerobic digestion system behind the facility, where it is broken down to generate a nutrient-rich liquid. This liquid compost is trucked back to the campus farms and applied there as a fertilizer to nourish the soil and help grow new crops. As KSU's first director of Culinary and Hospitality Services put it, "We go beyond farm-to-campus. We embrace farm-to-campus and back-to-farm operations."

Kennesaw State's ambitious program to mesh sustainable agriculture with campus dining got its start when the university committed to make sustainability a prime consideration in construction and operation of The Commons. KSU's architects and engineers designed a facility that minimizes energy use, water consumption, and waste generation. Illuminated with floor-to-ceiling windows and high-efficiency lighting, The Commons offers nine themed food stations and a rotating menu of 200–300 items. Foods are prepared to order or in small batches according to demand, drastically reducing the amount of leftover food. A "trayless" approach to

Student worker on a college farm ▲

food service further reduces food waste and the amount of water used in dishwashing, as diners without trays tend to eat less and use fewer plates. The Commons has special dishwashing systems designed for water and energy efficiency, napkin dispensers that reduce paper waste, hydration stations for refilling reusable water bottles, and a recycling and composting program that diverts 44,000 pounds of waste from the landfill each month. The facility even generates biodiesel from its used cooking oil (p. 393) to fuel university vehicles.

With an award-winning green building as the anchor of its program, KSU set about creating an agricultural system that could supply diners with fresh, healthy, local produce. The three campus farms grow dozens of items—everything from tomatoes to cucumbers to melons to apples—using practices that protect soil quality by minimizing the use of chemical pesticides and synthetic fertilizers. Indeed, soil quality is enhanced by the application of nutrient-rich compost and liquid from the digester. Sixty free-range chickens produce 300 organic eggs per week, and apiaries house 48 honeybee colonies that produce 30 gallons of honey each year.

On campus, herbs, lettuce, and shiitake mushrooms are grown in an herb garden and greenhouse behind The Commons and in 10 hydroponic stations scattered throughout the dining area. These are each watered with rainwater collected in barrels on the roof of The Commons. There is even a grist mill on site that grinds fresh grits and cornmeal. In addition, the university has cultivated relationships with local farms and meat producers, sourcing locally produced food whenever possible.

KSU's culinary program is receiving wide recognition, and in 2013 Kennesaw State became the first educational institution to win the National Restaurant Association's Innovator of the Year award, beating out competitors such as Walt Disney Parks and Resorts and the U.S. Air Force. Today at KSU, the links between farm and campus continue to grow. A farmers' market has been established on campus, and a student group—Students for Environmental Sustainability—has launched a permaculture project next to the science building. Student volunteers and interns work at the farms, and new classes in organic farming and beekeeping are using the farms as well.

Kennesaw State has enjoyed unusual success in linking campus and farm to enhance students' dining experiences, but food service operations on many campuses across North America today are embracing fresh locally farmed food as they pursue culinary sustainability in various ways. Michigan State University engages students in organic farming through teaching, research, and production. At California State University, Chico, more than 35 students are employed at the school's 800-acre farm, supplying fruits, vegetables, meat, and dairy products to the dining halls, farmers' markets, and local restaurants. Many dozens of schools—from Hampshire College to Clemson University to Dickinson College to Case Western Reserve University—provide their communities with food through community-supported agriculture (CSA) programs (p. 163).

Altogether, about 70% of America's largest colleges and universities now have a campus farm or garden from which their dining halls can source food directly, according to a recent survey of more than 300 major institutions. Meanwhile, more and more campus dining halls are gaining LEED certification as green buildings. Collectively, these efforts are reducing the ecological footprints of American colleges and universities and pointing the way to a more sustainable food future—all while supplying students with healthy and delicious meals.

Throughout this chapter, we'll see how principles of sustainability embraced at colleges and universities are being replicated at ever-larger scales, helping us chart a path toward ensuring adequate food for all people while minimizing the environmental impacts of agriculture.

The Race to Feed the World

Someone dining at The Commons, surrounded by a diversity of food options, might find it hard to imagine that many people in the world today struggle on a daily basis to get enough to eat. Unfortunately, such is the case, and as the human population continues to grow—with our numbers expected to swell to over 9 billion by the middle of this century—the challenge of feeding a growing world population will only become more dire. Feeding 2 billion more people than we do today while protecting the integrity of soil, water, and ecosystems will require the large-scale embrace of farming practices that are more sustainable than those we currently use. Providing **food security,** the guarantee of an adequate, safe, nutritious, and reliable food supply available to all people at all times, will be one of our greatest challenges in the coming decades.

There is good news, however. Over the past half-century, our ability to produce food has grown even faster than the global population (**FIGURE 7.1**). Improving people's quality of life by producing more food per person is a monumental achievement of which humanity can be proud. However, ensuring that our

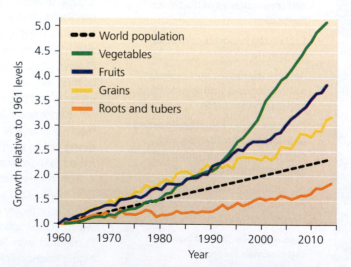

FIGURE 7.1 Global production of most foods has risen more quickly than the world population. This means that we have produced more food per person each year. The data lines show cumulative increases relative to 1961 levels (for example, a value of 2.0 means twice the 1961 amount). Food is measured by weight. *Data from UN Food and Agriculture Organization (FAO).*

food production can be sustained depends on conserving soil, water, and biodiversity by using careful agricultural practices. Today many of the world's soils are in decline, and most of the planet's arable land has already been claimed. Hence, even though agricultural production has outpaced population growth so far, we have no guarantee that this will continue.

We face undernutrition, overnutrition, and malnutrition

Despite our rising food production, nearly 800 million people worldwide do not have enough to eat. These people suffer from **undernutrition,** receiving fewer calories than the minimum dietary energy requirement. As a result, every 5 seconds, somewhere in the world, a child dies because he or she lacks access to a nutritious diet. In most cases, poverty limits the amount of food people can buy. One out of every seven of the world's people lives on less than $1.25 per day, and one out of three lives on less than $2 per day, according to World Bank estimates. Political obstacles, regional conflict and wars, and inefficiencies in distribution contribute significantly to hunger as well. Even our energy choices affect food supplies. For example, sizeable amounts of cropland are devoted to growing crops for the production of biofuels (pp. 393–394).

The good news is that globally, the number of people suffering from undernutrition has been falling since the 1960s. The percentage of people who are undernourished has fallen even more (**FIGURE 7.2**). We still have a long way to go to eliminate hunger, but these positive trends are encouraging.

Although nearly 1 billion people lack access to nutritious foods, many others consume too many calories each day. **Overnutrition** causes unhealthy weight gain, which leads to cardiovascular disease, diabetes, and other health problems. As a result, more than one in three adults in the United States are obese, according to the Centers for Disease Control and Prevention. Across the world as a whole, the World Health Organization estimates that 39% of adults are overweight, and that of these, one-third are obese. The growing availability of highly processed foods (which are often calorie-rich, nutrient-poor, and affordable to people of all incomes) suggests that overnutrition will remain a global nutritional problem, along with undernutrition, for the foreseeable future.

Just as the *quantity* of food a person eats is important for health, so is the *quality* of food. **Malnutrition,** a shortage of nutrients the body needs, occurs when a person fails to obtain a complete complement of proteins (p. 32), essential lipids (p. 33), vitamins, and minerals. Malnutrition can lead to disease (**FIGURE 7.3**). For example, people who eat a diet that is high in starch but deficient in protein can develop *kwashiorkor.* Children who have recently stopped breastfeeding and are no longer getting protein from breast milk are most at risk for developing kwashiorkor, which causes bloating of the abdomen, deterioration and discoloration of hair, mental disability, immune suppression, developmental delays, and reduced growth. Protein deficiency together with a lack of calories can lead to *marasmus,* which causes wasting or shriveling among millions of children in the developing world. Iron deficiency can result in anemia, which causes fatigue and developmental disabilities; iodine deficiency can produce swelling of the thyroid gland and brain damage; and vitamin A deficiency can lead to blindness.

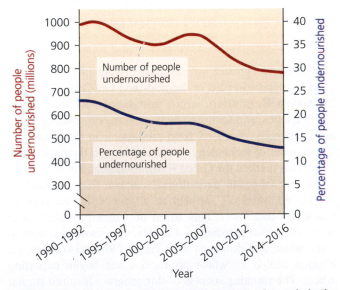

FIGURE 7.2 The number and the percentage of people in the developing world who suffer undernutrition have each been declining. *Data from Food and Agriculture Organization of the United Nations, 2015.* The state of food insecurity in the world, 2015. *Rome: FAO.*

DATA Q Explain how the percentage of undernourished people could have decreased between 2000–2002 and 2005–2007, while the number of undernourished people increased slightly during the same period.

Go to **Interpreting Graphs & Data** on Mastering Environmental Science

FIGURE 7.3 Millions of children suffer from forms of malnutrition, such as kwashiorkor and marasmus.

The Changing Face of Agriculture

If we are to enhance global food security, we will need to examine the ways we produce our food and how we can make them more sustainable. We can define **agriculture** as the practice of raising crops and livestock for human use and consumption. We obtain most of our food and fiber from **cropland,** land used to raise plants for human use, and from **rangeland,** land used for grazing livestock.

As the human population has grown, so have the amounts of land and resources we devote to agriculture. Agriculture is currently practiced on 38% of Earth's land surface, and uses more land area than any other human activity. Of this land, 26% is rangeland and 12% is cropland. The percentages are even greater in the United States, where nearly half the land is devoted to agriculture. Here, rangeland covers 27% and cropland covers 19% of our land area.

Industrial agriculture is a recent human invention

During most of the human species' 200,000-year existence, we were hunter-gatherers, depending on wild plants and animals for our food and fiber. Then about 10,000 years ago, as glaciers retreated and the climate warmed, people in some cultures began to raise plants from seed and to domesticate animals.

For thousands of years, the work of cultivating, harvesting, storing, and distributing crops was performed by human and animal muscle power, along with hand tools and simple machines—an approach known as **traditional agriculture.** Traditional farmers typically plant **polycultures** ("many types"), mixtures of different crops in small plots of farmland, such as the Native American farming systems that mixed maize, beans, squash, and peppers. Traditional agriculture is still practiced today but has rapidly been overtaken by newer methods of farming that increase crop yields.

Thousands of years after humans began practicing traditional agriculture, the industrial revolution (p. 5) introduced large-scale mechanization and fossil fuel combustion to agriculture, just as it did to industry. Farmers replaced horses and oxen with machinery that provided faster and more powerful means of cultivating, harvesting, transporting, and processing crops. Such **industrial agriculture** also boosted yields by intensifying irrigation and introducing synthetic fertilizers, while the advent of chemical pesticides reduced herbivory by crop pests and competition from weeds. Today, industrial agriculture is practiced on over 25% of the world's cropland and has been a major factor in reducing food prices worldwide due to its large-scale, intensive production model.

The use of machinery created a need for highly organized approaches to farming, and this led to the planting of vast areas with single crops in orderly, straight rows. Such **monocultures** ("one type") make farming more efficient, but they reduce biodiversity by eliminating habitats used by organisms in and around traditional farm fields. Moreover, when all plants in a field are genetically similar, they are equally susceptible to bacterial and viral diseases, fungal pathogens, or insect pests that can spread quickly.

Industrial agriculture's reliance on genetically similar crop varieties is a source of efficiency, but also an Achilles heel. Concern over potential crop failure has led to coordinated efforts to conserve the wild relatives of crop plants and crop varieties indigenous to various regions, because they contain genes we may one day need to introduce into our commercial crops. Seeds of these species are stored in some 1400 **seed banks,** institutions that preserve some 1–2 million different seed types in locations around the world. Many agricultural scientists feel that we also need to protect the genetic integrity of wild relatives of crop plants by preventing gene exchange between these species and crop plants that have been genetically modified with genes from other species (p. 158). These scientists want to avoid genetic "contamination" of wild populations so that we preserve the natural gene combinations of plants that are well adapted to their environments.

Our use of monocultures also contributes to a narrowing of the human diet. Globally, 90% of the food we consume now comes from just 15 crop species and eight livestock species—a drastic reduction in the diversity of food humans have historically eaten. For example, only 30% of the maize varieties that grew in Mexico in the 1930s exist today. The number of wheat varieties in China dropped from 10,000 in 1949 to 1000 by the 1970s. In the United States, apples and other fruit and vegetable crops have decreased in diversity by 90% in less than a century.

The Green Revolution boosted production—and exported industrial agriculture

The desire for greater quantity and quality of food for our growing population led in the mid- and late-20th century to the **Green Revolution,** which introduced new technology, crop varieties, and farming practices to the developing world and drastically increased food production in these nations. The transfer of technology and knowledge to the developing world that marked the Green Revolution began in the 1940s, when the American agricultural scientist Norman Borlaug introduced Mexico's farmers to a specially bred type of wheat (**FIGURE 7.4a**). This strain of wheat produced large seed heads, was resistant to diseases, was short in stature to resist wind, and produced high yields. Within two decades, Mexico tripled its wheat production and began exporting wheat. The stunning success of this program inspired similar projects around the world. Borlaug—who won the Nobel Peace Prize for his work—took his wheat to India and Pakistan and helped transform agriculture there.

Soon many developing countries were doubling, tripling, or quadrupling their yields using selectively bred strains of wheat, rice, corn, and other crops from industrialized nations (**FIGURE 7.4b**). These crops dramatically increased yields and helped millions avoid starvation. When Borlaug died in 2009 at age 95, he was widely celebrated as having "saved more lives than anyone in history."

(a) Norman Borlaug

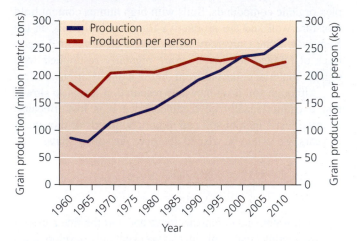

(b) Grain production in India

FIGURE 7.4 Norman Borlaug (a) helped launch the Green Revolution. The high-yielding, disease-resistant wheat that he bred helped boost agricultural productivity in many developing countries, such as India **(b)**.

The effects of industrial agriculture have been mixed

Along with the new grains, developing nations imported the methods of industrial agriculture. They began applying large amounts of synthetic fertilizers and chemical pesticides on their fields, irrigating crops generously with water, and using more machinery powered by fossil fuels. From 1900 to 2000, people increased energy inputs into agriculture by 80 times while expanding the world's cultivated area by just one-third.

The environmental and social impacts of these developments have been mixed. On the positive side, intensifying the use of already-cultivated land reduced pressures to convert additional lands for cultivation. Between 1961 and 2013, global food production more than tripled and per-person food production rose 48%, while area converted for agriculture increased only 11%. In this way, the Green Revolution helped preserve biodiversity and natural ecosystems by preventing a great deal of deforestation and habitat destruction. On the negative side, the intensified use of fossil fuels, water, inorganic fertilizers, and synthetic pesticides has worsened pollution, topsoil erosion, and soil and water quality (Chapter 9).

Sustainable agriculture reduces environmental impacts

We have achieved our impressive growth in food production by devoting more fossil fuel energy to agriculture; intensifying our use of irrigation, fertilizers, and pesticides; cultivating more land; planting and harvesting more frequently; and developing (through crossbreeding and genetic engineering) more productive varieties of crops and livestock. However, many of these practices have degraded soils, polluted waters, and affected biodiversity.

We cannot simply keep expanding agriculture into new areas, because land suitable and available for farming is running out. Instead, we must find ways to improve the efficiency of food production in areas already under cultivation. Industrial agriculture in some form seems necessary to feed our planet's more than 7 billion people, but many experts feel we will be better off in the long run by raising animals and crops in ways that are less polluting, are less resource-intensive, and cause less impact on natural systems. To this end, many farmers and agricultural scientists are creating agricultural systems that better mimic the way a natural ecosystem functions.

Sustainable agriculture describes agriculture that maintains the healthy soil, clean water, pollinators, and genetic diversity essential to long-term crop and livestock production. The best approach for making an agricultural system sustainable is to mimic the way a natural ecosystem functions. Ecosystems operate in cycles and are internally stabilized with negative feedback loops (p. 24). In this way they provide a useful model for agriculture. We will see that some farmers and ranchers have already adopted strategies and conservation methods to this end. Reducing fossil fuel inputs and the pollution these inputs cause is a key aim of sustainable agriculture, and to many this means moving away from the industrial model and toward more traditional organic and low-input models.

Low-input agriculture describes approaches to agriculture that use lesser amounts of fossil fuel energy, water, pesticides, fertilizers, growth hormones, and antibiotics than are used in industrial agriculture. Low-input approaches also seek to reduce food production costs by allowing nature to provide valuable ecosystem services—such as creating habitat for birds and other insect predators to provide pest control in farm fields—that farmers using industrial agricultural methods must pay for themselves.

In the sections that follow, we will examine the basic elements of agriculture—maintaining soil fertility, watering and fertilizing crops, and controlling agricultural pests—and how more sustainable methods can reduce these impacts while maintaining high crop yields. We will begin with soils, the foundation of agriculture.

Soils

Soil is not merely lifeless dirt; it is a complex system consisting of disintegrated rock, organic matter, water, gases, nutrients, and microorganisms. By volume, soil consists very roughly of 50% mineral matter and up to 5% organic matter.

The rest consists of the space between soil particles (pore space) taken up by air or water. The organic matter in soil includes living and dead microorganisms as well as decaying material derived from plants and animals. The soil ecosystem supports a diverse collection of bacteria, fungi, protists, worms, insects, and burrowing animals (**FIGURE 7.5**).

Soil forms slowly

Soil formation begins when the lithosphere's parent material is exposed to the effects of the atmosphere, hydrosphere, and biosphere. **Parent material** is the base geologic material in a particular location. It can be hardened lava or volcanic ash; rock or sediment deposited by glaciers; sediments deposited in riverbeds, floodplains, lakes, and the ocean; or **bedrock,** the continuous mass of solid rock that makes up Earth's crust. Parent material is broken down by **weathering,**

the physical, chemical, and biological processes that convert large rock particles into smaller particles.

Once weathering has produced fine particles, biological activity contributes to soil formation through the deposition, decomposition, and accumulation of organic matter. As plants, animals, and microbes die or deposit waste, this material is incorporated amid the weathered rock particles, mixing with minerals. For example, the deciduous trees of temperate forests drop their leaves each fall, making leaf litter available to the detritivores and decomposers (p. 74) that break it down and incorporate its nutrients into the soil. In decomposition, complex organic molecules are broken down into simpler ones, which plants can take up through their roots. Partial decomposition of organic matter creates **humus,** a dark, spongy, crumbly mass of material made up of complex organic compounds. Soils with high humus content hold moisture well and are productive for plant life. Soils on the Kennesaw State University farms have well-developed quantities of humus, because they are regularly resupplied with organic material collected from dining hall operations.

Although soil is a renewable resource, it forms so slowly that it cannot readily be regained once it is lost. Because forming just 1 inch of soil can require hundreds or thousands of years, we would be wise to conserve the soil we have.

A soil profile consists of layers known as horizons

As wind, water, and organisms move and sort the fine particles that weathering creates, distinct layers eventually develop. Each layer of soil is known as a **soil horizon,** and the cross-section as a whole, from surface to bedrock, is known as a **soil profile.**

Soil horizons can be generally categorized as A, B, and C horizons—or topsoil, subsoil, and parent material, respectively—but soil scientists often recognize at least three additional horizons, including an O horizon (litter layer) that consists primarily of organic matter (**FIGURE 7.6**). Soils from different locations vary, and few soil profiles contain all six horizons, but any given soil contains at least some of them.

Generally, the degree of weathering and the concentration of organic matter decrease as one moves downward in a soil profile. Minerals are transported downward by **leaching,** where solid particles suspended or dissolved in liquid are transported to another location. In some soils, minerals may be leached so rapidly that plants are deprived of nutrients. Leached minerals may enter groundwater, and some can pose human health risks when the water is extracted for drinking.

A crucial horizon for agriculture and ecosystems is the A horizon, or **topsoil.** Topsoil consists mostly of inorganic mineral components such as weathered substrate, with organic matter and humus from above mixed in. Topsoil is the portion of the soil that is most nutritive for plants, and it takes its loose texture, dark coloration, and strong water-holding capacity from its humus content. The O and A horizons are home to most of the countless organisms that give life to soil. Topsoil is vital for agriculture, but agriculture practiced unsustainably over time will deplete organic matter, reducing the soil's fertility and ability to hold water.

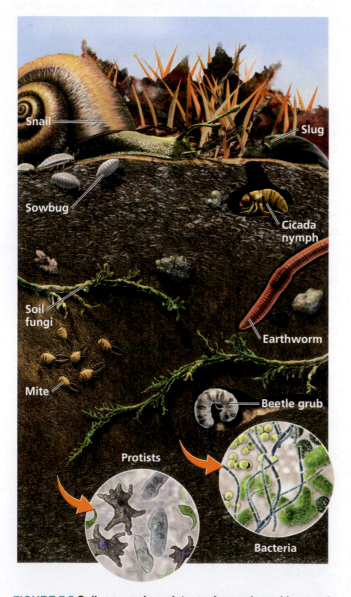

FIGURE 7.5 Soil, a complex mixture of organic and inorganic components, is full of living organisms. Most soil organisms decompose organic matter. Some, such as earthworms, also help to aerate the soil.

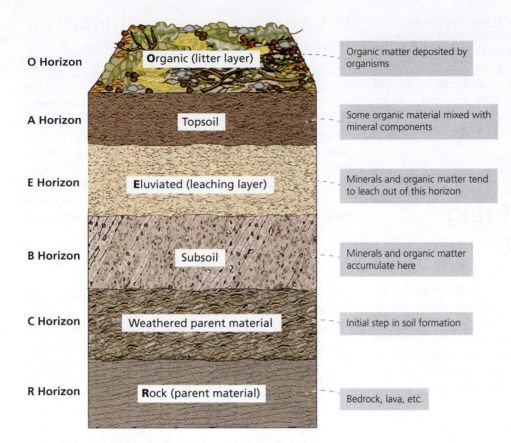

O Horizon — **O**rganic (litter layer) — Organic matter deposited by organisms

A Horizon — Topsoil — Some organic material mixed with mineral components

E Horizon — **E**luviated (leaching layer) — Minerals and organic matter tend to leach out of this horizon

B Horizon — Subsoil — Minerals and organic matter accumulate here

C Horizon — Weathered parent material — Initial step in soil formation

R Horizon — **R**ock (parent material) — Bedrock, lava, etc.

FIGURE 7.6 Mature soil consists of layers, or horizons, that have different attributes.

Regional differences in soil traits affect agriculture

Soil characteristics vary from place to place, and they are affected by climate and other variables. For example, it may surprise you to learn that soils of the Amazon rainforest are much less productive than soils in Iowa or Kansas. This is because the enormous amount of rain that falls in tropical regions such as the Amazon basin readily leaches minerals and nutrients out of the topsoil and E horizon, below the reach of plants' roots. At the same time, warm temperatures in the Amazon speed the decomposition of leaf litter and the uptake of nutrients by plants, so only small amounts of humus remain in the thin topsoil layer.

As a result, when tropical rainforest is cleared for farming, cultivation quickly depletes the soil's fertility. This is why the traditional form of agriculture in tropical forested areas is *swidden* agriculture, in which the farmer cultivates a plot for one to a few years and then moves on to clear another plot, leaving the first to grow back to forest. Plots are often burned before planting, in which case the practice is called **slash-and-burn** agriculture (**FIGURE 7.7a**). Cleared plots are increasingly being converted to pasture for ranching livestock instead of growing back as forest, and so agriculture has degraded the soils of many tropical areas.

In contrast, on the grasslands of North America, which have been almost entirely converted to agriculture (**FIGURE 7.7b**),

(a) Slash-and-burn agriculture on nutrient-poor soil in the tropics

(b) Industrial agriculture on rich topsoil in Iowa

FIGURE 7.7 Regional soil differences affect how people farm. In tropical forested areas such as Indonesia **(a)**, farmers pursue swidden agriculture by the slash-and-burn method because tropical rainforest soils (**inset**) are nutrient-poor and easily depleted. On American farmland **(b)**, less rainfall means fewer nutrients are leached from the topsoil, and more organic matter accumulates, forming a thick, dark topsoil layer (**inset**).

there is less rainfall and therefore less leaching, so nutrients remain within reach of plants' roots. Plants return nutrients to the topsoil as they die, maintaining its fertility. This creates the thick, rich topsoil of temperate grasslands, which can be farmed repeatedly with minimal loss of fertility as long as farmers guard against loss of soil.

FAQ

What is "slash-and-burn" agriculture?

Soils of tropical rainforests are not well suited for cultivating crops because they contain relatively low levels of plant nutrients. Instead, most nutrients are tied up in the forest's lush vegetation. When farmers cut tropical rainforest for agriculture, they enrich the soil by burning the plants on site. The nutrient-rich ash is tilled into the soil, providing sufficient fertility to grow crops. Unfortunately, the nutrients from the ash are usually depleted in just a few years. At this point, farmers move deeper into the forest and slash and burn another swath of land, causing additional impacts to these productive and biologically diverse ecosystems (p. 86).

Watering and Fertilizing Crops

Just as soil is a crucial resource for farming and ranching, so are water and nutrients. Plants require water and more than a dozen vital nutrients for growth, and crops and livestock are provided with supplemental water and nutrients when needed, to boost production.

Irrigation boosts productivity but can damage soil

Plants require water for optimum growth, and people have long supplemented the water that crops receive from rainfall. The artificial provision of water to support agriculture is known as **irrigation.** By irrigating crops, people maintain high yields in times of drought and turn previously dry and unproductive regions into fertile farmland. Some crops, such as rice and cotton, use large amounts of water and generally require irrigation, whereas others, such as beans and wheat, need relatively little water.

Worldwide, irrigated acreage has increased along with the adoption of industrial farming methods, and 70% of the fresh water withdrawn by people is applied to crops. In some cases, withdrawing water for irrigation has depleted aquifers and dried up rivers and lakes (pp. 269–271).

It's possible to overwater, however. **Waterlogging** occurs when overirrigation causes the water table to rise to the point that water drowns plant roots, depriving them of access to gases and essentially suffocating them. A more frequent problem is **salinization,** the buildup of salts in surface soil layers. In dryland areas where precipitation is minimal and evaporation rates are high, the evaporation of water from the soil's A horizon may pull water with dissolved salts up from lower horizons. When the water evaporates at the surface, those salts remain on the soil, often turning the soil surface white. Salinization now reduces productivity on 20% of all irrigated cropland, costing more than $11 billion each year.

Sustainable approaches to irrigation maximize efficiency

One of the most effective ways to reduce water use in agriculture is to better match crops and climate. Many arid regions have been converted into productive farmland through extensive irrigation, often with the support of government subsidies that make irrigation water artificially inexpensive. Some farmers in these areas cultivate crops that require large amounts of water, such as rice and cotton. This leads to extensive water loss from evaporation in the arid climate. Choosing other crops that require far less water could enable these areas to remain agriculturally productive while greatly reducing water use.

Another approach is to embrace technologies that improve efficiency in water use. Currently, irrigation efficiency worldwide is low, as plants end up using only about 40% of the water that we apply. The rest evaporates or soaks into the soil away from plant roots (**FIGURE 7.8a**). Drip irrigation systems that target water directly toward plant roots through hoses or tubes can increase efficiencies to more than 90% (**FIGURE 7.8b**). Such drip irrigation systems are used on the Kennesaw State University campus farms that supply The Commons with much of its produce, greatly reducing water use. In addition, rainwater is gathered in barrels atop the roof of The Commons; rainwater harvesting is another technique for making good use of water. As systems for drip irrigation and rainwater harvesting become more affordable, more farmers, gardeners, and homeowners are turning to them.

Fertilizers boost crop yields but can be overapplied

Along with water, crop plants require nitrogen, phosphorus, and potassium to grow, as well as smaller amounts of more than a dozen other nutrients. Leaching and uptake by plants removes these nutrients from soil, and if soils come to contain too few nutrients, crop yields decline. Therefore, we go to great lengths to enhance nutrient-limited soils by adding **fertilizers,** substances that contain essential nutrients for plant growth.

There are two main types of fertilizers. **Inorganic fertilizers** are mined or synthetically manufactured nutrient supplements. **Organic fertilizers** consist of the remains or wastes of organisms and include animal manure, crop residues, fresh vegetation (*green manure*), and **compost,** a mixture produced when decomposers break down organic matter, including food and crop waste, in a controlled environment.

One of the highlights of The Commons' sustainability efforts at KSU is its closed-loop system for recycling wastes. Uneaten food and scraps from food preparation are placed in a large "digester" tank outside the dining hall. Over time, the food items break down inside the digester, generating roughly 1900 L (500 gal) of nutrient-rich water daily. This liquid is trucked to the nearby campus farms, where it is used as an organic fertilizer.

(a) Flood-and-furrow irrigation

(b) Drip irrigation

FIGURE 7.8 Irrigation methods vary in their water use. Conventional methods **(a)** are inefficient, because most water is lost to evaporation and runoff. In drip irrigation systems **(b)**, hoses drip water directly into soil near plants' roots, so that much less is wasted.

Historically, people relied on organic fertilizers to replenish soil nutrients. But during the latter half of the 20th century, farmers in industrialized and Green Revolution regions widely embraced the use of inorganic fertilizers (**FIGURE 7.9**). This use has greatly boosted our global food production, but the overapplication of inorganic fertilizers is causing increasingly severe pollution problems. Nutrients from these fertilizers can also have impacts far beyond the boundaries of the fields. For instance, nitrogen and phosphorus runoff from farms and other sources spurs phytoplankton blooms in the Chesapeake Bay and creates an oxygen-depleted "dead zone" that kills animal and plant life (Chapter 2). Such eutrophication (pp. 28–30, 276) occurs

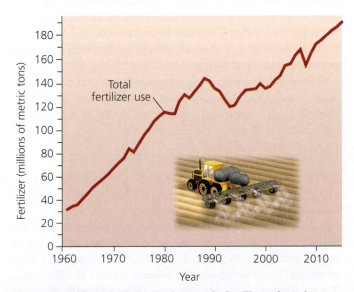

FIGURE 7.9 Use of synthetic, inorganic fertilizers has risen sharply over the past half-century. Today, usage stands at more than 190 million metric tons annually. *Data from International Fertilizer Industry Association and UN Food and Agriculture Organization (FAO, 2015).*

at countless river mouths, lakes, and ponds throughout the world. Components of some nitrogen fertilizers can even volatilize (evaporate) into the air, contributing to photochemical smog (p. 296) and acid deposition (p. 303).

Sustainable fertilizer use involves monitoring and targeting nutrients

Sustainable approaches to fertilizing crops with inorganic fertilizers target the delivery of nutrients to plant roots and avoid the overapplication of fertilizer. Farmers using drip irrigation systems can add fertilizer to irrigation water, thereby releasing it only above plant roots. Growers practicing no-till farming or conservation tillage (p. 150) often inject fertilizer along with seeds, concentrating it near the developing plant. Farmers can also avoid overapplication by regularly monitoring soil nutrient content and applying fertilizer only when nutrient levels are too low. These types of approaches are examples of **precision agriculture,** which involves using technology to precisely monitor crop conditions, crop needs, and resource use, to maximize production while minimizing waste of resources. In addition, by planting buffer strips of vegetation along field edges and watercourses, growers can help to capture nutrient runoff before it enters streams and rivers.

Sustainable agriculture embraces the use of organic fertilizers, because they can provide some benefits that inorganic fertilizers cannot. Organic fertilizers provide not only nutrients but also organic matter that improves soil structure, nutrient retention, and water-retaining capacity. When manure is applied in amounts needed to supply sufficient nitrogen for a crop, however, it may introduce excess phosphorus, which can run off into waterways. Accordingly, sustainable approaches do not rely solely on organic fertilizers but integrate them with the targeted use of inorganic fertilizer.

Conserving Agricultural Resources

Throughout the world, especially in drier regions, it has gotten more difficult to raise crops and graze livestock. Soils have deteriorated in quality and declined in productivity—a process termed **soil degradation.** Each year, our planet gains around 80 million people yet loses 5–7 million ha (12–17 million acres, about the size of West Virginia) of productive cropland to degradation. The common causes include soil erosion, nutrient depletion, water scarcity, salinization, waterlogging, chemical pollution, changes in soil structure and pH, and loss of organic matter from the soil. Over the past 50 years, scientists estimate that soil degradation has reduced potential rates of global grain production on cropland by 13%. This is a dangerous trend, given that we will need to maximize the productivity of our farmland if we are to feed an additional 2 billion people by 2050.

A major contributor to the degradation of soil ecosystems is **erosion,** the transfer of material from one place to another by the action of wind or water. **Deposition** occurs when eroded material is deposited at a new location. Erosion and deposition are natural processes that can help create soil. Flowing water can deposit freshly eroded nutrient-rich sediment across river valleys and deltas, producing fertile soils. However, erosion can be a problem locally because it generally occurs much more quickly than soil is formed. Erosion also tends to thin the biologically important topsoil layer. In general, steeper slopes, greater precipitation intensities, and sparser vegetation all lead to greater water erosion.

People have made land more vulnerable to erosion through three widespread practices: overcultivating fields through poor planning or excessive tilling (plowing); overgrazing rangeland with more livestock than the land can support; and clearing forests on steep slopes or with large clear-cuts (p. 200).

In today's world, humans are the primary cause of soil erosion, and we have accelerated it to unnaturally high rates. A 2004 study concluded that human activities move over 10 times more soil than all other natural processes on the surface of the planet combined. More than 19 billion ha (47 billion acres) of the world's croplands suffer from erosion and other forms of soil degradation. Farmlands in the United States, for example, lose roughly 5 tons of soil for every ton of grain harvested. A 2007 study found an even greater degree of human impact, but also pointed toward a solution, revealing that land farmed with sustainable approaches erodes at slower rates than land under industrial farming practices.

Desertification reduces productivity of arid lands

Much of the world's population lives and farms in *drylands*, arid and semi-arid environments that cover about 40% of Earth's land surface. These areas are prone to **desertification,** a form of land degradation in which more than 10% of productivity is lost as a result of erosion, soil compaction, forest removal, overgrazing, drought, salinization, climate change, water depletion, and other factors. Most such degradation results from wind and water erosion. By some estimates, desertification endangers the food supply or well-being of more than 1 billion people in over 100 countries and costs tens of billions of dollars in income each year through reduced productivity.

Grazing practices can contribute to soil degradation

We have focused in this chapter largely on the cultivation of crops, but raising livestock is a major component of agriculture and likewise exerts a toll on soils and ecosystems. People across the globe tend more than 3.4 billion cattle, sheep, and goats, most of which graze on grasses on the open range. As long as livestock populations do not exceed the rangeland's carrying capacity (p. 63) and do not consume grass faster than it can regrow, grazing can be sustainable. Moreover, human use of rangeland does not necessarily exclude its use by wildlife or its continued functioning as a grassland ecosystem. However, grazing too many livestock that destroy too much of the plant cover impedes plant regrowth. Without adequate regeneration of plant biomass, the result is **overgrazing.**

When livestock remove too much plant cover or churn up the soil with their hooves, soil is exposed and made vulnerable to erosion. In a positive feedback cycle, soil erosion makes it difficult for vegetation to regrow, a problem that perpetuates the lack of cover and gives rise to more erosion. Moreover, non-native weedy plants that are unpalatable or poisonous to livestock may invade and outcompete native vegetation in the new, modified environment. Too many livestock trampling the ground can also compact soil and alter its structure. Soil compaction makes it more difficult for water to infiltrate, for soil to be aerated, and for plants' roots to expand. All of these effects further decrease plant growth and survival.

Worldwide, overgrazing causes as much soil degradation as cropland agriculture does, and it causes more desertification. Degraded rangeland costs an estimated $23 billion per year in lost agricultural productivity.

The Dust Bowl prompted the United States to fight erosion

Prior to the large-scale cultivation of North America's Great Plains, native prairie grasses of this temperate grassland region held soils in place. In the late 19th and early 20th centuries, many homesteading settlers arrived in Oklahoma, Texas, Kansas, New Mexico, and Colorado with hopes of making a living there as farmers. Farmers grew abundant wheat, and ranchers grazed many thousands of cattle, contributing to erosion by removing native grasses and altering soil structure.

(a) Kansas dust storm, 1930s

FIGURE 7.10 Drought and poor agricultural practices devastated millions of U.S. farmers in the 1930s in the Dust Bowl. The photo **(a)** shows a dust storm approaching Rolla, Kansas. The map **(b)** shows the extent of the Dust Bowl region.

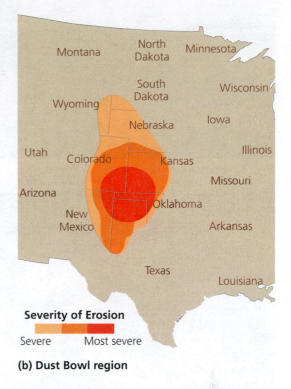

Severity of Erosion

Severe   Most severe

(b) Dust Bowl region

In the early 1930s, a drought exacerbated the ongoing human impacts, and the region's strong winds began to erode millions of tons of topsoil. Dust storms traveled up to 2000 km (1250 mi), blackening rain and snow as far away as New York and Washington, D.C. Some areas lost 10 cm (4 in.) of topsoil in a few years. The most affected region in the southern Great Plains became known as the **Dust Bowl,** a term now also used for the historical event itself. The "black blizzards" of the Dust Bowl ravaged towns in the heartland and forced thousands of farmers off their land (**FIGURE 7.10**).

In response, the U.S. government, along with state and local governments, increased its support for research on soil conservation practices. The U.S. Congress passed the Soil Conservation Act of 1935, establishing the Soil Conservation Service (SCS). This new agency worked closely with farmers to develop conservation plans for individual farms. The SCS (now the *Natural Resources Conservation Service*) served as a model for other nations that established their own soil conservation agencies to aid farmers in fighting soil erosion.

Sustainable agriculture begins with soil management

A number of farming techniques can reduce the impacts of conventional cultivation on soils and combat soil degradation (**FIGURE 7.11**). Not all these approaches are new, as some have been practiced since the dawn of agriculture.

Crop rotation In **crop rotation,** farmers alternate the type of crop grown in a given field from one season or year to the next (**FIGURE 7.11a**). Rotating crops can return nutrients to

the soil, break cycles of disease associated with continuous cropping, and minimize the erosion that can come from letting fields lie uncultivated. Many U.S. farmers rotate their fields between wheat or corn and soybeans from one year to the next. Soybeans are legumes, plants that have specialized bacteria on their roots that fix nitrogen (p. 42), revitalizing soil that the previous crop had partially depleted of nutrients. Crop rotation also reduces insect pests; if an insect is adapted to feed and lay eggs on one crop, planting a different type of crop will leave its offspring with nothing to eat.

In a practice similar to crop rotation, many farmers plant temporary cover crops, such as nitrogen-replenishing clover, to prevent erosion after crops have been harvested.

Contour farming Water running down a hillside with little vegetative cover can easily carry soil away, so farmers have developed several methods for cultivating slopes. **Contour farming** (**FIGURE 7.11b**) consists of plowing furrows sideways across a hillside, perpendicular to its slope and following the natural contours of the land. In contour farming, the side of each furrow acts as a small dam that slows runoff and captures eroding soil. Farmers also plant buffer strips of vegetation along the borders of their fields and along nearby streams, which further protect against erosion and water pollution.

Terracing On extremely steep terrain, the most effective method for preventing erosion is **terracing** (**FIGURE 7.11c**). Terraces are level platforms, sometimes with raised edges, that are cut into steep hillsides to contain water from irrigation and precipitation. Terracing transforms slopes into series of steps like a staircase, enabling farmers to cultivate hilly land without losing huge amounts of soil to water erosion.

(a) Crop rotation

(b) Contour farming

(c) Terracing

(d) Intercropping

(e) Shelterbelts

(f) No-till farming

FIGURE 7.11 Farmers have adopted various strategies to conserve soil. These include rotating crops **(a)**, contour farming **(b)**, terracing **(c)**, intercropping **(d)**, planting shelterbelts **(e)**, and no-till agriculture **(f)**.

Intercropping Farmers may also minimize erosion by **intercropping,** planting different types of crops in alternating bands (**FIGURE 7.11d**). Intercropping helps slow erosion by providing more ground cover than does a single crop. Like crop rotation, intercropping reduces vulnerability to insects and disease and, when a nitrogen-fixing legume is planted, replenishes the soil with nutrients.

Shelterbelts A technique to reduce erosion from wind is to establish **shelterbelts,** or *windbreaks* (**FIGURE 7.11e**). These are rows of trees or other tall plants that are planted along the edges of fields to slow the wind. On the Great Plains, fast-growing species such as poplars are often used. Shelterbelts

can be combined with intercropping; mixed crops are planted in rows surrounded by or interspersed with rows of trees that provide fruit, wood, and wildlife habitat, as well as protection from wind.

Conservation tillage **Conservation tillage** describes approaches that reduce the amount of tilling (plowing) relative to conventional farming. Turning the earth by tilling aerates the soil and works weeds and old crop residue into the soil to nourish it, but tilling also leaves the surface bare, allowing wind and water to erode away precious topsoil.

No-till farming is the ultimate form of conservation tillage (**FIGURE 7.11f**). Rather than plowing after each harvest,

When agriculture in South America vastly expanded in the 1970s, farmers largely used conventional soil tilling. This practice, however, led to massive losses of topsoil and organic matter due to erosion from torrential rainfall in wetter regions, and dry, windy winters in the more arid regions. Farms in the wetter portions of Brazil, for example, saw soil organic matter decrease by up to 50% after only a few years of conventional tilling.

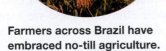

Farmers across Brazil have embraced no-till agriculture.

With their livelihoods threatened, farmers—with the vigorous support of national governments—enthusiastically embraced the emerging technology of no-till agriculture. Farmers, on farms big and small, were trained in no-till farming techniques, which were aided by the availability of inexpensive, locally produced herbicides for weed suppression. Farmers not only saw topsoil losses significantly decrease—by up to 97% in the wetter portions of Brazil and 80% in Argentina—but also enjoyed increased crop yields and higher profitability from reduced labor and fuel costs. Today, the use of no-till farming is widespread in South America, being used on 90% of cropland in Paraguay, 80% in Argentina and Uruguay, and 50% in Brazil (80% in southern Brazil), saving tons of soil each year from erosion in this major food-producing region.

EXPLORE THE DATA at **Mastering** Environmental Science

---

farmers leave crop residues atop their fields or plant cover crops, keeping the soil covered with plant material at all times. This reduces erosion by shielding the soil from wind and water. To plant the next crop, they cut a thin, shallow groove into the soil surface, drop in seeds, and cover them, using a machine called a no-till drill. By planting seeds of the new crop through the residue of the old, less soil erodes away, organic material accumulates, and the soil soaks up more water—all of which encourage better plant growth.

By increasing organic matter and soil biota while reducing erosion, conservation tillage can improve soil quality and combat global climate change by storing carbon in soils. Today across the United States, nearly one-quarter of farmland is under no-till cultivation, and more than 40% is farmed using conservation tillage. Forty percent of soybeans, 21% of corn, and 18% of cotton receive no-till treatment. All of these conservation approaches are having positive effects: According to U.S. government figures, soil erosion on cropland in the United States declined from 3.05 billion tons in 1982 to 1.72 billion tons in 2010.

Policy can promote conservation measures in agriculture

Many nations spend billions of dollars in government subsidies to support agriculture. In the United States, roughly one-fifth of the income of the average farmer comes from subsidies. Proponents stress that the uncertainties of weather make profits and losses from farming unpredictable from year to year. To persist, these proponents say, an agricultural system needs some way to compensate farmers for bad years. This may be the case, but subsidies can encourage people to cultivate land that would otherwise not be farmed; to produce more food than is needed, driving down prices for other producers; and to practice unsustainable farming methods that further degrade the land. They suggest that a better model is for farmers to buy insurance to protect against short-term production failures, an approach increasingly embraced by the U.S. government in its support for agriculture.

While government subsidies may promote soil degradation, the U.S. Congress has also enacted provisions promoting soil conservation through the farm bills it passes every five to six years. Many provisions require farmers to adopt soil conservation practices before they can receive subsidies.

The **Conservation Reserve Program (CRP),** established in the 1985 U.S. Farm Bill, pays farmers to stop cultivating highly erodible cropland and instead place it in conservation reserves planted with grasses and trees. Lands under the Conservation Reserve Program now cover an area nearly the size of Virginia, and the United States Department of Agriculture (USDA) estimates that each dollar invested in this program saves nearly 1 ton of topsoil. Besides reducing erosion, the CRP generates income for farmers, improves water quality, and provides habitat for wildlife. Recently, the federal government paid farmers about $2 billion per year for the conservation of 11 million ha (27 million acres) of land. The 2014 Farm Bill continued this program, but capped the amount of land it could cover by 2017 at 10 million ha (24 million acres). Internationally, the United States Nations promotes soil conservation and sustainable agriculture through a variety of programs led by its Food and Agriculture Organization (FAO).

Controlling Pests, Preserving Pollinators

Throughout the history of agriculture, the insects, fungi, viruses, rodents, and weeds that eat, infect, or compete with our crop plants have taken advantage of the ways we cluster food plants into agricultural fields. Pests pose an especially great threat to monocultures, where a pest adapted to specialize on the crop can move easily from plant to plant.

What people term a **pest** is any organism that damages crops that are valuable to us. What we term a **weed** is any plant that competes with our crops. There is nothing inherently malevolent in the behavior of a pest or a weed. These organisms are simply trying to survive and reproduce, like all living things, but they affect farm productivity when doing so.

We have developed thousands of chemical pesticides

Because industrial monocultures limit the ability of natural enemies to control pest populations, farmers have felt the need to introduce some type of pest control to produce food economically at a large scale. In the past half-century, most farmers have turned to chemicals to suppress pests and weeds. In that time we have developed thousands of synthetic chemicals to kill insects (*insecticides*), plants (*herbicides*), and fungi (*fungicides*). Such poisons are collectively termed **pesticides.**

All told, nearly 400 million kg (900 million lb) of active ingredients from conventional pesticides are applied in the United States each year—almost 3 pounds per person. Four-fifths of this total is applied on agricultural land. Since 1960, pesticide use has risen fourfold worldwide. Usage in industrialized nations has leveled off, but it continues to rise in the developing world. Today more than $44 billion is expended annually on pesticides across the world. Exposure to synthetic pesticides can have health impacts on people and other nontarget organisms (Chapter 10), so their use in agriculture has far-reaching effects.

Pests evolve resistance to pesticides

Despite the toxicity of pesticides, their effectiveness tends to decline with time as pests evolve resistance to them. Recall from our discussion of natural selection (pp. 50–57) that individuals within populations vary in their genetic makeup. Because most insects, weeds, and microbes can exist in huge numbers, it is likely that a small fraction of individuals may by chance already have genes that enable them to metabolize and detoxify a pesticide (p. 218). These individuals will survive exposure to the pesticide, whereas individuals without these genes will not. If an insect that is genetically resistant to an insecticide survives and mates with another resistant individual, the genes for pesticide resistance will be passed to their offspring. As resistant individuals become more prevalent in the pest population, insecticide applications will cease to be effective, the pest population will increase in size (**FIGURE 7.12**), and agricultural losses to pests will escalate.

In many cases, industrial chemists are caught up in an evolutionary arms race (p. 72) with the pests they battle, racing to increase or retarget the toxicity of their chemicals while the armies of pests evolve ever-stronger resistance to their efforts. Because we seem to be stuck in this cyclical process, it has been nicknamed the "pesticide treadmill." Currently, among arthropods (insects and their relatives) alone, there are more than 10,000 known cases of resistance by 600 species to more than 330 insecticides. Hundreds more weed species and plant diseases have evolved resistance to herbicides and other pesticides. Many species, including insects such as the green peach aphid, Colorado potato beetle, and diamondback moth, have evolved resistance to multiple chemicals.

1 Pests attack crops

2 Pesticide is applied

3 Most pests are killed. A few with innate resistance survive

4 Survivors breed and produce a pesticide-resistant population

5 Pesticide is applied again

6 Pesticide has little effect. New, more toxic, pesticides are developed

FIGURE 7.12 Through the process of natural selection, crop pests often evolve resistance to the poisons we apply to kill them.

Biological control pits one organism against another

Because of pesticide resistance, toxicity to nontarget organisms, and human health risks from some synthetic chemicals, agricultural scientists increasingly battle pests and weeds with organisms that eat or infect them. This more sustainable pest control strategy, called **biological control** or **biocontrol,** operates on the principle that "the enemy of one's enemy is one's friend." For example, parasitoid wasps (p. 72) are natural enemies of many caterpillars. These wasps lay eggs on a caterpillar, and the larvae that hatch from the eggs feed on the caterpillar, eventually killing it. Parasitoid wasps are frequently used as biocontrol agents and have often succeeded in controlling pests and reducing chemical pesticide use.

A widespread biocontrol effort is the use of *Bacillus thuringiensis* **(Bt),** a naturally occurring soil bacterium that produces a protein that kills many caterpillars and some fly and beetle larvae. Farmers spray Bt spores on their crops to protect against insects. In addition, scientists have isolated the gene responsible for the bacterium's poison, engineering it into crop plants so that the plants produce the poison (p. 160).

One classic case of successful biological control is the introduction of the cactus moth, *Cactoblastis cactorum*, from Argentina to Australia in the 1920s to control the invasive prickly pear cactus that was overrunning rangeland. Within just a few years, the moth managed to free millions of hectares of Australian rangeland from the cactus.

However, biocontrol approaches entail risks. Biocontrol organisms are sometimes more difficult to manage than chemical controls, because they cannot be "turned off" once they are initiated. Further, biocontrol organisms have in some cases become invasive and harmed nontarget organisms. Following the cactus moth's success in Australia, for example, it was introduced in other countries to control non-native prickly pear. Moths introduced to Caribbean islands spread to Florida on their own and are now eating their way through rare native cacti in the southeastern United States. If these moths reach Mexico and the southwestern United States, they could decimate many native and economically important species of prickly pear cacti. Because of concerns about unintended impacts, researchers study biocontrol proposals carefully before putting them into action.

Integrated pest management combines varied approaches to pest control

As it became clear that both chemical and biocontrol approaches pose risks, agricultural scientists and farmers began developing more sophisticated strategies, trying to combine the best attributes of each approach. **Integrated pest management (IPM)** incorporates numerous techniques, including close monitoring of pest populations, biocontrol approaches, use of synthetic chemicals when needed, habitat alteration, crop rotation, transgenic crops, alternative tillage methods, and mechanical pest removal.

IPM has become popular in many parts of the world that are embracing sustainable agriculture techniques. Indonesia stands as an exemplary case. This nation had subsidized pesticide use heavily for years, but its scientists came to understand that pesticides were actually making pest problems worse. They were killing the natural enemies of the brown planthopper, which began to devastate rice fields as its populations exploded. Concluding that pesticide subsidies were costing money, causing pollution, and apparently decreasing yields, the Indonesian government in 1986 banned the import of 57 pesticides, slashed pesticide subsidies, and promoted IPM. Within just four years, pesticide production fell by half, imports fell by two-thirds, and the government saved money by eliminating subsidy payments. Rice yields rose 13% with IPM, and since then the approach has spread to dozens of other nations, particularly in rice-growing regions in Asia. While the 11-year program was highly successful, its support by the Indonesian government has been inconsistent over the past 15 years. This has led pesticide manufacturers to once again aggressively market their products to farmers in the countryside, and some of the very same problems that gave rise to IPM in the first place are once again occurring.

Pollinators are beneficial "bugs" worth preserving

Managing insect pests is such a major issue in agriculture that it is easy to fall into a habit of thinking of all insects as somehow bad or threatening. But in fact, most insects are harmless to agriculture—and some are absolutely essential. The insects that pollinate crops are among the most vital factors in our food production. Pollinators are the unsung heroes of agriculture.

Pollination (p. 73) is the process by which male sex cells of a plant (pollen) fertilize female sex cells of a plant; it is the botanical version of sexual intercourse. Pollinators are animals that move pollen from one flower to another. Flowers are, in fact, evolutionary adaptations that function to attract pollinators. The sugary nectar and protein-rich pollen in flowers serve as rewards to lure pollinators, and the sweet smells and bright colors of flowers advertise these rewards.

Our staple grain crops are derived from grasses and are wind-pollinated, but 800 types of cultivated plants rely on bees, wasps, beetles, moths, butterflies, and other insects for pollination. Farmers in the United States alone gain an estimated $17 billion per year in pollination services from the introduced European honeybee that beekeepers have domesticated for use with crops, and more than $3 billion per year from many of the nation's 4000 species of wild native bees. However, scientific data indicate that populations of honeybees and of wild native bees are declining steeply across North America.

Scientists studying the pressures on bees, butterflies, and other pollinators are concluding that they are suffering a "perfect storm" of stresses—many of which result from industrial agriculture. A direct source of mortality is the vast arsenal of chemical insecticides we apply to crops, lawns, and gardens (see THE SCIENCE BEHIND THE STORY). Pollinators have also

What Role Do Pesticides Play in the Collapse of Bee Colonies?

**Dr. Dennis vanEngelsdorp,
University of Maryland**

Chemical pesticides have been implicated as one of the factors scientists believe may be contributing to the elevated rate of honeybee (*Apis mellifera*) colony collapses observed over the past decade (p. 156). Highly controlled laboratory studies have shown that honeybees suffer physiological harm, increased mortality, and altered behavioral patterns when exposed to elevated levels of some common agricultural pesticides. These studies, which typically test single pesticides at a time, have provided extremely valuable insight into the toxicity levels of some pesticides for bees, but do not capture the stresses faced by bees in natural settings, where they can be exposed to hundreds of potentially harmful chemicals.

A recent study—led by Dennis vanEngelsdorp of the University of Maryland and published in 2016 in *Nature Scientific Reports*—took on the challenge of determining how the multitude of pesticides in the environment may be affecting honeybee colonies. The researchers embraced an approach by which they treated the highly organized, communal beehive as a single "super-organism" and measured the cumulative pesticide exposures experienced by colonies during a typical agricultural season. They then compared the pesticide exposures of individual colonies with measures of colony health—such as whether the colony survived the season or if the colony lost its queen—to identify factors related to pesticide exposure that may predict bee colony collapses.

Treating a bee colony as a single organism makes logical sense, as a thriving honeybee colony depends on the collective efforts of hundreds or thousands of individual bees, so stressors that impair even one of the critical functions that sustain the hive can cause the entire colony to collapse. Hives typically have a single queen, a fertile female whose main function is to deposit eggs into honeycomb cells within the hive; the queen is the mother of most, if not all, bees in the hive. Worker bees, the most numerous class of bees in the hive, are sterile females that have many responsibilities. *Nurse bees* are workers that care for the queen and the larval bees that hatch from her eggs. *Foragers* leave the hive to forage for food, such as nectar and pollen, which they bring back to the hive. Other workers maintain, clean, and defend the hive. Chemicals secreted by the queen, called pheromones,

prevent the creation of a new queen while she is healthy and active. But when the queen becomes old or dies, her pheromones are no longer present and the workers create new queens by feeding several larval worker bees a high-nutrient food called *royal jelly*. The newly created queens then embark on a mating flight, during which they are fertilized by male bees, called *drones*, and they are then able to begin their life as the matron of a hive.

As part of the study, vanEngelsdorp's team surveyed a total of 91 bee colonies from March 2007 to January 2008. The colonies were part of three migratory commercial beekeeping operations that, starting in Florida, moved northward up the East Coast of the United States, pollinating farmers' crops (a service for which the farmers paid the beekeepers). Between pollination jobs, the bees foraged in natural areas for pollen to create honey, or rested in holding areas where the hives were maintained and prepared for their next job.

Pesticides can become concentrated in beehives because bees forage over a wide area in their search for pollen and nectar, often covering 100 km^2 (38 mi^2) around the hive. Forager bees bring pollen, some of it containing agricultural pesticides, back to the hive and store it in honeycomb cells near developing bee larvae. Nurse bees consume this pollen-rich substance, called "beebread," and use specialized glands to convert it to a protein-rich and calorically dense secretion that is then fed to the larvae (**FIGURE 1**). Nectar, which can also be contaminated with pesticides, is similarly deposited in honeycomb cells and is used to create honey, which sustains the worker bees and the queen throughout their adult lives. Thus, contaminated pollen and nectar from up to 100 km^2 can become concentrated inside a single beehive, potentially affecting all life stages of the colony.

FIGURE 1 Nurse bees tending to larvae in the hive. Nearby honeycomb cells are used to store beebread while others store honey.

This is akin to bioaccumulation (p. 221), where toxic substances attain higher concentrations in the tissues of organisms than are found in the surrounding environment.

To assess the colony's cumulative pesticide exposure, the researchers collected samples of beebread throughout the season, and took samples of honeycomb wax (in which larvae were brooded and honey and beebread were stored) at the beginning and end of the season. The samples were then sent to a laboratory and screened for the presence of 171 common agricultural pesticides. At the end of the season, the researchers noted which hives survived the entire season, as well as those that experienced a "queen event"—the death of the queen bee. The death of the queen can occur naturally, but is also seen in stressed hives where the worker bees, perceiving something is wrong with the queen, kill her and attempt to replace her with a new queen. These events are predictors of hive survival, as studies have shown that as the number of queen events increases in a hive, so does its likelihood of dying.

For each of the beebread and honeycomb wax samples collected, the researchers determined the total number of pesticides detected, the number of those pesticides that exceeded accepted toxicity levels, and the colony's "hazard quotient (HQ)," a measure of the cumulative toxicity the hive experienced from all the detected pesticides. These three measures were then subjected to correlation analysis with measures of hive health to determine how the bee colonies responded to the diverse collection of chemical pesticides in the modern agricultural environment.

In all, 93 different pesticides were detected in the colonies over the course of the season. More than half of the bee colonies perished during the study period, with hive mortality rates increasing as the season progressed. Correlation analysis revealed that the likelihood of a colony dying was positively correlated with the *total number* of pesticides in beebread and honeycomb wax but not with the *concentration* of the pesticides, as measured by HQ. This was an unexpected result; it

indicated that as the number of detected pesticides increased, so did the chances of the hive dying, regardless of the concentration of each pesticide in the hive. In the words of vanEngelsdorp, "Our results fly in the face of one of the basic tenets of toxicology—that the dose makes the poison" (p. 223). The researchers hypothesized that as the sheer number of toxic compounds increased in the hive, the bees experienced extreme physiological stress as they tried to detoxify such a diverse collection of contaminants within their bodies. In some cases this stress was simply too much to overcome, and the hive collapsed as more and more workers perished.

When examining queen events, the study determined that hives experiencing a queen event (1) contained more total pesticides in their beebread and wax and (2) had a higher HQ in wax than hives that maintained their queen throughout the season (**FIGURE 2**). So, while the concentration of pesticides in the hive, as measured by HQ, did not correlate with the survival of the hive, it did correlate with queen events, suggesting that hives with higher cumulative concentrations of pesticides were more likely to experience the loss of their queen.

Additionally, the team found that fungicides, which are sprayed on crops to reduce damage from fungal infections, correlated with increased risk of hive death. This result was unexpected. While laboratory studies had shown that bee larvae experience higher mortality when exposed to elevated concentrations of some fungicides, it was thought that the relatively lower concentrations of fungicides bees experience in agricultural settings were safe.

It is important to note that correlative studies (p. 11) such as this one do not establish causation. Rather, they highlight connections between factors, providing a roadmap of promising avenues for future research. This study, and others like it to come, may therefore play a valuable role in aiding scientists to one day unravel the mystery of what is killing one of the most important organisms in modern agriculture—and help us find solutions to save our bees.

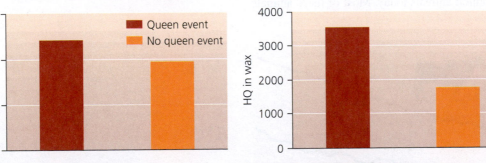

(a) Total number of pesticides in wax

(b) Hazard quotient (HQ) in wax

FIGURE 2 Hives that experienced a "queen event" had significantly more pesticides in their wax honeycomb and higher hazard quotient (HQ) values in their wax than did hives whose queen survived the season. Hazard quotient is calculated by dividing the total concentration of pesticides in the wax by the lethal dose of each detected pesticide required to kill 50% of the bees in a sample (the pesticide's LD_{50}; p. 223). *Figure from Traynor, K., et al., 2016. In-hive pesticide exposome: Assessing risks to migratory honey bees from in-hive pesticide contamination in the eastern United States.* Nature Scientific Reports, 6:33207, http://www.nature.com/articles/srep33207.

suffered losses of habitat and flower resources for decades, as natural areas are developed. Bees are also being attacked by invasive parasites and pathogens. In particular, two accidentally introduced parasitic mites have swept through honeybee populations in recent years, decimating hives.

Researchers are finding that these multiple sources of stress seem to interact and cause more damage than the sum of their parts. For example, pesticide exposure and difficulty finding food might weaken a bee's immune system, making it more vulnerable to parasites. Any or all of these factors may possibly play a role in **colony collapse disorder,** in which the majority of worker bees in a colony disappear, endangering the hive by depriving the queen and its developing bee larvae of their life-sustaining workers. For the past decade, this mysterious disorder has destroyed up to one-third of all honeybees in the United States each year.

Fortunately, we have a number of solutions at hand to help restore bee populations. By retaining or establishing wildflowers and flowering shrubs in or near farm fields, farmers can provide bees a refuge and a source of diverse food resources. Encouraging flowers on highway rights-of-way can provide resources to pollinators while beautifying roadsides. In addition, farmers can decrease their use of chemical insecticides by using biocontrol or integrated pest management. Homeowners can help pollinators by reducing or eliminating the use of pesticides, tending gardens of flowering plants, and providing nesting sites for bees.

Raising Animals for Food

Food from cropland agriculture makes up a large portion of the human diet, but most of us also eat animal products. Just as farming methods have changed over time, so have the ways we raise animals for food.

As wealth and global commerce have increased, so has humanity's production of meat, milk, eggs, and other animal products (**FIGURE 7.13**). The world population of domesticated animals raised for food rose from 7.3 billion animals in 1961 to over 30 billion animals today. Most of these animals are chickens. Global meat production has increased fivefold since 1950, and per capita meat consumption has doubled. The United Nations Food and Agriculture Organization (FAO) estimates that as more developing nations go through the demographic transition (pp. 127–129) and become wealthier, total meat consumption could nearly double again by the year 2050.

Our food choices are resource choices

What we choose to eat has ramifications for how we use resources that support agriculture. Every time that one organism consumes another, only about 10% of the energy moves from one trophic level up to the next; the great majority of energy is used up in cellular respiration (p. 35). For this reason, eating meat is far less energy-efficient than relying on a vegetarian diet and leaves a far greater ecological footprint.

Some animals convert grain feed into milk, eggs, or meat more efficiently than others (**FIGURE 7.14**). Scientists have calculated relative energy-conversion efficiencies for

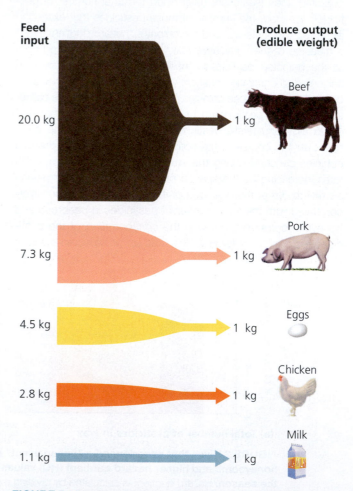

FIGURE 7.14 Producing different animal food products requires different amounts of animal feed. Twenty kilograms of feed must be provided to cattle to produce 1 kg of beef. *Data from Smil, V., 2001. Feeding the world: A challenge for the twenty-first century. Cambridge, MA: MIT Press.*

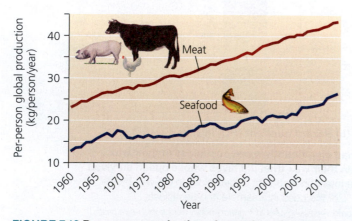

FIGURE 7.13 Per-person production of meat from farmed animals and of seafood has risen steadily worldwide. *Data from UN Food and Agriculture Organization (FAO).*

different types of animals. Such energy efficiencies have ramifications for land use: Land and water are required to raise food for the animals, and some animals require more than others. **FIGURE 7.15** shows the area of land and weight of water required to produce 1 kg (2.2 lb) of food protein for milk, eggs, chicken, pork, and beef, as well as the greenhouse gas emissions released in the process. Producing eggs and chicken meat requires the least space and water, whereas producing beef requires the most. Such differences make clear that when we choose what to eat, we are also indirectly choosing how to make use of resources such as land and water.

Feedlots have benefits and costs

In traditional agriculture, livestock are kept by farming families near their homes or are grazed on open grasslands by nomadic herders or sedentary ranchers. These traditions survive, but the advent of industrial agriculture has brought a new method. **Feedlots,** also known as *factory farms* or *concentrated animal feeding operations (CAFOs)*, are essentially huge warehouses or pens designed to deliver energy-rich food to animals living at extremely high densities. Today nearly half the world's pork and most of its poultry come from feedlots.

Feedlot operations allow for economic efficiency and increased production, which make meat affordable to more people. Concentrating cattle and other livestock in feedlots also removes them from rangeland, thereby reducing the grazing impacts these animals would exert across large portions of the landscape.

Intensified animal production through the industrial feedlot model has some negative consequences, though. Forty-five percent of our global grain production goes to feed livestock and poultry. This elevates the price of staple grains and endangers food security for the very poor. Livestock produce prodigious amounts of manure and urine, and their waste can pollute surface water and groundwater near feedlots. The crowded conditions under which animals are often kept necessitate heavy use of antibiotics to control disease. The overuse of antibiotics can cause microbes to become resistant to the antibiotics (just as pests become resistant to pesticides), making these drugs less effective.

Livestock are also a major source of greenhouse gases, such as methane, that lead to climate change (p. 313). The FAO reported in 2013 that livestock agriculture contributes over 14% of greenhouse gas emissions

weighing the
ISSUES

Feedlots and Animal Rights

Animal rights advocates denounce factory farming because they argue that it mistreats animals. Chickens, pigs, and cattle are crowded together in small pens their entire lives, fattened up, and slaughtered. Should we concern ourselves with the quality of life of the animals that constitute part of our diet? Do you think animal rights concerns are as important as environmental concerns? Are conditions at feedlots a good reason for being vegetarian?

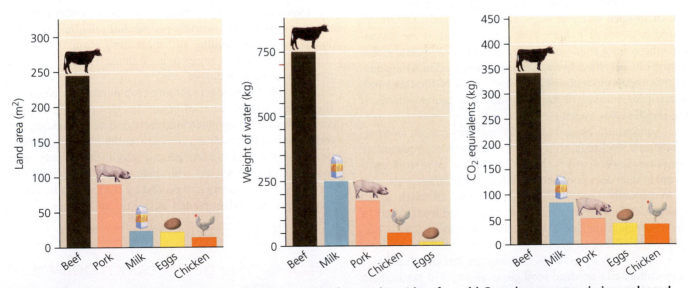

(a) Land required to produce 1 kg of protein

(b) Water required to produce 1 kg of protein

(c) Greenhouse gas emissions released in producing 1 kg of protein

FIGURE 7.15 Producing different types of animal products requires different amounts of (a) land and (b) water—and releases different amounts of (c) greenhouse gas emissions. Raising cattle for beef exerts the greatest impacts in all three ways. *Data (a, b) from Smil, V., 2001. Feeding the world: A challenge for the twenty-first century. Cambridge, MA: MIT Press; and (c) from FAO, 2015. Global Livestock Environmental Assessment Model (GLEAM).*

DATA Answer the following in terms of protein, pound for pound. • How many times more land does it take to produce beef than chicken? • How many times more water does beef require, compared with chicken? • How many times more greenhouse gas emissions does beef release, relative to chicken?

Go to **Interpreting Graphs & Data** on **Mastering** Environmental Science

worldwide. On a brighter note, the FAO judged that greenhouse gas emissions from livestock could be reduced by as much as 30% with the widespread adoption of best practices in livestock production.

Not all livestock are raised in intensive facilities, however. Meat, eggs, and dairy products are also produced from "free-range" animals that are not always confined indoors or in pens. The certification of an operation as free range is quite broad, though. For example, the only requirement the USDA imposes for certifying poultry operations as "free range" is that the birds be given access to the outdoors. The Kennesaw State University farms maintain some 60 free-range chickens, and these hens supply The Commons with 300 or more eggs weekly. Raising animals in this manner can be costlier than in intensive facilities, so prices for free-range meat and eggs are often higher than for conventionally raised animals.

We raise seafood with aquaculture

Besides growing plants as crops and raising animals on rangelands and in feedlots, we rely on aquatic organisms for food. Increased demand and new technologies have led us to overharvest most marine fisheries (pp. 281–283); as a result, wild fish populations are plummeting throughout the world's oceans. This means that raising fish and shellfish on "fish farms" may be the only way to meet our growing demand for these foods.

The cultivation of aquatic organisms for food in controlled environments, called **aquaculture,** is now being pursued with over 220 freshwater and marine species of fish, crustaceans, mollusks, and plants. Many aquatic species are grown in open water in large, floating net-pens. Others are raised in ponds or holding tanks. Aquaculture is the fastest-growing type of food production; in the past 20 years, global output has increased fivefold. Most widespread in Asia, aquaculture today produces $125 billion worth of food and provides three-quarters of the freshwater fish and two-thirds of the shellfish that we eat.

Aquaculture helps reduce fishing pressure on overharvested and declining wild stocks. Furthermore, aquaculture consumes fewer fossil fuels and provides a safer work environment than does commercial fishing. Fish farming can also be remarkably energy-efficient, producing as much as 10 times more fish per unit area than is harvested from waters of the continental shelf and up to 1000 times more than is harvested from the open ocean.

Along with its benefits, aquaculture has disadvantages. Aquaculture can produce prodigious amounts of waste, both from the fish and shellfish and from the feed that goes uneaten and decomposes in the water. Like feedlot animals, commercially farmed fish often are fed grain, and this affects food supplies for people. In other cases, farmed fish are fed fish meal made from wild ocean fish such as herring and anchovies, whose harvest from oceans may place additional stress on wild fish populations.

If farmed aquatic organisms escape into ecosystems where they are not native (as several carp species have done in U.S. waters), they may spread disease to native stocks or may outcompete native organisms for food or habitat. For

FIGURE 7.16 Transgenic salmon (top) grow faster than wild salmon of the same species. They often reach a larger size; these two fish are the same age.

example, if salmon that have been genetically engineered for rapid growth were to escape from aquaculture facilities, they could spread disease or introduce genes into wild salmon populations (**FIGURE 7.16**).

Genetically Modified Food

The Green Revolution enabled us to feed a greater number and proportion of the world's people, but relentless population growth is demanding still more innovation. A new set of potential solutions began to arise in the 1980s and 1990s as advances in genetics enabled scientists to directly alter the genes of organisms, including crop plants and livestock. The genetic modification of organisms that provide us food holds promise to enhance nutrition and the efficiency of agriculture while lessening impacts on the planet's environmental systems. However, genetic modification may also pose risks that are not yet well understood. This possibility has given rise to anxiety and protest by consumer advocates, small farmers, environmental activists, and critics of big business.

Foods can be genetically modified

Genetic engineering is any process whereby scientists directly manipulate an organism's genetic material in the laboratory by adding, deleting, or changing segments of its DNA (p. 32). **Genetically modified (GM) organisms** are organisms that have been genetically engineered using recombinant DNA, which is DNA that has been patched together from the DNA of multiple organisms (**FIGURE 7.17**). The goal is to place genes that code for certain desirable traits (such as rapid growth, disease resistance, or high nutritional content) into organisms lacking those traits. An organism that contains DNA from another species is called a **transgenic**

organism, and the genes that have moved between them are called **transgenes.**

The creation of transgenic organisms is one type of **biotechnology,** the material application of biological science to create products derived from organisms. Biotechnology has helped us develop medicines, clean up pollution, understand the causes of cancer, dissolve blood clots, and make better beer and cheese. **TABLE 7.1** shows several notable developments in GM foods. The stories behind them illustrate both the promises and pitfalls of food biotechnology.

The genetic alteration of plants and animals by people is nothing new; through artificial selection (p. 52), we have influenced the genetic makeup of our livestock and crop plants for thousands of years. However, as critics are quick to point out, the techniques geneticists use to create GM organisms differ from traditional selective breeding in several ways. For one, selective breeding mixes genes from individuals of the same or similar species, whereas scientists creating recombinant DNA routinely mix genes of organisms as different as bacteria and plants, or spiders and goats. For another, selective breeding deals with whole organisms living in the field, whereas genetic engineering works with genetic material in the lab. Third, traditional breeding selects from combinations of genes that come together naturally, whereas genetic engineering creates the novel combinations directly.

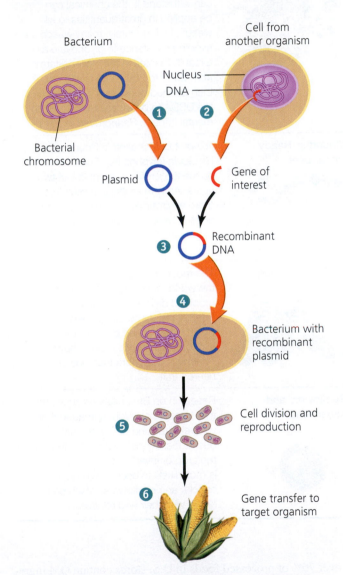

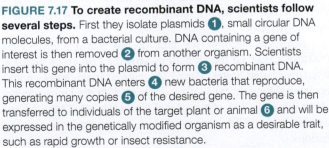

FIGURE 7.17 To create recombinant DNA, scientists follow several steps. First they isolate plasmids ❶, small circular DNA molecules, from a bacterial culture. DNA containing a gene of interest is then removed ❷ from another organism. Scientists insert this gene into the plasmid to form ❸ recombinant DNA. This recombinant DNA enters ❹ new bacteria that reproduce, generating many copies ❺ of the desired gene. The gene is then transferred to individuals of the target plant or animal ❻ and will be expressed in the genetically modified organism as a desirable trait, such as rapid growth or insect resistance.

Biotechnology is transforming the products around us

In just three decades, GM foods have gone from science fiction to mainstream business (**FIGURE 7.18**). Some GM crops today are engineered to resist herbicides, so that farmers can apply herbicides to kill weeds without having to worry about killing their crops. Other crops are engineered to resist insect attack. Some are modified for both types of resistance. Resistance to herbicides and insect pests enables large-scale commercial farmers to grow crops more efficiently. As a result, sales of GM seeds to these farmers in the United States and other countries have risen quickly.

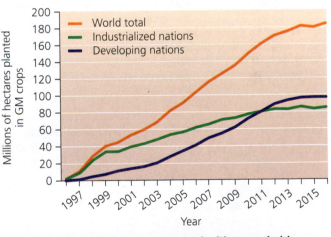

FIGURE 7.18 GM crops have spread with remarkable speed since their commercial introduction in 1996. They now are planted on more than 10% of the world's cropland. *Data from the International Service for the Acquisition of Agri-biotech Applications (ISAAA, 2017).*

DATA Compare the growth in the number of hectares planted with GM crops in developing nations from 2007 to 2011 with the growth from 2012 to 2016. • What are some possible explanations for the slowdown in the cultivation of GM crops in developing nations in recent years? • If current trends continue, which group of nations will have more GM crops in 2020? • Can you estimate how much more this group of nations might have?

Go to **Interpreting Graphs & Data** on **Mastering** Environmental Science

TABLE 7.1 Several Notable Examples of Genetically Modified Food Technologies

CROP	DESCRIPTION AND STATUS	CROP	DESCRIPTION AND STATUS
Golden rice 	Engineered to produce beta-carotene to fight vitamin A deficiency in Asia and the developing world. May offer only moderate nutritional enhancement despite years of work. Still undergoing research and development.	Bt cotton	Engineered with genes from bacterium *Bacillus thuringiensis* (Bt), which kills insects. Has increased yield, decreased insecticide use, and boosted income for 14 million small farmers in India, China, and other nations.
Virus-resistant papaya 	Resistant to ringspot virus and grown in Hawai'i. In 2011 became the first biotech crop approved for consumption in Japan.	Roundup-Ready alfalfa 	One of many crops engineered to tolerate Monsanto's Roundup herbicide (glyphosate). Because the crop can withstand it, the chemical can be applied in great quantities to kill weeds. Unfortunately, many weeds are evolving resistance to glyphosate as a result. Planted in the United States from 2005 to 2007, GM alfalfa was then banned because a lawsuit forced the USDA to better assess its environmental impact. Reapproved in 2011.
GM salmon 	Engineered for fast growth and large size. The first GM animal approved for sale as food. To prevent fish from breeding with wild salmon and spreading disease to them, the company AquaBounty promised to make their fish sterile and raise them in inland pens.	Roundup-Ready sugar beet 	Tolerant of Monsanto's Roundup herbicide (glyphosate). Swept to dominance (95% of U.S. crop) in just two years. As with alfalfa, a lawsuit forced more environmental review, after it had already become widespread. Reapproved in 2012.
Biotech potato 	Resistant to late blight, the pathogen that caused the 1845 Irish Potato Famine and that still destroys $7.5 billion of potatoes each year. Being developed by European scientists, but struggling with European Union (EU) regulations on research.	Biotech soybean 	The most common GM crop in the world, covering nearly half the cropland devoted to biotech crops. Engineered for herbicide tolerance, insecticidal properties, or both. Like other crops, soybeans may be "stacked" with more than one engineered trait.
Bt corn	Engineered with genes from bacterium *Bacillus thuringiensis* (Bt), which kills insects. One of many Bt crops developed.	Sunflowers and superweeds 	Research on Bt sunflowers suggests that transgenes might spread to their wild relatives and turn them into vigorous "superweeds" that compete with the crop or invade ecosystems. This is most likely to occur with squash, canola, and sunflowers, which can breed with their wild relatives.

Globally in 2017, more than 18 million farmers grew GM crops on 180 million ha (445 million acres) of farmland—nearly 12% of all cropland in the world. In the United States today, roughly 90% of corn, soybeans, cotton, and canola consist of genetically modified strains. This is incredible growth, considering that GM crop varieties have been commercially planted only since 1996. Worldwide, four of every five soybean and three of every four cotton plants are now transgenic, as are one of every three corn plants and one of every four canola plants. It is conservatively estimated that over 70% of processed foods in U.S. stores contain GM ingredients. Thus, it is highly likely that you consume GM foods on a daily basis.

Soybeans account for half of the world's GM crops (**FIGURE 7.19a**). Of the 28 nations growing GM crops in 2017, five (the United States, Brazil, Argentina, India, and Canada) accounted for over 90% of production, with the United States alone growing 40% of the global total (**FIGURE 7.19b**). Moreover, half of all GM crops worldwide are grown in developing nations.

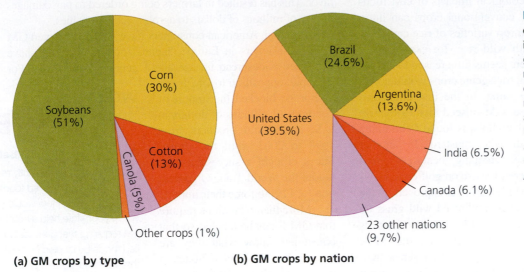

Brazil
(24.6%)

Argentina
(13.6%)

United States
(39.5%)

Corn
(30%)

Soybeans
(51%)

Cotton
(13%)

Canola (5%)

India (6.5%)

Canada (6.1%)

23 other nations
(9.7%)

Other crops (1%)

(a) GM crops by type

(b) GM crops by nation

FIGURE 7.19 So far, genetic engineering has mainly involved common crops grown in industrialized nations. Of the world's GM crops **(a)**, soybeans are the most common. Of global acreage planted in GM crops **(b)**, the United States devotes the most area. *Data from the International Service for the Acquisition of Agri-biotech Applications (ISAAA).*

What are the impacts of GM foods?

Genetic modification has the potential to advance agriculture by engineering crops with high drought tolerance for use in arid regions, and by developing high-yield crops that can feed our growing population on existing cropland. However, most of these noble intentions have not yet come to pass. This is largely because the corporations that develop GM varieties cannot easily profit from selling seed to small farmers in developing nations. Instead, most biotech crops have been engineered for insect resistance and herbicide tolerance, which improve efficiency for large-scale industrial farmers who can afford the technology.

Regardless, proponents of GM foods maintain that these foods bring environmental and social benefits and promote sustainable agriculture in several ways. Increased crop yields

enhance food security and reduce the need for new farmland, conserving natural areas. Crops engineered for drought resistance reduce the need for irrigation, and those engineered for better nutrition—such as golden rice—combat malnutrition. Herbicide-resistant crops promote no-till farming. Planting insect-resistant GM crops, proponents maintain, also reduces pesticide applications.

Although GM crops do appear to result in lower levels of insecticide use, studies find that the cultivation of these crops tends to result in *more* herbicide use. As weeds evolve resistance to herbicides, farmers apply ever-larger quantities of herbicide. Worldwide, over 250 weed species have evolved resistance to herbicides, and resistance to the weed-killer glyphosate is being documented more widely than in the past (**FIGURE 7.20**).

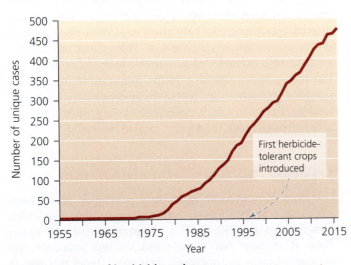

First herbicide-tolerant crops introduced

(a) Known cases of herbicide resistance

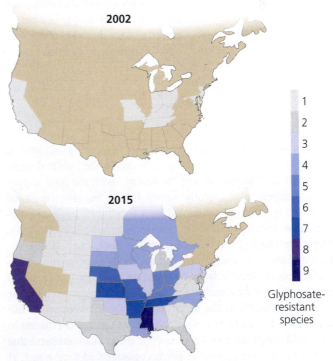

2002

2015

Glyphosate-resistant species

(b) Spread of glyphosate resistance

FIGURE 7.20 Weeds are evolving resistance to herbicides. Documented cases of herbicide resistance have surpassed **(a)** 475 biotypes involving 275 species of plants. In little over a decade, weed resistance to glyphosate **(b)** spread across North America. *Data from Heap, I. International Survey of Herbicide-Resistant Weeds. January, 2017. www.weedscience.com.*

In principle, there is nothing about the process of genetic engineering that should make genetically modified food any less safe to eat than food produced by conventional methods. The fact that technology is used to move a gene does not make the gene unsafe. Thus, to determine whether GM foods pose any health risks, studies must be done comparing those foods with conventional versions, one by one—just as researchers would study any other substance for health risks (Chapter 10). Thus far, no study has shown undeniable evidence of health impacts on humans from any GM food, but this lack of evidence does not guarantee that such foods pose no risk. A great deal of research is controlled by the companies that develop GM foods, and we will never be able to test all GM foods. Hence, the effects on humans of consuming GM foods will largely be examined with correlative studies (p. 11) over long time periods in coming decades.

Most scientists feel that ecological impacts of GM foods pose the greatest threat. Many conventional crops can interbreed with their wild relatives (crop varieties of rice can breed with wild rice, for example), so there seems little reason to believe that transgenic crops would not do the same. In the first confirmed case, GM oilseed rape (a relative of canola) was found hybridizing with wild mustard. In another case, creeping bentgrass engineered for use on golf courses—a GM plant not yet approved by the USDA—pollinated wild grass up to 21 km (13 mi) away from its experimental growing site. Most scientists think transgenes will inevitably make their way from GM crops into wild plants, but the ecological impacts of this are open to debate.

Because biotechnology is rapidly changing, and because the large-scale introduction of GMOs into our environment is recent, we cannot yet know everything about the consequences. Many experts feel we should proceed with caution, adopting the **precautionary principle,** the idea that one should not undertake a new action until its ramifications are well understood. Others feel that enough research has been done to allow for informed choices, and that it is time to double our bet on biotechnology because it appears to offer more benefits than risks.

Public debate over GM foods continues

Science helps inform us about genetic engineering, but ethical and economic concerns have largely driven the public debate. For many people, the idea of "tinkering" with the food supply seems dangerous or morally wrong. Others fear that the global food supply is being dominated by a handful of large corporations that develop GM technologies, among them Monsanto, Syngenta, Bayer CropScience, Dow, DuPont, and BASF. Critics say these multinational corporations threaten family farmers, and recent legal actions by these corporations against small farmers have stoked such fears. Agrobiotech corporations have taken out patents on the transgenes they've developed for use in GM crops—and some have sued farmers who are found to have GM crops on their land, even if the farmers contend that GM crops from neighboring fields contaminated their non-GM crops. Court cases have largely been decided in the companies'

favor. This has resulted in farmers being ordered to pay damages of tens of millions of dollars to agrobiotech companies.

Although American consumers have largely accepted GM foods, consumers in Europe, Japan, and other nations have expressed widespread unease about genetic engineering. For example, opposition to GM foods in Europe blocked the import of hundreds of millions of dollars in U.S. agricultural products from 1998 to 2003, until the United States brought a successful case before the World Trade Organization (p. 110) to force their import.

More than 60 nations require that GM foods be labeled so that consumers know what they are buying, but the United States is not one of them. Proponents of labeling argue that consumers have a right to know what's in the food they buy. Opponents argue that labeling implies the foods are dangerous and may result in lower sales.

weighing the
ISSUES

Do You Want Food Labeled?
Dozens of nations currently require that genetically modified foods be labeled as such. How would you vote if your state held a ballot initiative requiring that genetically modified foods be labeled? Consider that well over 70% of processed food now contains GM ingredients, such as corn syrup from GM corn. Would you personally choose among foods based on such labeling? Why or why not?

The Growth of Sustainable Agriculture

Industrial agriculture has allowed food production to keep pace with our growing population, but it involves many adverse environmental and social impacts. These range from the degradation of soils to reliance on fossil fuels to problems arising from pesticide use, genetic modification, and intensive feedlot and aquaculture operations. Although intensive commercial agriculture may help alleviate certain environmental pressures, it often worsens others. Throughout this chapter, we have seen examples of sustainable approaches to agriculture that maintain high crop yields, minimize resource inputs into food production, and lessen the environmental impacts of farming. Let's now take a closer look at the growth of sustainable agriculture and its adoption around the world.

Organic agriculture is booming

One type of sustainable agriculture is **organic agriculture,** which uses no synthetic fertilizers, insecticides, fungicides, or herbicides. In 1990, the U.S. Congress passed the Organic Food Production Act to establish national standards for organic products and facilitate their sale. Under this law, in 2000 the USDA issued criteria by which crops and livestock could be officially certified as organic, and these standards went into effect in 2002 as part of the National Organic Program. California, Washington, and Texas established stricter state guidelines for labeling foods organic, and today many U.S. states and over 80 nations have laws spelling out organic standards.

For farmers, organic farming can bring a number of benefits: lower input costs, enhanced income from higher-value produce, and reduced chemical pollution and soil degradation. Transitioning to organic agriculture does involve some risk, however, because farmers must use organic approaches for three years before their products can be certified as organic and sold at the higher prices commanded by organic foods. One of Kennesaw State University's campus farms that supplies The Commons with fruits and vegetables, for example, had to undergo such a waiting period before being able to label its produce as "organic."

The main obstacle for consumers to organic foods is price. Organic products tend to be 10–30% more expensive than conventional ones, and some (such as milk) can cost twice as much. However, because many consumers are willing to pay more for organic products, grocers and other businesses are making them more widely available.

Today, about 80% of Americans buy organic food at least occasionally, and most retail groceries offer it. U.S. consumers spent more than $35 billion on organic food in 2016, amounting to 5% of all food sales. This is up significantly from 1998, when organic food sales in the United States were only $5 billion and 1% of all food sales. Worldwide, sales of organic food grew similarly fast, increasing nearly fivefold between 2000 and 2016.

Production of organic food is increasing along with demand (FIGURE 7.21). Although organic agriculture takes up less than 1% of agricultural land worldwide, this area is rapidly expanding. Two-thirds of organic acreage is in developed nations. In the United States, the number of certified operations has more than tripled since the 1990s, while acreage devoted to organic crops and livestock has quadrupled. In the European Union, organic crops are grown on more than 5% of agricultural land.

Government policies have aided organic farming. In the United States, the 2014 Farm Bill had a number of provisions that directly aid organic agriculture, including funds to defray certification expenses. The European Union supports farmers financially during conversion to organic agriculture. Once conversion is complete, studies suggest that reduced inputs and higher market prices can make organic farming at least as profitable for the farmer as conventional methods.

Locally supported agriculture is growing

Apart from organic methods, another component of the move toward sustainable agriculture is an attempt to reduce the use of fossil fuels for the long-distance transport of food. The average food product sold in a U.S. supermarket travels at least 1600 km (1000 mi) between the farm and the grocery. Because of the travel time, supermarket produce is often chemically treated to preserve freshness and color.

In response, increasing numbers of farmers and consumers in developed nations are supporting local small-scale agriculture and adopting the motto "think global, eat local." Farmers' markets are springing up throughout North America as people rediscover the joys of fresh, locally grown produce. At **farmers' markets,** consumers buy meats and fresh fruits and vegetables in season from local producers. These markets generally offer a wide choice of organic items and unique local varieties not found in supermarkets. At Kennesaw State University, a weekly, on-campus farmers' market provides students, faculty, and staff with an opportunity to purchase fresh produce while supporting local agriculture.

Some consumers are even partnering with local farmers in a phenomenon called **community-supported agriculture (CSA).** In a CSA program, consumers pay farmers in advance for a share of their yield, usually a weekly delivery of produce. Consumers get fresh seasonal produce, and farmers get a guaranteed income stream up front to invest in their crops—a welcome alternative to taking out loans and being at the mercy of the weather.

Sustainable agriculture provides a roadmap for the future

A variety of approaches exist as pathways to sustainable agriculture. On one end of the spectrum is conventional industrial agriculture. It produces high yields, thanks to intensive inputs of fossil fuels, pesticides, and fertilizers applied to monocultures. At the other end of the spectrum is organic agriculture, which rejects chemical pesticides and fertilizers and takes a low-input approach, accepting lower yields but protecting natural resources that support agriculture in the long term. In between are a host of approaches that modify elements of conventional industrial agriculture and reduce environmental degradation from food production.

Sustainable agriculture, like sustainability itself, involves a triple bottom line of social, economic, and environmental dimensions (p. 115). Sustainable agriculture consists of agriculture that provides food security to society, that is profitable enough to provide farmers and ranchers a viable living, and that conserves resources adequate to support future agriculture. While many paths can be embraced to reach this goal, it is imperative to follow them if we are to leave a productive, ecologically healthy planet to future generations.

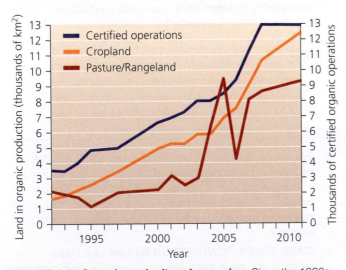

FIGURE 7.21 Organic agriculture is growing. Since the 1990s, U.S. acreage devoted to organic crops and livestock has quadrupled, and certified operations have more than tripled. *Data from USDA Economic Research Service.*

closing THE LOOP

Even if we have never set foot on a farm, milked a cow, or shorn a sheep, we depend on agriculture for our daily needs, including food and clothing. It is thus important for all of us to help ensure that the world's agricultural systems are sound and sustainable. This means safeguarding the quality and availability of resources on which crops and livestock rely, including soil, water, nutrients, and pollinators. When students at Kennesaw State University enter their dining hall, they actually know where a good portion of their food comes from, and how it has been grown. By sourcing produce locally at its campus farms, The Commons at KSU provides wholesome, delicious food produced in sustainable ways. With a minimum of fossil fuel consumption, chemical use, and processing, Kennesaw State supplies its students with food choices that have a light footprint, thus protecting the region's soil, water, and other agricultural resources.

Today many colleges and universities are managing campus farms, serving organic food, running trayless dining halls, composting food scraps, reusing waste oil as biodiesel, and even growing food for the communities around them. Chances are that you, as a college student, can take part in such activities and play a role in helping to reduce the ecological footprint of modern agriculture.

If our planet is to support well over 9 billion people by mid-century without further degrading the soil, water, pollinators, and other resources and ecosystem services that support our food production, we must find ways to shift to sustainable agriculture. Biological pest control, organic agriculture, pollinator conservation, preservation of native crop diversity, sustainable aquaculture and meat production, and likely some degree of careful and responsible genetic modification of food may all be parts of the game plan we will need to achieve a sustainable future.

TESTING Your Comprehension

1. Identify patterns in global food security from 1970 to the present, and describe the techniques people have used to increase agricultural production.

2. Compare and contrast the methods used in traditional and industrial agriculture. How does sustainable agriculture differ from industrial agriculture?

3. How are soil horizons created? List and describe the major horizons in a typical soil profile. How is organic matter distributed in a typical soil profile?

4. Explain how overirrigation can damage soils and reduce crop yields, and describe how irrigation processes can be made more efficient.

5. Explain the differences between inorganic and organic fertilizers. How do fertilizers boost crop growth? Describe approaches that act to reduce nutrient runoff into waterways.

6. Name three human activities that can promote soil erosion. Describe several farming techniques (such as terracing and no-till farming) that can help to reduce the risk of erosion.

7. Compare and contrast the approaches to controlling pests used in biological control, in integrated pest management, and in conventional industrial agriculture. Explain how agricultural pests can develop resistance to pesticides.

8. Name several positive and negative environmental consequences of raising animals for food in feedlot operations and aquaculture.

9. How is a transgenic organism created? How is genetic engineering different from traditional agricultural breeding? How is it similar?

10. Describe the recent growth of organic and locally supported agriculture in the United States.

SEEKING Solutions

1. Select two techniques or approaches described in this chapter that you think are especially effective in sustaining agricultural resources—for instance, in conserving soil or water, or in minimizing nutrient pollution—and describe how each sustains the resource.

2. Describe and assess several ways in which high-input industrial agriculture can be beneficial for the environment, and several ways in which it can be detrimental. Suggest several ways in which we might modify industrial agriculture to reduce its environmental impacts.

3. What factors make for an effective biological control strategy of pest management? What risks are involved in using a biocontrol approach? If you had to decide whether to use biocontrol against a particular pest, what questions would you want to have answered before you decide?

4. **CASE STUDY CONNECTION** As part of a class project, you have been asked to analyze your campus dining services and to provide a list of approaches that could increase the sustainability of the dining service. If your institution does not have on-campus

dining operations, assume such a facility is in the planning stages and that your input will affect its design and operation. Create a prioritized list of the following factors (and any others you'd like to offer) to present to the director of campus dining: the establishment of a campus farm; preferentially purchasing food from local farms; using "trayless" dining to reduce waste and conserve water; converting used cooking oil to biodiesel; composting organic wastes; and shortening the dining hall's hours of operation to conserve energy.

Provide a rationale for how you ranked the items in the list.

5. **THINK IT THROUGH** You are a USDA official and must decide whether to allow the planting of a new genetically modified strain of cabbage that produces its own pesticide and has twice the vitamin content of regular cabbage. What questions would you ask of scientists before deciding whether to approve the new crop? What scientific data would you want to see? Would you also consult nonscientists or consider ethical, economic, and social factors?

CALCULATING Ecological Footprints

Many people who want to reduce their ecological footprint have focused on how much energy is expended (and how many climate-warming greenhouse gases are emitted) in transporting food from its place of production to its place of sale. The typical grocery store item is shipped by truck, air, and/or sea for many hundreds of miles before reaching the shelves, and this transport consumes petroleum. This concern over "food-miles" has helped drive the "locavore" movement to buy and eat locally sourced food. However, food's

transport from producer to retailer, as measured by food-miles, is just one source of carbon emissions in the overall process of producing and delivering food.

In 2008, environmental scientists Christopher Weber and H. Scott Mathews conducted a thorough analysis of U.S. food production *and* delivery. By filling in the table below, you will get a better idea of how our dietary choices contribute to climate change.

FOOD TYPE	TOTAL EMISSIONS[1] ACROSS LIFE CYCLE	EMISSIONS[1] FROM DELIVERY[2]	PERCENTAGE EMISSIONS FROM DELIVERY[2,3]
Fruits and vegetables	0.85	0.10	11.8
Cereals and carbohydrates	0.90	0.07	
Dairy products	1.45	0.03	
Chicken/fish/eggs	0.75	0.03	
Red meat	2.45	0.03	
Beverages	0.50	0.04	

[1]"Emissions" are measured in metric tons of carbon dioxide equivalents, per household per year.
[2]"Delivery" means transport from producer to retailer.
[3]"Percentage emissions from delivery" is calculated by dividing emissions from delivery/total emissions across life cycle, and then multiplying by 100 (to convert proportion to percentage).

Source: Weber, C.L., and H.S. Mathews, 2008. Food-miles and the relative climate impacts of food choices in the United States. Environmental Science and Technology 42: 3508–3513.

1. Which type of food is responsible for the most greenhouse gas emissions across its whole life cycle? Which type is responsible for the least emissions?

2. What is the range of values for percentage of emissions from transport of food to retailers (delivery)? Weber and Mathews found that 83% of food's total emissions came from its production process on the farm or feedlot. What do these numbers tell you about how you might best reduce your own footprint with regard to food?

3. After measuring mass, energy content, and dollar value for each food type, the researchers calculated emissions per kilogram, calorie, and dollar. In every case, red meat produced the most emissions, followed by dairy products and chicken, fish, and eggs. They then calculated that shifting one's diet from meat and dairy to fruits, vegetables, and grains for just one day per week would reduce emissions as much as eating 100% locally (cutting food-miles to zero) all the time. Knowing all this, how would you choose to reduce your own food footprint? By how much do you think you could reduce it?

Mastering Environmental Science

CHAPTER 8

Biodiversity and Conservation Biology

Will We Slice through the Serengeti?

> " Construction of the road will be a huge relief for us. We will sell . . . maize and horticultural products to our colleagues in Arusha and they will bring us cows and goats.
> —Bizare Mzazi, a farmer outside Serengeti National Park

> If we construct this road, all our rhinos will disappear. . . . We should strive to conserve our heritage for future generations.
> —Sirili Akko, executive officer of the Tanzania Association of Tour Operators "

It's been called the greatest wildlife spectacle on Earth. Each year more than 1.2 million wildebeest migrate across the vast plains of the Serengeti in East Africa, along with more than 700,000 zebras and hundreds of thousands of antelope. The herds can stretch as far as the eye can see. Packs of lions track the procession and pick off the weak and the unwary, while hungry crocodiles wait in ambush at river crossings. After bearing their calves during the wet season, the wildebeest journey north to find fresh grass. The great herds spend the dry season at the northern end of the Serengeti ecosystem, and then head back south to complete their cyclical annual journey.

This epic migration, with its dramatic interplay of predators and prey, has cycled on for millennia. Yet today, the entire phenomenon may be threatened by a proposal to build a commercial highway across the Serengeti, slicing straight across the animals' migratory route.

Before examining the highway proposal, let's step back for a broad view of the Serengeti. The people native to this region, the Maasai, are semi-nomadic herders who have long raised cattle on the grasslands and savannas. Because the Maasai subsist on their cattle and have lived at low population densities, wildlife thrived here long after it had declined in other parts of Africa.

When East Africa was under colonial rule, the British created game reserves to conserve wildlife for their own hunting. Tanzania, Kenya, and other African nations gained independence in the mid-20th century, and the British reserves became the basis for today's national protected areas. Serengeti National Park was established in 1951, and the Maasai Mara National Reserve was later created just across the border in Kenya. These two protected areas, together with several adjacent ones, encompass the Serengeti ecosystem. This 30,000-km² (11,500-mi²) region is one of the last places on the planet where an ecosystem remains nearly intact and functional over a vast area.

Today 2 million people from around the world visit Tanzania and Kenya each year, most of them ecotourists who visit the parks and protected areas. Serengeti National Park alone receives 800,000 annual visitors. Tourism injects close to $3 billion into these nations' economies and creates jobs for tens of thousands of local people. Because the region's people see that functional ecosystems full of wildlife bring foreign dollars into their communities, many support the parks. Indeed, East Africa has been at the forefront of community-based conservation

A Maasai man herding his cattle ▲

Upon completing this chapter, you will be able to:

- Characterize the scope of biodiversity on Earth
- Specify the benefits that biodiversity brings us
- Discuss today's extinction crisis in geologic context
- Evaluate the primary causes of biodiversity loss
- Assess the science and practice of conservation biology
- Analyze efforts to conserve threatened and endangered species
- Compare and contrast conservation efforts above the species level

Wildebeest crossing the vast plains of the Serengeti

(p. 185), in which local people act as stewards managing their own natural resources, often in collaboration with international conservationists.

However, most people living in northern Tanzania remain desperately poor and live without electricity or medical care. Farmers, villagers, and townspeople along the shores of Lake Victoria feel isolated from the rest of Tanzania by a poor road system. Walled off by Serengeti National Park to their east (which does not allow commercial truck traffic on its few dirt roads), these people have little access to outside markets to buy and sell goods. In response, Tanzania's president Jakaya Kikwete promised to build a paved commercial highway across the Serengeti, connecting Lake Victoria communities with cities to the east and ports on the Indian Ocean. The World Bank and the German government offered to finance the $480-million project, and Chinese contractors stood ready to build it.

Around the world, conservationists reacted with alarm. The proposed highway would slice right through the middle of the wildebeest migration path (**FIGURE 8.1**). Scientists predicted that the road would block migration and that vehicles would kill countless animals in collisions. A highway would also provide access for poaching (the illegal killing of wildlife for meat or body parts) and would allow an entry corridor for exotic plant species that could invade the ecosystem. A highway would encourage human settlement right up to the park boundary, making the park an island of habitat hemmed in by agriculture, housing, and commerce. And by boosting development, a highway could spur the towns along Lake Victoria to grow into large cities, creating demand for still-larger transportation corridors in the future. For all these reasons, experts predicted that the highway would diminish animal populations and possibly destroy the migration spectacle.

Such an outcome could devastate tourism, so the region's tourism operators opposed the highway. So did most Kenyans, who feared that the highway would prevent migratory animals from reaching Kenya's Maasai Mara Reserve. In 2010 a Kenyan nongovernmental organization, the African Network for Animal Welfare, sought to stop the highway with a lawsuit in the East Africa Court of Justice, a body that adjudicates international matters in the region. After various hearings and appeals, this court in 2014 issued a final ruling prohibiting the project from going forward.

Meanwhile, international pressure rose on Tanzania to abandon its plans. Highway opponents proposed an alternative route that would wrap around the Serengeti's southern end, passing through more towns and serving five times as many people along the way. The World Bank and the German government offered to help fund this alternative route instead.

In 2016, John Magufuli took office as Tanzania's new president. Late that year he resurrected the original highway proposal and authorized construction of the highway through the park. Again, the diverse coalition of highway opponents raised an international outcry, and within weeks Magufuli backed down.

Meanwhile, a new plan kicked off debate. The nation of Uganda wanted to export oil from its recently developed oilfields eastward to ports on the Indian Ocean. In 2016, three routes for an oil pipeline were proposed, and the most direct option would cut straight across Serengeti National Park.

Such an oil pipeline (accompanied by a road) would create a barrier to migratory animals just as a highway would. It would bring all the impacts of a highway, along with risks of oil spills. So, when news of the pipeline project became public, scientists and conservationists again rushed to express opposition to a development corridor through the park. The non-profit group Serengeti Watch said it would sue to stop the proposal in the East Africa Court of Justice.

Faced with the opposition, Magufuli's government announced that any pipeline route would avoid passing through game reserves and national parks. Officials instead began studying routes around the Serengeti.

Today in Tanzania, poaching is on the rise, and animal populations are falling. The Serengeti is one of our planet's last intact large ecosystems, so impacts here have global ramifications for biological diversity on Earth. We would all be impoverished if the Serengeti's biodiversity were lost, so we must hope that Africans can find ways to improve their standard of living while conserving their wildlife and natural systems. East Africa has helped to pioneer win-win solutions in conservation thus far, so there is hope that it will show the way yet again.

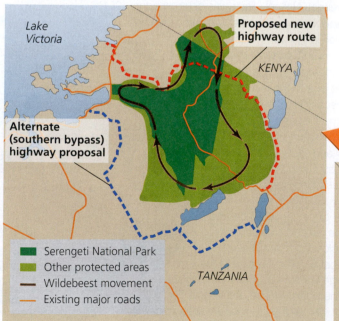

- 🟩 Serengeti National Park
- 🟩 Other protected areas
- ▬ Wildebeest movement
- ▬ Existing major roads

FIGURE 8.1 A proposed highway would slice through Serengeti National Park. It would increase commerce and connect Tanzanian people on each side, but would cut across the migration route for wildebeest and other animals. Highway opponents suggest an alternate route around the park's southern edge.

Life's Diversity on Earth

Rising human population and resource consumption are putting ever-greater pressure on the flora and fauna of our planet. We are diminishing the ultimate source of our civilization's wealth and happiness: Earth's diversity of life, the very quality that makes our planet unique. Thankfully, many people around the world are working tirelessly to save threatened animals, plants, and ecosystems in efforts to stop the loss of our planet's priceless biological diversity.

Biodiversity encompasses multiple levels

Biological diversity, or **biodiversity,** is the variety of life across all levels of biological organization. It includes diversity in species, genes, populations, communities, and ecosystems (**FIGURE 8.2**). The level that people find easiest to visualize and that we refer to most commonly is species diversity.

Species diversity A **species** is a distinct type of organism, a set of individuals that uniquely share certain characteristics and can breed with one another and produce fertile offspring (p. 50). Species form by the process of speciation (p. 53) and may disappear by extinction (p. 58). **Species diversity** describes the number or variety of species found in a particular area. One component of species diversity is *species richness,* the number of species inhabiting an area. Another is *evenness* or *relative abundance,* the degree to which species in a given area differ in numbers of individuals (greater evenness means they differ less).

Biodiversity exists below the species level in the form of *subspecies,* populations of a species that occur in different geographic areas and differ from one another in slight ways. Subspecies arise by the same processes that drive speciation but result when divergence stops short of forming separate species. As an example, the black rhinoceros diversified into about eight subspecies, each inhabiting a different part of Africa. The eastern black rhino, which is native to Kenya and Tanzania, differs in its attributes from each of the other subspecies.

Genetic diversity Scientists designate subspecies when they recognize substantial genetically based differences among individuals from different populations of a species. However, all species consist of individuals that vary genetically from one another to some degree, and this variation is an important component of biodiversity. **Genetic diversity** encompasses the differences in DNA composition (p. 32) among individuals, and these differences provide the raw material for adaptation to local conditions. In the long term, populations with more genetic diversity may be more likely to persist, because their variation better enables them to cope with environmental change.

Populations with little genetic diversity are vulnerable to environmental change if they lack genetic variants to help them adapt to changing conditions. Populations with low genetic diversity may also show less vigor, be more vulnerable to disease, and suffer *inbreeding depression,* which occurs when

Ecosystem diversity

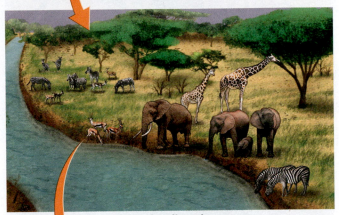

Species diversity

Genetic diversity

FIGURE 8.2 The concept of biodiversity encompasses multiple levels in the hierarchy of life.

genetically similar parents mate and produce weak or defective offspring. Scientists have sounded warnings over low genetic diversity in species that have dropped to low population sizes, including American bison, elephant seals, and the cheetahs of the East African plains. Diminished genetic diversity in our crop plants is a prime concern to humanity (p. 142).

Ecosystem diversity **Ecosystem diversity** refers to the number and variety of ecosystems (pp. 36, 59), but biologists may also refer to the diversity of communities (pp. 59, 73) or habitats (p. 60). Scientists may also consider the geographic arrangement of habitats, communities, or ecosystems across a landscape, including the sizes and shapes of patches and the connections among them (p. 38). Under any of these concepts, a seashore of beaches, forested cliffs, offshore coral reefs, and ocean waters would hold far more biodiversity than the same acreage of a monocultural cornfield. A mountain slope whose vegetation changes with elevation from desert to forest to alpine meadow would hold more biodiversity than a flat area the same size consisting of only desert, forest, or meadow.

The Serengeti region holds a diversity of habitats, including savanna (p. 87), grassland (p. 85), hilly woodlands, seasonal wetlands, and rock outcroppings. This habitat diversity contributes to the rich diversity of species in the region.

Biodiversity is unevenly distributed

In numbers of species, insects show a staggering predominance over all other forms of life (**FIGURE 8.3**). Among insects, about 40% are beetles, and beetle species alone outnumber all non-insect animal species and all plant species. No wonder the British biologist J.B.S. Haldane famously quipped that God must have had "an inordinate fondness for beetles."

Biodiversity is also greater in some places than others. Near the equator, greater amounts of solar energy, heat, and humidity spur plant growth, making tropical regions more productive than temperate regions and consequently better able to support larger numbers of organisms. Species diversity generally is higher near the equator, likely because the steady amount of sunlight year-round and the relatively stable climates of tropical regions allow numerous species to coexist. Whereas variable environmental conditions favor generalists (species that can tolerate a wide range of circumstances; p. 60), stable conditions favor specialists (species highly adapted to particular circumstances; p. 60).

Structurally diverse habitats tend to create more ecological niches (p. 60) and support greater species diversity. For instance, forests generally support more diversity than grasslands. For any given area, species diversity tends to increase with diversity of habitats, because each habitat supports a different mix of organisms.

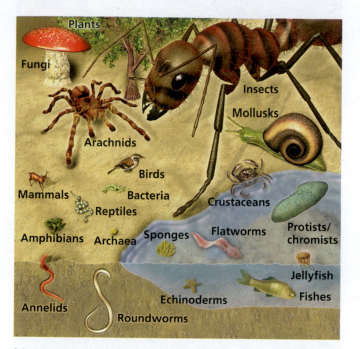

(a) Organisms scaled in size according to species richness

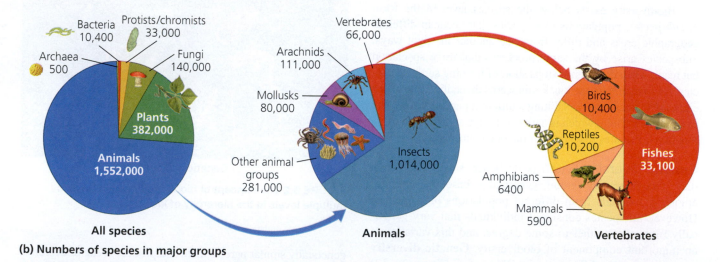

(b) Numbers of species in major groups

FIGURE 8.3 Some groups contain more species than others. The illustration **(a)** shows organisms scaled in size to the number of species known from each group, giving a visual sense of their species richness. The pie charts **(b)** show that most species are animals and that nearly two-thirds of animals are insects (whereas vertebrates make up only 4%). *Data from Roskov, Y., et al. (eds.), 2017. Species 2000 & ITIS catalogue of life, 30 June 2017. Digital resource at www.catalogueoflife.org/col. Leiden, the Netherlands: Species 2000: Naturalis.*

DATA Q • What percentage of vertebrate species do mammal species make up? • What percentage of animal species are mammals? • What percentage of the world's total species are mammals? • How many insect species exist for every mammal species?

Go to **Interpreting Graphs & Data** on **Mastering** Environmental Science

Human disturbance often creates patchwork combinations of habitats. This increases habitat diversity locally; so, in moderately disturbed areas, species diversity generally rises. However, at larger scales, human disturbance can decrease diversity if it replaces regionally unique habitats with homogenized disturbed habitats, causing many specialist species to disappear while widespread generalist species thrive. Moreover, species that rely on large expanses of habitat disappear when those habitats are fragmented by human disturbance.

Many species await discovery

We still are profoundly ignorant of the number of species that exist. So far, scientists have identified and described about 1.8 million species of plants, animals, fungi, and microorganisms. However, estimates for the total number that actually exist range from 3 million to 100 million, with the most widely accepted estimates in the neighborhood of 14 million.

Our knowledge of species numbers is incomplete for several reasons. First, many species are tiny and easily overlooked. These include bacteria, nematodes (roundworms), fungi, protists, and soil-dwelling arthropods. Second, many organisms are difficult to identify; sometimes, organisms that are thought to be of the same species turn out to be different species once biologists examine them more closely. Third, some areas of Earth remain little studied. We have barely sampled the ocean depths, hydrothermal vents (p. 35), or the tree canopies and soils of tropical forests. There remain many frontiers on our planet to explore!

Benefits of Biodiversity

These days most of us live in cities and suburbs, spend nearly all our time indoors, and pass hours each day staring at electronic screens. It's no wonder that we often fail to appreciate how biodiversity relates to our lives. Yet we benefit from biodiversity and it supports our society in fundamental ways. Indeed, our cities, homes, and technology would simply not exist without the resources and services that Earth's living species provide us.

Biodiversity enhances food security

Biodiversity provides the food we eat. Throughout our history, human beings have used at least 7000 plant species and several thousand animal species for food. Today industrial agriculture has narrowed our diet. Globally, we now get 90% of our food from just 15 crop species and 8 livestock species, and this lack of diversity leaves us vulnerable to crop failures. In a world where 800 million people go hungry, we can improve food security (the guarantee of an adequate, safe, nutritious, and reliable food supply; p. 140) by finding sustainable ways to harvest or farm wild species and rare crop varieties.

TABLE 8.1 shows a selection of promising wild food resources from just one region of the world—Central and South America. Plenty of additional new or underused food sources exist there and elsewhere worldwide. As examples, the babassu palm of the Amazon produces more vegetable

TABLE 8.1 Potential New Food Sources[*]

Amaranths
(three species of *Amaranthus*)

Grain and leafy vegetable; livestock feed; rapid growth, drought resistant

Capybara
(*Hydrochoeris hydrochaeris*)

World's largest rodent; meat esteemed; easily ranched in open habitats near water

Buriti palm
(*Mauritia exuosa*)

"Tree of life" to Amerindians; vitamin-rich fruit; pith as source for bread; palm heart from shoots

Vicuna
(*Lama vicugna*)

Threatened species related to llama; source of meat, fur, and hides; can be profitably ranched

Maca
(*Lepidium meyenii*)

Cold-resistant root vegetable resembling radish, with distinctive flavor; near extinction

Chachalacas
(*Ortalis*, many species)

Tropical birds; adaptable to human habitations; fast growing

*The wild species shown here—all native to Latin America—are just a few of the many plants and animals that could supplement our food supply.

Adapted from Wilson, E.O., 1992. The diversity of life. Cambridge, MA: Belknap Press.

oil than any other plant. The serendipity berry generates a sweetener 3000 times sweeter than table sugar. Some salt-tolerant grasses and trees are so hardy that farmers can irrigate them with saltwater to produce animal feed and other products.

Moreover, the wild relatives of our crop plants hold reservoirs of genetic diversity that can help protect the crops we grow in monocultures by providing helpful genes for crossbreeding or genetic engineering (p. 142). We have already received tens of billions of dollars' worth of disease resistance from the wild relatives of potatoes, wheat, corn, barley, and other crops.

Organisms provide drugs and medicines

People have made medicines from plants and animals for centuries, and about half of today's pharmaceuticals are derived from chemical compounds from wild plants (TABLE 8.2). A well-known example is aspirin, which was derived from compounds found in willows and meadowsweet from Europe and the Middle East.

Each year, pharmaceutical products owing their origin to wild species generate up to $150 billion in sales and save thousands of human lives. The world's biodiversity holds an even greater treasure chest of medicines still to be discovered. Yet with every species that goes extinct, we lose one more opportunity to find cures and treatments.

Biodiversity provides ecosystem services

Contrary to popular opinion, some things in life can indeed be free—as long as we protect the ecological systems that provide them. Forests cleanse air and water, while buffering us against floods. Native crop varieties insure us against disease and drought. Wildlife can attract tourism that generates income for people. Intact ecosystems provide these and other valuable processes, known as ecosystem services (pp. 4, 39), for all of us, free of charge. According to scientists, biodiversity helps to:

- Provide food, fuel, fiber, and shelter.
- Purify air and water.
- Detoxify and decompose wastes.
- Stabilize Earth's climate.
- Moderate floods, droughts, and temperatures.
- Cycle nutrients and renew soil fertility.
- Pollinate plants, including many crops.
- Control pests and diseases.
- Maintain genetic resources for crop varieties, livestock breeds, and medicines.
- Provide cultural and aesthetic benefits.

In these ways, organisms and ecosystems support vital processes that people cannot replicate or would need to pay

---

TABLE 8.2 Natural Plant Sources of Pharmaceuticals[*]

Pineapple
(*Ananas comosus*)

Drug: Bromelain
Application: Controls tissue inflammation

Pacific yew
(*Taxus brevifolia*)

Drug: Taxol
Application: Anticancer agent (especially ovarian cancer)

Autumn crocus
(*Colchicum autumnale*)

Drug: Colchicine
Application: Anticancer agent

Velvet bean
(*Mucuna pruriens*)

Drug: L-Dopa
Application: Parkinson's disease suppressant

Yellow cinchona
(several species of *Cinchona*)

Drug: Quinine
Application: Antimalarial agent

Common foxglove
(*Digitalis purpurea*)

Drug: Digitoxin
Application: Cardiac stimulant

[*]Shown are just a few of the many plants that provide chemical compounds of medical benefit.
Adapted from Wilson, E.O., 1992. The diversity of life. *Cambridge, MA: Belknap Press.*

for if nature did not provide them. The global economic value of just 17 ecosystem services has been estimated at more than $148 trillion per year (p. 99).

Biodiversity helps maintain functioning ecosystems

Ecological research demonstrates that biodiversity tends to enhance the stability of communities and ecosystems. Research has also found that biodiversity tends to increase the resilience (p. 77) of ecological systems—their ability to withstand disturbance, recover from stress, or adapt to change. Thus, the loss of biodiversity can diminish a natural system's ability to function and to provide services to our society.

Will the loss of a few species really make much difference in an ecosystem's ability to function? Consider a metaphor first offered by Paul and Anne Ehrlich (p. 122): The loss of one rivet from an airplane's wing—or two, or three—may not cause the plane to crash. But as rivets are removed the structure will be compromised, and eventually the loss of just one more rivet will cause it to fail.

Research shows that removing a keystone species (p. 76) such as a top predator can significantly alter an ecological system. Think of lions, leopards, and cheetahs on the Serengeti—or wolves, mountain lions, and grizzly bears at Yellowstone National Park (a place sometimes called "America's Serengeti"). These predators prey on herbivores that consume many plants. The removal of a top predator can have consequences that multiply as they cascade down the food chain.

Likewise, losing an "ecosystem engineer" (such as ants or earthworms; p. 77) can have major effects. For example, elephants eat and trample many plants, helping to maintain the open structure of Africa's savannas. Scientists have found that when elephants are removed (as by illegal hunting), the landscape fills in with scrubby vegetation, converting the savanna into a dense scrub forest and affecting countless other species.

Ecosystems are complex, however, and it is difficult to predict which species may be most influential. Thus, many people prefer to apply the precautionary principle (p. 161) in the spirit of Aldo Leopold (p. 16), who advised, "To keep every cog and wheel is the first precaution of intelligent tinkering."

Biodiversity boosts economies through tourism and recreation

When people travel to observe wildlife and explore natural areas, they create economic opportunities for area residents. Visitors spend money at local businesses, hire local people as guides, and support parks that employ local residents. The parks and wildlife of Kenya and Tanzania are prime examples. Ecotourism (p. 65) brings in fully a quarter of all foreign money entering Tanzania's economy each year. Leaders and citizens in both nations who recognize biodiversity's economic benefits have managed their parks and reserves diligently.

Ecotourism is a vital source of income for many nations, including Costa Rica, with its rainforests; Australia, with its Great Barrier Reef; and Belize, with its caves and coral reefs. The United States, too, benefits from ecotourism; its national parks draw millions of visitors from around the world. Although excessive development of infrastructure for ecotourism can damage the natural assets that draw people, ecotourism can provide a powerful financial incentive for nations, states, and local communities to preserve natural areas and reduce impacts on the landscape and on native species.

People value connections with nature

Not all of biodiversity's benefits to people can be expressed in the hard numbers of economics or the practicalities of food and medicine. Some scientists and philosophers argue that people find a deeper value in biodiversity. Harvard University biologist Edward O. Wilson has popularized the notion of **biophilia,** asserting that human beings share an instinctive love for nature and feel an emotional bond with other living things (**FIGURE 8.4**). As evidence of biophilia, Wilson and others cite our affinity for parks and wildlife, our love for pets, the high value of real estate with a view of natural landscapes, and our interest in hiking, bird-watching, fishing, hunting, backpacking, and other outdoor pursuits.

Indeed, a love for nature and biodiversity appears to be good for us: We thrive mentally and physically when we have access to nature, and we suffer when we don't. As children in recent years have been increasingly deprived of outdoor experiences and contact with wild organisms, writer Richard Louv argues that they suffer what he calls "nature-deficit disorder." In his 2005 book, *Last Child in the Woods,* Louv maintained that an alienation from nature and biodiversity damages childhood development and may lie behind much of the angst and anxiety felt by young people today. Since then, many researchers have begun to study this question scientifically. In a 2017 book, *The Nature Fix,* writer Florence Williams surveyed the growing scientific evidence that exposure to nature revitalizes our bodies and our brains. Interviewing scientists, assessing research, and presenting stories from around the world, her book shows how access to wildlife and green spaces can relieve stress, bring us happiness, make us mentally sharper, and improve our physical health.

FIGURE 8.4 An Indonesian girl gazes into a flower of *Rafflesia arnoldii*, the largest flower in the world. Biophilia holds that human beings have an instinctive love and fascination for nature and a deep-seated desire to affiliate with other living things.

Do we have ethical obligations toward other species?

Aside from all of biodiversity's pragmatic benefits, many people feel that living organisms simply have an inherent right to exist. Human beings are part of nature, and like any other animal we need to use resources and consume other organisms to survive. However, we also have conscious reasoning ability and can make deliberate decisions. Our ethical sense has developed from this intelligence and ability to choose. As our society's sphere of ethical consideration has widened over time, and as more of us take up biocentric or ecocentric worldviews (p. 15), more people have come to feel that other organisms have intrinsic value. In this view, the conservation of biodiversity is justified on ethical grounds alone.

Biodiversity Loss and Extinction

Despite our society's expanding ethical breadth and despite the many clear benefits that biodiversity brings us, the future of many species remains far from secure. In today's fast-changing world, every corner of our planet has been touched in some manner by human impact, and biological diversity is being rapidly lost.

Human disturbance creates winners and losers

We affect ecosystems and landscapes in many ways, and this creates both "winners" and "losers" among the world's plants and animals. In general, when we alter natural systems we tend to make each area more similar to other areas. This is because we spread into diverse natural environments and then shape them to our own species' particular tastes and needs. We tend to make landscapes more open in structure, by clearing vegetation to make room for farms, pastures, towns, and cities. And we frequently create pollution. Because the overall nature of our impacts is similar across regions and cultures, certain types of organisms tend to do well in our wake, whereas other types tend not to do well. As a result, the species that benefit from the changes we make—and the species that are harmed—each tend to show predictable sets of attributes (TABLE 8.3).

"Winning" species tend to be generalists able to fill many niches, tolerate disturbance, and use open habitats or edges. The house mouse (*Mus musculus*) is one example. This small, fast-reproducing mammal thrives by living and feeding in and near our buildings. In contrast, "losing" species tend to be those that specialize on certain resources, have trouble coping with change, and rely on mature and well-vegetated habitats. The tiger (*Panthera tigris*) is such a species. Large, slow-reproducing, and high in the food chain, it needs huge areas of mature habitat full of prey and free of people. Geographically widespread species stand a much better chance to succeed in a changing world undergoing human impact than species limited to small areas, and mainland species tend to do better than island species.

Many populations are declining

As the population size of a species shrinks, the species encounters two problems. First, it loses genetic diversity; and second, its geographic range tends to become smaller as it disappears from parts of its range. Both problems make a population vulnerable to further declines. Many species today are less numerous and occupy less area than they once did. For example, studies have documented significant population declines among large mammals of the Serengeti in recent years, due to a variety of human impacts.

To quantify such change globally, scientists at the World Wildlife Fund and the United Nations Environment Programme (UNEP) developed the *Living Planet Index*. This index expresses how large the average population size of a species is now, relative to its size in the year 1970. The most recent compilation summarized trends from 14,152 populations of 3706 vertebrate species that are sufficiently monitored. Between 1970 and 2012, the Living Planet Index fell by 58%—meaning that on average, population sizes became 58% smaller (FIGURE 8.5). This suggests that in the mere lifetimes of the authors of this textbook, Earth has lost the majority of its vertebrate animals.

Extinction is irreversible

When a population declines to a very low level, extinction becomes a possibility. **Extinction** (p. 58) occurs when the last member of a species dies and the species ceases to exist. The disappearance of a particular population from a given area, but not the entire species globally, is referred to as "local extinction" or **extirpation**. Extirpation is an erosive process that can, over time, lead to extinction. The black rhinoceros has been extirpated from most of its historic range across Africa (FIGURE 8.6), but as a species it is not yet extinct. However, at least three of its subspecies are extinct.

TABLE 8.3 Characteristics of Winning and Losing Species

WINNERS TEND TO BE	LOSERS TEND TO BE
• Generalists, using many resources or habitats • Geographically widespread • Users of open, early successional habitats • Able to cope with fast-changing conditions • Small and fast-reproducing • Low on the food chain • Not in need of large areas of habitat • Mainland species	• Specialists on certain resources or habitats • Limited to a small range • Users of mature, dense habitats • Needing stable conditions • Large and slow-reproducing • High on the food chain • Needing large areas of habitat • Island species

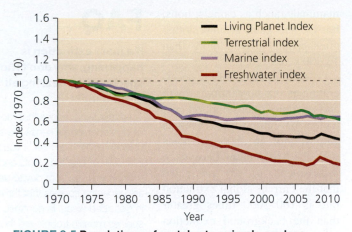

FIGURE 8.5 **Populations of vertebrate animals are less than one-half the size today that they were just 42 years ago.** Between 1970 and 2012, the Living Planet Index fell by 58%. The index for terrestrial species fell by 38%; for marine species, by 36%; and for freshwater species, by 81%. *Data from WWF, 2016.* Living planet report 2016. *Gland, Switzerland: WWF International.*

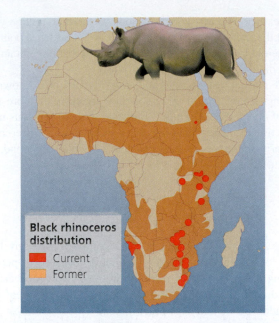

FIGURE 8.6 **The black rhinoceros has disappeared from most of its range across Africa.** Poaching is a major cause (p. 178). *Based on data from Deon Furstenburg, wildliferanching.com, and other sources.*

Human impact is responsible for most extirpation and extinction today, but these processes also occur naturally at a much slower rate. If species did not naturally go extinct, our world would be filled with dinosaurs, trilobites, ammonites, and millions of other creatures that vanished from our planet long before humans appeared. Paleontologists estimate that roughly 99% of all species that ever lived are now extinct.

Most extinctions preceding the appearance of human beings occurred singularly for independent reasons, at a pace referred to as the **background extinction rate** (p. 58). By studying traces of organisms preserved in the fossil record (p. 58), scientists infer that for mammals and marine animals, each year, on average, 1 species out of every 1–10 million has vanished.

Earth has experienced five mass extinction events

Extinction rates rose far above this background rate at several points in Earth's history. In the past 440 million years, our planet has experienced five **mass extinction events** (p. 58). Each event eliminated more than one-fifth of life's families (p. 55) and at least half its species (**FIGURE 8.7**). The most severe episode occurred at the end of the Permian period (see **APPENDIX E**). At this time, about 250 million years ago, close to 90% of all species

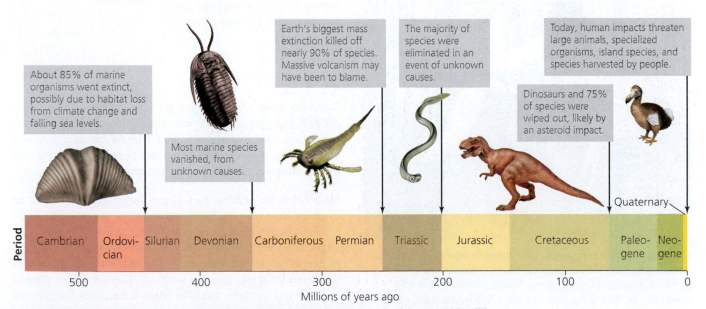

FIGURE 8.7 **Scientists have documented five mass extinction events in the past 500 million years.** Ongoing study of geology and the fossil record (p. 58) is revealing clues about the causes and consequences of each event. Today our impacts are threatening to cause a sixth mass extinction.

went extinct. The best-known episode occurred 66 million years ago at the end of the Cretaceous period, when evidence points to an asteroid impact (and possibly volcanism) that brought an end to the dinosaurs and many other groups.

If current trends continue, our modern era, known as the Quaternary period, may see the extinction of more than half of all species. Although similar in scale to previous mass extinctions, today's ongoing mass extinction is different in two primary respects. First, we are causing it. Second, we will suffer as a result.

We are setting the sixth mass extinction in motion

In the past few centuries alone, we have recorded hundreds of instances of species extinction caused by people. Among North American birds in the past two centuries, we have driven into extinction the Carolina parakeet, great auk, Labrador duck, and passenger pigeon (p. 61); almost certainly the Bachman's warbler and Eskimo curlew; and likely the ivory-billed woodpecker (**FIGURE 8.8**). Several more species, including the whooping crane, Kirtland's warbler, and California condor (p. 184), teeter on the brink of extinction.

People have been hunting species to extinction for thousands of years. Archaeological evidence shows that in case after case, a wave of extinction followed close on the heels of human arrival on islands and continents. After Polynesians reached Hawai'i, half its birds went extinct. Birds, mammals, and reptiles vanished following human arrival on many other oceanic islands, including huge landmasses such as New Zealand and Madagascar. Dozens of species of large vertebrates died off in Australia after people arrived roughly 50,000 years ago. North America lost 33 genera of large

FIGURE 8.8 The ivory-billed woodpecker was one of North America's most majestic birds. It lived in old-growth forests of the southeastern United States. Forest clearing and timber harvesting eliminated the mature trees it needed for food, shelter, and nesting, and this symbol of the South appeared to go extinct. Recent fleeting, controversial reports raised hopes that the species persists, but proof has been elusive.

mammals (such as camels, lions, horses, and giant ground sloths) after people arrived more than 13,000 years ago.

Today, species loss is accelerating as our population growth and resource consumption put increasing strain on habitats and wildlife. In 2005, scientists with the Millennium Ecosystem Assessment (p. 95) calculated that the current global extinction rate is 100–1000 times greater than the background extinction rate, and rising.

To monitor threatened and endangered species, the International Union for Conservation of Nature (IUCN) maintains the Red List, a regularly updated list of species facing high risks of extinction. As of 2017, the Red List reported that fully 20% of the 64,000 species with data adequate to assess were threatened with extinction. This included 21% (1194) of mammal species, 13% (1460) of bird species, 20% (1090) of reptile species, 32% (2067) of amphibian species, and 14% (2153) of fish species. In the United States alone during the past 500 years, 236 animal species and 38 plant species are known to have gone extinct. For all these figures, the actual numbers of species are without doubt greater than the known numbers.

FAQ

If a mass extinction is happening, why don't I notice species disappearing all around me?

There are two reasons that most of us don't personally sense the scale of biodiversity loss. First, if you live in a town or city, the plants and animals you see from day to day are generalist species that thrive in disturbed areas. In contrast, the species most in trouble are those that rely on less-disturbed habitats far from urban areas.

Second, a human lifetime is very short! The loss of populations and species may seem slow to us, but on Earth's timescale it is sudden. Because each of us is born into a world that has already lost species, we don't recognize what's already vanished. Likewise, our grandchildren won't appreciate what we lose in our lifetimes. Each human generation experiences just a portion of the overall phenomenon, so we have difficulty sensing the big picture. Nonetheless, researchers and naturalists who spend their time outdoors observing nature see biodiversity loss around them all the time—and that's precisely why they feel so passionate about preventing it.

Several major causes of biodiversity loss stand out

Scientists have identified five primary causes of population decline and species extinction: habitat loss, pollution, overharvesting, invasive species, and climate change. Each of these causes is intensified by human population growth and by our increasing per capita consumption of resources.

Habitat loss Habitat loss is the single greatest threat to biodiversity today. A species' habitat is the specific environment in which it lives (p. 60). Because organisms have adapted to their habitats over thousands or millions of years of evolution, any sudden, major change in their habitat will likely render it less suitable for them. Habitat is lost when it is destroyed outright, but also when it becomes fragmented or degraded.

Many human activities alter, degrade, or destroy habitat. Farming replaces diverse natural communities with simplified

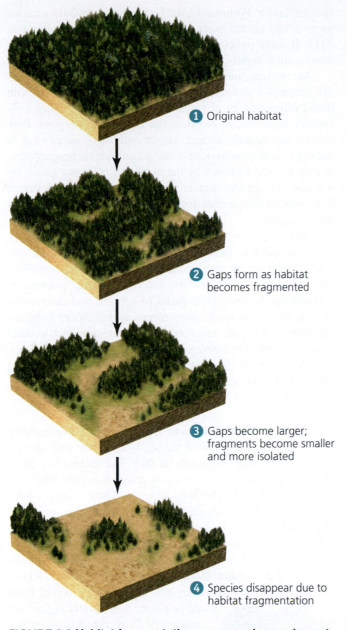

1 Original habitat

2 Gaps form as habitat becomes fragmented

3 Gaps become larger; fragments become smaller and more isolated

4 Species disappear due to habitat fragmentation

FIGURE 8.9 Habitat fragmentation occurs as human impact creates gaps that expand and eventually come to dominate the landscape, stranding islands of habitat. As habitat becomes fragmented, fewer populations can persist, and numbers of species in the fragments decline.

ones of just a few plant species. Grazing modifies grasslands and can lead to desertification (p. 148). Clearing forests removes the food and shelter that forest-dwelling organisms need to survive. Damming rivers creates reservoirs upstream while affecting water conditions and floodplain communities downstream. Urban sprawl supplants natural ecosystems, driving many species from their homes.

Habitat loss occurs most commonly by gradual, piecemeal degradation, such as **habitat fragmentation** (**FIGURE 8.9**). When farming, logging, road building, or development intrude into an unbroken expanse of forest or grassland, this breaks up a continuous area of habitat into fragments, or patches. As habitat fragmentation proceeds across a landscape, animals

and plants requiring the habitat disappear from one fragment after another. Fragmentation can also block animals from moving from place to place; this is the concern of opponents of the proposed highway through the Serengeti. In response to habitat fragmentation, conservationists try to link fragments together with corridors of habitat along which animals can travel. (In Chapter 9 we will learn more about fragmentation, the effects on wildlife, and potential solutions.)

Habitat loss affects all the world's biomes. More than half of the world's temperate forests, grasslands, and shrublands had been converted by 1950 (mostly for agriculture). Today, habitat is being lost most rapidly in tropical rainforests, tropical dry forests, and savannas. Within most biomes, wetlands (p. 260) are especially threatened. More than half the wetlands of the lower 48 U.S. states and Canada have been drained for agriculture. Habitat loss is the primary source of population declines for more than 80% of threatened mammals and birds, according to UNEP data. For example, the prairies of North America's Great Plains are today almost entirely converted to agriculture. Less than 1% of original prairie habitat remains. As a result, grassland bird populations have declined by an estimated 82–99%.

Of course, when we alter habitat, we benefit some species. Animals such as starlings, raccoons, house sparrows, pigeons, gray squirrels, rats, mosquitoes, and cockroaches thrive among us in farmland, towns, and cities. However, these "winning" species tend to be weedy generalists that are in little danger of disappearing. The concern is that far too many other species are "losing" as a result of the ways we alter habitats.

weighing the ISSUES

Habitat Fragmentation in Your Region

Examine satellite imagery of the region where you live using an online application such as Google Earth or Google Maps. How much of the area around you is developed with roads, buildings, and agriculture? How much is forested or in natural areas? Are the natural areas connected to one another, or are they fragmented? Discuss how easy or difficult it would be for an animal such as a deer or a bear to move from one habitat patch to another. Do you see ways in which natural areas could be better connected using corridors? Now zoom in or zoom out a bit on the map, and address these same questions at a different landscape scale.

Pollution Pollution can harm organisms in many ways. Air pollution degrades forest ecosystems and affects the atmosphere and climate. Noise and light pollution impinge on the behavior and habitat use of animals. Water pollution impairs fish and amphibians. Agricultural runoff containing fertilizers, pesticides, and sediments harms many terrestrial and aquatic species. Heavy metals, endocrine-disrupting compounds, and other toxic chemicals poison people and wildlife. Plastic garbage in the ocean can strangle, drown, or choke marine creatures. The effects of oil spills on wildlife are dramatic and well known. We examine all these impacts in other chapters of this book. However, although pollution can cause extensive damage to organisms and ecosystems, it tends to be less significant as a cause of population-wide decline than public perception holds it to be, and it is far less influential than habitat loss.

Overharvesting For most species, being hunted or harvested will not in itself pose a threat of extinction. However, with today's illegal global trade in wildlife products surpassing $20 billion per year, **poaching**—the unlawful killing of wildlife for meat or body parts—has led to steep population declines for many animals. Most vulnerable are large-bodied species that are long-lived and slow to reproduce, such as elephants, rhinoceroses, and other large mammals of the African savanna. Rhinoceros populations have crashed as poachers slaughter rhinos for their horns, which are ground into powder and sold illegally to ultra-wealthy Asian consumers as cancer cures, party drugs, and hangover treatments, even though the horns have no such properties and consist of the same material as our fingernails. In Central Africa, gorillas and other primates are killed for their "bush meat" and could soon face extinction. Across Asia, tigers are threatened by poaching as well as habitat loss; body parts from one tiger can fetch a poacher $15,000 on the black market, where they are sold as aphrodisiacs in China and other Asian countries. Today half the world's tiger subspecies are extinct, and most of the remaining animals are crowded onto just 1% of the land they occupied historically.

In the world's oceans, many fish stocks today are overharvested (p. 281). Whaling drove the Atlantic gray whale extinct and has left several other types of whales threatened or endangered. Thousands of sharks are killed each year merely for their fins, which are used in soup. Altogether, the oceans contain only 10% of the large animals they once did (p. 282).

Elephants have long been killed to extract their tusks for ivory (**FIGURE 8.10**). By 1989, 7% of African elephants were being slaughtered each year, so the world's nations enacted a global ban on the commercial trade of ivory. Elephant numbers recovered following the ban, but after 2007, poaching rose to all-time highs, driven by high black-market prices for ivory paid by wealthy overseas buyers. From 2011 through 2015, more than 170,000 African elephants were killed—enough to send populations downward and threaten the species' future. Fortunately, the United States in 2016 enacted its own ban on the ivory trade, and China followed suit in 2017. If these policies are enforced, it should help elephant conservation greatly.

Meanwhile, in much of Africa today, protecting wildlife remains a dangerous job. Poaching is conducted with brutal efficiency by organized crime syndicates using helicopters, night-vision goggles, and automatic weapons. Park rangers are heavily armed, yet are routinely outgunned in firefights with poachers, and many courageous rangers have lost their lives. In this way the demand for luxury goods by wealthy foreign consumers in Asia, Europe, and America has grave consequences for Africans living in regions like the Serengeti.

Today scientists are assisting efforts to curb poaching and save wildlife. Researchers in the field are tracking elephants and ivory by putting radio collars on animals, satellite-tracking tusks with microchips, and flying drones (unmanned surveillance aircraft) over parks to capture real-time video of poachers. Meanwhile, researchers in the lab are conducting genetic analyses to expose illegal hunting and wildlife trade. For instance, forensic DNA testing can reveal the geographic origins of elephant ivory, helping authorities to focus on hotspots of illegal activity (see **THE SCIENCE BEHIND THE STORY**, pp. 186–187).

Invasive species When non-native species are introduced to new environments, most perish, but the few that survive may do very well, especially if they find themselves freed from the predators, parasites, and competitors that had kept their populations in check back home. Once released from such limiting factors (p. 63), an introduced species may proliferate and become invasive (pp. 78–79), and may often displace native species (**TABLE 8.4**).

Some introductions are accidental. Examples include animals that escape from the pet trade; weeds whose seeds cling to our clothing and shoes as we travel from place to place; and aquatic organisms transported in the ballast water of ships. If a highway is built through the Serengeti, ecologists fear that passing vehicles would introduce weed seeds. Several American plants, such as datura, parthenium, and prickly poppy, have already spread through African grasslands and are toxic to native herbivores.

Other introductions are intentional. Throughout history, people have brought food crops and animals with them as they colonized new places, and today we continue global trade in exotic pets and ornamental plants. In Lake Victoria near the Serengeti, the Nile perch was introduced as a food fish to supply people much-needed protein (see Table 8.4). It soon spread throughout the vast lake, however, preying on and driving extinct dozens of native species of cichlid fish from

FIGURE 8.10 Poachers slaughter elephants to sell their tusks for ivory. Here Kenyan officials at Maasai Mara National Reserve prepare to set fire to tusks confiscated from poachers, in an effort to discourage the trade.

TABLE 8.4 Invasive Species

European gypsy moth
(*Lymantria dispar*)

Introduced to Massachusetts in the hope it could produce silk. The moth failed to do so, and instead spread across the eastern United States, where its outbreaks defoliate trees over large regions every few years.

Nile perch
(*Lates niloticus*)

A large fish from the Nile River. Introduced to Lake Victoria in the 1950s, it proceeded to eat its way through hundreds of species of native cichlid fish, driving a number of them to extinction. People value the perch as food, but it has radically altered the lake's ecology.

European starling
(*Sturnus vulgaris*)

Introduced to New York City in the 1800s by Shakespeare devotees intent on bringing every bird mentioned in Shakespeare's plays to America. Outcompeting native birds for nest holes, within 75 years starlings became one of North America's most abundant birds.

Emerald ash borer
(*Agrilus planipennis*)

Discovered in Michigan in 2002, this wood-boring insect reached 12 U.S. states and Canada by 2010, killing millions of ash trees in the upper Midwest. Billions of dollars will be spent in trying to control its spread.

Kudzu
(*Pueraria montana*)

A Japanese vine that can grow 30 m (100 ft) in a single season, the U.S. Soil Conservation Service introduced kudzu in the 1930s to help control erosion. Kudzu took over forests, fields, and roadsides throughout the southeastern United States.

Sudden oak death pathogen
(*Phytophthora ramorum*)

This disease has killed more than 1 million oak trees in California since the 1990s. The pathogen (a water mold) was likely introduced via infected nursery plants. Scientists are concerned about damage to eastern U.S. forests if it spreads to oaks there.

Brown tree snake
(*Boiga irregularis*)

Nearly every native forest bird on the South Pacific island of Guam has disappeared, eaten by these snakes, which arrived from Asia as stowaways on ships and planes after World War II. Guam's birds had not evolved with snakes, and had no defenses against them.

Polynesian rat
(*Rattus exulans*)

One of several rat species that have followed human migrations across the world. Polynesians transported this rat to islands across the Pacific, including Easter Island (pp. 8–9). On each island it caused ecological havoc, and has driven extinct birds, plants, and mammals.

---

one of the world's most spectacular evolutionary radiations of animals. The Nile perch is providing people food, but at significant ecological cost.

Species native to islands are especially vulnerable to introduced species. Island species have existed in isolation for millennia with relatively few parasites, predators, and competitors; as a result, they have not evolved the defenses necessary to resist invaders that are better adapted to these pressures. For instance, Hawaii's native plants and animals have been under siege from invasive organisms such as rats, pigs, and cats, and this has led to a number of extinctions (Chapter 3).

Some of the most devastating invasive species are microscopic pathogens that cause disease. In Hawai'i, malaria and avian pox transmitted by introduced mosquitoes are killing off the islands' native birds, which lack immunity to these foreign diseases. Some scientists classify disease separately as a major cause of biodiversity loss.

Experts debate the role of introduced species in today's world. For decades most biologists have focused on the negative impacts that invasive species exert on native ecosystems and the economic damage they cause. However, many introduced species, such as the European honeybee (p. 153), provide economic benefits. And in today's world there is no truly pristine ecosystem—all have been touched in some way by human impact, and many contain *novel communities* (p. 78), newly formed mixtures of native and non-native species. In some cases these communities host greater species diversity than the communities they replaced, and may function just as well in providing ecosystem services. Thus, while it is undeniably true that invasive species have driven extensive losses of native biodiversity in case after case, introduced species often increase overall biodiversity at local scales.

Climate change Our manipulation of Earth's climate (Chapter 14) is having global impacts on biodiversity. As we warm the atmosphere with emissions of greenhouse gases from fossil fuel combustion, we modify climate patterns and increase the frequency of extreme weather events (such as droughts and storms) that put stress on populations.

In the Arctic, melting sea ice is threatening polar bears and people alike (**FIGURE 8.11**). Across the world, warming temperatures are forcing organisms to shift toward the poles and upward in altitude. Mountaintop organisms cannot move farther upslope, so many may perish. Trees may not move toward the poles fast enough. As ranges shift, animals and plants encounter new prey, predators, and parasites to which they are not adapted. In a variety of ways, scientists predict that climate disruption will put many thousands of the world's plants and animals at increased risk of extinction.

FIGURE 8.11 The polar bear became the first species listed under the Endangered Species Act as a result of climate change. As Arctic warming melts the sea ice from which they hunt seals, polar bears must swim farther for food.

A mix of causes threatens many species

For many species, multiple factors are conspiring to cause declines. The monarch butterfly, once familiar to every American schoolchild, is today in precipitous decline (**FIGURE 8.12**). On its breeding grounds in the United States and Canada, industrial agriculture and chemical herbicides have eliminated most of the milkweed plants monarchs depend on. Our highly efficient monocultures leave no natural habitat remaining, while insecticides intended for crop pests also kill monarchs and other beneficial insects. In winter the entire monarch population of eastern and central North America migrates south and funnels into a single valley in Mexico, where the butterflies cluster by the millions in groves of tall trees. Here some people are illegally logging these forests while others fight to save the trees, the butterflies, and the ecotourism dollars they bring to the community.

Reasons for the decline of a population or species can be complex and difficult to determine. The worldwide collapse of amphibians provides an example of a "perfect storm" of bewildering factors. Today entire populations of frogs, toads, and salamanders are vanishing without a trace. More than 40% of the world's known species of amphibians are in decline, 30% are threatened, and at least 170 species studied just years or decades ago are thought to be extinct. As these

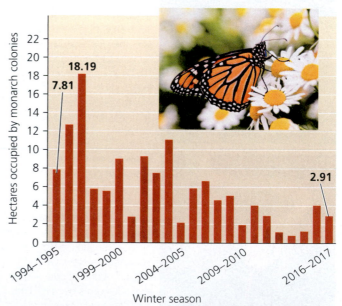

FIGURE 8.12 The once-abundant monarch butterfly has undergone alarming declines due to herbicides, pesticides, and habitat loss. Annual surveys on its Mexican wintering grounds show the population occupying many fewer hectares of forest than in the past. *Data from Monarch Watch, collected by the Monarch Butterfly Biosphere Reserve and World Wildlife Fund Mexico.*

DATA
• How much area did monarchs occupy in 2016–2017, as a proportion of the area occupied in 1994–1995?
• As a proportion of the area occupied in 1996–1997?

Go to **Interpreting Graphs & Data** on **Mastering** Environmental Science

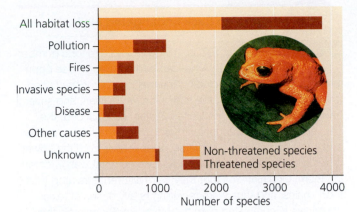

FIGURE 8.13 The world's amphibians are vanishing.
The golden toad is one of at least 170 species of amphibians that have suddenly gone extinct in recent years. This brilliant orange toad of Costa Rican cloud forests disappeared due to drought, climate change, and/or disease. Habitat loss is the main reason for amphibian declines, but many declines remain unexplained. *Data from IUCN, 2008.* Global amphibian assessment.

DATA Q • What is the second-greatest known cause of amphibian declines, after habitat loss? • What is the greatest cause for threatened species? • What is a greater cause for non-threatened species: fires or pollution?

Go to **Interpreting Graphs & Data** on **Mastering** Environmental Science

creatures disappear before our eyes, scientists are racing to discover why, and studies implicate a wide array of factors (**FIGURE 8.13**). These include habitat destruction, chemical pollution, invasive species, climate change, and a disease called chytridiomycosis caused by a fungal pathogen. Biologists suspect that multiple factors are interacting and multiplying the effects.

As researchers learn more, they are designing responses to amphibian declines. An IUCN conservation action plan recommends that we protect and restore habitat, crack down on illegal harvesting, enhance disease monitoring, and establish captive breeding programs.

Today many people are striving to save vanishing species. The search for solutions to our biodiversity crisis is dynamic and inspiring, and scientists are developing innovative approaches to sustain Earth's diversity of life.

Conservation Biology: The Search for Solutions

The urge to act as responsible stewards of natural systems, and to use science as a tool in this endeavor, sparked the rise of **conservation biology.** This scientific discipline is devoted to understanding the factors, forces, and processes that influence the loss, protection, and restoration of biological diversity. Conservation biologists aim to develop solutions to such problems as habitat degradation and species loss (**FIGURE 8.14**). Conservation biology is thus an applied and goal-oriented science, with implicit values and ethical standards.

Conservation biology responds to biodiversity loss

Conservation biologists integrate an understanding of evolution and ecology as they use field data, lab data, theory, and experiments to study our impacts on other organisms. They also design, test, and implement responses to these impacts.

At the genetic level, conservation geneticists ask how small a population can become and how much genetic variation

(a) Sampling insects in Madagascar

(b) Drawing blood from a Seychelles magpie robin

(c) Checking camera traps in Africa

FIGURE 8.14 Conservation biologists use many approaches to study the loss, protection, and restoration of biodiversity, seeking to develop scientifically sound solutions.

it can lose before running into problems such as inbreeding depression (p. 169). By determining a *minimum viable population size,* conservation geneticists help wildlife managers decide how vital it may be to increase a population. Studies of genes, populations, and species inform conservation efforts with habitats, communities, ecosystems, and landscapes. By examining how organisms disperse from one habitat patch to another, and how their genes flow among subpopulations, conservation biologists try to learn how likely a population is to persist or succumb in the face of environmental change.

Endangered species are a focus of conservation efforts

The primary legislation for protecting biodiversity in the United States is the **Endangered Species Act**. Enacted in 1973, the Endangered Species Act (ESA) offers protection to species that are judged to be **endangered** (in danger of becoming extinct in the near future) or **threatened** (vulnerable to becoming endangered soon). The ESA forbids the government and private citizens from taking actions that destroy individuals of these species or the habitats that are critical to their survival. The ESA also forbids trade in products made from threatened and endangered species. The aim is to prevent extinctions and enable declining populations to recover. As of 2017, there were 1276 species in the United States listed as endangered and 376 more listed as threatened. For most of these species, government agencies are running recovery plans to protect them and stabilize or increase their populations.

The ESA has had a number of successes. Species such as the bald eagle, peregrine falcon, and brown pelican have recovered and are no longer listed as endangered. Intensive management programs with species such as the red-cockaded woodpecker have held populations steady in the face of continued pressure on habitat. For every listed species that has gone extinct, three have recovered enough that they have been removed from the endangered species list.

These successes have come despite the fact that the U.S. Fish and Wildlife Service and the National Marine Fisheries Service, the agencies that administer the ESA, are perennially underfunded. Federal authorization for spending under the ESA expired in 1992, so Congress appropriates funds for its administration year by year. Today a number of species are judged by scientists to need protection but have not been added to the endangered species list because funding is inadequate to help recover them. Such species are said to be "warranted but precluded"—meaning that listing is warranted by scientific research but precluded by lack of resources. This has led some environmental advocacy groups to sue the government for failing to enforce the law. Well-meaning Fish and Wildlife Service staff are frequently caught in a no-win situation, battling lawsuits at the same time as they suffer budget cuts.

Polls repeatedly show that most Americans support protecting endangered species. Yet some opponents feel that the ESA can imperil people's livelihoods. This has been a common perception in the Pacific Northwest, where protection

Bringing Back Endangered Birds

In the early 1970s, things looked dire for a number of iconic North American bird species whose populations had collapsed, including the majestic peregrine falcon (the world's fastest bird), the stately brown pelican, and several hawks and owls. Even the United States' national bird, the bald eagle, was threatened with extinction. The falcon, pelican, and eagle had almost completely disappeared from the Lower 48 U.S. states before scientists discovered what was threatening them. One hint was that they were all predators atop their food chains, and thus were receiving heavy doses of toxic chemicals accumulated from the many smaller animals they ate over time (p. 222). Research eventually revealed that the chemical insecticide DDT (p. 216)—which had become widely used in the mid-20th century—was causing these birds' eggshells to become thin and break too early, killing the young. U.S. leaders banned DDT in 1973, the same year they enacted the Endangered Species Act. Together these two actions led to spectacular recoveries of the peregrine falcon, brown pelican, and bald eagle. Biologists began running programs to assist recovery, and the populations of these birds roared back. Today all three species are thriving across large portions of North America.

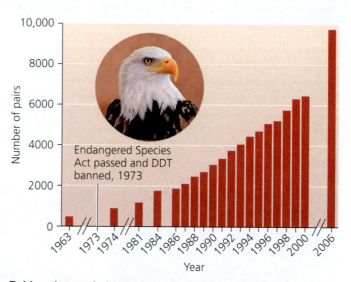

Endangered Species Act passed and DDT banned, 1973

Bald eagle populations rebounded following their protection as an endangered species.

EXPLORE THE DATA at Mastering Environmental Science

for the northern spotted owl and the marbled murrelet—birds that rely on old-growth forest—have slowed timber harvesting. In addition, many landowners worry that federal officials will restrict the use of private land on which threatened or endangered species are found. This has led to a practice described as "shoot, shovel, and shut up," among landowners who want to conceal the presence of such species on their land.

FIGURE 8.15 Populations of the greater sage grouse have declined steeply. Their listing as an endangered species would have complicated efforts to ranch and drill for oil and gas, so the U.S. government instead designed collaborative agreements with ranchers and the oil and gas industry to try to conserve the bird without listing it.

However, the ESA has stopped very few development projects—and a number of its provisions and amendments promote cooperation with landowners. A landowner and the government can agree to a *habitat conservation plan,* which grants the landowner an "incidental take permit" to harm some individuals of a species if he or she voluntarily improves habitat for the species. Likewise, in a *safe harbor agreement,* the government agrees not to mandate additional or different management requirements if the landowner acts to assist a species' recovery.

Recent efforts to conserve the greater sage grouse (**FIGURE 8.15**) exemplify the cooperative public-private approach. This species had been judged "warranted but precluded," but cattle ranchers and the oil and gas industry opposed listing the species as endangered because restrictions on land use across vast sagebrush regions of the West could complicate ranching and drilling activities. As a result, federal agencies embarked on a campaign with ranchers and the energy industry across 13 western states to design voluntary agreements to lessen impacts on sage grouse populations. In 2015 the bird was denied listing, as federal officials said the collaborative agreements were adequate to conserve the species. This produced celebration and relief in many quarters but also criticism both from development advocates, who felt the agreements were too restrictive, and from environmental advocates, who judged that the strategy would fail to save the species.

Treaties promote conservation

On the global stage, the United Nations (p. 110) has facilitated international treaties to protect biodiversity. The 1973 **Convention on International Trade in Endangered Species of Wild Fauna and Flora (CITES)** protects endangered species by banning the international transport of their body parts. The 1990 global ban on the ivory trade may be this treaty's biggest accomplishment so far. When nations enforce its provisions, CITES can protect rhinos, elephants, tigers, and other species whose body parts are traded.

In 1992, nations agreed to the **Convention on Biological Diversity,** aiming to help conserve biodiversity, use it in a sustainable manner, and ensure the fair distribution of its benefits. The treaty prompted nations to protect more areas, enhanced markets for sustainable crops such as shade-grown coffee, and helped Africans profit from ecotourism at their parks and preserves. Yet the overall goal— "to achieve, by 2010, a significant reduction of the current rate of biodiversity loss"—was not met.

Today the Convention's signatory nations aim to achieve 20 new biodiversity targets by 2020. Goals include:

- Cutting the loss of natural habitats in half—and, where feasible, bringing this loss close to zero
- Conserving 17% of land areas and 10% of marine and coastal areas
- Restoring at least 15% of degraded areas
- Alleviating pressures on coral reefs

Captive breeding, reintroduction, and cloning are being pursued

In the effort to save species at risk, zoos and botanical gardens have become centers for **captive breeding,** in which individuals are bred and raised in controlled conditions with the intent of reintroducing their progeny into the wild. The IUCN counts 65 plant and animal species that now exist *only* in captivity or cultivation. Reintroducing species into areas they used to inhabit is resource-intensive, but it can pay big dividends. In 2010 the first of 32 black rhinos were translocated from South Africa to Serengeti National Park to help restore a former population (**FIGURE 8.16**).

FIGURE 8.16 We can reestablish populations by reintroducing them to areas where they were extirpated. Black rhinos have been helicoptered in to Serengeti National Park.

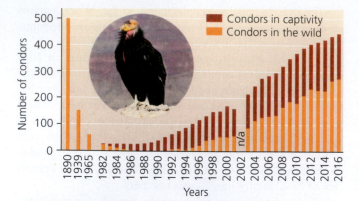

FIGURE 8.17 California condors are being bred in captivity and released to the wild, gradually rebuilding their population.

DATA • In 2016, were there more condors in the wild or in captivity? • Describe how the ratio of wild to captive condors has changed over the years. • Roughly how many condors are alive today compared with the number alive in the 1980s? • How large is today's condor population relative to the wild condor population in 1890?

Go to **Interpreting Graphs & Data** on **Mastering** Environmental Science

A prime example of captive breeding and reintroduction is the program to save the California condor, North America's largest bird (**FIGURE 8.17**). Condors are harmless scavengers of dead animals, yet people a century ago used to shoot them for sport. Condors also collided with electrical wires and succumbed to lead poisoning after scavenging carcasses of animals killed with lead shot. By 1982, only 22 condors remained, and biologists made the wrenching decision to take all the birds into captivity.

Today the collaborative program between the Fish and Wildlife Service and several zoos has boosted condor numbers. As of 2016, there were 170 birds in captivity and 276 birds living in the wild. Condors have been released at sites in California, Arizona, and Baja California (Mexico), where they thrill people lucky enough to spot the huge birds soaring through the skies. Unfortunately, many of these birds still die of lead poisoning, and wild populations will likely not become sustainable until hunters convert from lead shot to nontoxic shot made of copper or steel. California has now banned lead shot for all hunting, effective in 2019. This may give condors a fighting chance, while helping to stop the accumulation of a highly toxic substance in the environment.

One new idea for saving species from extinction is to create individuals by cloning them. In this technique, DNA from an endangered species is inserted into a cultured egg without a nucleus, and the egg is implanted into a female of a closely related species that acts as a surrogate mother. Several mammals have been cloned in this way, with mixed results. Some scientists even talk of recreating extinct species from DNA recovered from preserved body parts, and researchers may attempt this with the woolly mammoth, the passenger pigeon, and other recently extinct species. In 2009 a subspecies of Pyrenean ibex (a type of mountain goat) was cloned from cells taken from the last surviving individual, which had died in 2000. The cloned baby ibex died shortly after birth. Even if cloning can succeed from a technical standpoint, however, it is not an adequate response to biodiversity loss. Cloning does nothing to protect genetic diversity, and without ample habitat and protection in the wild, having cloned animals in a zoo does little good.

Forensics can help protect species

To counter poaching and illegal trade, scientists have a new tool at their disposal. **Forensic science,** or *forensics,* involves the scientific analysis of evidence to make an identification or answer a question relating to a crime or an accident. Conservation biologists are now using forensics to protect species at risk. By analyzing DNA from organisms or their tissues sold at market, researchers can often determine the species or subspecies—and sometimes the geographic origin. This can help detect illegal activity and enhance law enforcement.

One example is the analysis of illegal ivory shipments to determine the origin of African elephants that were killed to obtain their tusks (see **The Science behind the Story,** pp. 186–187). Forensic analysis has also helped researchers trace the origins of meat from whales sold in Asian markets, providing valuable data used to set policy for whaling and whale conservation.

Several strategies address habitats, communities, and ecosystems

Scientists know that protecting species does little good if the larger systems they rely on are not also sustained. Yet no law or treaty exists to protect communities or ecosystems. For these reasons, conservation biologists pursue several strategies for conserving ecological systems on broader scales.

Biodiversity hotspots To prioritize regions for conservation efforts, scientists have mapped biodiversity hotspots (**FIGURE 8.18a**). A **biodiversity hotspot** is a region that supports an especially great number of species that are endemic (p. 58), found nowhere else in the world (**FIGURE 8.18b**). To qualify as a hotspot, a region must harbor at least 1500 endemic plant species (0.5% of the world's total plant species). In addition, a hotspot must have already lost 70% of its habitat to human impact and be at risk of losing more.

The ecosystems of the world's biodiversity hotspots together once covered 15.7% of the planet's land surface. Today, because of habitat loss, they cover only 2.3%. This small amount of land is the exclusive home for half the world's plant species and 42% of terrestrial vertebrate species. The hotspot concept motivates us to focus on these valuable regions, where the greatest number of unique species can be protected.

Parks and protected areas A prime way to conserve habitats, communities, ecosystems, and landscapes is to set aside areas of undeveloped land in parks and preserves. Currently people have set aside nearly 15% of the world's land area in national parks, state parks, provincial parks, wilderness areas, biosphere reserves, and other protected areas. Many of these lands are managed for recreation, water

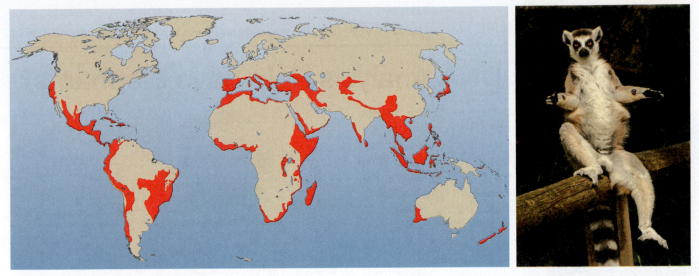

(a) The world's biodiversity hotspots

(b) Ring-tailed lemur

FIGURE 8.18 **Biodiversity hotspots are priority regions for habitat preservation.** Highlighted in red (a) are the 34 hotspots mapped by Conservation International, a nongovernmental organization. (Only 15% of the highlighted area is actually habitat; most is developed.) These regions are home to species such as the ring-tailed lemur (b), a primate endemic to Madagascar that has lost more than 90% of its forest habitat as a result of human population growth and resource extraction. *Data from Conservation International.*

quality protection, or other purposes (rather than for biodiversity), and many suffer from illegal logging, poaching, and resource extraction. Yet these areas offer animals and plants a degree of protection from human persecution, and some are large enough to preserve entire natural systems that otherwise would be fragmented, degraded, or destroyed.

Serengeti National Park and the adjacent Maasai Mara National Reserve are two of the world's largest and most celebrated parks, but Tanzania and Kenya have each set aside a number of other protected areas. Some of the best known include (in Kenya) Amboseli, Tsavo, Mount Kenya, and Lake Nakuru National Parks; and (in Tanzania) Ngorongoro Conservation Area, Selous Game Reserve, Kilimanjaro National Park, and Gombe Stream National Park. Altogether roughly 25% of Tanzania's land area and 12% of Kenya's land area is protected.

Alas, simply setting aside land is not always enough to ensure effective conservation. In Kenya and Tanzania, pressures from outside the reserves (settlements, hunting, competition with livestock, and habitat loss to farming) are reducing populations of migratory wildlife within the reserves. Similar situations occur in North America: Despite the large size of Yellowstone National Park, animals such as elk, bears, bison, and wolves roam seasonally in and out of the park, sometimes coming into conflict with ranchers. As a result, conservationists have tried to find ways to protect animals and habitats across the Greater Yellowstone Ecosystem, the larger region over which the animals roam.

As global warming (Chapter 14) drives species toward the poles and upward in elevation, this can force them out of protected areas. A major challenge today is to link protected areas across the landscape with corridors of habitat so that species like wildebeest or grizzly bears can move in response to climate change. (We will explore parks and protected areas more fully in Chapter 9).

Ecological restoration Protecting natural areas before they become degraded is the best way to safeguard native biodiversity and ecological systems. However, in some cases we can restore degraded natural systems to a semblance of their former condition through the practice of ecological restoration (p. 82). In Kenya, efforts are being made to restore the Mau Forest Complex, Kenya's largest remaining forested area and a watershed that provides water for the Maasai Mara Reserve and for the people of the region. Over the years so much forest in this densely populated area has been destroyed by agriculture, settlement, and timber extraction that the water supply for the Serengeti's wildlife and for millions of Kenyan people is now threatened. Kenya's government is working with international agencies and with U.S. funding to replant and protect areas of the forest.

Community-based conservation is growing

Helping people, wildlife, and ecosystems all at the same time is the focus of many current conservation efforts. In the past, conservationists from industrialized nations, in their zeal to preserve ecosystems in developing nations, often neglected the needs of people in the areas they wanted to protect. Developing nations came to view this as a kind of neocolonialism. Today, in contrast, many conservation biologists actively engage local people in efforts to protect land and wildlife—a cooperative approach called **community-based conservation.** A quarter of the world's protected areas are now being managed using community-based conservation. In several African nations, the African Wildlife Foundation funds community-based conservation programs to help communities conserve elephants, lions, rhinos, gorillas, and other animals.

Can Forensic DNA Analysis Help Save Elephants?

As any television buff knows, forensic science is a crucial tool in solving mysteries and fighting crime. In recent years, conservation biologists have been using forensics to unearth secrets and catch bad guys in the multibillion-dollar illegal global wildlife trade. One such detective story centers on the poaching of Africa's elephants for ivory.

Each year, tens of thousands of elephants are slaughtered illegally by poachers, simply for their tusks (**FIGURE 1**). Customs agents and law enforcement authorities manage to discover and confiscate tons of tusks being shipped internationally in the ivory trade. Yet only a small percentage of tusks are found and confiscated, and poachers are rarely apprehended, so the organized international crime syndicates that run these lucrative operations have been largely unhindered thus far.

Enter conservation biologist Samuel Wasser of the University of Washington in Seattle. By bringing the tools of genetic analysis, he and his colleagues have been shedding light on where elephants are being killed and where tusks are being shipped, thereby helping law enforcement efforts. In 2015, Wasser's team published a summary of nearly 20 years of work in the journal *Science,* revealing two major "poaching hotspots" in Africa.

The researchers began by accumulating 1350 reference samples of DNA from elephants at 71 locations across 29 African nations. Two subspecies of African elephant exist—savanna elephants, which live in open savannas, and forest elephants, which live in the forests of West and Central Africa. Of the 1350 genetic samples (from tissues or dung), 1001 came from savanna elephants and 349 came from forest elephants.

Wasser's teams sequenced DNA from the reference samples and compiled data on 16 highly variable stretches of DNA. For each of the 71 geographic locations, they measured

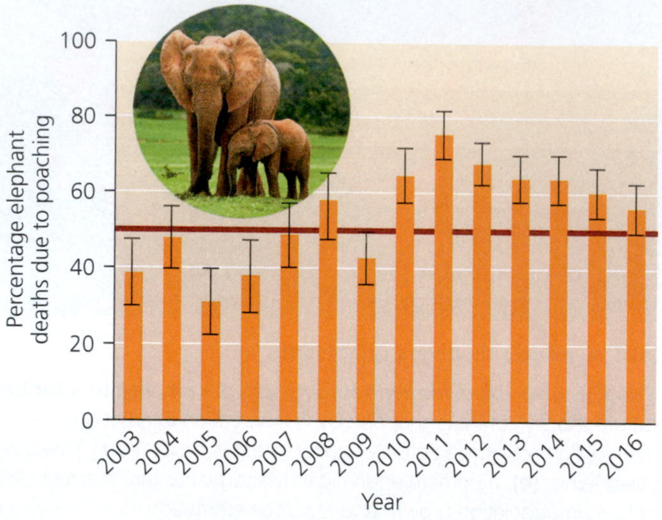

Confiscated tusks being destroyed in Kenya, to discourage poaching

FIGURE 1 Since 2010, more than half of African elephant deaths have been due to poaching, a level scientists conclude is unsustainable. Values above the horizontal line in this graph are thought to cause declines in the population. Data from *MIKE (Monitoring the Illegal Killing of Elephants), 2017.* Levels and trends of illegal killing of elephants in Africa to 31 December 2016—preliminary findings.

frequencies of alleles (different versions of genes) in these variable DNA stretches. By this process they created a map of allele frequencies for 71 geographic areas for both types of elephants; this would act as a kind of reference library with which they could compare any samples from tusks confiscated from the ivory trade.

Working with law enforcement officials, Wasser was able to access 20% of all ivory seizures made internationally between 1996 and 2005, 28% made between 2006 and 2011, and 61% made between 2012 and 2014 (**FIGURE 2**). Taking samples from these tusks and sequencing the DNA, Wasser's teams were then able to compare the results to their library of 71 locations and look for matches. In this way, with the help of sophisticated statistical techniques, they were able to determine the geographic origin of each of the tusks they tested, within an estimated distance of about 300–400 km (185–250 mi).

For instance, ivory seized in the Philippines from 1996 to 2005 all appeared to come from forest elephants in an area of

In East Africa, conservationists and scientists began working with the Maasai and other people of the region years ago, understanding that to conserve animals and ecosystems, local people need to be stewards of the land and feel invested in conservation. This has proven challenging because the parks and reserves were created on land historically used by local people. Residents were forcibly relocated; by some estimates 50,000 Maasai were evicted to create Serengeti National Park. In the view of many local people, the parks were a government

land grab, and laws against poaching deprive them of a right to kill wildlife. As human population grew in the region, conflicts between people and wildlife increased. Ranchers worried that wildebeest and buffaloes might spread disease to their cattle. Farmers lost produce when elephants ate their crops. And the economic benefits of ecotourism were not being shared with all people in the region.

In response, proponents of conservation have tried to reallocate tourist dollars to local villages and to transfer some

FIGURE 2 Dr. Samuel Wasser worked with law enforcement officials to obtain samples of confiscated ivory.

eastern Democratic Republic of Congo—just one small portion of their large geographic range. This region was difficult to patrol, however, due to its remoteness and to warfare occurring at the time, so little could be done with the information.

In contrast, when customs agents seized 6.5 tons of tusks in Singapore in 2002, Wasser's team determined that their DNA matched known samples from savanna elephants in Zambia, indicating that many more elephants were being killed there than Zambia's government had realized. The Zambian government responded; it replaced its wildlife director and began imposing harsher sentences on poachers and ivory smugglers.

For shipments seized between 2006 and 2014, the genetic research indicated a surprisingly clear pattern of two main poaching hotspots. Most forest elephant tusks seized originated from a small region of West-Central Africa where two protected areas overlap the boundaries of four nations (**FIGURE 3a**). As for savanna elephant tusks, most came from animals in southern Tanzania and northern Mozambique during the early portion of the period. Later in the period, tusks originated from throughout Tanzania (**FIGURE 3b**), pointing to a shift northward and an increase in poaching in the parks of central and northern Tanzania.

In most cases, shipments seized in ports such as Hong Kong, Malaysia, Taiwan, and Sri Lanka were labeled with their shipping origin (often a coastal port in Kenya, Tanzania, Togo, or other African countries). With the additional help of Wasser's research, authorities could learn the entire route of the ivory shipments, from where the elephants were killed to where the tusks were exported to where they were imported. Combined with other data on poaching collected by international survey efforts, the genetic information from DNA forensic studies is painting a clearer picture of the crime network threatening elephants, and is giving law enforcement authorities more and better information with which to work.

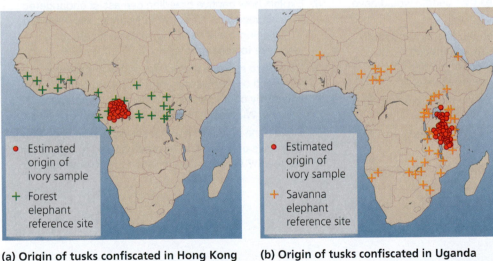

(a) Origin of tusks confiscated in Hong Kong

(b) Origin of tusks confiscated in Uganda

FIGURE 3 Genetic analysis of confiscated ivory reveals where elephants were killed. For example, analysis of a shipment confiscated in Hong Kong in 2013 **(a)** shows that it came from forest elephants killed in a small area of West-Central Africa. Likewise, tusks from a shipment confiscated in Uganda in 2013 **(b)** were found to have come from various areas within Tanzania. *Adapted from Wasser, S., et al., 2015. Genetic assignment of large seizures of elephant ivory reveals Africa's major poaching hotspots. Science 349: 84–87.*

authority over wildlife management to local people. In the regions around the Maasai Mara Reserve, the Kenya Wildlife Service and international non-profits have been helping farmers and ranchers build electric fences to keep wildlife away from their crops and livestock. These efforts are reducing conflicts between people and wildlife and are fostering more favorable attitudes toward conservation.

Working cooperatively to make conservation beneficial for local people requires patience, investment, and trust on all sides. Setting aside land for preservation may deprive local people of access to exploitable resources, but it also helps ensure that those resources can be sustainably managed and will not be used up or sold to foreign corporations. If tourism revenues are adequately distributed, people gain direct economic benefits from conserving wildlife. Community-based conservation has not always been successful, but in a world of rising human population, sustaining biodiversity will require locally based management that sustainably meets people's needs.

closing THE LOOP

Data from scientists worldwide confirm what any naturalist who has watched the habitat change in his or her hometown already knows: From amphibians to zebras, biological diversity is being lost rapidly within our lifetimes. This erosion of biodiversity threatens to result in a mass extinction event equivalent to those of the geologic past. Habitat alteration, pollution, overharvesting, invasive species, and climate change are the primary causes of biodiversity loss. This loss matters, because society cannot function without biodiversity's pragmatic benefits. Conservation biologists are conducting research that guides efforts to save endangered species, protect their habitats, recover populations, and preserve and restore natural ecosystems.

Areas such as East Africa have outsized importance globally for biodiversity and its conservation. Tanzania and Kenya are home to exceptionally rich species diversity. Both nations have invested heavily in protected areas to try to safeguard wildlife populations and functioning ecosystems, and both nations have been pioneers in community-based conservation. However, these countries also face severe economic and demographic challenges as people in their swelling populations attempt to rise up from poverty amid widespread land degradation in rural regions. East Africa's people desire economic development, and it can be challenging to achieve this while protecting natural resources. The debates over the Serengeti highway and pipeline proposals bring these issues into stark relief, but there is hope that if local people can obtain economic benefits from ecotourism, they will be inspired to help protect the Serengeti ecosystem and achieve a win-win solution for economic development and biodiversity conservation.

TESTING Your Comprehension

1. What is biodiversity? Describe three levels of biodiversity.

2. Define the term *ecosystem services*. Give three examples of ecosystem services that people would have a hard time replacing if these were lost.

3. What is the relationship between biodiversity and food security? Between biodiversity and pharmaceuticals? Give three examples of the benefits of biodiversity conservation for food supplies and medicine.

4. List three reasons why people suggest that biodiversity conservation is important.

5. What are the five primary causes of biodiversity loss? Give one specific example of each.

6. List three invasive species, and describe their impacts.

7. Describe one successful accomplishment of the U.S. Endangered Species Act. Now describe one reason some people have criticized it.

8. Explain how captive breeding can assist endangered species recovery, and give an example. Now explain why cloning could never be, in itself, an effective response to species loss.

9. Name two reasons that a large national park like Serengeti or Yellowstone might not be adequate to effectively conserve a population of a threatened species. What solutions exist to address these reasons?

10. Explain the notion of community-based conservation. Why have conservation advocates been turning to this approach? What challenges exist in implementing it?

SEEKING Solutions

1. Many arguments have been advanced for the importance of preserving biodiversity. Which argument do you find most compelling, and why? Which argument do you find least compelling, and why?

2. Some people declare that we shouldn't worry about endangered species because extinction has always occurred. How would you respond to this view?

3. Compare the approach of setting aside protected areas with the approach of community-based conservation. What are the advantages and disadvantages of each? Can we—and should we—follow both approaches?

4. **CASE STUDY CONNECTION** You are an adviser to the president of Tanzania, who is seeking to develop a formal policy on the question of potential highways and pipelines through Serengeti National Park. Given what you know from our Central Case Study, what would be your preliminary advice to the president, and why? To bolster your advice, what further information would you seek to learn from the region's residents? From scientists and conservationists? From tourism operators? From international sources that might help fund a project?

5. **THINK IT THROUGH** As a resident of your community and a parent of two young children, you attend a town meeting called to discuss the proposed development of a shopping mall and condominium complex. The development would eliminate a 100-acre stand of forest, the last sizeable forest stand in your town. The developers say the forest loss will not matter because plenty of 1-acre stands still exist scattered throughout the area. Consider the development's possible impacts on the community's biodiversity, children, and quality of life. What will you choose to tell your fellow citizens and the town's decision-makers at this meeting, and why?

CALCULATING Ecological Footprints

Research shows that much of humanity's footprint on biodiversity comes from our use of grasslands for grazing livestock and forests for timber and other resources. Grasslands and forests contribute different amounts to each nation's biocapacity (a region's natural capacity to provide resources and absorb our wastes), depending on how much of these habitats each nation has. Likewise, the per capita biocapacity and per capita ecological footprints of each nation vary further according to their populations. When footprints are equal to or below biocapacity, then resources are being used sustainably. When footprints surpass biocapacity, then resources are being used unsustainably.

In the table, fill in the proportion of each nation's per capita footprint accounted for by use of grazing land and forest land. Then fill in the proportion of each nation's per capita biocapacity provided by grazing land and forest land.

FOOTPRINTS AND BIOCAPACITIES (HECTARES PER PERSON)	KENYA	TANZANIA	UNITED STATES	CANADA
Footprint from grazing land	0.23	0.32	0.30	0.33
Footprint from forest land	0.26	0.22	0.82	1.11
Total ecological footprint	1.02	1.24	8.58	8.76
Percent footprint from grazing and forest	48			
Biocapacity of grazing land	0.24	0.32	0.28	0.29
Biocapacity of forest land	0.02	0.16	1.57	8.99
Total biocapacity	0.53	1.01	3.78	16.18
Percent biocapacity from grazing and forest				57

Data from Global Footprint Network, 2017.

1. In which nations is grazing land being used sustainably? In which nations is forest land being used sustainably?

2. Do forest use and grazing make up a larger part of the footprint for temperate-zone industrialized nations such as Canada and the United States, or for tropical developing nations such as Kenya and Tanzania? What do you think accounts for this difference between these two types of nations? What else besides use of forests and grasslands contributes to an ecological footprint?

3. The Living Planet Index (p. 174) declined 52% between 1970 and 2010, but its temperate and tropical components differed. During this period, the index for temperate regions decreased by 36%, whereas the index for tropical regions declined by 56%. Based on this information, do you predict that biodiversity loss has been steepest in Kenya and Tanzania or in Canada and the United States? Explain your answer.

Mastering Environmental Science

CHAPTER 9

Forests, Forest Management, and Protected Areas

Saving the World's Greatest Rainforest

Amazon rainforest

BRAZIL

> **Imagine a man without lungs. Imagine Earth without Amazon rainforest.**
> —Vinita Kinra, Indian-Canadian author

> **Destroying rainforest for economic gain is like burning a Renaissance painting to cook a meal.**
> —Edward O. Wilson, biologist and Pulitzer Prize–winning author

By any measure, the Amazon rainforest is enormous—and enormously important. Two-thirds the size of the contiguous United States, it encompasses most of the rainforest on Earth. It captures water, regulates climate, and absorbs much of our planet's carbon dioxide while releasing oxygen. Incredibly biodiverse, it hosts many thousands of plant and animal species—as well as tribes of indigenous people, some not yet contacted by outside civilization.

The Amazon plays all these vital roles in the health of our planet. And yet, we are losing this vast forest as people clear its trees for agriculture and settlement. Fully one-fifth of the immense forest that was standing just half a century ago is now gone.

The Amazon rainforest spreads across portions of nine South American nations, but nearly two-thirds lies within Brazil. Just as the United States in the 1800s pushed its frontier west to seek resources and expand its influence, Brazil today is pushing into the Amazon basin. Like the United States in the past, Brazil has sought to relieve crowding and poverty in its cities by encouraging urban-dwellers to settle on lands at its frontier.

For several decades, Brazil's government has provided incentives for settlement while foreign investors have helped fund development. In the 1970s, Brazil built the Trans-Amazonian highway across the forest and promised each settler 100 hectares (250 acres) of land, along with loans and 6 months' salary. People flooded into the region and cleared a great deal of forest for small farms, but most were unable to make a living at farming, and many plots were abandoned.

Once land is cleared for one purpose, however, it can be sold for others. Land speculators bought and sold parcels, and vast areas ended up going to wealthy landowners overseeing large-scale agriculture. Since that time, cattle ranching has accounted for three-quarters of the forest loss in the Brazilian Amazon (**FIGURE 9.1**).

Starting in the 1990s, rising global demand for soy, sugar, rice, corn, and palm oil fueled the destruction of Amazonian forest for industrial-scale farming. In particular, vast monocultures of soybeans spread across the land. Most soy was exported for biodiesel and animal feed.

Meanwhile, each new highway, road, or path cut into the forest allowed access not just for farmers, ranchers, and settlers, but also for miners, loggers, and poachers—many of whom were exploiting resources illegally. These pursuits in turn created more roads, which divided the forest into smaller and smaller fragments. Soon, dam construction

Member of the Xingu tribe from the Amazon ▲

Upon completing this chapter, you will be able to:

- Summarize the ecological and economic contributions of forests
- Outline the history and current scale of deforestation
- Assess approaches to resource management, describe methods of harvesting timber, and appraise aspects of forest management
- Identify federal land management agencies and the lands they manage
- Recognize types of parks and protected areas and evaluate issues involved in their design

Aerial view of a stretch of Amazon rainforest

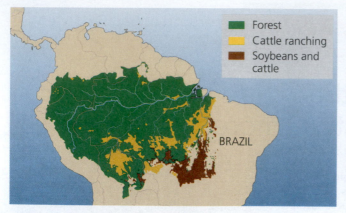

■	Forest
■	Cattle ranching
■	Soybeans and cattle

BRAZIL

(a) Amazon rainforest and regions of cleared forest

(b) An area of rainforest cleared for agriculture

FIGURE 9.1 Large areas of Amazon rainforest have been cleared for cattle ranching and soybean farming. *Map adapted from Nepstad, D., et al., 2014. Slowing Amazon deforestation through public policy and interventions in beef and soy supply chains. Science 344: 1118–1123.*

and oil and gas drilling caused more forest loss. As a result of all these activities, each year from the 1970s through 2008 an average of nearly 17,000 km² (6600 mi²) were cleared in Brazil (**FIGURE 9.2**)—an area larger than Connecticut. The Brazilian Amazon was losing more forest than anywhere else in the world.

In response, an international outcry arose from scientists and the public. People worldwide feared the loss of the region's biodiversity, and lamented the fate of the Amazon's indigenous tribespeople, who were dying from disease and conflict introduced by loggers, miners, and poachers trespassing on their land. As the threat of global climate change grew worse, people also dreaded the loss of one of Earth's biggest carbon sinks. For all these reasons, the Amazon rainforest came to be perceived as an international treasure—and its protection was regarded as vital for the good of people everywhere.

This new international scrutiny put pressure on Brazil to curb deforestation. Brazilian policymakers responded by strengthening the nation's Forest Code, a 1965 law governing the use of private land. After revision, the Forest Code mandated that landowners in the Amazon conserve 80% of their land as forest, and that rural landowners elsewhere in Brazil conserve 20%. The newly revised law also established large protected areas to protect tribes, conserve water, and prevent soil erosion. In addition, the government strengthened its historically weak enforcement of forest protections.

International pressure also convinced banks to begin requiring landowners to document compliance with environmental regulations in order to get loans. And it persuaded purchasers of beef, leather, and soy to stop buying Brazilian products or to buy only those certified as sustainable by third-party certifiers. In 2006, soy traders responded to an international campaign by agreeing not to buy Brazilian soybeans planted in recently deforested areas. This "soy moratorium" gave growers in Brazil incentive to plant on previously cleared land rather than cutting down forest. At the same time, global commodity prices were affecting Brazilian agriculture; for instance, a fall in beef prices made cattle ranching less profitable.

As a result, after 2004 deforestation rates in the Brazilian Amazon fell sharply. Since 2009 they have averaged 6300 km² (2400 mi²) per year (see Figure 9.2), an area the size of Delaware. Scientists and environmental advocates around the world held up Brazil as a model for slowing forest loss.

Today, however, deforestation is rising again as Brazil faces economic challenges and political instability amid corruption scandals engulfing its political leaders. Policymakers had begun revising the Forest Code again in 2012, granting an amnesty for people who had illegally cleared forest under the old law, and loosening the restrictions on forest clearance. Today's Forest Code offers much weaker protections than the earlier version.

As Brazil's current government acts to encourage development at the expense of forests, observers worry that the nation will be unable to keep promises it made at the Paris Climate Conference in 2015 to reduce forest loss. Brazil's decisions in coming years will shape the fate of the Amazon, which in turn bears significance for forests and societies around the world.

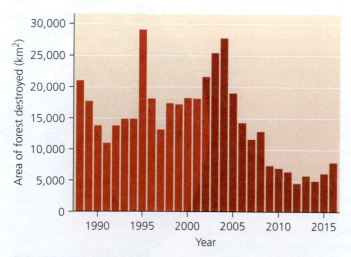

FIGURE 9.2 Annual forest loss has slowed in the Brazilian Amazon. Bars show area of forest cleared each year. *Data from Instituto Nacional de Pequisas Espaciais (INPE).*

Forest Ecosystems and Forest Resources

A **forest** is any ecosystem with a high density of trees. Forests provide habitat for countless organisms; help maintain the quality of soil, air, and water; and play key roles in our planet's biogeochemical cycles (pp. 39–44). Forests also provide us with wood for fuel, construction, paper production, and more.

There are many types of forests

Most of the world's forests occur as boreal forest, a biome that stretches across much of Canada, Scandinavia, and Russia; or as tropical rainforest, a biome in Latin America, equatorial Africa, Indonesia, and Southeast Asia. Temperate deciduous forests, temperate rainforests, and tropical dry forests also cover large regions (pp. 83–87).

Within each forest biome, the plant community varies from region to region because of differences in soil and climate. As a result, ecologists classify forests into **forest types**, categories defined by their predominant tree species (**FIGURE 9.3**). The eastern United States contains 10 forest types, ranging from spruce-fir to oak-hickory to longleaf–slash pine. The western United States has 13 forest types, ranging from Douglas fir and hemlock–sitka spruce forests of the moist Pacific Northwest to ponderosa pine and pinyon-juniper woodlands of the dry interior. In the Amazon rainforest, two of the main forest types are *várzea* (forest that gets flooded by swollen waters of the Amazon River or its tributaries for weeks or months each year) and *terra firme* (forest on land high enough to avoid seasonal flooding).

Altogether, forests currently cover 31% of Earth's land surface (**FIGURE 9.4**). Forests occur on all continents except Antarctica.

Forests are ecologically complex

Because of their structural complexity and their capacity to provide many niches for organisms, forests comprise some of the richest ecosystems for biodiversity. Insects, birds,

FIGURE 9.3 This maple-birch-beech forest from the Upper Peninsula of Michigan belongs to 1 of 23 forest types found in the continental United States.

mammals, and other animals subsist on the leaves, fruits, and seeds that trees produce, or find shelter in the cavities of tree trunks. Tree canopies are full of life, and understory shrubs and groundcover plants provide food and shelter for yet more organisms. Plants are colonized by an extensive array of fungi and microbes, in both parasitic and mutualistic relationships (pp. 72–73). And much of a forest's biodiversity resides on the forest floor, where the fallen leaves and branches of the leaf litter nourish the soil (**FIGURE 9.5**). A multitude of soil organisms helps to decompose plant material and cycle nutrients (p. 144).

Forests with a greater diversity of plants tend to host a greater diversity of organisms overall. As forests change by the process of succession (p. 77), their species composition changes along with their structure. In general, old-growth forests contain more structural diversity, microhabitats, and resources for more species. Old-growth forests also are home to more species that are threatened, endangered, or declining, because today old forests have become rare relative to young forests.

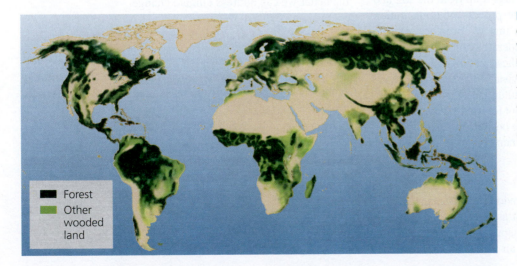

FIGURE 9.4 Forests cover 31% of Earth's land surface. Most widespread are boreal forests in the north and tropical forests near the equator. "Wooded land" supports trees at sparser densities. *Data from Food and Agriculture Organization of the United Nations, 2010.* Global forest resources assessment 2010.

Forest

Other wooded land

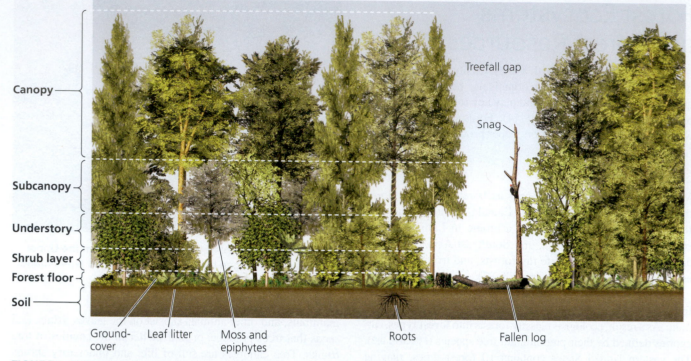

Canopy

Subcanopy

Understory

Shrub layer

Forest floor

Soil

Treefall gap

Snag

Ground-cover   Leaf litter   Moss and epiphytes   Roots   Fallen log

FIGURE 9.5 **A mature forest is complex in its structure.** Crowns of tall trees form the canopy, and trees beneath them form the subcanopy and understory. Shrubs and ground cover grow just above the forest floor, and vines, mosses, lichens, and epiphytes cover portions of trees and the forest floor. Snags (dead trees) provide food and nesting sites for woodpeckers and other animals, and logs nourish the soil. Fallen trees create openings called treefall gaps, letting light through and allowing early successional plants to grow.

Forests provide ecosystem services

Besides hosting biodiversity, a typical forest supplies us with many vital ecosystem services (pp. 4, 39, 172; **FIGURE 9.6**). As plants grow, their roots stabilize the soil and help to prevent erosion. Trees' roots draw minerals up from deep soil layers and deliver them to surface layers where other plants can use them. Plants also return organic material to the topsoil when they die or drop their leaves. When rain falls, leaves and leaf litter intercept water, slowing runoff. This helps water soak into the ground to nourish roots and recharge aquifers, thereby preventing flooding, reducing soil erosion, and helping keep streams and rivers clean. Forest plants also filter pollutants and purify water as they take it up from the soil and release it to the atmosphere by the process of transpiration (p. 41). Plants release the oxygen that we breathe, regulate moisture and precipitation, and moderate climate.

Because forests perform all these ecological functions, they are indispensable for our survival. Forests also enhance our health and quality of life with cultural, aesthetic, and recreational values (p. 98). People seek out forests for adventure and for spiritual solace alike—to admire beautiful trees, to observe wildlife, to enjoy clean air, and for many other reasons.

Carbon storage limits climate change

Of all the services that forests provide, their storage of carbon is eliciting the greatest interest as nations debate how to control global climate change (Chapter 14). Because plants absorb carbon dioxide from the air during photosynthesis (p. 34) and then store carbon in their tissues, forests serve as a major reservoir for carbon. Scientists estimate that the world's forests store about 296 billion metric tons of carbon in living tissue, which is more than the atmosphere contains. Each year forests absorb about 2.4 billion metric tons of carbon from the air, with the Amazon rainforest taking care of one-quarter of that total. Conversely, when plant matter is burned or when plants die and decompose, carbon dioxide is released—and thereafter less vegetation remains to soak it up. Carbon dioxide is the primary greenhouse gas driving climate change (p. 314). Therefore, when we destroy forests, we worsen climate change. The more forests we preserve or restore, the more carbon we keep out of the atmosphere, and the better we can address climate change.

Forests provide us valuable resources

Carbon storage and other ecosystem services alone make forests priceless to our society, but forests also provide many economically valuable resources. Among these are plants for medicines, dyes, and fibers; and animals, fruits, and nuts for food. Amazonian forests abound with resources ranging from Brazil nuts to açaí berries to latex for making rubber (from the sap of rubber trees). And, of course, forest trees provide us with wood. For millennia, wood has fueled the fires that cook our food and keep us warm. It has built the homes that keep us sheltered. It built the ships that carried people and cultures between continents. And it gave us paper, the medium of the first information revolution. Forest resources have helped us achieve the standard of living we now enjoy.

FIGURE 9.6 A forest provides us with a diversity of ecosystem services, as well as resources that we can harvest.

Labels in figure: Stores carbon; CO_2; Supports biodiversity; O_2; Produces oxygen; Provides fuel wood, lumber, paper, medicines, dyes, foods, fibers; H_2O; Purifies water, filters pollution; Promotes and provides health, beauty, recreation; Slows runoff, prevents flooding; Returns organic matter to soil; Transports minerals to soil surface; Stabilizes soil, prevents erosion

In recent decades, industrial harvesting has allowed us to extract more timber than ever before. Most commercial timber extraction today takes place in Canada, Russia, and other nations with large expanses of boreal forest; and in tropical nations with large areas of rainforest, such as Brazil and Indonesia. In the United States, most logging takes place in pine plantations of the South and conifer forests of the West.

Forest Loss

When trees are removed more quickly than they can regrow, the result is **deforestation**, the clearing and loss of forests. Deforestation has altered landscapes across much of our planet. In the time it takes you to read this sentence, 2 hectares (5 acres) of tropical forest will have been cleared. As we alter, fragment, and eliminate forests, we lose biodiversity, worsen climate change, and disrupt the ecosystem services that support our societies.

Agriculture and demand for wood put pressure on forests

To make way for agriculture and to extract wood products, people have been clearing forests for millennia. This has fed our civilization's growth, but the loss of forests today exerts damaging impacts as our population swells. Deforestation leads to biodiversity loss, soil degradation, and desertification (p. 148). It also releases carbon dioxide to the atmosphere, contributing to climate change.

In 2015, the UN Food and Agriculture Organization (FAO) released its latest *Global Forest Resources Assessment*, a periodic report based on remote sensing data from satellites, analysis from forest experts, questionnaire responses, and statistical modeling. The assessment concluded that we are eliminating 7.6 million hectares (ha; 18.8 million acres) of forest each year. Subtracting annual regrowth from this amount makes for an annual net loss of 3.3 million ha (8.2 million acres)—an area about the size of Maryland. This rate (for the period 2010–2015) is lower than deforestation rates for earlier years. In the 1990s, the world had been losing 1.8% of its forest each year; in 2010–2015 it lost just 0.8% each year.

While this is good news, some other research efforts have calculated higher rates of forest loss. In 2013, a large research team analyzed satellite data, published their work in the journal *Science*, and worked with staff from Google to create interactive maps of forest loss and gain across the world. They found much more deforestation than the FAO had: From 2000 to 2012, this team calculated annual losses of 19.2 million ha and annual gains of 6.7 million ha, for a net loss per year of 12.5 million ha (30.9 million acres—an area larger than New York State).

We deforested much of North America

Deforestation for timber and farmland propelled the expansion of the United States and Canada westward across the North American continent. The vast deciduous forests of the East were cleared by the mid-1800s, making way for countless small farms. Timber from these forests built the cities of the Atlantic Seaboard and the upper Midwest.

As a farming economy shifted to an industrial one, wood was used to fire the furnaces of industry. Logging operations moved south to the Ozarks of Missouri and Arkansas, and then to the pine woodlands and bottomland hardwood forests of the South, which were logged and converted to pine plantations. Once mature trees were removed from these areas, timber companies moved west, cutting the continent's biggest trees in the Rocky Mountains, the Sierra Nevada, the Cascade Mountains, and the Pacific Coast ranges. Exploiting forest resources fed the American economy, but we were depleting our store of renewable resources for the future.

By the 20th century, very little **primary forest**—natural forest uncut by people—remained in the lower 48 U.S. states, and today even less is left (**FIGURE 9.7**). Nearly all the largest trees in North America today are *second-growth* trees: trees that sprouted after old-growth trees were cut. Second-growth trees typify **secondary forest**, which contains smaller, younger trees. In terms of species composition, structure, and nutrient balance, a secondary forest differs markedly from the primary forest that it replaced.

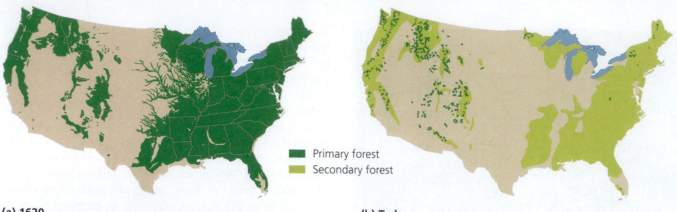

Primary forest
Secondary forest

(a) 1620

(b) Today

FIGURE 9.7 Areas of primary (uncut) forest have been dramatically reduced. When Europeans first colonized North America **(a)**, much of what is now the United States was covered in primary forest **(dark green)**. Today, nearly all this primary forest is gone **(b)**. Much of the landscape has become reforested with secondary forest **(pale green)**. *Adapted from (a) U.S. Forest Service; and (b) maps by Hansen, M.C., et al., 2013. High-resolution global maps of 21st-century forest cover change. Science 342: 850–853, and George Draffan, Endgame Research (www.endgame.org).*

Forests are being cleared most rapidly in developing nations

Uncut primary forests still remain in Brazil and in many developing countries. These nations are in the position the United States and Canada enjoyed a century or two ago, but today's powerful industrial technologies allow these nations to exploit their resources and push back their frontiers even faster than occurred in North America. As a result, deforestation is most rapid today in Indonesia, Africa, and Latin America **(FIGURE 9.8)**. In these regions, developing nations are striving to expand settlement for their burgeoning populations and to boost their economies by extracting natural resources.

As we have seen, Brazil was losing forests faster than any other country in the 1980s and 1990s **(FIGURE 9.9)**. Its reduction of deforestation in the Amazon since that time has been an inspiring success story, proving that a nation can save its forests while developing economically. Alas, that nation's recent political struggles are leading to an increase in forest loss in the Amazon, and to higher rates of clearing in other parts of the country. Moreover, most of Brazil's neighbors that share portions of the Amazon rainforest—Bolivia, Peru, Ecuador, Colombia, Venezuela, Guyana, Surinam, and French Guiana— have not yet managed to slow their deforestation rates.

In contrast, parts of Europe and North America are gaining forest as they recover from past deforestation. This is

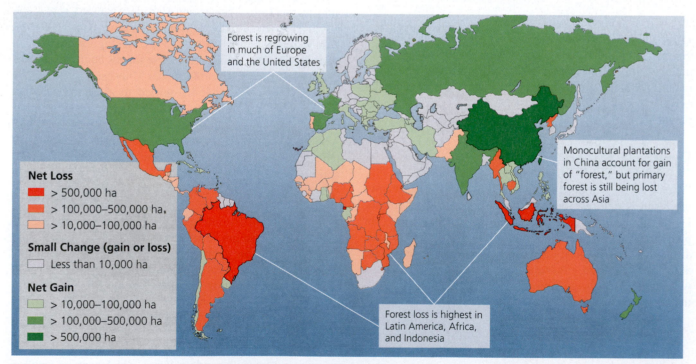

Forest is regrowing in much of Europe and the United States

Monocultural plantations in China account for gain of "forest," but primary forest is still being lost across Asia

Forest loss is highest in Latin America, Africa, and Indonesia

Net Loss
- ■ > 500,000 ha
- ■ > 100,000–500,000 ha
- ■ > 10,000–100,000 ha

Small Change (gain or loss)
- □ Less than 10,000 ha

Net Gain
- ■ > 10,000–100,000 ha
- ■ > 100,000–500,000 ha
- ■ > 500,000 ha

FIGURE 9.8 Tropical forests are being lost. Africa, Latin America, and Indonesia are losing the most forest, whereas Europe and North America are slowly gaining secondary forest. In Asia, tree plantations are increasing, but natural primary forests are still being lost. *Data from Food and Agriculture Organization of the United Nations, 2015. Global forest resources assessment 2015. By permission.*

FIGURE 9.9 Deforestation of Amazonian rainforest has been rapid. Satellite images of the state of Rondonia in Brazil show extensive clearing resulting from settlement in the region.

not compensating for the loss of tropical forests, however, because tropical forests are home to far more biodiversity than the temperate forests of North America and Europe.

Developing nations frequently are desperate enough for economic development and foreign capital that they impose few restrictions on logging. They often allow their timber to be extracted by foreign multinational corporations, which pay fees for a **concession,** or right to extract the resource. Once a concession is granted, the corporation has little or no incentive to manage resources sustainably.

FAQ

Is the Amazon all primary forest?

People long assumed that the entire Amazon rainforest was pristine, untouched primary forest, but recent scientific research is revealing that portions of the Amazon had been cleared for farms, orchards, and villages centuries ago and that the region evidently supported thousands, or perhaps millions, of people. Most of them died in epidemics that swept the Americas in the wake of early visits by Europeans, who sometimes unknowingly brought diseases against which Native American people had evolved no defenses. Once these Amazonian civilizations disappeared, the regions they had cleared regrew into the forest we see today.

Local people may receive temporary employment from the corporation, but once the timber is gone they no longer have the forest and the ecosystem services it provided. As a result, most economic benefits are short term and are reaped not by local residents but by the foreign corporation. Much of the wood extracted in developing nations is exported to Europe and North America. In this way, our consumption of high-end furniture and other wood products can drive forest destruction in poorer nations.

Throughout Southeast Asia and Indonesia today, vast areas of tropical rainforest are being cut to establish plantations of oil palms (**FIGURE 9.10**). Oil palm fruit produces palm oil, which we use in snack foods, soaps, cosmetics, and as a biofuel. In Indonesia, the world's largest palm oil

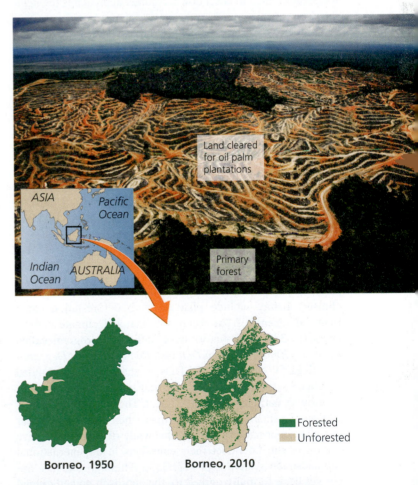

FIGURE 9.10 Oil palm plantations are replacing primary forest across Southeast Asia and Indonesia. Since 1950, the immense island of Borneo (maps at bottom) has lost most of its forest. *Data from Radday, M., WWF-Germany, 2007. Designed by Hugo Ahlenius, UNEP/GRID-Arendal. Extent of deforestation in Borneo 1950–2001, and projection towards 2020. http://www.grida.no/resources/8324.*

producer, oil palm plantations have displaced over 8 million ha (20 million acres) of rainforest. Clearing for plantations encourages further development and eases access for people to enter the forest and conduct logging illegally, leading to extensive loss of biodiversity.

weighing the ISSUES

Logging Here or There

Suppose you are an activist protesting the logging of old-growth trees near your hometown. Now let's say you know that if the protest is successful, the company will move to a developing country and cut its primary forest instead. Would you still protest the logging in your hometown? Would you pursue any other approaches?

Solutions are emerging

Many avenues are being pursued to address deforestation in developing nations. Some conservation proponents are running community-based conservation projects (p. 185) that empower local people to act as stewards of their forest resources. In other cases, conservation organizations, such as Conservation International, are buying concessions and using them to preserve forest rather than to cut it down. Another approach is the *debt-for-nature swap*, in which a conservation organization or a government offers to pay off a portion of a developing nation's international debt in exchange for a promise by the nation to conserve natural land.

Forest proponents are also trying to focus new agricultural development on lands that are already cleared. Brazil's 2006 soy moratorium, negotiated by soy producers, international conservationists, and Brazilian policymakers, successfully reduced clearing for soy by 30 times. People elsewhere in the world saw this as a model, and today are trying similar approaches. In Indonesia, the nonprofit World Resources Institute (WRI) began working with palm oil companies that own concessions to clear primary rainforest and steers them instead to plant their plantations on land that is already logged.

Because deforestation accounts for at least 12% of the world's greenhouse gas emissions—nearly as much as all the world's vehicles emit—international efforts to address global climate change include plans to curb deforestation using financial incentives. At recent international climate conferences (p. 335), negotiators have outlined a program called *Reducing Emissions from Deforestation and Forest Degradation* (REDD; changed to REDD+ as the program expanded in scope), whereby wealthy industrialized nations pay poorer developing nations to conserve forest. The aim is to make forests more valuable when saved than when cut down. Under this plan, poor nations gain income while rich nations receive carbon credits to offset their emissions in an international cap-and-trade system (pp. 113, 333). Although REDD+ has not yet been formally agreed to, the approach gained crucial momentum at the Paris Climate Conference in 2015. Brazil gained praise at the Paris conference by promising to reduce its emissions by 37% by 2025 and 43% by 2030 (compared with its 2005 levels) through aggressive steps to curb deforestation. Unfortunately, its policy shifts since then will make these goals more challenging.

Forest Management

As our demands on forests intensify, we need to manage forests with care. *Foresters* are professionals who manage forests through the practice of **forestry**. Foresters must balance our society's demand for forest products with the central importance of forests as ecosystems.

Debates over how to manage forest resources reflect broader questions about how to manage natural resources in general. Resources such as fossil fuels and many minerals are nonrenewable, whereas resources such as the sun's energy are perpetually renewable (p. 4). Between these extremes lie resources that are renewable if they are not exploited too rapidly. These include timber, as well as soils, fresh water, rangeland, wildlife, and fisheries.

Resource managers follow several strategies

Resource management describes our use of strategies to manage and regulate the harvest of renewable resources. Sustainable resource management involves harvesting these resources in ways that do not deplete them. A key question in managing resources is whether to focus strictly on the resource of interest or to look more broadly at the environmental system of which it is a part. Taking a broader view often helps avoid degrading the system and thereby helps to sustain the resource.

Maximum sustainable yield

A guiding principle in resource management has traditionally been **maximum sustainable yield.** Its aim is to achieve the maximum amount of resource extraction without depleting the resource from one harvest to the next. Recall the logistic growth curve (see Figure 3.14, p. 64), which shows how a population grows most quickly when it is at an intermediate size—specifically, at one-half of carrying capacity. A fisheries manager striving for maximum sustainable yield, for example, will aim to keep fish populations at intermediate levels so that they rebound quickly after each harvest. Doing so should result in the greatest amount of fish harvested over time while sustaining the population (**FIGURE 9.11**). This approach, however, keeps the fish population at only half its carrying capacity—well below the size it would attain in the absence of fishing. This will alter the food web dynamics of the community.

In forestry, maximum sustainable yield argues for cutting trees shortly after they go through their fastest stage of growth. Because trees increase in biomass most quickly at an intermediate age, they are cut long before growing as large as they would in the absence of harvesting. This practice maximizes timber production, but it alters forest ecology and eliminates habitat for species that depend on mature trees.

Ecosystem-based management

Because of these dilemmas, more and more managers espouse **ecosystem-based management,** which aims to minimize impact on the ecological processes that provide the resource. Under this approach, foresters may protect certain forested areas, restore ecologically important habitats, and consider patterns at the landscape level (p. 37), allowing timber harvesting while preserving the functional integrity of the forest ecosystem. Ecosystems are complex, however, so it can be challenging to determine how to implement this type of management.

Adaptive management

Some management actions will succeed, and some will fail. **Adaptive management** involves testing different approaches and trying

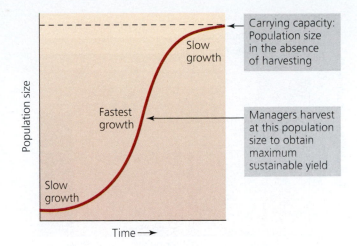

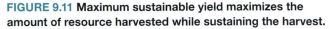

FIGURE 9.11 Maximum sustainable yield maximizes the amount of resource harvested while sustaining the harvest.

to improve methods through time. For managers, it entails monitoring the results of one's practices and adjusting them based upon what is learned.

We extract timber from private and public lands

The United States began formally managing forest resources a century ago, after depletion of eastern U.S. forests prompted fear of a "timber famine." This led the federal government to form forest reserves: public lands set aside to grow trees, produce timber, and protect water quality. Today's system of **national forests** consists of 77 million ha (190 million acres), managed by the U.S. Forest Service and covering more than 8% of the nation's land area (**FIGURE 9.12**). The Forest Service

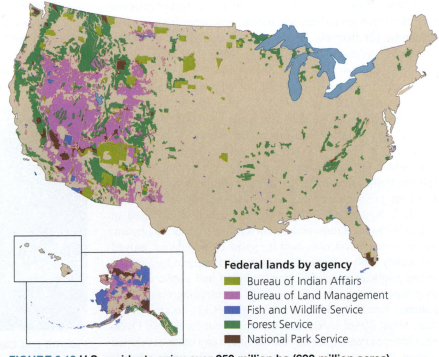

Federal lands by agency
- Bureau of Indian Affairs
- Bureau of Land Management
- Fish and Wildlife Service
- Forest Service
- National Park Service

FIGURE 9.12 U.S. residents enjoy over 250 million ha (600 million acres) of public lands. *Data from United States Geological Survey.*

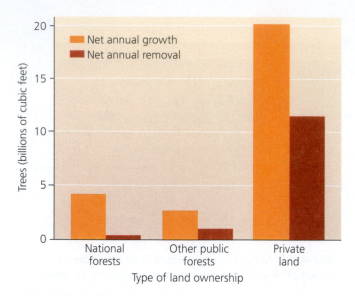

FIGURE 9.13 In the United States, trees are growing faster than they are being removed. However, forests that regrow after logging often differ greatly from the forests they replace. *"Private land" here combines land owned by the timber industry and by small landholders. Data from Oswalt, S.N., et al., 2014. Forest resources of the United States, 2012. For. Serv. Gen. Tech. Rep. WO-91. Washington, D.C.: USDA Forest Service.*

DATA For which of the three types of land is the ratio of growth to removal greatest?

Go to **Interpreting Graphs & Data** on *Mastering* Environmental Science

was established in 1905 under Gifford Pinchot, whose conservation ethic (p. 15) meant managing forests for "the greatest good of the greatest number in the long run." Pinchot believed that the nation should extract and use resources from its public lands, but that wise and careful management of timber resources was imperative.

Today almost 90% of timber harvesting in the United States takes place on private land owned by the timber industry or by small landowners (**FIGURE 9.13**). Timber companies generally pursue maximum sustainable yield on their land, as well as extracting timber from public forests. In the national forests, U.S. Forest Service staff conduct timber sales and build roads to provide access for logging companies, which sell the timber they harvest for profit. In this way, taxpayers subsidize private harvesting on public land (p. 113).

In U.S. national forests, timber extraction increased in the 1950s as the nation underwent a postwar economic boom, paper consumption rose, and the growing population moved into newly built suburban homes. Harvests began to decrease in the 1980s as economic trends shifted and public concern over logging grew. Today, regrowth is outpacing removal on national forests by 11 to 1 (see Figure 9.13); in an average year, about 2% of U.S. forest acreage is cut for timber. Overall, timber harvesting in the United States and other developed nations has remained stable for the past half-century, while it has more than doubled in developing nations.

Note, however, that even when the regrowth of trees outpaces their removal, the character of a forest may change. Once primary forest is replaced by secondary forest or single-species plantations, the resulting community may be very different and is generally less ecologically valuable.

Plantation forestry has grown

Today's timber industry focuses on production from plantations of fast-growing tree species planted in single-species monocultures (p. 142). All trees in a stand are planted at the same time, so the stands are said to be *even-aged*. Stands are cut after a certain number of years (the *rotation time*), and the land is replanted with seedlings. Plantation forestry is growing, and 7% of the world's forests are now plantations. One-quarter of these feature non-native tree species.

Ecologists and foresters view plantations as akin to crop agriculture. With few tree species and little variation in tree age, plantations do not offer habitat to many forest organisms. They lack the structural complexity of a mature natural forest (see Figure 9.5) and are vulnerable to outbreaks of pest species. For these reasons, some harvesting methods aim to maintain *uneven-aged* stands featuring a mix of trees of different ages and species.

We harvest timber in several ways

Timber companies choose from several methods to harvest trees. In **clear-cutting,** all trees in an area are cut at once (**FIGURE 9.14**). Clear-cutting is cost-efficient, and to some extent it can mimic natural disturbance events, such as fires or windstorms. However, the ecological impacts are severe. An entire ecological community is removed, soil erodes away, and sunlight penetrates to ground level, changing microclimatic conditions. As a result, new types of plants replace those of the original forest.

Concerns about clear-cutting led to alternative harvesting methods. In seed-tree systems and shelterwood systems, a few large trees are left standing in clear-cuts to reseed the area or to provide shelter for seedlings. In selection systems, a minority of trees is removed at any one time, while most are left

FIGURE 9.14 Clear-cutting is cost-efficient for timber companies but has ecological consequences. These include soil erosion, water pollution, and altered community composition.

standing. Selection systems preserve much of a forest's structural diversity, but are less cost-efficient for the industry.

All harvesting methods disturb soil, alter habitat, affect plants and animals, and modify forest structure and composition. Most methods speed runoff, raise flooding risk, and increase soil erosion, thereby degrading water quality. Finding ways to minimize these impacts is important, because timber harvesting is necessary to obtain the wood products that all of us use.

As people became more aware of the impacts of logging, many began to urge that public forests be managed for recreation, wildlife, and ecological integrity. In 1976 the U.S. Congress passed the National Forest Management Act, mandating that every national forest draw up plans for renewable resource management that assess the ecological impacts of logging and are subject to public input under the National Environmental Policy Act (p. 108). As a result, timber-harvesting methods were integrated with ecosystem-based management goals, and the Forest Service developed programs to manage wildlife and restore degraded ecosystems.

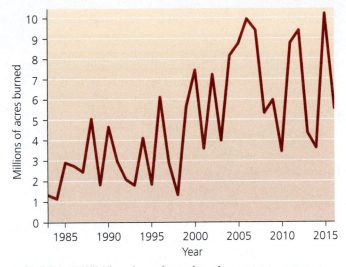

FIGURE 9.15 Wildfires have been burning more acreage across the United States. Fuel buildup from decades of fire suppression has contributed to this trend. *Data from National Interagency Fire Center.*

Fire can hurt or help forests

One element of management involves how to handle wildfire. For over a century, the Forest Service and other agencies suppressed fire whenever and wherever it broke out. Yet scientific research now shows that many species and ecological communities depend on fire. Some plants have seeds that germinate only in response to fire. When fire is suppressed, shrubs invade grasslands, pine woodlands become cluttered with hardwood understory, invasive plants move in, and animal diversity and abundance decline. Researchers studying tree rings have documented that historically, North America's grasslands and pine woodlands burned frequently. (Burn marks in a tree's growth rings reveal past fires, giving scientists an accurate history of fire events extending back hundreds or even thousands of years.)

Scientists also now know that putting out frequent low-intensity fires in forests increases the risk of occasional large catastrophic fires that can damage forests, destroy property, and threaten lives. When fire is suppressed, branches, logs, and leaf litter accumulate on the forest floor, producing kindling for a catastrophic fire. Severe fires have become more numerous in recent years (**FIGURE 9.15**), and fire control expenses now eat up most of the U.S. Forest Service's budget. Meanwhile, residential development alongside forested land—in the *wildland-urban interface*—is placing more homes in fire-prone areas.

FAQ

Aren't all forest fires bad?

No. Fire is a natural process that helps to maintain many forests and grasslands. When allowed to occur naturally, low-intensity fires generally burn moderate amounts of material, return nutrients to the soil, and promote lush growth of new vegetation. When we suppress fire, we allow unnaturally large amounts of dead wood, dried grass, and leaf litter to accumulate. This material becomes kindling that eventually can feed a truly damaging fire that grows too big and too hot to control. This is why many land managers today conduct carefully controlled prescribed burns and also allow some natural fires to run their course. By doing so, they aim to help return our fire-dependent ecosystems to a healthier and safer condition.

Once a catastrophic high-intensity fire burns a forest, it may leave many dead trees. Timber harvesters often try to remove dead trees, or snags, for their wood following a fire (or after a windstorm, insect damage, or disease)—a practice called **salvage logging.** From a short-term economic standpoint, salvage logging may seem to make good sense. However, ecologically, snags have immense value; the insects that decay them provide food for wildlife, and many animals depend on holes in snags for nesting and roosting. Running logging operations on recently burned land can also cause soil erosion, impede tree regeneration, and promote further wildfire.

To help prevent catastrophic fires, land management agencies now burn areas of forest deliberately with low-intensity fires under carefully controlled conditions (**FIGURE 9.16**). These **prescribed burns** clear away fuel loads, nourish the soil with ash, and encourage the vigorous growth of new vegetation.

FIGURE 9.16 Prescribed fire helps to promote forest health and prevent larger damaging fires. Forest Service staff are shown here conducting a carefully controlled, low-intensity burn.

Climate change and pest outbreaks are altering forests

Global climate change (Chapter 14) is now worsening wildfire risk in North America and other regions where it is bringing warmer and drier weather. In the Amazon in recent years, unprecedented fires have destroyed large areas of rainforest—wet forest ecosystems that are *not* adapted to fire. Tropical rainforests in Indonesia and Central America have also suffered severe fire damage during hot, dry spells resulting from strong El Niño conditions (p. 262) and climate change. Likewise, Canada's boreal forests have suffered vast fires in recent years.

Deforestation can worsen the impacts of climate change, because when forests are removed across a large region, humidity and transpiration are reduced and less precipitation falls, making the region drier. Many areas of the Amazon rainforest, particularly its edges, are much drier than they used to be, intensifying the risk of rainforest destruction by wildfire. In this way, deforestation and climate change create a dangerous positive feedback loop.

Climate change also is promoting outbreaks of pest insects. Bark beetles feed within the bark of conifer trees. They attract one another to weakened trees and attack *en masse,* eating tissue, laying eggs, and bringing with them a small army of fungi, bacteria, and other pathogens. Since the 1990s, infestations of bark beetles have devastated tens of millions of acres of forest in western North America (**FIGURE 9.17**), killing tens of *billions* of conifer trees and leaving them as fodder for fires.

Today's unprecedented outbreaks result from milder winters that allow beetles to overwinter farther north and from warmer summers that speed up their feeding and reproduction. In Alaska, beetles have switched from a two-year life-cycle to a one-year cycle. In parts of the Rocky Mountains, they now produce two broods per year instead of one. Moreover, drought has stressed and weakened trees, making them vulnerable to attack. On top of these climatic effects, past forest management has resulted in large areas of even-aged plantation forests dominated by single species that the beetles prefer, and many trees in these forests are now at a prime age for beetle infestation.

As climate change interacts with pests, diseases, fire, and deforestation, our forests are being altered in profound ways. Already, many dense, moist forests—from North America to the Amazon—have been replaced by drier woodlands, shrublands, or grasslands.

weighing the ISSUES

How to Handle Fire?

Can you suggest solutions to help protect people's homes from fire in the wildland-urban interface while improving the ecological condition of forests? Should people who choose to live in homes in fire-prone areas pay a premium for insurance against fire damage? Should homeowners in fire-prone areas be fined if they don't adhere to fire-prevention maintenance guidelines? Should new home construction be allowed in areas known to be fire-prone?

FIGURE 9.17 Climate change is enabling bark beetles to destroy vast numbers of trees in North America.

Sustainable forestry is gaining ground

Each of us can help address the challenges to forests by supporting certified sustainable forestry practices when we shop for wood or paper products. Certification organizations examine timber-harvesting practices and rate them against criteria for sustainability. These organizations then grant **sustainable forest certification** to forests, companies, and products produced using methods judged to be sustainable (**FIGURE 9.18**). As consumers, we can also choose to buy reclaimed or salvaged wood, which lowers the market demand for cutting trees.

Among certification organizations, the Forest Stewardship Council (FSC) is considered to have the strictest standards. FSC-certified timber-harvesting operations are required to protect rare species and sensitive habitats, safeguard water sources, control erosion, minimize pesticide use, and maintain the diversity of the forest and its ability to regenerate after harvesting.

FIGURE 9.18 Logs from trees harvested using certified sustainable practices are marked with the FSC logo.

As you turn the pages of this textbook, you are handling paper made from trees that were grown, managed, harvested, and processed using FSC-certified practices from sustainably managed forests. The paper for this textbook is "chain-of-custody certified," meaning that all steps in the life-cycle of the paper's production—from timber harvest to transport to pulping to production—have met strict standards.

More than 6% of the world's forests managed for timber production are FSC-certified, as of 2017—over 199 million ha (492 million acres) in 83 nations, and growing. The range of certification programs is also growing. In Brazil, for example, a new program by the Sustainable Agriculture Network certifies cattle ranches that meet sustainability guidelines that help protect forest.

Research indicates that certification can be effective. One recent study showed that FSC certification in Indonesia reduced deforestation there by 5% between 2000 and 2008, while also improving air quality, respiratory health, firewood availability, and nutrition for people living near timber-harvesting areas.

Pursuing sustainable forestry practices is often costlier for producers, but they recoup these costs when consumers pay more for certified products. If certification standards are kept strong, then we as consumers can help to promote sustainable forestry practices by exercising choice in the marketplace.

Parks and Protected Areas

As our world fills with more people consuming more resources, the sustainable management of forests and other ecosystems becomes ever more important. So does our need to preserve functional ecosystems by setting aside tracts of undisturbed land to remain forever undeveloped.

Why create parks and reserves?

People establish parks, reserves, and protected areas because these places offer us:

- Inspiration from their scenic beauty.
- Hiking, fishing, hunting, kayaking, bird-watching, and other recreation.
- Revenue from ecotourism (pp. 65, 173).
- Health, peace of mind, exploration, wonder, and solace.
- Utilitarian benefits and ecosystem services (such as how forested watersheds provide cities clean drinking water and a buffer against floods).
- Refuges for biodiversity, helping to maintain populations, habitats, communities, and ecosystems.

Parks and reserves were pioneered in the United States

The striking scenery of the American West persuaded U.S. leaders to create the world's first **national parks,** public lands protected from resource extraction and development

(a) Yosemite National Park, California

(b) Arches National Park, Utah

FIGURE 9.19 The awe-inspiring beauty of America's national parks draws millions of people for recreation and wildlife-watching.

but open to nature appreciation and recreation (**FIGURE 9.19**). Yellowstone National Park was established in 1872, followed by Sequoia, General Grant (now Kings Canyon), Yosemite, Mount Rainier, and Crater Lake National Parks.

The National Park Service was created in 1916 to administer the growing system of parks, monuments, historic sites, and recreation areas, which today numbers 417 units totaling more than 34 million ha (84 million acres; see Figure 9.12). Because America's national parks are open to everyone and showcase the nation's natural beauty in a democratic way, writer Wallace Stegner famously called them "the best idea we ever had."

Another type of protected area in the United States is the **national wildlife refuge.** The national wildlife refuge system, begun in 1903 by President Theodore Roosevelt, now totals more than 560 sites comprising nearly 60 million ha (150 million acres; see Figure 9.12). The U.S. Fish and Wildlife Service administers these refuges, which serve as havens for wildlife and encourage hunting, fishing, wildlife observation, photography, environmental education, and other public uses.

In response to the public's desire for undeveloped areas of land, in 1964 the U.S. Congress passed the Wilderness Act, which allowed some existing federal lands to be designated as **wilderness areas**. These areas are off-limits to development but are open to hiking, nature study, and other low-impact public recreation. Overall, the nation has 765 wilderness areas totaling 44 million ha (109 million acres), covering 4.5% of U.S. land area (2.7% if Alaska is excluded).

Efforts to set aside and manage public land at the national level are paralleled at state, regional, and local levels. Each U.S. state has agencies that manage resources and provide for recreation on public lands, as do many counties and municipalities.

Private nonprofit groups also preserve land. **Land trusts** are local or regional organizations that purchase ecologically important tracts of land in order to preserve them for future generations. The Nature Conservancy is the world's largest land trust, but nearly 1700 local land trusts in the United States together own 870,000 ha (2.1 million acres) and have helped preserve an additional 5.6 million ha (13.9 million acres), including scenic areas such as Big Sur on the California coast, Jackson Hole in Wyoming, and Maine's Mount Desert Island.

Protected areas are increasing internationally

Many nations have established protected areas and are benefiting from ecotourism as a result—from Kenya and Tanzania (Chapter 8) to Costa Rica (Chapter 5) to Belize to Ecuador to India to Australia. Today thousands of protected areas cover almost 15% of the planet's land area and 4% of its ocean area. However, parks often do not receive funding adequate to manage resources, provide recreation, or protect wildlife from poaching and trees from logging. As a result, many of the world's protected areas are merely *paper parks*—protected on paper but not in reality.

Brazil has designated more acreage for protection than any other nation. Most of this area lies deep in the Amazon rainforest, in large tracts set aside for indigenous tribes of people who live in the forest. Altogether about 30% of the Brazilian Amazon is afforded—on paper, at least—some degree of protection. A recent scientific study of the Brazilian Amazon found that protected areas suffered 10 times less deforestation than did non-protected areas. However, policymakers tend to locate protected areas in regions that are remote and that face fewer development pressures. Thus, to obtain a fair comparison, the researchers controlled for location variables and found that about half the difference in deforestation rates was due to protected status and half was due to location.

More and more protected areas today incorporate sustainable use to benefit local people. The United Nations oversees **biosphere reserves** (FIGURE 9.20), areas with exceptional biodiversity that couple preservation with sustainable development (p. 114). Each biosphere reserve consists of (1) a core area that preserves biodiversity; (2) a buffer zone that allows

Biodiversity protection

Sustainable agriculture, small settlements

Core area

Buffer zone

Transition zone

Limited development; research; education; ecotourism

FIGURE 9.20 Biosphere reserves couple preservation with sustainable development. Each biosphere reserve includes three zones. The photo above shows one example of sustainable use of forest products: Women process and sell Maya nuts harvested from rainforest trees in the Maya Biosphere Reserve in Guatemala. Here, management of forest concessions by local communities has helped to reduce illegal logging.

local activities and limited development; and (3) an outer transition zone where agriculture, settlement, and other land uses are pursued sustainably.

World heritage sites, another type of international protected area, are designated for their natural or cultural value. One such site is a transboundary reserve for mountain gorillas shared by three nations in Africa. Protected areas that overlap national borders sometimes function as "peace parks," acting as buffers between nations.

Habitat fragmentation makes preserves more vital

Protecting large areas of land has taken on new urgency now that scientists understand the risks posed by habitat fragmentation (see THE SCIENCE BEHIND THE STORY, pp. 206–207). Expanding agriculture, spreading cities, highways, logging, and other impacts routinely divide large expanses of habitat into small, disconnected ones (see Figure 8.9, p. 177). Even where forest cover is increasing, forests are becoming fragmented into ever-smaller parcels (FIGURE 9.21a, b).

When forests are fragmented, many species suffer. Bears, mountain lions, and other animals that need large areas of habitat may disappear. Birds that thrive in the interior of forests may fail to reproduce near the edge of a fragment

(FIGURE 9.21c). Their nests are attacked by predators and parasites that favor open habitats or that travel along habitat edges. Because of such **edge effects,** avian ecologists judge forest fragmentation to be a main reason why populations of many North American songbirds are declining.

Fragmentation is affecting our national parks, which are islands of habitat surrounded by farms, ranches, roads, and cities. In 1983 conservation biologist William Newmark examined historical records of mammal sightings in North American national parks. He found that most parks were missing species they had held previously. The red fox and river otter had vanished from Sequoia and Kings Canyon National Parks, for example, and the white-tailed jackrabbit and spotted skunk no longer lived in Bryce Canyon National Park. In all, 42 species had disappeared. As ecological theory predicted, smaller parks lost more species than larger parks. The parks were too small to sustain these populations, Newmark concluded, and they had become too isolated to be recolonized by new arrivals.

Because habitat fragmentation is a central issue in biodiversity conservation, and because there are limits on how much land can feasibly be set aside, conservation biologists have come to see value in **corridors** of protected land that allow animals to travel between islands of habitat (see Figure 2.17, p. 38). In theory, connecting fragments provides

(a) Fragmentation from clear-cuts in British Columbia

(c) Wood thrush

1831   1882   1950

(b) Fragmentation of wooded area (green) in Cadiz Township, Wisconsin

FIGURE 9.21 Forest fragmentation has ecological consequences. Fragmentation results from **(a)** clear-cutting and **(b)** agriculture and residential development. Fragmentation affects forest-dwelling species such as the **(c)** wood thrush, whose nests are parasitized by cowbirds that thrive in surrounding open country. *Source (b): Curtis, J.T., 1956. The modification of mid-latitude grasslands and forests by man. In W.L. Thomas Jr. (Ed.),* Man's role in changing the face of the earth. *©1956. Used by permission of the publisher, University of Chicago Press.*

Forest Fragmentation in the Amazon

What happens to animals, plants, and ecosystems when we fragment a forest? A massive research experiment in the middle of the Amazon rainforest is helping scientists learn the answers.

Stretching across 1000 km^2 (386 mi^2) of tropical rainforest north of the Brazilian city of Manaus, the Biological Dynamics of Forest Fragments Project (BDFFP) is the world's largest and longest-running experiment on forest fragmentation. For more than 30 years, hundreds of researchers have muddied their boots here, publishing over 700 scientific research papers, graduate theses, and books.

Dr. Thomas Lovejoy in the Amazon

The story begins back in the 1970s as conservation biologists debated how to apply ecological theory from ocean islands to forested landscapes that were being fragmented into islands of habitat. Biologist Thomas Lovejoy decided some good hard data were needed. He conceived a huge experiment to test ideas about forest fragmentation and established it in the heart of the biggest rainforest on the planet—South America's Amazon rainforest.

Farmers, ranchers, loggers, and miners were streaming into the Amazon then, and deforestation was rife. If scientists could learn how large fragments had to be in order to retain their species, it would help them work with policymakers to preserve forests in the face of development pressures.

Lovejoy's team of Brazilians and Americans worked out a deal with Brazil's government: Ranchers could clear some forest within the study area if they left square plots of forest standing as fragments within those clearings. By this process, 11 fragments of three sizes (1, 10, and 100 ha [2.5, 25, and 250 acres]) were left standing, isolated as "islands" of forest, surrounded by "seas" of cattle pasture (**FIGURE 1**). Each fragment was fenced to keep cattle out. Then, 12 study plots (of 1, 10, 100, and 1000 ha) were established within the large expanses of continuous forest still surrounding the pastures; these

would serve as control plots against which the fragments could be compared.

Besides comparing treatments (fragments) and controls (continuous forest), the project also surveyed populations in fragments before and after they were isolated. These data on trees, birds, mammals, amphibians, and invertebrates showed declines in diversity in most groups (**FIGURE 2**).

As researchers studied the plots over the years, they found that small fragments lost more species, and lost them faster, than large fragments—just as theory predicts. To slow down species loss by 10 times, researchers found that a fragment needs to be 1000 times bigger. Even 100-ha fragments were not large enough for some animals, and they lost half their species in less than 15 years. Monkeys died out because they need large ranges. So did colonies of army ants, as well as the birds that follow them to eat insects scared up as the ants swarm across the forest floor.

Fragments distant from continuous forest lost more species, but data revealed that even very small openings can stop organisms from dispersing to recolonize fragments. Many understory birds adapted to deep interior forest would not traverse cleared areas of only 30–80 m (100–260 ft). Distances of just 15–100 m (50–330 ft) were insurmountable for some bees, beetles, and tree-dwelling mammals.

FIGURE 1 Experimental forest fragments of 1, 10, and 100 ha were created in the BDFFP.

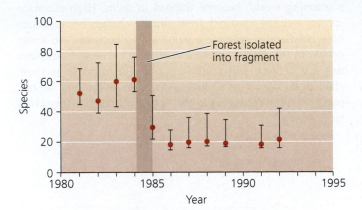

FIGURE 2 This graph shows how species richness of understory birds declined in a 1-ha parcel of forest after it was isolated as a fragment in 1984. Error bars show statistical uncertainty around mean estimates. *Data from Ferraz, G., et al., 2003. Rates of species loss from Amazonian forest fragments. Proc. Natl. Acad. Sci. 100: 14069–14073. © 2003 National Academy of Sciences. By permission.*

DATA Q
- On average, about how many bird species were present in this forest plot before fragmentation?
- How many were present after fragmentation?

Go to **Interpreting Graphs & Data** on **Mastering** Environmental Science

Soon, a complication ensued: Ranchers abandoned many of the pastures because the soil was unproductive, and these areas began filling in with young secondary forest, making the fragments less like islands. However, this led to new insights. Researchers learned that secondary forest can act as a corridor for some species, allowing them to disperse from mature forest and recolonize fragments where they'd disappeared. By documenting which species did this, scientists learned which may be more resilient to fragmentation.

The secondary forest habitat also introduced new species—generalists adapted to disturbed areas. Frogs, leaf-cutter ants, and small mammals and birds that thrive in second-growth

areas soon became common in adjacent fragments. Open-country butterfly species moved in, displacing interior-forest butterflies.

The invasion by open-country species illustrated one type of edge effect (p. 205). There were more: Edges receive more sunlight, heat, and wind than interior forest, which can kill trees adapted to the dark, moist interior. Tree death releases carbon dioxide, worsening climate change. Sunlight also promotes growth of vines and shrubs that create a thick, tangled understory along edges. As BDFFP researchers documented these impacts, they found that many edge effects extended deep into the forest (**FIGURE 3**). Small fragments essentially became "all edge."

The results on edge effects are relevant for all of Amazonia, because forest clearance and road construction create an immense amount of edge. A satellite image study in the 1990s estimated that for every 2 acres of land deforested, 3 acres were brought within 1 km of a road or pasture edge.

BDFFP scientists emphasize that impacts across the Amazon will be more severe than those revealed by their experiments. This is because most real-life fragments (1) are not protected from hunting, logging, mining, and fires; (2) do not have secondary forest to provide connectivity; (3) are not near large tracts of continuous forest that provide recolonizing species and maintain humidity and rainfall; and (4) are not square in shape, and thus feature more edge. Scientists say their years of data argue for preserving numerous tracts of Amazonian forest that are as large as possible.

The BDFFP has inspired other large-scale, long-term experiments on forest fragmentation elsewhere in the world; such studies are now running in Kansas, South Carolina, Australia, Borneo, Canada, France, and the United Kingdom. These projects are providing data showing how smaller area, greater isolation, and more edge are affecting species, communities, and ecosystems. Thus far, they indicate that fragmentation decreases biodiversity by 13–75% while altering nutrient cycles and other ecological processes—and that the effects magnify with time. Insights from these projects are helping scientists and policymakers strategize how best to conserve biodiversity amid continued pressures on forests.

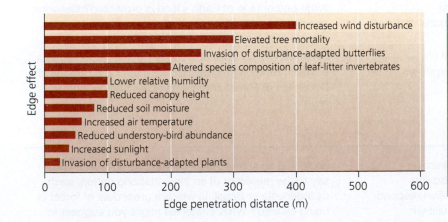

FIGURE 3 Edge effects can extend far into the interior of forest fragments. *Data from studies summarized in Laurance, W.F., et al., 2002. Ecosystem decay of Amazonian forest fragments: A 22-year investigation. Conservation Biology 16: 605–618, Fig. 3. Adapted by permission of John Wiley & Sons, Inc.*

DATA Q
- Would a tree inside a forest fragment, 275 meters in from the edge, be susceptible to edge effects?
- If so, which ones?

Go to **Interpreting Graphs & Data** on **Mastering Environmental Science**

animals access to more habitat and encourages gene flow to maintain populations in the long term. For these reasons, many land managers now try to establish corridors to join new reserves to existing reserves.

Climate change threatens protected areas

Today global climate change (Chapter 14) threatens our investments in protected areas. As temperatures become warmer, species ranges shift toward the poles and upward in elevation (p. 327). In a landscape of fragmented habitat, some organisms may be unable to move from one fragment to another. Species we had hoped to protect in parks may, in a warming world, become trapped in them. High-elevation species face the most risk from climate change, because they have nowhere to go once a mountaintop becomes too warm or dry. In this light, corridors to allow movement from place to place become still more important. In response to these challenges, conservation biologists are now looking beyond parks and protected areas as they explore strategies for saving biodiversity.

closing THE LOOP

Forests are ecologically vital and economically valuable, yet we continue to lose them around the world. The slowing of deforestation in the Brazilian Amazon has been encouraging, but renewed commitment will be needed to build on that trend. The condition of the Amazon rainforest today raises questions relevant for forests and people everywhere: How can we best safeguard areas of high biodiversity? How can economies transition from resource exploitation to conservation and sustainable use? How can protected areas function as intended? Can climate change policies and financial incentives protect and restore forests? And can sustainable certification programs help preserve forests while benefiting local people?

In North America, early emphasis on resource extraction evolved into policies of sustainable yield as land and resource availability began to decline. Public forests today are managed not only for timber production but also for recreation, wildlife habitat, and ecosystem integrity. Sustainable forest certification now provides economic incentives for conservation on forested lands.

Around the world, public support for the preservation of natural lands has led to the establishment of parks and protected areas. As development spreads across the landscape, fragmenting forests, scientists trying to conserve species, communities, and ecosystems are working more and more at the landscape level.

TESTING Your Comprehension

1. Name at least two reasons why natural primary forests contain more biodiversity than even-aged, single-species forestry plantations.

2. Describe three ecosystem services that forests provide.

3. Name several major causes of deforestation. Where is deforestation most severe today?

4. Compare and contrast maximum sustainable yield, ecosystem-based management, and adaptive management. How may pursuing maximum sustainable yield sometimes affect populations and communities?

5. Compare and contrast the major methods of timber harvesting. Which methods keep forest ecosystems most intact?

6. Describe several ecological impacts of logging. How has the U.S. Forest Service responded to public concern over these impacts?

7. Are forest fires a bad thing? Explain your answer. How do people use fire in a positive way?

8. Name at least three reasons that people have created parks and reserves. How do national parks differ from national wildlife refuges? What is a wilderness area?

9. What percentage of Earth's land is protected? Describe one type of protected area that has been established outside North America.

10. Give two examples of how forest fragmentation affects animals. How might research like that of the Biological Dynamics of Forest Fragments Project help us design reserves?

SEEKING Solutions

1. People in industrialized nations are fond of warning people in industrializing nations to stop destroying rainforest. People of industrializing nations often respond that this is hypocritical, because the industrialized nations became wealthy by deforesting their land and exploiting its resources in the past. What would you say to the president of an industrializing nation, such as Indonesia or Brazil, in which a great deal of forest is being cleared? What strategies might you suggest to help them achieve wealth while also conserving forests?

2. Do you think maximum sustainable yield represents a desirable policy for resource managers to follow? Why or why not? What might be some alternative approaches?

3. Consider the impacts that climate change may have on species' ranges. If you were trying to preserve an endangered mammal that occurs in a small area and you had generous funding to acquire land to help restore its population, how would you design a protected area for it? Would you use corridors? Would you include a diversity of elevations? Would you design a few large reserves or many small ones? Explain your answers.

4. **CASE STUDY CONNECTION** You run a non-profit environmental advocacy organization and are trying to save a tract of rainforest in the Brazilian Amazon. You have worked in this region for years and care for the local people, who want to save the forest and its animals but also need to make a living and use the forest's resources. Brazil's government plans to sell a concession to a multinational timber corporation to log the entire forest unless your group can work out some other solution. Describe what solution(s) you would try to arrange. Consider the range of options discussed in this chapter, including protected areas, land trusts, biosphere reserves, forest management techniques, carbon offsets, certified sustainable forestry, and more. Explain reasons for your choice(s).

5. **THINK IT THROUGH** You have just become supervisor of a national forest. Timber companies are asking to cut as many trees as you will let them, but environmental advocates want no logging at all. Ten percent of your forest is old-growth primary forest, and 90% is secondary forest. Your forest managers are split among preferring maximum sustainable yield, ecosystem-based management, and adaptive management. What management approach(es) will you take, and why? Will you allow logging of all old-growth trees, some, or none? Will you allow logging of secondary forest? If so, what harvesting strategies will you encourage? What would you ask your scientists before deciding on policies on fire management and salvage logging?

CALCULATING Ecological Footprints

We all rely on forest resources. The average North American consumes 225 kg (500 lb) of paper and paperboard each year. Using the estimates of paper and paperboard consumption for each region, calculate the per capita consumption for each using the population data in the table. Note: 1 metric ton = 2205 pounds.

REGION	POPULATION (MILLIONS)	TOTAL PAPER CONSUMED (MILLIONS OF METRIC TONS)	PER CAPITA PAPER CONSUMED (POUNDS)
Africa	1136	8	16
Asia	4351	190	
Europe	741	92	
Latin America	618	27	
North America	353	76	
Oceania	39	4	
World	7238	400	122

Data are for 2014, from Population Reference Bureau and UN Food and Agriculture Organization (FAO).

1. How much paper would be consumed if everyone in the world used as much paper as the average North American?

2. How much paper would North Americans save each year if they consumed paper at the rate of Europeans? At the rate of Africans?

3. North Americans have been reducing their per-person consumption of paper and paperboard by nearly 5% annually in recent years by recycling, shifting to online activity, and reducing packaging of some products. Name three specific things you personally could do to reduce your own consumption of paper products.

4. Describe three ways in which consuming FSC-certified paper rather than conventional paper can reduce the environmental impacts of paper consumption.

Mastering Environmental Science

CHAPTER 10

Environmental Health and Toxicology

Bisphenol A: Worldwide

Are We Being Poisoned by Our Food Packaging?

> " **This chemical is harming snails, insects, lobsters, fish, frogs, reptiles, birds, and rats, and the chemical industry is telling people that because you're human, unless there's human data, you can feel completely safe.**
> —Dr. Frederick vom Saal, BPA researcher

> **There is no basis for human health concerns from exposure to BPA.**
> —The American Chemistry Council "

Thanks to a lifetime of nutritional education, we are all aware that a diet high in sugars, fats, and processed foods can lead to harmful health impacts such as obesity, diabetes, and high blood pressure. Hence, we are taught to choose our foods wisely to avoid the dangers of an unhealthy diet. But, it turns out, the beverage bottles, food cans, and wrappers that surround our food may also pose a risk to our health by leaching harmful chemicals into the foods we eat and the liquids we drink.

Plastics are the predominant source of these chemicals. Plastics are polymers (p. 32) that contain a number of synthetic chemicals that can be released into foods and liquids—a process sped up by factors that promote the breakdown of plastics, such as extreme temperatures, exposure to ultraviolet light from the sun, or prolonged contact with highly acidic liquids, such as soda and fruit juices.

One chemical of concern is **bisphenol A** (**BPA** for short). Epoxy resins containing BPA are used to line the insides of metal food and drink cans and the insides of pipes for our water supply, as well as in enamels, varnishes, adhesives, and even dental sealants for our teeth. BPA is also present in polycarbonate plastic, a hard, clear type of plastic used in water bottles, food containers, CDs and DVDs, electronics, baby bottles, and children's toys.

Many plastic products also contain another class of hormone-disrupting chemical, called **phthalates.** Used to soften plastics, they are found in bottles and many types of food packaging, as well as in perfumes and children's toys.

Unfortunately, BPA and phthalates can leach out of their many products and into our food, water, air, and bodies. Over 90% of Americans carry detectable concentrations of PBA and phthalates in their urine, according to the Centers for Disease Control and Prevention (CDC). Because these chemicals pass through the body within hours, these data suggest that we are receiving almost continuous exposure.

What, if anything, are BPA and phthalates doing to us? To address such questions, scientists run experiments on laboratory animals, administering known doses of the substance and measuring the health impacts that result. Hundreds of studies with rats, mice, and other animals have shown many apparent health impacts from BPA and pthalates, including a wide range of reproductive abnormalities. Studies have also connected BPA and phthalates to impacts on human health. For example, a 2013 review of the scientific literature found 91 studies that examined the relationship between the level of BPA in subjects' urine or

Food packaging exposes people to BPA and phthalates. ▲

Upon completing this chapter, you will be able to:

- **Explain the goals of environmental health and identify major environmental health hazards**

- **Describe the types of toxic substances in the environment, the factors that affect their toxicity, and the defenses that organisms have against them**

- **Explain the movements of toxic substances and how they affect organisms and ecosystems**

- **Discuss the approaches used to study the effects of toxic chemicals on organisms**

- **Summarize risk assessment and risk management**

- **Compare philosophical approaches to risk and how they relate to regulatory policy**

Many consumers are embracing "BPA-free" water bottles due to concerns about the safety of bisphenol A.

FIGURE 10.1 **Studies have linked elevated blood/urine BPA concentrations to numerous health impacts in humans.** Although these correlative studies do not conclusively prove that BPA causes each observed ailment, they indicate topics for further research.

Labels on figure: Cardiovascular disease; Behavioral issues; Type 2 diabetes; Blood hormone levels; Hypertension; Obesity; Asthma; Kidney functioning; Sperm quality; Thyroid function; Male sexual performance; Immune function; Male genital abnormalities; Egg development and maturation

blood and a variety of health problems (**FIGURE 10.1**). Phthalates have also been associated with a range of adverse health impacts, such as birth defects, breast cancer, reduced sperm counts, diabetes, and cognitive impairment in children exposed to phthalates in the womb.

Many of these effects occur at extremely low doses—much lower than the exposure levels set so far by regulatory agencies for human safety. Scientists say this is because BPA and phthalates mimic certain hormones, such as male and female sex hormones; that is, they are structurally similar to sex hormones and can induce some of their effects in animals (see Figure 10.5, p. 218). Sex hormones function at minute concentrations, so when a synthetic chemical that is similar to the hormone reaches the body in a similarly low concentration, it can fool the body into responding. Other hormonal systems, such as the thyroid system that regulates growth and development, can also be affected by hormone-mimicking chemicals.

In reaction to research linking BPA and phthalates to health effects on humans, a growing number of researchers, doctors, and consumer advocates have called on governments to regulate BPA, phthalates, and other endocrine-disrupting chemicals and for manufacturers to stop using them. The chemical industry has long insisted that BPA and phthalates are safe, pointing to industry-sponsored research that finds no direct health impacts on people exposed to these chemicals.

Some governments have taken steps to regulate the use of BPA and phthalates in consumer products. Canada, for example, has banned BPA completely. In many other nations, including the United States, its use in products for babies and small children has been restricted. Accordingly, concerned parents can now more easily find BPA-free items for their infants and children, but the rest of us remain exposed through most food

cans, many drink containers, and thousands of other products. The European Union and nine other nations have banned some types of phthalates, and in 2008, the United States banned six types of phthalates in toys. Still, across North America, many routes of exposure remain.

In the face of mounting public concern about the safety of these chemicals, many companies are voluntarily choosing to remove them from their products, even in the absence of stringent regulation by the U.S. government. Walmart and Toys "R" Us, for example, decided to stop carrying children's products with BPA several years before the U.S. Food and Drug Administration (FDA) banned BPA use in baby bottles in 2012. Canned goods and containers that do not contain BPA are now available from large food companies such as ConAgra, Campbell's, and Tyson Foods, and are typically identified as BPA-free on their label. There is precedent for such efforts, because BPA was voluntarily phased out of can liners in Japan starting in the late 1990s.

To make matters worse, other classes of chemicals in packaging may be affecting our health. A 2017 study, for example, found that one-third of fast food packaging (mostly items with coatings designed for grease resistance) contained fluorinated chemicals that may also act as endocrine disruptors (see p. 218).

Although we don't yet know everything there is to know about BPA and phthalates, they aren't likely to be among our greatest environmental health threats. However, they provide a timely example of how we as a society assess health risks and decide how to manage them. As scientists and government regulators assess these chemicals' potential risks, their efforts give us a window on how hormone-disrupting chemicals are challenging the way we appraise and control the environmental health risks we face.

Environmental Health

Examining the impacts of human-made chemicals such as BPA is just one aspect of the broad field **environmental health,** which assesses environmental factors that influence our health and quality of life. These factors include both natural and anthropogenic (human-caused) factors. Practitioners of environmental health seek to prevent adverse effects on human health and on the ecological systems that are essential to our well-being.

We face four types of environmental hazards

We can categorize environmental health hazards into four main types: physical, chemical, biological, and cultural. Although some amount of risk is unavoidable, much of environmental health consists of taking steps to minimize the risks of encountering hazards and to lessen the impacts of the hazards we do encounter.

Physical hazards *Physical hazards* arise from processes that occur naturally in our environment and pose risks to human life or health. Some are ongoing natural phenomena, such as excessive exposure to ultraviolet (UV) radiation from sunlight, which damages DNA and has been tied to skin cancer, cataracts, and immune suppression (**FIGURE 10.2a**). We can reduce these risks by shielding our skin from intense sunlight with clothing and sunscreen.

Other physical hazards include discrete events such as earthquakes, volcanic eruptions, fires, floods, blizzards, landslides, hurricanes, and droughts. We cannot prevent many of these hazards, but we can minimize risk by preparing ourselves with emergency plans and avoiding practices that make us vulnerable to certain physical hazards. For example, scientists can map geologic faults to determine areas at risk of earthquakes, engineers can design buildings to resist damage, and governments and individuals can create emergency plans to prepare for a quake's aftermath.

Chemical hazards *Chemical hazards* include many of the synthetic chemicals that our society manufactures, such as pharmaceuticals, disinfectants, and pesticides (**FIGURE 10.2b**). Some substances produced naturally by organisms, such as venoms, can also be hazardous, as can many substances that are found in nature and then processed for our use, such as hydrocarbons, lead, and asbestos. Following our overview of environmental health, much of this chapter will focus on chemical health hazards and the ways we study and regulate them.

Biological hazards *Biological hazards* result from ecological interactions among organisms (**FIGURE 10.2c**). When we become sick from a virus, bacterial infection, or other pathogen, we are suffering parasitism (pp. 72–73). Some infectious diseases are spread when pathogenic microbes attack us directly. With others, infection occurs through a **vector,** an organism, such as a mosquito, that transfers the pathogen to the host. This is what we call **infectious disease.** Infectious diseases such as malaria, cholera, tuberculosis, and influenza (flu) are major environmental health hazards, especially in developing nations with widespread poverty and limited health care. As with physical and chemical hazards, it is impossible for us to avoid risk from biological

(a) Physical hazard **(b) Chemical hazard** **(c) Biological hazard** **(d) Cultural hazard**

FIGURE 10.2 Environmental health hazards come in four types. The sun's ultraviolet radiation is an example of a physical hazard **(a)**. Chemical hazards **(b)** include both synthetic and natural chemicals. Biological hazards **(c)** include diseases and the organisms that transmit them. Cultural or lifestyle hazards **(d)** include the behavioral decisions we make, such as smoking, as well as the socioeconomic constraints that may be forced upon us.

agents completely, but through monitoring, sanitation, and medical treatment we can reduce the likelihood and impacts of infection.

Cultural hazards Hazards that result from our place of residence, our occupation, the circumstances of our socioeconomic status, or our behavioral choices can be thought of as *cultural hazards*. We can minimize or prevent some of these cultural or lifestyle hazards, whereas others may be beyond our control. For instance, people can choose whether or not to smoke cigarettes (**FIGURE 10.2d**), but exposure to second-hand smoke at home or in the workplace may be beyond one's control. Much the same might be said for other cultural hazards such as diet and nutrition, workplace hazards, and drug use. Environmental justice advocates (pp. 16–17) argue that "forced" risks from cultural hazards, such as living near a hazardous waste site, are often higher for people with fewer economic resources or less political clout.

Disease is a major focus of environmental health

Despite all our technological advances, disease causes the vast majority of human deaths (**FIGURE 10.3a**). Over half the world's deaths result from **noninfectious diseases,** such as cancer and heart disease, which develop without the action of a foreign organism. You don't "catch" noninfectious diseases—people develop them through a combination of their genetics, coupled with environmental and lifestyle factors. For instance, whether a person develops lung cancer depends on his or her genes and on environmental conditions, such as the individual's exposure to airborne cancer-inducing chemicals, and to lifestyle choices, such as whether or not he or she chooses to smoke.

Although infectious disease accounts for fewer deaths than noninfectious disease, infectious disease robs society of more years of human life, because it strikes people at all ages, including the very young. Infectious diseases account for about 17% of deaths worldwide that occur each year (**FIGURE 10.3b**). Infectious disease is a greater problem in developing countries, where it accounts for nearly half of all deaths. Infectious disease causes many fewer deaths in developed nations because their wealth allows for better public sanitation and hygiene, which reduce the chance of contracting an infectious disease, and better access to medical care, which enables treatment if a disease is contracted.

In our world of global mobility and dense human populations, novel infectious diseases (or new strains of old diseases) that emerge in one location are more likely to spread quickly to other locations. Recent examples include the H5N1 avian flu ("bird flu"), which first appeared in humans in 1997; severe acute respiratory syndrome (SARS), where the first outbreak in humans was reported in 2003; the H1N1 swine flu that spread across the globe in 2009–2010; and the outbreak of Ebola in West Africa in 2014. Diseases like influenza are caused by pathogens that mutate readily, giving rise to a variety of strains of the disease with slightly

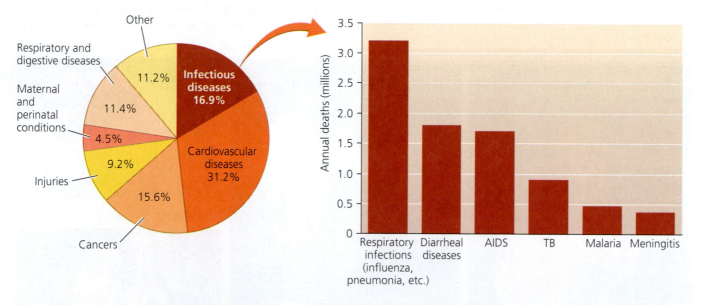

(a) Leading causes of death across the world

(b) Leading causes of death by infectious diseases

FIGURE 10.3 Infectious diseases are the second-leading cause of death worldwide. Six types of diseases account for 80% of all deaths from infectious disease. *Data from World Health Organization, 2015. Geneva, Switzerland: WHO, http://www.who.int.*

DATA AIDS is a well-known infectious disease, but respiratory and diarrheal diseases claim far more lives every year than AIDS. According to the figure, how many times more lives were lost to respiratory infections and diarrheal diseases than to AIDS?

Go to **Interpreting Graphs & Data** on **Mastering** Environmental Science

different genetics. As a pathogen's genes determine its virulence (a measure of how fast a disease spreads and the harm it does to infected individuals), these frequent mutations make it more likely that a highly virulent strain may arise and threaten a global pandemic at any time.

There are many ways to fight infectious disease. We can reduce the chance of infection through immunization, improved public sanitation, access to clean water, food security (p. 140), and public education campaigns that inform people of risks and how to avoid them. If infection occurs, the health effects and spread of infectious disease can be minimized by providing access to health care and medication.

SUCCESS STORY

Improving Access to Clean Drinking Water and Sanitation

Water-borne diseases, such as cholera and dysentery, have been a vexing problem for many of the world's developing countries. But thanks to the efforts of the United Nations (UN), national governments, and aid organizations, the threat from these infectious diseases—spread by the ingestion of food or water that has been contaminated with human or animal feces—is abating. Public health education campaigns have taught people how to sterilize their drinking water, as well as how to avoid contaminating the source of it. Infrastructure projects, such as the drilling of water wells and the construction of modern facilities to treat drinking water and wastewater, have provided 91% of people across the globe with access to safe drinking water—an increase of 2.6 billion people since 1990. In that same time period, some 2.1 billion people gained access to sanitary facilities as people in cities and villages embraced approaches to minimize the contamination of water supplies with fecal material, such as by installing composting toilets (as piped water is sometimes unavailable to the very poor) and through improved access to sewer systems, septic systems, and public latrines.

A young woman sampling water from her village's new well

Initiatives to provide people with clean drinking water have been very successful, enabling the United Nations to meet its Millennium Development Goal (enacted in 2000) of providing 88% of the world's population with clean water by 2015. While these efforts represent a significant improvement, many people in rural areas in poorer nations continue to lack access to safe drinking water and advanced sanitation. Efforts are currently underway to address these deficiencies, and the United Nations in 2017 announced the goal of providing all people access to clean drinking water by 2030.

EXPLORE THE DATA at **Mastering** Environmental Science

Toxicology is the study of chemical hazards

Although most indicators of human health are improving as the world's wealth increases, our modern society is exposing us to more and more synthetic chemicals. Some of these substances pose threats to human health, but figuring out which of them do—and how, and to what degree—is a complicated scientific endeavor. **Toxicology** examines how poisonous chemicals affect the health of humans and other organisms.

Toxicologists assess and compare substances to determine their **toxicity,** the degree of harm a chemical substance can inflict. A toxic substance, or poison, is called a **toxicant,** but any chemical substance may exert negative impacts if we expose ourselves to enough of it. Conversely, if the quantity is small enough, a toxicant may pose no health risk at all. These facts are often summarized in the catchphrase "The dose makes the poison." In other words, a substance's toxicity depends not only on its chemical properties but also on its quantity.

In recent decades, our ability to produce new chemicals has expanded, concentrations of chemical contaminants in the environment have increased, and public concern for health and the environment has grown. These trends have driven the rise of **environmental toxicology,** which deals specifically with toxic substances that come from or are discharged into the environment. Toxicologists generally focus on human health, using other organisms as models and test subjects. Environmental toxicologists study animals and plants to determine the ecological impacts of toxic substances and to see whether other organisms can serve as indicators of health threats that could soon affect people.

Many environmental health hazards exist indoors

Modern Americans spend roughly 90% of their lives indoors. Unfortunately, our homes and workplaces can be rife with environmental hazards.

Cigarette smoke and **radon,** a radioactive gas that seeps up from the ground, are leading indoor hazards and are the top two causes of lung cancer in developed nations. Homes and offices can have problems with toxic compounds produced by mold, which can flourish in wall spaces when moisture levels are high. **Asbestos,** used in the past as insulation in walls and other products, is dangerous when inhaled. **Lead poisoning** from water pipes or leaded paint can cause cognitive problems and behavioral abnormalities, damage to vital organs, and even death. Lead poisoning among U.S. children has greatly declined in recent years, however, as a result of education campaigns and the phaseout of lead-based paints and leaded gasoline (p. 19) in the 1970s.

One recently recognized indoor hazard is **polybrominated diphenyl ethers (PBDEs).** These compounds are used as fire retardants in computers, televisions, plastics, and furniture, and they may evaporate at very slow rates throughout the lifetime of the product. Like bisphenol A, PBDEs appear to act as endocrine disruptors. Animal testing suggests that

PBDEs may also cause cancer, affect thyroid hormones, and influence the development of the brain and nervous system. The U.S. government's National Health and Nutrition Examination Survey in 2009 tested the blood and urine of a cross-section of Americans and found one type of PBDE was detected in nearly every person in the study—showing how widespread exposure to PBDEs is in modern America. The European Union banned PBDEs in 2003, but in the United States there has so far been little movement on government regulation of these chemicals.

Risks must be balanced against rewards

It is important to keep in mind that artificially produced chemicals have played a crucial role in giving us the high standard of living we enjoy today. These chemicals have helped create the industrial agriculture that produces our food, the medical advances that protect our health and prolong our lives, and many of the modern materials and conveniences that we use every day. With most hazards, there is some tradeoff between risk and reward. Plastic bottles, can liners, and wrappers serve a useful purpose in safely containing and preserving our food, and their usefulness means that despite their health risks, we may as a society choose to continue including them in products when we judge their benefits to outweigh their impacts on human health. Thus, for BPA, phthalates, and numerous other toxic chemicals, it is appropriate to remember their benefits as we examine some of the unfortunate side effects these substances may elicit.

Toxic Substances and Their Effects on Organisms

Our environment contains countless natural substances that may pose health risks. These include oil oozing naturally from the ground; radon gas seeping up from bedrock; and **toxins,** toxic chemicals manufactured in the tissues of living organisms—for example, chemicals that plants use to ward off herbivores or that insects use to defend themselves from predators. In addition, we are exposed to many synthetic (human-made) chemicals, some of which have toxic properties.

Silent Spring began the public debate over synthetic chemicals

Synthetic chemicals surround us in our daily lives, and each year in the United States we manufacture or import around 113 kg (250 lb) of chemical substances for every man, woman, and child. Many of these substances, particularly the pesticides we use to control insects and weeds, find their way into the soil, air, and water—and into humans and other organisms (**FIGURE 10.4**).

It was not until the 1960s that most people began to learn about the risks of exposure to pesticides. The key event was the publication of Rachel Carson's 1962 book *Silent Spring* (p. 106), which brought the insecticide dichloro-diphenyl-trichloroethane (DDT) to the public's attention. The book was written at a time when large amounts of pesticides virtually untested for health impacts were indiscriminately sprayed, on the assumption that the chemicals would do no harm to people.

Carson synthesized scientific studies, medical case histories, and other data to contend that DDT in particular, and synthetic pesticides in general, were hazardous to people, wildlife, and ecosystems. The book became a best-seller and helped generate significant social change in views and actions toward the environment. The use of DDT was banned in the United States in 1973 and is now illegal in a number of nations.

Despite its damaging effects, DDT is still manufactured today because some developing countries with tropical climates use it to control disease vectors, such as mosquitoes that transmit malaria. In these countries, malaria represents a greater health threat than do the toxic effects of the pesticide. New research and technologies are showing promise for controlling mosquito populations without the use of DDT, however, giving hope that the chemical will soon no longer be needed for controlling disease vectors and could be phased out worldwide.

Not all toxic substances are synthetic

Although many toxicologists focus on synthetic chemicals, toxic substances also exist naturally in the environment around us and in the foods we eat. Thus, it would be a mistake to assume that all synthetic substances are unhealthy and that all natural substances are healthy. In fact, the plants and animals we eat contain many chemicals that can cause us harm. Recall that many plants produce toxins to ward off animals that eat them. In domesticating crop plants, we have selected (p. 52) for strains with reduced toxin content, but traces of naturally occurring plant toxins can still be found in many of the plants we eat. Furthermore, when we consume animal meat, we ingest toxins the animals obtained from plants or animals they ate. Scientists are actively debating just how much risk natural toxicants pose, and it is clear that more research is required on these questions.

Toxic substances come in different types

Toxic substances can be classified based on their particular impacts on health. The best-known toxicants are **carcinogens,** which are substances or types of radiation that cause cancer. In cancer, malignant cells grow uncontrollably, creating

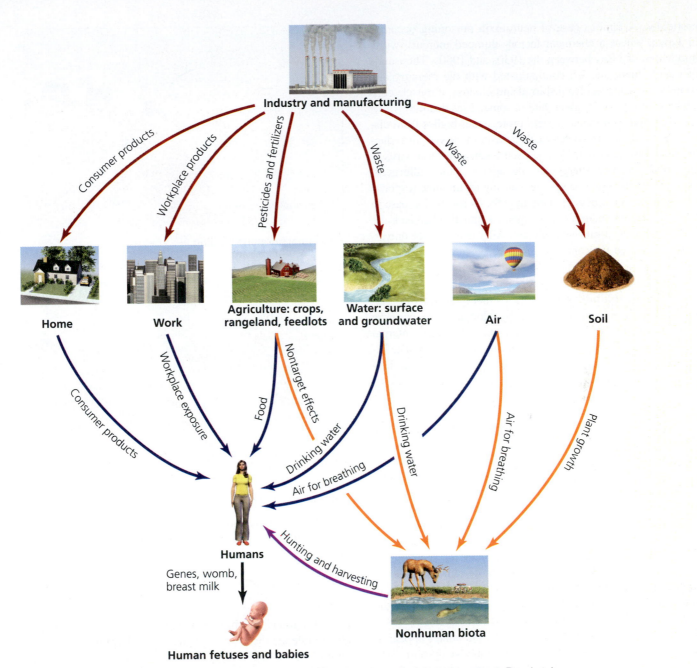

FIGURE 10.4 Synthetic chemicals take many routes in traveling through the environment. People take in only a tiny proportion of these compounds, and many compounds are harmless. However, people receive small amounts of toxicants from many sources, and developing fetuses and babies are particularly sensitive.

tumors, damaging the body, and often leading to death. Cancer frequently has a genetic component, but a wide variety of environmental factors are thought to raise the risk of cancer. In our society today, the greatest number of cancer cases is thought to result from carcinogens contained in cigarette smoke. Carcinogens can be difficult to identify because there may be a long lag time between exposure to the agent and the detectable onset of cancer—up to 15–30 years in the case of cigarette smoke—and because only a portion of the people exposed to a carcinogen eventually develop cancer.

Mutagens are substances that cause genetic mutations in the DNA of organisms (p. 32). Although most mutations have little or no effect, some can lead to severe problems, including cancer and other disorders. If mutations occur in an individual's sperm or egg cells, then the individual's offspring suffer the effects.

Chemicals that cause harm to the unborn are known as **teratogens.** Teratogens that affect development of human embryos in the womb can cause birth defects. One example is the drug thalidomide, which was developed in the 1950s to aid in sleeping and to prevent nausea during pregnancy. Tragically, the drug caused severe organ defects and limb deformities in thousands of babies whose mothers were prescribed this medication. Thalidomide was banned in the 1960s once scientists recognized its connection with birth defects.

Other chemical toxicants known as **neurotoxins** assault the nervous system. Neurotoxins include venoms produced by animals, heavy metals such as lead and mercury, and some pesticides. A famous case of neurotoxin poisoning occurred in Japan, where a chemical factory dumped mercury waste into Minamata Bay between the 1930s and 1960s. Thousands of people there ate fish contaminated with the mercury and soon began suffering from slurred speech, loss of muscle control, sudden fits of laughter, and in some cases death.

The human immune system protects our bodies from disease. Some toxic substances weaken the immune system, reducing the body's ability to defend itself against bacteria, viruses, and other threats. Allergy-causing agents, called **allergens,** overactivate the immune system, causing an immune response when one is not necessary. One hypothesis for the increase in asthma in recent years is that allergenic synthetic chemicals are more prevalent in our environment. Allergens are not universally considered toxicants, however, because they affect some people but not others and because one's response does not necessarily correlate with the degree of exposure.

Pathway inhibitors are toxicants that interrupt vital biochemical processes in organisms by blocking one or more steps in a pathway. Rat poisons, for example, cause internal hemorrhaging in rodents by interfering with the biochemical pathways that create blood-clotting proteins. Some herbicides, such as atrazine, kill plants by chemically blocking steps in photosynthesis (p. 34).

Most recently, scientists have recognized **endocrine disruptors,** toxic substances that interfere with the endocrine system. The endocrine system consists of chemical messengers, known as *hormones,* that travel through the bloodstream at extremely low concentrations and have many vital functions. They stimulate growth, development, and sexual maturity, and they regulate brain function, appetite, sex drive, and many other aspects of our physiology and behavior. Some hormone-disrupting toxicants affect an animal's endocrine system by blocking the action of hormones or accelerating their breakdown. Others are so similar to certain hormones in their molecular structure and chemistry that they "mimic" the hormone by interacting with receptor molecules just as the actual hormone would (**FIGURE 10.5**).

Among other effects, both BPA and phthalates appear to act as endocrine disruptors on the reproductive system. BPA is one of many chemicals that seem to mimic the female sex hormone estrogen and bind to estrogen receptors. Indeed, emerging research is indicating that bisphenol A might not be the only estrogen-mimicking compound in plastics, calling into question the safety of all the plastics that are ubiquitous in our lives. With their diverse impacts on human health, BPA and phthalates show how a substance can be a carcinogen, a mutagen, and an endocrine disruptor all at the same time.

Organisms have natural defenses against toxic substances

Although synthetic toxicants are new, organisms have long been exposed to natural toxicants. Mercury, cadmium, arsenic, and other harmful substances are found naturally in the environment. Some organisms produce biological toxins to deter predators or capture prey. Examples include venom in poisonous snakes and spiders, and the natural insecticide

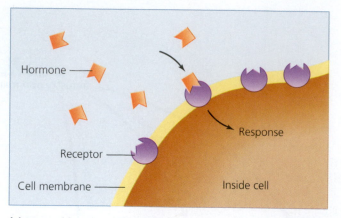

(a) Normal hormone binding

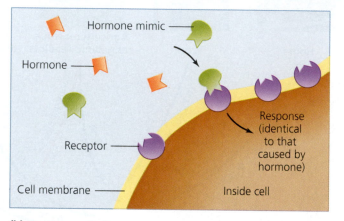

(b) Hormone mimicry

FIGURE 10.5 Many endocrine-disrupting substances mimic the chemical structure of hormone molecules. Like a key similar enough to fit into another key's lock, the hormone mimic binds to a cellular receptor for the hormone, causing the cell to react as though it had encountered the hormone.

pyrethrin found in chrysanthemums. Over time, organisms able to tolerate these harmful substances have gained an evolutionary advantage.

Skin, scales, and feathers are the first line of defense against toxic substances because they resist uptake from the surrounding environment. However, toxicants can circumvent these barriers and enter the body through eating, drinking, and breathing. Once inside the organism, they are distributed widely by the circulatory and lymph systems in animals, and by the vascular system in plants.

Organisms possess biochemical pathways that use enzymes to detoxify harmful chemicals that enter the body. Some pathways break down, or metabolize, toxic substances to render them inert. Other pathways make toxic substances water soluble so they are easier to excrete through the urinary system. In humans, many of these pathways are found in the liver. As a result, this organ is disproportionately affected by the intake of harmful substances, such as excessive alcohol.

Some toxic substances cannot be effectively detoxified or made water soluble by detoxification enzymes. Instead, the body sequesters these chemicals in fatty tissues and cell membranes to keep them away from vital organs. Heavy metals, dioxins, PBDEs, and some insecticides (including DDT) are stored in body tissue in this manner.

Defense mechanisms for natural toxins have evolved over millions of years. For the synthetic chemicals that are so prevalent in today's environment, however, organisms have not had long-term exposure, so the impacts of these toxic substances can be severe and unpredictable.

FAQ

Why do some insects survive exposure to a pesticide while others are killed by it?

When a population of organisms is exposed to a toxicant, such as a pesticide, a few individuals often survive while the vast majority of the population is killed. These individuals survive because they possess genes (which others in the population do not) that code for enzymes that counteract the toxic properties of the toxicant. Because the effects of these genes are expressed only when the pesticide is applied, many people think the toxicant "creates" detoxification genes by mutating the DNA of a small number of individuals. This is not the case. The genes for detoxifying enzymes were present in the DNA of resistant individuals from birth, but their effects were seen only when pesticide exposure caused selective pressure (p. 51) for resistance to the pesticide.

Individuals vary in their responses to hazards

Some of the defenses described above have a genetic basis. As a result, individuals may respond quite differently to identical exposures to hazards because they happen to have different combinations of genes. Poorer health also makes an individual more sensitive to biological and chemical hazards. Sensitivity also can vary with sex, age, and weight. Because of their smaller size and rapidly developing organ systems, younger organisms (for example, fetuses, infants, and young children) tend to be much more sensitive to toxicants than are adults. Regulatory agencies such as the U.S. Environmental Protection Agency (EPA) typically set human chemical exposure standards for adults and extrapolate downward for infants and children. However, many scientists contend that these linear extrapolations often do not offer adequate protection to fetuses, infants, and children.

The type of exposure can affect the response

The risk posed by a hazard often varies according to whether a person experiences high exposure for short periods of time, known as **acute exposure**, or low exposure over long periods of time, known as **chronic exposure**. Incidences of acute exposure are easier to recognize because they often stem from discrete events, such as accidental ingestion, an oil spill, a chemical spill, or a nuclear accident. Toxicity tests in laboratories generally reflect acute toxicity effects. However, chronic exposure is more common—and more difficult to detect and diagnose. Chronic exposure often affects organs gradually, as when smoking causes lung cancer or when alcohol abuse leads to liver damage. Because of the long time periods involved, relationships between cause and effect may not be readily apparent.

Toxic Substances and Their Effects on Ecosystems

When toxicants concentrate in environments and harm the health of many individuals, populations (p. 50) of the affected species become smaller. This decline in population size can then affect other species. For instance, species that are prey of the organism affected by toxicants could experience population growth because predation levels are lower. Predators of the poisoned species, however, would decline as their food source became less abundant. Cascading impacts can cause changes in the composition of the biological community (p. 73) and threaten ecosystem functioning. There are many ways toxicants can concentrate and persist in ecosystems and affect ecosystem services.

Airborne substances can travel widely

Toxic substances may sometimes be redistributed by air currents (Chapter 13), exerting impacts on ecosystems far from their site of release. Because so many substances are carried by the wind, synthetic chemicals are ubiquitous worldwide, even in seemingly pristine areas. Earth's polar regions are particularly contaminated, because natural patterns of global atmospheric circulation (p. 290) tend to move airborne chemicals toward the poles (**FIGURE 10.6**). Thus, although we

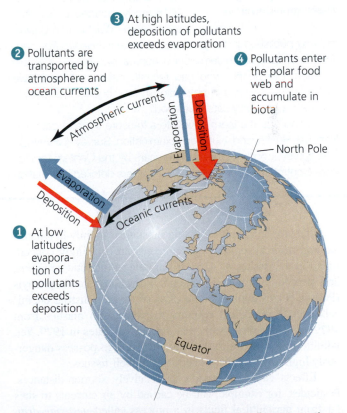

FIGURE 10.6 Air and water currents direct pollutants to the poles. In the process of "global distillation," pollutants that evaporate and rise high into the atmosphere at lower latitudes are carried toward the poles by atmospheric currents, while ocean currents carry pollutants deposited in the ocean toward the poles. This process exposes polar organisms to unusually concentrated levels of toxic substances.

Are Endocrine Disruptors Lurking in Your Fast Food?

Dr. Ami Zota, George Washington University

The plastics used in food packaging preserve the contents and protect against food pathogens, but can also expose us to endocrine-disrupting chemicals that can adversely affect our health. Increasingly, diet is being recognized as a major source of exposure to bisphenol A, which is used to line food cans and beverage containers, and phthalates, which can leach from food-processing equipment and into foods. But do some types of food pose a higher risk than others?

Researchers have hypothesized that fast food may expose people to higher levels of endocrine disruptors than other types of food. A 2016 study, headed by Dr. Ami Zota of George Washington University and published in the journal *Environmental Health Perspectives,* embraced an epidemiological approach to answer a simple question: Did people who had recently eaten fast food have higher levels of BPA and phthalates in their bodies than people who had not recently eaten fast food?

To find out, the team dove into a treasure trove of data, the National Health and Nutrition Examination Survey (NHANES). This survey, conducted every two years by the Centers for Disease Control and Prevention (CDC), gathers detailed information from people across the United States by asking participants to undergo a physical examination by a medical professional, providing blood and urine samples and completing a detailed questionnaire about their lifestyle, including a description of the foods they had recently consumed.

The team scoured the surveys from 2003 to 2010 and found that around one-third of participants reported having eaten fast food in the 24 hours preceding their examination. (For the sake of the study, fast food was defined as food from restaurants that lack wait service, carryout and delivery food options, and pizza.) The team then used a correlational approach to look for relationships in the subjects' reported consumption of fast food and their urinary concentrations of BPA and two types of phthalates used in food packaging and processing—di(2-ethylhexyl) phthalate (DEHP) and diisononyl phthalate (DiNP).

After analyzing more than 8800 individuals, Dr. Zota's team found a positive correlation between the consumption of fast food and urinary concentrations of both types of phthalates, demonstrating that people who had recently eaten fast food had measurably higher levels of phthalates than people who had not eaten fast food (**FIGURE 1**). The quantity of fast food eaten was also related to concentrations of the two phthalates in subjects. People who consumed less than 35% of their calories from fast food ("low consumers") had DEHP levels 15% higher than people who had not eaten fast food ("nonconsumers"), while those who consumed more than 35% of their calories from fast food ("high consumers") had DEHP levels 23% higher than people who had not eaten fast food. Similar results were seen for DiNP, where low consumers and high consumers had urinary concentrations

---

manufacture and apply synthetic substances mainly in temperate and tropical regions, contaminants are strikingly concentrated in the tissues of Arctic polar bears, Antarctic penguins, and people living in Greenland. Polychlorinated biphenyls (PCBs), for example, were a class of toxic industrial chemicals used in electrical equipment, paints, and plastics from 1929 until they were banned in the United States in 1979. Yet, wildlife in the Arctic to this day are found to possess dangerously high levels of these chemicals in their tissues.

Effects can also occur over relatively shorter distances. Pesticides, for example, can be carried by air currents to sites far from agricultural fields in a process called *pesticide drift.* The Central Valley of California is the world's most productive agricultural region, and the region's frequent winds often blow airborne pesticide spray—and dust particles containing pesticide residue—for long distances. In the nearby mountains of the Sierra Nevada, research has associated pesticide drift from the Central Valley with population declines in four species of frogs.

Toxic substances may concentrate in water

Water running off from land often transports toxicants from large areas and concentrates them in small volumes of surface water. Wastewater treatment plants also add toxins, pharmaceuticals, and detoxification products from humans to waterways. Many chemicals are soluble in water and enter organisms' tissues through drinking or absorption. For this reason, aquatic animals such as fish, frogs, and stream invertebrates are effective indicators of pollution. The contaminants that wash into streams and rivers also flow and seep into the water we drink. Once concentrated in water, toxicants can often move long distances through aquatic systems and affect a diversity of organisms and ecosystems. For example, in 2017 scientists reported that small crustaceans collected from some of the deepest trenches in the Pacific—depths up to 9.5 km (6 mi) below the surface—contained levels of PCBs

24% and 39% higher than nonconsumers, respectively. The researchers hypothesized that plastics in gloves from people preparing fast food meals and from food-processing equipment were releasing phthalates into foods, particularly hot foods and foods that contain high levels of fat to which phthalates can bind. Unlike phthalates, BPA did not show a statistically significant correlation with fast food consumption. The researchers hypothesize that BPA exposure from fast food consumption may be minor relative to exposure from other sources of BPA, such as eating canned foods and drinking beverages from cans and bottles, resulting in any added BPA from fast food items simply being "washed out" by larger exposures from other sources.

Industry groups, such as the National Restaurant Association and the American Chemistry Council, point out that the concentrations of phthalates detected in subjects was below the levels deemed dangerous for humans by the EPA, and so argue that phthalate exposures from fast food pose no risk to human health. Critics counter that the guidelines established by the EPA have not been revised since 1988, despite recent research that has shown impacts on reproductive and developmental systems at levels like those seen in the study. The regulation of chemicals that operate at extremely low concentrations, such as phthalates, will continue to pose regulatory challenges. Studies such as this one, however, can identify major sources of exposure people may experience and prompt further study and action by scientists, government regulators, and fast food companies to reduce such exposures—and safeguard human health from the far-reaching health impacts of endocrine-disrupting chemicals.

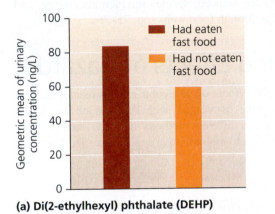

(a) Di(2-ethylhexyl) phthalate (DEHP)

(b) Diisononyl phthalate (DiNP)

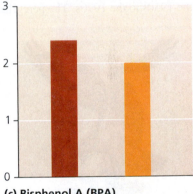

(c) Bisphenol A (BPA)

FIGURE 1 People who had recently eaten fast food showed significantly higher levels of (a) di(2-ethylhexyl) phthalate (DEHP) and (b) diisononyl phthalate (DiNP) in their urine (measured as nanograms of chemical per liter of urine) than people who had not eaten fast food. Urinary concentrations of **(c)** BPA were not significantly different in the two groups. Note the scale on the *y* axis differs in each of the three figure parts. *Data from Zota, A.R., et al., 2016. Recent fast food consumption and bisphenol A and phthalates exposures among the U.S. population in NHANES, 2003–2010.* Environmental Health Perspectives *124: 1521–1528.*

and PDBEs that were up to 50 times higher than concentrations found in crustaceans from some of the most polluted rivers on Earth.

Some toxicants persist in the environment

A toxic substance that is released into the environment may degrade quickly and become harmless, or it may remain unaltered and persist for many months, years, or decades. The rate at which a given substance degrades depends on its chemistry and on factors such as temperature, moisture, and sun exposure. The *Bt* toxin (p. 153) used in biocontrol and in genetically modified crops has a very short persistence time, whereas chemicals such as DDT and PCBs can persist for decades.

Persistent synthetic chemicals exist in our environment today because we have designed them to persist. The synthetic chemicals used in plastics, for instance, are used precisely because they resist breakdown. Sooner or later, however, most toxic substances degrade into simpler compounds called **breakdown products.** Often these are less harmful than the original substance, but sometimes they are just as toxic as the original chemical, or more so. For instance, DDT breaks down into DDE, a highly persistent and toxic compound in its own right.

Toxic substances may accumulate and move up the food chain

Within an organism's body, some toxic substances are quickly excreted, and some are degraded into harmless breakdown products. Others persist intact in the body. Substances that are fat soluble or oil soluble (including organic compounds such as DDT and DDE) are absorbed and stored in fatty tissues. Such persistent toxicants accumulate in an organism's body in a process termed **bioaccumulation,** such that the organism's tissues have a greater concentration of the substance than exists in the surrounding environment.

Toxic substances that bioaccumulate in an organism's tissues may be transferred to other organisms as predators consume prey, resulting in a process called **biomagnification** (**FIGURE 10.7**). When one organism consumes another, the predator takes in any stored toxicants and stores them in its own body. Thus, bioaccumulation takes place on all trophic levels. Moreover, each individual predator consumes many individuals from the trophic level beneath it, so with each step up the food chain, concentrations of toxicants become magnified.

The process of biomagnification occurred throughout North America with DDT. Top predators, such as birds of prey, ended up with high concentrations of the pesticide because concentrations became magnified as DDT moved from water to algae to plankton to small fish to larger fish and finally to fish-eating birds.

Biomagnification of DDT caused populations of many North American birds of prey to decline precipitously from the 1950s to the 1970s. The peregrine falcon was nearly wiped out in the eastern United States, and the bald eagle, the U.S. national bird, was virtually eliminated from the lower 48 states. Eventually scientists determined that DDT was causing these birds' eggshells to grow thinner, so that eggs were breaking in the nest and killing the embryos within. In a remarkable environmental success story, populations of all these birds have rebounded since the United States banned DDT.

Impacts from biomagnification still persist, though. Unfortunately, DDT continues to impair wildlife in parts of the world where it is still used. Mercury bioaccumulates in some commercially important fish species, such as tuna. Polar bears of Svalbard Island in Arctic Norway show extremely high levels of PCB contamination from biomagnification, and polar bear cubs suffer immune suppression, hormone disruption, and high mortality when they receive PCBs in their mothers' milk.

In all these cases, biomagnification affects ecosystem composition and the ecosystem services (p. 4) that nature provides. When populations of top predators such as eagles or polar bears are reduced, species interactions change, and effects cascade through food webs (pp. 75–76).

Studying Effects of Hazards

Determining the effects of particular environmental hazards on organisms and ecosystems is a challenging job, and scientists rely on several different methods to do this, ranging from correlative surveys to manipulative experiments (p. 11).

Wildlife studies integrate work in the field and lab

Scientists study the impacts of environmental hazards on wild animals to help conserve animal populations and to understand potential risks to people. Often, wildlife toxicologists work in the field with animals to take measurements, document patterns, and generate hypotheses before heading to the laboratory to run controlled manipulative experiments to test their hypotheses. The work of two of pioneers in the study of endocrine disruptors illustrates the approaches embraced in wildlife studies.

Biologist Louis Guillette studied alligators in Florida and discovered that many showed unusual reproductive problems. Females had trouble producing viable eggs; young alligators had abnormal gonads; and male hatchlings had too little of the male sex hormone testosterone, while female hatchlings had too much of the female sex hormone estrogen. Because certain lakes received agricultural runoff that included insecticides such as DDT and dicofol and herbicides such as atrazine, Guillette hypothesized that chemical contaminants were disrupting the endocrine systems of alligators during their development in the egg. Indeed, when Guillette and his team compared alligators in polluted lakes with those in cleaner lakes, they found the ones in polluted lakes to be suffering far more problems. Moving into the lab, the researchers found that several contaminants detected in alligator eggs and young could bind to receptors for estrogen and reverse the sex of male embryos. Their experiments showed that atrazine

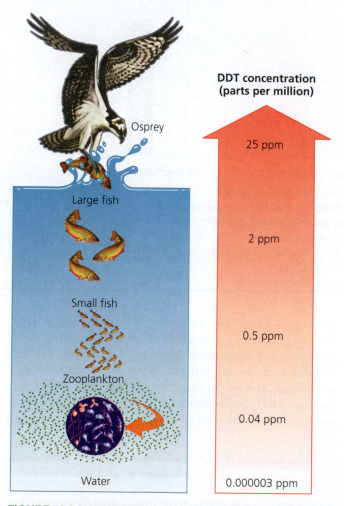

FIGURE 10.7 In a classic case of biomagnification, DDT becomes highly concentrated in fish-eating birds such as ospreys. Organisms at the lowest trophic level take in fat-soluble compounds such as DDT from water. As animals at higher trophic levels eat organisms lower on the food chain, each organism passes its load of toxicants up to its consumer, such that organisms on all trophic levels bioaccumulate the substance in their tissues.

DDT concentration (parts per million)

25 ppm

2 ppm

0.5 ppm

0.04 ppm

0.000003 ppm

Osprey

Large fish

Small fish

Zooplankton

Water

FIGURE 10.8 **Wildlife studies examine the effects of toxic substances in the environment.** Researcher Tyrone Hayes found that frogs show reproductive abnormalities that he attributes to endocrine disruption by pesticides.

appeared to disrupt hormones by inducing production of aromatase, an enzyme that converts testosterone to estrogen.

Following Guillette's work, researcher Tyrone Hayes (**FIGURE 10.8**) found similar reproductive problems in frogs and attributed them to atrazine. In lab experiments, male frogs raised in water containing very low doses of the herbicide became feminized and hermaphroditic, developing both testes and ovaries. Hayes then moved to the field to look for correlations between herbicide use and reproductive impacts in the wild. His field surveys showed that leopard frogs across North America experienced hormonal problems in areas of heavy atrazine usage. His work indicated that atrazine, which kills plants by blocking biochemical pathways in photosynthesis, can also act as an endocrine disruptor.

Human studies rely on case histories, epidemiology, and animal testing

In studies of human health, we gain much knowledge by directly studying sickened individuals. This process of observation and analysis of individual patients is known as a **case history** approach. Case histories have advanced our understanding of human illness, but they do not always help us infer the effects of rare hazards or chemicals that exist at low environmental concentrations and exert minor, long-term effects. Case histories also tell us little about probability and risk, such as how many extra deaths we might expect in a population due to a particular cause.

For such questions, which are common in environmental toxicology, we need **epidemiological studies,** large-scale comparisons among groups of people, usually contrasting a group known to have been exposed to some hazard against a group that has not. Epidemiologists track the fate of all people in the study for a long period of time (often years or decades) and

measure the rate at which deaths, cancers, or other health problems occur in each group. The epidemiologists then analyze the data, looking for observable differences between the groups, and statistically test hypotheses accounting for differences. When a group exposed to a hazard shows a significantly greater degree of harm, it suggests that the hazard may be responsible. The epidemiological process is akin to a natural experiment (p. 11), in which an event creates groups of subjects that researchers can study (for example, people exposed to carcinogenic compounds in their drinking water versus those not similarly exposed).

Epidemiological studies measure a statistical association between a health hazard and an effect, but they do not confirm that the hazard causes the effect. To establish causation, manipulative experiments are needed. However, subjecting people to massive doses of toxic substances in a lab experiment would clearly be unethical. This is why researchers have traditionally used animals—such as laboratory strains of rats, mice, and other mammals—as test subjects. Because of shared evolutionary history, substances that harm mice and rats are reasonably likely to harm us. Some people feel the use of animals for testing is unethical, but animal testing enables scientific and medical advances that would be impossible or far more difficult otherwise.

Dose-response analysis is a mainstay of toxicology

The standard method of testing with lab animals in toxicology is **dose-response analysis.** Scientists quantify the toxicity of a substance by measuring the strength of its effects or the number of animals affected at different doses. The **dose** is the amount of substance the test animal receives, and the **response** is the type or magnitude of negative effects the animal exhibits as a result. The response is generally quantified by measuring the proportion of animals exhibiting negative impacts. The data are plotted on a graph, with dose on the x axis and response on the y axis (**FIGURE 10.9a**). The resulting curve is called a **dose-response curve.**

Once they have plotted a dose-response curve, toxicologists can calculate a convenient shorthand gauge of a substance's toxicity: the amount of the substance it takes to kill half the population of study animals used. This lethal dose for 50% of individuals is termed the lethal dose–50%, or LD_{50}. A high LD_{50} indicates low toxicity for a substance, and a low LD_{50} indicates high toxicity.

If the experimenter is interested in nonlethal health impacts, he or she may want to document the level of toxicant at which 50% of a population of test animals is affected in some other way (for instance, the level of toxicant that causes 50% of lab mice to develop reproductive abnormalities). Such a level is called the effective dose–50%, or ED_{50}.

Some substances can elicit effects at any concentration, but for others, responses may occur only above a certain dose, or threshold. Such a **threshold dose** (**FIGURE 10.9b**) might be expected if the body's organs can fully metabolize or excrete a toxicant at low doses but become overwhelmed at higher concentrations.

Sometimes a response may *decrease* as a dose increases. Toxicologists are finding that some dose-response curves

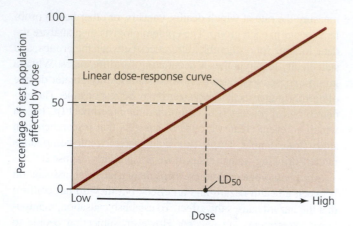

(a) Linear dose-response curve

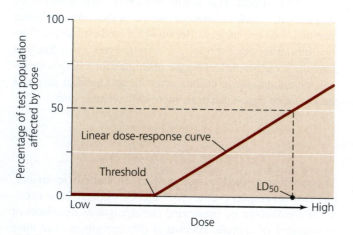

(b) Dose-response curve with threshold

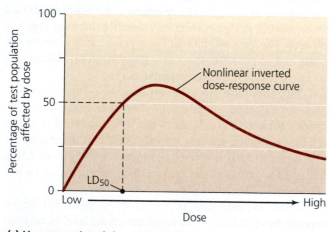

(c) Unconventional dose-response curve

FIGURE 10.9 Dose-response curves show that organisms' responses to toxicants may sometimes be complex. In a classic linear dose-response curve **(a)**, the percentage of animals killed or otherwise affected by a substance rises with the dose. The point at which 50% of the animals are killed is labeled the lethal dose–50, or LD$_{50}$. For some toxic substances, a threshold dose **(b)** exists, below which doses have no measurable effect. Some substances—in particular, endocrine disruptors—show unconventional, nonlinear dose-response curves **(c)** that are U-shaped, J-shaped, or shaped like an inverted U.

are U-shaped, J-shaped, or shaped like an inverted U (**FIGURE 10.9c**). Such counterintuitive curves contradict toxicology's traditional assumption that "the dose makes the poison." These unconventional dose-response curves often occur with endocrine disruptors, such as BPA and phthalates, likely because the hormone system is geared to respond to minute concentrations of substances (normally, hormones in the bloodstream). Because the endocrine system responds to minuscule amounts of chemicals, it may be vulnerable to disruption by contaminants that are dispersed through the environment and that reach our bodies in very low concentrations.

Researchers generally give lab animals much higher doses relative to body mass than people would receive in the environment. This is so that the response is great enough to be measured and so that differences between the effects of small and large doses are evident. Data from a range of doses give shape to the dose-response curve. Once the data from animal tests are plotted, researchers can extrapolate downward to estimate responses to still-lower doses from a hypothetically large population of animals. This way, they can come up with an estimate of, say, what dose causes cancer in 1 mouse in 1 million. A second extrapolation is required to estimate the effect on humans, with our greater body mass. Because these two extrapolations stretch beyond the actual data obtained, they introduce uncertainty into the interpretation of what doses are safe for people.

Chemical mixes may be more than the sum of their parts

It is difficult enough to determine the impact of a single hazard, but the task becomes astronomically more difficult when multiple hazards interact. Chemical substances, when mixed, may act together in ways that cannot be predicted from the effects of each in isolation. Mixed toxicants may sum each other's effects, cancel out each other's effects, or multiply each other's effects. Interactive impacts that are greater than the simple sum of their constituent effects are called **synergistic effects.**

With Florida's alligators, lab experiments have indicated that the DDT breakdown product DDE can either promote or inhibit sex reversal, depending on the presence of other chemicals. Mice exposed to a mixture of nitrate, atrazine, and the insecticide aldicarb have been found to show immune, endocrine, and nervous system effects that were not evident from exposure to each of these chemicals alone.

Traditionally, environmental health has tackled impacts of single hazards one at a time. In toxicology, the complex experimental designs required to test interactions, and the sheer number of chemical combinations, have meant that single-substance tests have received priority. This approach is changing, but the interactive effects of most chemicals are unknown.

Endocrine disruption poses challenges for toxicology

As today's emerging understanding of endocrine disruption leads toxicologists to question their assumptions, unconventional dose-response curves are presenting challenges for scientists studying toxic substances and for policymakers trying to set safety standards for them. Knowing the shape of a dose-response curve is crucial if one is using it to predict responses at doses below those that have been tested. Because so many novel synthetic chemicals exist in very low concentrations over wide areas, many scientists suspect that we may have underestimated the dangers of compounds that exert impacts at low concentrations.

Scientists first noted endocrine-disrupting effects decades ago, but the idea that synthetic chemicals might be altering the hormones of animals was not widely appreciated until the 1996 publication of the book *Our Stolen Future*, by Theo Colburn, Dianne Dumanoski, and J.P. Myers. Like *Silent Spring*, this book integrated scientific work from various fields and presented a unified view of the hazards posed by endocrine-disrupting chemicals.

Today, thousands of studies have linked hundreds of substances to effects on reproduction, development, immune function, brain and nervous system function, and other hormone-driven processes. Evidence is strongest so far in nonhuman animals, but many studies suggest impacts on humans. Some researchers argue that the sharp rise in breast cancer rates (one in eight U.S. women today develops breast cancer) may be due to hormone disruption, because an excess of estrogen appears to feed tumor development in older women. Other scientists attribute male reproductive problems to elevated BPA exposure. For example, studies found that workers in Chinese factories that manufactured BPA had elevated rates of erectile dysfunction and reduced sperm counts when compared to workers in factories manufacturing other products.

Much of the research into hormone disruption has brought about strident debate. This is partly because scientific uncertainty is inherent in any developing field. Another reason is that negative findings about chemicals pose an economic threat to the manufacturers of those chemicals, who stand to lose many millions of dollars in revenue if their products were to be banned or restricted in the United States.

Risk Assessment and Risk Management

Policy decisions on whether to ban chemicals or restrict their use generally follow years of rigorous testing for toxicity. Likewise, strategies for combating disease and other health threats are based on extensive scientific research. However, policy and management decisions also incorporate economics and ethics—and all too often the decision-making process is heavily influenced by pressure from powerful corporate and political interests. The steps between the collection and interpretation of scientific data and the formulation of policy involve assessing and managing risk.

We express risk in terms of probability

Exposure to an environmental health threat does not invariably produce a given consequence. Rather, it causes some probability of harm, a statistical chance that damage will result. To understand a health threat, a scientist must know more than just its identity and strength. He or she must also know the chance that one will encounter it, the frequency with which one may encounter it, the amount of substance or degree of threat to which one is exposed, and one's sensitivity to the threat. Such factors help determine the overall risk posed.

Risk can be measured in terms of *probability*, a quantitative description of the likelihood of a certain outcome. The probability that some harmful outcome (for instance, injury, death, environmental damage, or economic loss) will result from a given action, event, or substance expresses the overall risk posed by a particular threat.

Our perception of risk may not match reality

Every action we take and every decision we make involves some element of **risk**, some (generally small) probability that things will go wrong. We typically try to behave in ways that minimize risk, but our perceptions of risk do not always match statistical reality (**FIGURE 10.10**). People often worry unduly about small risks yet readily engage in activities that pose higher risks. For instance, most of us perceive flying in an airplane as a riskier activity than driving a car, but, statistically speaking, plane travel is much safer. Psychologists argue that this disconnect occurs because we feel more at risk when we are not controlling a situation and safer when we are "at the wheel"—regardless of the actual risk involved.

This psychology may help account for people's anxiety over exposure to bisphenol A, nuclear power, toxic waste, and

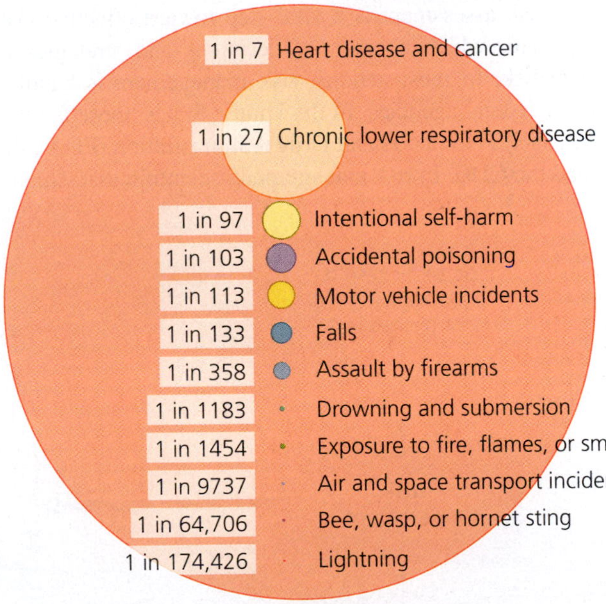

1 in 7	Heart disease and cancer
1 in 27	Chronic lower respiratory disease
1 in 97	Intentional self-harm
1 in 103	Accidental poisoning
1 in 113	Motor vehicle incidents
1 in 133	Falls
1 in 358	Assault by firearms
1 in 1183	Drowning and submersion
1 in 1454	Exposure to fire, flames, or smoke
1 in 9737	Air and space transport incidents
1 in 64,706	Bee, wasp, or hornet sting
1 in 174,426	Lightning

FIGURE 10.10 Our perceptions of risk do not always match the reality of risk. Listed here are several leading causes of death in the United States, along with a measure of the risk each poses. The larger the area of the circle in the figure, the greater the risk of dying from that cause. *Data are for 2013, from* Injury Facts, *2016.* Itasca, IL: National Safety Council.

DATA Q People tend to view car travel as being safer than airplane travel, but a person is how many times more likely to die from a car accident than from an airplane crash?

Go to **Interpreting Graphs & Data** on **Mastering** Environmental Science

pesticide residues on foods—environmental hazards that are invisible or little understood and whose presence in our lives is largely outside our personal control. In contrast, people are readier to accept and ignore the risks of smoking cigarettes, overeating, and not exercising, which are voluntary activities statistically shown to pose far greater risks to health.

Risk assessment analyzes risk quantitatively

The quantitative measurement of risk and the comparison of risks involved in different activities or substances together are termed **risk assessment.** Risk assessment is a way to identify and outline problems. In environmental health, it helps ascertain which substances and activities pose health threats to people or wildlife and which are largely safe.

Assessing risk for a chemical substance involves several steps. The first steps involve the scientific study of toxicity we examined above—determining whether a substance has toxic effects and, through dose-response analysis, measuring how effects vary with the degree of exposure. Subsequent steps involve assessing the individual's or population's likely extent of exposure to the substance, including the frequency of contact, the concentrations likely encountered, and the length of encounter.

Risk management combines science and other social factors

Accurate risk assessment is a vital step toward effective **risk management,** which consists of decisions and strategies to minimize risk. In most nations, risk management is handled largely by federal agencies. In the United States, these include agencies such as the Food and Drug Administration (FDA), the EPA, and the CDC. In risk management, scientific assessments

of risk are considered in light of economic, social, and political needs and values. Risk managers assess costs and benefits of addressing risk in various ways, with regard to both scientific and nonscientific concerns, before making decisions on whether and how to reduce or eliminate risk (**FIGURE 10.11**).

In environmental health and toxicology, comparing costs and benefits (p. 96) can be difficult because the benefits are often economic, whereas the costs often pertain to health. Moreover, economic benefits are generally known, easily quantified, and of a discrete and stable amount, whereas health risks are hard-to-measure probabilities, often involving a small percentage of people likely to suffer greatly and a large majority likely to experience little effect. Because of the lack of equivalence in the way costs and benefits are measured, risk management frequently tends to stir up debate.

In the case of BPA and phthalates, eliminating food packaging in the name of safety could do more harm than good. The plastic lining inside metal cans, for example, can release BPA into the food, but also helps prevent metal corrosion and the contamination of food by pathogens. Some alternative substances exist to those that expose users to BPA and phthalates, but replacing them with alternatives will entail economic costs to industry, and these costs get passed on to consumers in the prices of products. Such complex considerations can make risk management decisions difficult even if the science of risk assessment is fairly clear.

Two approaches exist for testing the safety of new products

Because we cannot know a substance's toxicity until we measure and test it, and because so many untested chemicals and combinations exist, science will never eliminate the many uncertainties that accompany risk assessment. In such a world

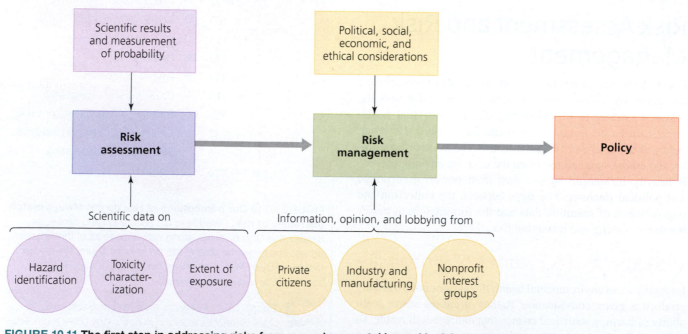

FIGURE 10.11 The first step in addressing risks from an environmental hazard is risk assessment. Once science identifies and measures risks, then risk management can proceed. In risk management, economic, political, social, and ethical issues are considered in light of the scientific data from risk assessment.

Sequence of events	"Innocent until proven guilty" approach	Precautionary principle approach
Industrial research and development	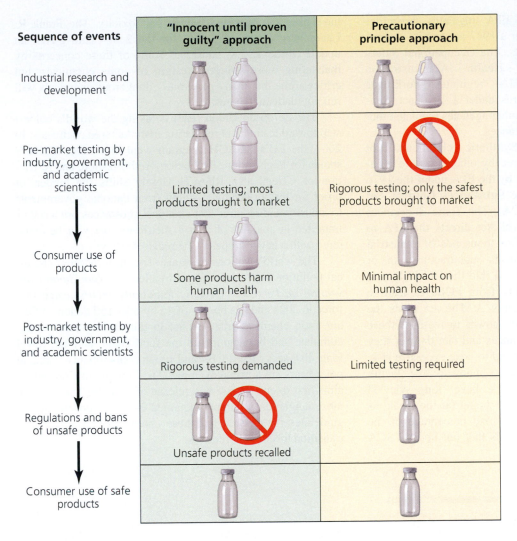	
Pre-market testing by industry, government, and academic scientists	Limited testing; most products brought to market	Rigorous testing; only the safest products brought to market
Consumer use of products	Some products harm human health	Minimal impact on human health
Post-market testing by industry, government, and academic scientists	Rigorous testing demanded	Limited testing required
Regulations and bans of unsafe products	Unsafe products recalled	
Consumer use of safe products		

FIGURE 10.12 Two main approaches can be taken to introduce new substances to the market. In one approach, substances are "innocent until proven guilty"; they are brought to market relatively quickly after limited testing. Products reach consumers more quickly, but some fraction of them **(white jug in the diagram)** may cause harm to some fraction of people. The other approach is to adopt the precautionary principle, bringing substances to market cautiously, only after extensive testing. Products that reach the market should be safe, but many perfectly safe products **(gray bottle in diagram)** will be delayed in reaching consumers.

of uncertainty, there are two basic philosophical approaches to categorizing substances as safe or dangerous (**FIGURE 10.12**).

One approach is to assume that substances are harmless until shown to be harmful. This is nicknamed the "innocent-until-proven-guilty" approach. Because thoroughly testing every existing substance (and combination of substances) for its effects is a hopelessly long, complicated, and expensive pursuit, the innocent-until-proven-guilty approach has the virtue of facilitating technological innovation and economic activity. However, it has the disadvantage of putting into wide use some substances that may later turn out to be dangerous.

The other approach is to assume that substances are harmful until shown to be harmless. This approach follows the precautionary principle (p. 162). This more cautious approach should enable us to identify troublesome toxicants before they are released into the environment, but it may also impede the pace of technological and economic advance.

These two approaches are actually two ends of a continuum of possible approaches. The two endpoints differ mainly in where they lay the burden of proof—specifically, whether product manufacturers are required to prove a product is safe or whether government, scientists, or citizens are required to prove a product is dangerous.

The choice of philosophical approach has direct implications for policy, and nations vary in how they blend the two approaches. European nations have recently embarked on a policy course that incorporates the precautionary principle. The United States, however, has long embraced an innocent-until-proven-guilty approach. This may be changing, however, as the passage of the **Frank R. Lautenberg Chemical Safety for the 21st Century Act** in 2016 directs the EPA to require more stringent testing of industrial chemicals before they are used in products.

Governments regulate industrial chemicals

In the United States, several federal agencies are assigned responsibility for tracking and regulating synthetic chemicals under various legislative acts. The FDA, under an act first passed in 1938, monitors foods and food additives, cosmetics, drugs, and medical devices.

weighing the
ISSUES

The Precautionary Principle

Industry's critics say chemical manufacturers should bear the burden of proof for the safety of their products before they hit the market. Industry's supporters say that mandating more safety research will hamper the introduction of products that consumers want and will increase the price of products as research costs are passed on to consumers. What do you think? Which approach should U.S. government regulators embrace?

FAQ

If the government allows a product to be sold in stores, isn't it safe?

Just because a product is available to the public doesn't mean it poses no risk to consumers. Medicines, cosmetics, and some types of food are tested for safety prior to release, but other potentially dangerous products, such as BPA in plastics, are not similarly tested. And with the federal agencies that oversee product toxicity tests, such as the EPA, now struggling due to extremely steep budget cuts, it would be a mistake to assume that the products you use have been thoroughly tested. The best approach is to educate yourself on the risks in products you use and exercise the famous sentiment of "*caveat emptor*"—Latin for "let the buyer beware."

The EPA regulates pesticides under a 1947 act and its amendments. The Occupational Safety and Health Administration (OSHA) regulates workplace hazards under a 1970 act. Several other agencies regulate other substances.

Synthetic chemicals not covered by other laws are regulated by the EPA under the 1976 **Toxic Substances Control Act (TSCA)**. The Toxic Substances Control Act directs the EPA to monitor thousands of industrial chemicals manufactured in or imported into the United States, ranging from PCBs to lead to bisphenol A. The act gives the agency power to regulate these substances and ban them if they are found to pose excessive risk. However, many public health advocates have long viewed TSCA as being far too weak, as only a small percentage of the chemicals that fall under TSCA have been thoroughly screened for toxicity. The Frank R. Lautenberg Chemical Safety for the 21st Century Act, an update to the TSCA, addresses some of these concerns by mandating more stringent testing of some chemicals, but critics argue that many potentially dangerous chemicals will remain inadequately tested.

The European Union (EU) is taking the world's boldest step toward testing and regulating manufactured chemicals. In 2007, the EU's **REACH** program went into effect (*REACH* stands for *R*egistration, *E*valuation, *A*uthorization, and restriction of *CH*emicals). REACH largely shifts the burden of proof for testing chemical safety from national governments to industry and requires that chemical substances produced or imported in amounts of over 1 metric ton per year be registered with a new European Chemicals Agency.

The world's nations have also sought to address chemical pollution with international treaties. The *Stockholm Convention on Persistent Organic Pollutants* (*POPs*) came into force in 2004 and has been ratified by over 150 nations. POPs are toxic chemicals that persist in the environment, bioaccumulate and biomagnify up the food chain, and can travel long distances. The PCBs and other contaminants found in polar bears are a prime example. The Stockholm Convention aims first to end the use and release of 12 POPs shown to be most dangerous, a group nicknamed the "dirty dozen". It sets guidelines for phasing out these chemicals and encourages transition to safer alternatives.

closing THE LOOP

Domestic regulation in the United States by the FDA and EPA, and international agreements such as REACH and the Stockholm Convention, indicate that governments may act to protect the world's people, wildlife, and ecosystems from toxic substances and other environmental hazards. At the same time, solutions often come more easily when they do not arise from government regulation alone. Consumer choice exercised through the market can often be an effective way to influence industry's decision making, but this requires consumers to have full information from scientific research regarding the risks involved. Once scientific results are in, a society's philosophical approach to risk management will determine what policy decisions are made.

All these factors have come into play regarding regulation of BPA, phthalates, and other harmful chemicals in food packaging and other consumer products. Research into the adverse effects of these chemicals is emerging, and while some nations have banned BPA and phthalates, some have only restricted their use in children's products and others have chosen not to restrict them at all. But growing consumer concern over the presence of harmful chemicals such as BPA and phthalates has spurred some companies to shift to safer alternatives, even in the absence of governmental regulation in the United States.

It is important to remember, however, that synthetic chemicals, while exposing people to some risk, have brought us innumerable modern conveniences, a larger food supply, and medical advances that save and extend human lives. A safer and happier future, one that safeguards the well-being of both people and the environment, therefore depends on knowing the risks that some hazards pose, assessing these risks, and having means in place to phase out harmful substances and replace them with safer ones whenever possible.

TESTING Your Comprehension

1. What four major types of health hazards are examined by practitioners of environmental health?
2. In what way is disease the greatest hazard that people face? What kinds of interrelationships must environmental health experts study to learn how diseases affect human health?
3. Where does most exposure to lead, asbestos, radon, and PBDEs occur?

4. List and describe the general categories of toxic substances covered in this chapter.

5. Explain the mechanisms within organisms that protect them from damage caused by toxic substances.

6. How do toxic substances travel through the environment? Describe and contrast the processes of bioaccumulation and biomagnification.

7. What are epidemiological studies, and how are they most often conducted?

8. Why are animals used in laboratory experiments in toxicology? Explain the dose-response curve. Why is a substance with a high LD_{50} considered safer than one with a low LD_{50}?

9. What factors may affect an individual's response to a toxic substance? What are synergistic effects, and why are they difficult to measure and diagnose?

10. How do scientists identify and assess risks from substances or activities that may pose health risks?

SEEKING Solutions

1. Describe some environmental health hazards you may be living with indoors. How may you have been affected by indoor or outdoor hazards in the past? How could you best deal with these hazards in the future?

2. Do you feel that laboratory animals should be used in experiments in toxicology? Why or why not?

3. Describe differences in the policies of the United States and the European Union regarding the study and management of the risks of synthetic chemicals. Which do you believe is more appropriate—the policies of the United States or those of the European Union? Why?

4. **CASE STUDY CONNECTION** You work for a public health organization. You have been asked to educate the public about BPA and to suggest ways that people can minimize their exposure to the chemical. You begin by examining your own lifestyle and finding ways to use alternatives to BPA-containing products. Create a list of five ways that you are exposed to BPA daily, and then list approaches you could take to help you avoid or minimize these exposures. What are some potential costs in terms of time and/or money in embracing these changes? What information about BPA could you provide to the public as it relates to human health?

5. **THINK IT THROUGH** You are the parent of two young children and want to minimize the environmental health risks they are exposed to. Name five steps you could take in your household and in your daily life that would minimize their exposure to environmental health hazards.

CALCULATING Ecological Footprints

In 2007, the last year the EPA gathered and reported data on pesticide use, Americans used 1.13 billion pounds of pesticide active ingredients, and global use totaled 5.21 billion pounds. In that same year, the U.S. population was 302 million, and the world's population was 6.63 billion. In the table, calculate your share of pesticide use as a U.S. citizen in 2007 and the amount used by (or on behalf of) the average citizen of the world.

Annual pesticide use

YOU	POUNDS OF ACTIVE INGREDIENTS
Your class	
Your state	
United States	1.13 billion
World (total)	5.21 billion
World (per capita)	

1. What is the ratio of your annual pesticide use to the world's per capita average?

2. In 2007, the average U.S. citizen had an ecological footprint of 8.0 hectares, and the average world citizen's footprint was 2.7 hectares. Compare the ratio of pesticide usage with the ratio of the overall ecological footprints. How do these differ, and how would you account for the difference?

3. Does the per capita pesticide use for a U.S. citizen seem reasonable for you personally? Why or why not? Do you find this figure alarming or of little concern? What else would you like to know to assess the risk associated with this level of pesticide use?

Mastering Environmental Science

CHAPTER 11

Geology, Minerals, and Mining

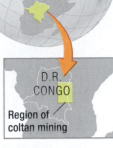

Region of coltan mining

central CASE STUDY

Mining for … Cell Phones?

> **The conflict in the Democratic Republic of the Congo has become mainly about access, control, and trade of five key mineral resources: coltan, diamonds, copper, cobalt, and gold.**
> —Report to the United Nations Security Council, April 2001

> **Coltan . . . is not helping the local people. In fact, it is the curse of the Congo.**
> —African journalist Kofi Akosah-Sarpong

You walk across your college campus to your next class. In the process, you pull out your cell phone and text a friend—and likely give very little thought to the technology that makes this text possible. What you probably don't know is that inside your phone is a little-known metal called tantalum—just a tiny amount—and without it, no cell phone can operate. Half a world away, a miner in the heart of Africa toils all day in a jungle streambed, sifting sediment for nuggets of coltan ore, which contain tantalum.

In bedeviling ways, tantalum links our glossy global high-tech economy with one of the most abused regions on Earth. Democratic Republic of Congo (D.R. Congo) has long been embroiled in a sprawling regional conflict fueled by ethnic tensions and access to valuable natural resources. This conflict has involved six nations and various rebel militias and has claimed more than 5 million lives since 1998. The current conflict is the latest chapter in the sad history of a nation rich in natural resources—copper, cobalt, gold, diamonds, uranium, and timber—whose impoverished people have been repeatedly robbed of the control of those resources.

Tantalum (Ta), element number 73 on the periodic table (**APPENDIX D**), is at the heart of the current battle over resources in D.R. Congo. As noted, we rely on this metal for our cell phones, but it is also vital for production of computer chips, DVD players, game consoles, and digital cameras. Tantalum powder is ideal for capacitors (the components that store energy and regulate current in miniature circuit boards) because it is highly heat resistant and readily conducts electricity.

Tantalum comes from a dull blackish mineral called tantalite, which often occurs with a mineral called columbite—so the ore is referred to as columbite–tantalite, or coltan for short. In eastern Congo, miners dig craters in rainforest streambeds, panning for coltan much as early California miners panned for gold.

As information technology boomed in the late 1990s, global demand for tantalum rose, and market prices for the metal shot up. High prices led some Congolese miners to mine coltan by choice, but many more were forced to work as miners.

Open war broke out in D.R. Congo in 1998 when rebel groups, supported by forces from neighboring Rwanda and Uganda, attempted to overthrow the government of President Laurent-Désiré Kabila. The country was fragmented by conflict between government forces and various rebel groups, with 11 other African nations

Mining is harming the endangered okapi in eastern Congo. ▲

Upon completing this chapter, you will be able to:

- Explain how plate tectonics and the rock cycle shape the landscape around us

- Identify major types of geologic hazards and describe ways to minimize their impacts

- Describe the types of mineral resources and their uses

- Describe the major methods of mining

- Discuss the environmental and social impacts of mining

- Explain reclamation efforts and mining policy

- Evaluate ways to encourage sustainable use of mineral resources

Coltan mining in eastern D.R. Congo

becoming involved in the hostilities. In mineral-rich eastern D.R. Congo, fighting was particularly intense. Local farmers were chased off their land; villages were burned; and civilians were raped, tortured, and killed. Soldiers and rebels seized control of mining operations. They forced farmers, refugees, prisoners, and children to work, and enriched themselves by selling coltan to traders, who in turn made profits by selling it to processing companies in the United States, Europe, and Asia. These companies refine and sell tantalum powder to capacitor manufacturers, which in turn sell capacitors to Nokia, Motorola, Sony, Intel, Compaq, Dell, and other high-tech corporations that use the capacitors in their products.

The conflict also caused ecological havoc as people streamed into national parks to avoid the fighting and locate new sources of minerals. This led to the clearing of rainforests for fuelwood and the killing of wildlife for food, including forest elephants; endangered gorillas; and okapi, a rare zebra-like relative of the giraffe. Miners disturbed streambeds in the search for coltan, increasing erosion rates and choking streams with sediments.

A grass-roots activist movement urged international action to stop the violence and exploitation in D.R. Congo and advanced the slogan, "No blood on my cell phone!" In 2001, an expert panel commissioned by the United Nations (UN) concluded that coltan riches were fueling, financing, and prolonging the war. The panel urged a UN embargo on coltan and other minerals from regions of D.R. Congo where conflict flourished.

Corporations rushed to assure consumers that they were not using tantalum from eastern D.R. Congo—noting that the region was producing less than 10% of the world's supply. Meanwhile, some observers felt an embargo could hurt the long-suffering Congolese people. The mining life may be miserable, they said, but it pays better than most jobs in a land where the average income is only 20 cents a day.

A 2002 peace treaty led to the withdrawal of foreign troops from D.R. Congo, but conflict with rebel groups within the country continues, bankrolled by the mineral riches of the region. Success by Congolese troops and an African-led UN intervention brigade against a major rebel group in eastern D.R. Congo has, however, helped to reduce conflict in the region.

Steps are now being taken to help support legitimate Congolese mines while preventing the exploitation that has defined mining in D.R. Congo in the recent past. Industry groups, working with national governments and nongovernmental aid organizations, are creating a certification system for conflict-free coltan to aid consumers in avoiding products that contain conflict minerals.

These efforts and others offer an opportunity to significantly reduce trade in conflict minerals while promoting trade of minerals sourced from legitimate mines in poor nations such as D.R. Congo. It is hoped that ongoing regulatory efforts to certify minerals will establish a framework that not only satisfies the world's demand for mineral resources but also protects the people and ecosystems that provide them.

Geology: The Physical Basis for Environmental Science

Coltan provides just one example of how we extract raw materials from beneath our planet's surface and turn them into products we use every day. To understand the environmental impacts of extracting resources from the earth, and the many ways we can make mineral extraction less damaging, we first need a working knowledge of some of the physical processes that shape our planet.

Our planet is dynamic, and this dynamism is what motivates **geology,** the study of Earth's physical features, processes, and history. A human lifetime is just a blink of an eye in the long course of geologic time, and the earth we experience is merely a snapshot in our changing planet's long history. We can begin to grasp this long-term dynamism as we consider two processes of fundamental importance to geology—plate tectonics and the rock cycle.

Earth consists of layers

Our planet consists of multiple layers (**FIGURE 11.1**). At Earth's center is a dense **core** consisting mostly of iron, solid in the inner core and molten in the outer core. Surrounding the core is a thick layer of less dense, elastic rock called the **mantle.** A portion of the upper mantle called the **asthenosphere** contains especially soft rock, melted in some areas. The harder rock above the asthenosphere is the **lithosphere.** The lithosphere includes both the uppermost mantle and the entirety of Earth's third major layer, the **crust,** the thin, brittle, low-density layer of rock that covers Earth's surface. The intense heat in the inner Earth rises from core to mantle to crust, and it eventually dissipates at the surface.

The heat from the inner layers of Earth also drives convection currents that flow in loops in the mantle, pushing the mantle's soft rock cyclically upward (as it warms) and downward (as it cools), like a gigantic conveyor belt system. As the mantle material moves, it drags large plates of lithosphere along its surface. This movement is known as **plate tectonics,** a process of extraordinary importance to our planet.

Plate tectonics shapes Earth's geography

Our planet's surface consists of about 15 major tectonic plates, which fit together like pieces of a jigsaw puzzle (**FIGURE 11.2**). Imagine peeling an orange and then placing the pieces back onto the fruit; the ragged pieces of peel are like the lithospheric plates riding atop Earth's surface. However, the plates are thinner relative to the planet's size, more like the skin of an apple. These plates move at rates of roughly 2–15 cm (1–6 in.) per year. This slow movement has influenced Earth's climate and life's evolution throughout our

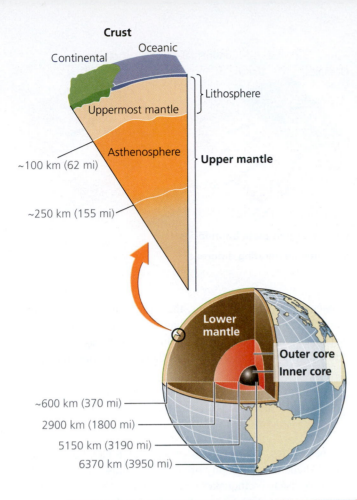

FIGURE 11.1 Earth's three primary layers—core, mantle, and crust—are themselves layered. The inner core of solid iron is surrounded by an outer core of molten iron, and the rocky mantle includes the molten asthenosphere near its upper edge. At Earth's surface, dense and thin oceanic crust abuts lighter, thicker continental crust. The lithosphere consists of the crust and the uppermost mantle above the asthenosphere.

planet's history as the continents combined, separated, and recombined in various configurations. By studying ancient rock formations throughout the world, geologists have determined that at least twice, all landmasses were joined together in a "supercontinent." Scientists have dubbed the landmass that resulted about 225 million years ago Pangaea (see inset in Figure 11.2).

There are three types of plate boundaries

The processes that occur at each type of plate boundary all have major consequences.

At **divergent plate boundaries,** tectonic plates push apart from one another as magma rises upward to the surface, creating new lithosphere as it cools (**FIGURE 11.3a**). An example is the Mid-Atlantic Ridge, part of a 74,000-km (46,000-mi) system of divergent plate boundaries slicing across the floors of the world's oceans.

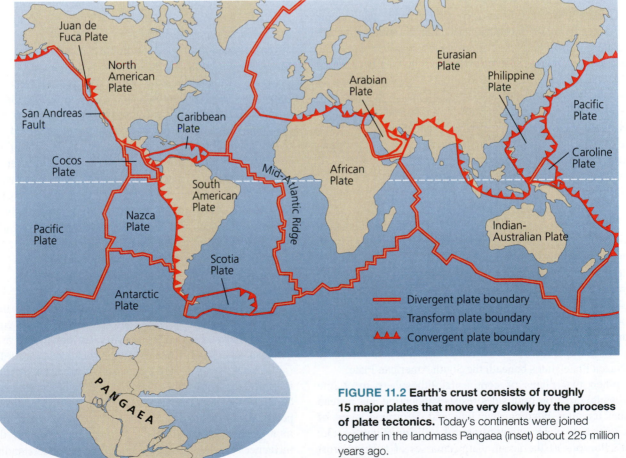

FIGURE 11.2 Earth's crust consists of roughly 15 major plates that move very slowly by the process of plate tectonics. Today's continents were joined together in the landmass Pangaea (inset) about 225 million years ago.

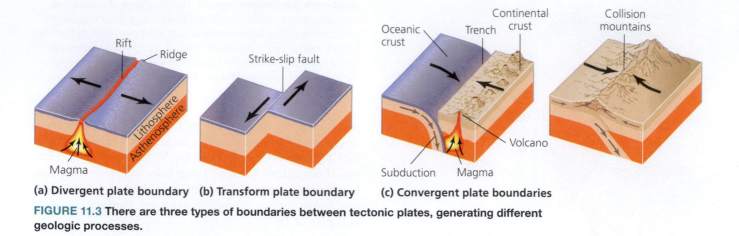

(a) Divergent plate boundary   (b) Transform plate boundary   (c) Convergent plate boundaries

FIGURE 11.3 There are three types of boundaries between tectonic plates, generating different geologic processes.

Where two plates meet, they may slip and grind alongside one another, forming a **transform plate boundary** (**FIGURE 11.3b**). This movement creates friction that generates earthquakes (p. 236) along strike-slip faults. The Tohuku earthquake, for example, occurred at such a fault off the coast of Japan. Faults are fractures in Earth's crust, and at strike-slip faults, each landmass moves horizontally in opposite directions. The Pacific Plate and the North American Plate, for example, slide past one another along California's San Andreas Fault. Southern California is slowly inching its way northward along this fault, and so the site of Los Angeles will eventually—in about 15 million years or so—reach that of modern-day San Francisco.

Convergent plate boundaries, where two plates come together, can give rise to different outcomes (**FIGURE 11.3c**). As plates of newly formed lithosphere push outward from divergent plate boundaries, this oceanic lithosphere gradually cools, becoming denser. After millions of years, it becomes denser than the asthenosphere beneath it and dives downward into the asthenosphere in a process called **subduction.** As the lithospheric plate descends, it slides beneath a neighboring plate that is less dense, forming a convergent plate boundary. The subducted plate is heated and pressurized as it sinks, and water vapor escapes, helping to melt rock (by lowering its melting temperature). The molten rock rises, and this magma may erupt through the surface via volcanoes (p. 236).

When one plate of oceanic lithosphere is subducted beneath another plate of oceanic lithosphere, the resulting volcanism may form arcs of islands, such as Japan and the Aleutian Islands of Alaska. Subduction zones may also create deep trenches, such as the Mariana Trench, our planet's deepest abyss, located in the western Pacific Ocean. When oceanic lithosphere slides beneath continental lithosphere, volcanic mountain ranges form that parallel coastlines (Figure 11.3c, left). An example is South America's Andes Mountains, where the Nazca Plate slides beneath the South American Plate.

When two plates of continental lithosphere meet, the continental crust on both sides resists subduction and instead crushes together, bending, buckling, and deforming layers of rock from both plates in a **continental collision** (Figure 11.3c, right). Portions of the accumulating masses of buckled crust are forced upward as they are pressed together, and mountain

ranges result. The Himalayas, the world's highest mountains, result from the Indian-Australian Plate's collision with the Eurasian Plate beginning 40–50 million years ago, and these mountains are still rising today as these plates converge.

Tectonics produces Earth's landforms

In these ways, the processes of plate tectonics build mountains; shape the geography of oceans, islands, and continents; and give rise to earthquakes and volcanoes. The coltan mining areas of eastern Congo are situated along the western edge of Africa's Great Rift Valley system, a region where the African plate is slowly pulling itself apart. Some of the world's largest lakes have formed in the immense valley floors, far below towering volcanoes such as Mount Kilimanjaro.

The topography created by tectonic processes, in turn, shapes climate by altering patterns of rainfall, wind, ocean currents, heating, and cooling—all of which affect rates of weathering and erosion and the ability of plants and animals to inhabit different regions. Thus, the locations of biomes (pp. 83–89) are influenced by plate tectonics. Moreover, tectonics has affected the history of life's evolution; the convergence of landmasses into supercontinents such as Pangaea is thought to have contributed to widespread extinctions by reducing the area of species-rich coastal regions and by creating an arid continental interior with extreme temperature swings.

The rock cycle alters rock

We tend to think of rock as pretty solid stuff. Yet in the long run, over geologic time, rocks and the minerals that compose them are heated, melted, cooled, broken down, and reassembled in a very slow process called the **rock cycle** (**FIGURE 11.4**).

A **rock** is any solid aggregation of minerals. A **mineral,** in turn, is any naturally occurring solid element or inorganic compound with a crystal structure, a specific chemical composition, and distinct physical properties. The type of rock in a given region affects soil characteristics and thereby influences the region's plant community. Understanding the rock cycle enables us to better appreciate the formation and

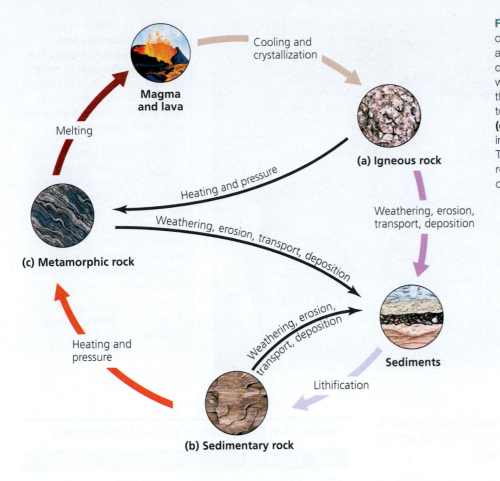

FIGURE 11.4 The rock cycle. Igneous rock **(a)** is formed when rock melts and the resulting magma or lava then cools. Sedimentary rock **(b)** is formed when rock is weathered and eroded and the resulting sediments are compressed to form new rock. Metamorphic rock **(c)** is formed when rock is subjected to intense heat and pressure underground. Through these processes, each type of rock can be converted into either of the other two types.

Magma and lava

Cooling and crystallization

Melting

(a) Igneous rock

Heating and pressure

Weathering, erosion, transport, deposition

Weathering, erosion, transport, deposition

(c) Metamorphic rock

Weathering, erosion, transport, deposition

Heating and pressure

Sediments

Lithification

(b) Sedimentary rock

conservation of soils, mineral resources, fossil fuels, and other natural resources.

Igneous rock All rocks can melt. At high enough temperatures, rock will enter a molten, liquid state called **magma.** If magma is released through the lithosphere (as in a volcanic eruption), it may flow or spatter across Earth's surface as **lava.** Geologists call the rock that forms when magma or lava cools **igneous rock** (from the Latin *ignis,* meaning "fire") (**FIGURE 11.4a**).

Sedimentary rock All exposed rock weathers away with time. The relentless forces of wind, water, freezing, and thawing eat away at rocks, stripping off one tiny grain (or large chunk) after another. Through weathering (p. 144) and erosion (p. 148), particles of rock come to rest downhill, downstream, or downwind from their sources, forming **sediments.** Alternatively, some sediments form chemically from the precipitation of substances out of solution.

Over time, deep layers of sediment accumulate, causing the weight and pressure on the layers below them to increase. **Sedimentary rock** (**FIGURE 11.4b**) is formed as sediments are physically pressed together and as dissolved minerals seep through sediments and act as a kind of glue, binding sediment particles (a process termed lithification).

These processes also create the fossils of organisms (p. 58) we use to learn about the history of life on Earth and the fossil fuels we use for energy (p. 343). Because sedimentary layers, or strata, pile up in chronological order, geologists

and paleontologists can assign relative dates to fossils they find in sedimentary rock.

Metamorphic rock Geologic forces may bend, uplift, compress, or stretch rock. When any type of rock is subjected to great heat or pressure, it may alter its form, becoming **metamorphic rock** (from the Greek for "changed form") (**FIGURE 11.4c**). The forces that metamorphose rock generally occur deep underground, at temperatures lower than the rock's melting point but high enough to change its appearance and physical properties.

Geologic processes occur at timescales that are difficult to conceptualize. But only by appreciating the long periods within which our planet's geologic forces operate can we realize how exceedingly slow processes such as plate tectonics or the formation of sedimentary rock can reshape our planet. This lengthy timescale is referred to as deep time, or geologic time.

Geologic and Natural Hazards

Plate tectonics shapes our planet, but the consequences of tectonic movement can also pose hazards to us. Earthquakes and volcanic eruptions are examples of such geologic hazards. We can see how such hazards relate to tectonic processes by examining a map of the *circum-Pacific belt*, or "ring of fire"

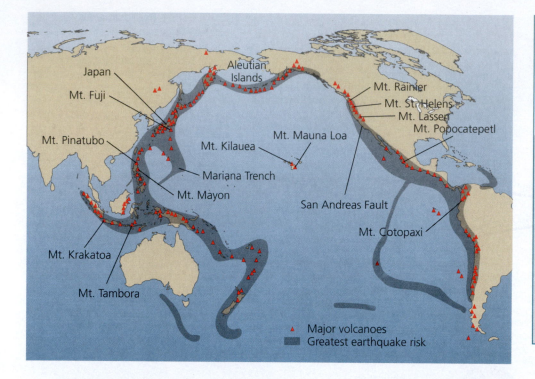

Japan
Mt. Fuji
Mt. Pinatubo
Mt. Krakatoa
Mt. Tambora
Mt. Mayon
Mariana Trench
Mt. Kilauea
Aleutian Islands
Mt. Mauna Loa
Mt. Rainier
Mt. St. Helens
Mt. Lassen
Mt. Popocatepetl
San Andreas Fault
Mt. Cotopaxi

▲ Major volcanoes
▬ Greatest earthquake risk

(**FIGURE 11.5**). Ninety percent of earthquakes and over half the world's volcanoes occur along this 40,000-km (25,000-mi) arc of subduction zones and fault systems.

Earthquakes result from movement at plate boundaries and faults

Along tectonic plate boundaries, and in other places where faults occur, the earth may relieve built-up pressure in fits and starts. Each release of energy causes what we know as an **earthquake.** Most earthquakes are barely perceptible, but occasionally they are powerful enough to do tremendous damage to human life and property (**TABLE 11.1**). Earthquakes can also occur in the interior tectonic plates, when faults are formed by continental plates being stretched and pulled apart by geologic forces within the earth. Such earthquakes are not only rare but poorly understood. The New Madrid seismic zone, which lies beneath the lower Mississippi River basin in the central United States, is one area where such an "intraplate" earthquake may occur (**FIGURE 11.6**). And human activities may also be inducing earthquakes in areas far from the boundaries of tectonic plates (see **THE SCIENCE BEHIND THE STORY**, pp. 238–239).

Volcanoes arise from rifts, subduction zones, or hotspots

Where molten rock, hot gas, or ash erupts through Earth's surface, a **volcano** is formed, often creating a mountain over time as cooled lava accumulates. As we have seen, lava can extrude in rift valleys and along mid-ocean ridges, or above subduction zones as one tectonic plate dives beneath another. Lava may also be emitted at *hotspots*, localized areas where plugs

TABLE 11.1 Examples of Large Earthquakes

YEAR	LOCATION	FATALITIES	MAGNITUDE[1]
1556	Shaanxi Province, China	830,000	~8
1755	Lisbon, Portugal	70,000[2]	8.7
1906	San Francisco, California	3,000	7.8
1923	Kwanto, Japan	143,000	7.9
1964	Anchorage, Alaska	128[2]	9.2
1976	Tangshan, China	255,000+	7.5
1985	Michoacan, Mexico	9,500	8.0
1989	Loma Prieta, California	63	6.9
1994	Northridge, California	60	6.7
1995	Kobe, Japan	5,502	6.9
2004	Northern Sumatra	228,000[2]	9.1
2005	Kashmir, Pakistan	86,000	7.6
2008	Sichuan Province, China	50,000+	7.9
2010	Port-au-Prince, Haiti	236,000	7.0
2010	Maule, Chile	500	8.8
2011	Northern Japan	18,000[2]	9.0
2015	Kathmandu, Nepal	8,900	7.8

[1]Measured by moment magnitude; each full unit is roughly 32 times as powerful as the preceding full unit.
[2]Includes deaths from the resulting tsunami.

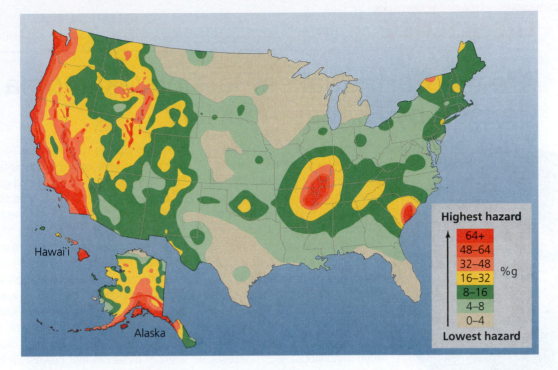

FIGURE 11.6 Many parts of the United States are at elevated risk for earthquakes. The West Coast faces threats from earthquakes due to its position at the boundary of tectonic plates. Portions of the continental interior have elevated risk due to naturally occurring intraplate earthquakes or human-induced earthquakes, typically from wastewater injection or hydraulic fracturing. The units for the figure are %g, a measure of acceleration related to the force of gravity. *Source: U.S. Geological Survey.*

Highest hazard

64+	
48–64	
32–48	%g
16–32	
8–16	
4–8	
0–4	

Lowest hazard

of molten rock from the mantle erupt through the crust. As a tectonic plate moves across a hotspot, repeated eruptions from this source may create a linear series of volcanoes. The Hawaiian Islands provide an example of this process (**FIGURE 11.7**).

At some volcanoes, lava flows slowly downhill and at other times, a volcano may let loose large amounts of ash and cinder in a sudden explosion. Sometimes a volcano can unleash a pyroclastic flow—a fast-moving cloud of toxic gas, ash, and rock fragments that races down the slopes, enveloping everything in its path. Such a flow buried the inhabitants of the ancient Roman cities of Pompeii and Herculaneum in A.D. 79, when Mount Vesuvius erupted.

Volcanic eruptions affect people as well as the environment. Ash blocks sunlight, and sulfur emissions lead to a

FIGURE 11.7 The Hawaiian Islands are the product of a hotspot on Earth's mantle. The Hawaiian Islands **(a)** have been formed by repeated eruptions from a hotspot of magma in the mantle as the Pacific Plate passes over the hotspot. The Big Island of Hawai'i is most recently formed, and it is still volcanically active. The other islands are older and have already begun eroding away. To their northwest stretches a long series of former islands, now submerged. The active volcano Kilauea **(b)**, on the Big Island's southeast coast, is currently located above the edge of the hotspot.

Are the Earthquakes Rattling Oklahoma Caused by Human Activity?

**Geophysicist
Katie Keranen,
Cornell University**

In November 2011, a series of earthquakes and aftershocks struck the small town of Prague, Oklahoma (population 2300), roughly 100 km (60 mi) east of Oklahoma City. The shaking damaged 14 homes, buckled the pavement of a local highway, and caused several injuries. One of the tremors measured 5.7 on the Richter scale—the largest earthquake ever recorded in the state.

Earthquakes are uncommon in Oklahoma, especially ones of such magnitude. But scientists were not completely surprised by the 2011 event, because they had been noting an increase in earthquake activity (**FIGURE 1**). For example, between 1978 and 2008, Oklahoma experienced a yearly average of only ~1.5 earthquakes of 3.0 magnitude or greater. In 2009, that number rose to 20, and by 2016 had jumped to 641. And scientists had even proposed an explanation for the increase—the injection of wastewater from oil and gas extraction into porous rock layers beneath the state.

Spurred by high energy prices, the extraction of crude oil and natural gas from conventional wells (p. 346) and hydraulic fracturing (p. 342) operations in Oklahoma increased from 2010 to 2013, with gas extraction rising by 17% and oil extraction by 65%. And as Oklahoma's oil and gas output increased, so did its disposal of wastewater by injection, increasing by 20% from 2010 to 2013. Oil and gas are the modified remains of ancient marine organisms, and deposits often contain briny water that is separated from the oil and gas after the fuels are extracted. This salty wastewater, which can also contain toxic and radioactive compounds, is then typically disposed of by trucking it to a facility far from the oil and gas wells. Once there, it is injected into porous rock formations thousands of feet underground, well below the more shallow rock layers that contain groundwater aquifers. This approach is designed to dispose of wastewater in a manner that prevents it from contaminating sources of drinking water, both aboveground and belowground.

Wastewater had been injected in this manner beneath Oklahoma for decades without any measurable increase in seismic activity, but scientists were aware that continued pumping of wastewater into rock formations could lead to earthquakes. As the pores within rocks become saturated with water, pressure grows in the underground rock layers, causing the rocks to expand. These expanding rock layers then push against existing faults in the earth, which "lubricates" them and causes them to slip and produce earthquakes. The rock formations beneath Oklahoma facilitate this process, as many of the porous layers of limestone into which wastewaters are injected are located near stressed rocks around faults.

OKLAHOMA

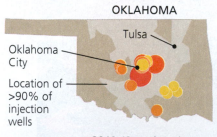

2010 43 earthquakes

2013 109 earthquakes

2016 641 earthquakes

FIGURE 1 The number of earthquakes in Oklahoma has increased greatly in recent years. Many of these earthquakes are thought to be related to the injection of wastewater from oil and gas extraction into underground rock layers. Each circle indicates an earthquake event. The higher the magnitude of the earthquake, the darker and wider the circle. *Source: Oklahoma Geological Survey, 2017, http://earthquakes.ok.gov/what-we-know/ earthquake-map/.*

Scientists grew more and more convinced that the increased seismic activity in Oklahoma was due to wastewater injections; however, persuading legislators, regulators, and the public proved challenging. Oil and gas extraction is big business in Oklahoma. By some estimates, one in five jobs in the state is connected to the industry. Landowners benefit from royalties earned by fossil fuel extractions on their land. Tax revenue from sales of oil and gas is the state's third-largest revenue source—behind only sales taxes and personal income taxes.

With such widespread economic benefits, the oil and gas industry enjoys high levels of support in Oklahoma government. When calls arose following the Prague quake to temporarily halt further wastewater injections, the government urged a cautious approach and echoed the industry's position that greater study was needed before decisive action should be taken.

Decisive action was eventually spurred in part by a scientific study published in 2013 that directly linked the Prague earthquake to nearby injections of wastewater from oil and gas extraction. Although the connection between underground fluid injection and earthquakes was well known, studies directly connecting specific events with injection sites were rare. The study, led by geophysicist Katie Keranen of the University of Oklahoma (now at Cornell University) and published in the journal *Geology,* measured the aftershocks produced by the 2011 earthquake to determine the location of the fault that produced the quake.

Reacting quickly to the initial earthquake, the researchers deployed seismic sensors near Prague, and gathered detailed readings on two major tremors and 1183 aftershocks that followed. Analysis of the patterns revealed that the tip of the fault that ruptured was within about 200 m (650 ft) of an active wastewater injection well and occurred at depths consistent with injected rock layers. Keranen's work also showed that it is possible for nearly two decades to pass between the initiation of wastewater injection and a subsequent seismic event, calling into question the safety of many other injection sites. A subsequent study led by Keranen found that earthquakes could be induced as far as 30 km (19 mi) from injection well fields, and concluded that up to 20% of the induced seismic activity over an area covering 2000 km^2 (770 mi^2) in the central United States could be traced to the activity of four high-volume wastewater disposal wells in Oklahoma.

As research continued into the connection between wastewater injection and seismic activity, and the issue gained public attention, the state incrementally increased its regulation of wastewater injection wells. The government mandated regular monitoring of well pressures in injection sites, and directed operators to slow injection rates or stop injections altogether if underground conditions were deemed conducive to initiating a seismic event. These actions led to lower levels of wastewater injection in 2016 versus 2015, and may have resulted in immediate benefit, as the number of earthquakes of magnitude 3.0 or greater in Oklahoma dropped from 890 in 2015 to 641 in 2016. But while the number of earthquakes declined in 2016, the state did experience the largest earthquake in its history—a magnitude 5.8 earthquake that struck near Pawnee, Oklahoma.

While recent events may be showing progress in reducing induced earthquakes in Oklahoma, research has shown that seismic activity is likely to occur long after wastewater injection has ceased, even in locations far from injection sites. Therefore, the south central United States, much like the U.S. West Coast, will be a hotbed for seismic study in coming decades (**FIGURE 2**).

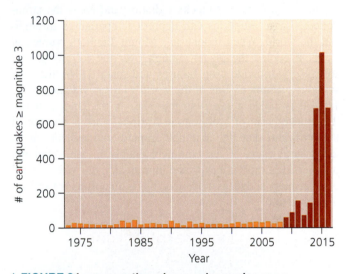

FIGURE 2 Larger earthquakes are becoming more common in the central and eastern United States. Much of the increase is centered in Oklahoma, where the local geology, coupled with the use of seismic-inducing activities such as oil extraction and wastewater injection, has led to more frequent tremors. *Source: Rubinstein, J.L., and A.B. Mahani, 2015. Myths and facts on wastewater injection, hydraulic fracturing, enhanced oil recovery, and induced seismicity. Seismol. Res. Lett. 86: 1–8 and U.S. Geological Survey, 2016.*

 • In which two sequential years was there the greatest change in the number of earthquakes? • How many times more earthquakes occurred in the later year of this time period than in the earlier year?

Go to **Interpreting Graphs & Data** on **Mastering** Environmental Science

FIGURE 11.8 Tsunami waves overtop a seawall following the Tohoku earthquake in 2011. The tsunami caused a greater loss of life and property than the earthquake that generated it and led to a meltdown at the Fukushima Daiichi nuclear power plant.

sulfuric acid haze that blocks radiation and cools the atmosphere. Large eruptions can depress temperatures throughout the world. When Indonesia's Mount Tambora erupted in 1815, it cooled the planet enough over the following year to cause worldwide crop failures and make 1816 "the year without a summer." One of the world's largest volcanoes—so large it is called a supervolcano—lies in the United States. The entire basin of Yellowstone National Park is an ancient supervolcano that has at times erupted so massively as to cover large parts of the continent deeply in ash. Although another eruption is not expected imminently, the region is still geothermally active, as evidenced by its numerous hot springs and geysers.

Landslides are a form of mass wasting

At a smaller scale than volcanoes or earthquakes, a **landslide** occurs when large amounts of rock or soil collapse and flow downhill. Landslides are severe, often sudden, manifestations of the phenomenon of **mass wasting,** the downslope movement of soil and rock due to gravity. Mass wasting occurs naturally, but often it is brought about by human land use practices that expose or loosen soil, making slopes more prone to collapse. Heavy rains may saturate soils and trigger mudslides of soil, rock, and water.

Most often, mass wasting eats away at unstable hillsides, damaging property one structure at a time. Occasionally mass wasting events can be colossal and deadly; mudslides that followed the torrential rainfall of Hurricane Mitch in Nicaragua and Honduras in 1998 killed over 11,000 people. Mudslides caused when volcanic eruptions melt snow and send huge volumes of destabilized mud racing downhill are called lahars, and they are particularly dangerous. A lahar following an eruption in 1985 buried the entire town of Armero, Colombia, killing 21,000 people.

Tsunamis can follow earthquakes, volcanoes, or landslides

Earthquakes, volcanic eruptions, and large coastal landslides can all displace huge volumes of ocean water instantaneously and trigger a **tsunami,** an immense swell, or wave, of water that can travel thousands of miles across oceans. In 2011, a tsunami generated by an offshore earthquake devastated large portions of northeastern Japan (**FIGURE 11.8**). The tsunami and earthquake killed more than 18,000 people, caused hundreds of billions of dollars in economic impacts, and contributed to the meltdown of the Fukushima Daiichi nuclear power plant (p. 369). In December 2004, a massive tsunami, triggered by an earthquake off Sumatra, devastated the coastlines of countries all around the Indian Ocean, including Indonesia, Thailand, Sri Lanka, India, and several African nations. Roughly 228,000 people were killed and 1–2 million were displaced. Since the 2004 tsunami, nations and international agencies have stepped up efforts to develop systems to give coastal residents advance warning of approaching tsunamis.

Those of us who live in the United States and Canada should not consider tsunamis to be something that occurs only in faraway places. Residents of the Pacific Northwest—such as the cities of Seattle, Washington, and Portland, Oregon—could be at risk if there is a slip in the Cascadia subduction zone that lies 1100 km (700 mi) offshore. The tsunami produced by such a slip would inundate 1.1 million km^2 (440,000 mi^2) of coastal land and cause massive destruction over an area that is currently home to 7 million people.

We can worsen or lessen the impacts of natural hazards

Aside from geologic hazards, people face other types of natural hazards. Heavy rains can lead to flooding that ravages low-lying areas near rivers and streams (p. 272). Coastal erosion can eat away at beaches (p. 255). Wildfire can threaten life and property in fire-prone areas. Tornadoes and hurricanes can cause extensive damage and loss of life.

Although we refer to such phenomena as "natural hazards," the magnitude of their impacts on us often depends on choices we make. We sometimes worsen the impacts of so-called natural hazards in various ways. For example, we live in areas that are prone to hazards, such as the floodplains of rivers or in coastal areas susceptible to flooding. People also use and engineer landscapes around us in ways that can increase the frequency or severity of natural hazards. Damming and diking rivers to control floods can sometimes lead to catastrophic flooding, and the clear-cutting of forests on slopes (p. 200) can induce mass wasting and increase water runoff. Human-induced climate change (Chapter 14) can cause sea levels to rise and promote coastal flooding, and can increase the risks of drought, fire, flooding, and mudslides by altering precipitation patterns.

We can often reduce or lessen the impacts of hazards through the thoughtful use of technology, engineering, and policy, informed by a solid understanding of geology and ecology. Examples include building earthquake-resistant structures; designing early warning systems for earthquakes, tsunamis, and volcanoes; and conserving reefs and shoreline vegetation to protect against tsunamis and coastal erosion. In addition, better forestry, agriculture, and mining practices can help prevent mass wasting. Zoning regulations, building codes, and insurance incentives that discourage development in areas prone to landslides, floods, fires, and storm surges can help keep us out of harm's way. Finally, addressing global climate change may help reduce the frequency of natural hazards in many regions.

Earth's Mineral Resources

Coltan provides just one example of how we extract raw materials from beneath our planet's surface and turn them into products we use every day. We mine and process a wide array of mineral resources in the modern world. Indeed, without these resources—which we use to make everything from building materials to fertilizers—civilization as we know it could not exist. Just consider a typical scene from a student lounge at a college or university (**FIGURE 11.9**) and note how many items are made with elements from the minerals we take from the earth.

We obtain minerals by mining

We obtain the minerals we use through the process of mining. The term *mining* in the broad sense describes the extraction of any resource that is nonrenewable on the timescale of our society. In this sense, we mine fossil fuels and groundwater, as well as minerals. When used specifically in relation to minerals, **mining** refers to the systematic removal of rock, soil, or other material for the purpose of extracting minerals of economic interest. Because most minerals of interest are widely

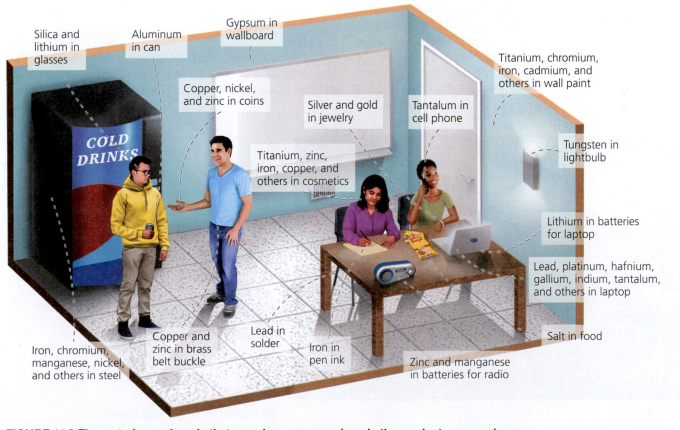

FIGURE 11.9 Elements from minerals that we mine are everywhere in the products we use in our everyday lives. This scene from a typical college student lounge points out just a few of the many elements from minerals that surround us.

How do geologists find mineral deposits that are deep in the earth?

Searching for reserves of underground minerals, also called prospecting, can be pursued in a number of ways. The earliest prospectors explored promising areas on foot, looking for exposed seams of mineral-containing rocks or for minerals carried into streams by runoff. Today, geologists direct vibrations into underground rock strata and capture the vibrations with sensors as the vibrations bounce off underground rock layers and reflect back to the surface. This process enables scientists to visualize the underlying rock layers and identify likely locations for reserves, just as they do for fossil fuel deposits (p. 346). Geologists also measure the magnetic fields in rock layers to look for metal ores, and they conduct chemical analyses of stream water to detect minerals of interest. If promising sites are located, cores can be drilled deep into the ground and inspected for the desired mineral before actual mining begins.

spread but in low concentrations, miners and mining geologists first try to locate concentrated sources of minerals before mining begins.

We use mined materials extensively

The average American consumes more than 17,900 kg (39,500 lb) of new minerals and fuels every year, according to 2016 estimates by the U.S. Minerals Education Coalition. At current rates of use, a child born in 2016 will use more than 1.4 million kg (3.1 million lb) of minerals and fuel during his or her lifetime (**FIGURE 11.10**).

More than half of the annual mineral and fuel use is from the coal, oil, and natural gas used to supply our intensive demands for energy. Much of the remaining mineral use is attributable to the sand, gravel, and stone used in constructing our buildings, roads, bridges, and parking lots. Metal use is dwarfed by these other two categories, but the average American will still use more than 2 tons of aluminum over his

1.59 Troy oz.
Gold

945 lb
Copper

37,587 lb
Salt

11,977 lb
Clays

15,287 lb
Phosphate rock

393,054 lb
Coal

552 lb
Zinc

1.31 million lb
Stone, sand, and gravel

4,885 lb
Bauxite (Aluminum)

72,983 gal
Petroleum

21,276 lb
Iron ore

886 lb
Lead

50,274 lb
Cement

7.08 million cu.ft.
Natural gas

+ 50,210 lb
Other minerals and metals

FIGURE 11.10 At current rates of use, an American baby born in 2016 is predicted to use more than 1.4 million kg (3.1 million lb) of minerals over his or her lifetime. *Data from Minerals Education Coalition, 2016.*

or her lifetime. This level of consumption clearly shows the potential of recycling and reuse (such as recycling stone and gravel from old highways into new construction) to make our modern, mineral-intensive lifestyle more sustainable.

Metals are extracted from ores

Some minerals can be mined for metals. As we have seen, the tantalum used in electronic components comes from the mineral tantalite (**FIGURE 11.11**). A **metal** is a type of chemical

(a) Coltan ore

(b) Capacitors containing tantalum

FIGURE 11.11 Tantalum is used to manufacture electronics. Coltan ore **(a)** is mined from the ground and then processed to extract the pure metal tantalum. This metal is used in capacitors **(b)** and other electronic components.

FIGURE 11.12 **A worker guides molten iron out of a blast furnace.** The metal is separated from the surrounding rock by melting ore at high temperatures and collecting the heavier metal.

FIGURE 11.13 **This surface impoundment at the Upper Big Branch mine in West Virginia holds coal tailings from a surface mining operation.** The impoundment, shown in the bottom left of the image—below the mined hillside above it—is like a large in-ground swimming pool holding potentially toxic liquids produced by the mining or processing of metals, minerals, or fuels.

element, or a mass of such an element, that typically is lustrous, opaque, and malleable, and can conduct heat and electricity. Most metals are not found in a pure state in Earth's crust but instead are present within **ore,** a mineral or grouping of minerals from which we extract metals. Copper, iron, lead, gold, and aluminum are among the many economically valuable metals we extract from mined ore.

We process metals after mining ore

Extracting minerals from the ground is the first step in putting them to use. However, most metals need to be processed in some way to become useful for our products. For example, after ores are mined, the metal-bearing rock is pulverized and washed, and the desired minerals are then isolated using chemical and/or physical means. For iron, this involves heating ore-bearing rocks to extremely high temperatures in a blast furnace and collecting the molten iron when it separates from the surrounding minerals, a process known as **smelting** (FIGURE 11.12). For aluminum, bauxite ore is first treated with chemicals to extract alumina (an aluminum oxide), and then an electrical current is used to generate pure aluminum from alumina. With coltan, processing facilities use acid solvents to separate tantalite from columbite. Other chemicals are then used to produce metallic tantalum powder. This powder can be consolidated by various melting techniques and can be shaped into wire, sheets, or other forms. Sometimes we mix, melt, and fuse a metal with another metal or a nonmetal substance to form an **alloy.** For example, steel is an alloy of the metal iron that has been fused with a small quantity of carbon.

Processing minerals exerts environmental impacts. Most methods are water-intensive and energy-intensive. Moreover, extracting metals from ore emits air pollution—smelting plants in particular can be hotspots of toxic air pollution. In addition, soil and water commonly become polluted by **tailings,** portions of ore left over after metals have been extracted. Tailings may leach heavy metals present in the ore waste as well as chemicals applied in the extraction process. For instance,

cyanide is used to extract gold from ore and sulfuric acid is used to extract copper. Mining operations often store toxic slurries of tailings in large reservoirs called **surface impoundments** (FIGURE 11.13). Their walls are designed to prevent leaks, but accidents can occur if the structural integrity of the impoundment is compromised. In 2000, a breach of an impoundment near Inez, Kentucky, released over 1 billion liters (250–300 million gal) of coal slurry, blackening 120 km (75 mi) of streams, killing aquatic wildlife, and affecting drinking water supplies. The failure of two impoundments at an iron ore mine in Brazil in 2015 buried nearby villages in a toxic slurry of water and mining waste, claiming 19 lives.

We also mine nonmetallic minerals and fuels

We mine and use many minerals that do not contain metals. FIGURE 11.14 illustrates the nation of origin and uses for some economically important mineral resources, both metallic and nonmetallic. As you can see, many geologic resources in the products we use are mined in faraway nations. Concentrated deposits of minerals form in several different ways, causing them to be unevenly distributed on Earth. In some cases, minerals are concentrated in magma and so accumulate in rock layers beneath magma chambers and in areas with large quantities of igneous rocks. This is true of many metals, including tantalum, iron, nickel, and platinum. In other cases, hot groundwater dissolves minerals from large areas of rock, and then concentrates them into one area when the water cools and the minerals precipitate out of solution. This occurs for minerals containing sulfur, zinc, and copper. Geologic processes, such as plate tectonics and rock weathering, also play a role in creating mineral reserves and making them accessible for mining.

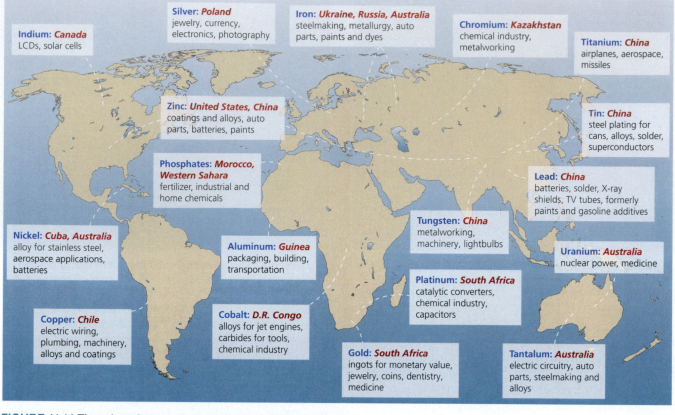

FIGURE 11.14 The minerals we use come from all over the world. Shown is a selection of economically important minerals (mostly metals, with several nonmetals), together with their major uses and their main nation of origin. Only a minority of minerals, uses, and origins is shown.

Sand and gravel (the most commonly mined mineral resources) provide fill and construction materials. Phosphates provide us with fertilizer. We mine limestone, salt, potash, and other minerals for many diverse purposes.

Gemstones are treasured for their rarity and beauty. For instance, diamonds have long been prized—and like coltan, they have fueled resource wars. Besides the conflict in eastern Congo, the diamond trade has acted to fund, prolong, and intensify wars in Angola, Sierra Leone, Liberia, and other nations. This is the origin of the term "blood diamonds," just as coltan has been called a "conflict mineral."

We also mine substances for fuel. Uranium ore is a mineral from which we extract the metal uranium, used in nuclear power (p. 366). One of the most common fuels we mine, coal (p. 346), is the modified remains of ancient swamp plants and is made up of the mineral carbon. Other fossil fuels—petroleum, natural gas, and alternative fossil fuels such as oil sands, oil shale, and methane hydrates—are also organic and are extracted from the earth (Chapter 15).

Mining Methods and Their Impacts

Mining for minerals is an important industry that provides us with raw materials for the countless products we use daily. However, the extraction of minerals—and the cleaning and processing of mined materials—often involves the degradation of large areas of land, thereby exerting severe impacts on the environment and on the people living near the mining locations.

If various methods are appropriate for a given resource, companies typically select a method based on its economic efficiency. In the sections that follow, we'll examine the primary mining techniques that are used throughout the world, and take note of the impacts of each as we proceed.

Strip mining removes surface layers of soil and rock

When a resource occurs in shallow horizontal deposits near the surface, the most effective mining method is often **strip mining,** whereby layers of surface soil and rock are removed from large areas to expose the resource. Heavy machinery removes the overlying soil and rock (termed *overburden*) from a strip of land, and the resource is extracted. This strip is then refilled with the overburden that had been removed, and miners proceed to an adjacent strip of land and repeat the process. Strip mining is commonly used for coal (**FIGURE 11.15a**) and oil sands (p. 347), and sometimes for sand and gravel.

Strip mining can be economically efficient, but it obliterates natural communities over large areas, and the soil in refilled areas can easily erode away. Strip mining also

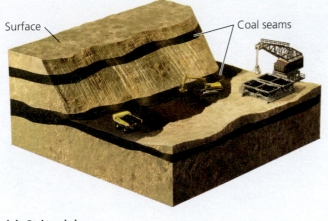

(a) Strip mining

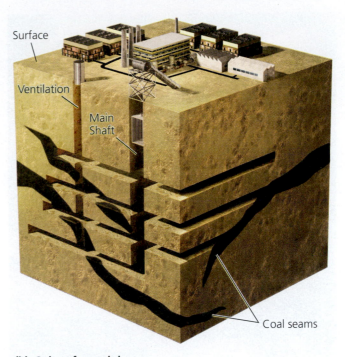

(b) Subsurface mining

FIGURE 11.15 Coal mining illustrates two types of mining approaches. In strip mining **(a)**, soil is removed from the surface in strips, exposing seams from which coal is mined. In subsurface mining **(b)**, miners work belowground in shafts and tunnels blasted through the rock. These passageways provide access to underground seams of coal or minerals.

FAQ

Why would anyone choose to work in a mine when it's such dangerous work?

Subsurface mining is the most dangerous form of mining and indeed one of society's most dangerous occupations. Besides risking injury or death from dynamite blasts, natural gas explosions, and collapsing shafts and tunnels, miners inhale toxic fumes and coal dust, which can lead to respiratory diseases, including fatal black lung disease.

Many of the people who work in mines do so because they have few other options. Underground mining often occurs in economically depressed areas, such as Appalachia in the United States, where mining is one of the few jobs that pay well. And for most mining jobs, people can begin working right out of high school. So, although the work is dangerous, many miners are willing to accept those risks to provide for themselves and their families because relatively fewer reliable, well-paying career opportunities are available.

pollutes waterways through the process of **acid mine drainage,** which occurs when sulfide minerals in newly exposed rock surfaces react with oxygen and rainwater to produce sulfuric acid (**FIGURE 11.16**). As the sulfuric acid runs off, it leaches metals from the rocks, and many of these metals are toxic to organisms. Acid drainage can affect fish and other aquatic organisms when it runs into streams and can pollute groundwater supplies people use for drinking water or irrigating crops. Although acid drainage is a natural phenomenon, mining greatly accelerates this process by exposing many new rock surfaces at once.

In subsurface mining, miners work underground

When a resource occurs in concentrated pockets or seams deep underground, and the earth allows for safe tunneling, then mining companies pursue **subsurface mining.** In this approach, shafts are excavated

deep into the ground, and networks of tunnels are dug or blasted out to follow deposits of the mineral (**FIGURE 11.15b**). Miners remove the resource systematically and ship it to the surface.

We use subsurface mining for metals such as zinc, lead, nickel, tin, gold, copper, and uranium, as well as for diamonds, phosphate, salt, and potash. In addition, a great deal of coal is mined using the subsurface technique.

FIGURE 11.16 Acidic drainage flows from a coal mine in Scotland. The orange color is due to iron from the drainage settling out on the soil surface and forming rust.

FIGURE 11.17 Smoldering mine fires beneath Centralia, Pennsylvania, have led to the creation of a "ghost town." A nearly 1-mile-long section of State Route 61 was closed, as subsidence due to mine fires caused the road to buckle and crack. The roadway has become a tourist attraction for people visiting the area.

Occasionally, subsurface mines can have impacts even long after the mines have been closed. For example, coal veins in abandoned mines underneath Centralia, Pennsylvania, caught fire in 1961. The once-thriving city became a ghost town as nearly all of its residents accepted buyouts in the 1980s and relocated when it became clear that the smoldering fires beneath them could not be contained; and indeed, the fires continue to burn to this day (**FIGURE 11.17**). Acidic drainage from subsurface mines can contaminate surface and groundwater, sometimes for centuries. Natural disasters or accidents can lead to catastrophic releases of toxin-laden groundwater that accumulates in abandoned mines, as occurred in 2015 when millions of gallons of heavy metal–laden mine drainage from an abandoned gold mine in Colorado were accidentally released into the nearby Animas River, killing aquatic wildlife and contaminating water supplies for communities in Colorado, Utah, New Mexico, and the Navajo nation.

Open pit mining creates immense holes in the ground

When a mineral is spread widely and evenly throughout a rock formation, or when the earth is unsuitable for tunneling, the method of choice is **open pit mining.** This essentially

FIGURE 11.18 The Bingham Canyon open pit mine outside Salt Lake City, Utah, is the world's largest human-made hole in the ground. This immense mine produces mostly copper.

involves digging a gigantic hole and removing the desired ore, along with waste rock that surrounds the ore. Some open pit mines are inconceivably enormous. The world's largest, the Bingham Canyon Mine near Salt Lake City, Utah, is 4 km (2.5 mi) across and 1.2 km (0.75 mi) deep (**FIGURE 11.18**). Conveyor systems and immense trucks with tires taller than a person carry out nearly half a million tons of copper ore and waste rock each day.

Open pit mines are terraced so that people and machinery can move about, and waste rock is left in massive heaps outside the pit. The pit is expanded until the resource runs out or becomes unprofitable to mine. Open pit mining is used to extract copper, iron, gold, diamonds, and coal, among other resources. This technique is also used to extract clay, gravel, sand, and stone (such as limestone, granite, marble, and slate), but such pits are generally called *quarries*.

Once mining is complete, abandoned pits generally fill up with groundwater, which soon becomes toxic as sulfides from the ore react and produce sulfuric acid. Acidic water from the pit can harm wildlife and can percolate into aquifers and spread through the region.

Mountaintop mining reshapes ridges and can fill valleys

When a resource occurs in underground seams near the tops of ridges or mountains, mining companies may practice **mountaintop removal mining,** in which several hundred vertical feet of mountaintop may be blasted off to allow recovery of entire seams of the resource (**FIGURE 11.19**). This method of mining is used primarily for coal in the Appalachian Mountains of the eastern United States. In mountaintop removal mining, a mountain's forests are clear-cut, the timber is sold, topsoil is removed, and then rock is repeatedly blasted away to expose the coal for extraction.

FIGURE 11.19 Mountaintop mining removes entire mountaintops to obtain the coal underneath. The rocks removed during the process are dumped into adjacent valleys, burying streams, promoting flooding, and contaminating drinking water supplies.

Afterward, overburden is placed back onto the mountaintop, but this waste rock is unstable and typically takes up more volume than the original rock, so generally a great deal of waste rock is dumped into adjacent valleys (a practice called "valley filling"). So far, mountaintop removal in the Appalachian Mountains has blasted away an area the size of Delaware and has buried nearly 3200 km (2000 mi) of streams.

Scientists are finding that dumping tons of debris into valleys degrades or destroys immense areas of habitat, clogs streams and rivers, and pollutes waterways with acid drainage. With slopes deforested and valleys filled with debris, erosion intensifies, mudslides become frequent, and flash floods ravage the lower valleys. Worsening the environmental impact of mountaintop removal mining is the fact that the Appalachian forests cleared in mountaintop mining are some of the richest forests for biodiversity in the nation.

People living in communities near the mining sites experience social and health impacts. Explosions that are a part of clearing the mountaintops crack house foundations and wells, loose rock tumbles down into yards and homes, and floods tear through properties when mining operations block or divert streams. Coal dust causes respiratory ailments, and contaminated water unleashes a variety of health problems. Studies have shown that people in mountaintop mining areas exhibit elevated levels of birth defects, lung cancer, heart disease, kidney disease, pulmonary disorders, hypertension, and early mortality.

Critics of mountaintop removal mining argue that valley filling violates the Clean Water Act (p. 107) because runoff flowing through waste rocks in valleys often contains high levels of salts and toxic metals that degrade water quality and affect aquatic organisms. Although hundreds of permits for mountaintop mining were issued during the Bill Clinton and George W. Bush administrations, in 2010 the U.S. Environmental Protection Agency (EPA) announced new guidelines that prohibit valley fills unless strict measures of water quality can be attained. Critics of the policy argue that these new guidelines will essentially end the practice of mountaintop mining. In 2011, the EPA revoked the permit of an existing mountaintop mining operation in West Virginia, citing these new guidelines. A 2014 court ruling backed the EPA's decision, and the agency continued to reexamine other permits across Appalachia.

Placer mining uses running water to isolate minerals

Some metals and gems accumulate in riverbed deposits, having been displaced from elsewhere and carried along by flowing water. To search for these metals and gems, miners sift through material in modern or ancient riverbed deposits, generally using running water to separate lightweight mud and gravel from heavier minerals of value (**FIGURE 11.20**). This technique is called **placer mining** (pronounced "plasser").

Placer mining is the method used by D.R. Congo's coltan miners, who wade through streambeds, sifting through large amounts of debris by hand with a pan or simple tools, searching for high-density tantalite that settles to the bottom while low-density material washes away. Today's African miners practice small-scale placer mining similar to the method used by American miners who ventured to California in the Gold Rush of 1849, and later to Alaska in the Klondike Gold Rush

FIGURE 11.20 Miners in eastern Congo find coltan by placer mining. Sediment is placed in plastic tubs, and water is run through them. A mixing motion allows the sediment to be poured off while the heavy coltan settles to the bottom.

of 1896–1899. Placer mining for gold is still practiced in areas of Alaska and Canada, although today, miners use large dredges and heavy machinery.

Placer mining is environmentally destructive because most methods wash large amounts of debris into streams, making them uninhabitable for fish and other life for many miles downstream. This type of mining also disturbs stream banks, causing erosion and harming ecologically important plant communities.

Solution mining dissolves and extracts resources in place

When a deposit is especially deep underground and the resource can be dissolved in a liquid, miners may use a technique called **solution mining.** In this technique, a narrow borehole is drilled to reach the deposit, and water, acid, or another liquid is injected down the borehole to leach the resource from the surrounding rock and dissolve it in the liquid. The resulting solution is then sucked out, and the desired resource is removed from solution. Sodium chloride (table salt), lithium, boron, bromine, magnesium, potash, copper, and uranium can be mined in this way.

Solution mining generally exerts less environmental impact than other mining techniques, because less area at the surface is disturbed. The primary potential impacts involve accidental leakage of acids into groundwater surrounding the borehole, and the contamination of aquifers with acids, heavy metals, or uranium leached from the rock.

weighing the ISSUES

Restoring Mined Areas

Mining has severe environmental impacts, but restoring mined sites to their pre-mining condition can be costly and difficult. How extensively should mining companies be required to restore a site after a mine is shut down, and what criteria should we use to guide restoration? Should we require nearly complete restoration? What should our priorities be—to minimize water pollution, impacts on human health, biodiversity loss, soil damage, or other factors? What measures should we use to evaluate the results of restoration? Should the amount of restoration we require depend on how much money the company made from the mine? Explain your recommendations.

Some mining occurs in the ocean

The oceans hold many minerals useful to our society. We extract some minerals from seawater, such as magnesium from salts held in solution. We extract other minerals from the ocean floor. For example, many minerals are concentrated in manganese nodules, small ball-shaped accretions that are scattered across parts of the ocean floor. More than 1.5 trillion tons of manganese nodules may exist in the Pacific Ocean alone, and their reserves of metal may exceed all terrestrial reserves. As land resources become scarcer and as undersea mining technology develops, mining companies may turn increasingly to the seas. The logistical difficulty of mining offshore resources, however, has limited their extraction so far.

FIGURE 11.21 More mine sites are now being restored. Here, bison graze on land reclaimed from a tar sands mining operation in Alberta, Canada.

Restoration helps to reclaim mine sites

Because of the environmental impacts of mining, governments of the United States and other developed nations now require that mining companies restore, or reclaim, surface-mined sites following mining. The aim of such restoration, or **reclamation,** is to restore the site to an approximation of its pre-mining condition.

To restore a site, companies are required to remove buildings and other structures used for mining, replace overburden, fill in shafts, and replant the area with vegetation (**FIGURE 11.21**). In the United States, the 1977 Surface Mining Control and Reclamation Act mandates restoration efforts, requiring companies to post bonds to cover reclamation costs before mining can be approved. This ensures that if the company fails to restore the land for any reason, the government will have the money to do so. Most other nations exercise less oversight regarding reclamation, and in nations such as D.R. Congo, there is essentially no regulation.

The mining industry has made great strides in reclaiming mined land, but even on sites that are restored, impacts from mining—such as soil and water damage from acid drainage—can be severe and long-lasting because the soil is often acidic and can contain high levels of metals that are toxic to native plant life. It is therefore often difficult to regain the same biotic communities that were naturally present before mining. Researchers are hybridizing varieties of wild grasses to create new strains of plants that can tolerate the soil conditions on reclaimed sites, with the hope that such plants will pave the way for the reintroduction of native vegetation. Establishing plant communities on reclaimed sites is key, because the plants stabilize the soil, prevent erosion, and can help create conditions that favor the reestablishment of native vegetation and functioning ecosystems.

An 1872 law still guides U.S. mining policy

The ways in which mining companies stake claims and use land in the United States is guided by a law that is well over a century old. The **General Mining Act of 1872** encourages people and companies to prospect for minerals on federally owned land by allowing any U.S. citizen, domestic company, or foreign company with permission to do business in the United States to stake a claim on any plot of public land open to mining. The person or company owning the claim gains the sole right to take minerals from the area. The claim-holder can also patent the claim (i.e., buy the land) for only about $5 per acre. Regardless of the profits they might make on the minerals they extract, the law requires no payments of any kind to the public and, until recently, no restoration of the land after mining. Supporters of the policy say that it is appropriate and desirable to continue encouraging the domestic mining industry, which must undertake substantial financial risk and investment to locate resources that are vital to our economy. Critics counter that the policy gives valuable public resources away to private interests nearly for free, and have tried unsuccessfully over the past two decades to repeal or modify this Act.

The General Mining Act of 1872 covers a wide variety of metals, gemstones, uranium, and minerals used for building materials. In contrast, fossil fuels, phosphates, sodium, and sulfur are governed by the Mineral Leasing Act of 1920. This law sets terms for leasing public lands that vary according to the resource being mined, but in all cases the terms include the payment of rents for the use of the land and the payment of royalties on profits.

Toward Sustainable Mineral Use

Mining exerts plenty of environmental impacts, but we also have another concern to keep in mind: Minerals are not inexhaustible resources (p. 96). As a result, it will benefit us to find ways to conserve the supplies we have left and to make them last.

Minerals are nonrenewable resources in limited supply

Some minerals we use are abundant and will likely never run out, but others are rare enough that they could soon become unavailable. For instance, geologists in 2017 calculated that the world's known reserves of tantalum will last about 129 more years at today's rate of consumption. If demand for tantalum increases, it could run out faster. And if everyone in the world began consuming tantalum at the rate of U.S. citizens, then it would last for only 31 more years!

Most pressing may be the dwindling supply of indium. This obscure metal, which is used in LCD screens and other electrical components, might last only another 30 years. A lack of indium and gallium would threaten the production of high-efficiency cells for solar power. Platinum is dwindling as well, and if this metal became unavailable, it would be harder to develop fuel cells and catalytic converters for vehicles. However, because of supply concerns and price volatility,

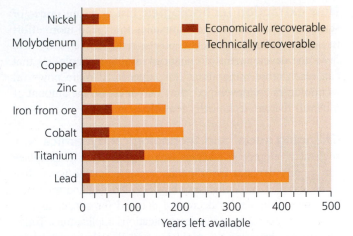

FIGURE 11.22 Minerals are nonrenewable resources, so supplies of metals are limited. Shown in red are the numbers of remaining years that certain metals are estimated to be economically recoverable at current prices. The entire lengths of the bars (red plus orange) show how long certain metals are estimated to be available at present rates of consumption, using current technology on all known deposits whether economically recoverable or not. *Data are for 2016, from U.S. Geological Survey, 2017. Mineral commodity summaries 2017. Reston, VA: USGS.*

DATA • Which metal has the highest proportion of its technically recoverable reserves that are currently economically recoverable? • Approximately what percentage is economically recoverable? • Which metal has the smallest proportion of its technically recoverable reserves that are currently economically recoverable, and what is this value?

Go to **Interpreting Graphs & Data** on Mastering Environmental Science

industries now are working intensely to develop ways of substituting other materials for indium, and platinum's high market price encourages recycling, which may keep it available, albeit as an expensive metal.

FIGURE 11.22 shows estimated years remaining for selected minerals at today's consumption rates. Calculating how long a given mineral resource will be available to us is beset by a great deal of uncertainty. There are several major reasons why such estimates may increase or decrease over time, each of which we'll discuss in turn.

Discovery of new reserves As we discover new deposits of a mineral, the known reserves—and thus the number of years this mineral is available to us—increase. For example, in 2010 geologists associated with the U.S. military discovered that Afghanistan holds immense mineral riches. The newly discovered reserves of iron, copper, niobium, lithium, and many other metals are estimated to be worth over $1 trillion—enough to realign the entire Afghan economy around mining. It is important to note, however, that these mineral riches are not guaranteed to make Afghanistan a wealthy nation; history teaches us that regions rich in nonrenewable resources, such as D.R. Congo and Appalachian regions of the United States, have often been unable to prosper from them.)

New extraction technologies Just as rising prices of scarce minerals encourage companies to expend more effort to access difficult-to-reach deposits, rising prices may also favor the development of enhanced mining technologies that can reach more minerals at less expense. If more powerful technologies are developed, they may increase the amount of minerals that are technically feasible to mine.

Changing social and technological dynamics New societal developments and new technologies in the marketplace can modify demand for minerals in unpredictable ways. Just as cell phones and computer chips boosted demand for tantalum, fiber-optic cables decreased demand for copper as they replaced copper wiring in communications applications. Today lithium–ion batteries are replacing nickel–cadmium batteries in many devices. Synthetically produced diamonds are driving down prices of natural diamonds and extending their availability. Additionally, health concerns sometimes motivate change: We have replaced toxic substances such as lead and mercury with safer materials in many applications, for example.

Changing consumption patterns Changes in the rates and patterns of consumption also alter the speed with which we exploit mineral resources. For instance, economic recession depressed demand and led to a decrease in production and consumption of most minerals from 2007 to 2009 after rising for long stretches. However, over the long term, demand has been rising. This is especially true today as China, India, and other major industrializing nations rapidly increase their consumption.

Recycling Advances in recycling technologies and the extent to which minerals are currently being recycled have helped to extend the lifetimes of many mineral resources, and will likely continue to do so as the technology and the demand for recycled materials grows. Considering Earth's finite supplies of mineral resources, conserving them through reuse and recycling can benefit us today—by preventing price hikes that result from reduced supply—while safeguarding these valuable resources for future generations.

We can make our mineral use more sustainable

We can address the challenges of a finite supply of mineral resources and the environmental damage associated with its extraction by encouraging the recycling of these resources. Electronic waste, or e-waste—from discarded computers, printers, cell phones, handheld devices, and other electronic products—is rising fast. Recycling old electronic devices helps to keep the hazardous substances associated with them out of landfills, while also assisting us to conserve valuable minerals such as tantalum, for which the mining industry estimates that recycling accounts for 20–25% of this metal's availability for use in manufacturing. Municipal recycling programs handle used items that consumers place in recycling bins, thus diverting metals from the waste stream (p. 400). Currently, around 35% of metals in the U.S. municipal solid waste stream are diverted for recycling. For example, 80% of the lead we consume today comes from recycled materials, in

---

SUCCESS STORY

Recycling Metals from E-Waste

Amplifier and receiver: Arsenic and gallium

Touch screen: Indium

Circuitry:
Copper
gold
palladium
platinum
silver
tungsten

Case: Petroleum and magnesium

Your cell phone contains a diversity of mined materials from around the world.

By some estimates, about 500 million old cell phones are currently lying inactive in people's homes and offices, and upgrades and improvements render more than 130 million additional cell phones obsolete each year in the United States alone. So, what can you do with your old cell phone? Aware of the problems posed by e-waste, more and more people are donating their retired cell phones to recycling programs, and by doing so, are giving the valuable metals and minerals within them a second life.

For example, when you turn in your old phone for recycling rather than discarding it, the phone may be dismantled and the various parts recycled for their metals. Alternatively, the phone may be refurbished and resold—people in African nations in particular readily buy used cell phones because they are inexpensive and land-line phone service does not always exist in poor and rural areas. Either way, by recycling your cell phone you're helping to extend the availability of our finite resources.

Today only about 10% of old cell phones are recycled, meaning that we have a long way to go! But as more of us recycle our phones, computers, and other electronic items, we'll be closer to sustainably reusing tantalum and other valuable metals, while decreasing the amount of e-waste that enters the waste stream.

EXPLORE THE DATA at **Mastering** Environmental Science

TABLE 11.2 Recycled minerals in the United States

MINERAL	U.S. RECYCLING RATE
Gold	Slightly less is recycled than is consumed
Iron and steel scrap	85% for autos, 82% for appliances, 72–98% for construction materials, 70% for cans
Lead	69% consumed comes from recycled post-consumer items
Tungsten	59% consumed is from recycled scrap
Nickel	45% consumed is from recycled nickel
Zinc	37% produced is recovered, mostly from recycled materials used in processing
Chromium	34% is recycled in stainless steel production
Copper	32% of U.S. supply comes from various recycled sources
Aluminum	30% produced comes from recycled post-consumer items
Tin	30% consumed is from recycled tin
Germanium	30% consumed worldwide is recycled. Optical device manufacturing recycles more than 60%
Molybdenum	About 30% gets recycled as part of steel scrap that is recycled
Cobalt	28% consumed comes from recycled scrap
Niobium (columbium)	Perhaps 20% gets recycled as part of steel scrap that is recycled
Silver	15% consumed is from recycled silver. U.S. recovers as much as it produces
Bismuth	All scrap metal containing bismuth is recycled, providing less than 10% of consumption
Diamond (industrial)	7% of production is from recycled diamond dust, grit, and stone

Data are for 2015, from U.S. Geological Survey, 2016. *Mineral commodity summaries 2016*. Reston, VA: USGS.

particular, recycled car batteries. Similarly, 33% of our copper comes from recycled copper sources such as pipes and wires. We recycle steel, iron, platinum, and other metals from auto parts. Altogether, we have found ways to recycle much of our gold, lead, iron and steel scrap, chromium, zinc, aluminum, and nickel. **TABLE 11.2** shows minerals that currently boast high recycling rates in the United States.

In many cases, recycling can decrease energy use substantially. For instance, making steel by re-melting recycled iron and steel scrap requires much less energy than producing steel from virgin iron ore. Because this practice saves money, the steel industry today is designed to make efficient use of iron and steel scrap. Over half its scrap comes from discarded consumer items such as cars, cans, and appliances. Similarly, more than 40% of the aluminum in the United States today is recycled. This is beneficial because it takes over 20 times more energy to extract virgin aluminum from ore (bauxite) than it does to obtain it from recycled sources.

closing THE LOOP

The physical processes of geology, such as plate tectonics and the rock cycle, are centrally important because they shape Earth's terrain and form the foundation for living systems. Geologic processes also generate phenomena that can threaten our lives and property, including earthquakes, volcanoes, landslides, and tsunamis. In addition, geologic processes also influence our access to the diversity of minerals and metals on which we depend through mining.

Economically efficient mining methods have greatly contributed to our material wealth, but they have also resulted in extensive environmental impacts, ranging from habitat loss to acid drainage.

As shown in the opening case study, which profiled the mining of tantalum in D.R. Congo, procuring these metals can have profound impacts on people and ecosystems, a realization that prompted the current movement toward the certification of minerals and gemstones as "conflict free." These initiatives are vital, as conflict minerals are an emerging problem in regions other than central Africa. A thriving black market in coltan is developing in remote portions of the northern Amazon jungle, and the recent discovery of vast mineral reserves in Afghanistan suggests that it, too, could become a significant source of conflict minerals.

Manufacturers, governments, and nongovernmental organizations are collaborating to make certification efforts practical and meaningful in the complex, multilayered global trade in metals and minerals. Consumers can also play a role in supporting these initiatives, by using online guides to choose products that avoid conflict minerals. Through these efforts, along with endeavors to make our mineral use more sustainable by maximizing recovery and recycling, we are working toward a future in which we can simultaneously meet our demand for these vital resources while minimizing environmental impacts and adverse effects on the people who extract and produce them.

TESTING Your Comprehension

1. Name the primary layers that make up our planet. Which portions does the lithosphere include?

2. Describe what occurs at a divergent plate boundary. What happens at a transform plate boundary? Compare and contrast the types of processes that can occur at a convergent plate boundary.

3. Name the three main types of rocks, and describe how each type may be converted to the others via the rock cycle.

4. Explain the processes that produce earthquakes, volcanoes, tsunamis, and mass wasting.

5. Define each of the following: (1) mineral, (2) metal, (3) ore, (4) alloy. Compare and contrast the terms.

6. A mining geologist locates a horizontal seam of coal very near the surface of the land. What type of mining method will the mining company likely use to extract it? What is one common environmental impact of this type of mining?

7. How does strip mining differ from subsurface mining? How does each of these approaches differ from open pit mining?

8. What is acid drainage, and where does it come from in a mining context? Why can such drainage be toxic to fish?

9. List five factors that can influence how long global supplies of a given mineral will last, and explain how each might increase or decrease the time span the mineral will be available to us.

10. Summarize the major factors that influence our estimates of reserves of valuable minerals and metals.

SEEKING Solutions

1. For each of the following natural hazards, describe one thing that we can do to minimize its impacts on our lives and property:

   - Earthquake
   - Tsunami
   - Mass wasting

2. List three impacts of mining on the natural environment, and describe how particular mining practices can lead to each of these impacts. How are these impacts being addressed? Can you think of additional solutions to prevent, reduce, or mitigate these impacts?

3. You have won a grant from the EPA to work with a mining company to develop a more effective way of restoring a mine site that is about to be closed. Describe a few preliminary ideas for carrying out restoration better than it is typically being done. Now describe a field experiment you would like to run to test one of your ideas.

4. **CASE STUDY CONNECTION** The story of coltan in D.R. Congo is just one example of how an abundance of exploitable resources can often worsen or prolong military conflicts in nations that are too poor or ineffectively governed to protect these resources. In such resource wars, civilians often suffer the most as civil society breaks down. Suppose you are the head of an international aid agency that has earmarked $10 million to help address conflicts related to mining in D.R. Congo. You have access to government and rebel leaders in D.R. Congo and neighboring countries, to ambassadors of the world's nations in the United Nations, and to representatives of international mining corporations. Based on what you know from this chapter, what steps would you consider taking to help improve the situation in this nation?

5. **THINK IT THROUGH** As you finish your college degree, you learn that the mountains behind your childhood home in the hills of Kentucky are slated to be mined for coal using the mountaintop removal method. Your parents, who still live there, are worried for their health and safety and do not want to lose the beautiful forested creek and ravine behind their property. However, your brother is out of work and could use a steady, well-paying mining job. What would you attempt to do in this situation?

CALCULATING Ecological Footprints

As we saw in Figure 11.22, the supplies of some metals are limited enough that, at today's prices, these metals could be available to us for only a few more decades. After that, prices will rise as they become scarcer. The number of years of total availability (at all prices) depends on a number of factors: On the one hand, metals will be available longer if new deposits are discovered, if new mining technologies are developed, and/or if recycling efforts improve. On the other hand, if our consumption of metals increases, the number of years we have left to use them will decrease.

Currently the United States consumes metals at a much higher per-person rate than the world does as a whole. If one goal of humanity is to lift the rest of the world up to U.S. living standards, then this will sharply increase pressures on mineral supplies.

The table below shows currently known economically recoverable global reserves for several metals, together with the amount used per year (each figure in thousands of metric tons). For each metal, calculate and enter (in the fourth column) the years of supply left at current prices by dividing the reserves by the amount used annually. The fifth column shows the amount the world would use if everyone in the world consumed the metal at the rate that Americans do. Now calculate the years of supply left at current prices for each metal if the world were to consume the metals at the U.S. rate, and enter these values in the sixth column.

METAL	KNOWN ECONOMIC RESERVES	AMOUNT USED PER YEAR	YEARS OF ECONOMIC SUPPLY LEFT	AMOUNT USED PER YEAR IF EVERYONE CONSUMED AT U.S. RATE	YEARS OF ECONOMIC SUPPLY LEFT IF EVERYONE CONSUMED AT U.S. RATE
Titanium	830,000	6600	125.8	24,692	33.6
Copper	720,000	19,400		41,308	
Nickel	78,000	2250		4846	
Tin	4700	280		932	
Tungsten	3100	86		337	
Antimony	1500	130		531	
Silver	570	27		167	
Gold	57	3.1		3.8	

Data are for 2016, from U.S. Geological Survey, 2017. Mineral commodity summaries 2017. USGS, Washington, D.C.
All numbers are in thousands of metric tons. World consumption data are assumed equal to world production data.
"Known economic reserves" include extractable amounts under current economic conditions. Additional reserves exist that could be mined at greater cost.

1. Which of these eight metals will last the longest under current economic conditions and at current rates of global consumption? For which of these metals will economic reserves be depleted fastest?

2. If the average citizen of the world consumed metals at the rate that the average U.S. citizen does, the economic reserves of which of these eight metals would last the longest? Which would be depleted fastest if everyone consumed at the U.S. rate?

3. In this chart, our calculations of years of supply left do not factor in population growth. How do you think population growth will affect these numbers?

4. Describe two general ways that we could increase the years of supply left for these metals. What do you think it will take to accomplish this?

Mastering Environmental Science

Students Go to **Mastering** Environmental Science for assignments, the etext, and the Study Area with practice tests, videos, current events, and activities.

Instructors Go to **Mastering** Environmental Science for automatically graded activities, current events, videos, and reading questions that you can assign to your students, plus Instructor Resources.

CHAPTER 12

Fresh Water, Oceans, and Coasts

Louisiana's vanishing coastal wetlands support a diversity of wildlife, such as these Roseate Spoonbills.

Starving the Louisiana Coast

> **The Louisiana and Mississippi coastal region is critical to the economic, cultural, and environmental integrity of the nation.**
> —Nancy Sutley, Chair of the White House Council on Environmental Quality (2010)

> **What really screwed up the marsh is when they put the levees on the river. They should take the levees out and let the water run; that's what built the land.**
> —Frank "Blackie" Campo, Resident of Shell Beach, Louisiana

The state of Louisiana is shrinking. Its coastal wetlands straddle the boundary between the land and the ocean, and these wetlands are disappearing beneath the waters of the Gulf of Mexico. Louisiana loses 75 km² (29 mi²) of coastal wetlands each year—about the size of Manhattan Island in New York City. Comparisons of wetland area from the mid-1800s to the early 1990s show a drastic decrease in wetlands (**FIGURE 12.1a**).

Louisiana's coastal wetlands transition from communities of salt-tolerant grasses at the ocean's edge to freshwater bald cypress swamps farther inland, supporting a diversity of animals such as shrimp, alligators, and black bears. The state's coastal wetlands also protect New Orleans, Baton Rouge, and other communities from damaging storms, acting as as buffer against strong winds and storm surges coming inland from the Gulf.

Louisiana's millions of acres of coastal wetlands formed over the past 7000 years as the Mississippi River deposited sediments at its delta before emptying into the Gulf of Mexico. The Mississippi River accumulates large quantities of sediment from water flowing over land and into streams in the river's 3.2-million-km² (1.2-million-mi²) watershed (**FIGURE 12.1b**).

The salt marshes in the river's delta naturally compact over time, lowering the level of the marsh bottom and submerging vegetation under increasingly deeper waters. When waters become too deep, the vegetation dies and soils are washed away by the Gulf of Mexico. The natural compaction is offset, however, by inputs of sediments from the river and from the deposition of organic matter from marsh grasses. These additions keep soil levels high, water depths relatively stable, and vegetation healthy.

So why are Louisiana's wetlands being swallowed by the sea? It's because people have modified the Mississippi River so extensively that much of its sediments no longer reach the wetlands that need them. The river's basin contains roughly 2000 dams, which slow river flow and allow sediments suspended in the water to settle in reservoirs. This not only prevents sediments from reaching the river's delta, but also slowly fills in each dam's reservoir, decreasing its volume and shortening its life span. In this way, dams throughout the Mississippi Basin affect the Louisiana coastline hundreds of miles downriver.

The Mississippi River is also lined with thousands of miles of *levees*—long raised mounds of earth—to prevent small-scale flooding. Levees at the

People fishing in a wetland in Louisiana. ▲

1839

1993

2020

(a) Coastal wetland area in 1839, 1993, and 2020

(b) Mississippi River watershed

Helena

Yellowstone River

Missouri River

Sioux Falls
Des Moines

Chicago

Pittsburgh

Columbus

Mississippi River
Watershed

Platte River

St. Louis

Louisville

Ohio

Oklahoma
City

Memphis

Arkansas River

Tennessee

Red River

Mississippi

New Orleans

Gulf of Mexico

(c) Sediment plumes from Mississippi River entering Gulf

FIGURE 12.1 Human modifications along the Mississippi River affect coastal wetlands at the river's mouth. Louisiana's coastal wetlands shrank **(a)** from 1839 to 1993, and are predicted to shrink even more by 2020 due to the construction of dams and levees along the river. The Mississippi River system **(b)** is the largest in the United States, draining over 40% of the land area of the lower 48 states. A satellite image of south Louisiana **(c)** shows the brown plumes of sediments being released into the Gulf of Mexico from the Mississippi River (plume on the right) and the Atchafalaya River (left). *(a) Adapted from Environmental Defense Fund.*

mouth of the Mississippi provide a deep river channel for shipping into the Gulf of Mexico and prevent the river from spilling into its delta, pouring sediments off the continental shelf and into the waters of the Gulf (**FIGURE 12.1c**).

Although oil and gas extraction has benefited Louisiana's economy, it has also promoted wetland losses. The extraction of oil, natural gas, and saline groundwater associated with oil deposits causes the land to compact, lowering soil levels. Additionally, engineers have cut nearly 13,000 km (8000 mi) of canals through coastal wetlands to facilitate shipping and oil and gas exploration, thus fragmenting wetlands and increasing erosion rates by enabling salty ocean water to penetrate inland, damaging vegetation and wildlife in freshwater marshes.

The need for rapid interventions to combat wetland losses was bolstered by a 2017 paper from researchers at Tulane University, reporting that the Louisiana coast was, on average, sinking by 9 mm (0.35 in.) a year. The team used sensors, installed along the Louisiana coast after Hurricane Katrina in 2005, to accurately measure the level of coastal subsidence, or sinking. Their findings were significant because previous studies had indicated that a "worst case scenario" would be the land sinking by 8–10 mm a year (0.31–0.39 in.), showing that current levels of coastal subsidence are already at alarming levels.

Proposed solutions for coastal erosion center on restoring the system to its natural state by diverting large quantities of water from the Mississippi River into coastal wetlands instead of shooting it out into the Gulf in the river's main channel. Proponents of this approach point to the Atchafalaya River, which currently diverts one-third of the lower Mississippi River's volume and carries it to the Gulf. The Atchafalaya delta, fed by this water and sediment, is actually gaining coastal land area. Additionally, the 2012 Resources and Ecosystems Sustainability, Tourist Opportunities and Revived Economies of the Gulf Coast States Act (RESTORE Act), legislation passed in the aftermath of the *Deepwater Horizon* spill (pp. 358–359), created a comprehensive ecosystem restoration plan for the Gulf Coast, financed by 80% of the fines paid for Clean Water Act (p. 107) violations associated with the spill.

Given the conflicting demands we put on waterways for water withdrawal, shipping, and flood control, there are no easy solutions to the problems faced in the Mississippi River and southern Louisiana. But how we tackle problems like those in Louisiana's coastal wetlands will help determine the long-term sustainability of one of our most precious natural resources—the aquatic ecosystems that provide us life-sustaining water.

Freshwater Systems

"Water, water, everywhere, nor any drop to drink." The well-known line from the poem *The Rime of the Ancient Mariner* is an apt description of the situation on our planet. Water may seem abundant, but water that we can drink is quite rare and limited (**FIGURE 12.2**). About 97.5% of Earth's water resides in the oceans and is too salty to drink or to use to water crops. Only 2.5% is considered **fresh water,** water that is relatively pure with few dissolved salts. Because most fresh water is tied up in glaciers, ice caps, and underground aquifers, just over 1 part in 10,000 of Earth's water is easily accessible for human use.

Water is renewed and recycled as it moves through the *water cycle* (pp. 40–41). The movement of water in the water cycle creates a web of interconnected aquatic systems that exchange water, organisms, sediments, pollutants, and other dissolved substances (**FIGURE 12.3**). What happens in one system therefore affects other systems—even those that are far away. Precipitation falling from the sky either sinks into the ground or flows off the land to form rivers, which carry water to the oceans or large inland lakes. As they flow, rivers can interact with ponds, wetlands, and coastal aquatic ecosystems. Underground aquifers exchange water with rivers, ponds, and lakes through the sediments on the bottoms of these water bodies. Let's examine the components of this interconnected system, beginning with groundwater.

Groundwater plays key roles in the water cycle

Liquid water occurs as either surface water or groundwater. **Surface water** is water located atop Earth's surface (such as a river or lake), and **groundwater** is water beneath the surface that resides within pores in soil or rock. Some of the precipitation reaching Earth's land surface infiltrates the surface to become groundwater. Groundwater flows slowly beneath the surface from areas of high pressure to areas of low pressure and can remain underground for long periods, in some cases for thousands of years. Groundwater makes up one-fifth of Earth's fresh water supply and plays a key role in meeting human water needs.

Groundwater is contained within **aquifers,** porous formations of rock, sand, or gravel that hold water (**FIGURE 12.4**). An aquifer's upper layer, or *zone of aeration*, contains pore spaces partly filled with water. In the lower layer, or *zone of saturation*, the spaces are completely filled with water. The boundary between these two zones is the **water table.** Any area where water infiltrates Earth's surface and reaches an aquifer below is known as a **recharge zone.** When a porous, water-bearing layer of rock, sand, or gravel is trapped between upper and lower layers of less permeable substrate (often clay), it is called a **confined aquifer,** or *artesian aquifer.* In such a situation, the water is under great pressure. In contrast, an **unconfined aquifer** has no impermeable upper layer to confine it, so its water is under less pressure and can be readily recharged by surface water.

FAQ

Is groundwater found in huge underground caverns?

As one of the "out of sight" elements of the water cycle, it's sometimes difficult for people to visualize how water exists underground. Many incorrectly assume that groundwater is always found in large underground caves—essentially lakes beneath Earth's surface. That is not the case. If you look at soil under a microscope, you'll see there are small pores between the particles of minerals and organic matter that compose the soil. Many types of rock, such as limestone and sandstone, have relatively large pores between the particles of minerals that make up the rock. So, when people extract groundwater with wells, we are simply sucking water out of the pores between soil particles or within rocks in the portion of the soil beneath the water table.

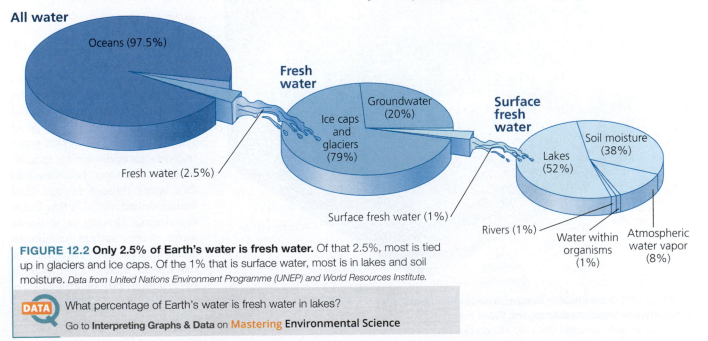

FIGURE 12.2 Only 2.5% of Earth's water is fresh water. Of that 2.5%, most is tied up in glaciers and ice caps. Of the 1% that is surface water, most is in lakes and soil moisture. *Data from United Nations Environment Programme (UNEP) and World Resources Institute.*

DATA What percentage of Earth's water is fresh water in lakes?

Go to **Interpreting Graphs & Data** on **Mastering** Environmental Science

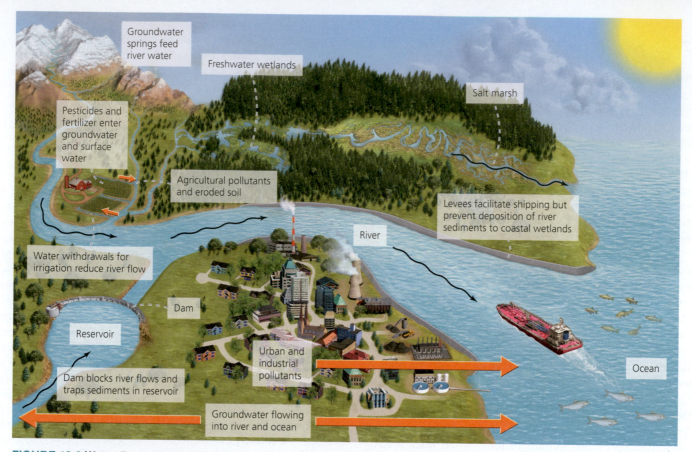

FIGURE 12.3 Water flows through freshwater systems and marine and coastal aquatic systems that interact extensively with one another. People affect the components of the system by constructing dams and levees, withdrawing water for human use, and introducing pollutants. Because the systems are closely connected, these impacts can cascade through the system and cause effects far from where they originated. In the figure, orange arrows indicate inputs into water bodies and black arrows indicate the direction of water flow.

Surface water converges in river and stream ecosystems

Surface water accounts for just 1% of fresh water, but it is vital for our survival and for the planet's ecological systems. Groundwater and surface water interact, and water can flow from one type of system to the other. Surface water becomes groundwater by infiltration. Groundwater becomes surface water through springs (and human-drilled wells), often keeping streams flowing or wetlands moist when surface conditions are otherwise dry. Each day in the United States, 1.9 trillion L (492 billion gal) of groundwater are released into surface waters—nearly as much as the daily flow of the Mississippi River.

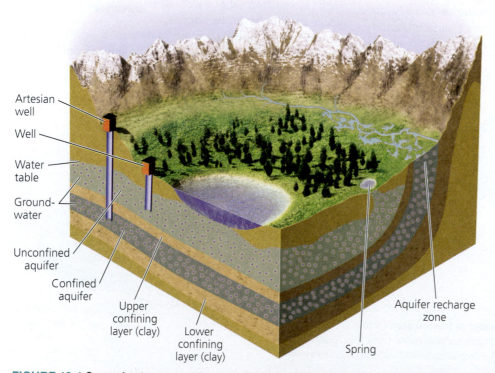

FIGURE 12.4 Groundwater occurs in unconfined aquifers above or in confined aquifers between impermeable layers. Water may rise to the surface at springs, in wetlands, and through wells. Artesian wells tap into confined aquifers to mine water under pressure.

Water that falls from the sky as rain, emerges from springs, or melts from snow or a glacier and then flows over the land surface, is called **runoff.** As it flows downhill, runoff converges where the land dips lowest, forming streams, creeks, or brooks. These small watercourses may merge into rivers, whose water eventually reaches a lake or ocean. A smaller river flowing into a larger one is called a *tributary*. The area of land drained by a *river system*—a river and all its tributaries—is that river's **drainage basin,** or **watershed** (p. 23). If you could trace every drop of water in the Mississippi River back to the spot where it first fell as precipitation, you would have delineated the river's watershed, the area shown in Figure 12.1.

Rivers shape the landscapes through which they run. Over thousands or millions of years, a meandering river may shift from one course to another, back and forth over a large area, carving out a flat valley and picking up sediment that is later deposited in coastal wetlands. Areas nearest to a river's course that are subject to periodic flooding are said to be within the river's **floodplain.** Frequent deposition of silt (eroded soil) from flooding makes floodplain soils especially fertile. As a result, agriculture thrives in floodplains, and *riparian* (riverside) forests are productive and species-rich. A river's meandering course is often driven by large-scale flooding events that scour new channels. However, extensive damming on the Mississippi and other rivers has reduced the rate of river meandering by 66–83% from its historic rate. Instead of coursing downriver, floodwaters are often trapped in reservoirs behind dams or are contained in river channels by levees.

Lakes and ponds are ecologically diverse systems

Lakes and ponds are bodies of standing surface water. The largest lakes, such as North America's Great Lakes, are sometimes known as inland seas. Although lakes and ponds can vary greatly in size, scientists have described several zones common to these waters (**FIGURE 12.5**).

Around the nutrient-rich edges of a water body, the water is shallow enough that aquatic plants grow from the mud and reach above the water's surface. This region, named the *littoral zone*, abounds in invertebrates—such as insect larvae, snails, and crayfish—that fish, birds, turtles, and amphibians feed on. The *benthic zone* extends along the bottom of the lake or pond, from the shore to the deepest point. Many invertebrates live in the mud, feeding on detritus or on one another. In the open portion of a lake or pond, far from shore, sunlight

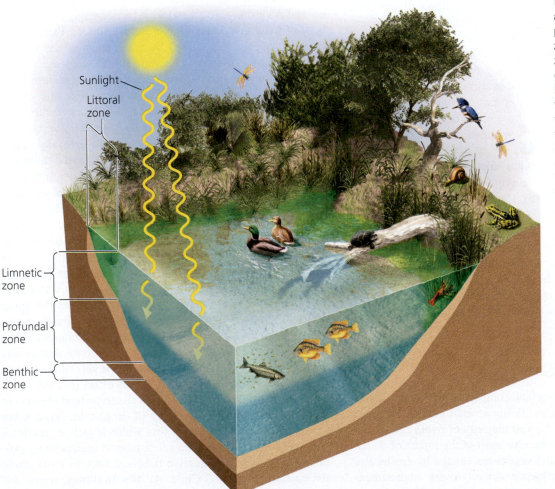

FIGURE 12.5 Lakes and ponds are composed of distinct zones. In the littoral zone, emergent plants grow along the shoreline. The limnetic zone is the layer of open, sunlit water, where photosynthesis takes place. Sunlight does not reach the deeper profundal zone. The benthic zone, at the bottom of the water body, often is muddy, rich in detritus and nutrients, and low in oxygen.

Sunlight

Littoral zone

Limnetic zone

Profundal zone

Benthic zone

penetrates the shallow waters of the *limnetic zone*. Because light enables photosynthesis, the limnetic zone supports phytoplankton (algae, protists, and cyanobacteria), which in turn support zooplankton (p. 28), both of which are eaten by fish. Below the limnetic zone lies the *profundal zone*, the volume of open water that sunlight does not reach. This zone lacks photosynthetic life and is lower in dissolved oxygen than are the upper waters.

Ponds and lakes change over time as streams and runoff bring them sediment and nutrients. **Oligotrophic** lakes and ponds, which are low in nutrients and high in oxygen, may slowly transition to the high-nutrient, low-oxygen conditions of **eutrophic** water bodies (pp. 28–30). Eventually, water bodies may fill in completely by the process of aquatic succession (p. 78). These changes occur naturally, but eutrophication can also result from human-caused nutrient pollution.

FIGURE 12.6 Freshwater wetlands, such as this bald cypress swamp in Louisiana, support biologically diverse and productive ecosystems.

Freshwater wetlands include marshes, swamps, bogs, and vernal pools

Wetlands are systems in which the soil is saturated with water, and generally feature shallow standing water with ample vegetation. There are many types of freshwater wetlands, and most are enormously rich and productive. In *freshwater marshes*, shallow water allows plants such as cattails and bulrushes to grow above the water surface. *Swamps* also consist of shallow water rich in vegetation, but they occur in forested areas (**FIGURE 12.6**). *Bogs* are ponds covered with thick floating mats of vegetation and can represent a stage in aquatic succession. *Vernal pools* are seasonal wetlands that form in early spring from rain and snowmelt, and dry up once the weather becomes warmer.

Wetlands are extremely valuable habitat for wildlife. Louisiana's coastal wetlands, for example, provide habitat for approximately 1.8 million migratory waterbirds each year. Wetlands also provide important ecosystem services by slowing runoff, reducing flooding, recharging aquifers, and filtering pollutants.

Despite the vital roles that wetlands play, people have drained and filled them extensively for agriculture. Many wetlands are lost when people divert and withdraw water, channelize rivers, and build dams. The United States and southern Canada, for example, have lost well over half their wetlands since European colonization.

The Oceans

The oceans are an important component of Earth's interconnected aquatic systems. While a small number of rivers empty into inland seas, the vast majority of rivers empty into oceans; thus, the oceans receive most of the inputs of water, sediments, pollutants, and organisms carried by freshwater systems. The oceans influence virtually every environmental system and every human endeavor, so even if you live in a landlocked region far from the coast, the oceans still affect you, and you affect the oceans.

The physical makeup of the ocean is complex

The world's five major oceans—Pacific, Atlantic, Indian, Arctic, and Antarctic—are all connected, comprising a single vast body of water that covers 71% of Earth's surface. Ocean water contains roughly 96.5% H_2O by mass; most of the remainder consists of ions from dissolved salts. Ocean water is salty primarily because ocean basins are the final repositories for runoff that collects salts from weathered rocks and carries them, along with sediments, to the ocean. Whereas the water in the ocean evaporates, the salts do not, and they accumulate in ocean basins (the salinity of ocean water generally ranges from 33,000 to 37,000 parts per million, whereas the salinity of freshwater runoff typically is less than 500 parts per million). If we were able to evaporate all the water from the oceans, a layer of dried salt 63 m (207 ft) thick would be left behind.

Sunlight warms the ocean's surface but does not penetrate deeply, so ocean water is warmest at the surface and becomes colder with depth. Deep below the surface, water is dense and sluggish, unaffected by winds and storms, sunlight, and daily temperature fluctuations. Ocean water travels in **currents,** vast riverlike flows that move in the upper 400 m (1300 ft) of water, horizontally and for great distances (**FIGURE 12.7**). These flows are driven by differences in the density of seawater (warmer water is less dense than cooler water, but water also becomes denser as it gets saltier), heating and cooling, wind, and the Coriolis effect (p. 290).

Surface winds and heating also create vertical currents in seawater. **Upwelling** is the rising of deep, cold, dense water toward the surface. Because this water is rich in nutrients from the bottom, upwellings often support high primary productivity (p. 36) and lucrative fisheries, such as those along the coasts of Peru and Chile. At **downwellings,** warm surface water rich in dissolved gases is displaced downward,

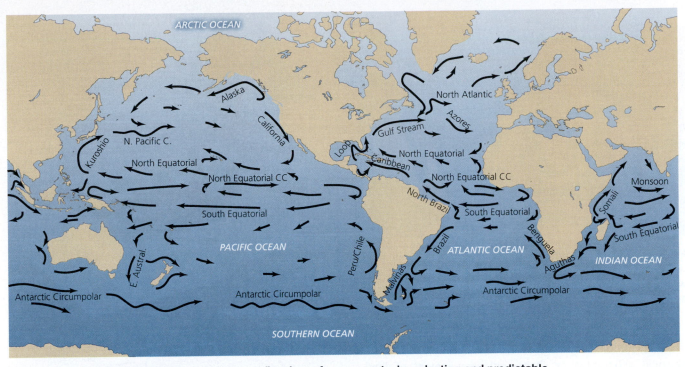

FIGURE 12.7 The upper waters of the oceans flow in surface currents, long-lasting and predictable global patterns of water movement. Warm- and cold-water currents interact with the planet's climate system, and people have used them for centuries to navigate the oceans. *Adapted from Rick Lumpkin (NOAA/AOML).*

DATA • If you released a special buoy that traveled on the surface ocean currents shown above into the Pacific Ocean from the southeastern coast of Japan, would it likely reach the United States or Australia first? • On what currents would it be carried?

Go to **Interpreting Graphs & Data** on **Mastering** Environmental Science

providing an influx of oxygen for deep-water life and "burying" CO_2 in ocean sediments.

Many parts of the ocean floor are rugged and complex. Underwater volcanoes shoot forth enough magma to build islands above sea level, such as the Hawaiian Islands. Our planet's longest mountain range is under water—the Mid-Atlantic Ridge (p. 233) that runs the length of the Atlantic Ocean. Stylized maps (**FIGURE 12.8**) that reflect bathymetry

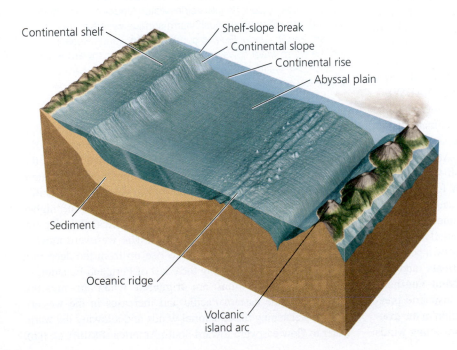

FIGURE 12.8 A stylized bathymetric profile shows key geologic features of the submarine environment. Shallow water exists around the edges of continents over the continental shelf, which drops off at the shelf-slope break. The steep continental slope gives way to the more gradual continental rise, all of which are underlain by sediments from the continents. Vast areas of seafloor are flat abyssal plain. Seafloor spreading occurs at oceanic ridges, and oceanic crust is subducted in trenches (p. 234). Volcanic activity along trenches may give rise to island chains such as the Aleutian Islands. Features on the left side of this diagram are more characteristic of the Atlantic Ocean, and features on the right side of the diagram are more characteristic of the Pacific Ocean.

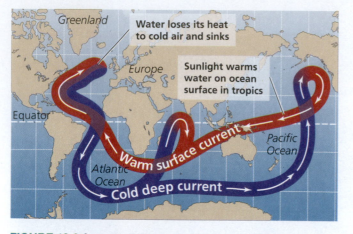

FIGURE 12.9 As part of the oceans' thermohaline circulation, warm surface currents carry heat from equatorial waters northward toward Europe, where they warm the atmosphere. The water then cools and sinks, forming the North Atlantic Deep Water (NADW).

(the measurement of ocean depths) and topography (the physical geography, or shape and arrangement of landforms) show that gently sloping *continental shelves* sit beneath the shallow waters bordering the continents. The continental shelf then drops off at the *shelf-slope break*, where the *continental slope* angles more steeply downward to the deep ocean basin below.

Ocean currents affect Earth's climate

The horizontal and vertical movements of ocean water can have far-reaching effects on climate globally and regionally. The **thermohaline circulation** is a worldwide current system in which warmer water with lower salt content moves along the surface and colder, saltier water (which is denser) moves deep beneath the surface (**FIGURE 12.9**). One segment of this worldwide conveyor-belt system includes the warm surface water in the Gulf Stream that flows across the Atlantic Ocean to Europe. Upon reaching Europe, this water releases heat to the air, keeping Europe warmer than it would otherwise be, given its latitude. The now-cooler water becomes saltier through evaporation, and thus becomes denser and sinks, creating a region of downwelling known as the **North Atlantic Deep Water (NADW).**

Scientists hypothesize that interrupting the thermohaline circulation could trigger rapid climate change. If climate change (Chapter 14) causes much of Greenland's ice sheet to melt, the resulting freshwater runoff into the North Atlantic would make surface waters less dense, because fresh water is less dense than saltwater. This could stop the NADW formation and shut down the northward flow of warm water, causing Europe to cool rapidly. A 2015 study indicated that climate change is already slowing the flow of this current, potentially affecting climate in Europe over the long term if this slowdown persists.

Another interaction between ocean currents and the atmosphere that influences climate is the **El Niño–Southern Oscillation (ENSO),** a systematic shift in atmospheric pressure, sea surface temperature, and ocean circulation in the tropical Pacific Ocean. Under normal conditions, prevailing winds

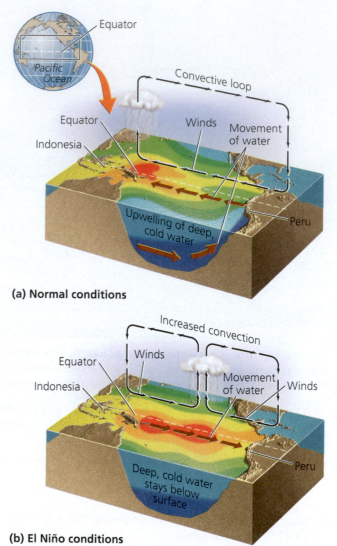

(a) Normal conditions

(b) El Niño conditions

FIGURE 12.10 El Niño conditions occur every 2 to 8 years, causing marked changes in weather patterns. In these diagrams, red and orange colors denote warmer water, and blue and green colors denote colder water. Under normal conditions **(a)**, prevailing winds push warm surface waters toward the western Pacific. Under El Niño conditions **(b)**, winds weaken, and the warm water flows back across the Pacific toward South America, like water sloshing in a bathtub. *Adapted from National Oceanic and Atmospheric Administration, Tropical Atmospheric Ocean Project.*

blow from east to west along the equator, from a region of high pressure in the eastern Pacific to one of low pressure in the western Pacific, forming a large-scale convective loop in the atmosphere (**FIGURE 12.10a**). The winds push surface waters westward, causing water to "pile up" in the western Pacific. As a result, water near Indonesia can be 50 cm (20 in.) higher and 8°C warmer than water near South America, elevating the risk of coastal flooding in the Pacific. The westward-moving surface waters allow cold water to rise up from the deep in a nutrient-rich upwelling along the coast of Peru and Ecuador.

El Niño conditions are triggered when air pressure decreases in the eastern Pacific and increases in the western Pacific, weakening the equatorial winds and allowing the warm water to flow eastward toward South America (**FIGURE 12.10b**).

This suppresses upwelling along the Pacific coast of the Americas, shutting down the delivery of nutrients that support marine life and fisheries. This phenomenon was called El Niño (Spanish for "little boy" or "Christ Child") by Peruvian fishermen because the arrival of warmer waters usually occurred shortly after Christmas. El Niño events alter weather patterns around the world, creating rainstorms and floods in areas that are generally dry, such as southern California, and causing drought and fire in regions that are typically moist, such as Indonesia. The El Niño event of 2015–2016 was especially strong, with massive heat waves afflicting India and drought ravaging Southeast Asia.

La Niña events are the opposite of El Niño events; in a La Niña event, unusually cold waters rise to the surface and extend westward in the equatorial Pacific when winds blowing to the west strengthen, and weather patterns are affected in opposite ways. ENSO cycles are periodic but irregular, occurring every 2–8 years. Scientists are exploring whether warming air and sea temperatures due to climate change may be increasing the frequency and strength of these cycles.

Marine and Coastal Ecosystems

With their variation in topography, temperature, salinity, nutrients, and sunlight, marine and coastal environments feature a variety of ecosystems. These systems may not give us the water we need for drinking and growing crops, but they teem with biodiversity and provide many other necessary resources.

Intertidal zones undergo constant change

Where the ocean meets the land, **intertidal**, or *littoral*, ecosystems (**FIGURE 12.11**) spread between the uppermost reach of the high tide and the lowest limit of the low tide. **Tides** are the periodic rising and falling of the ocean's height at a given location, caused by the gravitational pull of the moon and sun. Intertidal organisms spend part of each day submerged in water, part of the day exposed to air and sun, and part of the day being lashed by waves.

Life abounds in the crevices of rocky shorelines, which provide shelter and pools of water (tide pools) during low tides. Sessile (stationary) animals such as anemones, mussels, and barnacles live attached to rocks, filter-feeding on plankton in the water that washes over them. Urchins, sea slugs, chitons, and limpets eat intertidal algae or scrape food from the rocks. Sea stars (starfish) prey upon the filter-feeders and herbivores, while crabs scavenge detritus. The rocky intertidal zone is so diverse because environmental conditions such as temperature, salinity, and moisture change dramatically from the high to the low reaches.

FIGURE 12.11 The rocky intertidal zone stretches along rocky shorelines between the lowest and highest reaches of the tides. The intertidal zone provides niches for a diversity of organisms, including sea stars (starfish), crabs, sea anemones, corals, chitons, mussels, nudibranchs (sea slugs), and sea urchins. Areas higher on the shoreline are exposed to the air more frequently and for longer periods, so organisms that tolerate exposure best specialize in the upper intertidal zone. The lower intertidal zone is exposed less frequently and for shorter periods, so organisms less tolerant of exposure thrive in this zone.

Are We Destined for a Future of "Megadroughts" in the United States?

From 2012 to 2016, record-low levels of precipitation coupled with record-high heat kept California in severe drought conditions, prompting the state to act aggressively to implement a series of far-reaching measures to promote water conservation in agriculture, industry, and homes. While near-record levels of precipitation bathed the state in 2016 and 2017—refilling reservoirs and expanding snowpack in the mountains—state officials maintained many of the water conservation measures enacted during the drought, as climatic studies suggest that California may very well experience severe drought in the future.

Benjamin I. Cook, NASA

This was the sobering conclusion reached by a team of researchers led by Benjamin Cook, research scientist at NASA Goddard Institute for Space Studies, and advanced in a paper published in the journal *Science Advances* in 2015. In an effort to frame the current drought in the western United States in the proper context, the group compared the climate of the Central Plains and Southwest over the past 1000 years with predictions of their climates over the next 100 years based on computer simulations. "We are the first to do this kind of quantitative comparison between the projections and the distant past," said co-author Jason Smerdon, "and the story is a bit bleak."

Researchers used the North American Drought Atlas (NADA)—a reconstruction of climates based on data from tens of thousands of samples of tree rings from the United States, Canada, and Mexico—to estimate the past climates of the Central Plains and Southwest regions. Trees are a natural archive of climate data, because they grow at varying rates depending on moisture, temperature, and other factors. Using data from the recent past, scientists have determined the relationship between tree growth patterns and climatic factors, and they then use this relationship to reconstruct past climates for which they have data on tree growth from tree rings.

To predict the future climate of the southwestern United States and the Central Plains, the researchers used 17 climate models and ran simulations based on a "business as usual" scenario, in which the growth in greenhouse gas emissions followed current trends, as well as on a "moderate reduction" scenario, in which growth in greenhouse gas emissions was more modest.

The study also used three indicators of drought, measuring the level of soil moisture available to plants from the surface to 30-cm depth; from the surface to 2-m depth; and by using the Palmer Drought Severity Index (PDSI), a measure of the difference between soil moisture supply (from precipitation) and soil moisture demand (from evaporation and uptake by plants). In the PDSI, negative values mean drier conditions, or drought, and positive values mean wetter conditions.

When the researchers combined data from the past, present, and future, the results were stunning—and troubling. The climate models predicted unprecedented levels of drought in the Central Plains and Southwest regions through 2100, with high levels of agreement between the 17 climate models and the three measures of soil moisture (**FIGURE 1**). The study concluded that the drying of soils was not being driven by drastic reductions in precipitation but rather by increased evaporative pressure; that is, warmer temperatures leading to higher rates of evaporation and elevated rates of soil water uptake by plants.

The models also concluded that human-induced climate change, and not natural variability in climate, would be the driving force for drier soils in the Central Plains and Southwest. A comparison with the past indicated that these future conditions would be far worse than those seen during the "Medieval megadrought period" from A.D. 1100 to 1300. This extended drought is thought to have led to the decline of the ancient Pueblos, the Anasazi, who lived along the Colorado Plateau. Smerdon described their results as follows: "Even when selecting for the worst megadrought-dominated period, the 21st Century projections make the megadroughts seem like quaint walks through the Garden of Eden."

Comparing the probability of extended droughts in the latter half of the 20th century (1950–2000) with the probability of such droughts in the latter half of the 21st century (2050–2099) provided further cause for concern. According to the simulations, the probability that the two regions would experience a decade-long drought essentially doubled, and the chance that they would experience a multidecade drought increased from around 10% to more than 80% (**FIGURE 2**). If greenhouse gases are reduced, the chance of a multidecade drought drops to 60–70% for the Central Plains but remains above 80% for the Southwest, providing yet another incentive for combating global climate change.

Studies like these help us to prepare for the future by providing an idea of what to expect. They show that it would be wise to maintain indefinitely the water conservation strategies embraced by California during its historic drought—even in times of unusually high levels of precipitation. These studies also suggest that, given the future predictions of dire drought, implementing water conservation in other states in the Central Plains and Southwest would be a prudent course of action. In the words of study co-author Toby Ault, "The time to act is now. The time to start planning for adaptation is now. We need to assess what the rest of this century will look like for our children and grandchildren."

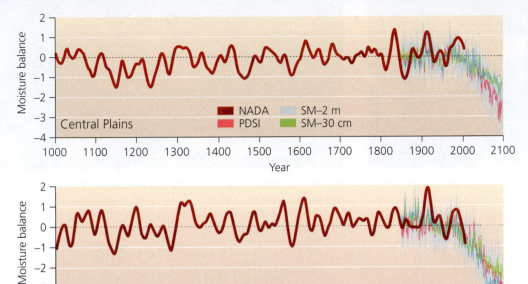

FIGURE 1 Soil moisture levels for the Central Plains and Southwest regions, as predicted by the North American Drought Atlas (NADA) and computer climate models. Negative values for moisture balance indicate drier soils (drought) and positive values indicate wetter soils. The predicted drought conditions in the late 21st century are unprecedented in the past 1000 years. (PDSI = Palmer Drought Severity Index, SM–30 cm = soil moisture to 30 cm depth, and SM–2 m = soil moisture to 2 m depth. The gray-shaded areas represent the variability in model PDSI values across computer climate models.) *Source: Cook, B.I., et al., 2015. Science Advances 1(1): e1400082.*

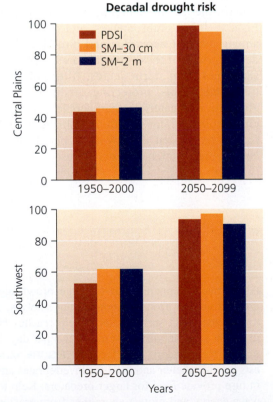

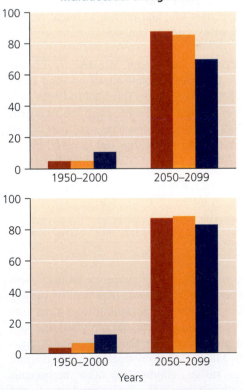

FIGURE 2 Risk of decadal (11-year) and multi-decadal (35-year) drought in the Southwest and Central Plains in the late 20th century and late 21st century for three measures of soil moisture. Due to human-induced climate change, both regions are far more likely to have long-term drought in the future than in the past. *Source: Cook, B.I., et al., 2015. Science Advances 1(1): e1400082.*

FIGURE 12.12 Salt marshes occur in temperate intertidal zones where the substrate is muddy. Tidal waters flow in channels called tidal creeks amid flat areas called benches, sometimes partially submerging the salt-adapted grasses.

FIGURE 12.13 Mangrove forests line tropical and subtropical coastlines. Mangrove trees, with their unique roots, are adapted for growing in saltwater and provide habitat for many fish, birds, crabs, and other animals.

Fresh water meets saltwater in estuaries

Water bodies where rivers flow into the ocean, mixing fresh water with saltwater, are called **estuaries**. Estuaries are biologically productive ecosystems that experience fluctuations in salinity with the daily and seasonal variations in tides and freshwater runoff. The shallow water of estuaries nurtures sea grass beds and other plant life, and provides critical habitat for shorebirds and many commercially important shellfish species.

Estuaries everywhere have been affected by coastal development, water pollution, habitat alteration, and overfishing. The Chesapeake Bay estuary (profiled in Chapter 2) is one such example. Estuaries and other coastal ecosystems have borne the brunt of human impact because two out of every three people choose to live within 160 km (100 mi) of the ocean.

Salt marshes line temperate shorelines

Along many of the world's coasts at temperate latitudes, **salt marshes** occur where the tides wash over gently sloping sandy or silty substrates. Rising and falling tides flow into and out of channels called *tidal creeks* and at highest tide spill over onto elevated marsh flats, like those in coastal Louisiana (**FIGURE 12.12**). Marsh flats grow thick with salt-tolerant grasses, as well as rushes, shrubs, and other herbaceous plants. Salt marshes boast very high primary productivity and provide critical habitat for shorebirds, waterfowl, and many fish and shellfish species. Salt marshes also filter pollution and stabilize shorelines against storm surges.

Mangrove forests line coasts in the tropics and subtropics

In tropical and subtropical latitudes, mangrove forests replace salt marshes along the coasts. **Mangroves** are salt tolerant, and they have unique roots that curve upward like snorkels to attain oxygen or downward like stilts to support the tree in changing water levels (**FIGURE 12.13**). Fish, shellfish, crabs, snakes, and other organisms thrive among the root networks, and birds feed and nest in the dense foliage of these coastal forests. Mangroves protect shorelines from storm surges, filter pollutants, and capture eroded soils, protecting offshore coral reefs. They also provide materials that people use for food, medicine, tools, and construction. Half the world's mangrove forests have been destroyed as people have developed coastal areas, often for tourist resorts and shrimp farms (p. 158).

Kelp forests harbor many organisms

Along many temperate coasts, large brown algae, or **kelp**, grow from the floor of continental shelves, reaching up toward the sunlit surface. Some kelp reaches 60 m (200 ft) in height and can grow 45 cm (18 in.) per day. Dense stands of kelp form underwater "forests" (**FIGURE 12.14**). Kelp forests provide shelter and food for invertebrates and fish, which in turn provide food for larger predators. Kelp forests absorb wave energy and protect shorelines from erosion. People eat some types of kelp, and kelp provides compounds that serve as thickeners in cosmetics, paints, ice cream, and other consumer products.

FIGURE 12.14 "**Forests" of tall brown algae known as kelp grow from the floor of the continental shelf.** Numerous fish and other creatures eat kelp or find refuge among its fronds.

Coral reefs are treasure troves of biodiversity

Shallow subtropical and tropical waters are home to coral reefs. A reef is an underwater outcrop of rock, sand, or other material. A **coral reef** is a mass of calcium carbonate composed of the shells of tiny marine animals known as **corals.** A coral reef may occur as an extension of a shoreline; along a *barrier island* paralleling a shoreline; or as an *atoll*, a ring around a submerged island.

Corals are tiny invertebrate animals related to sea anemones and jellyfish. They remain attached to rock or existing reef and capture passing food with stinging tentacles. Corals also derive nourishment from symbiotic algae known as **zooxanthellae,** which inhabit their bodies and produce food through photosynthesis—and provide the diversity of vibrant colors in reefs. Most corals are colonial, and the surface of a coral reef consists of millions of densely packed individuals. As corals die, their shells remain part of the reef and new corals grow atop them. This accumulation of coral shells enables the reef to persist and grow larger over time.

Like kelp forests, coral reefs protect shorelines by absorbing wave energy. They also host tremendous biodiversity (**FIGURE 12.15a**). This is because coral reefs provide complex physical structure (and thus many habitats) in shallow nearshore waters, which are regions of high primary productivity. If you have ever gone diving or snorkeling over a coral reef, you will have noticed the staggering diversity of anemones, sponges, hydroids, tubeworms, and other sessile invertebrates; the innumerable mollusks, flatworms, sea stars, and urchins; and the many fish species that find food and shelter in reef nooks and crannies.

The promotion of biodiversity by coral reefs makes their alarming decline worldwide particularly disturbing. Many reefs have fallen victim to "coral bleaching," which occurs when zooxanthellae die or abandon the coral, thereby depriving the coral of nutrition. Corals lacking zooxanthellae lose color and frequently die, leaving behind ghostly white patches in the reef (**FIGURE 12.15b**). The Great Barrier Reef, which stretches for

(a) Coral reef community

Bleaching is evident in the whitened regions of this coral

(b) Bleached coral

FIGURE 12.15 Coral reefs provide food and shelter for a tremendous diversity (a) of fish and other creatures. Today these reefs face multiple stresses from human impacts. Many corals have died as a result of coral bleaching **(b)**, in which corals lose their zooxanthellae.

2300 km (1400 mi) along the coast of Australia, has recently been severely damaged by bleaching. Major bleaching events in 2016 and 2017, attributed to rising sea temperatures due to climate change, affected some 1500 km (900 mi) of the reef and have decimated large stretches of its northern and middle sections.

Once large areas of coral die, species that hide within the reef are exposed to higher levels of predation, and their numbers decline. Without living coral—the foundation of the ecosystem on which so many species rely—biological diversity on reefs declines as organisms flee or perish. Coral bleaching is thought to occur when coral are strongly stressed, with common stressors including increased sea surface temperatures associated with global climate change (p. 313), and exposure to elevated levels of pollutants.

Another threat to coral comes from nutrient pollution in coastal waters, which promotes the growth of algae that are smothering reefs in the Florida Keys and in many other regions. Coral reefs also sustain damage when divers use cyanide to stun fish in capturing them for food or for the pet trade, a common practice in Indonesia and the Philippines.

A few coral species thrive in waters outside the tropics and build reefs on the ocean floor at depths of 200–500 m

(650–1650 ft). These little-known reefs—which occur in cold-water areas off the coasts of Norway, Spain, the British Isles, and elsewhere—are only now being studied by scientists.

Open-ocean ecosystems vary in their biodiversity

The uppermost 10 m (33 ft) of ocean water absorbs 80% of the solar energy that reaches its surface. For this reason, nearly all of the oceans' primary productivity occurs in the well-lit top layer, or *photic zone*. Generally, the warm, shallow waters of continental shelves are the most biologically productive and support the greatest species diversity. Habitats and ecosystems occurring between the ocean's surface and floor are classified as **pelagic,** whereas those that occur on the ocean floor are classified as **benthic.**

Biological diversity in pelagic regions of the open ocean is highly variable in its distribution. Primary production (p. 36) and animal life near the surface are concentrated in regions of nutrient-rich upwelling. Microscopic phytoplankton constitute the base of the marine food chain in the pelagic zone and are the prey of zooplankton which in turn become food for fish, jellyfish, whales, and other free-swimming animals (**FIGURE 12.16**). Predators at higher trophic levels include larger fish, sea turtles, and sharks. Fish-eating birds such as puffins and petrels also feed at the surface of the open ocean.

In the little-known deep-water ecosystems, animals have adapted to tolerate extreme water pressures and to live in the dark without food from autotrophs (p. 34). Many of these often bizarre-looking creatures scavenge carcasses or organic detritus that falls from above. Others are predators, and still others attain food from mutualistic (p. 73) bacteria. Ecosystems also form around hydrothermal vents, where heated water spurts from the seafloor, carrying minerals that precipitate to form rocky structures. Tubeworms, shrimp, and other creatures in these recently discovered deep-water systems use symbiotic

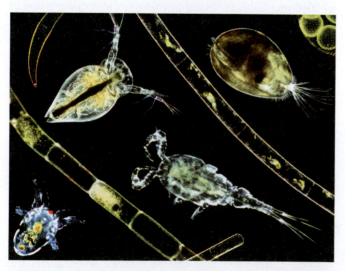

FIGURE 12.16 The uppermost reaches of ocean water contain billions upon billions of phytoplankton—tiny photosynthetic algae, protists, and bacteria that form the base of the marine food chain. This part of the ocean is also home to zooplankton, small animals and protists that dine on phytoplankton.

bacteria to derive their energy from chemicals in the heated water rather than from sunlight. They manage to thrive within narrow zones between scalding hot and icy-cold waters.

Aquatic systems are affected by human activities

Our tour of aquatic systems has shown the ecological and economic value of freshwater and marine ecosystems. We will now see how people affect these systems when we withdraw water for human use, build dams and levees, and introduce pollutants that alter water's chemical, biological, and physical properties.

Effects of Human Activities on Waterways

Although water is a limited resource, it is also a renewable resource as long as we manage our use sustainably. However, people are withdrawing water at unsustainable levels and are depleting many sources of surface water and groundwater. Already, one-third of the world's people are affected by water shortages.

Additionally, people have intensively engineered freshwater waterways with dams, levees, and diversion canals to satisfy demands for water supplies, transportation, and flood control. An estimated 60% of the world's 227 largest rivers, for example, have been strongly or moderately affected by human engineering. As we have seen in our Central Case Study, dams and channelization in the Mississippi River Basin have led to adverse impacts at the river's mouth. What we do in one part of the interconnected aquatic system affects many other parts, sometimes in significant ways.

Fresh water and human populations are unevenly distributed across Earth

World regions differ in the size of their human populations and, as a result of climate and other factors, possess varying amounts of groundwater, surface water, and precipitation. Hence, not every human around the world has equal access to fresh water (**FIGURE 12.17**). Because of the mismatched distribution of water and population, human societies have frequently struggled to transport fresh water from its source to where people need it.

Fresh water is distributed unevenly in time as well as space. For example, India's monsoon storms can dump half of a region's annual rain in just a few hours. Rivers have seasonal differences in flow because of the timing of rains and snowmelt. For this reason, people build dams to store water from wetter months that can be used in drier times of the year, when river flow is reduced.

As if the existing mismatches between water availability and human need were not enough, global climate change (Chapter 14) has and will continue to worsen conditions in many regions by altering precipitation patterns, melting glaciers, causing early-season runoff, and intensifying droughts and flooding.

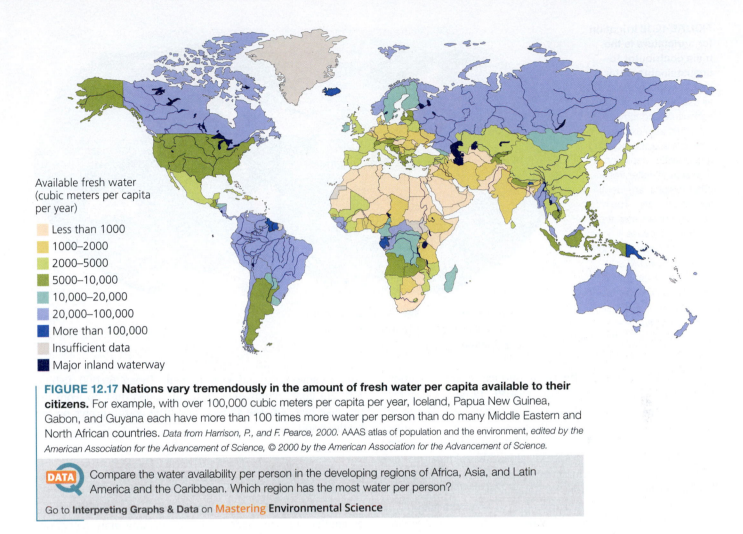

Available fresh water
(cubic meters per capita
per year)

- Less than 1000
- 1000–2000
- 2000–5000
- 5000–10,000
- 10,000–20,000
- 20,000–100,000
- More than 100,000
- Insufficient data
- Major inland waterway

FIGURE 12.17 Nations vary tremendously in the amount of fresh water per capita available to their citizens. For example, with over 100,000 cubic meters per capita per year, Iceland, Papua New Guinea, Gabon, and Guyana each have more than 100 times more water per person than do many Middle Eastern and North African countries. *Data from Harrison, P., and F. Pearce, 2000. AAAS atlas of population and the environment, edited by the American Association for the Advancement of Science, © 2000 by the American Association for the Advancement of Science.*

DATA Compare the water availability per person in the developing regions of Africa, Asia, and Latin America and the Caribbean. Which region has the most water per person?

Go to **Interpreting Graphs & Data** on **Mastering** Environmental Science

A 2009 study found that one-third of the world's 925 major rivers experienced reduced flow from 1948 to 2004, with the majority of the reduction attributed to effects of climate change.

Water supplies households, industry, and agriculture

We all use water at home for drinking, cooking, and cleaning. Most mining, industrial, and manufacturing processes require water. Farmers and ranchers use water to irrigate crops and water livestock. Globally, we allot about 70% of our annual fresh water use to agriculture; industry accounts for roughly 20%, and residential and municipal uses account for only 10%.

The removal of water from an aquifer or from a body of surface water without returning it to its source is called **consumptive use.** Our primary consumptive use of water is for agricultural irrigation (p. 146). **Nonconsumptive use** of water, in contrast, does not remove, or only temporarily removes, water from an aquifer or surface water body. Using water to generate electricity at hydroelectric dams is an example of nonconsumptive use; water is taken in, passed through dam machinery to turn turbines, and released downstream.

The large amount of water used for agriculture is due to our rapid population growth, which requires us to feed and clothe more and more people each year. Overall, we withdraw 70% more water for irrigation today than we did

50 years ago, and have doubled the amount of land under irrigation. This expansion of irrigated land has helped food and fiber production to keep up with population growth, but many irrigated areas are using water unsustainably, threatening their long-term productivity.

Excessive water withdrawals can drain rivers and lakes

In many places, we are withdrawing surface water at unsustainable rates and are drastically reducing river flows. Because of excessive water withdrawals, many major rivers—the Colorado River in North America, the Yellow River in China, and the Nile in Africa—regularly run dry before reaching the sea. This reduction in flow not only threatens the future of the cities and farms that depend on these rivers, but also drastically alters the ecology of the rivers and their deltas, changing plant communities, wiping out populations of fish and invertebrates, and devastating fisheries.

Worldwide, roughly 15–35% of water withdrawals for irrigation are thought to be unsustainable (**FIGURE 12.18**). In areas where agriculture is demanding more fresh water than can be sustainably supplied, *water mining*—withdrawing water faster than it can be replenished—is taking place. In these areas, aquifers are being depleted or surface water is being piped in from other regions to meet the demand for irrigation.

FIGURE 12.18 Irrigation for agriculture is the main contributor to unsustainable water use. Regions where over-all use of fresh water (for agriculture, industry, and domestic use) exceeds the available supply, requiring groundwater depletion or diversion of water from other regions, are identi-fied on this map. The map actually understates the problem, because it does not reflect seasonal short-ages. *Data from UNESCO, 2006.* Water: A shared respon-sibility. *World Water Develop-ment Report 2. UNESCO and Berghahn Books.*

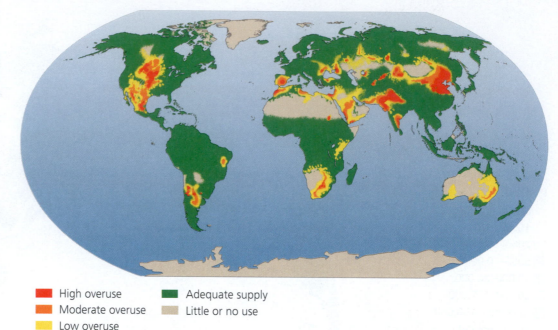

- 🟥 High overuse
- 🟧 Moderate overuse
- 🟨 Low overuse
- 🟩 Adequate supply
- ⬜ Little or no use

Nowhere are the effects of surface water depletion so evident as in the Aral Sea (**FIGURE 12.19a**). Once the fourth-largest lake on Earth, just larger than Lake Huron in the Great Lakes, it lost more than four-fifths of its volume in only 45 years (**FIGURE 12.19b**). This dying inland sea, on the border of present-day Uzbekistan and Kazakhstan, is the victim of poor irrigation practices. The former Soviet Union instituted industrial cotton farming in this dry region by flooding the land with water from the two rivers that supplied the Aral Sea its water. For a few decades this boosted Soviet cotton production, but it shrank the Aral Sea, and the irrigated

soil became salty and waterlogged. Today 60,000 fishing jobs are gone, winds blow pesticide-laden dust up from the dry lake bed (**FIGURE 12.19c**), and little cotton grows on the blighted soil.

Groundwater can also be depleted

Groundwater is more easily depleted than surface water because most aquifers recharge very slowly. If we compare an aquifer to a bank account, we are making more withdraw-als than deposits, and the balance is shrinking. Today we are mining groundwater, extracting 160 km³ (5.65 trillion ft³) more water each year than returns to the ground. This is a problem because one-third of Earth's human population—including

FIGURE 12.19 The Aral Sea in central Asia was once the world's fourth-largest lake. However, it has been shrinking **(a, b)** because so much water was withdrawn to irrigate cotton crops. Ships were stranded **(c)** along the former shoreline of the Aral Sea because the waters receded so far and so quickly. Today restoration efforts are beginning to reverse the decline in the northern portion of the sea, and waters there are slowly rising.

(a) Satellite image of Aral Sea, 1987

(b) Satellite image of Aral Sea, 2015

(c) A ship stranded by the Aral Sea's fast-receding waters

FAQ

When fresh water is scarce, why don't we irrigate crops with seawater?

Freshwater resources are often strained by agricultural irrigation. Some agriculture occurs in coastal areas, where huge volumes of seawater are available a short distance away, yet farmers do not use it to water their crops, even in times of extreme drought. Why?

The crops we grow for food, such as wheat and broccoli and strawberries, are terrestrial plants and are evolutionarily adapted to use fresh water to grow. If the cells of a terrestrial plant, such as corn, are placed in a solution of seawater, the water inside the cells will be "pulled out" into the surrounding saltwater by the process of diffusion. This occurs because there are fewer dissolved substances (such as sugars, proteins, and salts) within the cell than there are dissolved substances (such as salts) in the seawater. Accordingly, the concentration of water inside the cell is therefore greater than the concentration of water in the seawater.

Water always diffuses from areas of higher concentration to lower concentration, so it flows from inside the cell to outside the cell; this diffusion is called osmosis when it occurs across cell membranes. With continued exposure to saltwater, the plant will dehydrate and die. Marine organisms have evolved physiological mechanisms to combat or replace such water loss from their cells, so they are able to survive in seawater, unlike the terrestrial plants that supply us with so much of our food.

99% of the rural population of the United States—relies on groundwater for its needs.

As aquifers are mined, water tables drop deeper underground. This deprives freshwater wetlands of groundwater inputs, causing them to dry up. Groundwater also becomes more difficult and expensive to extract. In parts of Mexico, India, China, and multiple Asian and Middle Eastern nations, water tables are falling 1–3 m (3–10 ft) per year. In the United States, overpumping has drawn the Ogallala Aquifer in the Great Plains (**FIGURE 12.20**) down by more than 320 million liters (85 trillion gal), a volume equal to half the annual flow of the Mississippi River. Water from this massive aquifer has enabled American farmers to create the most productive grain-producing region in the world. However, unsustainable water withdrawals are threatening long-term use of the aquifer for agriculture.

As aquifers lose water, they become less able to support overlying strata, and the land surface above may sink, compact, or collapse, creating **sinkholes.** Venice, Bangkok, Beijing, and Mexico City are slowly sinking; overpumping is causing the streets to buckle and underground pipes to rupture. Once the ground sinks, soil and rock become compacted, losing the porosity that enabled them to hold water. Recharging a depleted aquifer thereafter becomes more difficult. Compaction of rock layers due to groundwater overextraction caused some agricultural regions in California to sink by 9 m (30 ft) from 1925 to 1977, and many areas continue to sink today as overextraction continues.

When groundwater is overextracted in coastal areas, saltwater from the ocean can intrude into inland aquifers. This causes coastal wells to draw up saline groundwater instead of the desired fresh groundwater. This problem has occurred in California, Florida, India, the Middle East, and many other locations where groundwater is heavily extracted.

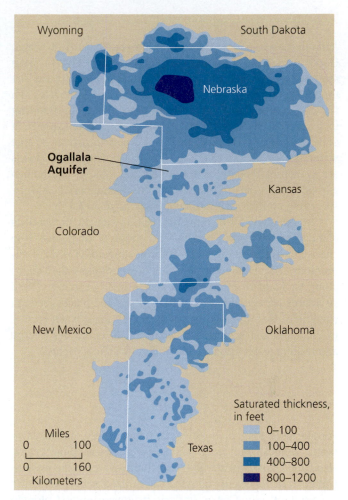

FIGURE 12.20 The Ogallala Aquifer is the world's largest aquifer, and it held 3700 km³ (881 mi³) of water before pumping began. This aquifer lies beneath 453,000 km² (175,000 mi²) of the Great Plains, stretching over eight U.S. states. Overpumping for irrigation is currently reducing the volume and extent of this aquifer.

Bottled water has ecological costs

These days, our groundwater is being withdrawn for a new purpose—to be packaged in plastic bottles and sold on supermarket shelves. The average American drinks over 136 L (36 gal) of bottled water a year, and sales top $14 billion annually in the United States and $160 billion worldwide.

Bottled water exerts substantial ecological impact. The energy costs of bottled water are estimated to be 1000–2000 times greater than the energy costs of tap water, mostly as a result of transportation costs. After their contents are consumed, at least three out of four bottles in the United States are thrown away, and not recycled. That's 30–40 billion containers per year (5 containers for every human being on Earth) and close to 1.5 million tons of plastic waste. Many people who buy bottled water do so because they believe it is superior to tap water. However, in blind taste tests people think tap water tastes just as good, and chemical analyses show that bottled water is no safer or healthier than tap water.

People build levees to control floods

Among the reasons we control the movement of fresh water, flood prevention ranks high. People have always been attracted to riverbanks for their water supply and for the flat topography and fertile soil of floodplains. But if one lives in a floodplain, one must be prepared to face flooding. **Flooding** is a normal, natural process that occurs when snowmelt or heavy rain swells the volume of water in a river so that water spills over the river's banks. In the long term, floods are immensely beneficial to both natural systems and human agriculture, because floodwaters build and enrich soil by spreading nutrient-rich sediments over large areas.

In the short term, however, floods can do tremendous damage to the farms, homes, and property of people who choose to live in floodplains. To protect against floods, communities and governments have built **levees** (also called *dikes*) along banks of rivers to hold water in main channels. These structures prevent flooding at most times and places, but can sometimes worsen flooding because they force water to stay in channels and accumulate, building up enormous energy and leading to occasional catastrophic overflow events.

We divert surface water to suit our needs

People have long diverted water from rivers and lakes to farms, homes, and cities with *aqueducts*—artificial rivers, also called *canals*. Water in the Colorado River in the western United States is heavily diverted and utilized as the river flows toward the Pacific Ocean. Early in its course, some Colorado River water is piped through a mountain tunnel and down the Rockies' eastern slope to supply the city of Denver. More is removed for Las Vegas and other cities, and for farmland as the water proceeds downriver. When the river reaches Parker Dam on the California–Arizona state line, large amounts of water are diverted into the Colorado River Aqueduct, which brings water to the Los Angeles and San Diego areas. Arizona also draws water from Parker Dam, transporting it in the canals of the Central Arizona Project. Farther south, water is diverted into the Coachella and All-American Canals, destined for agriculture, mostly in California's Imperial Valley.

The world's largest diversion project is underway in China. Three sets of massive aqueducts, totaling 2500 km (1550 mi) in length, are being built to move trillions of gallons of water from the Yangtze River in southern China, where water is plentiful, to northern China's Yellow River, which routinely dries up at its mouth due to the region's drier climate as well as withdrawal of much of its water for farms, factories, and homes. Many scientists say the $62 billion project won't transfer enough water to make a difference and will cause extensive environmental impacts, all while displacing hundreds of thousands of people.

We have erected thousands of dams

A **dam** is any obstruction placed in a river or stream to block its flow. Dams create **reservoirs**, artificial lakes that store water for human use. We build dams to prevent floods, provide drinking water, facilitate irrigation, and generate electricity.

Worldwide, we have erected more than 45,000 large dams (greater than 15 m, or 49 ft, high) across rivers in more than 140 nations, and have built tens of thousands of smaller dams. Only a few major rivers in the world remain undammed and free-flowing. These rivers run through the tundra and taiga of Canada, Alaska, and Russia, and in remote regions of Latin America and Africa.

Dams produce a mix of benefits and costs, as illustrated in **FIGURE 12.21**. As an example of this complex mix, we can consider the world's largest dam project. The Three Gorges Dam on China's Yangtze River, 186 m (610 ft) high and 2.3 km (1.4 mi) wide, was completed in 2008. Its reservoir stretches for 616 km (385 mi; as long as Lake Superior in the Great Lakes). This project provides flood control, enables boats and barges to travel farther upstream, and generates enough hydroelectric power to replace dozens of large coal or nuclear plants.

However, the Three Gorges Dam cost $39 billion to build and its reservoir flooded 22 cities, forcing the relocation of 1.24 million people. As the river slows upon entering the dam's reservoirs, sediments are deposited in the reservoir, eroding wetlands at the mouth of the river—just as in Louisiana. Many scientists also worry that the Yangtze's many pollutants will be trapped in the Three Gorges Dam reservoir, eventually making the water undrinkable.

People who feel that the costs of some dams outweigh their benefits are pushing for such dams to be dismantled. By removing dams and letting rivers flow freely, they say, we can restore ecosystems, reestablish economically valuable fisheries, and revive river recreation, such as fly-fishing and rafting. Many aging dams are in need of costly repairs or have outlived their economic usefulness, making them suitable candidates for removal. Some 400 dams have been dismantled in the United States in the past decade, and more will come down in the next 10 years, when the licenses of over 500 dams come up for renewal.

In 2014, the world's largest dam removal project was completed when the last section of the 64-m (210-ft) Glines Canyon Dam on the Elwah River in Washington State was demolished. Built in 1914 to supply power for local wood mills, the dam decimated local fisheries by preventing salmon from migrating upriver to spawn, imperiling the livelihood of Native Americans who had long harvested the river's salmon and shellfish. But since the dam's removal, the Elwah River system has rebounded. Sediments that were once held behind the dam have flooded downstream, rebuilding riverbanks, beaches, and estuaries. Habitats for shellfish and small fish are being restored in and around the river's mouth. And as salmon migrate upriver to spawn in the Elwah's tributaries, it is hoped that an entire functional ecosystem will emerge along with their return.

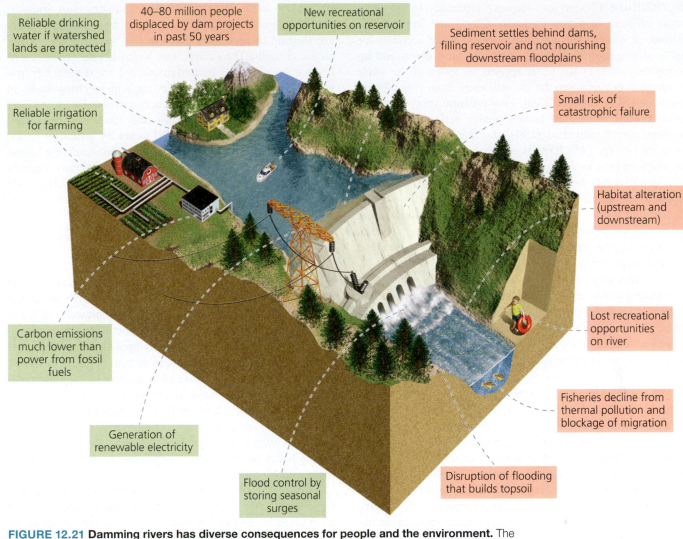

Reliable drinking water if watershed lands are protected

40–80 million people displaced by dam projects in past 50 years

New recreational opportunities on reservoir

Sediment settles behind dams, filling reservoir and not nourishing downstream floodplains

Reliable irrigation for farming

Small risk of catastrophic failure

Habitat alteration (upstream and downstream)

Carbon emissions much lower than power from fossil fuels

Lost recreational opportunities on river

Generation of renewable electricity

Fisheries decline from thermal pollution and blockage of migration

Flood control by storing seasonal surges

Disruption of flooding that builds topsoil

FIGURE 12.21 Damming rivers has diverse consequences for people and the environment. The generation of clean and renewable electricity is one of several major benefits **(green boxes)** of hydroelectric dams. Habitat alteration is one of several negative impacts **(orange boxes)**.

Solutions to Depletion of Fresh Water

To address the depletion of fresh water, we can aim to either increase supply or reduce demand. Increasing water supplies by constructing large dams was a common solution to water shortages in the past. However, large dams have already been constructed at sites most suitable for them, and most of the remaining locations are in regions that make such construction projects prohibitive. Building more dams, therefore, does not appear to be a viable solution to meet people's increasing demands for fresh water.

An alternate supply strategy is to generate fresh water through **desalination,** or *desalinization*—the removal of salt from seawater. Desalination can be accomplished by heating saltwater and condensing the water vapor that evaporates from it—essentially distilling fresh water. Over 20,000 desalination facilities are operating worldwide, but it is expensive, requires large inputs of fossil fuel energy, kills aquatic life at water intakes, and generates concentrated salty waste.

As a result, large-scale desalination is pursued mostly in wealthy oil-rich nations where energy is plentiful and water is extremely scarce. Saudi Arabia, for example, produces half of the nation's drinking water with desalination.

We can decrease our demand for water

Because supply-based strategies do not hold great promise for increasing water supplies, people are embracing demand-based solutions. Strategies for reducing fresh water demand include conservation and efficiency measures. Such strategies require changes in individual behaviors and can therefore be politically difficult, but they offer better economic returns and cause less ecological and social damage. Our existing shift from supply-based to demand-based solutions is already paying dividends. The United States, for example, decreased its water consumption by 16% from 1980 to 2010 thanks to conservation measures, even while its population grew 34%. Let's examine approaches that can conserve water in agriculture, households, industry, and municipalities.

Agriculture Farmers can improve efficiency by adopting more efficient irrigation methods. "Flood and furrow" irrigation, in which fields are flooded with water, accounts for 90% of irrigation worldwide. However, crop plants end up using only 40% of the water applied, as much of the water evaporates or seeps into the ground, away from crops. Other methods are far more efficient. Low-pressure spray irrigation squirts water downward toward plants, and drip irrigation systems target individual plants, introducing water directly onto the soil (p. 146). Experts estimate that drip irrigation—in which as little as 10% of water is wasted—could cut water use in half while raising crop yields, and could produce as much as $3 billion in extra annual income for farmers of the developing world. Researchers are currently experimenting with various materials and approaches to develop reliable, inexpensive drip irrigation systems that could convey these benefits to poorer farmers.

Another way to reduce agricultural water use would be to eliminate government subsidies for the irrigation of crops that require a great deal of water—such as cotton and rice—in arid areas. Biotechnology may play a role by producing crop varieties that require less water through selective breeding (p. 52) and genetic modification (pp. 158–162).

Households In our households, we can reduce water use by installing low-flow faucets, showerheads, washing machines, and toilets. Automatic dishwashers, studies show, use less water than does washing dishes by hand. Catching rain runoff from your roof in a barrel, or *rainwater harvesting,* will reduce the amount of water you need to use from the hose. Replacing exotic vegetation with native plants adapted to the region's natural precipitation patterns can also reduce water demand. For example, more and more residents in the U.S. Southwest are practicing **xeriscaping,** a type of landscaping that uses plants adapted to arid conditions.

Industry and municipalities Industry and municipalities can take water-saving steps as well. Manufacturers are shifting to processes that use less water, and in doing so are reducing their costs. Las Vegas is one of many cities that are recycling treated municipal wastewater for irrigation and industrial uses. Finding and patching leaks in pipes has saved some cities and companies large amounts of water—and money. Boston and its suburbs reduced water demand by 43% over 30 years by patching leaks, retrofitting homes with efficient plumbing, auditing industry, and promoting conservation. This program enabled Massachusetts to avoid an unpopular $500 million river diversion scheme.

Nations often cooperate to resolve water disputes

Population growth, expansion of irrigated agriculture, and industrial development doubled our annual fresh water use between 1960 and 2000. Increased withdrawals of fresh water can lead to shortages, and resource scarcity can lead to conflict. Many security analysts predict that water's role in regional conflicts will increase as human population continues to grow and as climate change alters precipitation patterns. A total of 261 major rivers—whose watersheds cover 45% of the world's land area—cross national borders, and transboundary disagreements are common (**FIGURE 12.22**). Access to water is already a key element in the disagreements among Israel, the Palestinian people, and neighboring nations in the arid Middle East.

Yet on the positive side, many nations have cooperated to resolve water disputes. India has struck agreements to co-manage transboundary rivers with Pakistan, Bangladesh, Bhutan, and Nepal. In Europe, nations along the Rhine and Danube rivers have signed water-sharing treaties.

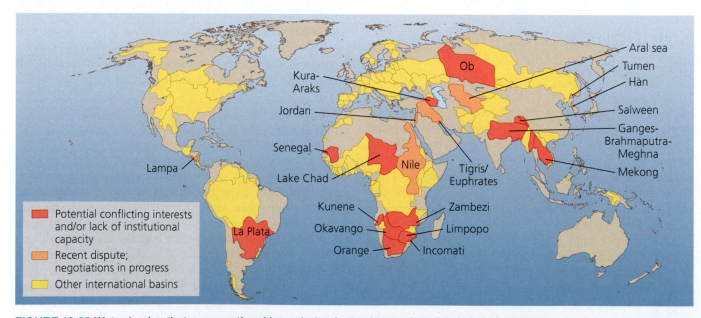

FIGURE 12.22 Water basins that cross national boundaries (yellow) have the potential for conflict if water supplies become scarce. Basins with higher potential for conflict (**red**) are found in regions with growing populations, but negotiations are underway on several international basins to prevent conflict (**orange**).

Water Pollution and Its Control

We have seen that people affect aquatic systems by withdrawing too much water and by erecting dams, diversions, and levees that alter natural processes in aquatic systems. Introducing toxic substances and disease-causing organisms into surface waters and groundwater is yet another way that people adversely affect aquatic ecosystems—and threaten human health.

Developed nations have made admirable advances in cleaning up water pollution over the past few decades. Still, the World Commission on Water, an organization that focuses on water issues, recently concluded that over half the world's major rivers remain "seriously depleted and polluted." In 2013, the U.S. Environmental Protection Agency (EPA) reported that 55% of U.S. streams and rivers are in poor condition to support aquatic life. The largely invisible pollution of groundwater, meanwhile, has been termed a "covert crisis." Preventing pollution is easier and more effective than treating it later. Many of our current solutions to pollution therefore embrace preventive strategies rather than "end-of-pipe" strategies such as treatment and cleanup.

Water pollution comes from point sources and non-point sources

Water pollution—changes in the chemical, physical, or biological properties of waters caused by human activities—comes in many forms and can have diverse impacts on aquatic ecosystems and human health. Most forms of water pollution are not conspicuous, so scientists and technicians measure water's chemical properties, such as pH, plant nutrient concentrations, and dissolved oxygen levels; physical characteristics, such as temperature and turbidity—the density of suspended particles in a water sample; and biological properties, such as the presence of harmful microorganisms or the species diversity in aquatic ecosystems.

Some water pollution is emitted from **point sources**—discrete locations, such as a factory, sewer pipe, or oil tanker (**FIGURE 12.23**). In contrast, pollution from

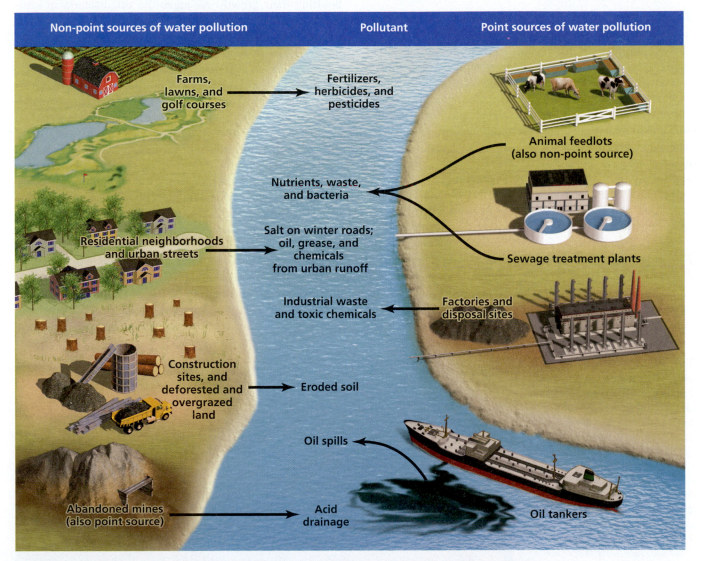

FIGURE 12.23 Point-source pollution (on right) comes from discrete facilities or locations, usually from single outflow pipes. Non-point-source pollution, such as runoff from streets, residential neighborhoods, lawns, and farms **(on left)**, originates from numerous sources spread over large areas.

non-point sources is cumulative, arising from multiple inputs over larger areas, such as farms, city streets, and residential neighborhoods. The U.S. Clean Water Act (p. 107) addressed point-source pollution with some success by targeting industrial discharges.

In the United States today, the greater threat to water quality comes from non-point-source pollution resulting from countless common activities, such as applying fertilizers and pesticides to farms and lawns, applying salt to roads in winter, and leaking automobile oil. To minimize non-point-source pollution of drinking water, governments often limit development on watershed land surrounding reservoirs.

Water pollution takes many forms

Water pollution comes in many forms that can impair waterways, and threaten people and organisms that drink the water or live in or near affected waters. Let's survey the major classes of water pollutants affecting waters in the world today.

Toxic chemicals Our waterways have become polluted with toxic organic substances of our own making, including pesticides, petroleum products, and other synthetic chemicals (p. 216). Many of these can poison animals and plants, alter aquatic ecosystems, and cause an array of human health problems, including cancer. Toxic metals such as arsenic, lead, and mercury also damage human health and the environment, as do acids from acid precipitation (p. 303) and acid drainage from mining sites (p. 245). Issuing and enforcing more stringent regulations on industry can help reduce releases of many toxic chemicals. We can also modify our industrial processes and our purchasing decisions to rely less on these substances.

Pathogens and waterborne diseases Disease-causing organisms (pathogenic viruses, protists, and bacteria) can enter drinking water supplies that become contaminated with human waste from inadequately treated sewage or animal waste from feedlots, chicken farms, or hog farms (p. 157).

Biological pollution by pathogens causes more human health problems than any other type of water pollution. In the United States, an estimated 20 million people fall ill each year from drinking water contaminated with pathogens. Worldwide, the United Nations estimates that 3800 children die every day from diseases associated with unsafe drinking water, such as cholera, dysentery, and typhoid fever.

We reduce the risks posed by waterborne pathogens by disinfecting drinking water (p. 279) and by treating wastewater (p. 279) prior to releasing it into waterways. Other measures to lessen health risks include public education to encourage personal hygiene and government enforcement of regulations to ensure the cleanliness of food production, processing, and distribution.

Nutrient pollution The Chesapeake Bay's dead zone shows how nutrient pollution causes eutrophication and hypoxia (low dissolved oxygen concentrations) in surface waters (p. 24). When excess nitrogen and/or phosphorus enters a water body, it fertilizes algae and aquatic plants, boosting their growth. As algae die off, bacteria in sediments consume them. Because this decomposition requires oxygen, dissolved oxygen levels decline. These levels can drop too low to support fish and shellfish, leading to dramatic changes in aquatic ecosystems.

A "dead zone" of very low dissolved oxygen levels appears annually in the northern Gulf of Mexico, fueled by nutrients from Midwest farms carried by the Mississippi and Atchafalaya rivers (**FIGURE 12.24**). The low-oxygen conditions have adversely affected marine life and greatly reduced catches of shrimp and fish.

Excessive nutrient concentrations sometimes give rise to population explosions among several species of marine algae that produce powerful toxins. Blooms of these algae are known as **harmful algal blooms.** Some toxic algal species produce a red pigment that discolors the water—hence the name **red tides.** Harmful algal blooms can cause illness and death in aquatic animals and people and adversely affect communities that rely on beach tourism and fishing.

Eutrophication is a natural process, but nutrient input from wastewater and fertilizer runoff from farms, golf courses,

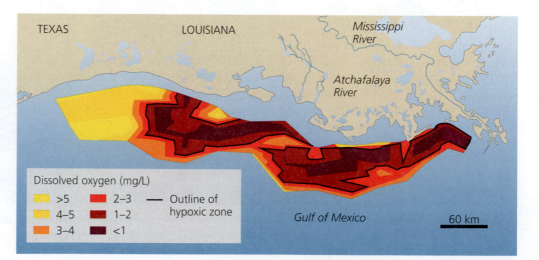

FIGURE 12.24 The map shows dissolved oxygen concentrations in bottom waters of the Gulf of Mexico off the Louisiana coast in 2015. The darkest areas indicate the lowest oxygen levels, with regions considered hypoxic (<2 mg/L) outlined in black. The dead zone forms to the west of the mouth of the Mississippi River because prevailing currents carry nutrients in that direction. *Hypoxic zone data from Nancy Rabalais, LUMICON, and R. Eugene Turner, LSU.*

Dissolved oxygen (mg/L)
- >5
- 4–5
- 3–4
- 2–3
- 1–2
- <1
- — Outline of hypoxic zone

TEXAS   LOUISIANA   Mississippi River   Atchafalaya River   Gulf of Mexico   60 km

lawns, and sewage can dramatically increase the rate at which it occurs. We can reduce nutrient pollution by specially treating municipal wastewater to remove nutrients, reducing fertilizer applications, using phosphate-free detergents, and planting vegetation and protecting natural areas around streams and rivers to reduce nutrient inputs into waterways.

Biodegradable wastes Introducing large quantities of biodegradable materials into waters also decreases dissolved oxygen levels. When human wastes, animal manure, paper pulp from paper mills, or yard wastes (grass clippings and leaves) enter waterways, bacterial decomposition escalates as organic material is metabolized, and this lowers dissolved oxygen levels in the water. **Wastewater** is water affected by human activities and is a source of biodegradable wastes. It includes water from toilets, showers, sinks, dishwashers, and washing machines; water used in manufacturing or industrial cleaning processes; and stormwater runoff. The widespread practice of treating wastewater to remove organic matter has greatly reduced impacts from biodegradable wastes in developed nations. Oxygen depletion remains a major problem in some developing nations, however, where wastewater treatment is less common.

Sediment Eroded soils, called **sediments,** can be carried to rivers by runoff and transported long distances by river currents (**FIGURE 12.25**). Clear-cutting, mining, clearing land for development, and cultivating farm fields all expose soil to wind and water erosion (p. 148). Some water bodies, such as the Colorado River and China's Yellow River, are naturally sediment-rich, but many others are not. When a clearwater river receives a heavy influx of eroded sediment, aquatic habitat changes dramatically, and fish adapted to clear water may be killed. We can reduce sediment pollution by better managing farms and forests and avoiding large-scale disturbance of vegetation.

Oil pollution Oil pollution is another important source of marine pollution. Although large oil spills occur infrequently, their impacts can be staggering near the spill site. The danger that oil spills pose to fisheries, economies, and ecosystems became clear in 1989, when the oil tanker *Exxon Valdez* ran aground in Prince William Sound, spilling hundreds of thousands of barrels of oil and causing an ecological disaster along the Alaskan coast.

The dangers of oil pollution made headlines again in 2010, when British Petroleum's *Deepwater Horizon* offshore drilling platform exploded and sank into the Gulf of Mexico off the Louisiana coast (pp. 358–359). Oil gushed from the platform's underwater well, was spread widely by ocean currents, and washed up on coastal areas across the northern Gulf of Mexico. Hundreds of miles of water, sediments, and shoreline along the coasts of Louisiana, Mississippi, Alabama, and Florida were impacted, both economically and ecologically. Five years after the accident, populations of oysters, crabs, and sea turtles remained at low levels, and large numbers of marine mammals were beaching themselves on Gulf shores, suggesting that the impacts of this accident may continue to be felt for some time.

FIGURE 12.25 Sediments wash into the Pacific Ocean from a river in Costa Rica. Farming, construction, and other human activities can cause elevated levels of soil to enter waterways affecting water quality and aquatic wildlife.

Much of the oil entering the ocean from sources other than seeps accumulates in waters from innumerable, widely spread, small non-point sources. Shipping vessels and recreational boats can leak oil as they ply ocean waters. Motor oil from vehicles on roads and parking lots is washed into streams by rains and carried to the sea. Spills from oil tankers (**FIGURE 12.26**) account for 12% of oil pollution in an average

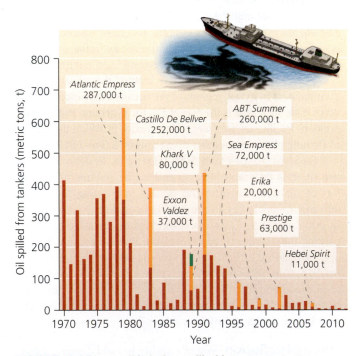

FIGURE 12.26 Less oil is being spilled into ocean waters today in large tanker spills, thanks in part to regulations imposed on the oil-shipping industry and improved spill-response techniques. The bar chart shows cumulative quantities of oil spilled worldwide from nonmilitary spills of more than 7 metric tons. Larger spills are identified by vessel name, and spill amounts from these events are indicated with orange or green bars. *Data from International Tanker Owners Pollution Federation Ltd.*

FIGURE 12.27 Aquatic life, such as this loggerhead turtle ensnared in a discarded fishing net, can suffer from their interactions with plastics in the ocean.

year, and 3% comes from leakage that occurs during the extraction of oil by offshore oil rigs. Although the *Exxon Valdez* spill was catastrophic, the good news is that the amount of oil spilled from tankers worldwide has decreased over the past three decades, in part because of an increased emphasis on spill prevention and response.

Nets and plastic debris Discarded fishing nets, fishing line, plastic bags and bottles, and other trash can harm aquatic organisms (**FIGURE 12.27**). Seabirds, fish, aquatic mammals, and sea turtles mistake floating plastic debris for food and can die as a result of ingesting material they cannot digest or expel. The chemicals that plastics contain also have toxic effects on organisms, and floating plastics can serve as "rafts" that facilitate the introduction of invasive species to new habitats.

In recent years, scientists have learned that plastic trash is accumulating in **gyres,** regions of the oceans where currents converge. One such area is the **Great Pacific Garbage Patch** in the northern Pacific, an area larger than Texas, in which tiny pieces of floating plastic outnumber organisms by a 6-to-1 margin.

Because plastics take 500–1000 years to degrade at sea and there is no viable way to collect the innumerable small bits of plastics that litter the oceans, preventing their entry into the oceans is key to remedying oceanic plastic pollution. In 2006, the U.S. Congress responded to ocean pollution by passing the Marine Debris Research, Prevention, and Reduction Act, aiding efforts to keep plastics out of marine waters. In 2015, Congress strengthened these efforts by passing a ban on the sale and distribution of products containing tiny plastic "microbeads," which were added to some toothpastes and shower gels to act as tiny "scrubbers."

Thermal pollution Water's ability to hold dissolved oxygen decreases as temperature rises, so some aquatic organisms may not survive when human activities raise water temperatures. When we withdraw water from a river and use it to cool an industrial facility, we transfer heat from the facility back into the river where the water is returned. People also raise water temperatures by removing streamside vegetation that shades water.

Too little heat can also cause problems. On the Mississippi and many other dammed rivers, water at the bottoms of reservoirs is colder than water at the surface. When dam operators release water from the depths of a reservoir, downstream water temperatures drop suddenly. In some river systems, these pulses of cold water have favored cold-loving invasive fish species over native species adapted to normal river temperatures.

Groundwater pollution is a difficult problem

Many pollutants that affect surface waters also affect groundwater. Groundwater pollution is difficult to detect. Moreover, pollutants may reside in groundwater longer than in surface waters. Because groundwater is not exposed to sunlight, contains fewer microbes and minerals, and holds less dissolved oxygen and organic matter than surface waters, pollutants decompose more slowly. For example, concentrations of the herbicide alachlor decline by half after 20 days in soil, but in groundwater this takes almost four years.

Some chemicals that are toxic at high concentrations—including aluminum, fluoride, nitrates, and sulfates—occur naturally in groundwater. However, groundwater pollution resulting from industrial, agricultural, and urban wastes—from heavy metals to petroleum products to solvents to pesticides—can leach through soil and seep into aquifers. For example, nitrate from fertilizers has leached into aquifers across the United States and Canada. Nitrate in drinking water has been linked to cancers, miscarriages, and "blue-baby" syndrome, a condition in which the oxygen-carrying capacity of infants' blood is reduced.

Leakage of carcinogenic pollutants from underground tanks—such as chlorinated solvents, gasoline, and industrial chemicals—also poses a threat to groundwater. The EPA manages a nationwide cleanup program to unearth and repair leaky tanks, and by 2017 the agency had confirmed leakage from over 535,000 tanks from across the United States, and had completed cleanup of more than 465,000 of them.

The leakage of radioactive compounds from underground tanks is also a source of groundwater pollution. Currently, 67 of the 177 underground storage tanks at the Hanford Nuclear Reservation in Washington State have been confirmed to be leaking radioactive waste into the soil. This site is the most radioactively contaminated area in the United States, storing 60% of the United States' high-level radioactive waste. Cleanup of the radioactive material stored at present in the site's aging underground tanks is not scheduled for completion until 2047.

Legislative and regulatory efforts have helped to reduce pollution

As numerous as our freshwater pollution problems may seem, it is important to remember that many were much worse in the United States a few decades ago, when, for example, the Cuyahoga River in Ohio repeatedly caught fire (p. 106). Citizen activism and government response during the 1960s and 1970s resulted in the Federal Water Pollution Control Act of 1972, amended and later renamed the Clean Water Act in

1977. These acts made it illegal to discharge pollution from a point source without a permit, set standards for industrial wastewater as well as contaminant levels in surface waters, and funded construction of sewage treatment plants. Thanks to such legislation, point-source pollution in the United States was reduced, and rivers and lakes became notably cleaner.

In the past several decades, however, enforcement of water quality laws has grown weaker, as underfunded and understaffed state and federal regulatory agencies have faced challenges when enforcing existing laws. A comprehensive investigation by the *New York Times* in 2009 revealed that violations of the Clean Water Act have risen, and that documented violations now number more than 100,000 per year—to say nothing of undocumented violations. The EPA and states act on only a tiny percentage of these violations, the *Times* found. As a result, 1 in 10 Americans have been exposed to unsafe drinking water—for the most part unknowingly, because many pollutants cannot be detected by smell, taste, or color.

The Great Lakes of Canada and the United States represent an encouraging success story in fighting water pollution. In the 1970s these lakes, which hold 18% of the world's surface fresh water, were badly polluted with wastewater, fertilizers, and toxic chemicals. Algal blooms fouled beaches, and Lake Erie was pronounced "dead." Today, efforts of the Canadian and U.S. governments have paid off. According to Environment Canada, releases of seven toxic chemicals are down by 71%, municipal phosphorus has decreased by 80%, and chlorinated pollutants from paper mills are down by 82%. Levels of PCBs and DDE are down by 78% and 91%, respectively. Bird populations are rebounding, and Lake Erie is now home to the world's largest walleye fishery. The Great Lakes' troubles are by no means over—sediment pollution is still heavy, algal blooms still plague Lake Erie, and fish are not always safe to eat. However, the progress so far shows how conditions can improve when citizens push their governments to take action.

We treat our drinking water

Technological advances as well as government regulation have improved drinking water quality. The treatment of drinking water is a widespread and successful practice in developed nations today. Before being sent to your tap, water from a reservoir or aquifer is treated with chemicals to remove particulate matter; is passed through filters of sand, gravel, and charcoal; and may be disinfected with small amounts of an agent such as chlorine to combat pathogenic bacteria. The U.S. EPA sets standards for more than 90 drinking water contaminants, which local governments and private water suppliers are obligated to meet.

We treat our wastewater

Wastewater treatment is also now a mainstream practice. Wastewater includes water that carries sewage; water from showers, sinks, washing machines, and dishwashers; water used in manufacturing or industrial cleaning processes; and stormwater runoff. Natural systems can process moderate amounts of wastewater, but the large and concentrated amounts that our densely populated areas generate can harm ecosystems and pose health threats. Thus, attempts are now widely made to treat wastewater before it is released into the environment.

In rural areas, **septic systems** are the most popular method of wastewater disposal. In a septic system, wastewater runs from the house to an underground septic tank, inside which solids and oils separate from water. The clarified water proceeds downhill to a drain field of perforated pipes laid horizontally in gravel-filled trenches underground. Microbes decompose pollutants in the wastewater these pipes emit. Periodically, solid waste from the septic tank is pumped out and taken to a landfill.

SUCCESS STORY | **Using Wetlands to Aid Wastewater Treatment**

Long before people built the first wastewater treatment plants, natural wetlands filtered and purified water. Recognizing this, engineers have begun manipulating wetlands—and have even constructed new wetlands— to employ as tools to cleanse wastewater. In this approach, wastewater that has gone through primary or secondary treatment at a conventional facility is pumped into the wetland, where microbes living amid the algae and aquatic plants decompose the remaining pollutants. Water cleansed in the wetland can then be released into waterways or allowed to percolate underground.

The Arcata Marsh and Wildlife Sanctuary

One of the first constructed wetlands was established in Arcata, a town on northern California's scenic Redwood Coast. This 35-hectare (86-acre) engineered wetland system was built in the 1980s after residents objected to a $50 million plan to build a large treatment plant that would have pumped treated wastewater into the ocean. The wetland, which cost just $7 million to build, not only treats wastewater but also serves as a haven for wildlife and human recreation. In fact, the Arcata Marsh and Wildlife Sanctuary has brought the town's waterfront back to life, with more than 200,000 people visiting each year. Additionally, wildlife has flourished in the area, with 100 species of plants, 300 species of birds, 6 species of fish, and a diversity of mammals and invertebrates populating the wetlands. The practice of treating wastewater with artificial wetlands is growing fast; today more than 500 artificially constructed or restored wetlands in the United States are performing this valuable service.

EXPLORE THE DATA at **Mastering** Environmental Science

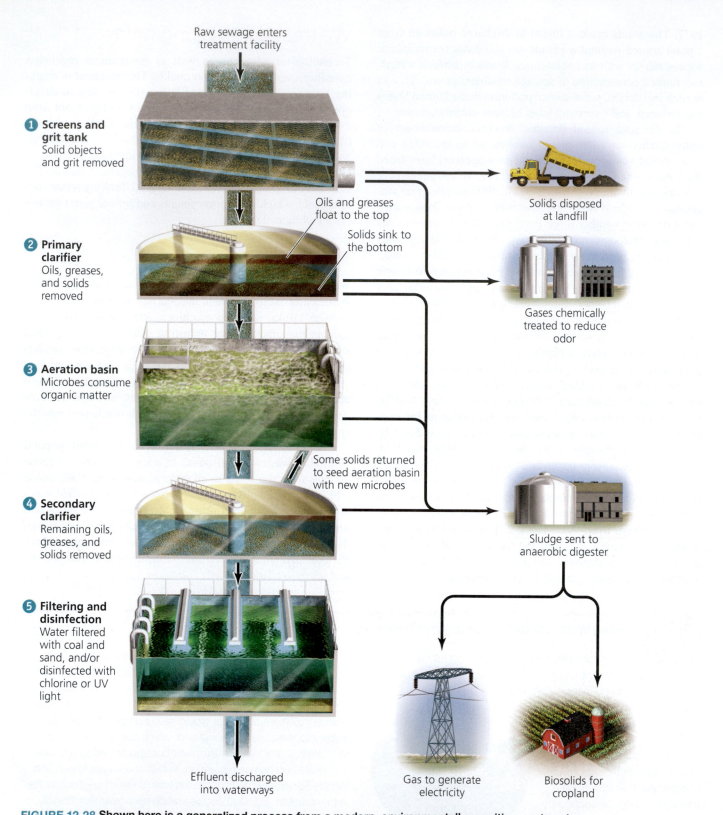

Raw sewage enters treatment facility

1 Screens and grit tank
Solid objects and grit removed

2 Primary clarifier
Oils, greases, and solids removed

Oils and greases float to the top

Solids sink to the bottom

3 Aeration basin
Microbes consume organic matter

4 Secondary clarifier
Remaining oils, greases, and solids removed

Some solids returned to seed aeration basin with new microbes

5 Filtering and disinfection
Water filtered with coal and sand, and/or disinfected with chlorine or UV light

Effluent discharged into waterways

Solids disposed at landfill

Gases chemically treated to reduce odor

Sludge sent to anaerobic digester

Gas to generate electricity

Biosolids for cropland

FIGURE 12.28 Shown here is a generalized process from a modern, environmentally sensitive wastewater treatment facility. Wastewater initially passes through screens to remove large debris and into grit tanks to let grit settle **1** It then enters tanks called *primary clarifiers* **2**, in which solids settle to the bottom and oils and greases float to the top for removal. Clarified water then proceeds to aeration basins **3** that oxygenate the water to encourage decomposition by aerobic bacteria. Water then passes into secondary clarifier tanks **4** for removal of further solids and oils. Next, the water may be purified **5** by chemical treatment with chlorine, passage through carbon filters, and/or exposure to ultraviolet light. The treated water (called *effluent*) may then be piped into natural water bodies, used for urban irrigation, flowed through a constructed wetland, or used to recharge groundwater. In addition, most treatment facilities use anaerobic bacteria to digest sludge removed from the wastewater. Biosolids from digesters may be sent to farm fields as fertilizer, and gas from digestion may be used to generate electric power.

In more densely populated areas, municipal sewer systems carry wastewater from homes and businesses to centralized treatment locations. There, pollutants are removed by physical, chemical, and biological means (**FIGURE 12.28**). At a treatment facility, **primary treatment,** the physical removal of contaminants in settling tanks or clarifiers, removes about 60% of suspended solids. Wastewater then proceeds to **secondary treatment,** in which water is stirred and aerated so that aerobic bacteria degrade organic pollutants. Roughly 90% of suspended solids may be removed after secondary treatment. Finally, the clarified water is treated with chlorine, and sometimes ultraviolet light, to kill bacteria. Most often, the treated water, called *effluent*, is piped into rivers or the ocean following primary and secondary treatment. However, many municipalities are recycling "reclaimed" water for lawns and golf courses, for irrigation, or for industrial purposes such as cooling water in power plants.

As water is purified throughout the treatment process, the solid material removed is termed *sludge*. Sludge is sent to digesting vats, where microorganisms decompose much of the matter. The result, a wet solution of "biosolids," is then dried and disposed of in a landfill, incinerated, or used as fertilizer on cropland. Methane-rich gas created by the decomposition process is sometimes burned to generate electricity, helping to offset the cost of treatment. Each year about 6 million dry tons of sludge are generated in the United States.

Emptying the Oceans

We affect the oceans and their biological resources by engineering waterways with dams and levees and by introducing water pollutants directly into the ocean or into rivers that empty into the sea. We are also putting pressure on oceans by overharvesting fish species and threatening the balance and functioning of marine and coastal ecosystems.

More than half the world's marine fish populations are fully exploited, meaning that we cannot harvest them more intensively without depleting them, according to the United Nations Food and Agriculture Organization (FAO). An additional 28% of marine fish populations are overexploited and already being driven toward extinction. Only one-fifth of the world's marine fish populations can yield more than they are already yielding without being driven into decline. If current trends continue, a comprehensive 2006 study in the journal *Science* predicted, populations of *all* ocean species that we fish for today will collapse by the year 2048.

If fisheries collapse as predicted, we will lose the ecosystem services they provide. Productivity will be reduced, ecosystems will become more sensitive to disturbance, and the filtering of water by vegetation and organisms (such as oysters) will decline, making harmful algal blooms, dead zones, fish kills, and beach closures more common. Aquaculture (p. 158) is booming—helping to relieve pressure on wild stocks—but it comes with its own set of environmental dilemmas. To prevent a collapse of the world's fisheries, it is vital that we turn immediately to more sustainable fishing practices.

Industrialized fishing facilitates overharvesting

Total global fisheries catch, after decades of increases, leveled off after the late 1980s (**FIGURE 12.29**). This occurred despite increased fishing effort, and aquaculture—humans raising fish as one raises livestock—now rivals captures of wild-caught fish in total seafood production. This seeming stability in catch since then can be explained by several factors that conceal population declines: Fishing fleets are exploiting increasingly remote fishing areas, are engaging in more intensive fishing, are capturing smaller fish than before, and are targeting less desirable fish species they formerly overlooked.

Today's industrialized commercial fishing fleets employ huge vessels and powerful new technologies to capture fish in large numbers using several methods. In *purse seining*, vessels

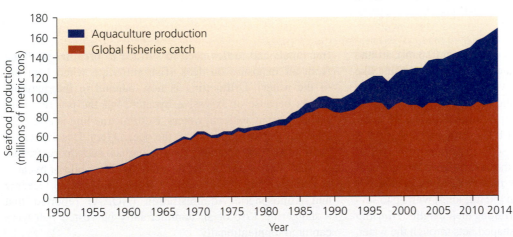

FIGURE 12.29 After rising for decades, global fisheries catch has stalled for the past 25 years. Many scientists fear that a global decline is imminent if conservation measures are not taken. Note how the growth in aquaculture has enabled seafood production to increase, despite flat hauls from capture fisheries in recent decades. *Data from the Food and Agriculture Organization of the United Nations, 2016.* The state of world fisheries and aquaculture 2016. *Fig. 1.*

DATA • Approximately what percentage of the total global production of fisheries in 2014 came from aquaculture? • How does this compare to 1980? • Offer an explanation for the difference in aquaculture production in these 2 years.

Go to **Interpreting Graphs & Data** on **Mastering** Environmental Science.

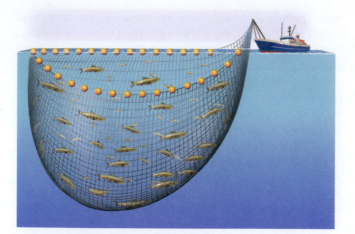

(a) Purse seining

(b) Driftnetting

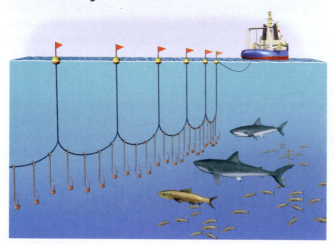

(c) Longlining

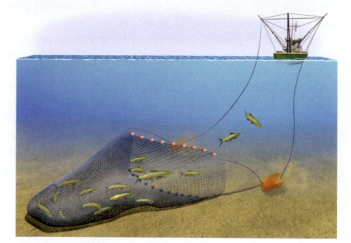

(d) Bottom-trawling

FIGURE 12.30 Commercial fishing fleets use several methods of capture. The illustrations above are schematic for clarity and do not portray the immense scale that these technologies can attain; for instance, industrial trawling nets can be large enough to engulf multiple jetliners.

deploy large nets, some as long as a kilometer (0.6 mi), around schools of fish near the surface and then draw the net shut like a laundry bag (or the hood on a sweatshirt) by the purse line, a rope running through the top of the net (**FIGURE 12.30a**). Some ships set out long *driftnets* that span large expanses of water (**FIGURE 12.30b**). These chains of transparent nylon mesh nets are arrayed to drift with currents so as to capture passing fish; floats at the top and weights at the bottom keep the nets vertical. *Longline fishing* (**FIGURE 12.30c**) involves setting out extremely long lines (up to 80 km [50 mi] long) with up to several thousand baited hooks spaced along their lengths. *Trawling* entails dragging immense cone-shaped nets through the water, with weights at the bottom and floats at the top. Trawling in open water captures pelagic fish, whereas *bottom-trawling* (**FIGURE 12.30d**) involves dragging weighted nets across the floor of the continental shelf to catch benthic organisms.

Unfortunately, these fishing practices catch more than just the species they target. **Bycatch,** the accidental capture of nontarget animals, accounts for the deaths of millions of animals each year. Purse seining and driftnetting capture dolphins, seals, and sea turtles, as well as countless nontargeted fish. Most of these creatures end up drowning

(mammals and turtles need to surface to breathe) or dying from air exposure on deck (fish suffocate when kept out of the water). Driftnetting is now banned in international waters because of excessive bycatch, but the practice continues in many national waters. Bottom-trawling is often likened to clear-cutting (p. 200) the ocean floor. It is especially destructive to structurally complex areas, such as reefs, that provide shelter and habitat for animals. The scale of bycatch can be stunning. A 2011 report from the National Oceanic and Atmospheric Administration (NOAA) reported that a staggering 17% of all commercially harvested fish were captured unintentionally.

Fisheries collapse quickly under intensive harvest

Throughout the world's oceans, today's industrialized fishing fleets are depleting marine populations. In a 2003 study, fisheries biologists Ransom Myers and Boris Worm analyzed fisheries data and concluded that the oceans today contain only one-tenth of the large-bodied fish and sharks they once did, largely because of the wide-scale use of industrialized fishing practices.

Many fisheries have collapsed in recent years. Groundfish (species that live in benthic habitats, such as Atlantic cod,

FAQ

I enjoy eating seafood, so how can I make sustainable choices?

To most of us, marine fishing practices may seem a distant phenomenon over which we have no control, especially given that more than 80% of seafood sold in the United States is imported. But although we don't have control over the seafood presented to us, we have full control over which items we buy. Finding out how seafood items were caught is difficult, however, because this information is not readily available to consumers in most cases. Thus, several organizations, such as the Environmental Defense Fund, have devised concise guides and smartphone apps to help consumers differentiate fish and shellfish that are overfished or whose capture is ecologically damaging from those that are harvested more sustainably.

haddock, halibut, and flounder) powered the economies of New England and Maritime Canada for 400 years. Yet fishing pressure became so intense that most stocks collapsed, bringing fishing economies down with them. The cod fishery of the northwestern Atlantic was decimated by more than a century of overfishing, but after the Canadian and American governments banned cod harvests in the 1990s, the fishery is now on a slow path to recovery.

Red snapper stocks in the Gulf of Mexico off the Louisiana coast have been similarly depleted by overfishing. Snapper have also suffered when they are captured as bycatch on shrimping vessels. The species was identified as severely overfished in 1989, and current populations are at a mere 3% of their historical abundance. A comprehensive recovery plan has been developed for red snapper, but the road to recovery will be a slow one as stocks remain exceedingly low.

Fishery catches have remained high, however, as fishing fleets have been shifting from large, desirable species to smaller, less desirable ones. Time and again, fleets have depleted popular food fish (such as cod) and shifted to species of lower value (such as capelin, a smaller fish eaten by cod). Because this often entails catching species at lower trophic levels, this phenomenon has been termed "fishing down the food chain."

Marine reserves protect ecosystems

Fisheries managers conduct surveys, study fish population biology, and monitor catches to determine the number of fish of a given species that can be harvested without reducing future catches—a concept called *maximum sustainable yield* (p. 199). Despite the use of this technique over several decades, a number of fish and shellfish stocks have plummeted. Thus, many scientists and managers feel it is time to shift the focus away from individual species and toward viewing marine resources as elements of larger ecological systems. One key aspect of such *ecosystem-based management* (p. 199) is to set aside areas of ocean where systems can function without human interference.

Hundreds of **marine protected areas (MPAs)** have been established worldwide, most of them along the coastlines of developed countries. However, despite their name, nearly all MPAs allow fishing or other extractive activities and so are not fully protected from impacts from people.

Because of the lack of true refuges from fishing pressure, many scientists want to establish areas where fishing is prohibited. Such "no-take" areas are called **marine reserves.** Designed to preserve ecosystems intact, marine reserves are also intended to improve fisheries. Scientists argue that marine reserves can act as production factories for fish for surrounding areas, because fish larvae produced inside reserves will disperse and stock other parts of the ocean. By serving both purposes, proponents maintain, marine reserves are a win-win proposition for environmentalists and fishers alike. However, many commercial and recreational fishers dislike the idea of no-take reserves and have opposed nearly every marine reserve that has been established.

Data from marine reserves around the world have indicated that reserves can work as win-win solutions that benefit ecosystems, fish populations, and fishing economies. A comprehensive review of data from marine reserves in 2001 revealed that just 1–2 years after their establishment, marine reserves, on average, increased species diversity by 23%; the density of organisms within the reserve by 91%; the total biomass of organisms by 192%; and the average size of each organism by 31%. From this and numerous other data sets, increasing numbers of scientists, fishers, and policymakers are advocating the establishment of fully protected marine reserves as a central management tool for marine biodiversity.

Climate change is altering ocean chemistry

Overfishing is not the only major threat to marine biodiversity. Elevated levels of carbon dioxide in the atmosphere from fossil fuel combustion can pollute ocean water and change its chemical properties—much in the way excess plant nutrients or toxic substances change the chemical properties of seawater and affect marine organisms. The oceans absorb carbon dioxide (CO_2) from the atmosphere, as we first saw in our study of the carbon cycle (p. 41). As our civilization pumps excess carbon dioxide into the atmosphere by burning fossil fuels for energy and removing vegetation from the land, the buildup of atmospheric CO_2 is causing the planet to grow warmer, setting in motion many changes and consequences (Chapter 16).

The oceans have soaked up roughly a third of the excess CO_2 that we've added to the atmosphere, and this has slowed global climate change. However, there are two concerns. The first concern is that the ocean's surface water may soon become saturated with as much CO_2 as it can hold. Once it reaches this limit, then climate change will accelerate as the oceans will no longer remove large amounts of carbon dioxide from the atmosphere.

The second concern is that as ocean water soaks up CO_2, it becomes more acidic. As **ocean acidification** proceeds, sea creatures such as coral, snails, and mussels have difficulty forming shells because the carbonate ions (CO_3^{2-}) they need to create them become less available with increasing acidity, and elevated acidity levels can cause the shells of sea creatures to even begin dissolving.

Chemistry tests in the lab show that coral shells, for example, begin to erode faster than they are built once the carbonate ion concentration falls below 200 micromoles/kg of seawater. Researchers studying coral reefs in the field are finding the same thing: Reefs are growing only in waters with greater than 200 micromoles/kg of carbonate ion availability. A 2007 study used historical data and computer simulations to model how the distribution of coral reefs around the world would vary at differing levels of atmospheric CO_2 and, hence, differing levels of ocean acidity. The results predicted that as atmospheric CO_2 levels rise, coral reefs will shrink in distribution, diversity, and density. By the time atmospheric CO_2 levels pass 500 ppm, little area of ocean will be left with conditions to support coral reefs.

Because of this threat to coral reefs, scientists are intensifying the study of coral responses to warmer and acidified ocean waters to better inform efforts to conserve ecologically important reef habitats. Research published in 2015, for example, revealed that individual coral colonies vary greatly in their responses to heat and acidity, suggesting some reefs may be more resilient than others to changes in water temperature and pH. In another study, scientists found that an endangered Caribbean coral was able to sustain growth under stressful conditions by increasing its feeding rate, indicating that some reefs may be able to better persist under increasingly stressful conditions as long as food is plentiful.

closing THE LOOP

The challenges faced in Louisiana's disappearing wetlands demonstrate how our planet's aquatic systems comprise an interconnected web of ecosystems where activities in one location can affect other locations far away. To meet our needs for water for farms, homes, and industries, we have overextracted surface water and groundwater in many locations, and have engineered waterways with canals, levees, and dams, altering their natural functions. Water pollutants threaten human health and ecosystem stability, and overharvesting of marine fish populations threatens the oceans' biodiversity. Climate change is altering the temperature and chemistry of the world's oceans, endangering ecologically important habitats such as coral reefs and the biodiversity those reefs support.

There is plenty of reason for optimism, however. Improvements in water use efficiency show promise for reducing demand for water, even with increasing human populations. Water quality in many freshwater bodies has improved in recent decades, thanks to legislative action from policymakers and the efforts of millions of concerned citizens. In the oceans, marine reserves give hope that we can restore ecosystems and commercial fisheries at the same time. Water is a vital need for people, so it is only with care and continued vigilance that we will be able to secure the water we need while maintaining the health of the aquatic ecosystems that provide us so many valuable ecosystem services.

TESTING Your Comprehension

1. Explain why the distribution of water on Earth makes it difficult for many people to access adequate fresh water.

2. Pick one of the aquatic systems profiled in this chapter, and provide three examples of ways it interacts with other aquatic systems.

3. Why are coral reefs biologically valuable? How are they being degraded by human impact? What is causing the disappearance of mangrove forests and salt marshes?

4. Why do the Colorado, Rio Grande, Nile, and Yellow rivers now slow to a trickle or run dry before reaching their deltas?

5. Describe three benefits and three costs of damming rivers. What are the costs and benefits of levees?

6. Name three major types of water pollutants, and provide an example of each. Explain which classes of pollutants you think are most important in your local area.

7. Define *groundwater*, and list some sources of groundwater pollution that come from human activities.

8. Describe and explain the major steps in the process of wastewater treatment. How can artificially constructed wetlands aid such treatment?

9. Name three industrial fishing practices, and explain how they result in bycatch and marine habitat degradation.

10. How does a marine reserve differ from a marine protected area? Why do many fishers oppose marine reserves? Explain why many scientists say no-take reserves will be good for fishers.

SEEKING Solutions

1. How can we lessen agricultural demand for water? Describe some ways we can reduce household water use.

2. Describe three ways in which your own actions contribute to water pollution. Now describe three ways in which you could diminish these impacts.

3. Describe the trends in global fish capture from 1950 to 1990 and from 1990 to 2016. Describe several factors that account for these trends.

4. **CASE STUDY CONNECTION** You are mayor of a coastal town in Louisiana that is slowly losing its land to the sea, and local residents are concerned about their homes and their community as the land disappears. Because this threat is due to actions upstream on the Mississippi River that rob it of land-nourishing sediments, your constituents want you to lobby your congressional representatives for federal action to restore sediment flows to the coastline. What actions, specifically, would you ask your representatives to propose to Congress in order to help save the coastline and your community?

5. **THINK IT THROUGH** Your state's governor has put you in charge of water policy for the state. The aquifer beneath your state has been overpumped, and many wells have run dry. Agriculture is a big part of the state's economy, but crop production recently declined for the first time in 40 years. Meanwhile, the state's largest city is growing so fast that more water is needed for its burgeoning urban population. What policies would you consider to restore your state's water supply? Would you try to take steps to increase supply, decrease demand, or both? Explain why you would choose such policies.

CALCULATING Ecological Footprints

One of the single greatest personal uses of water is for showering. Old-style showerheads that were standard in homes and apartments built before 1992 dispense at least 5 gallons (19 L) of water per minute, but low-flow showerheads produced after that year dispense just 2.5 gallons (9.5 L) per minute. Given an average daily shower time of 8 minutes, calculate the amounts of water used and saved over the course of a year with old standard versus low-flow showerheads, and record your results in the table.

	ANNUAL WATER USE WITH STANDARD SHOWERHEADS (GALLONS)	ANNUAL WATER USE WITH LOW-FLOW SHOWERHEADS (GALLONS)	ANNUAL WATER SAVINGS WITH LOW-FLOW SHOWERHEADS (GALLONS)
You			
Your class			
Your state			
United States			

1. In 2010, the EPA began promoting showerheads that produce still-lower flows of 2 gallons (7.5 L) per minute (gpm). How much water would you save per year by using a 2-gpm showerhead instead of a 2.5-gpm showerhead?

2. How much water would you be able to save annually by shortening your average shower time from 8 minutes to 6 minutes? Assume you use a 2.5-gpm showerhead.

3. Compare your answers to questions 1 and 2. Do you save more water by showering 8 minutes with a 2-gpm showerhead or 6 minutes with a 2.5-gpm showerhead?

4. Can you think of any factors that are *not* being considered in this scenario of water savings? Explain.

Mastering Environmental Science

CHAPTER 13

The Atmosphere, Air Quality, and Pollution Control

Clearing the Air in L.A. and in Mexico City

> 〝 **I left L.A. in 1970, and one of the reasons I left was the horrible smog. And then they cleaned it up. That was one of the greatest things the government has ever done for me. You have beautiful days now. It's a much, much nicer place to live.**
> —Actor and comedian Steve Martin, speaking to *Los Angeles Magazine*

> **This city can be a model for others.**
> — Mexico City Mayor Miguel Ángel Mancera 〞

Los Angeles has long symbolized air pollution. Smog—that unhealthy mix of air pollutants resulting from fossil fuel combustion—has blanketed the city for decades. Exhaust from millions of automobiles clogging L.A.'s freeways regularly becomes trapped by the mountains that surround the city, and the region's warm sunshine turns the pollutants to smog. In response, Los Angeles has used policy and technology to improve its air quality. Today L.A. still suffers the nation's worst smog, but its skies are clearer than in decades.

One city that looked to L.A.'s efforts as it planned its own responses to smog is Mexico City, the capital of Mexico and one of the world's largest metropolises. Not long ago, Mexico City suffered the most polluted air in the world. On days of poor air quality in the 1990s, residents wore face masks on the streets, teachers kept students inside at recess, and outdoor sports events were canceled. Children drawing pictures would use brown crayons to color the sky. Scientists documented severe health impacts of air pollution on the city's residents (see **THE SCIENCE BEHIND THE STORY**, pp. 298–299). Indeed, each year thousands of deaths and tens of thousands of hospital visits were blamed on pollution. Mexican novelist Carlos Fuentes called his nation's capital "Makesicko City."

As in Los Angeles, traffic generates most of the pollution in Mexico City, where motorists in nearly 7 million cars traverse miles of urban sprawl. And like L.A., Mexico City lies in a valley surrounded by mountains, vulnerable to temperature inversions that trap pollutants over the city. Moreover, at Mexico City's high altitude—2240 m (7350 ft) above sea level—solar radiation is intense, which worsens smog formed by the interaction of pollutants with sunlight. Mexico City environmental chemist Armando Retama likens his hometown to "a casserole dish with a lid on top."

Despite the challenges, Mexico City's 21 million people fought back and made notable improvements in air quality. City leaders took bold action to clean up the air, and in recent years Mexico City has been enjoying a renaissance. As the smog began to clear, revealing beautiful views of the snow-capped peaks that ring the valley, Mexico City became an international model for other cities seeking to fight pollution.

Efforts began in the 1990s, when city leaders shut down an oil refinery and pushed factories and power plants to shift to cleaner-burning natural gas. Policymakers mandated that lead be removed from gasoline, that the sulfur content of diesel fuel be reduced, and that catalytic converters (p. 294) be phased in for new vehicles.

Testing emissions at a vehicle inspection station in Los Angeles ▲

Upon completing this chapter, you will be able to:

- Describe the composition, structure, and function of Earth's atmosphere
- Relate weather and climate to atmospheric conditions
- Identify major outdoor air pollutants and outline the scope of air pollution
- Assess strategies and solutions for control of outdoor air pollution
- Explain stratospheric ozone depletion and identify steps taken to address it
- Describe acid deposition, discuss its consequences, and explain how we are addressing it
- Characterize the scope of indoor air pollution and assess solutions

Mexico City on a smoggy day

Levels of many pollutants fell, but as the population grew, smog from traffic persisted. In response, city officials stepped up vehicle emissions testing and upgraded taxis and city vehicles to cleaner models. To monitor air quality, 34 sampling stations were set up across the city, sending real-time data to city engineers.

In 2007 Mayor Marcelo Ebrard accelerated efforts as part of a 15-year sustainability plan he launched aiming to make Mexico City "the greenest city in the Americas." New lines were added to the subway system, and 800 exhaust-spewing minibuses were replaced with fuel-efficient buses. More than 450,000 people now use these buses each day, cutting carbon dioxide emissions by an estimated 80,000 tons per year.

In 2010 Ebrard introduced a bicycle-sharing program to free short-distance commuters from dependence on cars. With 1000 bikes at rental stations throughout the city, people can rent a bike cheaply at one location and drop it off at another. In the most popular initiative, every Sunday morning the city's main boulevard, the Paseo de la Reforma, is closed to car traffic, creating a safe and pleasant community space for pedestrians, bikers, joggers, and skateboarders.

After 2012, Ebrard's successor as mayor, Miguel Ángel Mancera, built on these efforts by expanding the bicycle program and putting electric buses and taxis on the roads. Mancera also rolled out a car-sharing program, aiming to remove 40,000 vehicles from circulation. Private entrepreneurs are now getting in on the act; recent university graduates have launched companies providing car-sharing and carpooling services, some using electric vehicles.

All these changes paid off with cleaner air. In 1991, Mexico City's air was deemed hazardous to breathe on all but 8 days of the year. By 2010–2015, most pollutants had been slashed by more than 75%, and the air was meeting health standards on one of every two days.

However, plenty of hurdles remain. Rampant development is challenging efforts to plan for sustainable growth. New highway construction is inducing more people to drive. The typical motorist still spends three hours a day stuck in traffic that averages just 13 mph. And smog still contributes to an estimated 4000 deaths each year. In 2016, a month-long spell of hot, windless weather brought back foul air conditions that the city had not suffered in over a decade. Angry residents complained that the city's politicians had become complacent and were failing to follow through on further reforms. In response, Mexico's president, Enrique Peña Nieto, urged leaders to buckle down anew and tighten emissions testing on vehicles. The hope is that Mexico City will continue to follow in the footsteps of Los Angeles, making steady progress toward cleaner air.

L.A. and Mexico City typify cities of developed and developing nations today. Nations that are industrializing as they try to build wealth for their citizens are confronting the same air quality challenges that plagued the United States and other wealthy nations a generation and more ago. We will examine the solutions sought in Los Angeles, Mexico City, and elsewhere as we learn about Earth's atmosphere and how to reduce the pollutants we release into it.

The Atmosphere

Every breath we take reaffirms our connection to the **atmosphere,** the layer of gases that envelops our planet. The atmosphere moderates our climate, provides oxygen, helps to shield us from meteors and hazardous solar radiation, and transports and recycles water and nutrients.

Earth's atmosphere consists of 78% nitrogen (N_2) and 21% oxygen (O_2), by volume of dry air. The remaining 1% is composed of argon (Ar) and minute concentrations of other gases (**FIGURE 13.1**). The atmosphere also contains water vapor (H_2O) in concentrations that vary with time and place from 0% to 4%. Today, human activity is altering the quantities of some atmospheric gases, such as carbon dioxide (CO_2), methane (CH_4), and ozone (O_3).

The atmosphere is layered

The atmosphere that stretches so high above us is actually just 1/100 of Earth's diameter, like the fuzzy skin of a peach. It consists of four layers that differ in temperature, density, and composition (**FIGURE 13.2**).

Within the bottommost layer, the **troposphere,** air movement drives the planet's weather. The troposphere is thin (averaging 11 km [7 mi] high) relative to the atmosphere's other layers, but it contains three-quarters of the atmosphere's mass. This is because gravity pulls mass downward, making

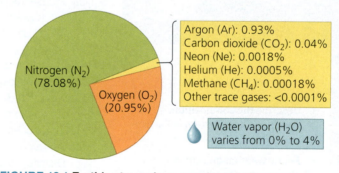

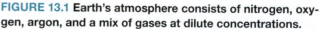

FIGURE 13.1 Earth's atmosphere consists of nitrogen, oxygen, argon, and a mix of gases at dilute concentrations.

air denser near Earth's surface. Tropospheric air gets colder with altitude, dropping to roughly –52°C (–62°F) at the top of the troposphere. At this point, temperatures stabilize, marking a boundary called the *tropopause*. The tropopause acts like a cap, limiting mixing between the troposphere and the atmospheric layer above it, the stratosphere.

The **stratosphere** extends 11–50 km (7–31 mi) above sea level. Similar in composition to the troposphere, the stratosphere is drier and less dense. Its gases experience little vertical mixing, so once substances (including pollutants) enter it, they tend to remain for a long time. The stratosphere warms with altitude, because its ozone and oxygen absorb the sun's ultraviolet (UV) radiation (p. 35). Most of the atmosphere's ozone concentrates in a portion of the

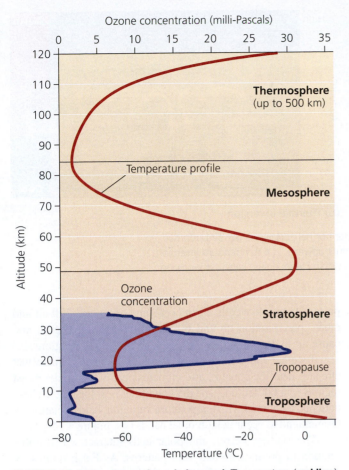

FIGURE 13.2 The atmosphere is layered. Temperature (**red line**) drops with altitude in the troposphere, rises with altitude in the stratosphere, drops in the mesosphere, and rises in the thermosphere. The tropopause separates the troposphere from the stratosphere. Ozone (**blue area**) is densest in the lower stratosphere. *Adapted from Jacobson, M.Z., 2002. Atmospheric pollution: History, science, and regulation. Cambridge: Cambridge University Press; and Parson, E.A., 2003. Protecting the ozone layer: Science and strategy. Oxford: Oxford University Press.*

stratosphere roughly 17–30 km (10–19 mi) above sea level, a region we call the **ozone layer.** By absorbing and scattering incoming UV radiation, the ozone layer greatly reduces the amount of this radiation that reaches Earth's surface. UV light can damage living tissue and induce genetic mutations, and life has evolved to rely on the protective presence of the ozone layer.

Above the stratosphere lies the mesosphere, where temperatures decrease with altitude and where incoming meteors burn up. Above this the thermosphere extends upward to an altitude of 500 km (300 mi). Still higher, the atmosphere merges into space in a region called the exosphere.

The sun influences weather and climate

An enormous amount of energy from the sun constantly bombards the upper atmosphere. Of this solar energy, about 70% is absorbed by the atmosphere and planetary surface, while the rest is reflected back into space (see Figure 14.2, p. 313).

Land and surface water absorb solar energy and then emit thermal infrared radiation, which warms the air and causes some water to evaporate. As a result, air near Earth's surface tends to be warmer and moister than air at higher altitudes. These differences set into motion a process of **convective circulation.** Warm air, being less dense, rises and creates vertical currents. As air rises into regions of lesser atmospheric pressure, it expands and cools, causing moisture to condense and fall as rain. Once the air cools, it descends and becomes denser, replacing warm air that is rising. The descending air picks up heat and moisture near ground level and prepares to rise again, continuing the process. Convective circulation influences both weather and climate.

Weather and climate each involve physical properties of the troposphere, such as temperature, pressure, humidity, cloudiness, and wind. **Weather** specifies atmospheric conditions in a location over minutes, hours, days, or weeks. In contrast, **climate** describes typical patterns of atmospheric conditions in a location over years, decades, centuries, or millennia. Mark Twain noted the distinction by remarking, "Climate is what we expect; weather is what we get." For example, Los Angeles has a climate characterized by warm dry summers and mild rainy winters, yet on some autumn days, dry Santa Ana winds blow in from the desert and bring extremely hot weather.

Inversions affect air quality

Under most conditions, air in the troposphere becomes cooler as altitude increases. Because warm air rises, vertical mixing results (**FIGURE 13.3a**). Occasionally, however, a layer of cool air may form beneath a layer of warmer air. This departure from the normal temperature profile is known as a **temperature inversion,** or thermal inversion (**FIGURE 13.3b**). The band of air in which temperature rises with altitude is called an **inversion layer** (because the normal direction of temperature change is inverted). The cooler air at the bottom of the inversion layer is denser than the warmer air above, so it resists vertical mixing and remains stable. Temperature inversions can occur in different ways, sometimes involving cool air at ground level and sometimes involving an inversion layer higher above the ground. One common type of inversion (shown in Figure 13.3b) occurs in mountain valleys where slopes block morning sunlight, keeping ground-level air within the valley shaded and cool.

Vertical mixing allows pollutants in the air to be carried upward and diluted, but temperature inversions trap pollutants near the ground. As a result, cities such as Los Angeles and Mexico City suffer their worst pollution when inversions prevent pollutants from being dispersed. Both metropolitan areas are encircled by mountains that promote inversions, interrupt air flow, and trap pollutants. Los Angeles experiences inversions most often when a "marine layer" of air cooled by the ocean moves inland. In Mexico City in 1996, a persistent temperature inversion sparked a five-day crisis in which air pollution killed at least 300 people and sent 400,000 to hospitals. Across the world, inversions frequently concentrate pollution over metropolitan areas in valleys ringed by mountains, from Tehran to Seoul to Río de Janeiro to São Paulo.

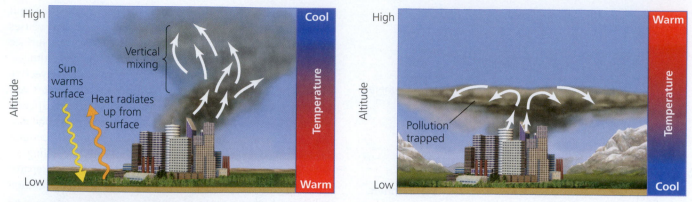

(a) Normal conditions　　　　　　　　**(b) Thermal inversion**

FIGURE 13.3 Temperature inversions trap air and pollutants. Under normal conditions **(a)**, air becomes cooler with altitude and air of different altitudes mixes, dispersing pollutants upward. In a temperature inversion **(b)**, dense cool air remains near the ground, and air warms with altitude within the inversion layer. Little mixing occurs, and pollutants are trapped.

Large-scale circulation systems produce global climate patterns

At large geographic scales, convective air currents contribute to broad climate patterns (**FIGURE 13.4**). Near the equator, solar radiation sets in motion a pair of convective cells known as *Hadley cells.* Here, where sunlight is most intense, surface air warms, rises, and expands. As it does so, it releases moisture, producing the heavy rainfall that gives rise to tropical rainforests. After releasing much of its moisture, this air diverges and moves in currents heading north and south. The air in these currents cools and descends at about 30 degrees latitude north and south. Because the descending air is now dry, the regions around 30° are quite arid, giving rise to deserts. Two additional

pairs of convective cells, *Ferrel cells* and *polar cells,* lift air and create precipitation around 60° latitude north and south and cause air to descend at 30° latitude and in the polar regions.

Together these three pairs of convective cells create wet climates near the equator, arid climates near 30° latitude, moist regions near 60° latitude, and dry conditions near the poles. These patterns, combined with temperature variation, help explain why biomes tend to occur in latitudinal bands (see Figure 4.14, p. 83).

The Hadley, Ferrel, and polar cells interact with Earth's rotation to produce global wind patterns. As Earth rotates on its axis, regions of the planet's surface near the equator move west to east more quickly than locations near the poles. As a result, from the perspective of an Earth-bound observer, air currents of the convective cells that flow north or south appear to be deflected from a straight path. This deflection is called the

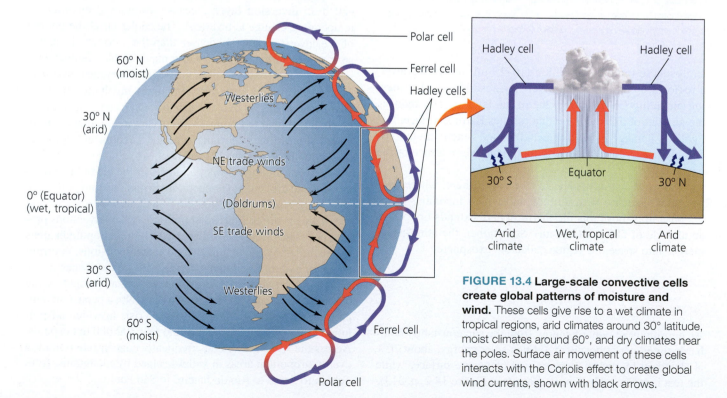

FIGURE 13.4 Large-scale convective cells create global patterns of moisture and wind. These cells give rise to a wet climate in tropical regions, arid climates around 30° latitude, moist climates around 60°, and dry climates near the poles. Surface air movement of these cells interacts with the Coriolis effect to create global wind currents, shown with black arrows.

Coriolis effect, and it results in the curving global wind patterns shown in Figure 13.4. For centuries, people made use of these patterns to facilitate ocean travel by wind-powered sailing ships.

Understanding how the atmosphere functions enables us to appreciate what drives Earth's climate and can help us predict our weather day by day. Such knowledge also helps us comprehend how our pollution of the atmosphere affects climate, ecological systems, economies, and human health.

Outdoor Air Quality

Throughout history, we have made the atmosphere a dumping ground for our airborne wastes. Whether from simple wood fires or modern coal-burning power plants, we have generated **air pollutants,** gases and particulate material added to the atmosphere that can affect climate or harm people or other living things. Fortunately, our efforts to control **air pollution,** the release of air pollutants, have brought some of our best successes in confronting environmental problems.

In recent decades, public policy and improved technologies have helped us reduce most types of **outdoor air pollution** (often called *ambient air pollution*) in industrialized nations. However, outdoor air pollution remains a problem, particularly in industrializing nations and in urban areas. Globally, the World Health Organization (WHO) estimates that each year 3.3 million people die prematurely as a result of health problems caused by outdoor air pollution. Moreover, we face an enormous air pollution challenge today in our emission of greenhouse gases (p. 314), which contribute to global climate change. (We discuss climate change separately and in depth in Chapter 14.)

Some pollution is from natural sources

When we think of outdoor air pollution, we tend to envision smokestacks belching smoke from industrial plants. However, natural processes also pollute the air (**FIGURE 13.5**).

Fires (p. 201) from burning vegetation generate soot and gases. Worldwide, more than 60 million hectares (ha; 150 million acres; an area the size of Texas) of forest and grassland burn in a typical year. Volcanic eruptions (p. 236) release particulate matter and sulfur dioxide that may spread over large regions. Ash from major eruptions can ground airplanes and pose respiratory health dangers. In 2012, Mexico City residents went on alert as the nearby volcano of Popocatepetl erupted, adding to the region's pollution challenges. Wind-blown dust is another natural pollution source. Dust storms in arid regions can sometimes blow dust from one continent to another.

Some natural impacts are made worse by human activity and land use policies. Farming and grazing practices that strip vegetation from the soil promote wind erosion and lead to dust storms, like those that devastated America's Dust Bowl in the 1930s (p. 149). In the tropics, many farmers set fires to clear forest (p. 145). In North American forests, the suppression of fire has allowed fuel to build up and eventually feed highly destructive fires (p. 201). And climate change (Chapter 14), driven by our use of fossil fuels, is leading to drought in many regions, worsening dust storms and fires as a result.

We create outdoor air pollution

Human activity produces many air pollutants. As with water pollution, anthropogenic (human-caused) air pollution can emanate from point sources or non-point sources (p. 275). A point source describes a specific location from which large quantities of pollutants are discharged (such as a coal-fired power plant). Non-point sources are more diffuse, consisting of many small, widely spread sources (such as millions of automobiles).

Pollutants released directly from a source are termed **primary pollutants.** Ash from a volcano, sulfur dioxide from a power plant, and carbon monoxide from an engine

(a) Natural fire in California   **(b) Mount Saint Helens eruption, 1980**   **(c) Satellite image of dust storm blowing dust from Africa to the Americas**

FIGURE 13.5 Wildfire, volcanoes, and dust storms are three natural sources of air pollution.

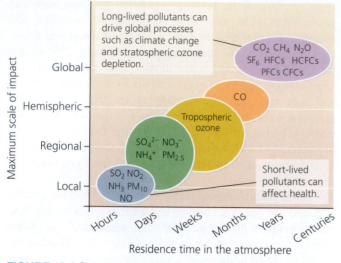

FIGURE 13.6 Substances with short residence times affect air quality locally, whereas those with long residence times affect air quality globally. *Source: United Nations Environment Programme, 2007. Global environmental outlook (GEO-4), Nairobi, Kenya.*

are all primary pollutants. Often primary pollutants react with one another, or with constituents of the atmosphere, and form other pollutants, called **secondary pollutants.** Examples include ozone formed near ground level (p. 294) from pollutants in urban smog, or the acids in acid rain, formed when certain primary pollutants react with water and oxygen.

Because substances differ in how readily they react in air and in how quickly they settle to the ground, they differ in their **residence time,** the amount of time a substance spends in the atmosphere. Pollutants with brief residence times exert localized impacts over short time periods. Most particulate matter and most pollutants from automobile exhaust stay aloft only hours or days, which is why air quality in a city like Mexico City or Los Angeles changes from day to day. In contrast, pollutants with long residence times can exert impacts regionally or globally for long periods, even centuries (**FIGURE 13.6**). The pollutants that drive climate change and those that deplete Earth's ozone layer (two separate phenomena!—see **FAQ,** p. 303) are each able to cause these long-lasting global impacts because they persist in the atmosphere for so long.

The Clean Air Act addresses pollution

To address air pollution in the United States, Congress has passed a series of laws, most notably the **Clean Air Act,** first enacted in 1963 and amended multiple times since, chiefly in 1970 and 1990. This body of legislation funds research into pollution control, sets standards for air quality, and encourages emissions standards for automobiles and for stationary point sources such as industrial plants. It also imposes limits on emissions from new sources, funds a nationwide air quality monitoring system, and enables citizens to sue parties violating the standards.

Under the Clean Air Act, the U.S. Environmental Protection Agency (EPA) sets nationwide standards for (1) emissions of several key pollutants and (2) concentrations of major pollutants in ambient air. It is largely up to the states to monitor emissions and air quality and to develop, implement, and enforce regulations within their borders. States submit implementation plans to the EPA for approval, and if a state's plans are not adequate, the EPA can take control of enforcement. If a region fails to clean up its air, the EPA can prevent it from receiving federal money for transportation projects.

Agencies monitor emissions

State and local agencies monitor and report to the EPA emissions of six major pollutants, profiled below. Across the United States in 2016, human activity polluted the air with roughly 80 million tons of these six pollutants.

Carbon monoxide Carbon monoxide (CO) is a colorless, odorless gas produced primarily by the incomplete combustion of fuel. Vehicles and engines account for most CO emissions in the United States. Other sources include industrial processes, waste combustion, and residential wood burning. Carbon monoxide is hazardous because it binds to hemoglobin in red blood cells, which in turn prevents hemoglobin from binding with oxygen. Carbon monoxide poisoning induces nausea, headaches, fatigue, heart and nervous system damage, and potentially death.

Sulfur dioxide Sulfur dioxide (SO_2) is a colorless gas with a pungent odor. Most emissions result from the combustion of coal for electricity generation and industry. During combustion, elemental sulfur (S), a contaminant in coal, reacts with oxygen (O_2) to form SO_2. Once in the atmosphere, SO_2 may react to form sulfur trioxide (SO_3) and sulfuric acid (H_2SO_4), which may settle back to Earth in acid deposition (p. 303).

Nitrogen oxides Nitrogen oxides (NO_X) are a family of compounds that include nitric oxide (NO) and nitrogen dioxide (NO_2). Most U.S. emissions of nitrogen oxides result when nitrogen and oxygen from the atmosphere react at high temperatures during combustion in vehicle engines. Fossil fuel combustion in industry and at electrical utilities accounts for most of the rest. NO_X emissions contribute to smog, acid deposition, and stratospheric ozone depletion.

Volatile organic compounds Volatile organic compounds (VOCs) are carbon-containing chemicals emitted by vehicle engines and a wide variety of solvents, industrial processes, household chemicals, paints, plastics, and consumer items. Examples range from benzene to acetone to formaldehyde. One group of anthropogenic VOCs consists of hydrocarbons (p. 32). Other VOCs are emitted naturally by plants. VOCs can react to produce secondary pollutants, as occurs in urban smog.

Particulate matter Particulate matter is composed of solid or liquid particles small enough to be suspended in the atmosphere. Particulate matter includes primary pollutants such as dust and soot, as well as secondary pollutants such as sulfates and nitrates. Scientists classify particulate matter by the size of the particles. Smaller particles are more likely to get deep into the lungs and to pass through tissues, causing damage to the

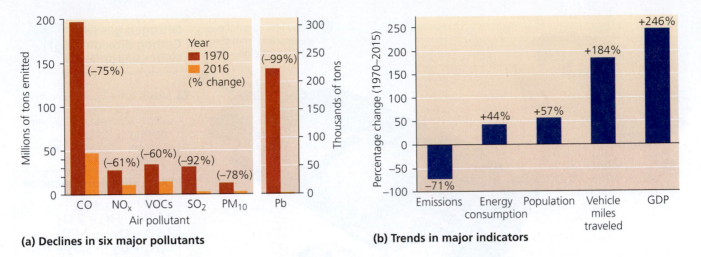

(a) Declines in six major pollutants

(b) Trends in major indicators

FIGURE 13.7 U.S. emissions have declined sharply since 1970. We have achieved reductions **(a)** in the six major pollutants tracked by the EPA, despite increases **(b)** in U.S. energy consumption, population, vehicle miles traveled, and gross domestic product. *Data from U.S. EPA.*

DATA • By what percentage has population increased since 1970? • By what percentage have emissions decreased? • Using these two percentages, calculate the change in emissions per person. (Hint: Begin by envisioning a population of 100 people and 100 units of emissions.)

Go to **Interpreting Graphs & Data** on **Mastering** Environmental Science

lungs, heart, and brain. PM$_{10}$ pollutants consist of particles less than 10 microns in diameter (one-seventh the width of a human hair). PM$_{2.5}$ pollutants consist of still-finer particles less than 2.5 microns in diameter. Most PM$_{10}$ pollution is from road dust, whereas most PM$_{2.5}$ pollution results from combustion.

Lead Lead (Pb) is a heavy metal that enters the atmosphere as a particulate pollutant. The lead-containing compounds tetraethyl lead and tetramethyl lead, when added to gasoline, improve engine performance. However, exhaust from the combustion of leaded gasoline emits airborne lead, which can be inhaled or can settle on land and water. When lead enters the food chain, it accumulates in body tissues and can cause central nervous system malfunction and many other ailments (p. 215). Since the 1980s the United States and many other nations have phased out leaded gasoline (p. 19), and now most developing nations are following suit. In developed nations today, the main source of atmospheric lead is industrial metal smelting.

We have reduced emissions

Since the Clean Air Act of 1970, the United States has reduced emissions of each of the six monitored pollutants substantially (**FIGURE 13.7a**). These dramatic reductions in emissions have occurred despite significant increases in the nation's population, energy consumption, miles traveled by vehicle, and gross domestic product (**FIGURE 13.7b**).

This success in controlling pollution has resulted from policy steps and technological advances, each motivated by grass-roots social demand for cleaner air. In factories, power plants, and refineries, technologies such as baghouse filters, electrostatic precipitators, and **scrubbers** (**FIGURE 13.8**) chemically convert or physically remove pollutants before they are emitted from smokestacks. Cleaner-burning motor

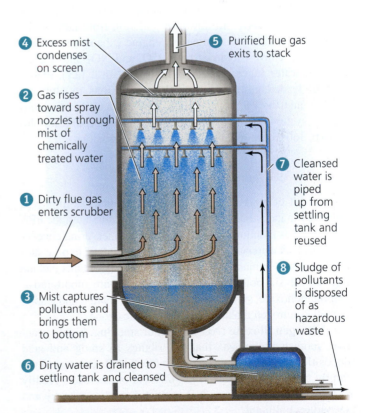

FIGURE 13.8 Scrubbers typically remove at least 90% of particulate matter and gases such as sulfur dioxide. Scrubbers and other pollution control devices come in many designs. In this diagram of a spray-tower wet scrubber, follow the numbers to see how polluted air rises through a chamber while nozzles spray a mist of water mixed with lime or other active chemicals to capture pollutants and wash them out of the air.

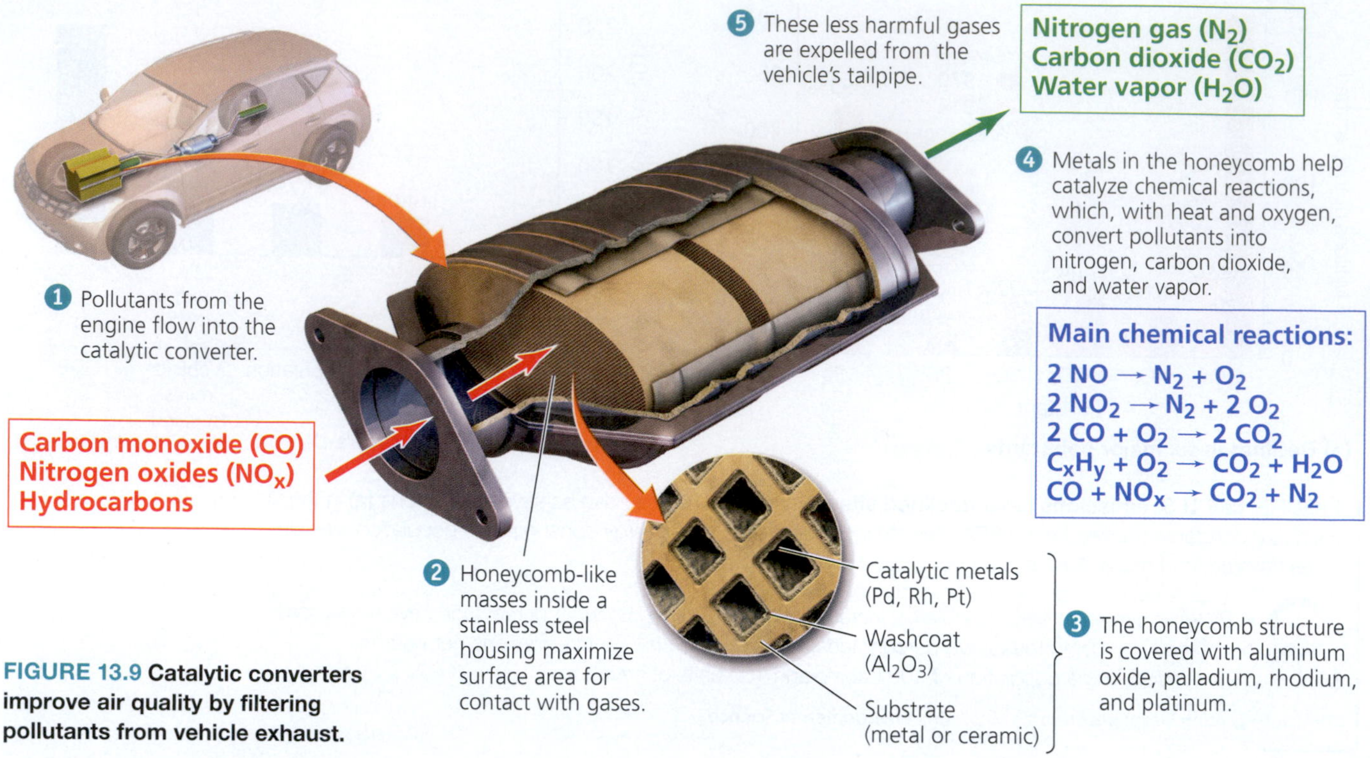

① Pollutants from the engine flow into the catalytic converter.

Carbon monoxide (CO)
Nitrogen oxides (NOₓ)
Hydrocarbons

② Honeycomb-like masses inside a stainless steel housing maximize surface area for contact with gases.

⑤ These less harmful gases are expelled from the vehicle's tailpipe.

Nitrogen gas (N₂)
Carbon dioxide (CO₂)
Water vapor (H₂O)

④ Metals in the honeycomb help catalyze chemical reactions, which, with heat and oxygen, convert pollutants into nitrogen, carbon dioxide, and water vapor.

Main chemical reactions:

$$2\,NO \rightarrow N_2 + O_2$$
$$2\,NO_2 \rightarrow N_2 + 2\,O_2$$
$$2\,CO + O_2 \rightarrow 2\,CO_2$$
$$C_xH_y + O_2 \rightarrow CO_2 + H_2O$$
$$CO + NO_x \rightarrow CO_2 + N_2$$

Catalytic metals (Pd, Rh, Pt)

Washcoat (Al_2O_3)

Substrate (metal or ceramic)

③ The honeycomb structure is covered with aluminum oxide, palladium, rhodium, and platinum.

FIGURE 13.9 Catalytic converters improve air quality by filtering pollutants from vehicle exhaust.

vehicle engines and automotive technologies such as catalytic converters have cut down on pollution from automobile exhaust. In a **catalytic converter,** engine exhaust reacts with metals that convert hydrocarbons, CO, and NO_X into carbon dioxide, water vapor, and nitrogen gas (**FIGURE 13.9**). Phaseouts of leaded gasoline caused lead emissions to plummet (p. 19), and the EPA's Acid Rain Program and its emissions trading system (pp. 304–305), along with clean coal technologies (p. 360), have reduced SO_2 and NO_X emissions.

Air quality has improved

As a result of emissions reductions, air quality has improved markedly in industrialized nations. In the United States, the EPA and the states monitor outdoor air quality by measuring concentrations of six **criteria pollutants,** pollutants judged to pose substantial risk to human health. The six criteria pollutants include four of the six pollutants whose emissions are monitored—carbon monoxide, sulfur dioxide, particulate matter, and lead—as well as nitrogen dioxide and tropospheric ozone.

Nitrogen dioxide (NO_2) is a foul-smelling, highly reactive, reddish brown gas that contributes to smog and acid deposition. Along with nitric oxide (NO), NO_2 belongs to the family of nitrogen oxides (NO_X). Nitric oxide reacts readily in the atmosphere to form NO_2, which is both a primary and secondary pollutant.

Although ozone in the stratosphere shields us from the dangers of UV radiation, ozone from human activity accumulates low in the troposphere. **Tropospheric ozone** (O_3; also called *ground-level ozone*) is a secondary pollutant, created by the reaction of nitrogen oxides and volatile carbon-containing chemicals in the presence of sunlight. A major component of photochemical smog (pp. 296–297), this colorless gas poses health risks because the O_3 molecule will readily split into a molecule of oxygen gas (O_2) and a free oxygen atom. The oxygen atom may then participate in reactions that injure living tissues and cause respiratory problems. Tropospheric ozone is the pollutant that most frequently exceeds its EPA standard.

Across the United States, more than 4000 monitoring stations take hourly or daily air samples to measure pollutant concentrations. The EPA compiles these data and calculates values for its Air Quality Index (AQI) for each site. Each of six pollutants—CO, SO_2, NO_2, O_3, PM_{10}, and $PM_{2.5}$—receives an AQI value from 0 to 500 that reflects its current concentration. AQI values below 100 indicate satisfactory air conditions, and values above 100 indicate unhealthy conditions. The highest AQI value from the pollutants on a particular day is reported as the overall AQI value for that day, and these values are made available online and reported in weather forecasts.

Thanks to the actions of scientists, policymakers, industrial leaders, and everyday people, outdoor air quality today is far better than it was a generation or two ago. However, there remains plenty of room for improvement. Concerns over new pollutants are emerging, greenhouse gas emissions are altering the climate, and people in low-income communities often suffer from hotspots of pollution. In fact, many Americans live in areas where pollution continues to reach unhealthy levels. For instance, despite great improvement over the past two decades, residents of Los Angeles County breathe air that violates health standards for five of the six criteria pollutants. As of 2015, 127 million Americans lived in counties that violated the national standard for tropospheric ozone. Still, L.A. and other metropolises are making perceptible headway toward cleaner air for their residents (**FIGURE 13.10**).

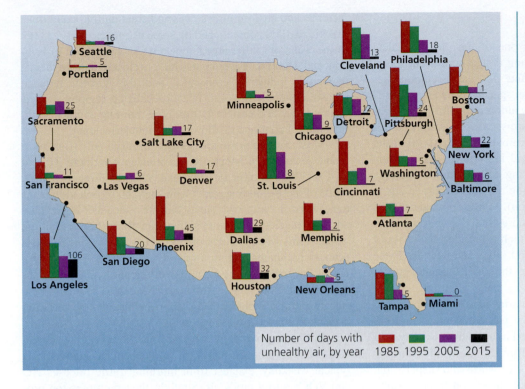

FIGURE 13.10 In most U.S. cities, air has become cleaner. This map shows numbers of days with unhealthy air from years spanning four decades for 29 metropolitan areas, according to the Air Quality Index (p. 294). *Data from U.S. EPA.*

| Number of days with unhealthy air, by year | 1985 | 1995 | 2005 | 2015 |

DATA Locate where you live on this map. • How does your city or the nearest major city to you compare to others in its air quality? • How has its air quality changed in recent years?

Now explore one of the EPA websites that lets you browse information on the air you breathe: www.airnow.gov, www.epa.gov/aircompare, or www.epa.gov/air/emissions/where.htm. • How does your region's air quality compare to that of the rest of the nation? • What factors influence the quality of your region's air? • Propose three steps for reducing air pollution in your region.

Go to **Interpreting Graphs & Data** on **Mastering** Environmental Science

---

Clearing the Air for Better Health Across America

With every breath we take, each of us alive today benefits from America's success in fighting outdoor air pollution. A generation or two ago, poor air quality was hampering people's health and well-being across the United States. The Clean Air Act changed that, by encouraging pollution reduction and spurring advances in technology. The resulting drop in outdoor air pollution since 1970 represents one of America's greatest accomplishments in safeguarding human health and environmental quality. As the figure shows, ambient concentrations of all criteria pollutants have declined below their health standards set by the EPA. Lead, not shown, has plummeted from 900% of its standard since 1980. The EPA estimates that between 1970 and 1990 alone, clean air regulations and the resulting technological advances in pollution control saved the lives of 200,000 Americans. This success demonstrates how seemingly intractable problems can be addressed when government and industry apply information from science and are responsive to the public's demands. However, progress gained can also be lost. To protect our hard-won gains and build on our success, we must insist that our policymakers uphold beneficial regulations and resist influence from polluting interests that threaten our health and safety. We also would do well to remember that outdoor air is cleaner than indoor air—so it pays to get off the couch, turn off the screen, and get outside!

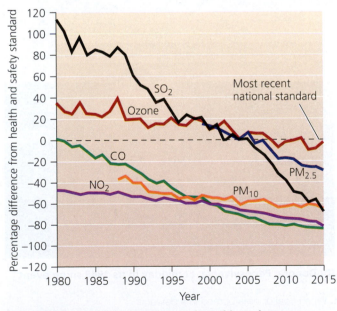

Concentrations of criteria pollutants in ambient air across the United States have fallen in the wake of Clean Air Act regulations. *Data from U.S. EPA.*

EXPLORE THE DATA at **Mastering** Environmental Science

Smog poses health risks

Now let's take a closer look at one of the most prevalent types of air pollution: smog. **Smog** is a general term for an unhealthy mixture of air pollutants that can accumulate as a result of fossil fuel combustion, generally over industrial regions or urban areas with heavy automobile traffic.

Since the onset of the industrial revolution, cities have suffered a type of smog known as **industrial smog** (**FIGURE 13.11a**). When coal or oil is burned, some portion is completely combusted, releasing CO_2; some is partially combusted, emitting CO; and some remains unburned and is released as soot (particles of carbon). Moreover, coal contains contaminants such as mercury and sulfur. Sulfur reacts with oxygen to form sulfur dioxide, which can undergo a series of reactions to form sulfuric acid and other compounds. These substances, along with soot, are the main components of industrial smog.

America's most severe industrial smog event occurred in Pennsylvania, in a small town named Donora, in 1948 (**FIGURE 13.11b**). Donora is located in a mountain valley, and one day after air had cooled during the night, the morning sun did not reach the valley floor to warm and disperse the cold air. The resulting temperature inversion trapped smog from a steel and wire factory. Twenty-one people died, and more than 6000 people—nearly half the town—became ill.

The world's worst industrial smog crisis occurred in London, England, in 1952, when a high-pressure system settled over the city for several days, trapping pollutants from factories and coal-burning stoves and creating foul conditions that killed 4000 people—and by some estimates up to 12,000. In the wake of "killer smog" episodes such as those in London and Donora, governments of developed nations began regulating industrial emissions, and this greatly reduced industrial smog. However, in industrializing regions such as China, India, and eastern Europe, coal burning and lax pollution control continue to allow severe industrial smog.

Most smog in urban areas today results largely from automobile exhaust. In Mexico City, vehicles contribute 31% of VOCs, 50% of sulfur dioxide, and 82% of nitrogen oxides. Some of these emissions react with sunlight, so pollution tends to be worst in cities with sunny climates, such as Mexico City and Los Angeles. Such cities suffer from **photochemical smog,** which forms when sunlight drives chemical reactions between primary pollutants and atmospheric compounds, producing a potent cocktail of more than 100 different chemicals, tropospheric ozone often being the most abundant (**FIGURE 13.12a**). Because it also includes NO_2, photochemical smog generally appears as a brownish haze (**FIGURE 13.12b**).

Hot, sunny, windless days in urban areas create perfect conditions for the formation of photochemical smog. On a typical weekday, exhaust from morning traffic releases NO and VOCs into a city's air. Sunlight then promotes the production of ozone and other secondary pollutants, leading pollution typically to peak in midafternoon. Photochemical smog irritates people's eyes, noses, and throats, and over time can lead to asthma, lung damage, heart problems, decreased resistance to infection, and even cancer.

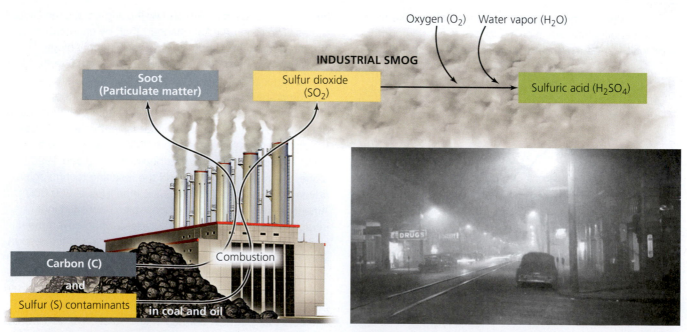

(a) Formation of industrial smog

(b) Donora, Pennsylvania, at midday in its 1948 smog event

FIGURE 13.11 Industrial smog results from fossil fuel combustion. When coal or oil is burned in a power plant or factory, soot (particulate matter of carbon) is released, and sulfur contaminants give rise to sulfur dioxide, which may react with atmospheric gases to produce further compounds **(a)**. Carbon monoxide and carbon dioxide are also emitted. Under certain weather conditions, industrial smog can blanket whole regions, as it did in Donora, Pennsylvania **(b)**, shown in the daytime during its deadly 1948 smog episode.

We can take steps to reduce smog

Photochemical smog afflicts countless American cities, from Atlanta to Newark to Baltimore to Houston to Salt Lake City. Mayors, city councilors, and state and federal regulators everywhere are trying to devise ways to clear their air.

Elsewhere in the world, many cities are tackling photochemical smog. In Tehran, Iran, city leaders now require vehicle inspections, regulate traffic into the city center, and pay drivers to turn in old, polluting cars for newer, cleaner ones. The government reduced sulfur in diesel fuel and converted buses to run on (cleaner-burning) natural gas. To raise public awareness, 22 electronic billboards were installed around the city, displaying current pollutant levels. All these efforts helped reduce pollution, yet so many people continued to stream into the city and buy cars that pollution soon grew worse again. In response, officials lowered gasoline subsidies, rationed fuel, and began expanding the subway system.

Los Angeles's struggle with air pollution began in 1943, when the city's first major smog episode cut visibility to three blocks. With the city's image as a clean and beautiful coastal haven at risk, civic leaders confronted the problem head-on.

Los Angeles's quest to solve its smog problem spurred much of the early research into photochemical smog and how automobiles might burn fuel more cleanly.

Los Angeles's city and county officials began by passing ordinances restricting emissions from power plants, oil refineries, and petrochemical facilities, then targeted emissions from motor vehicles. Because air pollution spreads from place to place, responsibility for pollution control soon moved from the city and county levels to the state and federal levels.

California took the lead among U.S. states in adopting pollution control technology and setting emissions standards for vehicles. In 1967 state leaders established the California Air Resources Board, the first state agency focused on regulating air quality. Today in California and 33 other states, drivers are required to have their vehicle exhaust inspected periodically. These inspection programs, which

weighing the ISSUES

Smog-Busting Solutions

Does the city you live in, or a major city near you, suffer from photochemical smog or other air pollution? How is this city responding? What policies do you think it should pursue? What benefits might your city enjoy from such policies? Might they cause any unintended consequences?

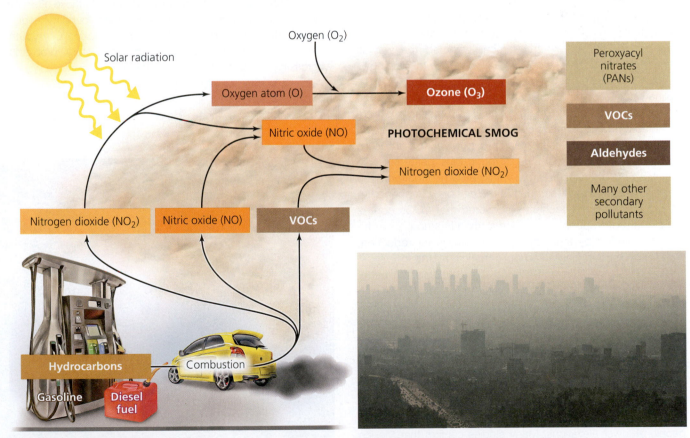

(a) Formation of photochemical smog

(b) Photochemical smog over Los Angeles

FIGURE 13.12 Photochemical smog results when pollutants from automobile exhaust react amid exposure to sunlight. Nitrogen dioxide, nitric oxide, and VOCs initiate a series of chemical reactions **(a)** that produce a toxic brew of secondary pollutants including ozone, peroxyacyl nitrates (PANs), aldehydes, and others. Photochemical smog is common over Los Angeles **(b)** and many other urban areas, especially those with hilly topography or frequent inversions.

Does Air Pollution Affect the Brain, as Well as the Lungs and Heart?

Dr. Lilian Calderón-Garcidueñas

"I know I'm inhaling poison," a 38-year-old candy vendor named Guadalupe told a reporter amid the fumes of a traffic-choked intersection in Mexico City. "But there is nothing I can do."

For as long as we have polluted our air, people have felt effects on their health. But quantifying those impacts poses a challenge for scientists. For researchers wanting to understand pollution's health impacts—and to design solutions for people like Guadalupe—what better place to go than Mexico City, long home to some of the world's worst air pollution?

A key first step is to determine what's in the air. One researcher who led the way is Mario Molina, a chemist who won a Nobel Prize for helping to discover the cause of stratospheric ozone depletion (p. 301). Molina organized hundreds of scientists to sample the air in Mexico City, his hometown. The nearly 200 research papers spawned by these efforts clarified many aspects of the city's air quality. One study used machines to identify and record individual particles in real time. It found that metal-rich particulates from trash incinerators were peaking in the morning, whereas smoke from fires outside the city blew in during the afternoon. Other researchers discovered that VOCs control the amount of tropospheric ozone formed in smog. City officials responded by targeting VOC emissions for reduction, while also discouraging automobile traffic (**FIGURE 1**).

Few people understand Mexico City's air pollution better than Armando Retama, the city's director of atmospheric monitoring. But he may grasp its impacts best when he leaves town. "I can breathe better. I'm not all dry. My eyes aren't irritated. My skin doesn't crack," he says. "We have chronic symptoms that we aren't aware of."

Most known health impacts of urban pollution affect the respiratory system. At high altitudes like Mexico City's, the "thin air" forces people to breathe deeply to obtain enough oxygen. This means they pull more air pollutants into their lungs than people at lower elevations. Many studies confirm that Mexico City residents show poorer lung function than people from less-polluted areas and that respiratory problems and emergency room visits become more numerous when pollution is severe.

In 2007 a research team led by Isabelle Romieu of Mexico's National Institute of Public Health examined the effects of growing up amid polluted air. Her team measured lung function in 3170 eight-year-old children from 39 Mexico City schools across 3 years and correlated this with their exposure to tropospheric ozone, nitrogen dioxide, and particulate matter. The team found that children from more-polluted neighborhoods lagged behind those from cleaner ones in the ability to inhale and exhale deeply—indicating smaller, weaker lungs. Romieu and her colleagues also showed that the city's pollution worsens asthma in children. Analyzing data from 200 asthmatic and healthy children, her team found that children in areas with more traffic and pollutants coughed, wheezed, and used medication more often.

Another Mexican researcher, Lilian Calderón-Garcidueñas—now at the University of Montana—compared chest X-ray films and medical records of Mexico City children with those of similar children from less-polluted locations. Her team found hyperinflation and other problems with the lungs of Mexico City youth. Mexico City children reported many respiratory problems, whereas rural children did not (**FIGURE 2**).

Air pollution also harms the heart and the cardiovascular system, affecting heart rate, blood pressure, blood clotting, blood vessels, and atherosclerosis. Epidemiological studies

FIGURE 1 During a resurgence of smog in Mexico City in 2016, commuters wore face masks and took advantage of free mass transit once authorities restricted car traffic.

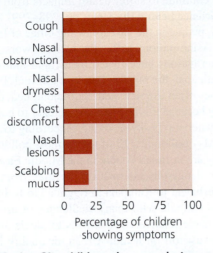

FIGURE 2 Mexico City children show respiratory symptoms from air pollution. These data are from 174 Mexico City children. Of 27 similar children from less-polluted areas outside Mexico City, none showed any of these conditions. *Data from Calderón-Garcidueñas, L., et al., 2003. Respiratory damage in children exposed to urban pollution. Pediatric Pulmonology 36: 148–161.*

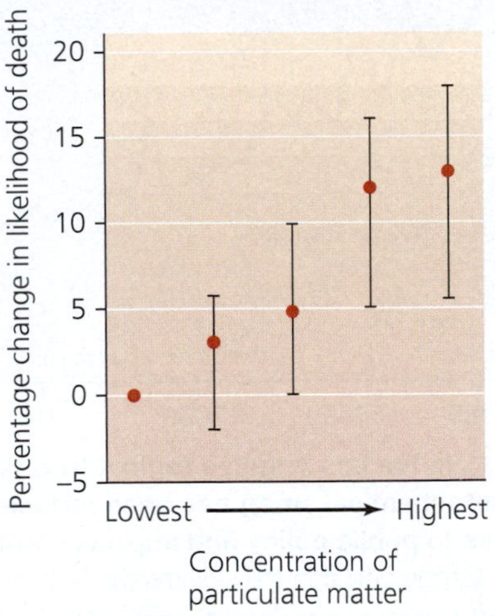

FIGURE 3 Rates of death increase in Mexico City with exposure to air pollution. *Data from Borja-Aburto, V., et al., 1997. Ozone, suspended particulates, and daily mortality in Mexico City. Am. J. Epidemiology 145: 258–268.*

(p. 223) show that pollution correlates with emergency room admissions for heart attacks, chest pain, and heart failure, as well as death from heart-related causes. This is because tiny particulates can work their way into the bloodstream, causing the heart to reduce blood flow or go out of rhythm. The heart mounts an inflammatory response against pollutant particles laden with dead bacteria in the blood, but if pollution is persistent, the inflammation becomes chronic and stresses the heart. Even young people are at risk. One Mexican research team analyzed the hearts of 21 Mexico City residents who had died at an early age, and found that pollution exacts a toll before age 18.

All these impacts of air pollution on the lungs and heart can lead to higher rates of death. Studies by one research team in Mexico City compared death certificate records against air pollution measurements. The team found that death rates rose immediately after severe pollution episodes, especially in response to particulate matter (**FIGURE 3**).

Today researchers are learning that air pollution also affects our brains. Calderón-Garcidueñas was one of the first to recognize this. She noticed that older dogs in polluted Mexico City neighborhoods often seemed lethargic, disoriented, and senile. Deciding to test her observations scientifically, she examined the brains of such dogs after they died and compared them to brains of dogs from less-polluted areas. The brains of dogs from polluted parts of the city showed deposits of the protein amyloid ß, the "plaques" that signal Alzheimer's disease.

Subsequent research of hers and others has indicated that pollution also appears to damage children's brain tissue

in ways similar to Alzheimer's disease. In one study, Calderón-Garcidueñas used brain scans and found that 56% of Mexico City youth had lesions on the prefrontal cortex, whereas fewer than 8% did in a region with clean air. In another study, her team compared 20 children from Mexico City with 10 similar children from a Mexican city with clean air, measuring their cognitive skills and scanning their brains with magnetic resonance imagery (MRI). The Mexico City children performed more poorly on most cognitive tests of reasoning, knowledge, and memory. The differences in cognition were consistent with differences in volume of white matter in key portions of the brain, as revealed by the MRIs.

Such findings led researchers at the University of Southern California to run rigorously controlled experimental tests with lab mice. These USC scientists are collecting polluted air alongside Los Angeles freeways and pumping the pollutants into the air that lab mice breathe. They are then comparing the brains of these mice after death to those of mice that breathed clean air. Results thus far are showing that tiny $PM_{2.5}$ pollutants do indeed pass through tissue layers and harm the brains of the mice.

Today a number of scientists are conducting long-term epidemiological research to better assess the impacts of air pollution on the human brain. A 2016 review of 18 such studies from six nations found that all but one showed some correlation between air pollution and dementia. Continued research could reveal promising new avenues to fight dementia and Alzheimer's disease.

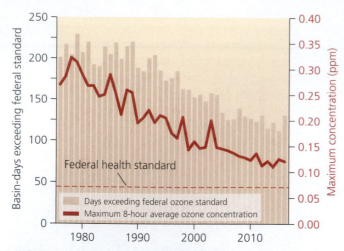

FIGURE 13.13 In the Los Angeles region, tropospheric ozone in photochemical smog has been reduced since the 1970s, thanks to public policy and improved automotive technology. Ozone pollution still violates the federal health standard, however. *Data from South Coast Air Quality Management District.*

DATA • By roughly what percentage has Los Angeles reduced its ozone pollution since the late 1970s? • Calculate changes in each data set shown. • Do the two data sets show similar patterns?

Go to **Interpreting Graphs & Data** on *Mastering* Environmental Science

require owners to repair cars that pollute excessively, have cut vehicle emissions by 30%.

California's demands also helped lead the auto industry to develop less-polluting cars. A study by the non-profit group Environment California concluded that a new car today generates just 1% of the smog-forming emissions of a 1960s-era car. For this reason, the air is cleaner, even with more vehicles on the road. In Los Angeles, VOC pollution has declined by 98% since 1960, even though the city's drivers now burn 2.7 times more gasoline. L.A.'s peak smog levels have decreased substantially since 1980 (**FIGURE 13.13**).

Despite its progress, Los Angeles still suffers the worst tropospheric ozone pollution of any U.S. metropolitan area, according to annual rankings by the American Lung Association. L.A. residents breathe air exceeding California's health standard for ozone on more than 90 days per year. One recent study calculated that air pollution in the L.A. basin and the nearby San Joaquin Valley each year causes nearly 3900 premature deaths and costs society $28 billion (due to hospital admissions, lost workdays, etc.).

Air pollution remains severe in industrializing nations

Although the United States and other industrialized nations have improved their air quality, outdoor air pollution is growing worse in many industrializing countries. In these societies, more and more people are driving automobiles while proliferating factories and power plants are emitting more pollutants. At the same time, many people continue to burn traditional

sources of fuel, such as wood, charcoal, and coal, for cooking and home heating. Mexico embodies these trends, and despite the progress in its capital, residents of many Mexican cities and towns continue to suffer health impacts from polluted air (see **The Science behind the Story,** pp. 298–299).

Today, people in the vast sprawling cities of India, China, and other Asian nations suffer the world's worst air quality. Many Asian nations have fueled their rapid industrial development with coal, the most-polluting fossil fuel. Power plants and factories often use outdated, inefficient, heavily polluting technology because it is cheaper and quicker to build. In addition, car ownership is skyrocketing. According to 2016 WHO data for $PM_{2.5}$ pollutants, 12 of the 25 most polluted cities on Earth today are in India. People in the Middle East and Africa also suffer dangerous air quality, in cities from Riyadh to Cairo to Peshawar to Kampala to Baghdad to Kabul. Because these cities struggle with many challenges of urban poverty, they have made relatively few efforts to tackle air pollution, despite its vast toll on health. Urban air quality is nearly as bad in eastern European nations such as Poland and Bulgaria, where old Soviet-era factories still pollute the air and where many people burn coal for home heating.

In China's capital of Beijing in the winter of 2013, pollution became so severe that airplane flights were canceled and people wore face masks to breathe (**FIGURE 13.14**). Thousands of people suffered ill health as pollution soared 30 times past the WHO's safe limits. Each winter since then, Beijing's "airmageddon" has returned.

Across China, the health impacts of outdoor air pollution are enormous. Breathing the air in some Chinese cities is like smoking two packs of cigarettes a day. Recent research has blamed outdoor air pollution for 1.2 million premature deaths in China each year, and one study found that residents of polluted northern China die on average five years earlier than residents of southern China, where the air is cleaner. Moreover, prevailing westerly winds carry some of China's pollution across the Pacific Ocean to North America! A 2014 study calculated that

FIGURE 13.14 At Tiananmen Square in Beijing, China, children wear face masks during an "airpocalypse" gripping the city.

Chinese pollution reaching the U.S. West Coast adds at least one extra smoggy day per year to Los Angeles's total.

In 2015, China's citizens were transfixed by an online video that went viral (and was soon banned by the Chinese government), titled *Under the Dome*. This powerful 104-minute documentary, produced and narrated by a Chinese investigative reporter, Chai Jing, confirmed for China's people what they already knew: They spend their daily lives trapped under a dome of dangerously polluted air.

China's government is now striving to reduce pollution. It has closed down some heavily polluting factories and mines, phased out some subsidies for polluting industries, and installed pollution controls in power plants. It subsidizes efficient electric heaters for homes to replace dirty, inefficient coal stoves. It has mandated cleaner gasoline and diesel and has raised standards for fuel efficiency and emissions for cars beyond what the United States requires. In Beijing, mass transit is being expanded and many buses run on natural gas. China is also aggressively developing wind, solar, and nuclear power to substitute for coal-fired power.

Should we regulate greenhouse gases as air pollutants?

Because humanity continues to release vast quantities of carbon dioxide and other greenhouse gases that warm the lower atmosphere and drive global climate change (Chapter 14), this is arguably today's biggest air pollution problem. Industry and utilities generate many of these emissions, but all of us contribute by living carbon-intensive lifestyles. Each year the average U.S. vehicle driver releases close to 6 metric tons of carbon dioxide, 275 kg (605 lb) of methane, and 19 kg (41 lb) of nitrous oxide, all of them greenhouse gases that drive climate change.

In 2007 the U.S. Supreme Court ruled that the EPA has legal authority under the Clean Air Act to regulate carbon dioxide and other greenhouse gases as air pollutants. President Barack Obama urged Congress to address greenhouse gas emissions through bipartisan legislation. When Congress failed to do so, Obama instructed the EPA to develop regulations for these emissions. In 2011, the EPA introduced moderate carbon emission standards for cars and light trucks, and in 2012 it announced that it would begin phasing in limits on carbon emissions for new coal-fired power plants and cement factories (but not existing ones).

The coal-mining and petrochemical industries objected, and several states joined them in suing the EPA, but a court of appeals unanimously upheld the EPA's regulations. The automotive industry supported these regulations. U.S. automakers had begun investing in fuel-efficient vehicles, and preferred one set of federal emissions standards to avoid having to worry about meeting many different state standards. The public also voiced strong support; 2.1 million Americans sent comments to the EPA in favor of its actions—a record number of public comments for any federal regulation.

In 2015, the EPA launched a regulatory plan for existing power plants, the Clean Power Plan. Under the plan, states were allowed to choose how to reduce their plants' emissions—by upgrading technology, switching from coal to natural gas, enhancing efficiency, promoting renewable energy, or through carbon taxes or cap-and-trade programs. The plan aimed to cut CO_2 emissions from power plants by 32% below 2005 levels by the year 2030. The EPA estimated that SO_2 and NO_X would be reduced by 20%, that cleaner air would save 3600 lives, and that by 2030 the plan would bring public health and climate benefits worth $54 billion each year.

A number of states and industries challenged the plan in court, and in 2016 the Supreme Court issued a stay of the plan in a controversial 5 to 4 ruling, preventing the EPA from enforcing the plan until the lawsuits were resolved. Subsequently, the rise of Donald Trump to the presidency, his appointment of longtime EPA foe Scott Pruitt as the agency's head, and the opposition of many Republicans in Congress to EPA policy all suggest the Clean Power Plan will likely be dismantled. A Trump executive order in spring 2017 initiated this process.

The EPA will no doubt continue to face formidable opposition from emitting industries and from policymakers who fear that regulations will hamper economic growth. Yet if we were able to reduce emissions of other pollutants sharply since 1970 while advancing our economy, we can hope to achieve similar results in reducing greenhouse gas emissions. Indeed, although U.S. carbon dioxide emissions have risen significantly since 1970, they fell by 14% from 2007 to 2016 even as the economy grew. This decrease in emissions resulted from a shift from coal to cleaner-burning natural gas, and from improved fuel efficiency in automobiles and other technologies.

Ozone Depletion and Recovery

Although ozone in the troposphere is a pollutant in photochemical smog, ozone in the stratosphere (p. 289) protects life on Earth by absorbing the sun's ultraviolet (UV) radiation, which can damage tissues and DNA. When scientists discovered that our planet's stratospheric ozone was being depleted, they realized this posed a major threat to human health and the environment. Years of research by hundreds of scientists revealed that certain airborne chemicals destroy ozone and that most of these **ozone-depleting substances** were human-made. Our subsequent campaign to halt degradation of the stratospheric ozone layer stands as one of society's most successful efforts to address a major environmental problem.

Synthetic chemicals deplete ozone

Researchers identifying ozone-depleting substances pinpointed primarily **halocarbons**—human-made compounds derived from simple hydrocarbons (p. 32) in which hydrogen atoms are replaced by halogen atoms such as chlorine, bromine, or fluorine. In the 1970s, industry was producing more than 1 million tons per year of one type of halocarbon, **chlorofluorocarbons (CFCs).** CFCs were useful as refrigerants,

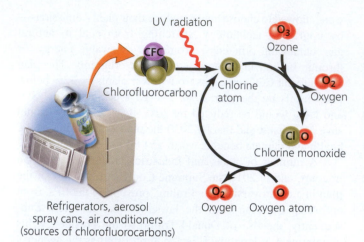

UV radiation

CFC
Chlorofluorocarbon

Refrigerators, aerosol
spray cans, air conditioners
(sources of chlorofluorocarbons)

O_3
Ozone

Cl
Chlorine
atom

O_2
Oxygen

Cl O
Chlorine monoxide

O_2
Oxygen

O
Oxygen atom

FIGURE 13.15 CFCs destroy ozone in a multistep process, repeated many times. A chlorine atom released from a CFC molecule in the presence of UV radiation reacts with an ozone molecule, forming one molecule of oxygen gas and one chlorine monoxide (ClO) molecule. The oxygen atom of the ClO molecule then binds with a stray oxygen atom to form oxygen gas, leaving the chlorine atom to begin the destructive cycle anew.

as fire extinguishers, as propellants for aerosol spray cans, as cleaners for electronics, and for making polystyrene foam.

Unfortunately, CFCs can linger in the stratosphere for a century or more. There, intense UV radiation from the sun breaks bonds in CFC molecules, releasing their chlorine atoms. In a two-step chemical reaction (**FIGURE 13.15**), each newly freed chlorine atom can split an ozone molecule and then ready itself to split more. During its long residence time in the stratosphere, each free chlorine atom can catalyze the destruction of as many as 100,000 ozone molecules!

The ozone hole appears each year

In 1985, researchers shocked the world when they announced that stratospheric ozone levels over Antarctica in the southern springtime had declined by nearly half during just the previous decade, leaving a thinned ozone concentration that was soon named the **ozone hole** (**FIGURE 13.16**). During each Southern Hemisphere spring since then, ozone concentrations over this immense region have dipped to roughly half their historic levels.

Extensive scientific detective work has revealed why seasonal ozone depletion is focused over Antarctica (and to a lesser extent, the Arctic). During the dark and frigid Antarctic winter (June to August), icy clouds form in the stratosphere containing condensed nitric acid, which splits chlorine atoms off from compounds such as CFCs. The freed chlorine atoms accumulate in the clouds, trapped over Antarctica by circular wind currents. In the Antarctic spring (starting in September), sunshine returns and dissipates the clouds. This releases the chlorine atoms, which begin destroying ozone. The solar radiation also catalyzes chemical reactions, speeding up ozone depletion as temperatures warm. The ozone hole lingers over Antarctica until December, when warmth weakens the circular currents, allowing ozone-depleted air to diffuse away and ozone-rich air from elsewhere to stream in. The ozone hole vanishes until the following spring.

By the time scientists had worked this out, plummeting ozone levels had become a serious international concern. Ozone depletion was occurring globally, year-round—not just in Antarctica. Scientists worried that intensified UV exposure at Earth's surface would promote skin cancer, damage crops, and kill off ocean phytoplankton, the base of the marine food chain.

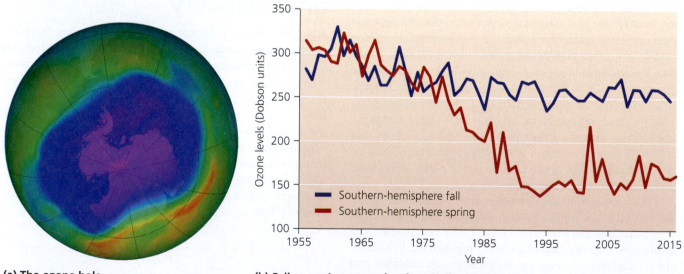

(a) The ozone hole

(b) Fall vs. spring ozone levels at Halley, Antarctica

FIGURE 13.16 The "ozone hole" is a vast area of thinned ozone density in the stratosphere over the Antarctic region. It has reappeared seasonally each September in recent decades. Colorized satellite imagery of Earth's Southern Hemisphere from September 24, 2006, **(a)** shows the ozone hole (**purple/blue**) at its maximal recorded extent to date. Data from Halley, Antarctica, **(b)** show a drop in springtime stratospheric ozone concentrations from the 1950s to 1990. Once ozone-depleting substances began to be phased out, ozone concentrations stopped declining. *Data from (a) NASA and (b) British Antarctic Survey.*

We addressed ozone depletion with the Montreal Protocol

Policymakers responded to the scientific concerns, and international efforts to restrict production of CFCs bore fruit in 1987 with the **Montreal Protocol.** In this treaty, the world's nations agreed to cut CFC production in half by 1998.

FAQ

Is the ozone hole related to global warming?

This is a common misconception held by many people. In reality, stratospheric ozone depletion and global warming are completely different issues. Ozone depletion allows excess ultraviolet radiation from the sun to penetrate the atmosphere, but this does not significantly warm or cool the atmosphere. Conversely, global warming does not appreciably affect ozone loss. However, by coincidence many ozone-depleting substances banned by the Montreal Protocol also happen to be greenhouse gases that warm the atmosphere. Thus, although the Montreal Protocol was designed to combat ozone depletion, it is also helping us slow down climate change.

Follow-up agreements deepened the cuts, advanced timetables for compliance, and added nearly 100 further ozone-depleting substances—most of which have now been phased out—with industry making the shift to alternative chemicals. As a result, we have evidently halted the advance of ozone depletion and stopped the Antarctic ozone hole from growing larger (**FIGURE 13.17**). This is a remarkable success that all humanity can celebrate.

Earth's ozone layer is not expected to recover completely until after 2060 (**FIGURE 13.18**). Much of the 5.5 million tons of CFCs emitted into the troposphere has not yet diffused into the stratosphere, so concentrations may not peak there until 2020. Because of this time lag and the long residence times of many halocarbons, many years will pass before our policies have the desired result—one reason that scientists often argue for proactive policy guided by the precautionary principle (pp. 162, 227).

Because of its success in addressing ozone depletion, the Montreal Protocol is widely viewed as an inspiring model for international cooperation on other global challenges, from biodiversity loss (p. 183) to persistent organic pollutants (p. 228) to climate change (p. 335).

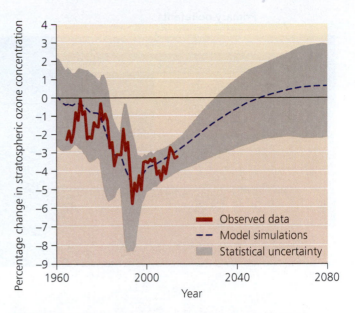

FIGURE 13.18 Globally, stratospheric ozone is starting to recover. Data averaged from multiple satellite and ground-based sources (**red line**) show a global decrease in observed ozone concentrations, followed by a gradual increase. Simulations from many models (**blue dashed line** shows mean; **gray shading** shows statistical uncertainty) indicate that recovery should be complete later this century. *Adapted from World Meteorological Organization, 2014. Scientific assessment of ozone depletion: 2014. Geneva, Switzerland: WMO, Global Ozone Research and Monitoring Project—Report #55.*

Addressing Acid Deposition

Just as stratospheric ozone depletion crosses political boundaries, so does **acid deposition,** the deposition of acidic (p. 32) or acid-forming pollutants from the atmosphere onto Earth's surface. As with ozone depletion, we are enjoying some success in addressing this challenge.

Fossil fuel combustion spreads acidic pollutants far and wide

Acid deposition often takes place by precipitation (commonly referred to as **acid rain**) but also may occur by fog, gases, or the deposition of dry particles. Acid deposition is one type of **atmospheric deposition,** which refers broadly to the wet or dry deposition of a variety of pollutants, including mercury, nitrates, organochlorines, and others, from the atmosphere onto Earth's surface.

Acid deposition originates primarily with the emission of sulfur dioxide and nitrogen oxides, largely through fossil fuel combustion by automobiles, electric utilities, and industrial facilities. Once airborne, these pollutants react with water, oxygen, and oxidants to produce compounds of low pH (p. 32), primarily sulfuric acid and nitric acid. Suspended in the troposphere, these acids may travel days or weeks for hundreds of kilometers (**FIGURE 13.19**).

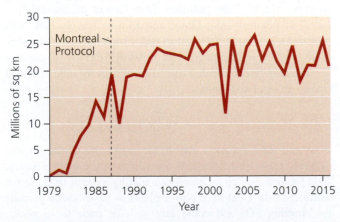

FIGURE 13.17 Phase-outs of ozone-depleting substances since 1987 have halted the growth of the Antarctic ozone hole. *Data from NASA, reflecting averages from 7 Sept. to 13 Oct. each year.*

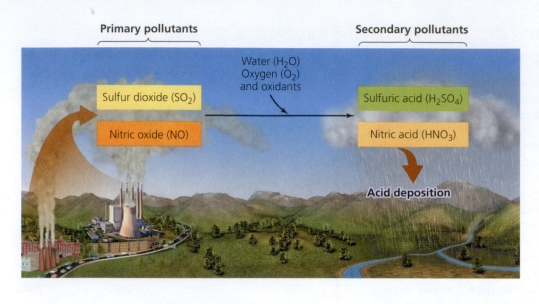

Water (H_2O)
Oxygen (O_2)
and oxidants

Sulfur dioxide (SO_2)

Nitric oxide (NO)

Secondary pollutants

Sulfuric acid (H_2SO_4)

Nitric acid (HNO_3)

Acid deposition

FIGURE 13.19 Acid deposition has impacts far downwind from where pollutants are released. Sulfur dioxide and nitric oxide emitted by industries, utilities, and vehicles react in the atmosphere to form sulfuric acid and nitric acid. These acidic compounds descend to Earth's surface in rain, snow, fog, and dry deposition.

Acid deposition has many impacts

Acid deposition has wide-ranging detrimental effects on ecosystems (**TABLE 13.1**). Acids leach nutrients such as calcium, magnesium, and potassium ions out of the topsoil, altering soil chemistry and harming plants and soil organisms. This occurs because hydrogen ions from acid deposition take the place of calcium, magnesium, and potassium ions in soil compounds, and these valuable nutrients leach into the subsoil, where they become inaccessible to plant roots.

Acid deposition also "mobilizes" toxic metal ions such as aluminum, zinc, mercury, and copper by chemically converting them from insoluble forms to soluble forms. Elevated soil concentrations of metal ions such as aluminum damage the root tissue of plants, hindering their uptake of water and nutrients. In some areas, acid fog with a pH of 2.3 (equivalent to vinegar, and over 1000 times more acidic than normal rainwater) has enveloped forests, killing trees. Animals are affected by acid deposition, too; populations of snails and other invertebrates typically decline, and this reduces the food supply for birds.

When acidic water runs off from land, it affects streams, rivers, and lakes. Thousands of lakes in Canada, Europe, the United States, and elsewhere have lost their fish because acid precipitation leaches aluminum ions out of soil and rock and into waterways. These ions damage the gills of fish and disrupt their salt balance, water balance, breathing, and circulation.

Besides altering ecosystems, acid deposition damages crops, erodes stone buildings, corrodes vehicles, and erases the writing from tombstones. Ancient cathedrals in Europe, sacred temples in Asia, and revered monuments in Washington, D.C., are experiencing unrecoverable damage as their features dissolve away (**FIGURE 13.20**).

Because the pollutants leading to acid deposition can travel long distances, their effects may be felt far from their sources. Much of the pollution from power plants and factories in Pennsylvania, Ohio, and Illinois travels east with prevailing winds and falls out in states such as New York, Vermont, and New Hampshire. As a result, regions of greatest acidification tend to be downwind from heavily industrialized source areas of pollution.

TABLE 13.1 Ecological Impacts of Acid Deposition

ACID DEPOSITION IN NORTHEASTERN U.S. FORESTS HAS ...

- accelerated leaching of base cations (ions such as Ca^{2+}, Mg^{2+}, NA^+, and K^+, which counteract acid deposition) from soil

- allowed sulfur and nitrogen to accumulate in soil, where excess N can encourage weeds

- increased dissolved inorganic aluminum in soil, hindering plant uptake of water and nutrients

- leached calcium from needles of red spruce, causing trees to die from wintertime freezing

- increased mortality of sugar maples due to leaching of base cations from soil and leaves

- acidified hundreds of lakes and diminished their capacity to neutralize further acids

- elevated aluminum levels in surface waters

- reduced species diversity and abundance of aquatic life, affecting entire food webs

Adapted from Driscoll, C.T., et al., 2001. Acid rain revisited. Hubbard Brook Research Foundation. © 2001 C. T. Driscoll. Used with permission.

We are addressing acid deposition

The Acid Rain Program established under the Clean Air Act of 1990 has helped fight acid deposition in the United States. This program set up an emissions trading system (p. 113) for sulfur dioxide. Coal-fired power plants were allocated permits for emitting SO_2 and could buy, sell, or trade these allowances. Each year the overall amounts of allowed pollution were decreased. (See p. 333 for further explanation of how such a system works.) The economic incentives created by this

(a) Before acid rain damage

(b) After acid rain damage

FIGURE 13.20 Acid deposition corrodes statues and buildings. Shown is an Egyptian obelisk known as Cleopatra's Needle, in Central Park, New York City, **(a)** before and **(b)** after significant acid deposition.

cap-and-trade program encouraged polluters to switch to low-sulfur coal, invest in technologies such as scrubbers (p. 293), and devise other ways to become cleaner and more efficient. During the course of the cap-and-trade program, SO_2 emissions across the United States fell by 67% (**FIGURE 13.21**). As a result, average sulfate loads in precipitation across the eastern United States were 51% lower in 2008–2010 than in 1989–1991, and are still lower today.

The Acid Rain Program also required power plants to reduce nitrogen oxide emissions, with the EPA allowing plants flexibility in how they did so. Emissions of NO_X fell significantly as a result, and wet nitrogen deposition declined as well. Thanks to the declines in SO_2 and NO_X, air and water quality improved throughout the eastern United States (**FIGURE 13.22**). This market-based program spawned similar

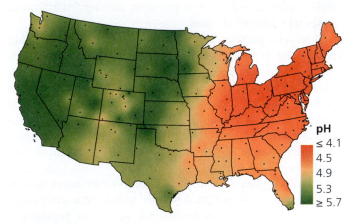

(a) pH of precipitation in 1990

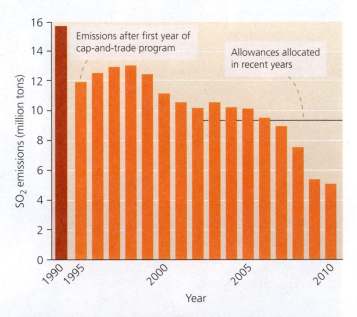

FIGURE 13.21 Sulfur dioxide emissions fell 67% in the wake of an emissions trading system. By 2010, emissions from U.S. power plants participating in this EPA program mandated by the 1990 Clean Air Act had dropped well below the amount allocated in permits (**black line**). *Data from U.S. EPA.*

(b) pH of precipitation in 2015

FIGURE 13.22 Precipitation has become less acidic as air quality has improved under the Clean Air Act. Average pH values for precipitation rose between **(a)** 1990 and **(b)** 2015. Precipitation remains most acidic in regions near and downwind from areas of heavy industry. *Data from the National Atmospheric Deposition Program.*

DATA • In the area where you live, how did the pH of precipitation change between 1990 and 2015? • Has precipitation become more acidic or less acidic?

Go to **Interpreting Graphs & Data** on **Mastering** Environmental Science

cap-and-trade programs for other pollutants, including greenhouse gases (p. 333). The Los Angeles region adopted its own cap-and-trade program in 1994. The RECLAIM (Regional Clean Air Incentives Market) program has helped the L.A. basin reduce emissions of sulfur oxides and nitrogen oxides by more than 70%.

Many have attributed the success in reducing acid deposition nationwide to the Acid Rain Program, and the EPA has calculated that the program's economic benefits (in health care expenses avoided, for instance) outweighed its costs by 40 to 1. However, some experts maintain that pollution declined because cleaner fuels became less expensive and because simultaneous conventional regulation mandated emissions cuts. Indeed, during this time period European nations using command-and-control regulation (p. 111) instead of emissions trading reduced their SO_2 emissions by even more than the United States did. In 2011, emissions trading ended once the EPA issued its Cross-State Air Pollution Rule, which aimed to limit pollution drifting from upwind states into downwind states.

As with recovery of the ozone layer, there is a time lag before the positive consequences of emissions cuts kick in, so it will take time for acidified ecosystems to recover. Meanwhile, in industrializing nations, the problem is becoming worse. Today China emits the most sulfur dioxide of any nation, as a result of coal combustion in power plants and factories that lack effective pollution control equipment. Not surprisingly, China has the world's worst acid rain problem.

Overall, data on acid deposition show advances in controlling outdoor air pollution but also indicate that more could be done. The same can be said for indoor air pollution, a source of human health threats that is less familiar to most of us, but statistically more dangerous.

Indoor Air Quality

Indoor air generally contains higher concentrations of pollutants than does outdoor air. As a result, the health impacts from **indoor air pollution** in workplaces, schools, and homes outweigh those from outdoor air pollution. The World Health Organization (WHO) attributes nearly 3.5 million premature deaths each year to indoor air pollution (compared with 3.3 million for outdoor air pollution). Indoor air pollution takes nearly 10,000 lives each day.

If this seems surprising, consider that the average American spends at least 90% of his or her time indoors. Then consider the dizzying array of consumer products in our homes and offices. Many of these products are made of synthetic materials, and novel synthetic substances are not comprehensively tested for health effects before being brought to market (Chapter 10). Furniture, carpeting, cleaning fluids, insecticides, and plastics all exude volatile chemicals into the air.

Ironically, some well-meaning attempts to enhance the energy efficiency of buildings have ended up worsening indoor air quality. To reduce heat loss, building managers often seal off ventilation, while designers often construct new buildings with limited ventilation and with windows that do not open. These steps save energy, but they also trap stable, unmixed air—and pollutants—inside. There is good news, however. In both developing and developed nations, we have feasible ways to address the primary causes of indoor air pollution.

Risks differ in developing and developed nations

Indoor air pollution exerts most impact in the developing world, where poverty forces millions of people to burn wood, charcoal, animal dung, or crop waste inside their homes for cooking and heating, with little or no ventilation (**FIGURE 13.23**). As a result, people inhale soot, carbon monoxide, and other pollutants on a regular basis. This increases the risks of premature death by pneumonia, bronchitis, and lung cancer, as well as allergies, sinus infections, cataracts, asthma, emphysema, and heart disease.

In industrialized nations, the primary indoor air health risks are cigarette smoke and radon. Smoking cigarettes irritates the eyes, nose, and throat; worsens asthma and other respiratory ailments; and greatly increases the risk of lung cancer and heart disease. Inhaling secondhand smoke from a nearby smoker causes the same problems. Tobacco smoke is a brew of more than 4000 chemical compounds, over 250 of which are known or suspected to be toxic or carcinogenic. Although smoking has become less popular in developed nations in recent years, it is still estimated in the United States alone to cause 160,000 lung cancer deaths per year.

FIGURE 13.23 In the developing world, many people build fires indoors for cooking and heating, as in this Maasai home in Kenya. Indoor fires expose people to severe pollution from particulate matter and carbon monoxide.

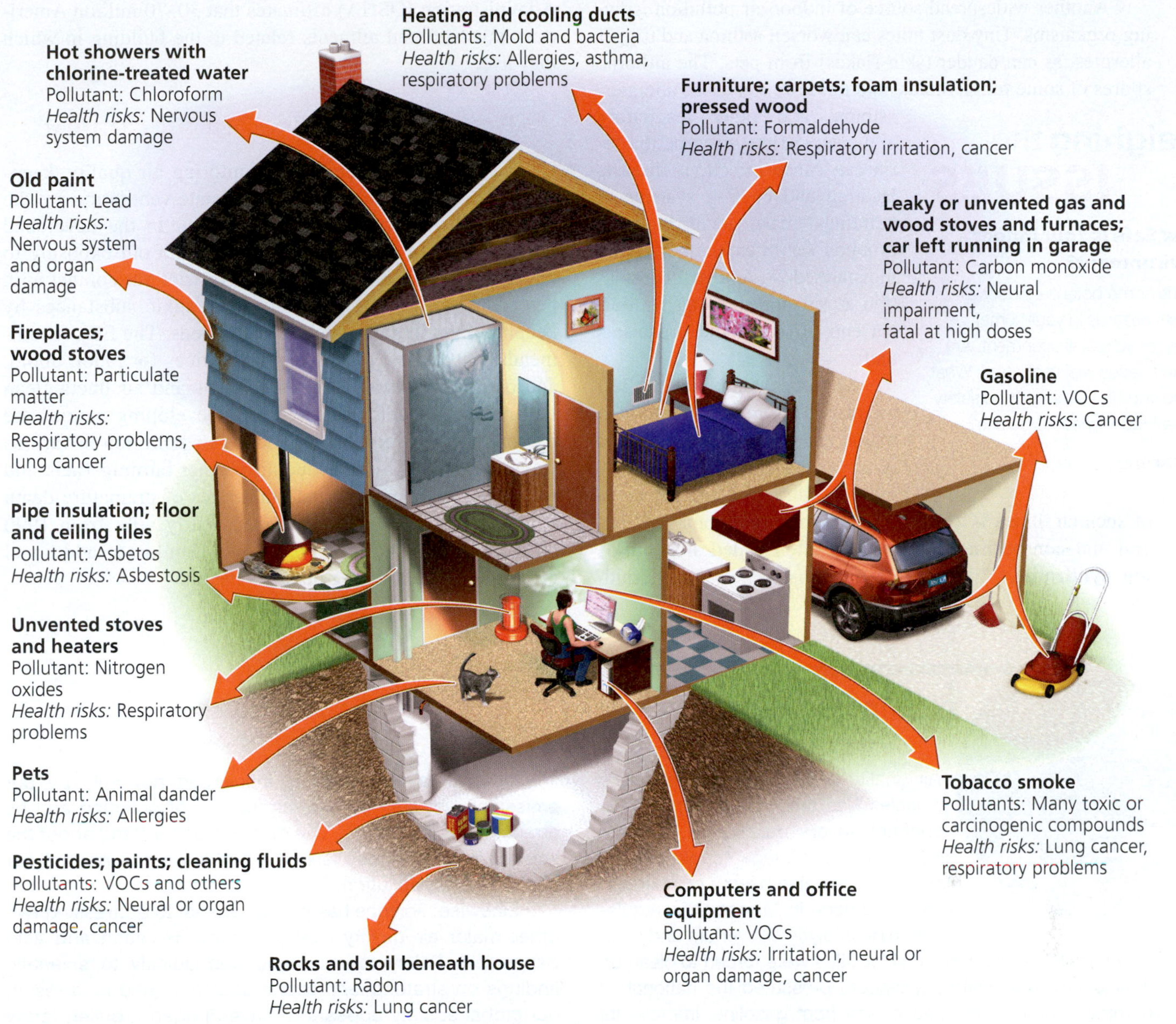

Hot showers with chlorine-treated water
Pollutant: Chloroform
Health risks: Nervous system damage

Old paint
Pollutant: Lead
Health risks: Nervous system and organ damage

Fireplaces; wood stoves
Pollutant: Particulate matter
Health risks: Respiratory problems, lung cancer

Pipe insulation; floor and ceiling tiles
Pollutant: Asbetos
Health risks: Asbestosis

Unvented stoves and heaters
Pollutant: Nitrogen oxides
Health risks: Respiratory problems

Pets
Pollutant: Animal dander
Health risks: Allergies

Pesticides; paints; cleaning fluids
Pollutants: VOCs and others
Health risks: Neural or organ damage, cancer

Rocks and soil beneath house
Pollutant: Radon
Health risks: Lung cancer

Heating and cooling ducts
Pollutants: Mold and bacteria
Health risks: Allergies, asthma, respiratory problems

Furniture; carpets; foam insulation; pressed wood
Pollutant: Formaldehyde
Health risks: Respiratory irritation, cancer

Leaky or unvented gas and wood stoves and furnaces; car left running in garage
Pollutant: Carbon monoxide
Health risks: Neural impairment, fatal at high doses

Gasoline
Pollutant: VOCs
Health risks: Cancer

Tobacco smoke
Pollutants: Many toxic or carcinogenic compounds
Health risks: Lung cancer, respiratory problems

Computers and office equipment
Pollutant: VOCs
Health risks: Irritation, neural or organ damage, cancer

FIGURE 13.24 The typical home contains many sources of indoor air pollution. Shown are common sources, the major pollutants they emit, and some of the health risks they pose.

Radon is the second-leading cause of lung cancer in the developed world, responsible for an estimated 21,000 deaths per year in the United States and 15% of lung cancer cases worldwide. Radon (p. 215) is a colorless, odorless, radioactive gas resulting from the natural decay of uranium in soil, rock, or water. It seeps up from the ground and can penetrate buildings. The only way to tell if radon is entering a building is to sample air with a test kit. The EPA estimates that 6% of U.S. homes exceed its safety standard for radon. More than a million homes have undergone mitigation, and today new homes are being built with radon-resistant features.

Many substances pollute indoor air

In our daily lives at home, we are exposed to many indoor air pollutants (**FIGURE 13.24**). The most diverse are volatile organic compounds (p. 292), airborne carbon-containing compounds released by plastics, oils, perfumes, paints, cleaning fluids, adhesives, and pesticides. VOCs evaporate from furnishings, building materials, carpets, laser printers, and fax machines. Some products, such as chemically treated furniture, release many VOCs when new and progressively fewer as they age. Others, such as photocopying machines, emit VOCs each time they are used. Formaldehyde—a VOC used in pressed wood, insulation, and other products—irritates mucous membranes, induces skin allergies, and causes other ailments. The "new car smell" that fills the interiors of new automobiles comes from a complex mix of dozens of VOCs as they outgas from the newly manufactured plastic, metal, and leather components of the car. Some scientific studies warn of health risks from this brew and recommend that you keep a new car well ventilated.

Another widespread source of indoor air pollution is living organisms. Tiny dust mites can worsen asthma and trigger allergies, as can dander (skin flakes) from pets. The airborne spores of some fungi, molds, and mildews can cause allergies, asthma, and other respiratory ailments. Some airborne bacteria can cause infectious disease. Heating and cooling systems in buildings make ideal breeding grounds for microbes, providing moisture, dust, and foam insulation as substrates, along with air currents to carry the organisms aloft.

Microbes that induce allergic responses are thought to be a major cause of *building-related illness*, a sickness produced by indoor pollution. When the cause of such an illness is a mystery, and when symptoms are general and nonspecific, the illness is often called **sick building syndrome.** The U.S. Occupational Safety and Health Administration (OSHA) estimates that 30–70 million Americans have suffered ailments related to the building in which they live.

weighing the
ISSUES

How Safe Is Your Indoor Environment?

Name some potential indoor air quality hazards in your home, work, or school environment. Are these spaces well ventilated? What could you do to improve the safety of the indoor spaces you use?

We can enhance indoor air quality

Using low-toxicity materials, monitoring air quality, keeping rooms clean, and providing adequate ventilation are the keys to alleviating indoor air pollution. In the developed world, we can avoid cigarette smoke, limit our exposure to new plastics and treated wood, and restrict our contact with pesticides, cleaning fluids, and other toxic substances by keeping them in garages or outdoor sheds. The EPA recommends that we test our homes and offices for radon, mold, and carbon monoxide. Keeping rooms and air ducts clean will reduce irritants and allergens. Developing nations are making progress in reducing indoor air pollution in various ways, especially by introducing cleaner-burning fuels and stoves. Researchers calculate that rates of premature death from indoor air pollution dropped nearly 40% from 1990 to 2010. With continued efforts, we should see additional progress in safeguarding people's health.

closing THE LOOP

Air quality is vital for our health. Los Angeles was among the first cities to confront severe outdoor air pollution and take major steps to alleviate it. Mexico City has also been a pioneer and a model for others. In Mexico City, regulations now require catalytic converters and vehicle emissions tests. Industrial facilities were forced to clean up their operations, while city leaders pressured the national oil company, Pemex, to remove lead from gasoline, improve its refineries, and sell cleaner-burning petroleum products. As a result of such efforts—alongside an expanded subway system, a fleet of low-emission buses, and bike- and car-sharing programs—smog in Mexico City has been reduced since 1990 by more than half. Particulate matter is down by 70%, carbon monoxide by 74%, sulfur dioxide by 86%, and lead by 95%.

Across the world, industrializing nations such as China and India are confronting the need to make similar progress, while in all nations the growing awareness of indoor air pollution is encouraging action and solutions.

Outdoor air pollution is influenced not only by our emissions but also by natural sources of pollution and by atmospheric conditions. The more we understand about the science of the atmosphere, the better we can protect our health against pollution.

Likewise, science has proven crucial to addressing two other major air quality issues—ozone depletion and acid deposition. Policymakers responded quickly to scientific findings on stratospheric ozone depletion, and as a result, our global society appears to have dodged a bullet; today our planet's ozone layer is on the mend. With acid deposition, we responded to scientific research by launching policies to reduce emissions of acidic pollutants, and ecosystems are now beginning to recover.

As the world's less-wealthy nations industrialize, continued integration of science, policy, economics, and technology can help achieve cleaner air. However, people in all nations and societies will need to remain vigilant and keep pressure on their policymakers to protect public health by reducing pollution and improving air quality.

TESTING Your Comprehension

1. About how thick is Earth's atmosphere? Name one characteristic of the troposphere and one characteristic of the stratosphere.

2. Where is the "ozone layer" located? Describe how and why stratospheric ozone is beneficial for people, whereas tropospheric ozone is harmful.

3. How does solar energy influence weather and climate? Describe how Hadley, Ferrel, and polar cells help to determine climate patterns and the location of biomes.

4. Describe a temperature inversion. Explain how inversions contribute to severe smog episodes such as the ones in London, England, and in Donora, Pennsylvania.

5. How does a primary pollutant differ from a secondary pollutant? Give an example of each.

6. What has happened with the emissions of major pollutants in the United States in recent decades? What has happened with concentrations of "criteria pollutants" in U.S. ambient air?

7. How does photochemical smog differ from industrial smog? Give three examples of the health risks posed by the outdoor air pollutants in smog.

8. Explain how chlorofluorocarbons (CFCs) deplete stratospheric ozone. Why is this depletion considered a long-term international problem? What was done to address this problem?

9. Why are the effects of acid deposition often felt in areas far from where the primary pollutants are produced? List three impacts of acid deposition.

10. Name three common sources of indoor pollution and their associated health risks. For each pollution source, describe one way to reduce exposure to the source.

SEEKING Solutions

1. Name one type of natural air pollution, and discuss how human activity can sometimes worsen it. What potential solutions can you think of to minimize this human impact?

2. Explain how and why emissions of major pollutants have been reduced by well over 50% in the United States since 1970, despite increases in population, energy use, and economic activity. Describe at least two ways you think air quality might be further improved.

3. International action through a treaty helped to halt stratospheric ozone depletion, but other transboundary pollution issues—such as acid deposition and greenhouse gas pollution—have not been addressed as effectively. What types of actions do you feel are appropriate for pollutants that cross political boundaries?

4. **CASE STUDY CONNECTION** Describe at least three ways in which Los Angeles or Mexico City has responded to its air pollution challenges. What results have each of these responses produced? Now consider your city or a major city near where you live. Describe at least one approach used by L.A. or Mexico City that you feel would help address air pollution in your city, and explain why.

5. **THINK IT THROUGH** You have just taken a job at a medical clinic in your hometown. The nursing staff has asked you to develop a brochure for patients featuring tips on how to minimize health impacts from air pollution (both indoor and outdoor) in their daily lives. List the top five tips you will feature, and explain for each why you will include it in your brochure.

CALCULATING Ecological Footprints

"While only some motorists contribute to traffic fatalities, all motorists contribute to air pollution fatalities." So stated a writer for the Earth Policy Institute, pointing out that air pollution kills far more people than vehicle accidents. According to EPA data, emissions of nitrogen oxides in the United States in 2016 totaled 10.4 million tons (20.8 billion lbs). Nitrogen oxides come from fuel combustion in power plants and various industrial, commercial, and residential sources, but fully 6.0 million tons (12.0 billion lbs) of the 2016 total came from motor vehicles. The U.S. Census Bureau estimates the nation's population to have been 323.1 million in 2016 and projects that it will reach 359.4 million in 2030. Considering these data, calculate the missing values in the table.

	TOTAL NO$_X$ EMISSIONS (lb)	NO$_X$ EMISSIONS FROM VEHICLES (lb)
You		
Your class		
Your state		
United States	20.8 billion	12.0 billion

Data from U.S. EPA.

1. By what percentage is the U.S. population projected to increase between 2016 and 2030? Do you think that NO$_X$ emissions will increase, decrease, or remain the same over that period of time? Why? (You may wish to refer to Figure 13.7.)

2. Assume you are an average American. Using the 2016 emissions totals, how many pounds of total NO$_X$ emissions are you responsible for creating? How many pounds of NO$_X$ emissions from vehicles are you responsible for creating? What percentage of your total NO$_X$ emissions would that be?

3. Describe three ways in which an American driver could reduce his or her NO$_X$ emissions from vehicular travel. What steps could you take to reduce the overall NO$_X$ emissions for which you are responsible?

Mastering Environmental Science

CHAPTER 14

Global
Climate Change

Rising Seas Threaten South Florida

> ❝ **Miami, as we know it today, is doomed. It's not a question of if. It's a question of when.**
> —Dr. Harold Wanless, University of Miami geologist

> **Miami Beach is not going to sit back and go underwater.**
> —Philip Levine, mayor of Miami Beach ❞

It happens now in Miami at least six times a year. Salty water bubbles up from drains, seeps up from the ground, fills the streets, and spills across lawns and sidewalks. Under the dazzling sun of a South Florida sky, floodwaters stall car traffic, creep into doorways, force businesses to close, and keep people from crossing the street. Employees struggle to get to work while tourists stand around, baffled.

The flooding is most severe in Miami Beach, the celebrated strip of glamorous hotels, clubs, shops, and restaurants that rises from a 7-mile barrier island just offshore from Miami. The carefree affluent image of Miami Beach, with its sun and fun, is increasingly jeopardized by the grimy reality of these unwelcome saltwater intrusions. By 2030, flooding is predicted to strike Miami and Miami Beach about 45 times per year—becoming no longer a curious inconvenience, but an existential threat.

These mysterious floods that seem to come out of nowhere are a recent phenomenon, so Miami-area residents are just now realizing that their coastal metropolis is slowly being swallowed by the ocean. The cause? Rising sea levels driven by global climate change.

The world's oceans rose 20 cm (8 in.) in the 20th century as warming temperatures expanded the volume of seawater and caused glaciers and ice sheets to melt, discharging water into the oceans. These processes are accelerating today, and scientists predict that sea level will rise another 26–98 cm (10–39 in.) or more in this century as climate change intensifies.

As sea levels rise, coastal cities across the globe—from Venice to Amsterdam to New York to San Francisco—are facing challenges. In the United States, scientists find that the Atlantic Seaboard and the Gulf Coast are especially vulnerable. The hurricane-prone shores of Florida, Louisiana, Texas, and the Carolinas are at risk, as are coastal cities such as Houston and New Orleans. From Cape Cod to Corpus Christi, millions of Americans who live in shoreline communities are beginning to suffer significant expense, disruption to daily life, and property damage as beaches erode, neighborhoods flood, aquifers are fouled, and storms strike with more force.

Perhaps nowhere in America is more vulnerable to sea level rise than Miami and its surrounding communities in South Florida. Six million people live in this region, and three-quarters of them inhabit low-lying coastal areas that also hold most of the region's wealth and property. Experts calculate that Miami alone has more than $400 billion in assets at risk from sea level rise—more than any other city in the

Flooding in Miami after Hurricane Irma in 2017 ▲

Upon completing this chapter, you will be able to:

- **Describe Earth's climate system and explain the factors that influence global climate**

- **Identify greenhouse gases, and characterize human influences on the atmosphere and on climate**

- **Summarize how researchers study climate**

- **Outline current and expected future trends and impacts of climate change in the United States and across the world**

- **Suggest and assess ways we may respond to climate change**

Bulldozing beach sand off a Fort Lauderdale boulevard after a storm surge

FIGURE 14.1 In Miami, property and infrastructure valued at billions of dollars are located within just meters of the ocean.

world (**FIGURE 14.1**). Hurricanes strike the area frequently, and every centimeter of sea level rise makes the surge of seawater from a storm more expansive, costly, and dangerous. In 2017, Hurricane Irma sent floodwaters coursing through the streets of Miami, Miami Beach, Jacksonville, and other Florida cities. Miami's mayor Tomás Regalado, a Republican, pleaded with federal leaders to recognize that climate change was fueling stronger storms and hurting cities like his. "If this isn't climate change, I don't know what is," Regalado declared. "This is truly [a] poster child for what is to come."

South Florida is highly sensitive to sea level change because its landscape is exceptionally flat; just 1 m (3.3 ft) of sea level rise would inundate more than a third of the region. A 4-m (13-ft) rise in sea level would submerge Miami and reduce the region to a handful of small islands.

The porous limestone bedrock that underlies South Florida also poses a challenge. Pockmarked with holes like Swiss cheese, this permeable rock lets water percolate through. This is why Miami's floods seem to arise out of nowhere; during the highest tides of the year, ocean water is forced inland, where it mixes with fresh water underground and is pushed up as a briny mixture through the limestone directly into yards and streets. As a result, Miami and its neighboring cities cannot simply wall themselves off from a rising ocean, because seawalls won't stop water from seeping up from below.

Moreover, as saltwater moves inland, it contaminates the fresh drinking water of South Florida's Biscayne Aquifer. Fort Lauderdale and several other communities are already struggling with saltwater incursion. Florida is building desalination plants to convert seawater to drinking water, but desalination (p. 273) is expensive and consumes large amounts of energy.

Some of Florida's top state-level politicians have long been in denial about climate change, but today Miami-area leaders and citizens are taking action to safeguard their region's future. In 2010, commissioners of Broward, Miami-Dade, Monroe, and Palm Beach Counties adopted an agreement to work together on strategies to combat climate change and its effects in the region. This agreement, the Southeast Florida Regional Climate Change Compact, is garnering wide recognition as a model for regional cooperation on climate issues. Despite a lack of money from the state and federal governments, these policymakers are helping to build up dunes, raise building foundations, shift development inland, and stop subsidizing insurance for development in low-lying coastal areas.

In Miami Beach, Mayor Philip Levine won election to office in 2013 after a campaign ad showed him paddling a kayak through the streets of the South Beach neighborhood, promising to address flooding. "I wasn't swept into office," Levine is fond of saying. "I floated in." Under Levine, the city has raised some roadways 3 feet, and businesses are being urged to remodel their first floors. The city raised stormwater charges on residents and is spending $400 million installing a system of massive pumps to extract floodwater. Engineers expect these measures to get the city through the next couple of decades, but they recognize that more interventions will be needed later.

Only time will tell whether South Florida's communities will overcome their challenges and provide a shining example for other regions. In Miami and many other coastal cities, vast sums will be spent on pumps, drains, pipes, seawalls, and other engineering solutions, but ultimately these are only temporary fixes; they may buy time, but they cannot stop the water forever. In the long term, only reducing our emissions of greenhouse gases will halt sea level rise and the many other imminent consequences of global climate change.

Our Dynamic Climate

Climate influences virtually everything around us, from the day's weather to major storms, from crop success to human health, and from national security to the ecosystems that support our economies. If you are a student in your teens or twenties, the accelerating change in our climate today may well be *the* major event of your lifetime and the phenomenon that most shapes your future.

Climate change is also the fastest-developing area of environmental science. New scientific studies that refine our understanding of climate are published every week, and policymakers and businesspeople respond just as quickly.

Because new developments are occurring so rapidly, we urge you to explore beyond this book, and with your instructor, the most recent information on climate change and the consequences it may have for your future.

What is climate change?

Climate describes an area's long-term atmospheric conditions, including temperature, precipitation, wind, humidity, barometric pressure, solar radiation, and other characteristics. *Climate* differs from *weather* (p. 289) in that weather specifies conditions over hours or days, whereas climate summarizes conditions over years, decades, or centuries.

Global climate change—generally referred to simply as **climate change**—describes an array of changes in aspects of Earth's climate, such as temperature, precipitation, and the frequency and intensity of storms. People often use the term *global warming* synonymously in casual conversation, but **global warming** refers specifically to an increase in Earth's average surface temperature. Global warming is only one aspect of global climate change, but warming does in turn drive other components of climate change.

Over the long term, our planet's climate varies naturally. However, today's climatic changes are unfolding at an exceedingly rapid rate, and they are creating conditions humanity has never experienced. Scientists agree that human activities, notably fossil fuel combustion and deforestation, are largely responsible. Some researchers point out that the term "climate change" is so mild-sounding as to be misleading, and that a more accurate term would be "climate disruption."

Three factors influence climate

Three natural factors exert the most influence on Earth's climate. The first is the sun. Without the sun, Earth would be dark and frozen. The second is the atmosphere. Without this protective layer of gases, Earth would be as much as 33°C (59°F) colder on average, and temperature differences between night and day would be far greater than they are. The third is the oceans, which store and transport heat and moisture.

The sun supplies most of our planet's energy. Earth's atmosphere, clouds, land, ice, and water together absorb about 70% of incoming solar radiation and reflect the remaining 30% back into space (**FIGURE 14.2**). The 70% that is absorbed warms the surface and the atmosphere, and powers everything from wind to waves to evaporation to photosynthesis.

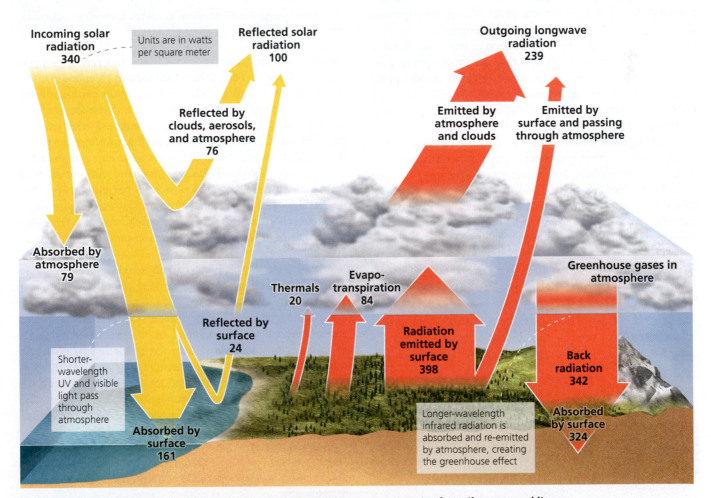

FIGURE 14.2 Our planet receives about 340 watts of energy per square meter from the sun, and it naturally reflects and emits this same amount. Earth absorbs nearly 70% of the solar radiation it receives, and reflects the rest back into space **(yellow arrows)**. The radiation absorbed is then re-emitted **(orange arrows)** as infrared radiation, which has longer wavelengths. Greenhouse gases in the atmosphere absorb a portion of this long-wavelength radiation and then re-emit it, sending some back downward to warm the atmosphere and the surface by the greenhouse effect. *Data from Intergovernmental Panel on Climate Change (IPCC); Stocker, T.F., et al. (Eds.), 2013.* Climate change 2013: The physical science basis. Contribution of Working Group I to the fifth assessment report of the IPCC. *Cambridge, UK, and New York, NY: Cambridge University Press.*

Greenhouse gases warm the lower atmosphere

As Earth's surface absorbs solar radiation, the surface increases in temperature and emits infrared radiation (p. 35), radiation with wavelengths longer than those of visible light. Atmospheric gases having three or more atoms in their molecules tend to absorb infrared radiation. These include water vapor (H_2O), ozone (O_3), carbon dioxide (CO_2), nitrous oxide (N_2O), and methane (CH_4), as well as halocarbons, a diverse group of mostly human-made gases (p. 301). All these gases are known as **greenhouse gases**. After absorbing radiation emitted from the surface, greenhouse gases re-emit infrared radiation. Some of this re-emitted energy is lost to space, but much of it travels back downward, warming the lower atmosphere (specifically the troposphere; p. 288) and the surface, in a phenomenon known as the **greenhouse effect**.

Greenhouse gases differ in their capacity to warm the troposphere and surface. *Global warming potential* refers to the relative ability of a molecule of a given greenhouse gas to contribute to warming. **TABLE 14.1** shows global warming potentials for several greenhouse gases. Values are expressed in relation to carbon dioxide, which is assigned a value of 1. For example, at a 20-year time horizon, a molecule of methane is 84 times more potent than a molecule of carbon dioxide. Yet because a methane molecule typically resides in the atmosphere for less time than a carbon dioxide molecule, methane's global warming potential is reduced at longer time horizons (it is 28 at a 100-year horizon).

Although carbon dioxide is less potent on a per-molecule basis than most other greenhouse gases, it is far more abundant in the atmosphere. Moreover, greenhouse gas emissions from human activity consist mostly of carbon dioxide; for this reason, carbon dioxide has caused nearly twice as much warming since the industrial revolution as have methane, nitrous oxide, and halocarbons combined.

Greenhouse gas concentrations are rising fast

The greenhouse effect is a natural phenomenon, and greenhouse gases have been present in our atmosphere throughout Earth's history. That's a good thing: Without the natural

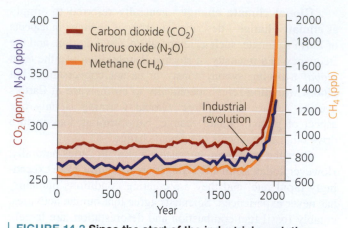

FIGURE 14.3 Since the start of the industrial revolution around 1750, global concentrations of carbon dioxide, methane, and nitrous oxide in the atmosphere have increased markedly. *Data from IPCC, 2013. Fifth assessment report.*

DATA Q By about what percentage has atmospheric carbon dioxide concentration increased since 1750?

Go to **Interpreting Graphs & Data** on **Mastering** Environmental Science

greenhouse effect, our planet would be too cold to support life as we know it. Thus, it is not the natural greenhouse effect that concerns scientists today, but rather the *anthropogenic* (human-generated) intensification of the greenhouse effect. By increasing the concentrations of greenhouse gases over the past 250 years (**FIGURE 14.3**), we are intensifying the greenhouse effect beyond what our species has ever experienced.

We have boosted Earth's atmospheric concentration of carbon dioxide from roughly 278 parts per million (ppm) in the late 1700s to more than 400 ppm today (see Figure 14.3). The concentration of CO_2 in our atmosphere now is far higher than it has been in over 800,000 years, and likely in the past 20 million years.

Why have atmospheric carbon dioxide levels risen so much? Most carbon is stored for long periods in the upper layers of the lithosphere (p. 232). The deposition, partial decay, and compression of organic matter (mostly plants and phytoplankton) in wetland or marine areas hundreds of millions of years ago led to the formation of coal, oil, and natural gas in buried sediments (p. 346). Over the past two centuries, we have extracted these fossil fuels from the ground and burned them in our homes, power plants, and automobiles, transferring large amounts of carbon from one reservoir (underground deposits that stored the carbon for millions of years) to another (the atmosphere). This sudden flux of carbon from the lithosphere to the atmosphere is the main reason atmospheric CO_2 concentrations have risen so dramatically.

At the same time, people have cleared and burned forests to make room for crops, pastures, villages, and cities. Forests serve as a reservoir for carbon as plants conduct photosynthesis (p. 34) and store carbon in their tissues. When we clear forests, we reduce the biosphere's ability to remove carbon dioxide from the atmosphere. In this way, deforestation (p. 195) contributes to rising atmospheric CO_2 concentrations.

TABLE 14.1 Global Warming Potentials of Four Greenhouse Gases

GREENHOUSE GAS	RELATIVE HEAT-TRAPPING ABILITY (IN CO_2 EQUIVALENTS)	
	OVER 20 YEARS	OVER 100 YEARS
Carbon dioxide	1	1
Methane	84	28
Nitrous oxide	264	265
Hydrochlorofluoro-carbon HFC-23	10,800	12,400

Data from IPCC, 2013. Fifth assessment report.

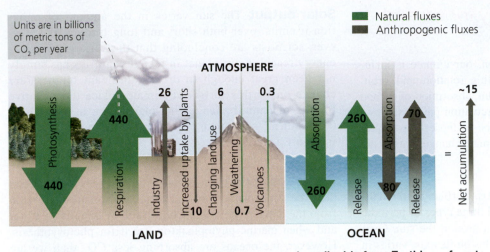

Units are in billions of metric tons of CO_2 per year

Natural fluxes
Anthropogenic fluxes

ATMOSPHERE

LAND

OCEAN

Photosynthesis 440 / 440

Respiration 440

Industry

Increased uptake by plants 26 / 10

Changing land use 6

Weathering 0.7

Volcanoes 0.3

Absorption 260 / 260

Release 260 / 80

Absorption 70

Release 70

Net accumulation ~15 =

FIGURE 14.4 Human activities are sending more carbon dioxide from Earth's surface to its atmosphere than is moving from the atmosphere to the surface. Shown are all current fluxes of CO_2, with arrows sized according to mass. Green arrows indicate natural fluxes, and gray arrows indicate anthropogenic fluxes. *Adapted from IPCC, 2007.* Fourth assessment report.

DATA For every metric ton of carbon dioxide we emit due to changes in land use (e.g., deforestation), how much do we emit from industry?

Go to **Interpreting Graphs & Data** on **Mastering** Environmental Science

FIGURE 14.4 summarizes scientists' understanding of the fluxes (natural and anthropogenic) of carbon dioxide among the atmosphere, land, and oceans.

Methane concentrations are also rising—150% since 1750 (see Figure 14.3)—and today's atmospheric concentration is the highest by far in over 800,000 years. We release methane by tapping into fossil fuel deposits, raising livestock that emit methane as a metabolic waste product, growing crops such as rice, and disposing of organic matter in landfills.

We have also elevated atmospheric concentrations of nitrous oxide. This greenhouse gas, a by-product of feedlots, chemical manufacturing plants, auto emissions, and synthetic nitrogen fertilizers, has risen by 20% since 1750 (see Figure 14.3).

Among other greenhouse gases, ozone concentrations in the troposphere have risen roughly 42% since 1750, a result of photochemical smog (p. 296). The contribution of halocarbons to global warming has begun to slow because of the Montreal Protocol and subsequent controls on their production and use (p. 303). Water vapor is the most abundant greenhouse gas in our atmosphere and contributes most to the natural greenhouse effect. Its concentrations vary locally, but because its global concentration has not changed, it is not thought to have driven industrial-age climate change.

Other factors warm or cool the surface

Whereas greenhouse gases warm the atmosphere, **aerosols,** microscopic droplets and particles, can have either a warming or a cooling effect. Soot particles, or "black carbon aerosols," generally cause warming by absorbing solar energy, but most other aerosols cool the atmosphere by reflecting the sun's rays. When sulfur dioxide from fossil fuel combustion enters the

atmosphere, it undergoes various reactions, some of which lead to acid deposition (p. 303) and form a sulfur-rich aerosol haze that blocks sunlight. Sulfate aerosols released by major volcanic eruptions can cool Earth's climate for up to several years. This occurred in 1991 with the eruption of Mount Pinatubo, a volcano in the Philippines.

To quantify the impact that a given factor exerts on Earth's temperature, scientists calculate its **radiative forcing,** the amount of change in thermal energy that the factor causes. Positive forcing warms the surface, whereas negative forcing cools it. When scientists sum up the effects of all factors, they find that Earth is now experiencing radiative forcing of about 2.3 watts/m^2 (**FIGURE 14.5**). This means that our planet today is receiving and retaining roughly 2.3 watts/m^2 more thermal energy than it is emitting into space. (By contrast, the pre-industrial Earth of 1750 was in balance, emitting as much radiation as it was receiving.) This extra amount is equivalent to the power converted into heat and light by 200 incandescent lightbulbs (or more than 900 compact fluorescent lamps) across a football field. As Figure 14.2 shows, Earth naturally receives and gives off about 340 watts/m^2 of energy. Although 2.3 may seem like a small proportion of 340, heat from this imbalance accumulates, and over time it is enough to alter climate significantly.

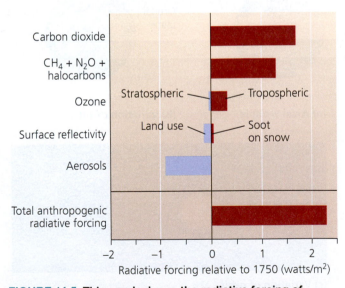

Carbon dioxide

CH$_4$ + N$_2$O + halocarbons

Ozone
Stratospheric / Tropospheric

Surface reflectivity
Land use / Soot on snow

Aerosols

Total anthropogenic radiative forcing

Radiative forcing relative to 1750 (watts/m^2)

FIGURE 14.5 This graph shows the radiative forcing of major factors that warm or cool our planet. Radiative forcing is expressed as the effect each factor has on temperature today relative to 1750, in watts/m^2. Red bars indicate positive forcing (warming), and blue bars indicate negative forcing (cooling). *Data from IPCC, 2013.* Fifth assessment report.

Climate varies naturally for several reasons

Aside from atmospheric composition, our climate is influenced by cyclical changes in Earth's rotation and orbit, variation in energy released by the sun, absorption of carbon dioxide by the oceans, and ocean circulation patterns. However, scientific data indicate that none of these four natural factors can fully explain the rapid climate change that we are experiencing today.

Milankovitch cycles In the 1920s, Serbian mathematician Milutin Milankovitch described three types of periodic changes in Earth's rotation and orbit around the sun. Over thousands of years, our planet wobbles on its axis, varies in the tilt of its axis, and experiences change in the shape of its orbit, all in regular long-term cycles of different lengths. These variations, known as **Milankovitch cycles,** alter the way solar radiation is distributed over Earth's surface (**FIGURE 14.6**). By modifying patterns of atmospheric heating, these cycles trigger long-term climate variation, including periodic episodes of *glaciation* during which global surface temperatures drop and ice sheets expand outward from the poles. These cycles are highly influential in the very long term, but science shows that they do not account for the very rapid, extreme climate disruption we are experiencing today.

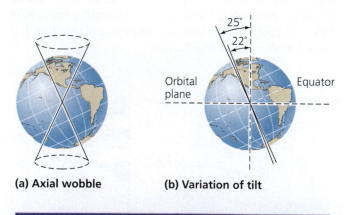

(a) Axial wobble   **(b) Variation of tilt**

(c) Variation of orbit

FIGURE 14.6 **There are three types of Milankovitch cycles:** **(a)** an axial wobble that occurs on a 19,000- to 23,000-year cycle; **(b)** a 3-degree shift in the tilt of Earth's axis that occurs on a 41,000-year cycle; and **(c)** a variation in Earth's orbit from almost circular to more elliptical, which repeats every 100,000 years.

Solar output The sun varies in the amount of radiation it emits, over both short and long timescales. However, scientists are concluding that the variation in solar energy reaching our planet in recent centuries has simply not been great enough to drive significant temperature change on Earth's surface. Estimates place the radiative forcing of natural changes in solar output at only about 0.05 watt/m^2—less than any of the anthropogenic causes in Figure 14.5.

Ocean absorption The oceans hold 50 times more carbon than the atmosphere holds. Oceans absorb carbon dioxide from the atmosphere when CO_2 dissolves directly in water and when marine phytoplankton use it for photosynthesis. However, the oceans are absorbing less CO_2 than we are adding to the atmosphere (see Figure 2.21, p. 42). Thus, carbon absorption by the oceans is slowing global warming but is not preventing it. Moreover, as ocean water warms, it absorbs less CO_2 because gases are less soluble in warmer water—a positive feedback effect (p. 25) that accelerates warming of the atmosphere.

Ocean circulation Ocean water exchanges heat with the atmosphere, and ocean currents move energy from place to place. For example, the oceans' thermohaline circulation system (p. 262) moves warm tropical water northward, providing Europe a far milder climate than it would otherwise have. Scientists are studying whether freshwater input from Greenland's melting ice sheet might shut down this warm-water flow—an occurrence that could plunge Europe into much colder conditions.

Multiyear climate variability results from what is known as the El Niño–Southern Oscillation (p. 262), which involves systematic shifts in atmospheric pressure, sea surface temperature, and ocean circulation in the tropical Pacific Ocean. These shifts overlie longer-term variability from the Pacific Decadal Oscillation. El Niño and La Niña events alter weather patterns in diverse ways, often leading to rainstorms and floods in dry regions and drought and fire in moist regions. This leads to impacts on wildlife, agriculture, and fisheries.

FAQ

The climate changes naturally, so why worry about climate change?

Earth's climate does indeed change naturally over very long periods of time, but there is nothing "natural" about today's sudden climate disruption. We know that human activity is directly causing the unnaturally rapid changes we are now witnessing. Moreover, humanity has never before experienced the sheer amount of change predicted for this century. In fact, the quantity by which the world's temperature is forecast to rise is greater than the amount of cooling needed to bring on an ice age! Greenhouse gas concentrations are already higher than they've been in more than 800,000 years, and they are rising. The human species, *Homo sapiens*, has existed for just 200,000 years, and our civilization arose only in the past few thousand years during an exceptionally stable period in Earth's climate history. Unless we reduce our emissions, we will soon be challenged by climate conditions our species has never lived through.

Studying Climate Change

To comprehend any phenomenon that is changing, we must study its past, present, and future. Scientists monitor present-day climate, but they also have devised clever means of inferring past change and sophisticated methods to predict future conditions.

Proxy indicators tell us of the past

To understand past climate, scientists decipher clues from thousands or millions of years ago by taking advantage of the record-keeping capacity of the natural world. **Proxy indicators** are types of indirect evidence that serve as proxies, or substitutes, for direct measurement.

For example, Earth's ice caps, ice sheets, and glaciers hold clues to climate history. In frigid regions near the poles and atop high mountains, snow falling year after year compresses into ice. Over millennia, this ice accumulates to great depths, preserving within its layers tiny bubbles of the ancient atmosphere. Scientists can examine the trapped air bubbles by drilling into the ice and extracting long columns, or cores. The layered ice, accumulating season after season for thousands of years, provides a timescale. By studying the chemistry of the bubbles in each layer, scientists can determine atmospheric composition, greenhouse gas concentrations, temperature, snowfall, solar activity, and even (from trapped soot particles) the frequency of forest fires and volcanic eruptions during each time period.

Recently, researchers analyzed the deepest ice core ever (**FIGURE 14.7**). At a remote site in Antarctica, they drilled down 3270 m (10,728 ft) and pulled out more than 800,000 years' worth of ice! This core chronicles Earth's history across eight glacial cycles. By analyzing air bubbles trapped in the ice, researchers discovered that over the past 800,000 years, atmospheric concentrations of carbon dioxide, methane, and nitrous oxide have never been as high as they are today. The ice core results also confirm that temperature swings in the past were tightly correlated with greenhouse gas concentrations. This bolsters the scientific consensus that greenhouse gas emissions are causing Earth to warm today.

Researchers also drill cores into beds of sediment beneath bodies of water. Sediments often preserve pollen grains and other remnants from plants that grew in the past (as described in the study of Easter Island; pp. 8–9). Because climate influences the kinds of plants that grow in an area, knowing what plants were present can tell us a great deal about the climate at that place and time. Other types of proxy indicators include tree rings (which reveal year-by-year histories of precipitation and fire), pack-rat middens (rodent dens in which plant parts may be preserved for centuries in arid regions), and coral reefs (p. 267), which reveal aspects of ocean chemistry.

Direct measurements tell us about the present

Today we measure temperature with thermometers, rainfall with rain gauges, wind speed with anemometers, and air pressure with barometers, using computer programs to integrate and analyze this information in real time. With these technologies and more, we document fluctuations in weather day-by-day and hour-by-hour across the globe.

We also measure the chemistry of the atmosphere and the oceans. Direct measurements of carbon dioxide concentrations in the atmosphere reach back to 1958, when scientist Charles Keeling began analyzing hourly air samples from a monitoring station at Hawaii's Mauna Loa Observatory. These data show that atmospheric CO_2 concentrations have increased from 315 ppm in 1958 to more than 400 ppm today.

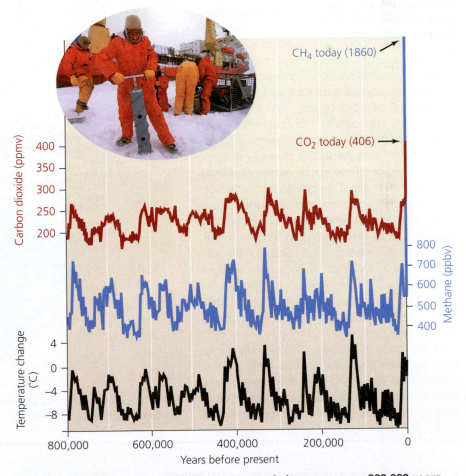

FIGURE 14.7 Data from the EPICA ice core reveal changes across 800,000 years. Shown are surface temperature (**black line**), atmospheric methane concentration (**blue line**), and atmospheric carbon dioxide concentration (**red line**). Concentrations of CO_2 and methane rise and fall in tight correlation with temperature. Today's current values are included at the top right of the graph, for comparison. *Adapted by permission of Macmillan Publishers Ltd: Brook, E. 2008. Paleoclimate: Windows on the greenhouse. Nature 453: 291–292, Fig. 1a. www.nature.com.*

How Do Climate Models Work?

Models are indispensable for scientists studying climate today—and they are increasingly vital for our society because they help us predict what conditions will confront us in the future. Yet to most of us, a climate model is a mysterious black box. So how *do* scientists create a climate model?

The colorful maps and data-rich graphs that scientists generate from a climate model are the end result, but the process begins when they put into the model a long series of mathematical equations. These equations describe how various components of Earth's systems function. Some equations are derived from physical laws such as those on the conservation of mass, energy, and momentum (p. 30). Others are derived from observational and experimental data on physics, chemistry, and biology, gathered from the field. Converted into computing language, these equations are integrated with information about Earth's landforms, hydrology, vegetation, and atmosphere (**FIGURE 1**).

Earth's climate system is mind-bogglingly complex, but as computers grow more powerful, sophisticated models are incorporating more and more of the factors that influence climate. Most models consist of submodels, each handling a different component—ocean water, sea ice, glaciers, forests, deserts, troposphere, stratosphere, and so on.

For a model to function, all its building blocks must be given equations to make them behave realistically in space and time. In the real climate system, time is continuous and spatial effects reach down to the level of molecules interacting

A researcher works with satellite data used to model climate change.

with one another. But the virtual reality of climate models cannot be so detailed—there is simply not enough computer power available. Instead, modelers approximate reality by dividing time into periods (called time steps) and by dividing Earth's surface into cells or boxes in a grid (called grid boxes) (**FIGURE 2**).

Each grid box contains land, ocean, or atmosphere, much like a digital photograph is made up of discrete pixels of certain colors. The grid boxes are arrayed in a three-dimensional layer by latitude and longitude, or in equal-sized polygons. The finer the scale of the grid, the greater resolution the model will have, and the better it will be able to predict results region by region. However, more resolution requires more computing power, and climate models already strain the most powerful

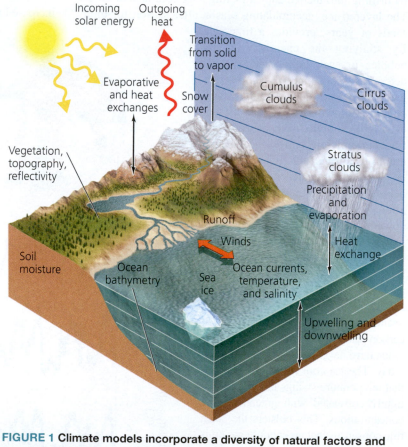

FIGURE 1 Climate models incorporate a diversity of natural factors and processes. Anthropogenic factors can then be added in.

Models help us predict the future

To understand how climate systems function and to predict future climate change, scientists simulate climate processes with sophisticated computer programs. **Climate models** are programs that combine what is known about atmospheric circulation, ocean circulation, atmosphere–ocean interactions, and feedback cycles to simulate climate dynamics (see **THE SCIENCE BEHIND THE STORY**). This requires manipulating vast amounts of data with complex mathematical

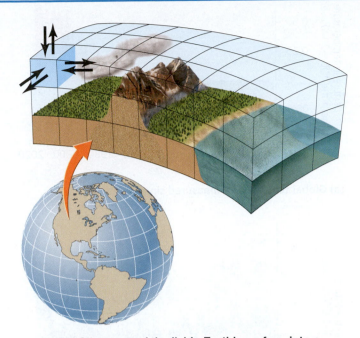

FIGURE 2 Climate models divide Earth's surface into a layered grid. Each grid box represents land, air, or water, and interacts with adjacent grid boxes via the flux of materials and energy. *Adapted from Bloom, Arnold J., 2010.* Global climate change: Convergence of disciplines. *Sunderland, MA: Sinauer Associates.*

supercomputing networks. Today's best climate models feature dozens of grid boxes piled up from the bottom of the ocean to the top of the atmosphere, with each grid box measuring a few dozen miles wide, and time measured in periods of just minutes.

Once the grid is established, the processes that drive climate are assigned to each grid box, with their rates parceled out among the time steps. The model lets the grid boxes interact through time by means of the flux of materials and energy into and out of each grid box.

Once modelers have input all this information, learned from our study of Earth and the climate system, they let the model run through time and simulate climate, from the past into the future. If the computer simulation accurately reconstructs past and present climate, then that inspires confidence in the model's ability to predict future climate accurately as well.

A number of studies have compared model runs that include only natural processes, model runs that include only human-generated processes, and model runs that combine both. These studies have found repeatedly that the model runs incorporating both human and natural processes are the ones

that fit real-world climate observations the best (**FIGURE 3**). This supports the idea that human activities, as well as natural processes, are influencing our climate.

The major human influence on climate is our emission of greenhouse gases, and modelers need to select values to enter for future emissions if they want to predict future climate. Generally, they will run their simulations multiple times, each time with a different future emission rate according to a specified scenario. Differences between the results from such scenarios tell us what influence these different emission rates would have.

Researchers are constantly testing and evaluating their models. They improve them by incorporating what is learned from new research and by taking advantage of what new computing technologies allow. As their work proceeds, we can expect increasingly precise and accurate predictions about future climate conditions.

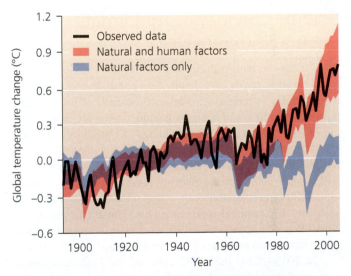

FIGURE 3 Models that incorporate both natural and anthropogenic factors predict observed climate trends best. *Adapted from Melillo, J.M., et al. (Eds.), 2014.* Climate change impacts in the United States: The third national climate assessment. *U.S. Global Change Research Program.*

DATA Q • Why do you think the red-shaded area diverges from the blue-shaded area in the right-hand portion of the graph? • Why does the observed data (black line) track with the red-shaded area rather than with the blue-shaded area?

Go to **Interpreting Graphs & Data** on **Mastering** Environmental Science

equations—a task not possible until the advent of modern computers.

Climate modelers provide starting information to the model based on real data from the field, set up rules for the simulation, and then let the program run. Researchers test

the efficacy of a model by entering past climate data and running the model toward the present. If a model accurately reconstructs current climate, then we have reason to believe that it simulates climate mechanisms realistically and may accurately predict future climate.

Plenty of challenges remain for climate modelers, because Earth's climate system is so complex. Yet as scientific knowledge builds and computing power intensifies, climate models continue to improve in resolution and are allowing us to make predictions region-by-region across the world.

Impacts of Climate Change

Virtually everyone is noticing changes in the climate these days. Miami-area residents suffer flooding. Texas ranchers and California farmers endure multiyear droughts. Coastal homeowners struggle to obtain insurance against hurricanes and storm surges. People from New York to Atlanta to Chicago to Los Angeles face unprecedented heat waves and cold snaps. Climate change has already exerted a multiplicity of impacts on our planet and on our society. If we continue to emit greenhouse gases, the consequences of climate change will grow more severe.

Scientific evidence for climate change is extensive

For decades, scientists have studied climate change in enormous breadth, depth, and detail. As a result, the scientific literature today is replete with many thousands of independent published studies, and we have gained a rigorous understanding of most aspects of climate change. To make this vast and growing research knowledge accessible to policymakers and the public, the Intergovernmental Panel on Climate Change (IPCC) has taken up the task of reviewing and summarizing it. This international body consists of many hundreds of scientists and national representatives. The IPCC shared the Nobel Peace Prize in 2007 for its work in informing the world of the trends and impacts of climate change.

In 2013–2014, the IPCC released its *Fifth Assessment Report*. Summarizing thousands of scientific studies, this report documents observed trends in surface temperature, precipitation patterns, snow and ice cover, sea level, storm intensity, and other factors. It also predicts future trends after considering a range of scenarios for future greenhouse gas emissions. The report addresses impacts of climate disruption on wildlife, ecosystems, and society. Finally, it discusses strategies we might pursue in response. To learn more, you may wish to download the *Fifth Assessment Report* yourself. Three working group reports and a synthesis report are accessible online at the IPCC's website.

For the United States, impacts have been assessed by the U.S. Global Change Research Program, which Congress created to coordinate federal climate research. Its 2014 *National Climate Assessment* summarized current research, observed trends, and predicted future impacts of climate change on the United States. You can explore scientists' predictions for the nation and for your own region by consulting this publicly accessible report online. An updated report was due out in

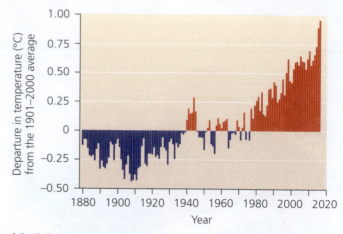

(a) Global temperature measured since 1880

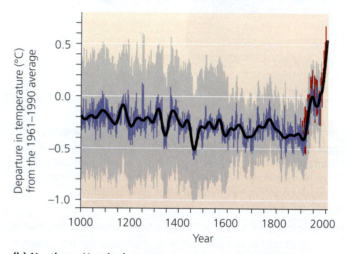

(b) Northern Hemisphere temperature, past 1000 years

FIGURE 14.8 Global temperatures have risen sharply in the past century. Data from thermometers **(a)** show changes in Earth's average surface temperature since 1880. Since 1976, every single year has been warmer than average. In **(b)**, proxy indicators **(blue line)** and thermometer data **(red line)** together show average temperatures in the Northern Hemisphere over the past 1000 years. The gray-shaded zone represents the 95% confidence range. *Data from (a) NOAA National Climatic Data Center; and (b) IPCC, 2001. Third assessment report.*

2018, but was thrown into limbo when President Trump disbanded its scientific advisory panel.

Temperatures continue to rise

Average surface temperatures on Earth have risen by about 1.1°C (2.0°F) in the past 100 years (**FIGURE 14.8**). Most of this increase has occurred since 1975—and just since 2000, we have experienced 17 of the 18 warmest years since global measurements began 140 years ago! Since the 1960s, each decade has been warmer than the last. If you were born after 1976, you have never in your life lived through a year with average global temperatures lower than the 20th-century average. If you were born after 1985, you have never even lived through a *month* with cooler-than-average global temperatures.

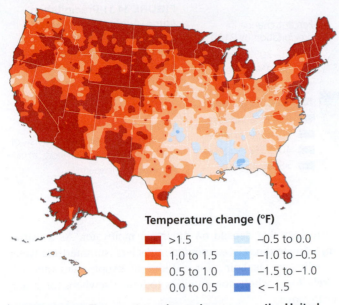

Temperature change (°F)

■ >1.5	■ −0.5 to 0.0
■ 1.0 to 1.5	■ −1.0 to −0.5
■ 0.5 to 1.0	■ −1.5 to −1.0
■ 0.0 to 0.5	■ < −1.5

FIGURE 14.9 Temperatures have risen across the United States. Most of the nation has warmed by more than 1°F (0.6°C) when the average from the period 1991–2012 is compared to the average from 1901–1960. *Data from NOAA National Climatic Data Center as presented in Melillo, J.M., et al. (Eds.), 2014.* Climate change impacts in the United States: The third national climate assessment. *U.S. Global Change Research Program.*

DATA
- By how much did the average temperature rise or fall where you live during the timeframe illustrated?
- How does this compare with other parts of the country?

Go to **Interpreting Graphs & Data** on **Mastering** Environmental Science

In just the past two decades, temperatures in most areas of the United States have risen by more than 1 full degree Fahrenheit (**FIGURE 14.9**).

We can expect global surface temperatures to continue rising because we are still emitting greenhouse gases and because the greenhouse gases already in the atmosphere will continue warming the globe for decades to come. At the end of the 21st century, the IPCC predicts global temperatures will be 1.0–3.7°C (1.8–6.7°F) higher than today's, depending on how well we control our emissions. Unusually hot days and heat waves will become more frequent. Future changes in temperature (**FIGURE 14.10**) are predicted to vary from region to region in ways that they already have. For example, polar regions will continue to experience the most intense warming.

Precipitation is changing

A warmer atmosphere speeds evaporation and holds more water vapor, and precipitation has increased worldwide by 2% over the past century. Yet some regions of the world are receiving less rain and snow than usual while others receive more. In the western United States, droughts have become more frequent and severe, harming agriculture, worsening soil erosion, reducing water supplies, and triggering wildfire. In parts of the eastern United States, heavy rain events have increased, leading to floods that have killed dozens of people and left thousands homeless.

Future changes in precipitation (**FIGURE 14.11**) are predicted to intensify regional changes that have already occurred. Many wet regions will receive more rainfall, increasing flooding risks, while many dry regions will become drier, worsening water shortages.

Extreme weather is becoming "the new normal"

The sheer number of extreme weather events in recent years—droughts, floods, hurricanes, snowstorms, cold snaps, heat waves—has caught everyone's attention, and weather records are falling left and right. In the United States in 2012 alone, the nation experienced a freakish March heat wave, a drought that devastated agriculture across three-fifths of the country, and Superstorm Sandy, which inflicted over $65 billion in damage along the Atlantic Coast. In 2017, Hurricane Harvey flooded Houston while Hurricane Irma devastated the Caribbean and drenched Florida and the Southeast, even as wildfires raged out of control in the West.

Scientific data summarized by the U.S. Climate Extremes Index confirm that the frequency of extreme weather events in the United States has doubled since 1970. Scientists are not the only ones to notice this trend. The insurance industry is finely attuned to such patterns, because insurers pay out money each time a major storm, drought, or flood hits. A major German insurer, Munich Re, calculated that since

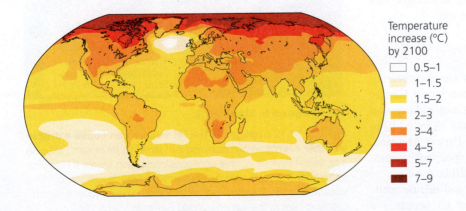

Temperature increase (°C) by 2100

□ 0.5–1	
▨ 1–1.5	
▨ 1.5–2	
▨ 2–3	
▨ 3–4	
▨ 4–5	
▨ 5–7	
■ 7–9	

FIGURE 14.10 Surface temperatures are projected to rise for the years 2081–2100, relative to 1986–2005. Landmasses are expected to warm more than oceans, and the Arctic will warm the most. This map was generated using an intermediate emissions scenario involving an average global temperature rise of 2.2°C (4.0°F). *Data from IPCC, 2013.* Fifth assessment report.

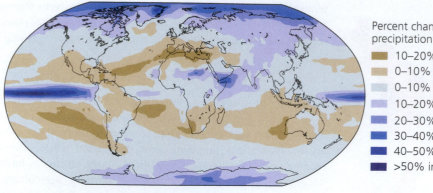

Percent change in precipitation by 2100

- 10–20% decrease
- 0–10% decrease
- 0–10% increase
- 10–20% increase
- 20–30% increase
- 30–40% increase
- 40–50% increase
- >50% increase

FIGURE 14.11 Precipitation (June–August) is projected to change for the years 2081–2100, relative to 1986–2005. Browner shades indicate less precipitation; bluer shades indicate more. This map was generated using an intermediate emissions scenario involving an average global temperature rise of 2.2°C (4.0°F). *Data from IPCC, 2013. Fifth assessment report.*

1980, extreme weather events causing losses have doubled in Europe and have risen by 2.5 times in Africa, 4 times in Asia, and 5 times in North America.

For years, researchers conservatively stated that although climate trends influence the probability of what the weather may be like on any given day, no single particular weather event can be directly attributed to climate change. In the aftermath of Superstorm Sandy, a metaphor spread across the Internet: When a baseball player takes artificial steroids and starts hitting more home runs, you can't attribute any one particular home run to the steroids, but you *can* conclude that the steroids were responsible for the increase in home runs. Our greenhouse gas emissions are like steroids that are supercharging our climate and increasing the instance of extreme weather events.

In 2012, research by Jennifer Francis of Rutgers University and Stephen Vavrus of the University of Wisconsin revealed a mechanism that may explain how and why global warming leads to more extreme weather. Warming has been greatest in the Arctic, and this has weakened the intensity of the Northern Hemisphere's polar jet stream, a high-altitude air current that blows west-to-east and meanders north and south, influencing the weather across North America and Eurasia. As the jet stream slows down, its meandering loops become longer, and may get stuck in what meteorologists call an *atmospheric blocking pattern* whereby the eastward movement of weather systems is blocked (**FIGURE 14.12**). When this happens, a rainy system that might normally move past a city in a day can instead be held in place for several days, causing flooding. Or dry conditions over a farming region might last two weeks instead of two days, resulting in drought. Hot spells last longer, and cold spells last longer, too.

Melting ice has far-reaching effects

As the world warms, mountaintop glaciers are disappearing (**FIGURE 14.13**). Between 1980 and 2016, the World Glacier Monitoring Service estimates that the world's major glaciers on average each lost mass equivalent to more than 19 m (62 ft) in vertical height of water. In Glacier National Park in Montana, only 25 of 150 glaciers present at the park's inception remain. Scientists estimate that by 2030 even these will be gone.

Mountains accumulate snow in winter and release meltwater gradually during summer. One out of six people live in regions that depend on mountain meltwater. As a warming climate diminishes mountain glaciers, summertime water supplies are declining for millions of people, and this will likely force whole communities to look elsewhere for water, or to move.

Warming temperatures are also melting vast amounts of polar ice. In Antarctica, coastal ice shelves the size of Rhode Island have disintegrated as a result of contact with warmer ocean water, and research now suggests that the entire West

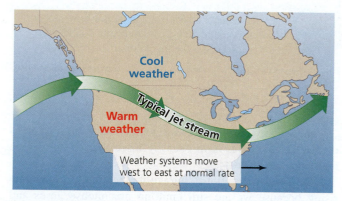

(a) Normal jet stream

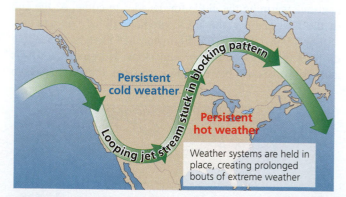

(b) Jet stream in March 2012

FIGURE 14.12 Changes in the jet stream can cause extreme weather events. Arctic warming can slow the jet stream, causing it to depart from its normal configuration **(a)** and create a blocking pattern **(b)** that stalls weather systems in place, leading to extreme weather events. The blocking pattern shown here brought record-breaking heat to the eastern United States in March 2012.

(a) Jackson Glacier in 1911

(b) Jackson Glacier in 2009

FIGURE 14.13 Glaciers are melting rapidly as global warming proceeds. The Jackson Glacier in Glacier National Park, Montana, retreated substantially between **(a)** 1911 and **(b)** 2009.

Antarctic ice shelf may be on its way to collapse, which would create a 3-m (10-ft) rise in sea level.

In the Arctic, where temperatures have warmed more than anywhere else, the immense ice sheet that covers Greenland is melting faster and faster. Sea ice is also thinning (**FIGURE 14.14**), and as this ice melts earlier in the season, freezes later, and recedes from shore, it becomes harder for Inuit people and for polar bears alike to hunt the seals they each rely on for food. As Arctic sea ice disappears, new shipping lanes open up for commerce, and governments and companies rush to exploit newly accessible underwater oil and mineral reserves. Russia, Canada, the United States, and other nations are jockeying for position, trying to lay claim to regions of the Arctic seafloor as the ice melts.

One reason warming is accelerating in the Arctic is that as snow and ice melt, darker, less reflective surfaces (such as bare ground and pools of meltwater) are exposed. As a result, more of the sun's rays are absorbed, and the surface warms. In a process of positive feedback, this warming causes more ice and snow to melt, which in turn causes more absorption and warming (see Figure 2.1b, p. 25).

Warming Arctic temperatures are also causing *permafrost* (permanently frozen ground) to thaw and settle, destabilizing buildings, pipelines, roads, and bridges. A recent study estimates Alaska will suffer $5.6–7.6 billion in damage to public infrastructure by 2080 due to climate change. Moreover, when permafrost thaws, it can release methane that has been stored for thousands of years. Because methane is a potent greenhouse gas, its release acts as positive feedback that intensifies climate change.

Rising sea levels may affect hundreds of millions of people

As glaciers and ice sheets melt, the runoff causes sea levels to rise. Sea levels also are rising because ocean water is warming, and water expands in volume as it warms. In addition, as we extract groundwater from aquifers (for drinking and to apply to farmland), the wastewater that enters rivers and the excess irrigation water that runs off farmland eventually reach the ocean, adding to sea level rise.

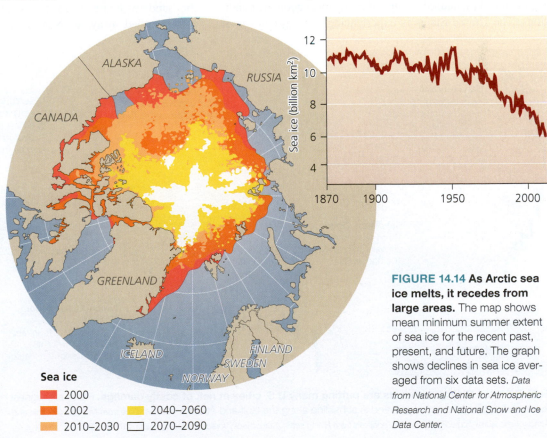

Sea ice
- 2000
- 2002
- 2010–2030
- 2040–2060
- 2070–2090

FIGURE 14.14 As Arctic sea ice melts, it recedes from large areas. The map shows mean minimum summer extent of sea ice for the recent past, present, and future. The graph shows declines in sea ice averaged from six data sets. *Data from National Center for Atmospheric Research and National Snow and Ice Data Center.*

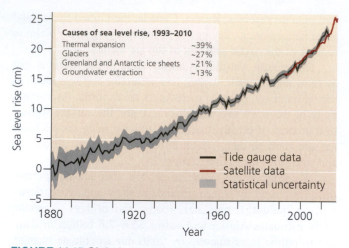

FIGURE 14.15 Global average sea level has risen more than 24 cm (9.5 in.) since 1880. Sea levels rise because water expands as it warms, glaciers and ice sheets are melting, and groundwater we extract eventually reaches the ocean. *Data from IPCC, 2013. Fifth assessment report; CSIRO; and NASA.*

Worldwide, average sea levels have risen 24.1 cm (9.5 in.) in the past 135 years (**FIGURE 14.15**), reaching a rate of 3.4 mm/yr from 1993 to 2016. These numbers represent vertical rises in water level, and on most coastlines a vertical rise of a few inches translates into a great many feet of incursion inland. Higher sea levels lead to beach erosion; coastal flooding; intrusion of saltwater into aquifers; and greater impacts from *storm surges*, temporary local rises in sea level generated by storms.

Regions experience differing amounts of sea level change due to patterns of ocean currents or because land may be rising or subsiding naturally, depending on local geologic conditions. The United States is experiencing varying degrees of

sea level rise (**FIGURE 14.16**), with the East Coast and the Gulf Coast most at risk.

In the southern Pacific Ocean, small island nations have seen sea levels rise as quickly as 9 mm/yr. Rising seas threaten the very existence of countries like the Maldives, a nation of 1200 islands in the Indian Ocean. In the Maldives, four-fifths of the land lies less than 1 m (39 in.) above sea level (**FIGURE 14.17a**). Saltwater is contaminating drinking supplies, and storms are eroding beaches and damaging the coral reefs that support the Maldives' tourism and fishing industries. Residents have already evacuated several low-lying islands. For these reasons, leaders of the Maldives have played a prominent role in international efforts to fight global warming. In 2009, Maldives President Mohamed Nasheed and his cabinet donned scuba gear and dove into a coastal lagoon, where they held the world's first underwater cabinet meeting—part of a campaign to draw global attention to the impacts of climate change (**FIGURE 14.17b**).

In the United States, several major hurricanes have demonstrated the impact that storm surges can inflict on highly developed metropolitan areas. In 2017, flooding from Hurricane Harvey devastated Houston, the nation's fourth-largest city, along with large regions of Texas and Louisiana. This was followed by Hurricane Irma, which flooded coastal cities from Miami to Jacksonville to Charleston, after ravaging the Florida Keys and many Caribbean islands. Then Hurricane Maria brought ruin to Puerto Rico. In 2012, Superstorm Sandy—a hurricane until just before it made landfall—battered the eastern part of the nation, leaving 160 people dead and thousands homeless (**FIGURE 14.18**). In New Jersey, coastal communities were inundated with salt water and sand, thousands of homes were destroyed, and iconic boardwalks were washed away. In New York City, economic activity

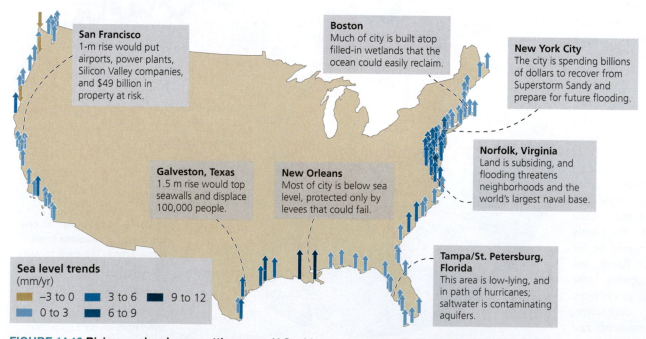

FIGURE 14.16 Rising sea levels are putting many U.S. cities at risk of costly damage. Rates are highest (taller darker blue arrows) where land is subsiding along the Gulf and Atlantic coasts. *Data from National Oceanic and Atmospheric Administration; city profiles adapted from Rising seas: A city-by-city forecast. Rolling Stone, 20 June, 2013.*

(a) Malé, capital of the Maldives

(b) The "underwater cabinet meeting"

FIGURE 14.17 Rising sea levels threaten island nations. The capital of the Maldives **(a)** is crowded onto an island averaging just 1.5 m (5 ft) above sea level. In 2009, the Maldives' president led his cabinet in an underwater meeting **(b)** to focus international attention on the plight of island nations vulnerable to sea level rise.

ground to a halt as tunnels, subway stations, vehicles, and buildings flooded.

Although not directly and solely caused by climate change, these storms were facilitated and strengthened by it. Warmer ocean water boosts the chances of large and powerful hurricanes. A warmer atmosphere retains more moisture, which a hurricane can dump onto land. A blocking pattern in the jet stream contributed to Sandy's energy. And higher sea levels magnify the damage caused by storm surges.

In 2005, Hurricane Katrina slammed into New Orleans and the Gulf Coast, killing more than 1800 people and inflicting $80 billion in damage. Outside New Orleans today, marshes of the Mississippi River delta continue to disappear under rising seas (pp. 255–256), weakening

protection against future storm surges. Around the world, rising seas are eating away at the salt marshes, dunes, mangrove forests, and coral reefs that serve as barriers protecting our coasts.

In its *Fifth Assessment Report*, the IPCC predicted that mean global sea level will rise 26–82 cm (10–32 in.) higher by 2100, depending on our level of emissions. However, research since then finds that Greenland's ice is melting faster and faster, which would lead seas to rise more quickly—perhaps over 1 m (3.3 ft) by 2100. More than half of the U.S. population lives in coastal counties, and 3.7 million Americans live within 1 vertical meter of the high tide line. It is estimated that a 1-m rise threatens 180 U.S. cities with losing an average of 9% of their land area. South Florida is judged most at risk

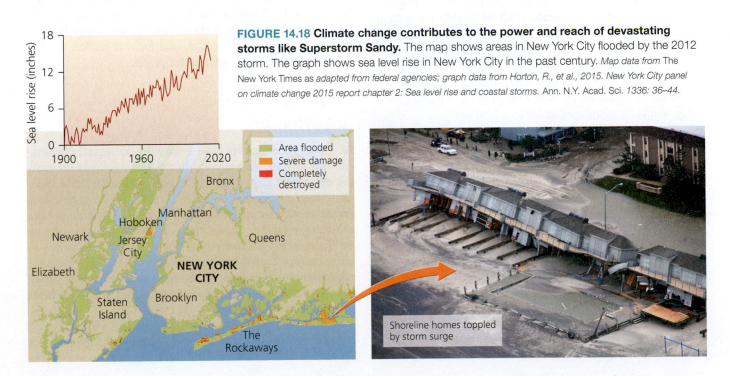

FIGURE 14.18 Climate change contributes to the power and reach of devastating storms like Superstorm Sandy. The map shows areas in New York City flooded by the 2012 storm. The graph shows sea level rise in New York City in the past century. *Map data from The New York Times as adapted from federal agencies; graph data from Horton, R., et al., 2015. New York City panel on climate change 2015 report chapter 2: Sea level rise and coastal storms. Ann. N.Y. Acad. Sci. 1336: 36–44.*

Shoreline homes toppled by storm surge

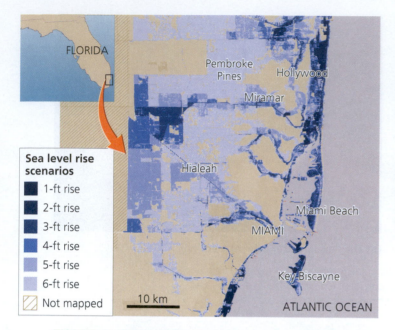

Sea level rise scenarios

- 1-ft rise
- 2-ft rise
- 3-ft rise
- 4-ft rise
- 5-ft rise
- 6-ft rise
- Not mapped

10 km

ATLANTIC OCEAN

FIGURE 14.19 Miami, Florida, is one of many U.S. cities vulnerable to sea level rise. Shown are areas of the Miami region that would be flooded by rises in sea level of 1–6 feet. *Data from NOAA Coastal Flood Exposure Mapper, www.coast.noaa.gov/floodexposure/#/map.*

(**FIGURE 14.19**). Here 2.4 million people, 1.3 million homes, and 1.8 million acres are vulnerable, according to experts with the Surging Seas project of Climate Central. In more than 100 South Florida towns, fully half the population is at risk.

Whether sea levels this century rise 26 cm, 1 m, or more, hundreds of millions of people will be displaced or will need to invest in costly efforts to protect against high tides and storm surges. A 2015 study calculated that if we burn all the fossil fuels remaining in the world, this would melt all the ice on Antarctica (half of it in just 1000 years), raising sea levels by about a foot per decade. If all the world's ice melted, the oceans would be 65 m (216 ft) higher. This would put the entire state of Florida under water, along with the rest of the Eastern Seaboard and Gulf Coast. The lower Mississippi River all the way up to Memphis would become a gigantic bay, and California's Central Valley, where much of our food is grown today, would be submerged under an inland sea.

Acidifying oceans imperil marine life

The oceans have absorbed roughly one-quarter of the carbon dioxide we have added to the atmosphere. This is altering ocean chemistry, making seawater more acidic—a phenomenon referred to as **ocean acidification** (p. 284). Ocean acidification threatens marine animals such as corals, clams, oysters, mussels, and crabs, which pull carbonate ions out of seawater to build their exoskeletons of calcium carbonate. As seawater becomes more acidic, carbonate ions become less available, and calcium carbonate begins to dissolve, jeopardizing the existence of these animals.

So far, global ocean chemistry has decreased by 0.1 pH unit, which corresponds to a 26% rise in acidity (hydrogen ion

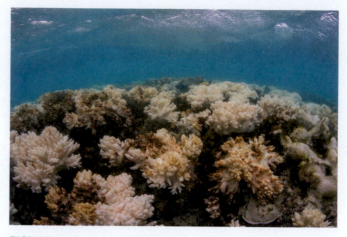

FIGURE 14.20 Australia's Great Barrier Reef suffered severe coral bleaching in 2016–2017. Coral bleaching can occur when warmed ocean waters cause corals to expel the symbiotic algae they rely on for food. The world's coral reefs are also vulnerable to ocean acidification.

concentration). Scientists are witnessing initial impacts on a variety of marine animals, and acidified seawater has already killed billions of larval oysters in Washington and Oregon, jeopardizing the region's once-thriving industry. By 2100, scientists predict that seawater will decline in pH by another 0.06–0.32 units—possibly enough to destroy most of our planet's living coral reefs (p. 267). Such destruction could be catastrophic for marine biodiversity and fisheries, because so many organisms depend on coral reefs for food and shelter. Indeed, ocean acidification, with the potential loss of marine life, threatens to become one of the most far-reaching impacts of global climate change.

Coral reefs face two additional risks from climate change: Warmer waters contribute to deadly coral bleaching (**FIGURE 14.20**; p. 267), and stronger storms physically damage reefs. All these factors concern residents of places like South Florida. South of Miami in the Florida Keys, coral reefs protect coastlines from erosion, offer snorkeling and scuba diving sites for tourism, and provide habitat for fish consumed locally and exported for profit.

Organisms and ecosystems are affected

As global warming proceeds, it modifies biological phenomena that rely on temperature. In the spring, plants are leafing out earlier and insects are hatching earlier. These shifts can create mismatches in seasonal timing with phenomena driven mostly by day length, such as bird migration. For example, in eastern North America, forests are greening up earlier in spring but many migratory songbirds are failing to arrive early enough to keep up with the change. This can have consequences; in Europe, some birds are raising fewer young because the insects they eat are peaking earlier and the birds have been unable to adjust.

Biologists are also recording spatial shifts in the ranges of species, as plants and animals move toward the poles or upward in elevation (toward cooler areas) as temperatures

FIGURE 14.21 Animal populations are shifting toward the poles and upward in elevation. Mountain-dwelling animals such as the pika, a unique mammal of western North America, are being forced upslope (into more limited habitat) as temperatures warm. Many pika populations in the Great Basin have disappeared from mountains already.

Pikas are disappearing from mountains after being forced upslope.

warm (**FIGURE 14.21**). Organisms that cannot adjust could face extinction. Many trees may not be able to shift their distributions fast enough. Rare species may be forced out of preserves and into developed areas where they cannot survive. Organisms adapted to mountainous environments may be forced uphill until there is nowhere left to go.

Changes in precipitation also have consequences. In regions where heavy rainstorms are increasing, erosion and flooding can pollute and alter aquatic systems. Where rain and snow are decreasing, lakes, ponds, wetlands, and streams are drying up. Overall, the many impacts of climate change on ecological systems will tend to diminish the ecosystem goods and services on which people depend.

Crops and forests experience a mix of consequences

Understanding the effects of climate change on plant communities is vital, especially in the ecosystems that we manage for food and resources. Plants draw in carbon dioxide for photosynthesis, so it stands to reason that an atmosphere richer in CO_2 might enhance plant growth, resulting in more CO_2 being removed from the air. On the other hand, if plant growth is inhibited by drought, fire, or disease, then more CO_2 gets released to the air. Scientists studying these questions are finding complex answers, and large-scale outdoor experiments show that extra CO_2 can both augment and diminish plant growth.

In the forests that provide our timber and paper products, enriched atmospheric CO_2 can spur growth, but drought, fire, and disease often eliminate these gains. Foresters increasingly find themselves battling catastrophic fires, invasive species, and insect and disease outbreaks, all of which can worsen with longer periods of warm and dry weather. For instance, milder winters and hotter, drier summers are promoting outbreaks of bark beetles that are destroying millions of acres of trees in western North America (p. 202).

For some agricultural crops in the temperate zones, moderate warming may slightly increase production because growing seasons become longer. Research shows that enriched atmospheric carbon dioxide may or may not boost yields, and also that crops can become less nutritious when supplied with more CO_2. If rainfall continues to shift in space and time, intensified droughts and floods will likely cut into agricultural productivity. Considering all factors, the IPCC has predicted that global crop yields will increase somewhat—but that beyond a rise of 3°C (5.4°F), they will decline, worsening hunger in developing nations.

Climate change affects our health, wealth, and security

Droughts, floods, storm surges, and other aspects of climate change are already taking a toll on the lives and livelihoods of millions of people. Ultimately, these impacts have consequences for our health, wealth, and national security.

Health As climate change proceeds, we are facing more heat waves—and heat stress can cause death. A 1995 heat wave in Chicago killed at least 485 people, and a 2003 heat wave in Europe killed 35,000 people. A warming climate also exposes us to other health risks:

- Respiratory ailments from air pollution as hotter temperatures promote photochemical smog (pp. 296–297)

- Expansion of tropical diseases, such as malaria and dengue fever, into temperate regions as disease vectors (such as mosquitoes) spread toward the poles

- Disease and sanitation problems when floods overcome sewage treatment systems

- Injuries and drowning from worsened storms

Wealth People are experiencing a variety of economic costs and benefits from the impacts of climate change, but researchers predict that costs will outweigh benefits. They also expect climate change to widen the gap between rich and poor, both within and among nations. Poorer people have less wealth and technology with which to adapt to climate change, and they rely more on resources (such as local food and water) that are sensitive to climate disruption.

Economists have tried to quantify damages from climate change by totaling up its various external costs (pp. 96, 104). Their estimates for the **social cost of carbon,** the economic cost of damages resulting from each ton of carbon dioxide we emit, run the gamut from $10 to $350 per ton, depending on what costs are included and what discount rate (p. 97) is used. The U.S. government has applied a formal estimate of roughly $40 per ton ($37 in 2007 dollars, rising with inflation) to decide when and how to regulate emissions. Other nations and many large corporations use their own estimates.

In terms of overall cost to society, the IPCC has estimated that climate change may impose costs of 1–5% of GDP globally, with poor nations losing proportionally more than rich nations. The *Stern Review on the Economics of Climate Change*, commissioned by the British government, concluded that climate change could cost us 5–20% of GDP by the year 2200. The U.S. Environmental Protection Agency (EPA) calculated in 2015 that reducing greenhouse gas emissions would save the United

FIGURE 14.22 Climate change can contribute to humanitarian, geopolitical, and national security problems. Prolonged drought associated with climate change weakened agriculture and helped spark the civil war in Syria and the rise of the Islamic State (ISIS), many experts have concluded. The resulting flow of refugees, in turn, put social and political strains on European nations that received them.

States $235–334 billion by the year 2050, and $1.3–1.5 trillion by 2100. Regardless of the precise numbers, most economists have concluded that investing money now to fight climate change will spare us a great deal of these costs in the future.

National security The many costs and impacts of climate disruption are beginning to endanger the ability of nations to ensure social stability and protect their citizens from harm. The Pentagon, the White House, the U.S. Navy, the Council on Foreign Relations, and the Central Intelligence Agency have all concluded and publicly reported that climate change is contributing to political violence, war and revolution, humanitarian disasters, and refugee crises (**FIGURE 14.22**). "Climate change will affect the Department of Defense's ability to defend the Nation and poses immediate risks to U.S. national security," the U.S Defense Department stated bluntly in a landmark 2014 report. The report described how storms, rising seas, and other impacts are "threat-multipliers," making small problems larger. Already, extreme weather events have damaged military installations, weakened the economies and infrastructure of allies and trading partners, disrupted flows of oil and gas, and strained emergency response abilities, while Arctic melting has set off a competitive race among nations to claim polar resources.

When environmental conditions worsen and people suffer as a result, some of them may leave their homes and become refugees, while others may turn to radical political ideologies or even terrorism. This is why national security experts have linked climate change and drought to the origins of the war in Syria, to conflicts elsewhere in the Middle East and Africa, and to the resulting refugee crisis in Europe. In the years ahead, the world's militaries and emergency responders will be devoting more and more of their efforts toward problems created or made worse by climate disruption.

What is behind the debate over climate change?

Scientists agree that today's global warming is due to the well-documented recent increase in greenhouse gas concentrations in our atmosphere. They agree that this rise results primarily from our combustion of fossil fuels for energy and secondarily from the loss of carbon-absorbing vegetation due to deforestation. They have documented a wide diversity of impacts on the physical properties of our planet, on organisms and ecosystems, and on human well-being (**FIGURE 14.23**).

Yet despite the overwhelming evidence for climate disruption and its impacts, many people, especially in the United States, have long tried to deny that it is happening. Most of these "climate skeptics" or "climate change deniers" now accept that the climate is changing but still express doubt that we are the cause. Public debate over climate change has been fanned by corporate interests, ideological think tanks, and a handful of scientists funded by fossil fuel industries, all of whom have aimed to cast doubt on the scientific consensus.

For instance, the oil corporation Exxon-Mobil funded attacks on climate science for years, methodically sowing doubt in the public discourse—even after its own in-house scientists had done cutting-edge research back in the 1980s documenting climate change! In the 2010 book *Merchants of Doubt*, science historians Naomi Oreskes and Erik Conway reveal how some of the ideologically motivated individuals who cast doubt on climate science had previously done the same against widely accepted scientific conclusions on the risks of tobacco smoke, DDT, ozone depletion, and acid rain.

The views of climate change deniers have long been amplified by the mainstream American news media, which traditionally have sought to present two sides to every issue, even when the arguments of the two sides were not equally supported by evidence. In today's more fragmented media environment, climate change has become a victim of political partisanship, and many of us receive information—and misinformation—from social media and partisan sources online.

The rise of Donald Trump to the U.S. presidency brought opposition to addressing climate change into the White House. Yet, as data have mounted over the decades and as the social and economic costs of climate disruption have grown clearer, more and more policymakers, business executives, military leaders, national security experts, and everyday people have concluded that climate change is escalating and is causing impacts to which we must begin to respond.

Responding to Climate Change

From this point onward, our society will be focusing on how best to respond to the challenges of climate change. The good news is that everyone can play a part in this all-important search for solutions—not just leaders in government and business, but people from all walks of life, and especially today's youth. Already we have made progress: Greenhouse

Average surface temperature has risen 1°C (1.8°F) since 1900. By 2100 it could rise 1.0–3.7°C (1.8–6.7°F) more.

Precipitation has varied by region, making wet areas wetter and dry ones drier.

Ocean surface waters are warming, melting Arctic sea ice and affecting circulation.

Polar ice and mountain glaciers are melting, worsening sea level rise and reducing drinking water supplies.

Storms, heavy rain, and flooding have increased, threatening crops, property, and human lives.

Ocean water is becoming more acidic, endangering marine life, coral reefs, and fisheries.

Most organisms and ecosystems are being affected. Some species could go extinct.

Heat waves and drought are intensifying in many regions, affecting farms, forests, and human health.

Sea level rose 21 cm (8 in.) since 1900 and could rise much more by 2100, displacing people and causing escalating expense.

FIGURE 14.23 Current trends and future impacts of climate change are extensive. Shown are major physical, biological, and social trends and impacts (both observed and predicted) as reported by the IPCC. (Mean estimates are shown; the IPCC reports ranges and statistical probabilities as well.) *Data from IPCC, 2013. Fifth assessment report.*

gas emissions globally may be finally reaching a peak, and in the United States they have begun to decline.

We can respond in two ways

We can respond to climate change in two fundamental ways. One is to pursue actions that reduce the magnitude of climate change. This strategy is called **mitigation** because the aim is to mitigate the problem; that is, to alleviate it or lessen its severity. To mitigate climate change, we need to reduce greenhouse gas emissions. Examples of mitigation include improving energy efficiency, switching to clean renewable energy sources, preserving and restoring forests, recovering landfill gas, and promoting farm practices that protect soil quality.

In the second type of response, we seek to cushion ourselves from the impacts of climate change. This strategy is called **adaptation** because the goal is to adapt to change. Installing elaborate pump systems, as Miami Beach is doing to pump out its floodwaters, is an example of adaptation (**FIGURE 14.24**). Other examples include erecting seawalls; restricting coastal development; adjusting farming practices to cope with drought; and modifying water management practices to deal with reduced river flows, glacial outburst floods, or salt contamination of groundwater.

Adaptation aims to address the impacts we suffer from climate change, whereas mitigation addresses the root cause

of the problem. We need to pursue adaptation because even if we could halt all our emissions tomorrow, the greenhouse gas pollution already in the atmosphere would continue driving global warming for years, with the temperature rising an estimated 0.6°C (1.0°F) more by the end of the century. Because this change is already locked in, we would be wise

FIGURE 14.24 Miami Beach is trying to adapt to sea level rise by constructing an elaborate system of pumps and drainage pipes. Adaptation is a more costly and less effective response to climate change than mitigation, however.

to find ways to adapt to its impacts. But we also need to pursue mitigation, because otherwise climate change will eventually overwhelm any efforts we might make to adapt. We will spend the remainder of our chapter examining approaches for the mitigation of climate change.

What can you do personally?

Although climate change requires action from leaders in policy, economics, business, and engineering, every one of us as individuals can play a role in reducing emissions. Just as we each have an ecological footprint (p. 5), we each have a **carbon footprint** that expresses the amount of carbon we are responsible for emitting. We can reduce our own carbon footprints by taking simple steps in our everyday lives. Taken together, individual decisions and actions scale up to make a difference at the societal level.

How you get around Transportation accounts for more than 35% of U.S. carbon dioxide emissions, largely because we rely on motor vehicles that run on gasoline. If you can drive less—or live without a car—you will cut down greatly on your carbon footprint. Today many people are reducing their reliance on cars. Some people are choosing to live nearer to their workplaces. Others use mass transit such as buses, subway trains, and light rail. Still others bike or walk to work.

If you drive, selecting a fuel-efficient vehicle makes a huge difference. The typical automobile is highly inefficient; only 14% of the energy from fuel in our gas tanks actually moves our cars down the road; most of the rest escapes as waste heat. Better engineering for more aerodynamic designs, increased engine efficiency, and improved tire design can improve fuel efficiency (p. 364). New technology is also bringing us alternatives to the traditional combustion-engine automobile. These include electric vehicles, gasoline-electric hybrids (p. 364), and vehicles that use hydrogen fuel cells (p. 395) or fuels such as compressed natural gas or biodiesel (p. 393). Finally, simple steps like driving the speed limit and keeping your tires inflated can boost your fuel efficiency by a third.

How you power up From cooking to heating to lighting to surfing the internet, much of what we do each day depends on electricity, which accounts for 35% of U.S. emissions. One easy way to reduce electricity consumption—and save money at the same time—is to purchase energy-saving products. You can consult EnergyGuide labels and labels from the EPA's Energy Star program when shopping for electronics, appliances, and home and office equipment to judge which brands and models will save energy and money (pp. 364–365). In recent years U.S. consumers, businesses, and industries have saved hundreds of billions of dollars while reducing emissions by adopting energy-efficient appliances, lighting, windows, doors, ducts, insulation, electronics, and heating and cooling systems.

As individuals and as a society we also can reduce emissions by switching to cleaner energy sources. Two-thirds of our electricity comes from burning fossil fuels, and coal accounts for most of the resulting emissions. Natural gas generates the same amount of energy as coal, but with half the emissions. In recent years the United States has reduced its emissions largely by switching from coal to natural gas at its power plants (see Figure 15.13, p. 353). Cleaner still are nuclear power (Chapter 15) and renewable energy sources such as solar power, wind power, geothermal power, bioenergy, and hydropower (Chapter 16). As a homeowner, you can install solar panels or ground-source heat pumps. As a student on campus, you can help persuade your college or university to purchase renewable power or to generate solar or wind power on campus.

What you eat One of the most effective ways to lower your carbon footprint is through your diet. Eating animal products such as meat, eggs, and milk is far more energy-intensive than eating lower on the food chain—so a vegetarian diet will reduce your emissions greatly. The type of meat one eats also makes a difference: Pound for pound, beef leads to eight times more emissions than chicken (see Figure 7.15, p. 157)! Scientists estimate that animal agriculture accounts for fully 14.5% of the world's greenhouse gas emissions. For many of us, reducing the amount of red meat we eat while shifting to more fruits and vegetables can reduce our carbon footprint more than any other single action we could take, while also enhancing our health.

Eating locally grown food (or buying any locally made product)—as opposed to items shipped in from far away—cuts down on fuel use, and thus emissions, from long-distance transportation. Studies show that the type of food you eat tends to have more influence than where it comes from (for example, eating less meat makes more difference than buying local produce), but keeping both things in mind will help make a dent in emissions.

Finally, being careful not to waste food will make a difference. About 40% of U.S. food goes to waste, adding greatly to the nation's emissions. Taking smaller portions (encouraged by trayless dining at campus dining halls), ordering wisely at restaurants, and cleaning out your refrigerator all help to reduce food waste.

What you do We can take many other steps to minimize our carbon footprints. Here are just a few:

- Choose energy-efficient products when shopping, thereby encouraging manufacturers and retailers to produce and sell low-emission products.

- Cut back on waste by buying only what you need, reusing items whenever possible, and recycling and composting what you no longer need (Chapter 17).

- Get involved in sustainability efforts on campus (pp. 19, 435), such as running recycling programs; finding ways to enhance energy and water efficiency; managing gardens and sustainable dining halls; or pressing administrators to build green buildings, divest from fossil fuel stocks, or invest in renewable energy (**FIGURE 14.25**).

- Get engaged politically by communicating with your representatives in government, supporting candidates for

FIGURE 14.25 You can help reduce the carbon footprint of your college or university. Pressing for divestment from fossil fuel stocks is just one of many ways students are working to reduce campus emissions.

office who will tackle the challenges of climate disruption, and advocating for policies to support clean energy and energy efficiency.

- Be a model for others in your daily life. Perhaps you can encourage resource conservation at your workplace. Maybe you will pursue a career in a profession that promotes energy efficiency or renewable energy. Perhaps as a parent you will pass on to your children what you've learned about addressing climate change. Our actions reverberate with the people around us—so through what you say and do, you can multiply your own efforts.

Multiple strategies can help us reduce emissions

There is no single "magic bullet" for stopping climate change—but there are many ways that together can help. Waste managers are cutting emissions by generating energy from waste in incinerators (p. 407); capturing methane seeping from landfills (p. 408); and encouraging recycling, reuse, and composting (pp. 403–405). Power producers are capturing excess heat from electricity generation and putting it to use by cogeneration (p. 363). Sustainably managing cropland and rangeland enables soil to store more carbon, while new techniques reduce methane emission from rice farming and cattle. Preserving forests, reforesting cleared areas, and pursuing sustainable forestry practices (pp. 202–203) all help to absorb carbon dioxide from the air.

As our society transitions to clean energy, we are also trying to capture emissions before they leak to the atmosphere. Billions of dollars are being spent to develop *carbon capture and storage*, by which carbon dioxide is removed from emissions and stored belowground under pressure in deep salt mines, depleted oil and gas deposits, or other underground reservoirs (see Figure 15.19, p. 360).

Back in 2004, environmental scientists Stephen Pacala and Robert Socolow advised that we follow some age-old wisdom: When the job is big, break it into smaller parts. Pacala and Socolow identified 15 strategies (**FIGURE 14.26**) that could each

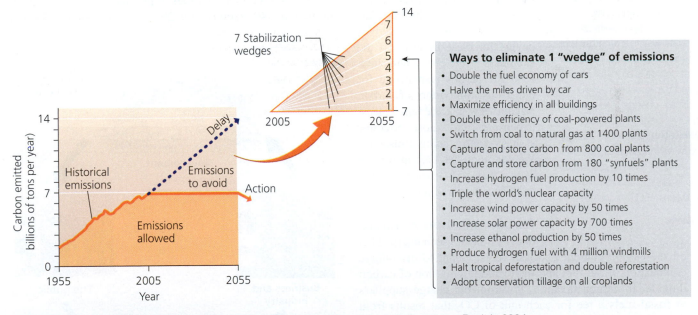

FIGURE 14.26 To stabilize emissions, we can break the job down into smaller steps. Back in 2004, environmental scientists Stephen Pacala and Robert Socolow created this graph showing the doubling of CO_2 emissions that scientists expected to occur from 2005 to 2055. They added a flat line to represent the trend if emissions were held constant and then separated the graph into emissions allowed (below the line) and emissions to be avoided (the triangular area above the line). They then divided this "stabilization triangle" into seven equal-sized wedges. Each "stabilization wedge" represents 1 billion tons of CO_2 emissions in 2055 to be avoided. Finally, they identified a series of strategies, each of which could take care of one wedge. Accomplishing just 7 of these strategies could halt our growth in emissions for the next half-century. Since then, society has made enough progress on multiple fronts (renewable energy, efficiency measures, converting from coal to gas, and more) that global emissions held steady in 2014–2016. *Adapted from Pacala, S., and R. Socolow, 2004. Stabilization wedges: Solving the climate problem for the next 50 years with current technologies. Science 305: 968–972.*

eliminate 1 billion tons of carbon per year by 2050 if deployed at a large scale. Achieving just 7 of these 15 aims would stabilize our emissions. If we were to achieve more, we would reduce emissions. Since that time, no one of these 15 strategies has been fully achieved, yet we have made progress with enough of them, and others, that global emissions at last held steady from 2014 to 2016, even as economic growth continued. With further progress on multiple fronts, global emissions may peak and begin to decline.

weighing the
ISSUES

Taking Climate Change to Court

People desiring stronger action against climate change are now taking their grievances to court. In the Netherlands in 2015, a group named Urgenda and 900 citizens sued the Dutch government for "knowingly contributing" to global warming—and won. The court agreed that Dutch policy would not hold warming to the internationally set 2°C goal, and ordered the government to deepen emission cuts. In a U.S. lawsuit in 2016, the nonprofit Our Children's Trust and 21 young people demanded action on climate change, saying the federal government had "willfully ignored this impending harm." In Peru, a farmer sued a fossil fuel power company, asking monetary compensation for impacts of glacier melt on his community.

Do you think lawsuits are an appropriate way to strengthen policy responses to climate change? To what degree is a government obligated to protect its citizens from climate change? Do you favor suing fossil fuel companies for compensation for climate change impacts? What ethical or human rights issues, if any, do you think climate change presents? How could these best be resolved?

We can put a price on carbon

To encourage efforts to reduce emissions, today there is a growing call for "putting a price on carbon"—using economic incentives as tools to motivate voluntary reductions in emissions. **Carbon pricing** is intended to compensate the public for the external costs (pp. 96, 104) we all suffer from fossil fuel emissions and climate change. Carbon pricing lifts the burden of paying for these impacts off the shoulders of the public and shifts it to the parties responsible for emissions. Supporters of carbon pricing view it as the fairest, least expensive, and most effective strategy for reducing emissions. We can approach carbon pricing in two ways: (1) carbon taxation and (2) carbon trading.

Carbon taxation A **carbon tax** is a type of green tax (p. 112) on the carbon content of fossil fuels, meant to discourage combustion of the fuel and the resulting emission of carbon dioxide. In carbon taxation, governments charge suppliers of fossil fuels a fee for each unit of CO_2 that results from their product. For instance, a government might tax firms that mine, process, sell, or import fossil fuels—such as coal-mining companies, natural gas suppliers, or oil refineries. These parties will pass the cost of the tax along to distributors and retailers, such as gas stations or the utilities that generate power from fossil fuels, and their costs in turn get passed along to consumers in the form of higher prices for products like gasoline or electricity.

Higher prices give consumers motivation to reduce fossil fuel use by switching to cheaper alternatives. For instance, a driver might opt to buy a more fuel-efficient vehicle, a homeowner might choose to live in a city near a bus route rather than in a remote suburb, or a factory might relocate to a site along a railroad line. Likewise, businesses, industries, and utilities gain motivation to switch to less carbon-intensive products as consumer demand changes—a coal-fired power plant might switch to cleaner-burning natural gas, or an oil company might boost its investments in solar and wind energy. As the cost of the tax is passed along through transactions, carbon-intensive products become more expensive and every party in the economy gains incentives to buy, sell, and use products that cause less carbon pollution.

Although costs are passed along to consumers, consumers end up paying nothing extra if carbon taxation takes a **fee-and-dividend** approach. In this approach, funds the government receives from fossil fuel suppliers through the carbon tax (the "fee") are transferred back to taxpayers as a tax cut or a tax refund (the "dividend"). This way, any costs of a carbon tax that get passed along to consumers will be reimbursed to them (**FIGURE 14.27**) in the form of a reduction in their income taxes. In theory, such a system gives everyone a financial incentive to reduce emissions while imposing no financial burden on taxpayers and no drag on the economy. The fee-and-dividend approach is a type of **revenue-neutral carbon tax,** because there is no net transfer of revenue from taxpayers to the government. For this reason, the approach is gaining broad appeal across the political spectrum.

Carbon taxes of various types have been introduced in roughly 40 nations and 20 cities and states. Sweden

Emitters pay carbon tax

Government reduces taxes for taxpayers

Government

Business and industry

The public

Consumers pay higher prices for products

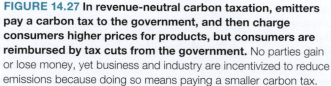

FIGURE 14.27 In revenue-neutral carbon taxation, emitters pay a carbon tax to the government, and then charge consumers higher prices for products, but consumers are reimbursed by tax cuts from the government. No parties gain or lose money, yet business and industry are incentivized to reduce emissions because doing so means paying a smaller carbon tax.

has one of the world's steepest and longest-running carbon taxes and has reduced fossil fuel use greatly as a result (see Figure 15.29, p. 367), while expanding its economy and providing its citizens one of the world's highest standards of living. The Canadian province of British Columbia has embarked on revenue-neutral carbon taxation that is also showing success. In the United States, Boulder, Colorado, taxes electricity consumption; Montgomery County, Maryland, taxes power plants; and San Francisco Bay Area counties tax businesses for emissions.

Carbon trading The second means of pricing carbon is carbon trading. In an emissions trading system (p. 113), a government sets up a market in permits for the emission of pollutants, and companies, utilities, or industries buy and sell the permits among themselves (**FIGURE 14.28**). A **carbon trading** system is one in which permits are traded for the emission of carbon dioxide or greenhouse gases in general. In such a market, the price of permits fluctuates freely according to supply and demand. In the approach known as **cap-and-trade** (p. 113), the government sets a cap on the amount of pollution it will allow, then gives, sells, or auctions permits to emitters that allow them to emit a certain fraction of the total amount. Over time the government lowers the cap to ensure that total emissions decrease. Emitters with too few permits to cover their pollution must reduce their emissions, buy permits from other emitters, or pay for carbon offset credits (p. 337).

As with carbon taxation, carbon trading harnesses the financial incentives of free-market capitalism while granting emitters the freedom to decide how they can best reduce emissions. In theory, once polluters are charged a price for polluting, market forces do the work of reducing pollution in an economically efficient way by allowing business, industries, or utilities flexibility in how to respond.

The world's largest cap-and-trade program is the European Union Emission Trading Scheme. This market began in 2005, but investors soon realized that national governments had allocated too many permits to their industries. The overallocation gave companies little incentive to reduce emissions, so permits lost their value and market prices collapsed. Europeans addressed these problems by making emitters pay for permits, and today the market runs more effectively. Similar difficulties befell the world's first emissions trading program for greenhouse gas reduction, the Chicago Climate Exchange, which operated from 2003 to 2010 and involved several hundred corporations, institutions, and municipalities.

Today, California is running a cap-and-trade program that seems to be succeeding. It is on track to reduce emissions to 1990 levels by the year 2020, while the state's economy continues to thrive. The Canadian province of Quebec joined California's market in 2017 and Ontario plans to join in 2018. Meanwhile, nine northeastern states are participating in the Regional Greenhouse Gas Initiative. In this effort, Connecticut, Delaware, Maine, Maryland, Massachusetts, New Hampshire, New York, Rhode Island, and Vermont run a cap-and-trade program for power plant emissions. From 2005 to 2014, these states cut their CO_2 emissions from power plants by 45%, even as their economies grew. It is estimated that investment of the auction proceeds will save consumers $4.7 billion in energy costs and eliminate 15 million tons of CO_2 emissions.

Internationally, China is setting up a national cap-and-trade program, and Mexico may do so as well. Many observers are skeptical that China will be able to ensure transparency, prevent corruption, and minimize government interference in its market—but if China succeeds, this could represent a major step toward controlling emissions globally. All these early experiments are providing lessons for how to set up effective and sustainable trading systems.

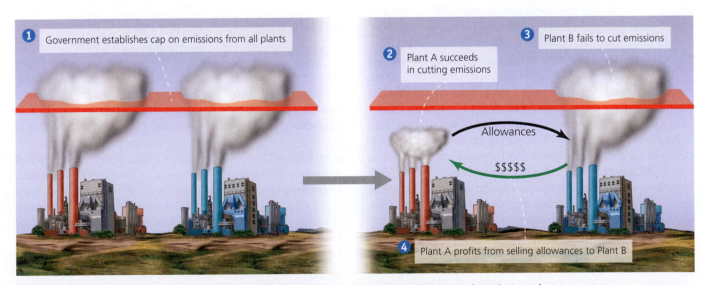

FIGURE 14.28 A cap-and-trade emissions trading system harnesses the efficiency of market capitalism to reduce emissions. In the diagram, Plant A succeeds in reducing its emissions below the cap and Plant B does not. As a result, Plant B pays money to Plant A to purchase allowances that Plant A no longer needs to use. Plant A profits from this sale, and the government cap is met, reducing pollution overall. Over time, the cap can be lowered to achieve further emissions cuts.

Taxing Carbon in British Columbia

The Canadian province of British Columbia introduced a revenue-neutral carbon tax in 2008. Established by a politically conservative government, the tax began at $10 per ton of CO_2 equivalent and gradually rose to $30, garnering more than a billion dollars in revenue per year. Monies raised from the carbon tax have been replacing revenues from corporate and personal income taxes, which were lowered. So far, the policy appears to be working: British Columbia's fuel consumption and greenhouse gas emissions have declined, the province's economy has remained strong, and its taxpayers are enjoying lower income taxes. According to calculations by the nonprofit Carbon Tax Center, British Columbia reduced its per capita greenhouse gas emissions by 12.9% in the five years following introduction of the carbon tax, while elsewhere in Canada per capita emissions fell by only 3.7%. Interpretations of the data vary, but perceived success in British Columbia led Canada's national government to propose nationwide carbon pricing, slated to begin in 2018. Under the plan, each of Canada's provinces would either adopt a national revenue-neutral carbon tax (starting at $10/ton and rising to $50/ton in 2022) or join a cap-and-trade market.

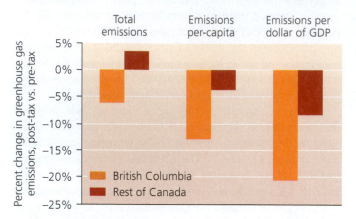

Relative to the rest of Canada, British Columbia released fewer emissions after introducing its carbon tax (2008–2013) than before the tax (2000–2007).

EXPLORE THE DATA at **Mastering** Environmental Science

What role should government play?

Amid all the promising approaches, strategies, and technologies for reducing greenhouse gas emissions, people often disagree on what role government should play to encourage such efforts: Should it mandate change through laws and regulations? Should it design policies that give private entities financial incentives to reduce emissions? Should it impose no policies at all and hope that private enterprise will develop solutions on its own? This debate has been vigorous in the United States, where many business leaders and politicians have opposed all government action to address climate change, fearful that emissions reductions will impose economic costs on industry and consumers.

In 2007, the U.S. Supreme Court ruled that carbon dioxide was a pollutant that the Environmental Protection Agency (EPA) could regulate under the Clean Air Act (p. 292). When Barack Obama became president, he instead urged that Congress craft laws to address emissions. In 2009, the House of Representatives passed legislation to create a nationwide cap-and-trade system in which industries and utilities would compete to reduce emissions for financial gain, and under which emissions were mandated to decrease 17% by 2020. However, legislation did not pass in the Senate. As a result, responsibility for addressing emissions passed to the EPA, which began phasing in emissions regulations on industry and utilities, hoping to spur energy efficiency retrofits and renewable energy use.

In 2013, President Obama announced that because of legislative gridlock, he would take steps to address climate change using the president's executive authority. His Climate Action Plan aimed to jumpstart renewable energy development, modernize the electrical grid, finance clean coal and carbon storage efforts, improve automotive fuel economy, protect and restore forests, and encourage energy efficiency. It also led to the Clean Power Plan (p. 301), under which the EPA proposed to regulate existing power plants.

The rise of Donald Trump to the presidency, together with Republican control of Congress, resulted in the sudden reversal of most Obama-era federal policies designed to reduce greenhouse gas emissions. By executive order, Trump halted the Clean Power Plan. His actions and policies raised questions as to whether and how the United States would continue to engage with other nations in international efforts to curb climate change.

International climate negotiations have sought to limit emissions

Disruption of the climate is a global problem, so global cooperation is needed to forge effective solutions. In 1992, most nations signed the **U.N. Framework Convention on Climate Change.** This treaty outlined a plan for reducing greenhouse gas emissions to 1990 levels by the year 2000 through a voluntary approach. Emissions kept rising, however, so nations forged a binding treaty to *require* emissions reductions. Drafted in 1997 in Kyoto, Japan, the **Kyoto Protocol** mandated signatory nations, by the period 2008–2012, to reduce emissions of six greenhouse gases to levels below those of 1990. The treaty took effect in 2005 after Russia became the 127th nation to ratify it.

The United States was the only developed nation not to ratify the Kyoto Protocol. U.S. leaders objected to how it required industrialized nations to reduce emissions but did not require the same of rapidly industrializing nations such as China and India. Proponents of the Kyoto Protocol countered that this was justified because industrialized nations created the climate problem and therefore should take the lead in resolving it.

As of 2015, nations that signed the Kyoto Protocol had decreased their emissions (**FIGURE 14.29**) by 12.0% from 1990 levels. However, much of this reduction was due to economic contraction in Russia and nations of the former Soviet Bloc following the breakup of the Soviet Union. When these nations are factored out, the remaining signatories showed only a 0.7% decrease in emissions. Nations not parties to the accord, including China, India, and the United States, increased their emissions.

All along, representatives of the world's nations were meeting at a series of annual conferences, trying to design a treaty to succeed the Kyoto Protocol. Delegates from European nations and small island nations generally took the lead, while China, India, and the United States were reluctant to commit to emissions cuts. At a contentious 2009 conference in Copenhagen, Denmark, nations endorsed a goal of limiting climate change to 2°C of warming, but failed to agree on commitments. In Cancun, Mexico, in 2010, nations made progress on a plan, nicknamed *REDD* (p. 198), to help tropical nations reduce forest loss, and developed nations promised to help fund mitigation and adaptation efforts for developing nations. In Doha, Qatar, in 2012, negotiators extended the Kyoto Protocol until 2020, but a number of nations backed out, and the protocol now applies to only about 15% of the world's emissions.

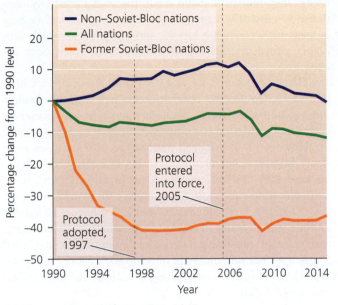

(a) Changes in emissions, 1990–2015

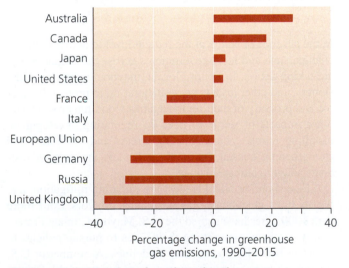

(b) Changes in emissions for selected nations

FIGURE 14.29 The Kyoto Protocol produced mixed results. Nations ratifying it reduced their emissions of six greenhouse gases by 12.0% by 2015 **(a)**, but this was largely due to unrelated economic contraction in the former Soviet-Bloc countries. A selection of major nations **(b)** shows varied outcomes in reducing emissions. The United States did not ratify the Protocol, Australia joined it late, and Canada left early. Values do not include influences of land use and forest cover. *Data from U.N. Framework Convention on Climate Change.*

DATA Q In part (b), compare the nations whose emissions increased with those whose emissions decreased. What difference(s) do you note between these two groups that might explain why their emission trends differ?

Go to **Interpreting Graphs & Data** on **Mastering** Environmental Science

The Paris Accord produced global consensus

At the 2015 climate conference in Paris, France, the world's nations made more far-reaching commitments than ever before. European nations announced plans to further reduce their emissions. China pledged to cut back on coal-fired power and to establish a cap-and-trade program. Brazil promised to halt deforestation. India agreed to slow its emissions growth and reforest its land while aggressively developing renewable energy. And the United States committed to emissions cuts it hoped would result from new regulations on coal-fired power plants and from switching from coal to natural gas. All these national commitments were voluntary and unenforceable, but a strong global consensus was forged, creating powerful diplomatic peer pressure on all parties to live up to their pledges.

The impressive progress at the Paris conference resulted because all nations were encouraged to bring their own particular solutions to the table. Success at Paris was also facilitated by an agreement a year earlier between U.S. President Obama and Chinese President Xi Jinping, which broke the impasse between these two largest polluting nations. In their joint announcement, Obama had pledged the United States would reduce carbon emissions by 28% by 2025, and Xi Jinping had promised China would halt its emissions growth by 2030.

In 2017, President Trump announced that his administration would withdraw the United States from the Paris Accord, breaking with 194 other nations and joining Nicaragua and Syria as the only nations not party to the agreement. Trump felt the accord would hurt America's economy. His decision drew scorn and outrage from around the world and

FIGURE 14.30 Commitments made in the 2015 Paris Accord would limit the global temperature increase to 3.3°C (6.0°F) by 2100. This is less than the 4.2°C (7.6°F) rise predicted in the absence of those commitments, but exceeds the target of limiting temperature rise to 2.0°C (3.6°F). *Data from Climate Interactive.*

was opposed in polls by the majority of the American people. What impact a U.S. withdrawal—which by the terms of the accord cannot formally take effect until 2020—might have is unclear. Many people feared that the decision would cause other nations to abandon the commitments they made in Paris. Yet the resolute unity nations displayed in the face of Trump's announcement suggested that the rest of the world might instead redouble its commitment to fighting climate change.

Ultimately, the historic conference in Paris will be judged by how well nations live up to their pledges. Yet even if all nations fully meet their Paris pledges, calculations indicate that the average global temperature by 2100 would still rise about 3.3°C (6.0°F) (**FIGURE 14.30**). Indeed, many experts now predict that success in mitigating climate change will rest on technological advances, economic incentives, and business investments in renewable energy and energy efficiency, as well as on government initiatives at the national, regional, state, and local levels.

Will emissions cuts hurt the economy?

Like President Trump, many U.S. policymakers have opposed reducing emissions because they fear this will hamper economic growth. China and India long resisted emissions cuts under the same assumption. This is understandable, given that our economies have relied so heavily on fossil fuels. Yet nations such as Germany, France, and the United Kingdom have reduced their emissions greatly since 1990 while enhancing their economies and providing their citizens very high standards of living. The per-person emission of greenhouse gases in wealthy nations from Denmark to New Zealand to Hong Kong to Switzerland to Sweden is less than half that in the United States.

Indeed, the United States was able to reduce its carbon dioxide emissions by 14% from 2007 to 2016. This occurred

as a result of numerous efficiency measures and a shift from coal to natural gas in power plants. The U.S. economy grew during this period, suggesting that cutting emissions need not hinder economic growth. Moreover, from 2014 through 2016, for the first time in history, global carbon emissions stopped rising; they remained stable while the world economy continued to grow. These encouraging data suggest that we may perhaps have reached a historic turning point at which economic growth has become decoupled from greenhouse gas emissions.

Because resource use and per capita emissions are high in the United States, policymakers and industry leaders often assume the United States has more to lose economically from restrictions on emissions than developing nations do. However, industrialized nations are also the ones most likely to *gain* economically from major energy transitions, because they are best positioned to invent, develop, fund, and market new technologies to power the world in a post-fossil-fuel era. Germany, Japan, and China have realized this and are now leading the world in production, deployment, and sales of renewable energy technology. If the United States does not act quickly to develop such energy technologies, then the future could belong to nations like China, Germany, and Japan.

States, cities, and businesses are advancing climate change efforts

In the absence of action at the federal level to address climate change, state and local governments across the United States are taking action. In South Florida, mayors, county commissioners, and other leaders are collaborating to protect the people and property of their region against flooding, erosion, and other impacts of sea level rise. By regulating development, installing pump systems, raising streets and buildings, and strengthening coastal dunes, these leaders are helping residents adapt to climate change.

Elsewhere, political leaders are pursuing mitigation, trying to limit greenhouse gas emissions. Mayors from more than 1000 cities have signed the U.S. Mayors Climate Protection Agreement, committing their cities to pursue policies to "meet or beat" Kyoto Protocol guidelines. A number of U.S. states have enacted targets or mandates for renewable energy production, seeking to boost alternatives to fossil fuels. The boldest state-level action so far has come in California. In 2006 that state's legislature worked with then-Governor Arnold Schwarzenegger to pass the Global Warming Solutions Act, which sought to cut California's greenhouse gas emissions 25% by the year 2020. This law established the state's cap-and-trade program and followed earlier efforts to mandate higher fuel efficiency for automobiles.

In the wake of President Trump's decision to leave the Paris Accord, state, regional, and local efforts went into overdrive. Former New York City Mayor Michael Bloomberg established a group that brought together dozens of governors, mayors, university presidents, and corporate leaders who promised action to support the U.S. emissions reduction pledge under the Paris Accord. California governor Jerry

Brown flew to Beijing and met with Chinese President Xi Jinping to discuss climate cooperation between China and California.

Moreover, individuals from the private sector have become involved as never before. At the Paris conference, Microsoft founder and billionaire philanthropist Bill Gates led an effort to attract private investment into renewable energy from some of the wealthiest people in the world. And as Trump mulled over his decision, CEOs of many major corporations urged him to stay in the Paris Accord. Business thrives on certainty and stability of policy direction, and most corporations today have already invested a great deal in energy efficiency and emissions reductions—and have found these efforts to help their bottom line.

Ultimately, many businesses, utilities, universities, governments, and individuals are aiming to achieve **carbon-neutrality**, a condition in which no net greenhouse gases are emitted. In practice, this generally requires buying **carbon offsets**, voluntary payments intended to enable another entity to help reduce the emissions that one is unable to reduce. The payment thus offsets one's own emissions. For example, a coal-burning power plant could fund a reforestation project to plant trees that will soak up as much carbon dioxide as the coal plant emits. Or a university could fund clean renewable energy projects to make up for fossil fuel energy the university uses. In principle, carbon offsets are a powerful idea, but rigorous oversight is needed to make sure that offset funds achieve what they are intended for—and that offsets fund only emissions cuts that would not occur otherwise.

Should we engineer the climate?

What if all our efforts to reduce emissions are not adequate to rein in climate change? As climate disruption becomes more severe, some scientists and engineers are reluctantly considering drastic, assertive steps to alter Earth's climate in a last-ditch attempt to reverse global warming—an approach called **geoengineering** (**FIGURE 14.31**).

One geoengineering approach would be to suck carbon dioxide out of the air. To achieve this, we might enhance photosynthesis by planting trees at large scales or by fertilizing ocean phytoplankton with nutrients such as iron. A more high-tech method might be to design "artificial trees," structures that chemically filter CO_2 from the air. A different approach would be to block sunlight before it reaches Earth, thereby cooling the planet. We might deflect sunlight by injecting sulfates or other fine dust particles into the stratosphere; by seeding clouds with seawater; or by deploying fleets of reflecting mirrors on land, at sea, or in orbit in space.

Scientists were long reluctant even to discuss the notion of geoengineering. The potential methods are technically daunting, would take years or decades to develop, and might pose unforeseen risks. Blocking sunlight does not reduce greenhouse gases, so ocean acidification would continue. And any method would work only as long as society has the capacity to maintain it. Moreover, many experts are wary of promulgating hope for easy technological fixes, lest politicians lose incentive to try to reduce emissions through policy.

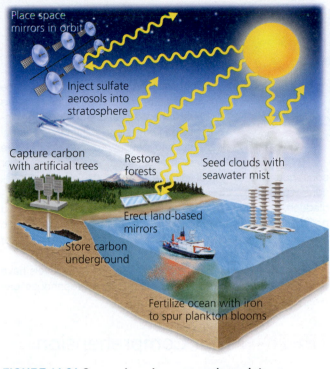

FIGURE 14.31 Geoengineering proposals seek to use technology to remove carbon dioxide from the air or reflect sunlight away from Earth. However, most geoengineering ideas would take years to develop, may not work well, or might cause undesirable side effects. Thus, they are not a substitute for reducing emissions.

However, as climate change intensifies, scientists are beginning to assess the risks and benefits of geoengineering, so that we can be ready to take well-informed action if climate change becomes severe enough to justify it.

We all can address climate change

Government policies, corporate actions, international treaties, carbon pricing, technological innovations—and perhaps even geoengineering—all have roles to play in addressing climate change. But in the end, the most influential factor may be the collective actions of millions of regular people. To help reduce emissions, each of us can take steps in our everyday lives, from shifting our diets to choosing energy-efficient appliances to deciding where to live and how to get to work.

College students are vital to driving personal and societal changes to reduce carbon footprints and address climate change—through common practices, educating others, and political engagement. Today a groundswell of interest is sweeping across campuses, and many students are pressing their administrations to seek carbon-neutrality.

Global climate change may be the biggest challenge we face, but halting it would be our greatest victory. With concerted action, there is still time to avert the most severe impacts. Through outreach, education, innovation, and lifestyle choices, we have the power to turn the tables on climate change and help bring about a bright future for humanity and our planet.

closing THE LOOP

Many factors influence Earth's climate, and human activities have come to play a major role. Climate change is well underway, and additional greenhouse gas emissions will intensify global warming and cause progressively severe and diverse impacts. Sea level rise and other consequences of global climate change are affecting locations worldwide—from Miami to the Maldives, Alaska to Bangladesh, and New York to the Netherlands. As scientists and political leaders come to better understand anthropogenic climate disruption and its consequences, more and more of them are urging immediate action.

Policymakers at the international and national levels have struggled to take meaningful steps to slow greenhouse gas emissions, so increasingly, people at local and regional levels are the ones making a difference. In South Florida, citizens and local leaders are investing time, thought, money, and creativity into finding solutions to rising sea levels. They are seeking to mitigate climate change by reducing greenhouse gas emissions and to adapt to climate change by building pumping systems, raising streets and foundations, and tailoring financial and insurance incentives to guide development toward upland areas. Like people anywhere who love their homes, residents of South Florida are girding themselves for a long battle to protect their land, communities, and quality of life while our global society inches its way toward emissions reductions. For all of us across the globe, taking steps to mitigate and adapt to climate change represents the foremost challenge for our future.

TESTING Your Comprehension

1. What happens to solar radiation after it reaches Earth? How do greenhouse gases warm the lower atmosphere?

2. Why is carbon dioxide considered the main greenhouse gas? Why are carbon dioxide concentrations increasing in the atmosphere?

3. What evidence do scientists use to study the ancient atmosphere? Describe what a proxy indicator is, and give two examples.

4. Has simulating climate change with computer programs been effective in helping us predict climate? Briefly describe how these programs work.

5. List three major trends in climate that scientists have documented so far. Now list three future trends that scientists predict, along with potential consequences.

6. Describe how rising sea levels, caused by global warming, can create problems for people. How is climate change affecting marine ecosystems?

7. How might a warmer climate affect agriculture? How is it affecting distributions of plants and animals? How might it affect human health?

8. What are the two largest sources of greenhouse gas emissions in the United States? How can we reduce emissions from these sources?

9. What roles have international treaties played in addressing climate change? Give two specific examples.

10. What is meant by "putting a price on carbon?" Describe the two major approaches to carbon pricing. Discuss advantages and disadvantages of each approach.

SEEKING Solutions

1. Some people argue that we need "more proof" or "better science" before we commit to changes in our energy economy. How much certainty do you think we need before we take action regarding climate change? How much certainty do you need in your own life before you make a major decision? Should nations and elected officials follow a different standard? Do you feel that the precautionary principle (pp. 162, 227) is an appropriate standard in the case of global climate change? Why or why not?

2. Suppose that you would like to make your own lifestyle carbon-neutral. You plan to begin by reducing the emissions you are responsible for by 25%. What three actions would you take first to achieve your goal?

3. How might your campus reduce its greenhouse gas emissions? Develop three concrete proposals for ways to reduce emissions on your campus that you feel would be effective and feasible. How would you present these proposals to campus administrators to gain their support?

4. **CASE STUDY CONNECTION** You are the city manager for a coastal U.S. city that scientists predict will be hit hard by sea level rise, with risks and impacts trailing those in Miami by just a few years. You have recently returned from a professional conference in Florida, where you toured Miami Beach and learned of the efforts being made there to adapt to climate change. What steps would you take to help your own city prepare for rising sea level? How would you explain the risks and impacts of climate disruption to your fellow city leaders to gain their support? Of the measures being taken in Florida communities, which would you

choose to study closely, which would you want to begin right away, and which would be highest priority in the long run? Explain your choices.

5. **THINK IT THROUGH** You have just been elected governor of a medium-sized U.S. state. Polls show that the public wants you to take action to reduce greenhouse gas emissions, but does not want prices of gasoline or electricity to rise. Your state legislature will support you in your efforts as long as you remain popular with voters. One neighboring state has just passed legislation mandating that one-third of its energy come from renewable sources within 10 years. Another neighboring state has recently joined a regional emissions trading consortium. A third neighboring state has just enacted a revenue-neutral carbon tax. What actions will you take in your first year as governor, and why? What effects would you expect each action to have?

CALCULATING Ecological Footprints

We all contribute to global climate change, because fossil fuel combustion plays such a large role in supporting the lifestyles we lead. Likewise, as individuals, each one of us can help to address climate change through personal decisions and actions in how we live our lives. Several online calculators enable you to calculate your own personal carbon footprint, the amount of carbon emissions for which you are responsible. Go to http://www.nature.org/greenliving/carboncalculator/ or to www.carbonfootprint.com/calculator.aspx, take the quiz, and enter the relevant data in the table.

	CARBON FOOTPRINT (TONS PER PERSON PER YEAR)
World average	
U.S. average	
Your footprint	
Your footprint with three changes (see Question 3)	

1. How does your personal carbon footprint compare to that of the average U.S. resident? How does it compare to that of the average person in the world? Why do you think your footprint differs in the ways it does?

2. As you took the quiz and noted the impacts of various choices and activities, which one surprised you most?

3. Think of three changes you could make in your lifestyle that would lower your carbon footprint. Now take the footprint quiz again, incorporating these three changes. Enter your resulting footprint in the table. By how much did you reduce your yearly emissions?

4. What do you think would be an admirable yet realistic goal for you to set as a target value for your own footprint? Would you choose to purchase carbon offsets to help reduce your impact? Why or why not?

Mastering Environmental Science

Students Go to **Mastering** Environmental Science for assignments, the etext, and the Study Area with practice tests, videos, current events, and activities.

Instructors Go to **Mastering** Environmental Science for automatically graded activities, current events, videos, and reading questions that you can assign to your students, plus Instructor Resources.

CHAPTER 15

Nonrenewable Energy Sources, Their Impacts, and Energy Conservation

Fracking the Marcellus Shale

> " **Hydraulic fracturing and horizontal drilling have opened up a new era of energy security, job growth, and economic strength.**
> —American Petroleum Institute director Erik Milito

> **There's no safe way to put toxic chemicals into the ground and control them.**
> —New York schoolteacher Elizabeth Bouiss "

Dimock, Pennsylvania

UNITED STATES

When the men from Cabot Oil and Gas Corporation came to the small town of Dimock in rural Pennsylvania, many of Dimock's 1500 residents were happy to sign on to the contracts the company offered. In exchange for the right to drill for natural gas on their land, Cabot would pay them royalties on sales of the gas extracted from their property. To some in this small, rural community, the gas payments and the potential for jobs seemed like a ticket to economic security.

Soon the new drilling sites around Dimock were producing the most natural gas from anywhere in the Marcellus Shale, the vast gas-bearing rock formation that underlies portions of Pennsylvania, New York, West Virginia, and Ohio (**FIGURE 15.1**). Money and jobs from the gas boom kept Dimock economically afloat, even as other towns in the region reeled from recession and cut funding for schools and basic services.

Yet despite the economic gains, some Dimock residents began having second thoughts about drilling. Their once-tranquil community was now experiencing noise, nighttime light, and air pollution; heavy truck traffic; and toxic wastewater spills. Soon, many people's drinking water began to turn brown, gray, or cloudy with sediment, and chemical smells began wafting from their wells. On New Year's Day, 2009, Norma Fiorentino's well exploded. Methane had built up in her well water, and a spark from a motorized pump set off a potentially lethal blast.

Residents blamed the drilling technique that Cabot Oil and Gas was using: hydraulic fracturing. Townspeople who could no longer drink their own well water appealed for help. Retired schoolteacher Victoria Switzer approached Cabot, local political leaders, and the Pennsylvania Department of Environmental Protection (DEP) but was turned away by them all. Then she went to the news media, and the story got out. Documentary filmmaker Josh Fox came to town and filmed residents setting their methane-contaminated tap water on fire. His 2010 film *Gasland* won numerous awards, and Dimock became Ground Zero in the burgeoning national debate over hydraulic fracturing.

Across the United States, virtually all the easily accessible oil and natural gas has already been discovered and extracted. To extract more, we've needed to develop ever more powerful technology to reach petroleum deposits deeper underground, deeper underwater, and at lower concentrations. In formations such as the Marcellus Shale, natural gas is locked up in tiny bubbles dispersed throughout the shale rock. Hydraulic fracturing is now making this **shale gas** accessible.

Hauling fracking wastewater—one of many jobs created by shale gas extraction ▲

Upon completing this chapter, you will be able to:

- Identify the energy sources that we use

- Understand the value of the EROI concept

- Describe the formation of coal, natural gas, and crude oil, and evaluate how we extract, process, and use these fossil fuels

- Assess concerns over the future decline of conventional oil supplies

- Outline ways in which we are extending our reach for fossil fuels and exploring unconventional new fossil fuel sources

- Examine and assess environmental, political, social, and economic impacts of fossil fuel use, and explore potential solutions

- Specify strategies for enhancing energy efficiency and conserving energy

- Describe nuclear energy and how we harness it

- Assess the benefits and drawbacks of nuclear power, and outline the societal debate over this energy source

Actor and anti-fracking activist Mark Ruffalo holds up water from a Dimock, Pennsylvania, well during a protest at New York's capitol building in Albany.

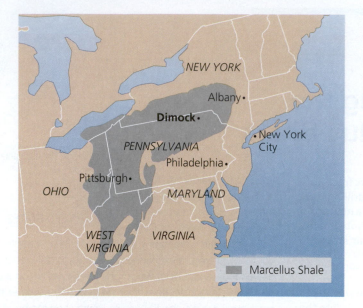

FIGURE 15.1 The Marcellus Shale formation underlies large portions of Pennsylvania, New York, Ohio, and West Virginia.

Hydraulic fracturing (commonly known as **fracking**) involves drilling deep into the earth and angling the drill horizontally once a shale formation is reached. An electrical charge sets off targeted explosions that perforate the drilling pipe and create fractures in the shale. Drillers then pump a slurry of water, sand, and chemicals down the pipe under great pressure. The sand lodges in the fractures and holds them open, while some of the liquids return to the surface. Natural gas trapped in the shale migrates into the fractures and rises through the drilling pipe to the surface, where it is collected (**FIGURE 15.2**).

Fracking has boosted the amount of natural gas extracted from the Marcellus Shale, the Barnett Shale in Texas, the Haynesville Shale in Louisiana, and elsewhere in recent years, igniting a boom in U.S. natural gas production. This has employed thousands of people and has driven down the price of natural gas. Expanded use of domestically produced natural gas has reduced the United States' reliance on coal for electricity. Because natural gas is cleaner-burning than coal, burning it in place of coal reduces the greenhouse gas emissions that drive climate change.

For all these reasons—and also because of the political influence of oil and gas corporations—policymakers have encouraged fracking. They have freed it from many regulatory constraints that would normally apply; fracking has been exempted from seven major federal environmental laws that protect public health, including the National Environmental Policy Act and the Safe Drinking Water Act.

As a result, gas companies do not need to report the chemical additives they plan to use in fracking. Nor do they need to test for chemical compounds drawn up in fracking wastewater (much of which is radioactive because drillers add radioisotopes [p. 31] as tracers to the fracking fluids they inject, and because naturally occurring radioisotopes are brought up from deep underground). Consequently, neither regulators nor policymakers nor scientists nor homeowners have access to data to make fully informed judgments about potential health or environmental effects.

In such a climate of uncertainty, people in places like Dimock are left to wonder, worry, and argue. After recognizing that for some Dimock families the water was undrinkable, Pennsylvania's DEP fined Cabot and required the company to haul in clean drinking water from outside town. However, the water shipments ended once Tom Corbett, an avid supporter of fracking, was elected Pennsylvania's governor. As publicity built and activists claimed Pennsylvania was not safeguarding its citizens, the U.S. Environmental Protection Agency (EPA) stepped in, sending federal researchers in 2012 to test Dimock's water. After sampling water from 64 wells, the EPA stated that the results did not warrant its taking action. In 2016, however, reanalysis of the results by a different federal agency showed levels of at least 10 chemicals high enough to threaten health in 27 of the wells tested, along with widespread methane contamination and other water quality issues.

In 2015 the EPA issued a draft of its 6-year nationwide study of the impacts of hydraulic fracturing on drinking water, but was forced to revise its conclusions after criticism from its own scientific advisory board. The study's final version, issued in 2016, admitted greater impacts on drinking water and detailed problems in Dimock and other locations such as Pavilion, Wyoming, and Parker County, Texas. However, research in Dimock and nationwide has suffered from a lack of data comparing water quality before and after drilling, so it has been difficult to prove that drilling has been the cause of water problems.

Meanwhile, in Dimock, 44 families brought lawsuits against Cabot. After years of legal struggles, all but two settled with the company and signed non-disclosure agreements binding them to silence. In 2016, a jury found Cabot negligent and awarded the remaining two families $4.24 million in damages. But in 2017, the judge reversed the jury's decision, canceling the award and ordering a new trial.

Debates like those in Dimock have been occurring across the nation as fracking has spread. People from Ohio to California are experiencing impacts on their water, land, and air, and are trying to weigh these against fracking's economic benefits. State governments have reacted in various ways. The neighboring states of Pennsylvania and New York offer a study in contrast. Pennsylvania's political leaders welcomed the gas industry with open arms, exempting it from regulations. New York's leaders, in contrast, chose to ban fracking entirely. Vermont and Maryland have also banned fracking, whereas Ohio, Texas, Louisiana, Wyoming, and other states have encouraged it.

Today many policymakers, scientists, and engineers are seeking ways to minimize the impacts of hydraulic fracturing, so that we can continue to use it to boost shale gas production. Their hope is that natural gas can bring economic and national security benefits and that this fossil fuel can serve as a "bridge" from the more-polluting fossil fuels of the past to the cleaner renewable energy sources we will need to develop now and for the future.

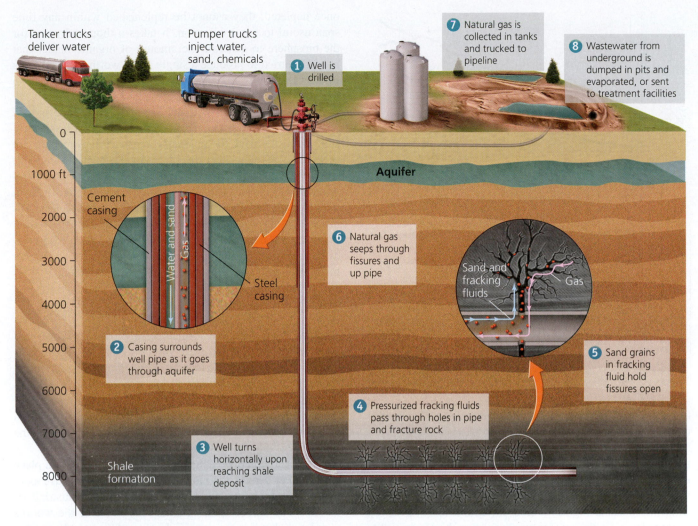

FIGURE 15.2 Hydraulic fracturing is used to extract natural gas trapped in shale deposits deep underground.

The following labels appear in the figure:

Tanker trucks deliver water

Pumper trucks inject water, sand, chemicals

7 Natural gas is collected in tanks and trucked to pipeline

8 Wastewater from underground is dumped in pits and evaporated, or sent to treatment facilities

1 Well is drilled

Aquifer

Cement casing

Water and sand

Gas

Steel casing

6 Natural gas seeps through fissures and up pipe

Sand and fracking fluids

Gas

2 Casing surrounds well pipe as it goes through aquifer

5 Sand grains in fracking fluid hold fissures open

4 Pressurized fracking fluids pass through holes in pipe and fracture rock

3 Well turns horizontally upon reaching shale deposit

Shale formation

0
1000 ft
2000
3000
4000
5000
6000
7000
8000

Sources of Energy

Humanity has devised many ways to harness the various forms of energy available on Earth (**TABLE 15.1**). We have taken most of our energy from nonrenewable sources—fossil fuels and nuclear power—but various renewable sources are also available. We use our planet's energy sources to heat and light our homes; power our machinery; fuel our vehicles; produce plastics, pharmaceuticals, and synthetic fibers; and provide the comforts and conveniences to which we've grown accustomed.

We still rely mostly on fossil fuels

Of all the energy sources shown in Table 15.1, we have relied the most heavily on **fossil fuels,** highly combustible substances formed underground over millions of years from the buried remains of ancient organisms. We use three main fossil fuels, in solid form (coal), liquid form (oil), and gas form (natural gas). Fossil fuels provide most of the energy that we buy, sell, and consume because their high energy content makes them efficient to ship, store, and burn. A single gallon

TABLE 15.1 Energy Sources We Use

ENERGY SOURCE	DESCRIPTION	TYPE OF ENERGY
Coal	Fossil fuel (solid)	Nonrenewable
Oil	Fossil fuel (liquid)	Nonrenewable
Natural gas	Fossil fuel (gas)	Nonrenewable
Nuclear energy	Energy from atomic nuclei	Nonrenewable
Biomass energy	Energy stored in plant matter	Renewable
Hydropower	Energy from running water	Renewable
Solar energy	Energy from sunlight directly	Renewable
Wind energy	Energy from wind	Renewable
Geothermal energy	Earth's internal heat	Renewable
Ocean energy	Energy from tides and waves	Renewable

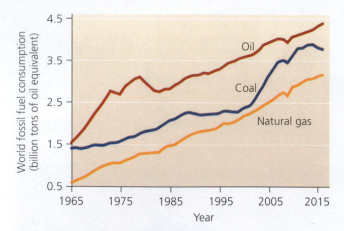

FIGURE 15.3 Annual global consumption of fossil fuels has risen greatly over the past half-century. Oil remains our leading energy source. *Data from BP p.l.c., 2017.* Statistical review of world energy 2017.

DATA Q
- By roughly what percentage has the annual consumption of oil risen since the year you were born?
- Which fuel has risen fastest since you were born—oil, coal, or natural gas?

Go to **Interpreting Graphs & Data** on **Mastering** Environmental Science

of oil contains as much energy as a person would expend in nearly 600 hours of human labor.

We use fossil fuels for transportation, manufacturing, heating, and cooking, and also to generate **electricity,** a secondary form of energy that we can transfer over long distances and apply to many uses. Global consumption of coal, oil, and natural gas is now at its highest level ever (**FIGURE 15.3**). Across the world today, over 80% of our energy and two-thirds of our electricity come from these three main fossil fuels (**FIGURE 15.4**).

Given our accelerating consumption, we risk using up Earth's reserves of coal, oil, and natural gas. These fuels are considered nonrenewable: They take so long to form that,

once depleted, they cannot be replenished within any time span useful to our civilization. It takes a thousand years for the biosphere to generate the amount of organic matter that must be buried to produce a single day's worth of fossil fuels for our society. To replenish the fossil fuels we have used so far would take many millions of years.

For this reason, fossil fuel companies have developed approaches such as hydraulic fracturing to exploit less-accessible deposits. And for this same reason, many people are seeking to develop other energy sources—such as sunlight, geothermal energy, and tidal energy—that are perpetually renewable because natural processes readily replenish them (p. 4). Such renewable energy sources also cause less pollution than do fossil fuels. While our society embarks on a transition from fossil fuels to renewable energy, measures to extend the availability of conventional fossil fuels by conserving energy and improving energy efficiency are vital.

Energy is unevenly distributed

Because fossil fuels are formed by geologic processes and because regions vary in their geologic histories and conditions, some areas of the globe have ended up with substantial reserves of oil, coal, or natural gas, whereas others possess very few. For example, half the world's proven reserves of crude oil lie in the Middle East. The Middle East is also rich in natural gas, as is Russia. The United States possesses the most coal of any nation (**TABLE 15.2**).

Rates at which we consume energy also vary from place to place. Per person, the most industrialized nations use more than 50 times more energy than do the least industrialized nations. The United States claims only 4.4% of the world's population, but it consumes nearly 18% of the world's energy.

Societies differ in how they use energy, as well. Industrialized nations apportion roughly one-third to transportation, one-third to industry, and one-third to all other uses. In contrast, industrializing nations devote more energy to

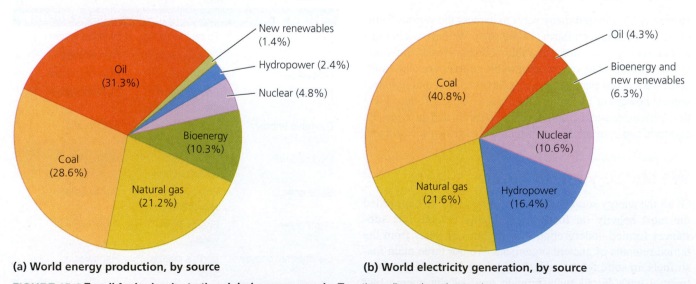

(a) World energy production, by source

(b) World electricity generation, by source

FIGURE 15.4 Fossil fuels dominate the global energy supply. Together, oil, coal, and natural gas account for **(a)** 81% of the world's energy production and **(b)** two-thirds of global electricity. *Data from International Energy Agency, 2016.* Key world energy statistics 2016. *Paris: IEA.*

TABLE 15.2 Nations with the Largest Proven Reserves of Fossil Fuels

OIL (% world reserves)		NATURAL GAS (% world reserves)		COAL (% world reserves)	
Venezuela*	17.6	Iran	18.0	United States	22.1
Saudi Arabia	15.6	Russia	17.3	China	21.4
Canada*	10.0	Qatar	13.0	Russia	14.1
Iran	9.3	Turkmenistan	9.4	Australia	12.7
Iraq	9.0	United States	4.7	India	8.3

*Most reserves in Venezuela and Canada consist of oil sands, which are included in these figures. *Data from BP p.l.c., 2017.* Statistical review of world energy 2017.

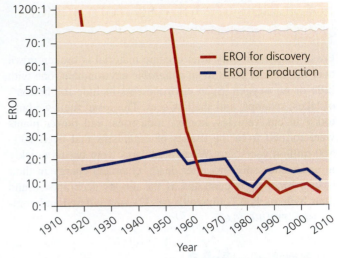

FIGURE 15.5 EROI values for the discovery and production of oil and gas in the United States have declined over the past century. *Data from Guilford, M., et al., 2011. A new long term assessment of energy return on investment (EROI) for U.S. oil and gas discovery and production. Pp. 115–136 in Hall, C., and D. Hansen (Eds.),* Sustainability, Special Issue, 2011, New studies in EROI (energy return on investment).

subsistence activities such as growing and preparing food and heating homes. Because industrialized nations rely more on mechanized equipment and technology, they use more fossil fuels. In the United States, oil, coal, and natural gas supply 82% of energy demand.

It takes energy to make energy

We do not simply get energy for free. To harness, extract, process, and deliver energy, we need to invest substantial inputs of energy. For instance, fracturing shale layers deep underground requires the use of powerful machinery, large quantities of water, and specialized chemicals, as well as vehicles, storage tanks, pipelines, waste ponds, processing facilities, equipment for workers, and more—all requiring the consumption of energy. Thus, when evaluating an energy source, it is important to subtract costs in energy invested from the benefits in energy received. **Net energy** expresses the difference between energy returned and energy invested:

$$\text{Net energy} = \text{Energy returned} - \text{Energy invested}$$

When we assess energy sources, it is also helpful to use a ratio known as **EROI (energy returned on investment)**. EROI ratios are calculated as follows:

$$\text{EROI} = \text{Energy returned} / \text{Energy invested}$$

Higher EROI ratios mean that we receive more energy from each unit of energy we invest. Fossil fuels are widely used because their EROI ratios have historically been high. However, EROI ratios can change over time. Ratios rise as technologies to extract and process fuels become more efficient, and they fall as resources are depleted and become harder to extract.

For example, EROI ratios for "production" (extraction and processing) of conventional oil and natural gas in the United States declined from roughly 24:1 in the 1950s to about 11:1 in recent years (**FIGURE 15.5**). This means we used to gain 24 units of energy for every unit of energy expended, but now we gain only 11. EROI ratios for both the discovery and the production of oil and gas have declined because we found and extracted the easiest deposits first and now must work harder and harder to find and extract the remaining amounts.

Where will we turn for energy?

Since the onset of the industrial revolution, coal, oil, and natural gas have powered the astonishing advances of our civilization. These extraordinarily rich sources of energy have helped to bring us a standard of living our ancestors could scarcely have imagined. Yet because fossil fuel deposits are finite and nonrenewable, easily accessible supplies of the three main fossil fuels have dwindled, and EROI ratios have fallen.

In response, today we are devoting enormous amounts of money, energy, and technology to extend our reach for fossil fuels. We are using potent extraction methods such as hydraulic fracturing to free gas and oil tightly bound in rock layers. We are deploying powerful new machinery and techniques to squeeze more fuel from sites that were already extracted. We are drilling deeper underground, farther offshore, and into the Arctic seabed. And we are pursuing new types of fossil fuels.

There is, however, a different way we can respond to the depletion of conventional fossil fuel resources: We can hasten the development of renewable energy sources to replace them (Chapter 16). By transitioning to clean and renewable alternatives, we can gain energy that is sustainable in the long term while reducing pollution and its health impacts and the emission of greenhouse gases that drive climate change.

Fossil Fuels: Their Formation, Extraction, and Use

To grapple effectively with the energy issues we face, it is important to understand how fossil fuels are formed, how we locate deposits, how we extract these resources, and how our society puts them to use.

Fossil fuels are formed from ancient organic matter

Fossil fuels form only after organic material is broken down over millions of years in an *anaerobic* environment, one with little or no oxygen. Such environments include the bottoms of lakes, swamps, and shallow seas. The fossil fuels we burn today in our vehicles, homes, industries, and power plants were formed from the tissues of organisms that lived 100–500 million years ago. When organisms were buried quickly in anaerobic sediments after death, chemical energy in their tissues became concentrated as the tissues decomposed and their hydrocarbon compounds (p. 32) were altered and compressed (**FIGURE 15.6**).

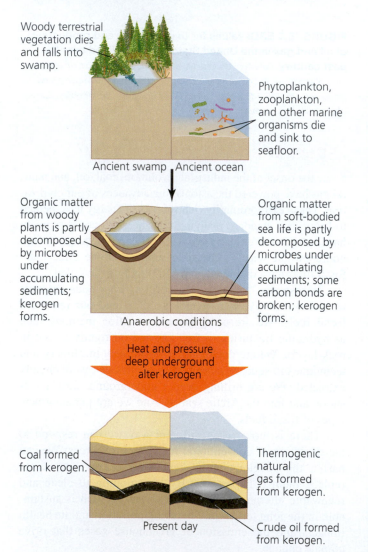

FIGURE 15.6 Fossil fuels begin to form when organisms die and end up in oxygen-poor conditions. This can occur when trees fall into lakes and are buried by sediment, or when phytoplankton and zooplankton drift to the seafloor and are buried **(top)**. Organic matter that undergoes slow anaerobic decomposition deep under sediments forms kerogen **(middle)**. Coal results when plant matter is compacted so tightly that there is little decomposition **(bottom left)**. The action of geothermal heating on kerogen may create crude oil and natural gas **(bottom right)**, which come to reside in porous rock layers beneath dense, impervious layers.

Because fossil fuels form only under certain conditions, they occur in isolated deposits. For example, the Marcellus Shale holds rich reserves of natural gas that other nearby rock formations do not. Geologists searching for fossil fuels drill cores and conduct ground, air, and seismic surveys to map underground rock formations and predict where fossil fuel deposits might occur.

Coal The world's most abundant fossil fuel is **coal,** a hard blackish substance formed from organic matter (generally woody plant material) compressed under very high pressure, creating dense, solid carbon structures. Coal typically results when water is squeezed out of such material as pressure and heat increase over time, and when little decomposition takes place. The proliferation 300–400 million years ago of swamps where organic material was buried created coal deposits in many regions of the world.

To extract coal from deposits near the surface, we use *strip mining*, in which heavy machinery scrapes away huge amounts of earth. For deposits deep underground, we use *subsurface mining*, digging vertical shafts and blasting out networks of horizontal tunnels to follow seams, or layers, of coal. (Strip mining and subsurface mining are illustrated in Figure 11.15, p. 245.) We are also mining coal on immense scales in the Appalachian Mountains, blasting away entire mountaintops in a process called *mountaintop removal mining* (pp. 246, 352).

Oil and natural gas The thick blackish liquid we know as **oil** consists of a mix of many types of hydrocarbon molecules (p. 32). The term **crude oil** refers specifically to oil extracted from the ground before it is refined. **Natural gas** is a gas consisting primarily of methane (CH_4) and lesser, variable, amounts of other volatile hydrocarbons. Oil is also known as **petroleum,** although this term is commonly used to refer to oil and natural gas collectively.

Both oil and natural gas are formed from organic material (especially dead plankton) that drifted down through coastal marine waters millions of years ago and became buried in sediments on the ocean floor. This organic material was transformed by time, heat, and pressure into today's natural gas and crude oil. Natural gas may form directly, or it may form from coal or oil altered by heating. As a result, natural gas is often found above deposits of oil or seams of coal, and is often extracted along with those fuels.

Underground pressure tends to drive oil and natural gas upward through cracks and fissures in porous rock until they become trapped under a dense, impermeable rock layer. Oil and gas companies employ geologists to study rock formations to identify promising locations. When such a location is identified, a company typically conducts *exploratory drilling*, drilling small holes to great depths. If enough oil or gas is encountered, extraction may begin. Because oil and gas are under pressure while in the ground, they rise to the surface when a deposit is tapped. Once pressure is relieved and some portion has risen to the surface, the remainder will need to be pumped out.

Unconventional fossil fuels Besides the three conventional fossil fuels—coal, oil, and natural gas—other types of fossil fuels exist, often called "unconventional" because we are not (yet) using them as widely. Three examples of unconventional fossil fuels are oil sands, oil shale, and methane hydrate.

Oil sands (also called **tar sands**) consist of moist sand and clay containing 1–20% bitumen, a thick and heavy form of petroleum. Oil sands result from crude oil deposits degraded and chemically altered by water erosion and bacterial decomposition.

Oil from oil sands is extracted by two main methods. For deposits near the surface (**FIGURE 15.7a**), a process akin to strip mining for coal or open-pit mining for minerals (pp. 244–246) is used. Shovel-trucks peel back layers of soil and dig out vast quantities of bitumen-soaked sand or clay. This is mixed with hot water and solvents at an extraction facility to purify the bitumen. Oil sands deeper underground (**FIGURE 15.7b**) are extracted by injecting steam and solvents down a drilling shaft to liquefy and isolate the bitumen, then pumping it out. Bitumen from either process must then be chemically refined and processed to create synthetic crude oil (called *syncrude*). Three barrels of water are required to extract each barrel of oil, and the toxic wastewater that results is discharged into vast reservoirs.

The second type of unconventional fossil fuel, **oil shale,** is sedimentary rock (p. 235) filled with organic matter that can be processed into a liquid form of petroleum called **shale oil.** Oil shale is formed by the same processes that form crude oil but occurs when the organic matter was not buried deeply enough or subjected to enough heat and pressure to form oil.

Oil shale is extracted using strip mines or subsurface mines. It can be burned directly like coal, or it can be processed in several ways. One way is to bake it in the presence of hydrogen and in the absence of air to extract liquid petroleum (a process called *pyrolysis*). The world's known deposits of oil shale may contain 3 trillion barrels of petroleum (more than all the world's conventional crude oil), but oil shale is costly to extract, and its EROI is low, ranging from 4:1 down to just 1.1:1.

The third unconventional fossil fuel, **methane hydrate,** is an ice-like solid consisting of molecules of methane embedded in a crystal lattice of water molecules. Methane hydrate occurs in sediments in the Arctic and on the ocean floor because it is stable at temperature and pressure conditions found there.

Scientists estimate there are enormous amounts of methane hydrate on Earth, holding perhaps twice as much carbon as all known deposits of oil, coal, and natural gas combined. Japan recently extracted methane hydrate from the seafloor by sending down a pipe and lowering pressure within it so that the methane turned to gas and rose to the surface. However, we do not yet know whether extraction is safe and reliable. If extraction were to destabilize a methane hydrate deposit on the seafloor, this could cause a landslide and tsunami and lead to a sudden release of methane, a potent greenhouse gas, into the atmosphere.

Economics determines how much will be extracted

As we develop more powerful technologies for locating and extracting fossil fuels, the proportions of these fuels that are physically accessible to us—the "technically recoverable"

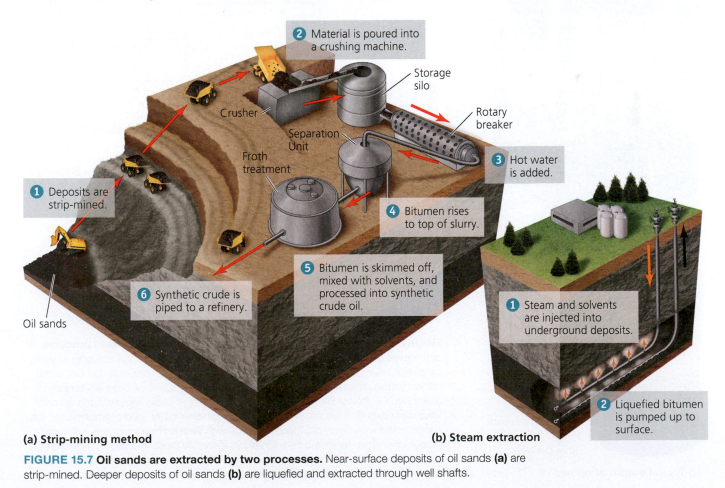

(a) Strip-mining method

2 Material is poured into a crushing machine.

Storage silo

Crusher

Rotary breaker

Separation Unit

Froth treatment

3 Hot water is added.

1 Deposits are strip-mined.

4 Bitumen rises to top of slurry.

6 Synthetic crude is piped to a refinery.

5 Bitumen is skimmed off, mixed with solvents, and processed into synthetic crude oil.

Oil sands

(b) Steam extraction

1 Steam and solvents are injected into underground deposits.

2 Liquefied bitumen is pumped up to surface.

FIGURE 15.7 Oil sands are extracted by two processes. Near-surface deposits of oil sands **(a)** are strip-mined. Deeper deposits of oil sands **(b)** are liquefied and extracted through well shafts.

portions—tend to increase. However, whereas technology determines how much fuel *can* be extracted, economics determines how much *will* be extracted. This is because extraction becomes increasingly expensive as a resource is removed, so companies rarely find it profitable to extract the entire amount. Instead, a company will consider the costs of extraction (and other expenses) and balance these against the income it expects from sale of the fuel. Because market prices of fuel fluctuate, the portion of fuel from a given deposit that is "economically recoverable" fluctuates as well. As market prices rise, economically recoverable amounts approach technically recoverable amounts.

The amount of a fossil fuel that is technologically *and* economically feasible to remove under current conditions is termed its **proven recoverable reserve.** Proven recoverable reserves increase as extraction technology improves or as market prices of the fuel rise. Proven recoverable reserves decrease as fuel deposits are depleted or as market prices fall (making extraction unprofitable). Some examples of proven recoverable reserves are shown in Table 15.2 (p. 345).

The amounts of a fossil fuel "produced" (extracted and processed) from a nation's reserves depends on many factors. **TABLE 15.3** shows amounts produced and amounts consumed (as a percentage of global production and consumption) by leading nations.

Refining gives us a diversity of fuels

Once we extract oil or gas, it must be processed and refined (**FIGURE 15.8**). Crude oil is a mix of hundreds of types of hydrocarbon molecules characterized by carbon chains of different lengths (p. 32). Chain length affects a substance's chemical properties, and this has consequences for human use, such as whether a given fuel burns cleanly in a car engine. Through the process of **refining** at a refinery, hydrocarbon molecules are separated by size and are chemically transformed to create specialized fuels for heating, cooking, and transportation and to create lubricating oils, asphalts, and the precursors of plastics and other petrochemical products.

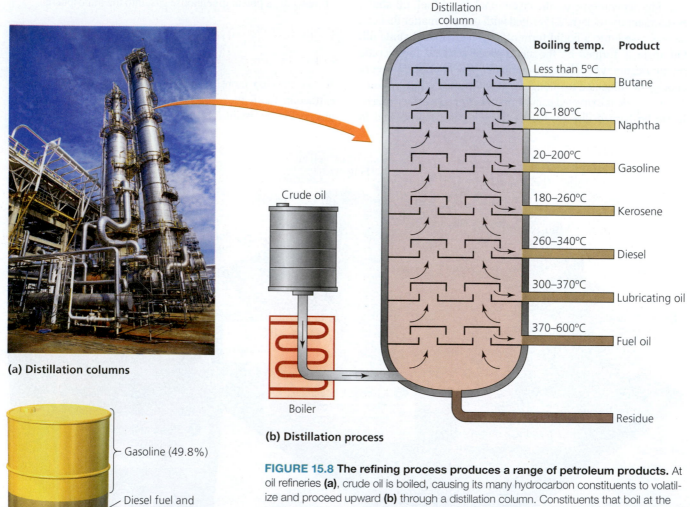

(a) Distillation columns

(b) Distillation process

(c) Typical output of refined oil

- Gasoline (49.8%)
- Diesel fuel and heating oil (24.1%)
- Jet fuel (8.2%)
- Liquefied petroleum gases (3.1%)
- Other (14.8%)

Distillation column

Boiling temp.	Product
Less than 5°C	Butane
20–180°C	Naphtha
20–200°C	Gasoline
180–260°C	Kerosene
260–340°C	Diesel
300–370°C	Lubricating oil
370–600°C	Fuel oil
	Residue

Crude oil

Boiler

FIGURE 15.8 The refining process produces a range of petroleum products. At oil refineries **(a)**, crude oil is boiled, causing its many hydrocarbon constituents to volatilize and proceed upward **(b)** through a distillation column. Constituents that boil at the hottest temperatures and condense readily once the temperature cools will condense at low levels in the column. Constituents that volatilize at cooler temperatures will continue rising through the column and condense at higher levels, where temperatures are cooler. In this way, heavy oils (generally those with hydrocarbon molecules with long carbon chains) are separated from lighter oils (generally those with short-chain hydrocarbon molecules). Shown in **(c)** are percentages of each major category of product typically generated from a barrel of crude oil. *Data (c) from U.S. Energy Information Administration.*

TABLE 15.3 Top Producers and Consumers of Fossil Fuels

PRODUCTION (% world production)		CONSUMPTION (% world consumption)	
Coal			
China	46.1	China	50.6
United States	10.0	India	11.0
Australia	8.2	United States	9.6
India	7.9	Japan	3.2
Indonesia	7.0	Russia	2.3
Oil			
United States	13.4	United States	20.3
Saudi Arabia	13.4	China	12.8
Russia	12.2	India	4.6
Iran	5.0	Japan	4.2
Iraq	4.8	Saudi Arabia	4.0
Natural Gas			
United States	21.1	United States	22.0
Russia	16.3	Russia	11.0
Iran	5.7	China	5.9
Qatar	5.1	Iran	5.7
Canada	4.3	Japan	3.1

Data from BP p.l.c., 2017. Statistical review of world energy 2017.

Fossil fuels have many uses

Each major type of fossil fuel has its own mix of uses.

Coal People have burned coal to cook food, heat homes, and fire pottery for thousands of years. Coal-fired steam engines helped drive the industrial revolution by powering factories, trains, and ships, and coal fueled the furnaces of the steel industry. Today we burn coal largely to generate electricity. In coal-fired power plants, coal combustion converts water to steam, which turns turbines to create electricity (**FIGURE 15.9**). Coal provides 40% of the electrical generating capacity of the United States, and it has powered China's surging economy.

Natural gas We use natural gas to generate electricity in power plants, to heat and cook in our homes, and for many other purposes. Converted to a liquid at low temperatures (*liquefied natural gas*, or *LNG*), it can be shipped long distances in refrigerated tankers. Versatile and clean-burning, natural gas emits just half as much carbon dioxide per unit of energy released as coal and two-thirds as much as oil. For this reason, many experts view natural gas as a climate-friendly "bridge fuel" that can help us transition from today's polluting fossil fuel economy toward a clean renewable energy economy. However, many other experts worry that investing in natural gas will simply delay our transition to renewables and instead deepen our reliance on fossil fuels.

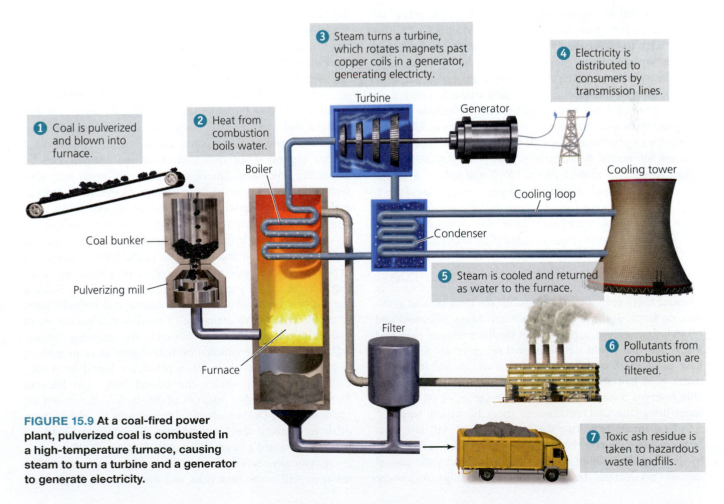

❸ Steam turns a turbine, which rotates magnets past copper coils in a generator, generating electricty.

❹ Electricity is distributed to consumers by transmission lines.

❶ Coal is pulverized and blown into furnace.

❷ Heat from combustion boils water.

Turbine

Generator

Boiler

Cooling tower

Cooling loop

Coal bunker

Condenser

Pulverizing mill

❺ Steam is cooled and returned as water to the furnace.

Filter

❻ Pollutants from combustion are filtered.

Furnace

❼ Toxic ash residue is taken to hazardous waste landfills.

FIGURE 15.9 At a coal-fired power plant, pulverized coal is combusted in a high-temperature furnace, causing steam to turn a turbine and a generator to generate electricity.

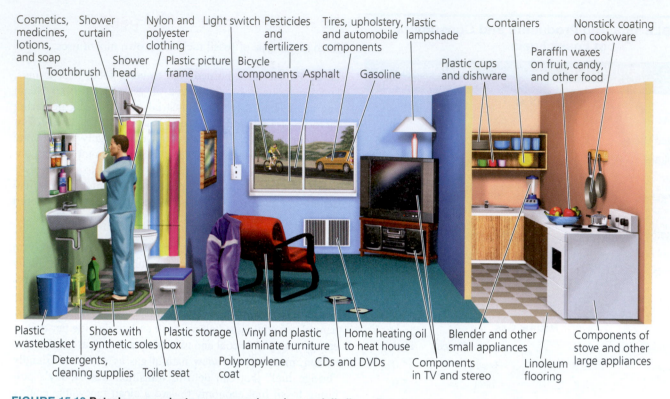

Cosmetics, medicines, lotions, and soap
Shower curtain
Nylon and polyester clothing
Light switch
Pesticides and fertilizers
Tires, upholstery, and automobile components
Plastic lampshade
Containers
Nonstick coating on cookware

Toothbrush
Shower head
Plastic picture frame
Bicycle components
Asphalt
Gasoline
Plastic cups and dishware
Paraffin waxes on fruit, candy, and other food

Plastic wastebasket
Shoes with synthetic soles
Plastic storage box
Vinyl and plastic laminate furniture
Home heating oil to heat house
Blender and other small appliances
Components of stove and other large appliances

Detergents, cleaning supplies
Toilet seat
Polypropylene coat
CDs and DVDs
Components in TV and stereo
Linoleum flooring

FIGURE 15.10 Petroleum products are everywhere in our daily lives. Petroleum helps to make many of the fabrics we wear, the materials we consume, and the plastics in countless items we use every day.

Oil Our global society consumes nearly 750 L (200 gal) of oil each year for every man, woman, and child. Most is used as fuel for vehicles, including gasoline for cars, diesel for trucks, and jet fuel for airplanes. Fewer homes burn oil for heating these days, but industry and manufacturing continue to use a great deal of it.

Refining techniques and chemical manufacturing have greatly expanded our uses of petroleum to include a wide array of products and applications, from plastics to lubricants to fabrics to pharmaceuticals. Today, petroleum-based products are all around us in our everyday lives (**FIGURE 15.10**). Take a moment to explore Figure 15.10, and reflect on all the conveniences in your own life that depend on petroleum products. The fact that we use petroleum to help create so many items and materials we rely on day by day makes it vital that we take care to conserve our remaining oil reserves.

We are depleting fossil fuel reserves

Because fossil fuels are nonrenewable, the total amount available on Earth declines as we use them. Many scientists and oil industry analysts calculate that we have already extracted nearly half the world's conventional oil reserves. So far, we have used up about 1.2 trillion barrels of oil, and most estimates hold that about 1.2 trillion barrels of proven recoverable reserves remain. Adding proven reserves of oil from oil sands brings the total remaining to about 1.7 trillion barrels.

To estimate how long this remaining oil will last, analysts calculate the **reserves-to-production ratio,** or R/P ratio, by dividing the amount of remaining reserves by the annual rate of "production" (extraction and processing). At current levels of production (33.6 billion barrels globally per year), 1.7 trillion barrels would last about 51 more years. Applying the R/P ratio to natural gas, we find that the world's proven reserves of this resource would last 53 more years. For coal, the latest R/P ratio estimate is 153 years.

The actual number of years remaining for these fuels could turn out to be less than these figures suggest if our demand and production continue to increase. Alternatively, the actual number of years may end up being *more* than the figures suggest if we reduce demand and consumption by enhancing efficiency. The actual number of years may also turn out to be greater because proven recoverable reserves increase as new deposits are discovered, as extraction technology becomes more powerful, and as market prices rise.

For instance, hydraulic fracturing for natural gas in the Marcellus Shale and elsewhere in the United States has expanded the nation's proven reserves of natural gas considerably in recent years. Likewise, fracking of the Bakken Formation—layers of shale and dolomite that underlie parts of North Dakota, Montana, and Canada—is allowing us to extract oil trapped tightly in this rock. By accessing this so-called *tight oil,* conventional oil held tightly in or near shale (which differs from shale oil, a petroleum liquid from specially processed oil shale), the United States has boosted its proven recoverable reserves of oil. In fact, the recent oil boom in North Dakota, along with increases in drilling deep offshore, enabled the United States, beginning in 2014, to become the world's largest extractor of oil.

Eventually, however, extraction of any nonrenewable resource will come to a peak and then decline. In general, extraction tends to decline once reserves are depleted halfway. If demand for the resource holds steady or rises while extraction declines, a shortage will result. With oil, this scenario has come to be nicknamed **peak oil**.

Peak oil will pose challenges

To understand concerns about peak oil, let's turn back the clock to 1956. In that year, Shell Oil geologist M. King Hubbert calculated that U.S. oil extraction would peak around 1970. His prediction was ridiculed at first, but it proved to be accurate; U.S. extraction peaked in that very year (**FIGURE 15.11a**). This peak in extraction came to be known as *Hubbert's peak*. Today, however, U.S. oil extraction has risen close to this

amount again, as fracking has enabled us to extract formerly inaccessible oil and as we have pursued various unconventional petroleum sources.

For the world as a whole, many scientists today are calculating that global extraction of oil will soon begin to decline (**FIGURE 15.11b**). Since about 2005, extraction of conventional oil has in fact been declining, and we have had to seek out a variety of less-conventional petroleum sources to compensate for this decline. Predicting an exact date for global peak oil is difficult, however. Many companies and governments do not reveal their data on oil reserves, and estimates differ as to how much oil we can continue extracting from existing deposits. For these reasons, estimates vary for the timing of an oil extraction peak, although most studies predict dates before 2035.

Whenever an oil peak occurs, a divergence of supply and demand could have momentous consequences that profoundly affect our lives. Writer James Howard Kunstler has sketched a frightening scenario of our post-peak world during what he calls "the long emergency": Lacking cheap oil with which to transport goods long distances, today's globalized

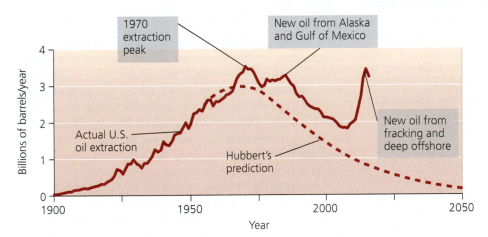

(a) Hubbert's prediction of peak in U.S. crude oil extraction, along with actual data

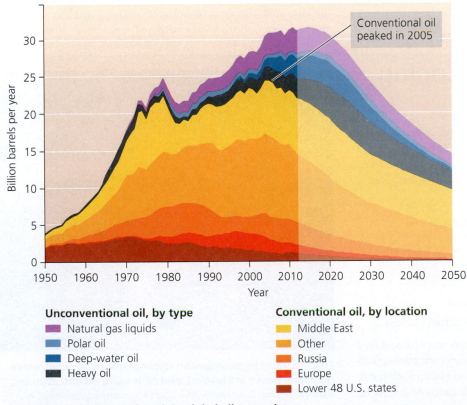

Unconventional oil, by type
- Natural gas liquids
- Polar oil
- Deep-water oil
- Heavy oil

Conventional oil, by location
- Middle East
- Other
- Russia
- Europe
- Lower 48 U.S. states

(b) Modern prediction of peak in global oil extraction

FIGURE 15.11 Peak oil describes a peak in extraction. U.S. extraction of crude oil peaked in 1970 **(a),** just as geologist M. King Hubbert had predicted. Drilling in Alaska and the Gulf of Mexico enhanced extraction during the subsequent decline, and today extraction has risen sharply as deep offshore drilling and hydraulic fracturing have made new deposits accessible. As for global oil extraction, many analysts calculate that it may soon peak. Shown in **(b)** is one projection, from a recent analysis by scientists at the Association for the Study of Peak Oil. These data show that conventional sources of oil peaked around 2005, and that since then unconventional sources have driven the continued overall rise. Paler shades indicate future data predicted by the researchers. *Data from (a) Hubbert, M.K., 1956. Nuclear energy and the fossil fuels. Shell Development Co. Publ. No. 95, Houston, TX; and U.S. Energy Information Administration; and (b) Campbell, C.J., and Association for the Study of Peak Oil. By permission of Dr. Colin Campbell.*

Hasn't "peak oil" been debunked? We're extracting more oil all the time, right?

An eventual peak is inevitable with any nonrenewable resource, so the notion of peak oil is sound. Its timing is a major question, though—and much depends on how one defines "oil." If one counts only conventional crude oil, then data indicate that we passed the global extraction peak around 2005. If one also includes "unconventional" oil from difficult-to-access sources such as oil sands, deep-water offshore oil, polar oil, and "tight oil" freed by fracking, then extraction has been roughly flat since 2005. If one also lumps in various additional petroleum sources (such as liquids condensed from natural gas), then extraction is still rising. In the big picture, conventional oil has already been declining for more than a decade, and today we increasingly rely on a host of petroleum sources that are more difficult and expensive to access.

economy would collapse into isolated local economies. Large cities would need to run urban farms to feed their residents, and with less mechanized farming and fewer petroleum-based fertilizers and pesticides, we might feed only a fraction of the world's people. The American suburbs would be hit particularly hard because of their dependence on the automobile.

More-optimistic observers argue that as oil supplies dwindle, rising prices will create powerful incentives for businesses, governments, and individuals to conserve energy and to develop alternative energy sources—and that these efforts will save us from major disruptions.

If we discover and exploit enough new deposits to continue extracting more and more oil, we might postpone our day of reckoning for decades. If we do so, however, we will find ourselves wrestling with another concern: trying to avoid runaway climate change driven by greenhouse gas emissions from the combustion of all that additional oil!

Reaching Further for Fossil Fuels . . . and Coping with the Impacts

To stave off the day when supplies of oil, gas, and coal begin to decline, we are investing more and more money, effort, and technology into locating and extracting new fossil fuel deposits. We are extending our reach for fossil fuels in several ways:

- Mountaintop mining for coal
- Secondary extraction from existing wells
- Directional drilling
- Hydraulic fracturing for oil and gas
- Offshore drilling in deep waters
- Moving into ice-free waters of the Arctic
- Exploiting new "unconventional" fossil fuel sources

All these pursuits are expanding the amount of fossil fuel energy available to us. However, as we extend our mining and drilling efforts into less-accessible places to obtain fuel that is harder to extract, we also reduce the EROI ratios of our fuels, intensify pollution, and worsen climate change.

Our society's love affair with fossil fuels has helped to ease constraints on travel, lengthen our life spans, and boost our material standard of living dramatically. Yet continued reliance on fossil fuels poses growing risks to human health; environmental quality; and social, political, and economic stability. Indeed, the risks from climate change alone are great enough that the International Energy Agency's chief economist recently joined a growing chorus of scientists in concluding that we will need to leave most fossil fuels in the ground if we are to avoid dangerous climate disruption. As we survey the ways we are expanding our reach for fossil fuel energy, we will also examine the impacts of fossil fuel use and assess ways to minimize these impacts.

Mountaintop mining extends our reach for coal

Coal mining has long been an economic mainstay of the Appalachian region, but **mountaintop removal mining** has brought coal extraction—and its impacts—to a whole new level (**FIGURE 15.12** and pp. 246–247). In this method, entire mountaintops are blasted away to access seams of coal. The massive scale of mountaintop removal mining makes it economically efficient, but it can cause staggering volumes of rock and soil to slide downslope, polluting or burying streams and disrupting life for people living nearby.

Mountaintop mining magnifies many of the impacts of traditional strip mining for coal, which erodes soil and destroys large areas of habitat. These mining methods also send chemical runoff into waterways in the form of acid drainage (p. 245), whereby sulfide minerals in newly exposed rock surfaces react with oxygen and rainwater to produce sulfuric acid. In most developed nations, mining companies are required to restore affected areas after mining, but

FIGURE 15.12 In mountaintop removal mining for coal, entire mountain peaks are leveled, and fill is dumped into adjacent valleys. Shown is an aerial view spanning many square miles in West Virginia.

reclamation is rarely able to re-create the ecological communities that preceded mining (p. 248). Subsurface coal mining, for its part, has long posed health risks to miners. They are at risk of accidents, and they breathe coal dust and toxic gases in confined spaces, which can lead to black lung disease and other respiratory ailments. (We explore coal mining and its impacts further in Chapter 11.)

Once mined, coal is transported by rail, and this can release coal dust into the air. In the Pacific Northwest, clean energy advocates are opposing the transport of coal by train from the interior West, where it is mined, to coastal terminals, to be shipped to Asia. Pollution along the route is one concern; another is that we are facilitating China's heavy reliance on coal, a driver of global climate change.

Secondary extraction yields additional fuel

At a typical oil or gas well, as much as two-thirds of a deposit may remain in the ground after primary extraction, the initial drilling and pumping of oil or gas. So, companies may return and conduct secondary extraction using new technology or approaches to force the remaining oil or gas out by pressure. In secondary extraction for oil, solvents are injected, underground rocks are flushed with water or steam, or hydraulic fracturing may be used. Because secondary extraction is more expensive than primary extraction, most deposits undergo secondary extraction only when market prices of oil and gas are high enough to make the process profitable.

Directional drilling reaches more fuel with less impact

Drilling for oil or gas typically requires building networks of access roads, housing for workers, transport pipelines, waste piles for removed soil, and ponds to collect toxic sludge. All this tends to pollute soil, air, and water; fragment habitat; and disturb wildlife and people. Today's **directional drilling** technology helps to lessen some of these impacts by allowing drillers to bore down vertically and then curve to drill horizontally. This enables them to follow horizontal layered deposits such as the Marcellus Shale or the Bakken Formation. By allowing access to a large underground area (up to several thousand meters in radius) around each drill pad, fewer drill pads are needed, and the surface footprint of drilling is smaller.

Hydraulic fracturing expands our access to oil and gas

For oil and natural gas trapped tightly in shale or other rock, petroleum companies now use hydraulic fracturing (see Figure 15.2). Chemically treated water under high pressure is pumped into layers of rock to crack them, and sand or small glass beads hold the cracks open as the water is withdrawn. Gas or oil then travels upward through the system of fractures.

By unlocking formerly inaccessible deposits of shale gas and tight oil, fracking ignited a boom in extraction in the

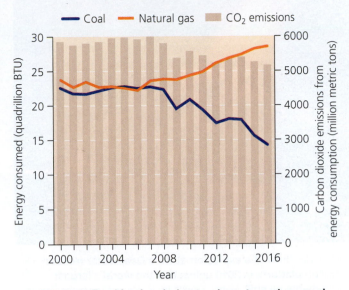

FIGURE 15.13 Fracking has led natural gas to replace coal at many U.S. power plants. Americans' consumption of coal and natural gas used to be roughly equal, but from 2007 to 2016, use of natural gas rose by more than 20% while use of coal fell by nearly 38%. This sudden divergence reduced greenhouse gas emissions. U.S. emissions of carbon dioxide from energy consumption (**gray bars** in the graph) peaked in 2007, and have since dropped nearly 14%.

United States. The new flow of oil reduced U.S. dependence on foreign imports and, by 2016, led to a glut on the world market that brought oil prices down to less than one-quarter of their 2008 high. The new flow of natural gas lowered the price of that fuel and has enabled many U.S. power plants to switch from coal to gas. Because natural gas is much cleaner-burning than coal, this has reduced carbon dioxide emissions from electricity generation substantially (**FIGURE 15.13**). Considering that during this period the American economy and population each grew, this drop in emissions is a notable achievement in the fight against global warming.

Yet despite the benefits it has delivered, fracking has sparked debate wherever it has occurred. Above the Marcellus Shale, it is affecting the landscapes, economies, politics, and everyday lives of people in Pennsylvania, Ohio, New York, and neighboring states. The choices people face between financial gain and protecting their health, drinking water, and environment have been dramatized in popular films such as *Promised Land* and *Gasland*. In North Dakota, fracking for oil supercharged the economy and drew young people from around the nation for high-paying jobs, but it also polluted the landscape, drew down water resources, and left many workers jobless once oil prices fell and rigs shut down.

One pollution risk from hydraulic fracturing is that fracking fluids may leak out of drilling shafts and into aquifers that people rely on for drinking water. Another concern is that methane may contaminate groundwater used for drinking if it travels up fractures or leaks through a shaft. Fracking also gives rise to air pollution as methane and volatile toxic components of fracking fluids seep up from drilling sites. Some of the unhealthiest air in the United States has been found in a remote region of Wyoming near fracking operations.

FIGURE 15.14 The explosion at BP's *Deepwater Horizon* drilling platform in 2010 unleashed the world's largest accidental oil spill. Here, vessels try to put out the blaze.

Like the people of Dimock, Pennsylvania, residents of areas near fracking sites across North America have experienced polluted air and fouled drinking water.

Fracking also consumes immense volumes of fresh water. Injected water often returns to the surface laced with salts, radioactive elements such as radium, and toxic chemicals such as benzene from deep underground. This wastewater may be sent to sewage treatment plants that are not designed to handle all the contaminants and that do not test for radioactivity. In Pennsylvania, millions of gallons of drilling waste from Marcellus Shale fracking sites have been sent to treatment plants, which then release their water into rivers that supply drinking water for people in Pittsburgh, Harrisburg, and other cities. In response to public outcry and government pressure, the oil and gas industry is beginning to reduce its water use by reusing wastewater in multiple injections.

In addition, fracking is now known to cause earthquakes (see **The Science behind the Story,** Chapter 11, pp. 238–239). Most have been minor, but this does raise questions about whether all social and economic costs have been considered.

We are drilling farther offshore

Roughly 35% of the oil and 10% of the natural gas extracted in the United States today come from offshore sites, primarily in the Gulf of Mexico and off southern California. Geologists estimate that most U.S. gas and oil remaining is found offshore. As oil and gas are depleted at shallow-water sites and as drilling technology improves, the industry is moving into deeper and deeper water.

Deep offshore drilling is boosting oil and gas production, but it poses risks. In the *Deepwater Horizon* oil spill of 2010 (p. 277), faulty equipment allowed natural gas accompanying an oil deposit to shoot up a well shaft and ignite on a British Petroleum platform off the Louisiana coast, killing 11 workers (**FIGURE 15.14**). The platform sank, emergency shut-off systems failed, and oil began gushing out of a broken pipe on the ocean floor at a rate of 30 gallons per second. British Petroleum's engineers tried one solution after another, but the flow of oil and gas continued out of control for three months, spilling roughly 4.9 million barrels (206 million gallons) of oil. BP's Macondo well, where the accident took place, lay beneath 1500 m (5000 ft) of water. The deepest wells in the Gulf of Mexico are now twice that depth.

As oil from the Macondo well spread through the Gulf and washed ashore, the region suffered a wide array of impacts (**FIGURE 15.15**). Unknown numbers of fish, birds, shrimp, corals, and other marine animals were killed, affecting coastal and ocean ecosystems in complex ways. Plants in coastal marshes died, causing erosion that put New Orleans and other cities at greater risk from storm surges and flooding. Beach tourism suffered, as did Gulf Coast fisheries, which supply much of the nation's seafood. Thousands of

(a) Brown pelican coated in oil

(b) Beach cleanup

FIGURE 15.15 Impacts of the *Deepwater Horizon* spill were many. This brown pelican, coated in oil **(a)**, was one of countless animals killed. For months, volunteers and workers labored **(b)** to clean oil from the Gulf's beaches.

fishermen and shrimpers were put out of work. Throughout this process, scientists studied the spill's many impacts on the region (see THE SCIENCE BEHIND THE STORY, pp. 358–359).

The *Deepwater Horizon* spill was the largest accidental oil spill in world history, far eclipsing the spill that resulted when the *Exxon Valdez* tanker ran aground in 1989 and damaged ecosystems and economies in Alaska's Prince William Sound. The *Exxon Valdez* event had led U.S. policymakers to tighten regulation and improve spill response capacity. But in 2008, responding to rising gasoline prices and a desire to reduce dependence on foreign oil, Congress lifted a long-standing moratorium on offshore drilling along much of the nation's coast. The Obama administration then opened vast areas for drilling, including most waters from Delaware to Florida, more of the Gulf of Mexico, and most waters off Alaska's North Slope. Once the *Deepwater Horizon* spill occurred, however, public reaction forced Obama's administration to backtrack. Later, the administration sought a middle path, opening access to areas holding 75% of technically recoverable offshore oil and gas reserves while banning drilling offshore from states that did not want it. Drilling leases were expanded off Alaska and in the Gulf of Mexico, but not along the East and West Coasts.

Globally, pollution from large oil spills has declined in recent decades (p. 277), thanks to government regulations (such as requirements for double-hulled ships) and improved spill response efforts. Most water pollution from oil today results from innumerable small non-point sources to which we all contribute (p. 275). Oil from automobiles, homes, gas stations, and businesses runs off roadways and enters rivers and wastewater facilities, being discharged eventually into the ocean.

Melting ice is opening up the Arctic

Today all eyes are on the Arctic. As climate change melts the sea ice that covers the Arctic Ocean (p. 323), new shipping lanes are opening and nations and companies are jockeying for position, scrambling to stake claim to areas of ocean where oil and gas deposits might lie beneath the seafloor. However, offshore drilling in Arctic waters poses severe pollution and safety risks. Frigid temperatures, ice floes, winds, waves, and brutal storms make conditions challenging and accidents likely. If a spill were to occur, icebergs, pack ice, storms, cold, and wintertime darkness would hamper response efforts, while frigid water temperatures would slow the natural breakdown of oil.

So far, Royal Dutch Shell has been the only company to pursue offshore drilling in Alaska's stormy waters—and it met with one mishap after another. One drilling rig ran aground while being towed during a storm. Another drilling ship nearly met the same fate in the Aleutian Islands. A containment dome intended to control leaks was crushed during testing. An icebreaker ran aground and had to be towed for repairs all the way to Portland, Oregon, where it faced media-savvy demonstrators protesting its arrival by rowing in kayaks and dangling from a bridge. In 2015, Shell gave up and withdrew from the Arctic after spending $7 billion in efforts to drill there.

We are exploiting new fossil fuel sources, such as oil sands

Three sources of "unconventional" fossil fuels—oil sands, oil shale, and methane hydrate—are abundant, and together could theoretically supply our civilization for centuries. However, they are difficult and expensive to extract and process, and their net energy values and EROI ratios are very low (p. 393). Extracting oil sands and oil shale consumes large volumes of water, devastates landscapes, and pollutes waterways. Burning these fossil fuels would likely emit more greenhouse gases than our use of coal, oil, and natural gas currently does, worsening air pollution and climate change.

Oil sands are becoming a major fuel source—and a focus of debate. Much of the world's oil sands underlie a vast region of boreal forest in northern Alberta in Canada. These tar-like deposits produce a low-quality fuel that requires a great deal of energy to extract and process. Most scientific estimates for the EROI ratio of Alberta's oil sands range from around 3:1 to 5:1. Yet when oil prices are high, mining oil sands becomes profitable, and in some recent years companies have produced nearly 2 million barrels of oil from them per day.

To extract these resources, companies clear vast areas of forest and dig enormous open pits miles wide and hundreds of feet deep (FIGURE 15.16a). The immense volumes of water used become polluted and are piped to gigantic reservoirs, where the toxic oily water kills waterfowl. The Syncrude company's wastewater reservoir near Fort McMurray, Alberta, is so massive that it is held back by the world's second-largest dam.

To sell its oil, Canada looked south to the United States. TransCanada Corporation built the Keystone Pipeline to pipe diluted bitumen 4700 km (2900 mi) to Illinois and to Texas refineries for export overseas. TransCanada then proposed a pipeline segment cutting across the Great Plains to shave off distance and add capacity to the line (FIGURE 15.16b). This Keystone XL pipeline proposal met opposition from people living along the route who were concerned about health, water quality, and property rights. It also faced nationwide opposition from advocates of action to address global climate change.

Pipeline proponents argued that the Keystone XL project would create jobs and guarantee a dependable oil supply for decades. They stressed that buying oil from Canada—a stable, friendly, democratic neighbor—could help reduce U.S. reliance on oil-producing nations with authoritarian governments and poor human rights records, such as Saudi Arabia and Venezuela.

Opponents of the pipeline extension expressed dismay at forest destruction in Alberta and anxiety about transporting oil over the Ogallala Aquifer (p. 271), where spills might contaminate drinking water for millions of people and irrigation water for America's breadbasket. They also sought to prevent extraction of a vast new source of fossil fuels whose combustion would emit huge amounts of greenhouse gases. By buying a source of oil that is energy-intensive to extract and that burns 14–20% less cleanly than conventional oil, they held, the United States would prolong fossil fuel dependence and worsen climate change.

(a) Massive oil sands mine in Alberta

(c) Protest at the White House

(b) Keystone pipeline and Keystone XL extension

FIGURE 15.16 The Keystone XL pipeline project has been a focus of debate. In Alberta **(a)**, oil sands are mined from enormous pits. The Keystone pipeline brings oil into the United States **(b)**, where its proposed extension set off a complex discussion, including large protests **(c)** in front of the U.S. White House.

The Keystone XL project required a permit from the U.S. State Department. Starting in 2008, this spurred an escalating political drama that came to include lawsuits, conflict-of-interest charges, high-stakes quarrels between President Obama and Congress, and street protests at the White House (**FIGURE 15.16c**). In 2015, Obama decided against approving the project, telling the nation that Keystone XL "would not serve the national interest of the United States" because its construction would not contribute meaningfully to the U.S. economy, it would not lower gas prices for consumers, and it would not enhance America's energy security. Moreover, he noted, approving it on the eve of global climate talks in Paris (p. 335) would undercut U.S. leadership just as America sought to gather nations together to address climate change.

In 2017 President Donald Trump reversed Obama's decision and signed an executive order approving the Keystone XL project. TransCanada will still need to obtain local permits, fight legal battles, and acquire funding to build the pipeline, however, and it may choose to do so only if oil prices rise high enough. We will leave it to you and your instructor to flesh out the rest of this evolving story.

Fuel can leak during transport

Another high-profile pipeline project is the Dakota Access Pipeline, a 1900-km (1200-mi) underground pipeline proposed to bring oil from Bakken Formation drilling sites in North Dakota to a tank farm in Illinois. In 2016–2017, thousands of people joined Native American protests against the pipeline on the Standing Rock Indian Reservation (**FIGURE 15.17a**). The Standing Rock Sioux objected to the pipeline's potential harm to sacred burial grounds where it would cross their land, and to the risk of water pollution where it would cross under the Missouri River. After months of protests, legal wrangling, and international media attention, the U.S. Army Corps of Engineers in the final weeks of President Obama's administration denied an easement for pipeline construction. The next month, however, President Trump ordered expedited construction of the pipeline.

Concerns about pipeline spills are justified. Oil from Canada's oil sands has leaked out of pipelines, causing spills along the Kalamazoo River in Michigan, in a residential neighborhood of Mayflower, Arkansas, and elsewhere. However, because there is

(a) Protest against the Dakota Access Pipeline

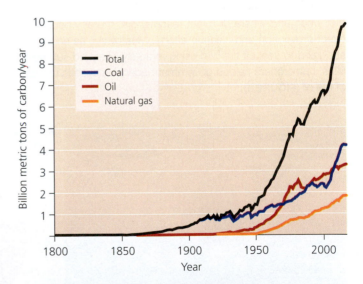

(b) Explosive derailment of an oil train in Lac-Mégantic, Quebec

FIGURE 15.17 Spills and pollution are always risks, however oil is transported. In 2016–2017, thousands of demonstrators protested the Dakota Access Pipeline on the Standing Rock Sioux Reservation **(a)**, contending that spills could contaminate water supplies and degrade sacred land. However, transporting oil by rail poses its own set of risks, as shown by the tragic derailment and explosion in Lac-Mégantic, Quebec **(b)**, in 2014.

not enough pipeline capacity to serve the Bakken oil fields, most Bakken oil is transported by rail, in pressurized tank cars. Tragically, a series of explosive derailments of trains carrying North Dakota crude oil has illustrated the risks of carrying oil by train (**FIGURE 15.17b**). Worst was the 2013 explosion in Lac-Mégantic, Quebec, which killed 47 people and destroyed the town's center. In the two following years the United States saw 10 major explosions, and in 2014 there were 141 tanker spills. The Obama administration responded with regulations to upgrade the safety of tanker cars; industry complained of additional cost while safety advocates said the steps were not strong enough.

Emissions pose health risks and drive climate change

As our society extends its reach for fossil fuel energy in so many ways, scientists, engineers, and policymakers are seeking solutions for air pollution and climate change. When we burn fossil fuels, we alter fluxes in Earth's carbon cycle (pp. 41–42). We essentially remove carbon from a long-term reservoir underground and release it into the air (**FIGURE 15.18**). This occurs as carbon from the hydrocarbon molecules of fossil fuels unites with oxygen from the atmosphere during combustion, producing carbon dioxide (CO_2). Carbon dioxide is a greenhouse gas (p. 314), and CO_2 released from fossil fuel combustion warms our planet (Chapter 14). Because climate change is beginning to have diverse and severe impacts, carbon dioxide pollution is becoming recognized as the single biggest negative consequence of fossil fuel use. Methane is also a potent greenhouse gas that drives climate warming.

Fossil fuel emissions affect our health, as well. Combusting coal can emit mercury that bioaccumulates in organisms' tissues, poisoning animals as it moves up food chains (pp. 221–222) and posing health risks to people. Gasoline combustion in automobiles releases cancer-causing pollutants such as benzene and toluene. Workers at drilling operations,

refineries, and in other jobs that entail frequent exposure to oil pollutants such as hydrogen sulfide, lead, and arsenic can develop serious health problems, including cancer.

The combustion of oil in vehicles and coal in power plants releases sulfur dioxide and nitrogen oxides, which contribute to smog (p. 296) and acid deposition (pp. 303–304). Air pollution from fossil fuel combustion is intensifying in

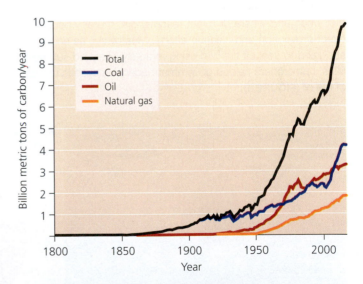

FIGURE 15.18 Emissions from fossil fuel combustion have risen dramatically as nations have industrialized and as population and consumption have grown. In this graph, global emissions of carbon from carbon dioxide are subdivided by source (oil, coal, or natural gas). *Data from Carbon Dioxide Information Analysis Center, Oak Ridge National Laboratory, U.S. Department of Energy, Oak Ridge, TN.*

 By what percentage have carbon emissions increased since the year your mother or father was born?

Go to **Interpreting Graphs & Data** on **Mastering** Environmental Science

Discovering Impacts of the Gulf Oil Spill

President Barack Obama echoed the perceptions of many Americans when he called the *Deepwater Horizon* oil spill "the worst environmental disaster America has ever faced." But what has scientific research told us about the actual impacts of the Gulf oil spill?

We will never have all the answers, because the deep waters affected by the spill have been difficult for scientists to study. Yet the intense and focused scientific response to the spill demonstrates the dynamic way in which science can assist society.

As the spill was taking place, government agencies called on scientists to help determine how much oil was leaking. Researchers eventually determined the rate reached 62,000 barrels per day. Using underwater imaging, aerial surveys, and shipboard water samples, researchers tracked the movement of oil up through the water column and across the Gulf. These data helped predict when and where oil might reach shore, thereby helping to direct prevention and cleanup efforts. Meanwhile, as engineers struggled to seal off the well using remotely operated submersibles, researchers helped government agencies assess the fate of the oil (**FIGURE 1**).

A scientist rescues an oiled Kemp's ridley sea turtle.

University of Georgia biochemist Mandy Joye, who had studied natural seeps in the Gulf for years, documented that the leaking wellhead was creating a plume of oil the size of Manhattan. She also found evidence of low oxygen concentrations, or hypoxia (p. 24), because some bacteria consume oil and gas, depleting oxygen from the water and making it uninhabitable for fish and other creatures.

Joye and other researchers feared that the thinly dispersed oil might devastate plankton (the base of the marine food chain) and the tiny larvae of shrimp, fish, and oysters (the pillars of the fishing industry). Scientists taking water samples documented sharp drops in plankton during the spill, but it will take years to learn whether the impact on larvae will diminish populations of adult fish and shellfish. Studies on the condition of living fish in the region have shown gill damage, tail rot, lesions, and reproductive problems at much higher levels than is typical.

What was happening to life on the seafloor was a mystery, because only a handful of submersible vehicles in the world are able to travel to the crushing pressures of the deep sea. Luckily, a team of researchers led by Charles Fisher of Penn State University had been scheduled to embark on a regular survey of deepwater coral across the Gulf of Mexico in late 2010—shortly after the spill occurred, as it turned out. Using the three-person submersible *Alvin* and the robotic vehicles *Jason* and *Sentry*, the team found healthy coral communities at sites far away from the Macondo well but found dying corals and brittlestars covered in a brown material at a site 11 km from the Macondo well.

Eager to determine whether this community was contaminated by the BP oil spill, the research team added chemist

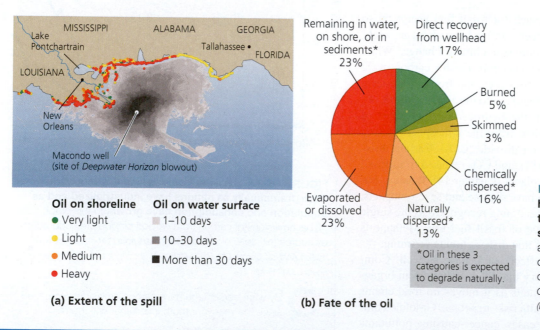

Oil on shoreline
- Very light
- Light
- Medium
- Heavy

Oil on water surface
- 1–10 days
- 10–30 days
- More than 30 days

(a) Extent of the spill

Remaining in water, on shore, or in sediments* 23%
Direct recovery from wellhead 17%
Burned 5%
Skimmed 3%
Chemically dispersed* 16%
Naturally dispersed* 13%
Evaporated or dissolved 23%

*Oil in these 3 categories is expected to degrade naturally.

(b) Fate of the oil

FIGURE 1 Scientists helped track oil from the *Deepwater Horizon* spill. The map **(a)** shows areas polluted by oil. The pie chart **(b)** gives a breakdown of the oil's fate. *Source: (a) NOAA Office of Response and Restoration; (b) NOAA*

Helen White of Haverford College and returned a month later, thanks to a National Science Foundation program that funds rapid response research. On this trip, chemical analysis of the brown material showed it to match oil from the BP spill.

Other questions revolve around the chemical dispersant that BP used to break up the oil, called Corexit 9500. Work by biologist Philippe Bodin following the *Amoco Cadiz* oil spill in France in 1978 had found that Corexit 9500 appeared more toxic to marine life than the oil itself. BP threw an unprecedented amount of this chemical at the *Deepwater Horizon* spill, injecting a great deal directly into the path of the oil at the wellhead. This caused the oil to dissociate into trillions of tiny droplets that dispersed across large regions. Many scientists worried that this caused the oil to affect far more plankton, larvae, and fish.

Impacts of the oil on birds, sea turtles, and marine mammals were less difficult to assess, and hundreds of these animals were cleaned and saved by wildlife rescue teams. Officially confirmed deaths numbered 6104 birds, 605 turtles, and 97 mammals, but a much larger, unknown, number surely succumbed to the oil and were never found. What impacts this mortality may have on populations in coming years is unclear. (After the *Exxon Valdez* spill in Alaska in 1989, populations of some species rebounded, but populations of others have never come back.) Researchers have been following the movements of marine animals in the Gulf with radio transmitters to try to learn what effects the oil may have had.

As images of oil-coated marshes saturated the media, researchers worried that the death of marsh grass would leave the shoreline vulnerable to severe erosion by waves. Louisiana has already lost many coastal wetlands to subsidence, dredging, sea level rise, and silt capture by dams on the Mississippi River (pp. 255–256). Fortunately, researchers found that oil did not penetrate to the roots of most plants and that oiled grasses were sending up new growth. Indeed, Louisiana State University researcher Eugene Turner said that loss of marshland from the oil "pales in comparison" with marshland lost each year due to other factors.

The ecological damage caused by the spill had measurable consequences for people. The region's mighty fisheries were shut down, forcing thousands of fishermen out of work. The government tested fish and shellfish for contamination and reopened fishing once they were found to be safe, but consumers balked at buying Gulf seafood. Beach tourism remained low all summer as visitors avoided the region. Together, losses in fishing and tourism totaled billions of dollars.

Scientists expect some consequences of the Gulf spill to be long-lasting. Oil from the similar *Ixtoc* blowout off Mexico's coast in 1979 continues to lie in sediments near dead coral reefs, and fishermen there say it took 15–20 years for catches to return to normal. After the *Amoco Cadiz* spill, it took seven years for oysters and other marine species to recover. In Alaska, oil from the 1989 *Exxon Valdez* spill remains embedded in beach sand today.

However, many researchers are hopeful about the Gulf of Mexico's recovery from the *Deepwater Horizon* spill. The Gulf's warm waters and sunny climate speed the natural breakdown of oil. In hot sunlight, volatile components of oil evaporate from the surface and degrade in the water, so that fewer toxic compounds affect marine life. In addition, bacteria that consume hydrocarbons thrive in the Gulf because some oil has always seeped naturally from the seafloor and because leakage from platforms, tankers, and pipelines is common. These microbes give the region a natural self-cleaning capacity.

Researchers continue to conduct a wide range of scientific studies (**FIGURE 2**). A consortium of federal and state agencies has been coordinating research and restoration efforts in the largest ever Natural Resource Damage Assessment, a process mandated under the Oil Pollution Act of 1990. Answers to questions will come in gradually as long-term impacts become clear.

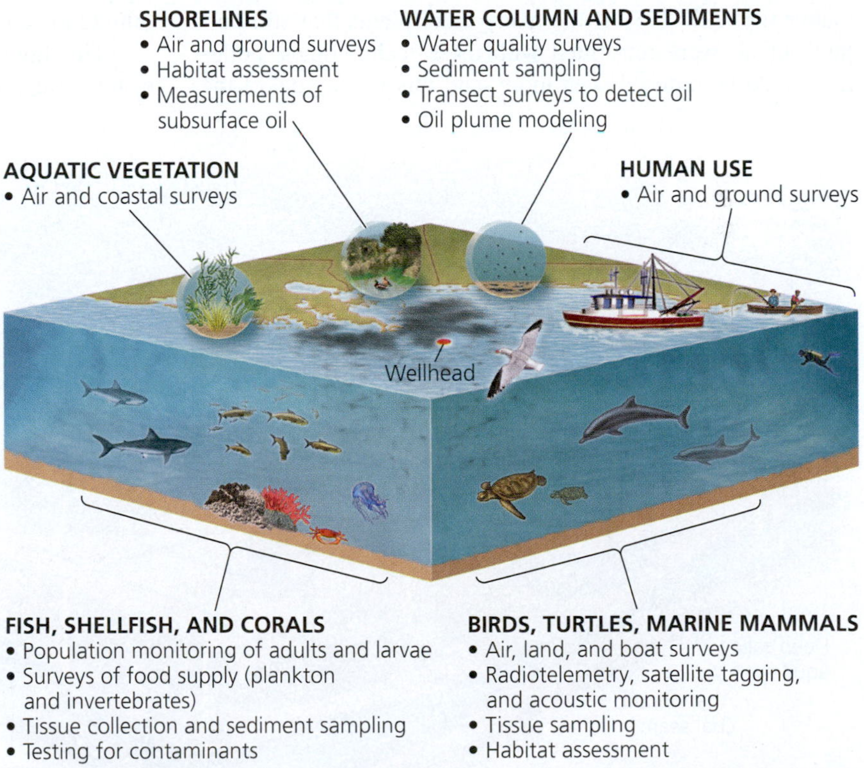

SHORELINES
• Air and ground surveys
• Habitat assessment
• Measurements of subsurface oil

WATER COLUMN AND SEDIMENTS
• Water quality surveys
• Sediment sampling
• Transect surveys to detect oil
• Oil plume modeling

AQUATIC VEGETATION
• Air and coastal surveys

HUMAN USE
• Air and ground surveys

Wellhead

FISH, SHELLFISH, AND CORALS
• Population monitoring of adults and larvae
• Surveys of food supply (plankton and invertebrates)
• Tissue collection and sediment sampling
• Testing for contaminants

BIRDS, TURTLES, MARINE MAMMALS
• Air, land, and boat surveys
• Radiotelemetry, satellite tagging, and acoustic monitoring
• Tissue sampling
• Habitat assessment

FIGURE 2 Thousands of researchers continue to help assess damage to natural resources from the *Deepwater Horizon* oil spill. They have been surveying habitats, collecting samples and testing them in the lab, tracking wildlife, monitoring populations, and more.

developing nations that are industrializing, but it has been reduced in developed nations as a result of laws and regulations to protect public health (Chapter 13). In these nations, public policy has encouraged industry to develop and install technologies that reduce pollution, such as catalytic converters that cleanse vehicle exhaust (see Figure 13.9, p. 294).

Clean coal technologies aim to reduce air pollution from coal

At coal-fired power plants, scientists and engineers are seeking ways to cleanse coal exhaust of sulfur, mercury, arsenic, and other impurities. **Clean coal technologies** refer to techniques, equipment, and approaches that aim to remove chemical contaminants during the generation of electricity from coal. Among these technologies are scrubbers, devices that chemically convert or physically remove pollutants (see Figure 13.8, p. 293). Another approach is to dry coal that has high water content, to make it cleaner-burning. We can also gain more power from coal with less pollution through *gasification*, in which coal is converted into a cleaner synthesis gas, or *syngas*, by reacting it with oxygen and steam at a high temperature. Syngas from coal can be used to turn a gas turbine or to heat water to turn a steam turbine.

The U.S. government and the coal industry have each invested billions of dollars in clean coal technologies for new power plants. These efforts have helped to reduce air pollution from sulfates, nitrogen oxides, mercury, and particulate matter (pp. 292–295). If the many older plants that still pollute our air were retrofitted with these technologies, pollution could be reduced even more, and this was a goal of the Obama administration's Clean Power Plan (pp. 301, 334). However, the coal industry spends a great deal of money fighting regulations on its practices. As a result, many power plants have little in the way of pollution control technologies. Many energy analysts argue that clean coal technologies will never result in energy that is completely clean and that coal should be replaced outright with cleaner energy sources.

Can we capture and store carbon?

Even if clean coal technologies were able to remove every last contaminant from power plant emissions, coal combustion would still pump huge amounts of carbon dioxide into the air, intensifying the greenhouse effect and worsening climate change. This is why many current efforts focus on **carbon capture and storage** (CCS; p. 331). This approach involves capturing CO_2 emissions, converting the gas to a liquid, and then sequestering (storing) it in the ocean or underground in a geologically stable rock formation (**FIGURE 15.19**).

Carbon capture and storage is being attempted at a variety of facilities. The world's first coal-fired power plant to approach zero emissions opened in 2008 in Germany. This plant removes its sulfate pollutants and captures its carbon dioxide, then compresses the CO_2 into liquid form, trucks it away, and injects it 900 m (3000 ft) underground into a depleted natural gas field. In North Dakota, the Great Plains Synfuels Plant gasifies its coal and sends half the CO_2 through a pipeline to Canada, where an oil company injects it into an oilfield to help it pump out the remaining oil.

The highest-profile CCS effort has suffered a rocky history. Beginning in 2003, the U.S. Department of Energy teamed up with seven energy companies to build a prototype of a near-zero-emissions coal-fired power plant. The *FutureGen* project (and its successor, the $1.65-billion *FutureGen 2.0*) aimed to design, construct, and operate a plant that captures 90% of its CO_2 emissions and sequesters the CO_2 deep underground beneath layers of impermeable rock. It was hoped that this showcase project, located in downstate Illinois, could be a model for a new generation of power plants across the world, but financial challenges plagued it, and in 2015 the project was suspended.

At present, carbon capture and storage remains unproven. We do not know how to ensure that carbon dioxide will stay underground once injected there. Injection might in some cases contaminate groundwater supplies or trigger earthquakes. Injecting carbon dioxide into the ocean would

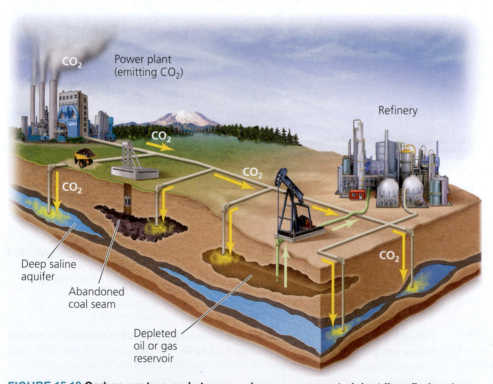

FIGURE 15.19 Carbon capture and storage schemes propose to inject liquefied carbon dioxide emissions underground. The CO_2 may be injected into depleted fossil fuel deposits, deep saline aquifers, or oil or gas deposits undergoing secondary extraction.

further acidify its waters (pp. 284, 326). Moreover, CCS is energy-intensive and decreases the EROI of coal, adding to its cost and the amount we consume. Finally, many renewable energy advocates fear that CCS takes the burden off emitters and prolongs our dependence on fossil fuels rather than facilitating a shift to renewables.

weighing the
ISSUES

Clean Coal and Carbon Capture

Do you think we should spend billions of dollars to try finding ways to burn coal more cleanly and to sequester carbon emissions from fossil fuels? Or is our money better spent on developing new clean and renewable energy sources, even if they don't yet have enough infrastructure to generate power at the scale that coal can? What pros and cons do you see in each approach?

We all pay external costs

The costs of addressing the many health and environmental impacts of fossil fuel extraction and use are generally not internalized in the market prices of fossil fuels. Instead, we all pay these external costs (pp. 96, 104) through medical expenses, costs of environmental cleanup, and impacts on our quality of life. The prices we pay at the gas pump or on our monthly utility bill have been kept inexpensive as a result of government subsidies to extraction companies. The profitable fossil fuel industries receive far more financial support from taxpayers than do the emerging renewable energy sources (see Figure 5.12, p. 113, and Figure 16.6, p. 379). In this way, we all pay extra for fossil fuel energy through our taxes, generally without realizing it.

Fossil fuel extraction has mixed consequences for local people

Wherever fossil fuels are extracted, people living nearby must weigh the environmental, health, and social drawbacks of extractive development against the financial benefits they may gain. In North America, communities where fossil fuel extraction takes place often experience a flush of high-paying jobs and economic activity. However, economic booms often prove temporary, whereas residents may be left with a polluted environment for generations to come.

In recent years debate over this type of trade-off has roiled inhabitants of Pennsylvania, New York, and other states above the Marcellus Shale. Many working-class residents of small New York towns have viewed jobs and economic activity across the border in Pennsylvania with envy and wish their state were encouraging fracking for shale gas as well. Other New Yorkers fear drinking water contamination and feel the short-term economic benefits are not worth the long-term health and environmental impacts (**FIGURE 15.20**).

People are wrestling with similar dilemmas in North Dakota and in parts of the South and West in response to new oil and gas drilling. In Appalachia, the debate has gone on for years over mountaintop removal mining. And along the Gulf of Mexico, the oil and gas industry employs 100,000 people and helps fund local economies—yet far more people are employed in tourism, fishing, and service industries, all of which were hurt by the *Deepwater Horizon* spill.

In Alaska, the oil industry maintains public support for drilling by paying the Alaskan government a portion of its revenues. Since the 1970s, the state of Alaska has received more than $70 billion in oil revenues. One-quarter of these revenues are placed in the Permanent Fund, an investment fund that pays yearly dividends to all citizens. Since 1982, each Alaska resident has received annual payouts ranging from $331 to $2072.

In contrast, in most parts of the world where fossil fuels are extracted, local residents suffer pollution without compensation. When multinational corporations pay governments of developing nations for access to oil or gas, the money generally does not trickle down to the people who live where the extraction takes place. Moreover, oil-rich developing nations such as Venezuela and Nigeria tend to have few environmental regulations, and governments may not enforce regulations if doing so would jeopardize the large sums of money associated with oil development.

In Ecuador, local people brought suit against Chevron for environmental and health impacts from years of oil extraction in the nation's rainforests. An Ecuadorian court in 2011 found the oil company guilty and ordered it to pay $9.5 billion for cleanup—the largest-ever such judgment. Chevron refused, and the court battle proceeded to the United States, where a judge threw out the ruling. The ongoing legal battle has now moved to other nations.

In Nigeria, the Shell Oil Company extracted $30 billion of oil from land of the native Ogoni people. Oil spills, noise, and gas flares caused chronic illness among them, but oil profits went to Shell and to the military dictatorships of Nigeria, while the Ogoni remained in poverty with no running water or electricity. Ogoni activist and leader Ken Saro-Wiwa worked for fair compensation to the Ogoni. After 30 years of persecution by the Nigerian government, he was arrested in 1994, given a trial universally regarded as a sham, and put to death by military tribunal.

FIGURE 15.20 Pollution from shale gas drilling creates external costs. This Pennsylvania homeowner can set fire to her tap water because it is contaminated with methane.

Dependence on foreign energy affects the economies of nations

Putting all your eggs in one basket is always a risky strategy. Because virtually all of our modern technologies and services depend in some way on fossil fuels, we are susceptible to supplies becoming costly or unavailable. Nations with few fossil fuel reserves of their own are especially vulnerable (**FIGURE 15.21**). In the wake of its 1970 oil extraction peak, the United States began relying more on foreign supplies.

Such reliance means that seller nations can control energy prices, forcing buyer nations to pay more as supplies dwindle. This became clear in 1973, when the Organization of Petroleum Exporting Countries (OPEC) resolved to stop selling oil to the United States. The predominantly Arab nations of OPEC opposed U.S. support for Israel in the Arab–Israeli Yom Kippur War and sought to raise prices by restricting supply. OPEC's embargo created panic in the West and caused oil prices to skyrocket, spurring inflation. Fear of oil shortages drove Americans to wait in long lines at gas pumps. A similar supply shock occurred in 1979 in response to the Iranian revolution.

With the majority of world oil reserves located in the politically volatile Middle East, crises in this region have dramatically affected oil supplies and prices time and again. For many U.S. leaders, this has enhanced the allure of using fracking to boost domestic oil and gas extraction. By supplying more of its own energy, the United States becomes less susceptible to foreign entanglements. Indeed, the United States now imports just one-quarter of its oil and

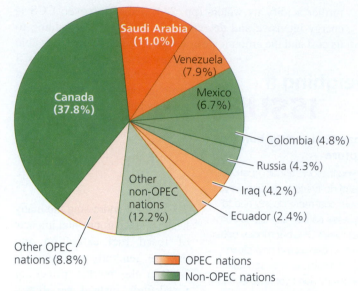

FIGURE 15.22 The United States now receives most of its imported oil from non-OPEC nations and from non-Middle-Eastern nations. *Data from U.S. Energy Information Administration.*

has diversified its sources of imported petroleum considerably (**FIGURE 15.22**).

While diversifying sources of foreign energy in response to the supply shocks of the 1970s, U.S. leaders also enacted conservation measures, funded research on renewable energy sources, and established an emergency stockpile (which today stores one month's supply of oil) deep underground in salt caverns in Louisiana, called the Strategic Petroleum Reserve. They also called for secondary extraction and the development of more domestic sources.

Since then, the desire to reduce reliance on foreign oil by boosting domestic production has driven the expansion of offshore drilling into deeper water. It has repeatedly driven a proposal to open the Arctic National Wildlife Refuge on Alaska's North Slope to oil extraction, despite arguments that drilling there would spoil America's last true wilderness while adding little to the nation's oil supply. Today it is driving the push to drill for oil in Arctic waters, despite the risks. As domestic oil and gas production rises with enhanced drilling, the United States becomes freer to make geopolitical decisions without being hamstrung by dependence on foreign energy. At the same time, however, climate change is posing new national security concerns. As we reach further for fossil fuels, our society will continue to debate the complex mix of social, political, economic, and environmental costs and benefits.

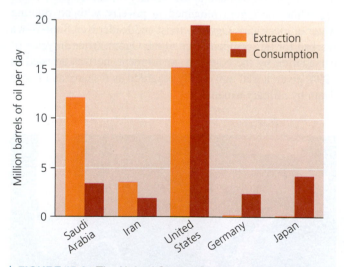

FIGURE 15.21 The United States, Germany, and Japan are among nations that consume more oil than they extract. Saudi Arabia and Iran extract more oil than they consume and are able to export oil to high-consumption countries. *Data from U.S. Energy Information Administration.*

DATA For every barrel of oil produced in the United States, how many barrels are consumed in the United States?

Go to **Interpreting Graphs & Data** on **Mastering** Environmental Science

weighing the ISSUES

Drill, Baby, Drill?

Do you think the United States should encourage hydraulic fracturing by subsidizing it and loosening regulations? Do you think the United States should open more of its offshore waters to oil and gas extraction? In each case, what benefits and costs do you foresee? Would the benefits likely exceed the costs? How strictly should government regulate oil and gas extraction once drilling begins? Give reasons for your answers.

Energy Efficiency and Conservation

Fossil fuels are limited in supply, and their use has health, environmental, political, and socioeconomic consequences. For these reasons, many people have concluded that fossil fuels are not a sustainable long-term solution to our energy needs. They see a need to shift to clean and renewable sources of energy that exert less impact on climate and human health. As our society transitions to renewable energy, it will benefit us to extend the availability of fossil fuels. We can do so by conserving energy and by improving energy efficiency.

Efficiency and conservation bring benefits

Energy efficiency describes the ability to obtain a given amount of output while using less energy input. **Energy conservation** describes the practice of reducing wasteful or unnecessary energy use. In general, efficiency results from technological improvements, whereas conservation stems from behavioral choices. Because greater efficiency allows us to reduce energy use, efficiency is a primary means of conservation.

Efficiency and conservation help us to waste less and to reduce our environmental impact. In addition, by extending the lifetimes of our nonrenewable energy supplies, efficiency and conservation help to alleviate many of the difficult individual choices and divisive societal debates related to fossil fuels.

Americans use far more energy per person than people in most other nations (**FIGURE 15.23a**). Residents of many European nations enjoy standards of living similar to those of U.S. residents yet use less energy per capita. This indicates that Americans could reduce their energy consumption considerably without diminishing their quality of life. Indeed, per-person energy consumption *has* declined slightly in the United States over the past three decades (see Figure 15.23a)—and this has occurred during a period of sustained economic growth.

During this period, the United States also cut in half its **energy intensity,** or energy use per dollar of Gross Domestic Product (GDP) (**FIGURE 15.23b**). Lower energy intensity indicates greater efficiency, and these data show that the United States now gets twice as much economic bang for its energy buck. Thus, although the United States continues to burn through more energy per dollar of GDP than most other industrialized nations, Americans have achieved tremendous gains in efficiency already and should be able to make further progress.

Personal actions and efficient technologies are two routes to conservation

As individuals, we can make conscious choices to reduce our energy consumption by driving less, dialing down thermostats, turning off lights when rooms are not in use, and investing in energy-efficient devices and appliances. For any given

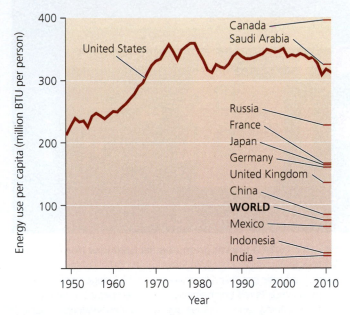

(a) Per capita energy consumption

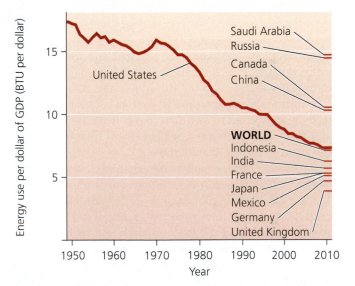

(b) Energy intensity

FIGURE 15.23 The United States trails other developed nations in energy efficiency but has made much progress. U.S. per-person energy use **(a)** has fallen slightly since 1979 but remains greater than that of most other nations. U.S. energy intensity **(b)** has fallen steeply and now approaches that of other developed nations. Energy intensity is energy use per inflation-adjusted 2005 dollars of GDP, using purchasing power parities, which control for differences among nations in purchasing power. *Data from U.S. Energy Information Administration.*

individual or business, reducing energy consumption saves money while helping to conserve resources.

As a society, we can conserve energy by developing technologies and strategies to make devices and processes more efficient. Currently, more than two-thirds of the fossil fuel energy we use is simply lost, as waste heat, in automobiles and power plants.

One way we can improve the efficiency of power plants is through **cogeneration,** in which excess heat produced

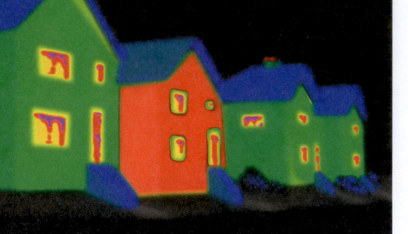

FIGURE 15.24 A thermogram reveals heat loss from buildings. It records energy in the infrared portion of the electromagnetic spectrum (pp. 34–35). In this image, one house is uninsulated; its red color signifies warm temperatures where heat is escaping. Green shades signify cool temperatures, where heat is being conserved. Also note that in all houses, more heat is escaping from windows than from walls.

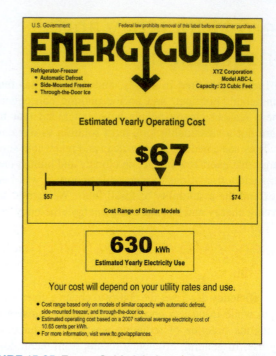

FIGURE 15.25 EnergyGuide labels provide consumers information on energy efficiency. These yellow-and-black labels on ovens, water heaters, and other appliances allow us to compare the performance of one brand or model versus others. In the example shown here, a refrigerator uses an estimated 630 kilowatt-hours of electricity per year. It has an estimated yearly operating cost of $67, which is intermediate compared with similar brands and models.

during electricity generation is captured and used to heat nearby workplaces and homes and to produce other kinds of power. Cogeneration can almost double the efficiency of a power plant. The same is true of coal gasification and combined cycle generation, in which coal is treated to create hot gases that turn a gas turbine, while exhaust from this turbine heats water to drive a steam turbine.

In homes, offices, and public buildings, a significant amount of heat is needlessly lost in winter and gained in summer because of poor design and inadequate insulation (**FIGURE 15.24**). Improvements in design can reduce the energy required to heat and cool buildings. Such improvements can involve passive solar design (p. 382), better insulation, a building's location, the vegetation around it, and the color of its roof (lighter colors keep buildings cooler by reflecting sunlight).

Many consumer products, from lightbulbs to computers to dishwashers, have been reengineered through the years to enhance efficiency. Energy-efficient lighting, for example, can reduce energy use by 80%. Compact fluorescent bulbs are more efficient than traditional incandescent lightbulbs, which is why the United States and many other nations are phasing out incandescent bulbs.

Federal standards for energy-efficient appliances have already reduced per-person home electricity use below what it was in the 1970s. The U.S. EPA's Energy Star program certifies appliances, electronics, doors and windows, and other products that surpass efficiency standards. The federal government also requires manufacturers of certain types of appliances to post test results for energy efficiency on yellow-and-black "EnergyGuide" labels on the products (**FIGURE 15.25**). These two labeling programs enable consumers to take energy use into account when shopping, and each program has been credited with reducing energy consumption and carbon emissions by enormous amounts. These programs also

save American consumers many billions of dollars; studies show in case after case that savings on utility bills more than offset the slightly higher prices of energy-efficient products.

Automobile fuel efficiency is a key to conservation

Automotive technology represents perhaps our best opportunity to conserve large amounts of fossil fuels. We can accomplish this with electric cars, electric/gasoline hybrids, plug-in hybrids, or vehicles that use hydrogen fuel cells (p. 395). Among electric/gasoline hybrids, current models obtain fuel-economy ratings of up to 50 miles per gallon (mpg)—twice that of the average American car. Many fully electric vehicles now obtain fuel-economy ratings of over 100 mpg. Automakers also can enhance fuel efficiency for gasoline-powered vehicles by using lightweight materials, continuously variable transmissions, and more efficient engines.

One of the ways in which U.S. leaders responded to the OPEC embargo of 1973 was to mandate an increase in the fuel efficiency of automobiles. Automakers responded by boosting fuel efficiency more than 60% between 1975 and 1982 (**FIGURE 15.26**). Over the next three decades, however, as market prices for oil fell, many of the conservation initiatives of this time were abandoned. Without high market prices and a threat of shortages, people lost the economic motivation to conserve, and U.S. policymakers repeatedly

Improving Energy Efficiency

When shopping for electronics, appliances, or home and office equipment, most of us would like to choose energy-efficient models. But how can we know how efficient a given TV or microwave oven or computer printer or refrigerator is? Luckily, American consumers can look for the Energy Star label. This familiar blue-and-white label tells us that a product has been independently certified as performing above federal energy-efficiency standards. By helping consumers identify and choose energy-efficient brands and models, the Energy Star program is reducing electricity demand and saving consumers money. For each extra dollar we spend on an Energy Star product, we save an average of $4.50 on energy costs, while preventing more than 35 pounds of greenhouse gas emissions, according to the Environmental Protection Agency, which runs the program. From 1992 through 2014, the Energy Star program helped Americans save $362 billion in utility bills, and reduced greenhouse gas emissions by 2.5 billion tons. Armed with information from this simple label, each and every one of us is empowered to make our own decisions about conserving energy through our purchasing behavior.

The Energy Star label certifies products as energy-efficient.

EXPLORE THE DATA at **Mastering** Environmental Science

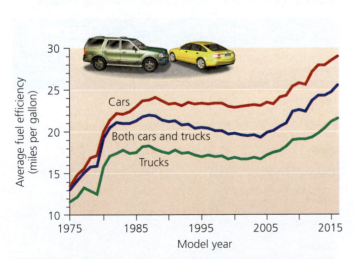

FIGURE 15.26 Automotive fuel efficiencies have responded to public policy. Fuel efficiency for automobiles in the United States rose dramatically in the late 1970s as a result of legislative mandates but then stagnated once no further laws were enacted to improve fuel economy. Recent legislation has now once again improved fuel efficiency. *Data from U.S. Environmental Protection Agency, 2016.* Light-duty automotive technology, carbon dioxide emissions, and fuel economy trends: 1975 through 2016.

failed to raise the corporate average fuel efficiency (CAFE) standards, which set benchmarks for auto manufacturers to meet. The average fuel efficiency of new vehicles fell from 22.0 mpg in 1987 to 19.3 mpg in 2004, as sales of sport-utility vehicles increased relative to sales of cars.

In 2007, Congress mandated that automakers raise average fuel efficiency to 35 mpg by the year 2020. Then when automakers requested a government bailout during the 2008–2009 recession, President Obama persuaded them to agree to boost average fuel economies to 54.5 mpg by 2025. The required technologies would add more than $2000 to the average price of a car, but drivers would save perhaps $6000 in fuel costs over the car's lifetime. These policies resulted in a substantial increase in average fuel efficiencies (see Figure 15.26). However, the 54.5-mpg goal may not come to pass, as President Trump in 2017 signaled to automakers that he would let them renege on the agreement.

U.S. policymakers could do more to encourage oil conservation. The United States has kept taxes on gasoline extremely low, relative to most other nations. Americans pay two to three times *less* per gallon of gas than drivers in many European countries. In fact, gasoline in the United States is sold more cheaply than bottled water! As a result, U.S. gasoline prices do not account for the substantial external costs (pp. 96–97, 104) that oil production and consumption impose on society. Some experts estimate that if all costs to society were taken into account, the price of gasoline would exceed $13 per gallon. Instead, our artificially low gas prices diminish our economic incentive to conserve.

The rebound effect cuts into efficiency gains

Energy efficiency is a vital pursuit, but it may not always save as much energy as we expect. This is because gains in efficiency from better technology may be partly offset if people engage in more energy-consuming behavior as a result. For instance, a person who buys a fuel-efficient car may choose to drive more because he or she feels it's okay to do so now that less gas is being used per mile. This phenomenon is called the **rebound effect,** and studies indicate that it is widespread. In some instances, the rebound effect may completely erase efficiency gains.

Nonetheless, efficiency will play a necessary role in the conservation efforts we make toward reducing energy use. It is often said that reducing energy use is equivalent to finding

weighing the
ISSUES

More Miles, Less Gas

If you drive an automobile, what gas mileage does it get? How does it compare to the vehicle averages in Figure 15.26? If your vehicle's fuel efficiency were 10 mpg greater, and if you drove the same amount, how many gallons of gasoline would you no longer need to purchase each year? How much money would you save?

Do you think U.S. leaders should mandate further increases in the CAFE standards? Should the government raise taxes on gasoline sales as an incentive for consumers to conserve energy? What effects on economics, on health, and on environmental quality might each of these steps have?

a new oil reserve. Some estimates hold that energy conservation and efficiency in the United States could save 6 million barrels of oil a day—nearly the amount gained from all offshore drilling, and considerably more than would be gained from Canada's oil sands. In fact, conserving energy is *better* than finding a new reserve because it alleviates health and environmental impacts while at the same time extending our future access to fossil fuels. Yet regardless of how much we conserve, we will still need energy. Among the alternatives to fossil fuels for our energy economy is nuclear power.

Nuclear Power

Nuclear power is free of the air pollution produced by fossil fuel combustion and thereby offers a powerful means of combating climate change. Yet nuclear power's promise has been clouded by nuclear weaponry, the thorny dilemma of radioactive waste disposal, and the long shadow of accidents at Chernobyl and Fukushima. As a result, public safety concerns and the costs of addressing them have constrained nuclear power's expansion.

Fission releases nuclear energy in reactors to generate electricity

Nuclear energy is the energy that holds together protons and neutrons (p. 30) in the nucleus of an atom. We can harness this energy by converting it to thermal energy inside **nuclear reactors,** facilities contained within nuclear power plants. This thermal energy is then used to generate electricity by heating water to produce steam that turns turbines. The generation of electricity using nuclear energy in this way is what we call **nuclear power.**

The reaction that drives the release of nuclear energy inside nuclear reactors is **nuclear fission,** the splitting apart of atomic nuclei (**FIGURE 15.27**). In fission, the nuclei of large, heavy atoms, such as uranium or plutonium, are bombarded with neutrons. Ordinarily, neutrons move too quickly to split nuclei when they collide with them, but if neutrons are slowed down, they can break apart nuclei. In a nuclear reactor, the neutrons bombarding uranium are slowed down with a substance called a *moderator*, most often water or graphite. Each split nucleus emits energy in the form of heat, light, and radiation, and it also releases neutrons. These neutrons (two to three in the case of uranium-235) can in turn bombard other uranium-235 (^{235}U) atoms, resulting in a self-sustaining chain reaction.

If not controlled, this chain reaction becomes a runaway process of positive feedback (p. 25)—the process that creates the explosive power of a nuclear bomb. Inside a power plant, however, fission is controlled so that, on average, only one of the two or three neutrons emitted with each fission event goes on to induce another fission event. To soak up the excess neutrons produced when uranium nuclei divide, *control rods* made of a metallic alloy that absorbs neutrons are placed among the water-bathed fuel rods of uranium. Engineers move the control rods in and out of the water to maintain fission at the desired rate. In this way, the chain reaction maintains a constant output of energy. All this takes place within the reactor core and is the first step in the electricity-generating process of a nuclear power plant (**FIGURE 15.28**).

First developed commercially in the 1950s, nuclear power experienced most of its growth during the 1970s and 1980s. The United States generates over a quarter of the world's nuclear power yet receives less than 20% of its electricity from this energy source. A number of other nations rely more heavily on nuclear power (**TABLE 15.4**). Today 452 nuclear power plants operate in 30 nations.

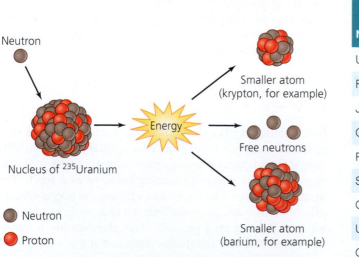

FIGURE 15.27 Nuclear fission drives modern nuclear power. In nuclear fission, the nucleus of an atom of uranium-235 is bombarded with a neutron. The collision splits the uranium atom into smaller atoms and releases two or three neutrons, along with heat, light, and radiation.

TABLE 15.4 Top Producers of Nuclear Power

NATION	NUCLEAR POWER CAPACITY (gigawatts)	NUMBER OF REACTORS	PERCENTAGE ELECTRICITY FROM NUCLEAR POWER
United States	100.4	100	19.7
France	63.1	58	72.3
Japan	40.3	43	2.2
China	31.4	36	3.6
Russia	26.5	36	17.1
South Korea	23.1	25	30.3
Canada	13.6	19	15.6
Ukraine	13.1	15	52.3
Germany	10.8	8	13.1
Sweden	9.7	10	40.0
United Kingdom	8.9	15	20.4

Data from International Atomic Energy Agency.

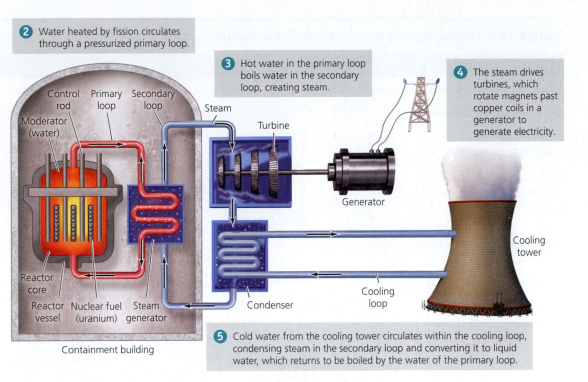

① Fission occurs in the reactor core, where fuel rods are submerged in water. Control rods absorb excess neutrons to regulate the reaction rate.

② Water heated by fission circulates through a pressurized primary loop.

③ Hot water in the primary loop boils water in the secondary loop, creating steam.

④ The steam drives turbines, which rotate magnets past copper coils in a generator to generate electricity.

⑤ Cold water from the cooling tower circulates within the cooling loop, condensing steam in the secondary loop and converting it to liquid water, which returns to be boiled by the water of the primary loop.

Control rod   Primary loop   Secondary loop   Steam
Moderator (water)
Turbine
Generator
Cooling tower
Reactor core
Reactor vessel   Nuclear fuel (uranium)   Steam generator   Condenser   Cooling loop
Containment building

FIGURE 15.28 In a pressurized light water reactor (the most common type of nuclear reactor), radioactive uranium fuel rods heat water, and steam turns turbines and generators to generate electricity.

Nuclear energy comes from processed and enriched uranium

We use the element uranium for nuclear power because its atoms are radioactive, emitting subatomic particles and high-energy radiation as they decay into a series of lighter isotopes (p. 31). We obtain uranium by mining. Uranium-containing minerals are uncommon and in finite supply, so nuclear power is generally considered a nonrenewable energy source.

More than 99% of the uranium in nature occurs as the isotope uranium-238. Uranium-235 (with three fewer neutrons) makes up less than 1% of the total. Because ^{238}U does not emit enough neutrons to maintain a chain reaction, we use ^{235}U for commercial nuclear power. Therefore, we must process the ore we mine to enrich the concentration of ^{235}U to at least 3%. This enriched uranium is formed into pellets of uranium dioxide, which are used in fuel rods.

After several years in a reactor, the decayed uranium fuel no longer generates adequate energy, so it must be replaced with new fuel. In some countries, the spent fuel is reprocessed to recover the remaining energy. However, this is costly relative to the market price of uranium, so most spent fuel is disposed of as radioactive waste.

Nuclear power delivers energy more cleanly than fossil fuels

Using fission, nuclear power plants generate electricity without creating the air pollution that fossil fuels do. Of course, the construction of plants and equipment has a large carbon footprint, but the actual nuclear power–generating process is essentially emission-free. All told, scientists estimate that using nuclear power in place of fossil fuels helps the world avoid emissions of 2.5 billion metric tons of carbon dioxide per year, about 7% of global CO_2 emissions.

In the nation of Sweden, nuclear power took the place of coal and natural gas to such an extent that Sweden was able to slash its use of fossil fuels in half (**FIGURE 15.29**). One recent study calculated that the replacement of coal-fired power with nuclear power in Sweden over the years had prevented 2.1 billion metric tons of CO_2 emissions and had saved 61,000 lives by cutting down on air pollution.

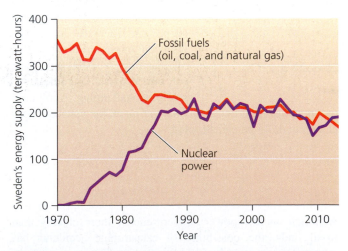

FIGURE 15.29 Sweden has cut its fossil fuel consumption in half since 1970, largely by replacing fossil fuels with nuclear power. *Data from Swedish Energy Agency.*

TABLE 15.5 Risks and Impacts of Coal-Fired versus Nuclear Power Plants

TYPE OF IMPACT	COAL	NUCLEAR
Land and ecosystem disturbance from mining	Extensive, on surface or underground	Less extensive
Greenhouse gas emissions	Considerable emissions	None from plant operation; much less than coal over the entire life-cycle
Other air pollutants	Sulfur dioxide, nitrogen oxides, particulate matter, and other pollutants	No pollutant emissions
Radioactive emissions	No appreciable emissions	No appreciable emissions during normal operation; possibility of emissions if severe accident occurs
Occupational health among workers	More known health problems and fatalities	Fewer known health problems and fatalities
Health impacts on nearby residents	Air pollution impairs health	No appreciable known health impacts under normal operation
Consequences of accident or sabotage	No widespread effects	Potentially catastrophic widespread effects
Solid waste	More generated	Less generated
Radioactive waste	None	Radioactive waste generated
Fuel supplies remaining	Should last several hundred more years	Uncertain; supplies could last longer or shorter than coal supplies

For each type of impact, the more severe impact is shown in red.

Nuclear power offers a mix of advantages and disadvantages compared with fossil fuels—coal in particular (TABLE 15.5). For residents living downwind from power plants, scientists calculate that nuclear power poses far fewer chronic health risks from pollutants (such as nitrogen oxides and sulfur dioxide). And because uranium generates far more power than coal by weight or volume, less of it needs to be mined, so uranium mining causes less damage to landscapes and generates less waste than coal mining. A drawback of nuclear power is that arranging for safe disposal of radioactive waste is challenging. Another is that if an accident occurs at a power plant, the consequences can potentially be catastrophic.

weighing the ISSUES

Choose Your Risk

Examine Table 15.5. Given the choice of living next to a nuclear power plant or living next to a coal-fired power plant, which would you choose? What would concern you most about each option?

Nuclear power poses small risks of large accidents

Although nuclear power delivers energy more cleanly than fossil fuels, the possibility of catastrophic accidents has spawned a great deal of public anxiety. Three events have been influential in shaping public opinion about nuclear power: Three Mile Island, Chernobyl, and Fukushima.

Three Mile Island In Pennsylvania in 1979, a combination of mechanical failure and human error at the **Three Mile Island** plant caused coolant water to drain from the reactor vessel, temperatures to rise inside the reactor core, and metal surrounding the fuel rods to melt, releasing radiation. This process is termed a **meltdown,** and at Three Mile Island it proceeded through half of one reactor core. Area residents stood ready to be evacuated as the nation held its breath, but fortunately most radiation remained inside the containment building.

Once this accident was brought under control, the damaged reactor was shut down, and multi-billion-dollar cleanup efforts stretched on for years. Three Mile Island is best regarded as a near miss; the emergency could have been far worse had the meltdown proceeded through the entire stock of uranium fuel or had the containment building not contained the radiation.

Chernobyl In 1986 the **Chernobyl** plant in Ukraine (part of the Soviet Union at the time) suffered the most severe nuclear accident yet (FIGURE 15.30a). Engineers had turned off safety systems to conduct tests, and human error, along with unsafe reactor design, led to explosions that destroyed the reactor and sent clouds of radioactive debris billowing into the atmosphere. Winds carried radioactive fallout across much of the Northern Hemisphere, particularly Ukraine, Belarus, and parts of Russia and Europe. For 10 days, radiation escaped while emergency crews risked their lives putting out fires. The Soviet government evacuated more than 100,000 residents of the area.

(a) The destroyed reactor at Chernobyl, 1986

(b) The confinement dome under construction, 2015

FIGURE 15.30 **The world's worst nuclear accident unfolded in 1986 at Chernobyl.** The destroyed reactor **(a)** was later encased in a massive concrete sarcophagus to contain radiation leakage. Today an international team has built a huge new confinement structure **(b)** that has been slid into place to encase the deteriorating sarcophagus.

The accident killed 31 people directly and sickened thousands more. Exact numbers are uncertain because of inadequate data and the difficulty of determining long-term radiation effects. Health authorities estimate that most of the 6000-plus cases of thyroid cancer diagnosed in people who were children at the time resulted from radioactive iodine. An international consensus effort 20 years after the event estimated that radiation raised cancer rates among exposed people by as much as a few percentage points, resulting in up to several thousand fatal cancer cases.

Following the catastrophe at Chernobyl, workers erected a gigantic concrete sarcophagus around the demolished reactor, scrubbed buildings and roads, and removed irradiated materials. However, the landscape for at least 30 km (19 mi) around the plant remains contaminated, the demolished reactor is still full of dangerous fuel and debris, and radioactivity leaks from the hastily built, deteriorating sarcophagus. In 2016 an international team finished building an enormous confinement structure (**FIGURE 15.30b**) and slid it on rails into place around the old sarcophagus to prevent a re-release of radiation.

Fukushima Daiichi On March 11, 2011, a magnitude 9.0 earthquake struck eastern Japan and sent an immense tsunami roaring onshore (p. 240). More than 18,000 people were killed and many thousands of buildings were destroyed. This natural disaster affected the operation of several of Japan's nuclear plants, most notably **Fukushima Daiichi.** Here, the earthquake shut down power, and the tsunami flooded the plant's emergency generators (**FIGURE 15.31a**). Without electricity, workers could not use moderators and control rods, and the fuel began to overheat as fission proceeded, uncontrolled.

Amid the chaos across the region, help was slow to arrive, so workers flooded the reactors with seawater in a desperate effort to prevent meltdowns. Several explosions and fires occurred. Three reactors experienced full meltdowns,

and three others were seriously damaged. Parts of the plant remained inaccessible for months because of radioactive water. It will likely require decades to fully clean up the site.

Radioactivity was released during and after these events at levels about one-tenth of those from Chernobyl. Thousands of area residents were evacuated and screened for radiation effects (**FIGURE 15.31b**), and restrictions were placed on food and water from the region. Much of the radiation spread by air or water into the Pacific Ocean, and trace amounts were detected around the world (**FIGURE 15.31c**). In the years following the event, a slow flow of radioactive groundwater from beneath the plant has continued to leak into the ocean. Japan's government has dedicated $1.2 billion to monitor the region's people for signs of any long-term health effects.

In the aftermath of the disaster, the Japanese government idled all 50 of the nation's nuclear reactors and embarked on safety inspections. Efforts to restart them were met with public debate and street protests. Across the world, many nations reassessed their nuclear programs. Germany reacted most strongly, shutting down half of its nuclear power plants and deciding to phase out the rest by 2022.

The calamity at Fukushima could likely have been avoided had the emergency generators not been located in the basement where a tsunami could flood them. And the design of most modern reactors is safer than Chernobyl's. Yet natural disasters and human error will always pose risks—and as plants age, they require more maintenance and become less safe. Moreover, radioactive material could be stolen from plants and used in terrorist attacks. This possibility has been especially worrisome in the cash-strapped nations of the former Soviet Union, where hundreds of former nuclear sites have gone without adequate security for years.

To address concerns about stolen fuel and to reduce the world's nuclear weapons stockpiles, the United States and Russia embarked on a remarkably successful program nicknamed "Megatons to Megawatts." Between 1993 and 2013,

(a) The tsunami barrels toward the Fukushima reactors

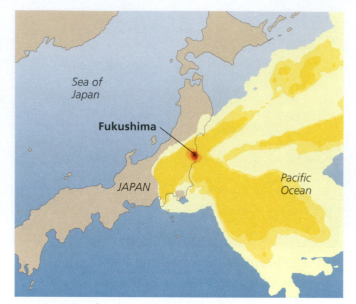

Intensity of cesium-137 radiation in fallout relative to intensity at the plant site (1 = 100%, 0.9 = 90%, etc)

▉ 0.9–1	▉ 0.4–0.6	▉ 0.01–0.1
▉ 0.7–0.9	▉ 0.3–0.4	▉ 0.005–0.01
▉ 0.6–0.7	▉ 0.1–0.3	▉ 0.001–0.005

(c) Most radiation drifted eastward over the ocean

(b) A Japanese child is screened for radiation

FIGURE 15.31 The Fukushima Daiichi crisis was unleashed after an earthquake generated a massive tsunami. The tsunami **(a)** tore through a seawall and inundated the plant's nuclear reactors. Children evacuated from the region **(b)** were screened for radiation exposure. Most of the radiation that escaped from the plant drifted over the ocean, as shown in this map **(c)** of cesium-137 isotopes in the 9 days following the accident. *Data in (c) from Yasunari, T.J., et al. 2011. Cesium-137 deposition and contamination of Japanese soils due to the Fukushima nuclear accident. Proc. Natl. Acad. Sci. USA 108: 19530–19534.*

the United States purchased weapons-grade uranium and plutonium from Russia, let Russia process them into lower-enriched fuel, and diverted the fuel to peaceful use in power generation. As a result, in recent years fully 10% of America's electricity has been generated from fuel recycled from Russian warheads that used to be atop missiles pointed at American cities!

Waste disposal remains a challenge

Even if nuclear power could be made completely safe, we would still be left with the conundrum of what to do with spent fuel rods and other radioactive waste. This waste will continue emitting radiation for as long as our civilization exists—uranium-235 has a half-life (p. 31; the time it takes for half the atoms to decay and give off radiation) of 700 million years.

Currently, such waste is held in temporary storage at nuclear power plants. To minimize leakage of radiation, spent fuel rods are sunken in pools of cooling water (**FIGURE 15.32a**) or encased in thick casks of steel, lead, and concrete (**FIGURE 15.32b**). In total, U.S. power plants are storing more than 70,000 metric tons of high-level radioactive waste (such as spent fuel)—enough to fill a football field to the depth of 7 m (21 ft)—as well as much more low-level radioactive waste (such as contaminated clothing and equipment). This waste is held at more than 120 sites spread across 39 states (**FIGURE 15.33**). The majority of Americans live within 125 km (75 mi) of temporarily stored waste.

Because storing waste at many dispersed sites creates a large number of potential hazards, nuclear waste managers would prefer to send all waste to a single, central repository that can be heavily guarded. In the United States, government scientists selected Yucca Mountain, a remote site in the Nevada desert. Choice of this site followed extensive study,

(a) Wet storage

(b) Dry storage

FIGURE 15.32 Nuclear waste is stored at nuclear power plants, because no central repository yet exists. Spent fuel rods are kept in wet storage **(a)** in pools of water, which keep them cool and reduce radiation release, or in dry storage **(b)** in thick-walled casks layered with lead, concrete, and steel.

and $13 billion was spent on its development. Most Nevadans opposed the choice, however, and concerns were raised about seismic activity. In 2010 the Obama administration ended support for the project, although lawsuits have challenged this decision and the Trump administration may attempt to revive it.

Until the United States establishes a central repository for radioactive waste, this waste will remain spread among many locations. However, one concern with a centralized repository is that waste would need to be transported there from the many current storage sites and from each nuclear plant in the future. Because this would involve many thousands of shipments by rail and truck across hundreds of public highways through almost every state of the union, some people worry that the risk of an accident is unacceptably high.

Nuclear power's growth has slowed

Dogged by concerns over waste disposal, safety, and cost overruns, nuclear power's growth has slowed. Public anxiety in the wake of Chernobyl made utilities less willing to

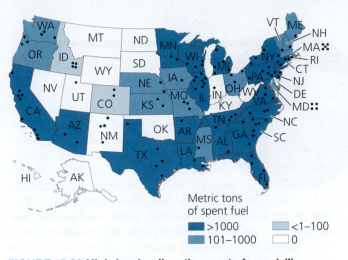

Metric tons of spent fuel

- >1000
- 101–1000
- <1–100
- 0

FIGURE 15.33 High-level radioactive waste from civilian reactors is currently stored at more than 120 sites in 39 states across the United States. In this map, dots indicate storage sites, and colors indicate the amount of waste stored in each state. *Data from Nuclear Energy Institute.*

invest in new plants, and reaction to Fukushima stalled a resurgence in the industry. Building, maintaining, operating, and ensuring the safety of nuclear facilities is enormously expensive, and almost every nuclear plant has overrun its budget. In addition, plants have aged more quickly than expected because of problems that were underestimated, such as corrosion in coolant pipes. The plants that have been shut down—well over 100 to date—have served on average less than half their expected lifetimes. Moreover, shutting down, or decommissioning, a plant is sometimes more expensive than the original construction. As a result of these financial issues, electricity from nuclear power remains more expensive than electricity from fossil fuels. Governments are still subsidizing nuclear power to keep electricity prices down for ratepayers, but many private investors lost interest long ago.

Nonetheless, nuclear power remains one of the few currently viable alternatives to fossil fuels with which we can generate large amounts of electricity in short order. This is why more and more environmental advocates propose expanding nuclear capacity using a new generation of reactors designed to be safer and less expensive. For a nation wishing to cut its pollution and greenhouse gas emissions quickly and substantially, nuclear power is in many respects the leading option. Yet with future growth for nuclear power uncertain, fossil fuels in limited supply, and climate change worsening, our society must determine where we will turn for clean and sustainable energy. Increasingly, people are turning to renewable energy sources (Chapter 16), those that cannot be depleted by our use.

weighing the
ISSUES

More Nuclear Power?

A number of European nations have reduced their carbon emissions by expanding their use of nuclear power to replace fossil fuels. Do you think the United States should expand its nuclear power program? Why or why not?

closing THE LOOP

Over the past two centuries, fossil fuels have helped us build the complex industrialized societies we enjoy today. Yet our supplies of conventional fossil fuels are declining. We can respond to this challenge by expanding our search for new sources of fossil fuels—and paying ever-higher economic, health, and environmental costs. Or, we can encourage conservation and efficiency while developing alternative energy sources that are clean and renewable. Nuclear power provides a climate-friendly alternative to fossil fuels, but high costs and public fears over safety in the wake of accidents have stalled its growth.

The paths we choose to meet our global energy needs will have far-reaching consequences for human health and well-being, for Earth's climate, and for the stability and progress of our civilization.

Today's debate over hydraulic fracturing for shale gas is a microcosm of the larger conversation about our energy future. A key question is whether natural gas will be a bridge fuel to renewables or an anchor that keeps us in the fossil fuel age. Fortunately, as renewable energy sources become increasingly well developed and economical, it becomes easier to envision freeing ourselves from a reliance on fossil fuels and charting a bright future for humanity and the planet with renewable energy.

TESTING Your Comprehension

1. Why are fossil fuels our most prevalent source of energy today? How are fossil fuels formed? Why are they considered nonrenewable?

2. Describe how *net energy* differs from *energy returned on investment (EROI)*. Why are these concepts important when evaluating energy sources?

3. Describe how coal is used to generate electricity. Now, describe how we create petroleum products. Provide examples of several of these products.

4. Summarize two fundamental ways by which we can respond to a global peak in the extraction of oil. What are the pros and cons of each type of response?

5. Describe three ways in which we are now extending our reach for fossil fuels. List several impacts (positive or negative) that these actions might have.

6. Give an example of a clean coal technology. Now describe how carbon capture and storage is intended to work.

7. Describe two specific examples of how technological advances can improve energy efficiency. Now describe one specific action you could take to conserve energy.

8. Based on data in this chapter, give two explanations for why it is reasonable to expect that Americans should be able to use energy more efficiently in the future as the U.S. economy expands.

9. In what ways did the events at Three Mile Island, Chernobyl, and Fukushima Daiichi differ? What consequences resulted from each incident?

10. List several concerns about the disposal of radioactive waste. What has been done so far about its disposal?

SEEKING Solutions

1. Summarize the main arguments for and against hydraulic fracturing for natural gas in the Marcellus Shale. What problems is this extraction helping to address? What problems is it creating? If a gas company offered you money to drill for gas in your backyard, how would you respond?

2. Describe three specific health or environmental impacts resulting from fossil fuel extraction or consumption. For each impact, what steps could governments, industries, or individuals take to alleviate the impact? What might prevent them from taking such steps? What could encourage them to take such steps?

3. Contrast the experiences of the Ogoni people of Nigeria with those of the citizens of Alaska. How have they been similar and different? Do you think businesses or governments should take steps to ensure that local people benefit from oil drilling operations? How could they do so?

4. **CASE STUDY CONNECTION** You are the mayor of a rural Pennsylvania town above the Marcellus Shale, and a gas company would like to drill in your town. Some of your town's residents are eager to have jobs that hydraulic fracturing for shale gas would bring. Others are fearful that leaks of methane and fracking fluids from drilling shafts will contaminate the water supply. Some of your town's landowners are excited to receive payments from the gas company for use of their land, whereas others dread the prospect of noise and pollution. If the company receives too much opposition in your town, it says it will drill elsewhere instead. What information would you seek from the gas company, from your state regulators, and from scientists and engineers before deciding whether support for fracking is in the best interest of your town? How would you make your decision? How might you try to address the diverse preferences of your town's residents?

5. **THINK IT THROUGH** You are elected governor of the state of Florida as the federal government is debating opening new waters to offshore drilling for oil and natural gas. Drilling in Florida waters would create jobs for Florida citizens and revenue for the state in the form of royalty payments from oil and gas companies. However, there is always the risk of a catastrophic oil spill, with its ecological, social, and economic impacts. Would you support or oppose offshore drilling off the Florida coast? Why? What, if any, regulations would you insist be imposed on such development? What questions would you ask of scientists before making your decision? What factors would you consider in making your decision?

CALCULATING Ecological Footprints

Scientists at the Global Footprint Network calculate the energy component of our ecological footprint by estimating the amount of ecologically productive land and sea required to absorb the carbon released from fossil fuel combustion. This translates into 6.1 hectares (ha) of the average American's 8.6-ha ecological footprint. Another way to think about our footprint, however, is to estimate how much land would be needed to grow biomass with an energy content equal to that of the fossil fuel we burn.

	HECTARES OF LAND FOR FUEL PRODUCTION
You	1649
Your class	
Your state	
United States	

Assume that you are an average American who burns about 6.1 metric tons of oil-equivalent in fossil fuels each year and that average terrestrial net primary productivity (p. 36) can be expressed as 0.0037 metric ton/ha/year. Calculate how many hectares of land it would take to supply our fuel use by present-day photosynthetic production.

1. Compare the energy component of your ecological footprint calculated in this way with the 6.1 ha calculated using the method of the Global Footprint Network. Explain why results from the two methods may differ.

2. Earth's total land area is approximately 15 billion hectares. Compare this to the hectares of land for fuel production from the table.

3. In the absence of stored energy from fossil fuels, how large a human population could Earth support at the level of consumption of the average American, if all of Earth's area were devoted to fuel production? Do you consider this realistic? Provide two reasons why or why not.

Mastering Environmental Science

Students Go to **Mastering** Environmental Science for assignments, the etext, and the Study Area with practice tests, videos, current events, and activities.

Instructors Go to **Mastering** Environmental Science for automatically graded activities, current events, videos, and reading questions that you can assign to your students, plus Instructor Resources.

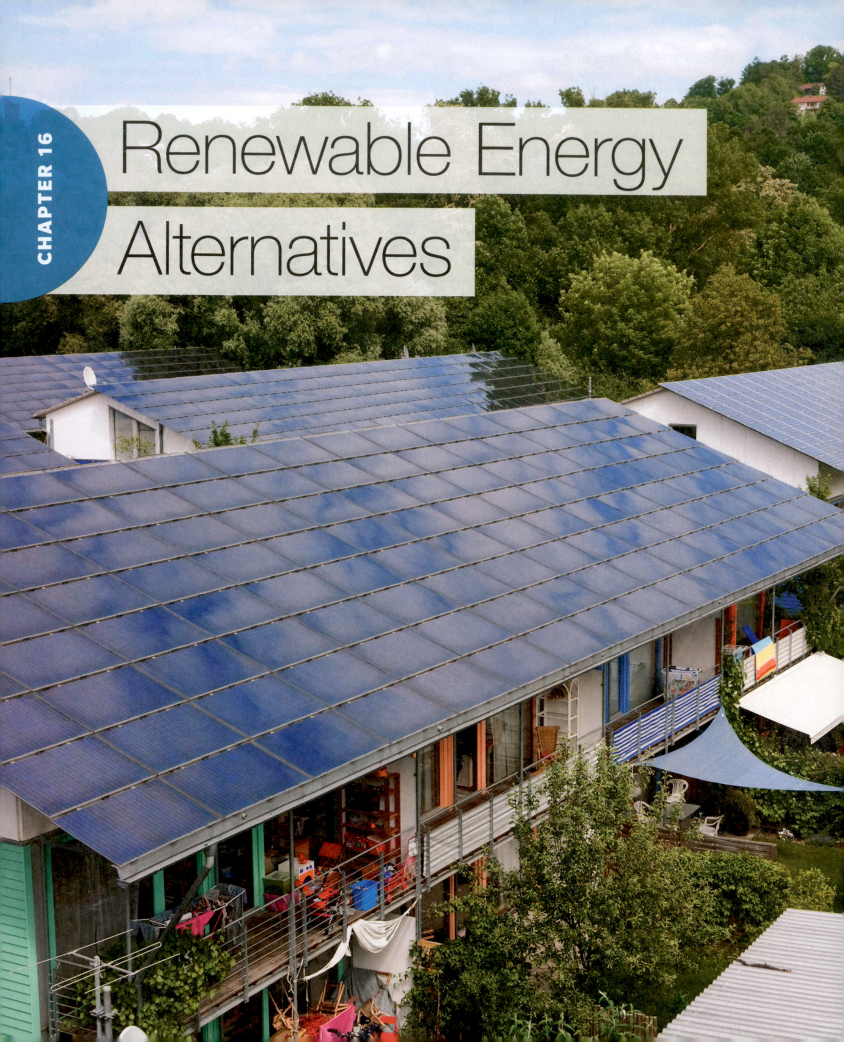

CHAPTER 16

Renewable Energy Alternatives

Germany Reaches for the Sun

> [Renewable energy] will provide millions of new jobs. It will halt global warming. It will create a more fair and just world. It will clean our environment and make our lives healthier.
> —Hermann Scheer, energy expert and member of the German parliament, 2009

> The nation that leads the clean energy economy will be the nation that leads the global economy.
> —U.S. President Barack Obama, 2010

When we think of solar energy, most of us envision a warm sunny place such as Arizona or southern California. Yet the country that produces the most solar power per person is Germany, a northern European nation that receives less sun than Alaska! In recent years Germany has been among the world's top installers and users of photovoltaic (PV) solar technology, which produces electricity from sunshine. Germany now obtains nearly 7% of its electricity from solar power—one of the highest rates in the world.

How is this possible in such a cool and cloudy country? A bold federal policy has used economic incentives to promote solar power and other forms of renewable energy. Germany's **feed-in tariff** system has required utilities to buy power (at guaranteed premium prices under long-term contract) from anyone who can generate power from renewable energy sources and feed it into the electrical grid. In response, German homeowners and businesses rushed to install PV panels and began selling their excess solar power to utilities at a profit.

The feed-in tariffs have applied to all forms of renewable energy. As a result, Germany became the world leader in wind power in the 1990s, until being overtaken by the United States and China. Today Germany ranks third or fourth in the world in solar water heating, biomass-generated power, electrical power capacity from renewable sources, and renewable energy generated per person. The nation gets fully one-third of its electricity from renewable sources.

Germany's push for renewable energy dates back to 1990. In the wake of the disaster at the Chernobyl nuclear power plant in Ukraine (part of the Soviet Union at the time) (p. 368), Germany decided to phase out its own nuclear power plants. However, if these were shut down, the nation would lose virtually all its clean energy and would become utterly dependent on oil, gas, and coal imported from Russia and the Middle East.

Enter Hermann Scheer, a German parliament member and an expert on renewable energy. While everyone else assumed that solar, wind, and geothermal energy were costly and risky, Scheer saw them as a great economic opportunity—and as the only long-term answer. In 1990, Scheer helped push through a landmark law establishing feed-in tariffs. Ten years later, the law was revised and strengthened: The Renewable Energy Sources Act of 2000 aimed to promote renewable energy production and

Upon completing this chapter, you will be able to:

- Discuss reasons for seeking alternatives to fossil fuels

- Identify the major sources of renewable energy, and assess their recent growth and future potential

- Describe solar energy and how we harness it, and evaluate its advantages and disadvantages

- Describe wind power and how we harness it, and evaluate its advantages and disadvantages

- Describe geothermal energy and how we harness it, and evaluate its advantages and disadvantages

- List the various ocean energy sources and describe their potential

- Outline the scale, methods, and impacts of hydroelectric power

- Describe established and emerging sources and techniques involved in harnessing bioenergy, and assess bioenergy's benefits and shortcomings

- Explain hydrogen fuel cells, and weigh options for energy storage and transportation

Installing PV solar panels ▲

Rooftop photovoltaic solar panels on German homes

use, enhance the security of the energy supply, reduce carbon emissions, and lessen the many external costs (pp. 96, 104) of fossil fuel use.

Under the law, each renewable source was assigned its own payment rate. These rates have since been lowered year by year to encourage increasingly efficient means of producing power. However, utilities pass along to consumers their costs of paying the feed-in tariffs, and these costs soon grew to 15% of the average German citizen's electric bill. To ease this burden on ratepayers, the German government in 2010 decided to slash PV solar tariff rates.

When German consumers heard that the solar tariff rate would be reduced, sales of PV modules skyrocketed as people rushed to lock in contracts at the existing rate. Further tariff reductions the next two years also spurred rushes to install PV systems before the rates dropped. In 2010, 2011, and 2012, Germans installed more than 7 gigawatts of PV solar capacity each year—an amount surpassing the total capacity of solar power in the United States at the time. Then from 2013 on, solar installations in Germany slowed as tariff rates became low (**FIGURE 16.1**).

By reducing the subsidies, Germany's leaders aimed to encourage technological innovation for efficiency within the solar industry, thereby creating a stronger industry that can sustain growth over the long term and outcompete foreign companies for international business. Indeed, boosted by domestic demand, German renewable energy industries have become global leaders, designing and selling technologies around the world while employing nearly 400,000 people. Germany is second in PV production behind China, leads the world in production of biodiesel, and has built several cellulosic ethanol (p. 394) facilities.

By 2050, Germany aims to obtain 80% of its electricity from renewable sources. To achieve this historic energy transition—or *Energiewende*, in German—the government has been allotting more public money to renewable energy than any other nation—over $25 billion annually in recent years.

As Germany replaces fossil fuels with renewable energy, it is improving its air quality and is helping to fight climate change. Since 1990, carbon dioxide emissions from German energy sources have fallen by 25%, and emissions of seven other major pollutants (CH_4, N_2O, SO_2, NO_X, CO, VOCs, and dust) have been reduced by 12–95%. At least half of these reductions are attributed to renewable energy paid for under the feed-in tariff system.

The planned transition from fossil fuels in Germany has not been completely smooth, however. In response to public protests following the Fukushima nuclear disaster in Japan in 2011 (p. 369), the German government shut down 7 of its 15 nuclear power plants. As a result, the electricity supply from nuclear power fell sharply, causing rates to rise and leading to an increase in the combustion of coal for power. At the same time, the rise of solar and wind, two intermittent power sources, meant that the German grid occasionally was flooded with excess electricity. Germany's *Energiewende* thus faced the challenge of keeping renewables growing at a steady and predictable rate without causing instability in supplies.

As a result, in 2016 Germany's leaders revised the national energy policy, replacing feed-in tariffs for large-scale producers with an auction system. Under the new system, the government decides how much new capacity of each type of renewable energy it will allow each year, and then auctions off permits for this development to the lowest bidders. Proponents of this policy change predict that it will lower costs for ratepayers, strengthen industries through competition, and allow time for the grid to expand along with the new mix of energy sources. Critics of the auction system say it favors powerful firms over small start-ups, creates risk and uncertainty that will discourage private investment, and threatens to stifle Germany's progress toward its renewable energy targets.

Small-scale PV solar installations are still eligible for feed-in tariffs under the revised law, so ordinary homeowners wishing to "go solar" are unaffected, but the impact of this major policy change on Germany's overall energy situation remains to be seen.

Germany's experience has served as a model for other countries. As of 2017, more than 100 nations, states, and provinces had implemented some sort of feed-in tariff for renewable energy. Nations with high tariff rates such as Spain and Italy ignited their wind and solar development as a result. In North America, Vermont and Ontario established feed-in tariff systems similar to Germany's, while California, Hawai'i, Maine, New York, Oregon, Rhode Island, Washington, and utilities in several additional states conduct more-limited programs. In 2010, Gainesville, Florida, became the first U.S. city to establish feed-in tariffs, and solar power grew quickly there as a result. Moreover, utilities in 46 U.S. states now offer *net metering*, in which utilities credit customers who produce renewable power and feed it into the grid. As more nations, states, and cities encourage renewable energy, we may soon experience a historic transition in the way we meet our energy demands.

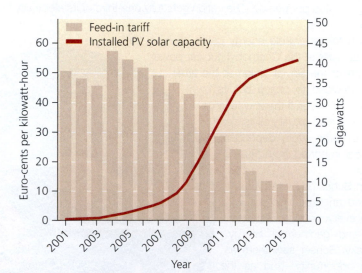

FIGURE 16.1 PV solar power rose steeply in Germany, thanks to feed-in tariffs. The rapid rise in installed capacity of PV solar then slowed as the feed-in tariffs were reduced. *Data from International Energy Agency, 2016.* Photovoltaic power systems programme annual report 2015; *and AGEE-Stat, Federal Ministry for Economic Affairs and Energy, Germany.*

Renewable Energy Sources

Germany's bold federal policy is just one facet of a global shift toward renewable energy sources. Across the world, nations are seeking to move away from fossil fuels while ensuring a reliable, affordable, and sustainable supply of energy.

We have alternatives to fossil fuels

Fossil fuels drove the industrial revolution and helped to create the unprecedented material prosperity we enjoy today. Our global economy is still powered by coal, oil, and natural gas, which together provide over four-fifths of the world's energy and two-thirds of our electricity (see Figure 15.4, p. 344). However, easily extractable supplies of these nonrenewable energy sources are in decline, while the economic and social costs, security risks, and health and environmental impacts of fossil fuel dependence continue to intensify (Chapter 15). For these reasons, most energy experts accept that we will need to shift to energy sources that are less easily depleted and gentler on our health and environment.

We have developed a range of alternatives to fossil fuels (see Table 15.1, p. 343). These include nuclear power (nonrenewable because it relies on uranium, a mineral in finite supply) and a diverse array of renewable energy sources. Of our renewable sources, hydroelectric power and energy from biomass are well established and already play substantial roles in our energy and electricity budgets. Although hydropower and bioenergy are renewable, supplies of water and biomass can be locally depleted if overharvested.

Perpetually renewable sources include energy from the sun, wind, Earth's geothermal heat, and ocean water. These energy sources are often called "new renewables" because (1) they are just beginning to be used on a wide scale in our modern industrial society, (2) they are harnessed using technologies still in a rapid phase of development, and (3) they will likely play much larger roles in the future.

Thus far, renewable sources are being used more for generating electricity than for fueling transportation. In the United States, renewable sources account for 10.5% of total energy use (**FIGURE 16.2a**) and 15.4% of electricity generated (**FIGURE 16.2b**), but for just 4.8% of transportation needs.

Renewable sources are growing fast

Over the past four decades, energy production from solar, wind, and geothermal sources has grown far faster than energy production from other sources. However, because these "new renewable" sources started from such low levels of use, it will take them some time to replace fossil fuels. One hurdle constraining their growth is that so far we lack infrastructure to transfer huge volumes of power from them inexpensively on a continent-wide scale. However, rapid growth in renewable energy seems likely to continue as technology improves, prices fall, conventional fossil fuel supplies decline, and people demand cleaner environments.

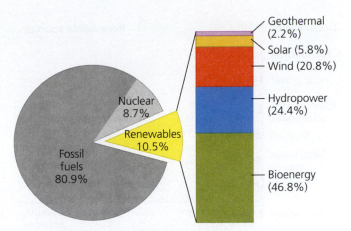

(a) U.S. consumption of energy, by source

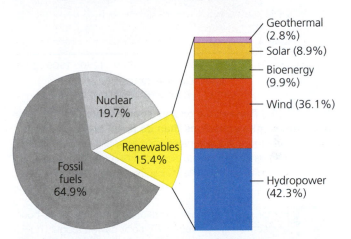

(b) U.S. electricity generation, by energy source

FIGURE 16.2 Renewable energy sources contribute a small but growing portion of the energy we consume. In the United States, about 10.5% of total energy consumption (transport, heating, electricity, etc.) **(a)** is from renewable energy, mostly bioenergy and hydropower. Of electricity generated in the United States **(b)**, about 15.4% comes from renewable sources, predominantly hydropower and wind. *Data are for 2016, from Energy Information Administration.*

DATA • What percentage of overall U.S. energy consumption does solar power contribute? • What percentage of overall U.S. electricity generation does wind power contribute?

Go to **Interpreting Graphs & Data** on Mastering **Environmental Science**

Renewable energy offers advantages

Renewable energy offers a large number of substantial benefits for individuals and for society. Unlike fossil fuels, most renewable sources are inexhaustible on timescales relevant to our society, thus providing us long-term security. Renewable alternatives also help us economically, by diversifying an economy's energy mix and thereby reducing price volatility and reliance on imported fuels. Some renewable sources generate income and property tax for rural communities, and some help people in developing regions of the world to produce their own energy. And of course, replacing fossil

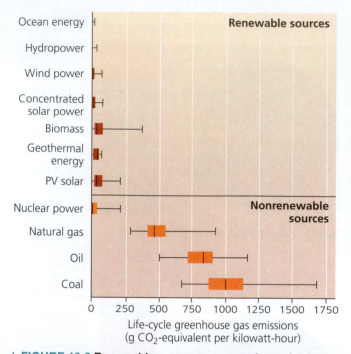

FIGURE 16.3 Renewable energy sources release far fewer greenhouse gas emissions than do fossil fuels. Shown are medians and ranges of estimates from scientific studies of each source when used to generate electricity. *Data from Intergovernmental Panel on Climate Change, 2012.* Renewable energy sources and climate change mitigation. Special report. *New York, NY: Cambridge University Press.*

DATA Q For every unit of emissions released from 1 kilowatt-hour of electricity generated by PV solar power, roughly how many units would be released from 1 kilowatt-hour of electricity generated by coal combustion?

Go to **Interpreting Graphs & Data** on **Mastering** Environmental Science

FAQ

Isn't renewable energy too expensive and untested to rely on?

Not at all. For decades now, renewable energy sources and technologies have been supplying power to many millions of people. This goes for hydropower and bioenergy, but also for most "new renewable" sources. These range from geothermal power and ground-source heat pumps to solar water heating and PV solar cells to onshore and offshore wind power. All have become far more affordable, and in many places electricity from wind is now cheaper than electricity from fossil fuels.

fuels with clean, renewable energy benefits our health by reducing air pollution (Chapter 13). Perhaps most importantly, replacing fossil fuels with renewable energy is the prime way to slow the greenhouse gas emissions (**FIGURE 16.3**) that drive global climate change (Chapter 14). In all likelihood, stopping climate change will require a full transition to clean renewable energy.

Shifting to renewable energy also creates employment opportunities. The design, installation, maintenance, and management required to develop technologies and to rebuild and operate our energy infrastructure are becoming major sources of employment today, through **green-collar jobs** (**FIGURE 16.4**). Already, nearly 10 million people work in renewable energy jobs around the world.

JOBS CREATED
PV solar: 3,095,000
Hydropower: 1,730,000
Biofuels: 1,724,000
Wind power: 1,155,000
Solar heating: 828,000
Biomass: 723,000
Other: 538,000

FIGURE 16.4 Renewable energy creates green-collar jobs. As of 2016, more than 9.8 million people worldwide were employed in jobs directly or indirectly connected to renewable energy. *Data from REN21, 2017.* Renewables 2017: Global status report. *REN21, UNEP, Paris.*

Germany and many other industrialized nations are gradually replacing fossil fuels with alternative sources while continuing to raise living standards for their citizens. For example, since 1970, Sweden has decreased its fossil fuel use from 81% to 38% of its national energy budget (p. 367). Bioenergy, hydropower, and nuclear power now provide Sweden with over 60% of its energy and virtually all of its electricity, and the nation's economy is as strong as ever. Worldwide, recent research makes a strong case that we can transition quickly to renewable energy and gain substantial benefits by doing so (see **THE SCIENCE BEHIND THE STORY**, pp. 380–381).

Policy and investment can accelerate our transition

Most renewable energy remains more expensive than fossil fuel energy (**FIGURE 16.5**), but prices are falling fast and renewable energy has already become cost-competitive with fossil fuel energy in many places. Today electricity from wind power is cheaper than electricity from fossil fuels across large portions of the United States. Where renewable energy is given political support, its market prices become still cheaper and it spreads faster. Feed-in tariffs like Germany's hasten the spread of renewable energy by creating financial incentives for businesses and individuals. Governments are also setting goals or mandating that certain percentages of power come from renewable sources. As of 2017, nearly all the world's nations and 30 U.S. states had set official targets for renewable energy use. Governments also invest in research and development of technologies, lend money to renewable energy businesses as they start up, and offer tax credits and tax rebates to companies and individuals who produce or buy renewable energy.

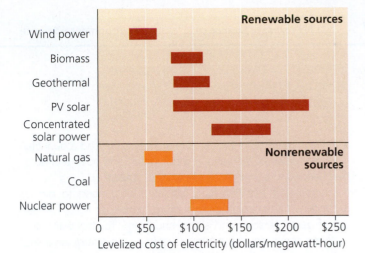

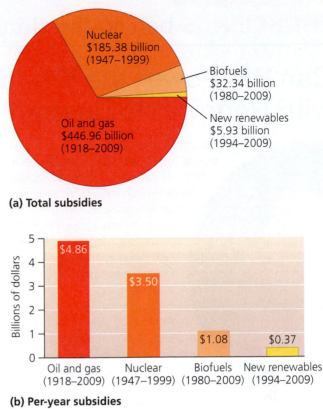

(a) Total subsidies

FIGURE 16.5 Renewable energy is becoming cost-competitive with nonrenewable energy. Shown are ranges for the price of electricity from major sources, as of 2016. Prices for renewable sources are lower than shown here if subsidies are included or if external costs (pp. 96, 104) are considered. Moreover, 2017 data from the U.S. Energy Information Administration forecast that prices for renewables will be still cheaper for plants that begin operation in the near future. *Data from Lazard, 2016.* Lazard's levelized cost of energy analysis—Version 10.0. *New York, NY: Lazard.*

(b) Per-year subsidies

FIGURE 16.6 Fossil fuels and nuclear power have received far more in U.S. government subsidies (mostly tax breaks) than have renewable energy sources. This is true both for **(a)** total amounts over the past century and for **(b)** average amounts per year. *Data are in 2010 dollars, from Pfund, N., and B. Healey, 2011. What would Jefferson do? The historical role of federal subsidies in shaping America's energy future. DBL Investors.*

When a government boosts an industry with such policies, the private sector often responds with investment of its own as investors recognize an enhanced chance of profit. Indeed, global investment (public plus private) in renewable energy has risen to exceed $200 billion each year since 2010. Yet the economics of renewable energy have been erratic. Technologies evolve quickly, and policies vary from place to place and can change unpredictably. As a result, renewable markets have been somewhat volatile.

Critics of public subsidies for renewable energy complain that funneling taxpayer money to particular energy sources is inefficient and skews the market. Instead, they propose, we should let energy sources compete freely. However, proponents of renewable energy point out that governments have long subsidized fossil fuels and nuclear power far more than they now subsidize renewables. As a result, there has never been a level playing field, nor a truly free market.

In one recent study, researchers dug into data on the U.S. government's many energy subsidies and tax breaks (p. 113) over the

weighing the ISSUES

How to Transition to Renewable Energy?

Many approaches are being pursued to help our society transition from fossil fuels to renewable energy. Reducing subsidies for fossil fuels is one approach. Increasing subsidies for renewables is another. Using feed-in tariffs or net metering are two specific ways of boosting support for renewables. Governments can also pass laws mandating greater use of renewable energy by utilities. Or they can establish carbon taxes (p. 332) or cap-and-trade programs (pp. 113, 333). Describe an advantage and a disadvantage of each of these approaches. Can you propose any additional approaches? What steps do you think our society should take to transition to renewable energy?

past century. Their report revealed that in total, oil and gas have received 75 times more subsidies—and nuclear power has received 31 times more—than new renewable sources (**FIGURE 16.6a**). Moreover, on a per-year basis (controlling for the differing amounts of time each source has been used), oil and gas have received 13 times more subsidies—and nuclear power over 9 times more—than solar, wind, and geothermal power (**FIGURE 16.6b**). Other research concludes that across the world today, for every $1 in taxpayer money that goes toward renewable energy, $4 continues to go toward fossil fuels.

Many feel that the subsidies showered on fossil fuels have helped to enhance America's economy, national security, and international influence by supporting thriving global energy industries dominated by U.S. firms. Yet by this logic, if America wants to be a global leader to rival nations such as China and Germany in the transition to renewable energy, then political and financial support will likely need to be redirected toward renewable energy sources.

Let's now begin our tour of today's renewable energy sources. We'll start with the new renewables (solar, wind, geothermal, and ocean energy), proceed through the well-established hydropower and bioenergy sources, and then briefly examine hydrogen fuel cells.

Can We Power the World with Renewable Energy?

Mark Jacobson of Stanford University

Despite the rapid growth of renewable energy, many experts remain skeptical that we will ever be able to replace fossil fuels entirely. Yet some recent scientific research has aimed to outline in detail how we might power our society completely with clean renewable energy—without fossil fuels, biofuels, or nuclear power.

For years, Mark Jacobson, director of the Atmosphere/ Energy Program at Stanford University, has been assessing energy sources with the goal of finding solutions to pollution, climate change, and energy insecurity. In 2009 he published a full life-cycle analysis (p. 410) of the social, health, and environmental impacts of all major energy sources and found that new renewable sources based on wind, water, and solar power were lower-impact than biofuels, nuclear power, and fossil fuels.

Jacobson then teamed up with Mark Delucchi of the University of California–Davis to examine whether and how the world could meet 100 percent of its energy needs with clean renewable energy from the sun, wind, and water. In 2011 these researchers published a pair of scientific papers in the journal *Energy Policy*. Jacobson and Delucchi first calculated the likely global demand for energy in the years 2030 and 2050, using government projections. They then examined the current outputs and limitations of renewable energy technologies and selected those that were technically and commercially proven and established. Because electrical power is more energy-efficient than fuel combustion, they chose to propose only electrical technologies (for example, battery-electric vehicles rather than gasoline-powered vehicles).

The researchers then calculated what it would take to manufacture these technologies at the needed scale and to build infrastructure for renewable energy storage and transmission throughout the world. Because sources such as solar power and wind power are intermittent (varying from hour to hour and day to day), Jacobson and Delucchi judged what balance of sources was needed to compensate for intermittency and ensure a consistent, reliable energy supply.

Once all the math was done, Jacobson and Delucchi concluded that the world *can*, in fact, fully replace fossil fuels, nuclear power, and biofuels and meet all its energy demands with clean renewable sources alone. They proposed a quantitative breakdown of the various wind, water, and solar technologies needed to power the world in 2030 (**TABLE 1**).

To achieve this transition, society would need to greatly expand its transmission infrastructure, construct fleets of fuel-cell-powered vehicles and ships, and more. To deal with intermittency of sun and wind and prevent gaps in energy supplies, we would need to link complementary combinations of sources across large regions and, when there is oversupply, use the extra energy to produce hydrogen fuel. Achieving all this in a few decades would clearly be ambitious, but Jacobson and Delucchi argued that most barriers to achieving a transition to renewables are social and political, not technological or economic.

With an eye toward impacts, Jacobson and Delucchi also added up the total area of land that all of this new renewable energy infrastructure would require. The researchers calculated that 0.74% of Earth's land surface would be occupied directly by energy infrastructure, with 0.41% being newly required beyond what is already taken up. An additional 1.18% of area would be needed for spacing between structures (mostly wind turbines), but half of this area could be over water if half our wind power is offshore, and most land between turbines could be used for farming or grazing.

Jacobson and Delucchi estimated that the overall economic cost of energy in their proposed scenario would be roughly the same as the cost of energy today. One main challenge, however,

TABLE 1 Renewable Energy Infrastructure Needed to Power the World in 2030

TECHNOLOGY	NUMBER OF PLANTS OR DEVICES NEEDED	PERCENTAGE OF GLOBAL DEMAND SATISFIED
Wind turbines	3,800,000	50
Wave devices	720,000	1
Geothermal plants	5350	4
Hydroelectric plants	900	4
Tidal turbines	490,000	1
Rooftop PV systems	1,700,000,000	6
Solar PV plants	40,000	14
CSP plants	49,000	20

Data from Jacobson, M.Z., and M.A. Delucchi, 2011. Providing all global energy with wind, water, and solar power, Part I: Technologies, energy resources, quantities and areas of infrastructure, and materials. Energy Policy 39: 1154–1169.

is the limited availability of a handful of rare-earth metals (such as platinum, lithium, indium, tellurium, and neodymium) used in certain materials and equipment for wind, water, and solar technologies. While additional reserves of these metals may be discovered and mined, we would likely need to enhance efforts to recycle them.

Jacobson and Delucchi's work drew both praise and criticism. Critics, such as Australian energy expert Ted Trainer, felt that they underestimated the costs of their proposal, overestimated efficiency gains from electrification, and failed to offer persuasive quantitative evidence that intermittency can be overcome.

More recently, Jacobson and Delucchi worked with eight colleagues to design plans for how each of the 50 U.S. states might power itself entirely with wind, water, and solar sources. This research team evaluated each state's resources and energy demands, and in 2015 published a study in the journal *Energy and Environmental Science* that summarized its state-specific "roadmaps." The researchers determined a mix of renewable sources and technologies that could feasibly meet the energy demand forecast for the United States in 2050. **FIGURE 1** shows their overall plan for transitioning to 100% renewable energy across the United States. Footprints of the infrastructure required would be 0.47% of land area (0.42% new), with spacing area of 2.4% (1.6% new).

Across all 50 states, the team estimated that a full conversion to renewable energy would raise the number of energy jobs from 3.9 million to 5.9 million and would eliminate up to 62,000 premature deaths caused by pollution each year. The proposal would also save an estimated $3.3 trillion in costs from predicted climate change impacts worldwide due to U.S emissions in 2050. As a result, they calculated, the average American in 2050 would save $260 in energy costs, $1500 in health costs, and $8300 in climate change costs each year.

In a related paper in 2015 in *Proceedings of the National Academy of Sciences*, Jacobson and his colleagues ran simulation models, aiming to demonstrate how hydropower and energy storage systems could solve the challenge of peaks and troughs from intermittent wind and solar energy production. Their analysis purported to show that this challenge could be overcome, enabling a full transition to an energy economy run purely on wind, water, and solar.

Many researchers disagreed with this conclusion. In 2017, a team of 21 scientists headed by energy expert Christopher Clack authored a paper in the same journal vehemently rebutting many aspects of the Jacobson team's analysis. They found fault with its methods and felt many assumptions made were unrealistic. Clack's team felt that energy research overall supports the idea that we can transition to an economy dominated by renewable energy, but that we will need to continue using nuclear power and fossil fuels to some degree.

Jacobson's team responded to its critics, and a lengthy debate between the two camps has ensued. Such open debate—and the further research it spurs—is precisely how science as a whole moves forward. As a result, we can look forward to better and better scientific guidance as our society accelerates its transition to renewable energy.

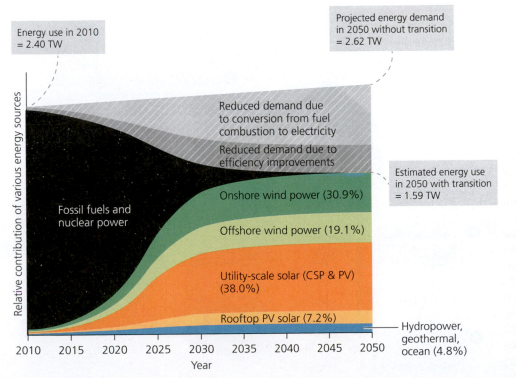

FIGURE 1 Fossil fuels and nuclear power can be phased out and replaced entirely by renewable energy by 2050 in the United States, research indicates. This graph summarizes the transition to renewables proposed by Jacobson's team. Heights of each colored band at a given year show percentages of total U.S. energy demand to be met by each energy source in that year. *Data from Jacobson, M., et al., 2015. 100% clean and renewable wind, water, and sunlight (WWS) all-sector energy roadmaps for the 50 United States. Energy Environ. Sci. 8: 2093–2117.*

Solar Energy

The sun releases astounding amounts of energy by converting hydrogen to helium through nuclear fusion. The tiny proportion of this energy that reaches Earth is enough to drive most processes in the biosphere, helping to make life possible on our planet. Each day, Earth receives enough **solar energy,** or energy from sunlight, to power human consumption for a quarter of a century. On average, each square meter of Earth's surface receives about 1 kilowatt of solar energy—17 times the energy of a lightbulb. As a result, a typical home has enough roof area to meet all its power needs with rooftop panels that harness solar energy.

We can collect solar energy using passive or active methods

The simplest way to harness solar energy is by **passive solar energy collection.** In this approach, buildings are designed to maximize absorption of sunlight in winter yet to minimize it in summer (**FIGURE 16.7**). South-facing windows maximize the capture of winter sunlight. Overhangs shade windows in summer, when the sun is high in the sky and cooling is desired. Planting vegetation around a building buffers it from temperature swings. Passive solar techniques also use materials that absorb heat, store it, and release it later. Such *thermal mass* (of straw, brick, concrete, or other materials) can make up floors, roofs, and walls, or can be used in portable blocks. All these approaches reduce energy costs.

   Active solar energy collection makes use of technology to focus, move, or store solar energy. For instance, *flat plate solar collectors* (**FIGURE 16.8**) heat water and air for homes or businesses. These panels generally consist of heat-absorbing metal plates mounted on rooftops in flat glass-covered boxes. Water, air, or antifreeze runs through tubes that pass through the collectors, transferring heat from them to the building or its water tank. More than 300 million households and businesses worldwide heat water with solar collectors. China is the world's leader in this technology, but Germans motivated by feed-in tariffs installed 200,000 new systems in one year (2008) alone. Many remote rural communities in the developing world also take advantage of active solar technology for heating, cooling, and water purification.

FIGURE 16.7 Passive solar design elements can be seen in this house built by college students. Stanford University students and staff designed and constructed this solar-powered house as part of the sixth biannual Solar Decathlon, in which college and university teams compete to build the best houses fully powered by solar energy.

Overhang shades summer sun from above.

Plants buffer house from temperature swings.

Thermal mass in floors and walls absorbs heat and then slowly releases it.

South-facing porch lets in low-angle sunlight in winter.

Concentrating sunlight focuses energy

We can intensify solar energy by gathering sunlight from a wide area and focusing it on a single point. This is the

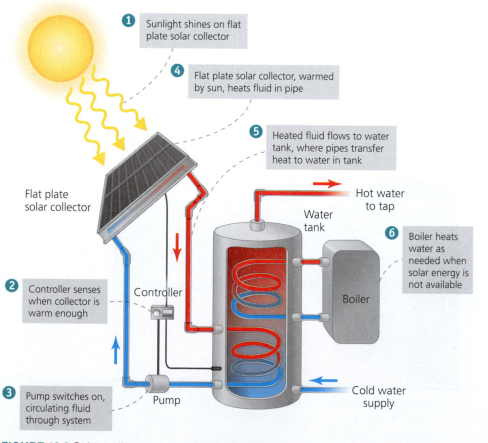

1 Sunlight shines on flat plate solar collector

4 Flat plate solar collector, warmed by sun, heats fluid in pipe

5 Heated fluid flows to water tank, where pipes transfer heat to water in tank

Flat plate solar collector

Hot water to tap

Water tank

6 Boiler heats water as needed when solar energy is not available

2 Controller senses when collector is warm enough

Controller

Boiler

3 Pump switches on, circulating fluid through system

Pump

Cold water supply

FIGURE 16.8 Solar collectors use sunlight to heat water for homes and businesses.

FIGURE 16.9 **Utilities concentrate solar power to generate electricity at large scales.** At this CSP plant in Spain, 624 mirrors reflect sunlight onto a central receiver atop a 115-m (375-ft) power tower.

principle behind *solar cookers*, simple portable ovens that use reflectors to cook food and that are proving useful in the developing world.

At much larger scales, utilities are using the principle behind solar cookers to generate electricity at large centralized facilities and transmit power to homes and businesses via the electrical grid. Such **concentrated solar power (CSP)** is being harnessed by several methods in sunny arid regions in Spain, the U.S. Southwest, and elsewhere. In one method, curved mirrors focus sunlight onto synthetic oil in pipes. The superheated oil is piped to a facility where it heats water, creating steam that drives turbines to generate electricity. In another method, hundreds of mirrors focus sunlight onto a central receiver atop a tall "power tower" (**FIGURE 16.9**). From here, air or fluids carry heat through pipes to a steam-driven generator.

Concentrated solar power holds promise for producing tremendous amounts of energy, but CSP developments (which typically sprawl across huge stretches of arid land) pose considerable environmental impacts. All land beneath the mirrors is cleared and graded, destroying sensitive habitat for desert species, while maintenance requires enormous amounts of water, which is scarce and precious in desert regions. In recent years, global electricity generation by CSP plants has grown almost as fast as electricity generation from photovoltaic rooftop panels.

PV cells generate electricity

The most direct way to generate electricity from sunlight involves photovoltaic (PV) systems. **Photovoltaic (PV) cells** convert sunlight to electrical energy when light strikes one of a pair of plates made primarily of silicon, a semiconductor that conducts electricity (**FIGURE 16.10**). The light causes one plate to release electrons, which are attracted by electrostatic forces to the opposing plate. Connecting the two plates with wires enables the electrons to flow back to the original plate, creating an electrical current (direct current, DC), which can

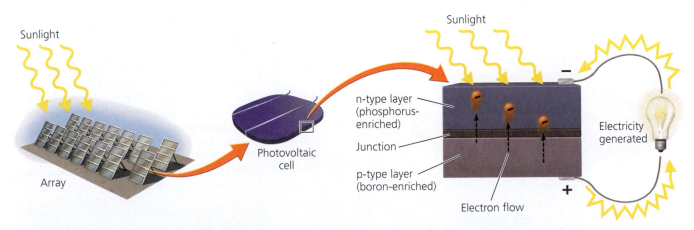

FIGURE 16.10 A photovoltaic (PV) cell converts sunlight to electrical energy. When sunlight hits the silicon layers of the cell, electrons are knocked loose from some of the silicon atoms and tend to move from the boron-enriched "p-type" layer toward the phosphorus-enriched "n-type" layer. Connecting the two layers with wiring remedies this imbalance as electrical current flows from the n-type layer back to the p-type layer. This direct current (DC) is converted to alternating current (AC) to produce usable electricity.

be converted into alternating current (AC) and used for residential and commercial electrical power. Small PV cells may power your watch or your calculator. Arrays of PV panels can be seen on the roofs of the German houses in the photo that opens this chapter (p. 374).

Researchers are experimenting with variations on PV technology, including **thin-film solar cells,** photovoltaic materials compressed into ultra-thin sheets. Although less efficient at converting sunlight to electricity, they are cheaper to produce. Thin-film technologies can be incorporated into roofing shingles and potentially many other surfaces, even highways!

Photovoltaic cells of all types can be connected to batteries that store the accumulated charge until needed. Or, producers of PV electricity can sell power to their local utility if they are connected to the regional electrical grid. In parts of 46 U.S. states, homeowners can sell power to their utility in a process called **net metering,** in which the value of the power the homeowner provides is subtracted from the homeowner's utility bill. Feed-in tariff systems like Germany's go a step further by paying producers more than the market price of the power, offering producers the hope of turning a profit.

Solar energy offers many benefits

The fact that the sun will continue shining for another 4–5 billion years makes it inexhaustible as an energy source for human civilization. Moreover, the amount of solar energy reaching Earth should be enough to power our society once we deploy technology adequate to harness it. These advantages of solar energy are clear, but the technologies themselves also provide benefits. PV cells and other solar technologies use no fuel, are quiet and safe, contain no moving parts, and require little maintenance. An average unit can produce energy for 20–30 years.

Solar systems allow for local, decentralized control over power. Homes, businesses, rural communities, and isolated areas in developing nations can use solar power to produce electricity without being near a power plant or connected to a grid. In places where PV systems are connected to the regional electrical grid, homeowners can sell excess solar power to their utility through feed-in tariffs or net metering.

Developing, manufacturing, and deploying solar technology also creates green-collar jobs; PV-related jobs now employ nearly 3.1 million people worldwide. Finally, a major advantage of solar energy over fossil fuels is that it does not emit greenhouse gases and other air pollutants (see Figure 16.3). The manufacture of equipment currently requires fossil fuels, but once up and running, a solar system produces no emissions.

Location, timing, and cost can be drawbacks

Solar energy currently has three main disadvantages—yet all three are being resolved. One limitation is that not all regions are equally sunny (**FIGURE 16.11**). People in Seattle or Anchorage will find it more challenging to rely on solar energy than people in Phoenix or San Diego. However, observe in Figure 16.11 that Germany receives less sunlight than Alaska, and yet it is the world leader in solar power!

A second limitation is that solar energy is an intermittent resource. Power is only produced when the sun shines, so daily or seasonal variation in sunlight can limit stand-alone solar systems if storage capacity in batteries or fuel cells is not adequate or if backup power is not available from a municipal electrical grid. However, at the utility scale, nonrenewable energy sources or renewable pumped-storage hydropower (p. 391) can help compensate for periods of low solar production. And for individuals, battery storage is fast becoming efficient and affordable; indeed, the majority of German homeowners installing solar panels are also now storing excess energy in batteries to use later.

The third drawback of current solar technology is the up-front cost of the equipment. However, recent declines in price and improvements in efficiency have been impressive. Solar systems continue to become more affordable and now can sometimes pay for themselves in less than 10 years. After that time, they provide energy virtually for free as long as the equipment lasts.

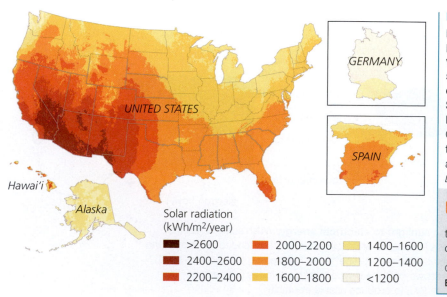

Solar radiation
(kWh/m²/year)

■ >2600	■ 2000–2200	1400–1600
■ 2400–2600	■ 1800–2000	1200–1400
■ 2200–2400	■ 1600–1800	<1200

FIGURE 16.11 Solar radiation varies from place to place. Harnessing solar energy is more profitable in sunny regions such as the southwestern United States than in cloudier regions such as Alaska and the Pacific Northwest. However, compare solar power leaders Germany and Spain with the United States. Spain is similar to Kansas in the amount of sunlight it receives, and Germany is cloudier than Alaska! This suggests that solar power can be used with success just about anywhere. *Data from National Renewable Energy Laboratory, U.S. Department of Energy.*

DATA Q • Roughly how much more solar radiation does southern Arizona receive than Germany? • How does your own state compare to Germany in its solar radiation?

Go to **Interpreting Graphs & Data** on **Mastering Environmental Science**

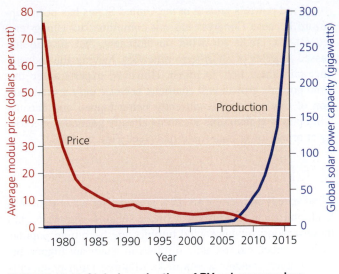

FIGURE 16.12 Global production of PV solar power has grown rapidly, while prices have fallen sharply. *Data from REN21, 2017. Renewables 2017: Global status report. REN21, UNEP, Paris; Bloomberg New Energy Finance; and pvXchange.*

Solar energy is expanding

Solar energy contributes just 0.61%—61 parts in 10,000—of the U.S. energy supply, and just 1.4% of U.S. electricity generation. Even in Germany, which gets more of its electricity from solar than any other nation, the percentage is only 7%. However, solar energy use has grown by nearly a third each year worldwide over the past four decades. Solar is proving especially attractive in developing countries rich in sun but poor in infrastructure, where hundreds of millions of people still live without electricity.

PV technology is the fastest-growing power source today, having recently doubled every two years. Germany led the world in installation of PV technology until China (with its much larger economy) passed it in 2015. Today China leads the world in production of PV cells, followed by Germany and Japan, while U.S. firms account for only 2% of the industry. The Chinese government's support of its solar industry led to so much production that supply has outstripped global demand in recent years. Highly subsidized Chinese firms have been selling solar products abroad at low prices (often at a loss), driving American and European solar manufacturers out of business. In response, the United States and European nations have slapped tariffs on Chinese imports, while both sides have filed complaints with the World Trade Organization.

As production of PV cells rises, prices fall (**FIGURE 16.12**). At the same time, efficiencies are increasing, making each unit more powerful. Use of solar technology should continue to expand as prices fall, technologies improve, and governments enact economic incentives to spur investment. Similar trends are apparent for wind power, another major and growing source of renewable energy.

Wind Power

As the sun heats the atmosphere, it causes air to move, producing wind. Today we are harnessing **wind power** by using the energy of wind to generate electricity.

Wind turbines convert kinetic energy to electrical energy

To produce wind power, we use **wind turbines,** mechanical assemblies that convert wind's kinetic energy (p. 33), or energy of motion, into electrical energy. Wind blowing into a turbine turns the blades of the rotor, which rotate machinery atop a tower (**FIGURE 16.13**). Today's towers average 80 m (260 ft) in height, and the largest are taller than a football field is long. Higher is generally better, to minimize turbulence (and potential damage) while maximizing wind speed. Engineers design turbines to rotate in response to changes in wind direction, so that the motor faces into the wind at all times. They also design them to begin turning at specified wind speeds to harness energy efficiently. Turbines are often erected in groups called **wind farms.** Today's largest wind farms contain hundreds of turbines.

FIGURE 16.13 A wind turbine converts wind's energy of motion into electrical energy. Wind spins a turbine's blades, turning a shaft that extends into the nacelle. Inside the nacelle, a gearbox converts the rotational speed of the blades into much higher speeds, providing adequate motion for the generator to produce electricity.

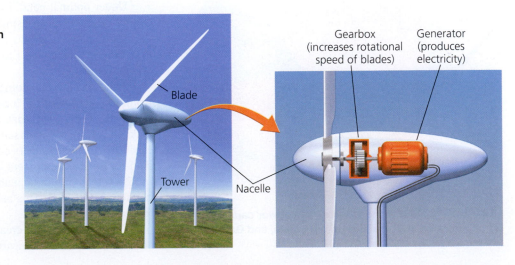

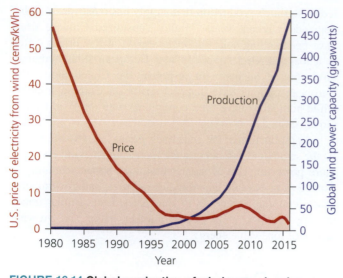

FIGURE 16.14 Global production of wind power has been doubling every three years in recent years, and prices have fallen. *Data from Global Wind Energy Council; and U.S. Department of Energy, EERE, 2016. 2015 Wind technologies market report.*

Wind power is growing fast

Like solar energy, wind currently provides just a small proportion of the world's power needs, but wind power is growing fast (**FIGURE 16.14**). Five nations account for three-quarters of the world's wind power output (**FIGURE 16.15**), but dozens of nations now produce wind power.

Denmark leads the world in obtaining the highest percentage of its energy from wind power. In this small European nation, wind supplies nearly 40% of electricity needs. Texas generates the most wind power of all U.S. states—fully one-quarter of the United States' wind power. Iowa obtains more than 30% of its electricity from wind, while South Dakota and Kansas get roughly 25% of their electricity from wind.

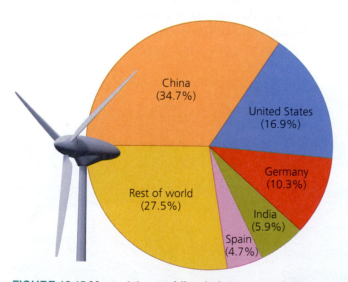

FIGURE 16.15 Most of the world's wind power capacity is concentrated in China, the United States, and Germany. *Data are for 2016, from Global Wind Energy Council.*

Wind power's growth in the United States has been haphazard because Congress has not committed to a long-term federal tax credit for wind development, but instead has passed a series of short-term renewals, leaving the industry uncertain about how to invest. However, experts agree that wind power's growth will continue, because only a small portion of this resource is currently being tapped and because wind power at favorable locations already generates electricity at prices lower than fossil fuels (see Figure 16.5).

Offshore sites are productive

Wind speeds on average are 20% greater over water than over land, and air is less turbulent (more steady) over water. For these reasons, offshore wind turbines are becoming popular. Costs to erect and maintain turbines in water are higher, but the stronger, steadier winds make offshore wind potentially more profitable.

Denmark erected the first offshore wind farm in 1991, and soon more came into operation across northern Europe. Germany raised its feed-in tariff rate for offshore wind sharply in 2009, and within just five years 14 new wind farms were operating. By 2016, more than 3500 wind turbines were powering 80 wind farms in the waters of 11 European nations. In the United States, a small project in Rhode Island waters is America's first commercial offshore wind farm. As of 2017, roughly 20 more U.S. offshore wind developments were in the planning stages, mostly off the North Atlantic coast.

Wind power has many benefits

Like solar power, wind power produces no emissions once the equipment is manufactured and installed. As a replacement for fossil fuel combustion in the average U.S. power plant, running a 1-megawatt wind turbine for 1 year prevents the release of more than 1500 tons of carbon dioxide, 6.5 tons of sulfur dioxide, 3.2 tons of nitrogen oxides, and 60 lb of mercury, according to the U.S. Environmental Protection Agency. The amount of carbon pollution that all U.S. wind turbines together prevent from entering the atmosphere is equal to the emissions from 24 million cars or from combusting the cargo of 17 hundred-car freight trains of coal each and every day.

Under optimal conditions, wind power is efficient in its energy returned on investment (EROI; p. 345). Studies find that wind turbines produce roughly 20 times more energy than they consume—an EROI value superior to that of most energy sources.

Wind turbine technology can be used on many scales, from a single tower for local use to large farms that supply whole regions. Small-scale turbine development can help make local areas more self-sufficient, just as solar energy can. Farmers and ranchers can also make money leasing their land for wind development. A single large turbine can bring in $2000 to $4500 in annual royalties while occupying just a quarter-acre of land. This also increases property tax income for rural communities.

Lastly, wind power creates job opportunities. More than 100,000 Americans and nearly 1.2 million people globally are employed in jobs related to wind power. More than 100 colleges and universities now offer programs that train people in the skills needed for jobs in wind power and other renewable energy fields.

Wind power has limitations

Wind is an intermittent resource; we have no control over when it will occur. This limitation is alleviated, though, if wind is one of several sources contributing to a utility's power generation. Pumped-storage hydropower (p. 391) can help to compensate during windless times, and batteries or hydrogen fuel (p. 395) can store energy generated by wind and release it later when needed.

Just as wind varies from time to time, it varies from place to place. Resource planners and wind power companies study wind patterns revealed by meteorological research before planning a wind farm. A map of average wind speeds across the United States (**FIGURE 16.16a**) shows that mountainous regions, offshore sites, and areas of the Great Plains are best. Based on such data, the wind industry has installed much of its generating capacity in states with high wind speeds (**FIGURE 16.16b**). However, most Americans live near the coasts, far from the Great Plains and mountain regions. Thus, continent-wide transmission networks would need to be enhanced to send wind-powered electricity to these population centers, or numerous offshore wind farms would need to be developed. When wind farms *are* proposed near population centers, though, local residents often oppose them for aesthetic reasons.

Turbines also pose a hazard to birds and bats, which are killed when they fly into the blades. Large open-country raptors such as golden eagles are known to be at risk, and turbines located on ridges along migratory flyways are likely most damaging. One strategy for protecting birds and bats is to select sites that are not on migratory flyways or amid prime habitat for species likely to fly into the blades, but more research on impacts on wildlife and how to prevent them is urgently needed.

Geothermal Energy

Geothermal energy is thermal energy that arises from beneath Earth's surface. The radioactive decay of elements (p. 31) amid high pressures deep in the interior of our planet generates heat that rises to the surface in molten rock and through cracks and fissures. Where this energy heats groundwater, spurts of water and steam rise from below and may erupt through the surface as geysers or as submarine hydrothermal vents (p. 35).

We harness geothermal energy for heating and electricity

Geothermal energy can be harnessed directly from geysers at the surface, but most often wells must be drilled down hundreds or thousands of meters toward heated groundwater.

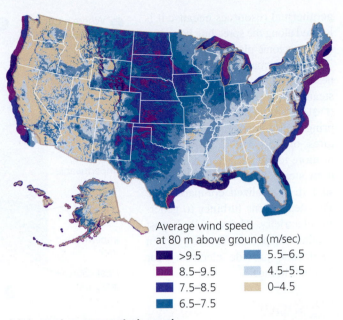

Average wind speed at 80 m above ground (m/sec)

>9.5	5.5–6.5
8.5–9.5	4.5–5.5
7.5–8.5	0–4.5
6.5–7.5	

(a) Annual average wind speed

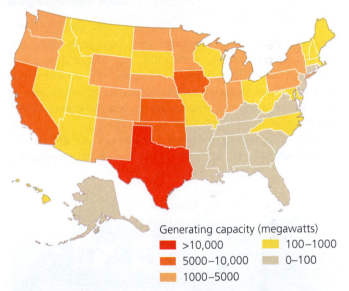

Generating capacity (megawatts)

>10,000	100–1000
5000–10,000	0–100
1000–5000	

(b) Installed wind power through 2017

FIGURE 16.16 Wind speed varies from place to place. Maps of average wind speeds **(a)** help guide the placement of wind farms. Another map **(b)** shows wind power generating capacity installed in each U.S. state through 2017. *Data from (a) U.S. National Renewable Energy Laboratory; and (b) American Wind Energy Association, 2017. Second quarter 2017 market report.*

DATA Compare parts **(a)** and **(b).** Which states or regions have high wind speeds but are not yet heavily developed with commercial wind power?

Go to **Interpreting Graphs & Data** on **Mastering** Environmental Science

Hot groundwater can be piped up and used directly for heating buildings and for industrial processes. The nation of Iceland heats nearly 90% of its homes in this way. Such direct heating is efficient and inexpensive, but it is feasible only where geothermal energy is readily available. Iceland has a wealth of geothermal resources because it is located along the spreading boundary of two tectonic plates (pp. 232–234).

Geothermal power plants harness naturally heated water and steam to generate electricity (**FIGURE 16.17**). A power plant brings groundwater at temperatures of 150–370°C (300–700°F) or more to the surface and converts it to steam by lowering the pressure in specialized compartments. The steam turns turbines to generate electricity. The world's largest geothermal plants, The Geysers in California, provide electricity for 725,000 homes.

Heat pumps make use of temperature differences

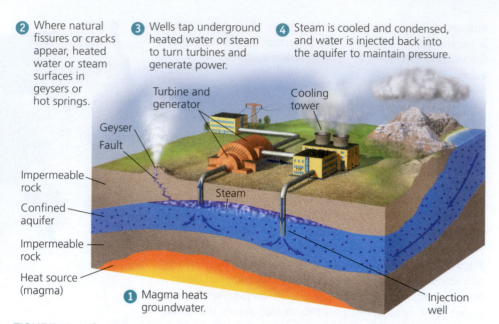

② Where natural fissures or cracks appear, heated water or steam surfaces in geysers or hot springs.

③ Wells tap underground heated water or steam to turn turbines and generate power.

④ Steam is cooled and condensed, and water is injected back into the aquifer to maintain pressure.

① Magma heats groundwater.

Turbine and generator
Cooling tower
Geyser
Fault
Impermeable rock
Confined aquifer
Impermeable rock
Heat source (magma)
Steam
Injection well

FIGURE 16.17 Geothermal power plants generate electricity using naturally heated water from underground.

Heated groundwater is available only in certain areas, but we can take advantage of the mild temperature differences that exist naturally between the soil and the air just about anywhere. Soil varies in temperature from season to season less than air does, and **ground-source heat pumps** (GSHPs) make use of this fact (**FIGURE 16.18**). These pumps provide heating in the winter by transferring heat from the ground into buildings, and they provide cooling in the summer by transferring heat from buildings into the ground. This heat transfer is accomplished with a network of underground plastic pipes that circulate water and antifreeze. More than 600,000 U.S. homes use GSHPs. Compared to conventional electric heating and cooling systems, GSHPs heat spaces 50–70% more efficiently, cool them 20–40% more efficiently, can reduce electricity use by 25–60%, and can reduce emissions by up to 70%.

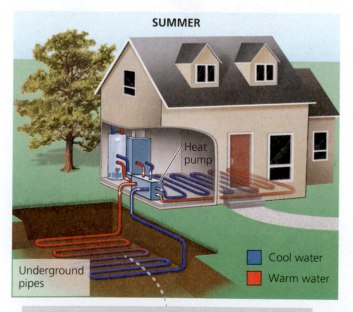

SUMMER

Heat pump

Underground pipes

■ Cool water
■ Warm water

In summer, soil underground is **cooler** than surface air. Water flowing through the pipes transfers heat from the house to the ground, cooling air in ducts or radiant cooling system under floor.

WINTER

In winter, soil underground is **warmer** than surface air. Water flowing through the pipes transfers heat from the ground to the house, warming air in ducts, water in tank, or radiant heating system under floor.

FIGURE 16.18 Ground-source heat pumps provide an efficient way to heat and cool air and water in a home. A network of pipes filled with water and antifreeze extends underground. Soil is cooler than air in the summer (**left**), and warmer than air in the winter (**right**), so by running fluid between the house and the ground, these systems adjust temperatures inside.

Geothermal power has pros and cons

All forms of geothermal energy—direct heating, electrical power, and ground-source heat pumps—greatly reduce emissions relative to fossil fuel combustion. Geothermal energy is renewable, but not every geothermal power plant will be able to operate indefinitely. If a plant uses heated water more quickly than groundwater is recharged, it will eventually run out of water. This was occurring at The Geysers in California, so operators began injecting municipal wastewater into the ground to replenish the supply! Moreover, patterns of geothermal activity in Earth's crust shift naturally over time, so an area that produces hot water may not always do so. In addition, some hot groundwater is laced with salts and minerals that corrode equipment and pollute the air. These factors may shorten the lifetime of plants, increase maintenance costs, and add to pollution.

The greatest limitation of geothermal power is that it is restricted to regions where we can tap energy from naturally heated groundwater. Places such as Iceland, northern California, and Yellowstone National Park are rich in naturally heated groundwater, but most areas of the world are not. Engineers are trying to overcome this limitation by developing **enhanced geothermal systems** (EGS), in which we drill deeply into dry rock, fracture the rock, and pump in cold water. The water becomes heated deep underground and is then drawn back up and used to generate power. In theory, we could use EGS widely in many locations. Germany, for instance, has little heated groundwater, but feed-in-tariffs enabled an EGS facility to operate profitably there. However, EGS appears to trigger occasional minor earthquakes. Unless we can develop ways to use EGS safely and reliably, our use of geothermal power will remain localized.

Ocean Energy Sources

The oceans are home to several underexploited sources of energy resulting from natural patterns of motion and temperature.

We can harness energy from tides, waves, and currents

Kinetic energy from the natural motion of ocean water can be used to generate electrical power. The rise and fall of ocean tides (p. 263) twice each day moves large volumes of water past any given point on the world's coastlines. Differences in height between low and high tides are especially great in long, narrow bays such as Alaska's Cook Inlet or the Bay of Fundy between New Brunswick and Nova Scotia. Such locations are best for harnessing **tidal energy**, by erecting dams across the outlets of tidal basins. As tidal currents pass through the dam, water turns turbines to generate electricity (**FIGURE 16.19**).

The world's largest tidal generating station is South Korea's Sihwa Lake facility, which opened in 2011. The oldest is the La Rance facility in France, which has operated for nearly 50 years. The first U.S. tidal station began operating in 2012 in Maine, and one is now being installed in New York City's East River. Tidal stations release few or no pollutant

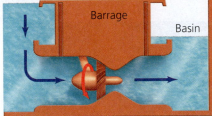

FIGURE 16.19 We can harness tidal energy by allowing ocean water to spin turbines. At the world's largest tidal power facility at Sihwa Lake, South Korea, water flows from the ocean into a huge enclosed basin as the tide rises, spinning turbines to generate electricity.

emissions, but they can affect the ecology of estuaries and tidal basins.

Scientists and engineers are also working to harness the motion of ocean waves and convert their energy into electricity. Many designs for machinery to harness **wave energy** have been invented. Some designs for offshore facilities involve floating devices that move up and down with the waves. Some designs for onshore facilities funnel waves from large areas into narrow channels and elevated reservoirs, from which water then flows out, generating electricity. Other coastal designs use rising and falling waves to push air into and out of chambers, turning turbines. The first commercial wave energy facility began operating in 2011 in Spain.

A third way to harness marine kinetic energy is to use the motion of ocean currents (p. 260), such as the Gulf Stream. Underwater turbines have been erected to test this approach.

The ocean stores thermal energy

Each day the tropical oceans absorb solar radiation with the heat content of 250 billion barrels of oil—enough to provide 20,000 times the electricity used daily in the United States. The ocean's sun-warmed surface is warmer than its deep water, and **ocean thermal energy conversion** (OTEC) relies on this gradient in temperature. In one approach, warm surface water is piped into a facility to evaporate chemicals, such as ammonia, that boil at low temperatures. The evaporated gases spin turbines to generate electricity. Cold water piped up from ocean depths then condenses the gases so they can be reused. In another approach, warm surface water is evaporated in a vacuum, and its steam turns turbines and then is condensed by cold water. So far, no OTEC facility operates commercially, but research is being conducted in Hawai'i and elsewhere.

Hydroelectric Power

In **hydroelectric power,** or **hydropower,** we use the kinetic energy of flowing river water to generate electricity. Hydroelectric power provides nearly one-sixth of the world's electricity. We examined hydropower and its environmental impacts in our discussion of freshwater resources (pp. 272–273). Now we will take a closer look at hydropower as an energy source.

Hydropower uses three approaches

Most hydroelectric power today comes from impounding water in reservoirs behind concrete dams that block the flow of river water and then letting that water pass through the dam. Because water is stored behind dams, this is called the **storage** technique. As reservoir water passes through a dam, it turns the blades of turbines, which cause a generator to generate electricity (**FIGURE 16.20**). This electricity is transmitted by transmission lines to the electrical grid that serves consumers, while the water flows into the riverbed below the dam and continues downriver. By storing water in reservoirs, dam operators can ensure a steady

(a) Ice Harbor Dam, Snake River, Washington

(b) Generators inside McNary Dam, Columbia River

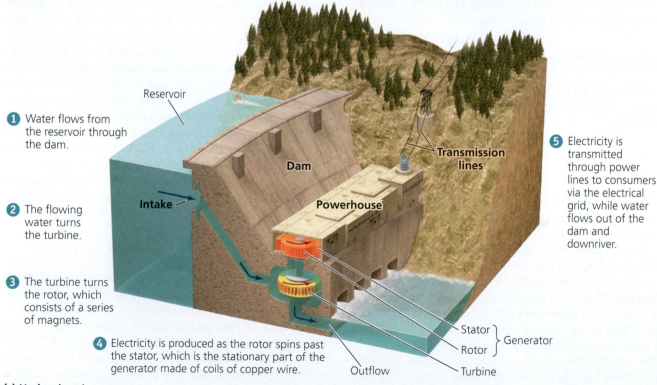

1 Water flows from the reservoir through the dam.

2 The flowing water turns the turbine.

3 The turbine turns the rotor, which consists of a series of magnets.

4 Electricity is produced as the rotor spins past the stator, which is the stationary part of the generator made of coils of copper wire.

5 Electricity is transmitted through power lines to consumers via the electrical grid, while water flows out of the dam and downriver.

Reservoir

Dam

Intake

Powerhouse

Transmission lines

Stator
Rotor
} Generator

Outflow

Turbine

(c) Hydroelectric power

FIGURE 16.20 We generate hydroelectric power with large dams. Inside these dams **(a)**, flowing water turns massive turbines **(b)** to generate electricity. Water is funneled from the reservoir through the dam and its powerhouse **(c)** and into the riverbed below.

and predictable supply of electricity, even during periods of naturally low river flow.

An alternative approach is the **run-of-river** technique, which generates electricity without greatly disrupting a river's flow. One method is to divert a portion of a river's flow through a pipe or channel, passing it through a powerhouse and returning it to the river. Run-of-river systems are useful in remote areas far from electrical grids and in regions without the economic resources to build and maintain large dams. This approach cannot guarantee reliable water flow in all seasons, but it minimizes many impacts of the storage technique by leaving most water in the river channel.

To better control the timing of flow, pumped-storage hydropower can be used. In **pumped storage,** water is pumped from a lower reservoir to a higher reservoir at times when demand for power is weak and prices are low. When demand is strong and prices are high, water is sent downhill through a turbine, generating electricity. Although energy must be input to pump the water, pumped storage can be profitable. When paired with intermittent sources such as solar and wind power, it can help balance a region's power supply by compensating for dips in power availability.

Hydropower is clean and renewable, yet has impacts

Hydroelectric power has three clear advantages over fossil fuels. First, it is renewable; as long as precipitation falls from the sky and fills rivers and reservoirs, we can use water to turn turbines. Second, hydropower is efficient. It has been shown to have an EROI ratio of more than 80:1, higher than any other modern-day energy source. Third, hydropower emits no carbon dioxide or other pollutants into the atmosphere. Fossil fuels *are* used in constructing and maintaining dams, and large reservoirs may release the greenhouse gas methane as a result of anaerobic decay in deep water. But overall, hydropower produces only a small fraction of the greenhouse gas emissions typical of fossil fuel combustion.

Damming rivers (p. 272), however, destroys habitat for wildlife as ecologically rich riparian areas above dam sites are submerged and those below often are starved of water. Because water discharge is regulated to optimize electricity generation, the natural flooding cycles of rivers are disrupted. Suppressing flooding prevents river floodplains from receiving fresh nutrient-laden sediments. Instead, sediments become trapped behind dams, where they begin filling the reservoir. Dams also modify water temperatures and generally block the passage of fish and other aquatic creatures. Along with habitat alteration, this has diminished biodiversity in many dammed waterways. These ecological impacts generally translate into negative social and economic impacts on local communities.

Hydroelectric power is widely used, but it may not expand much more

Hydropower accounts for one-sixth of the world's electricity production (see Figure 15.4b, p. 344). For nations with large amounts of river water and the economic resources to build dams, hydroelectric power has been a keystone of their development and wealth. Canada, Brazil, Norway, Venezuela, and certain other nations obtain the majority of their electricity from hydropower.

Today the world is witnessing some gargantuan hydroelectric projects. China's Three Gorges Dam (p. 272) is the world's largest. However, hydropower may not expand much more, because most of the world's large rivers are already dammed. In addition, as people have grown aware of the ecological impacts of dams, residents of some regions are resisting dam construction. In the United States, 98% of rivers appropriate for dam construction already are dammed, while many of the remaining 2% are protected under the Wild and Scenic Rivers Act. Some dams that are no longer functional are being dismantled so that river habitats can be restored (p. 272).

Bioenergy

Like hydropower, bioenergy is well established and has long provided a great deal of humanity's energy. **Bioenergy** is energy obtained from **biomass** (p. 74), which consists of organic material derived from living or recently living organisms. Biomass contains chemical energy that originated with sunlight and photosynthesis. We harness bioenergy by burning biomass for heating, using biomass to generate electricity, and processing biomass to create liquid fuels for transportation (**TABLE 16.1**).

TABLE 16.1 Sources and Uses of Bioenergy

BIOMASS FOR DIRECT COMBUSTION FOR HEATING

- Wood cut from trees (fuelwood)
- Charcoal
- Manure from farm animals

BIOMASS FOR GENERATING ELECTRICITY (BIOPOWER)

- Crop residues (such as cornstalks) burned at power plants
- Forestry residues (wood waste from logging) burned at power plants
- Processing wastes (from sawmills, pulp mills, paper mills, etc.) burned at power plants
- "Landfill gas" burned at power plants
- Livestock waste from feedlots for gas from anaerobic digesters

BIOFUELS FOR POWERING VEHICLES

- Corn grown for ethanol
- Bagasse (sugarcane residue) grown for ethanol
- Soybeans, rapeseed, and other crops grown for biodiesel
- Used cooking oil for biodiesel
- Algae grown for biofuels
- Plant matter treated with enzymes to produce cellulosic ethanol

FIGURE 16.21 More than a billion people in developing countries rely on wood from trees for heating and cooking. In principle, biomass is renewable, but in practice it may not be if forests are overharvested.

Fuelwood is used widely in the developing world

More than 1 billion people use wood from trees as their principal energy source. In rural regions of developing nations, people (generally women) gather fuelwood to burn in their homes for heating, cooking, and lighting (**FIGURE 16.21**). Although fossil fuels and electricity are replacing traditional energy sources as developing nations industrialize, fuelwood, charcoal, and livestock manure constitute almost half of all renewable energy used worldwide.

These traditional biomass sources are renewable only if they are not overharvested. Harvesting fuelwood at unsustainably rapid rates can lead to deforestation, soil erosion, and desertification (pp. 195, 148). Burning fuelwood for cooking and heating also poses health hazards from indoor air pollution (see Figure 13.23, p. 306).

We can generate electricity using biomass

Biomass can be burned to generate electricity. Such **biopower** can be produced in many ways. A variety of waste products are being combusted for biopower, including woody debris from logging, liquid waste from pulp mills, organic waste from landfills or feedlots, and residue from crops (such as cornstalks and corn husks). We are also starting to use certain fast-growing plants as crops to burn for biopower, including grasses such as bamboo, fescue, and switchgrass, and trees such as specially bred willows and poplars.

At small scales, farmers, ranchers, or villagers can operate modular biopower systems that use livestock manure to generate electricity. Small household biodigesters provide portable, decentralized energy production for remote rural areas. At large scales, power plants can be built or retrofitted to burn biomass to generate electricity. At some coal-fired power plants, wood chips or pellets are combined with coal in a high-efficiency boiler in a process called *co-firing*. At other power plants,

biomass is vaporized at high temperatures in the absence of oxygen (a process called *gasification*), creating gases that turn a turbine to propel a generator. Another method of heating biomass in the absence of oxygen results in *pyrolysis* (p. 347), producing a liquid fuel that can be burned to generate electricity.

By enhancing energy efficiency and putting waste products to use, biopower conserves resources and helps reduce greenhouse gas emissions. When biomass replaces coal in co-firing and direct combustion, we reduce emissions of sulfur dioxide as well, because plant matter, unlike coal, contains no appreciable sulfur. One disadvantage is that when we burn plant matter for power, we deprive the soil of nutrients and organic matter it would have gained from the plant matter's decomposition, thus requiring other actions to restore soil fertility.

 SUCCESS STORY   **Turning Waste into Energy**

The Swedish city of Kristianstad is best known as the home of Absolut Vodka. But Kristianstad is now gaining attention for its capacity to produce energy from waste. Back in 1999, this city of 81,000 people—the hub of an agricultural and food-processing region—aimed to free itself of a dependence on fossil fuels. After building a power plant that burns forestry waste, it constructed a facility to turn waste into *biogas* (a mix of methane and other gases that results from breaking down organic matter amid a lack of oxygen). Kristianstad's biogas plant receives household garbage, crop waste, food industry waste, and animal manure and uses anaerobic digestion to turn this waste into biogas. In addition, the city's landfill and wastewater treatment plant each collect methane, adding to the city's biogas supply. The biogas is burned to generate electricity and to heat homes by district heating. It is also refined and used to fuel hundreds of cars, trucks, and buses

Kristianstad's biogas plant uses organic waste to produce electricity, home heating, vehicle fuel, and fertilizer.

engineered to run on biogas. In total, this waste-to-energy system replaces about 7% of Kristianstad's gasoline and diesel fuel each year, all of its district heating, and much of its electricity, while excess biogas is sold to neighboring communities. A by-product of biogas production is liquid organic fertilizer, and nearly 100,000 tons per year are sold to area farms.

***EXPLORE THE DATA* at Mastering Environmental Science**

Biofuels can power vehicles

Some biomass sources can be converted into **biofuels,** liquid fuels used primarily to power automobiles. The two primary biofuels developed so far are ethanol (for gasoline engines) and biodiesel (for diesel engines).

Ethanol **Ethanol** is the alcohol in beer, wine, and liquor. It is produced as a biofuel by fermenting biomass, generally from carbohydrate-rich crops, in a process similar to brewing beer. In fermentation, carbohydrates are converted to sugars and then to ethanol. Spurred by the 1990 Clean Air Act amendments, a 2007 congressional mandate, and generous subsidies, ethanol is widely added to gasoline in the United States to conserve oil and reduce automotive emissions. In 2016 in the United States, over 58 billion L (15.3 billion gal) of ethanol were produced—47 gallons for every American— mostly from corn (**FIGURE 16.22**). Fully 40% of today's U.S. corn crop is used to make ethanol. (Some by-products of ethanol production are used in livestock feed; with this accounted for, 28% of U.S. corn goes toward ethanol.)

Any vehicle with a gasoline engine runs well on gasoline blended with up to 10% ethanol, but automakers are also producing *flexible-fuel* vehicles that run on E-85, a mix of 85% ethanol and 15% gasoline. More than 17 million such cars are on U.S. roads today. In Brazil, almost all new cars are flexible-fuel vehicles, and ethanol from crushed sugarcane residue (called *bagasse*) accounts for half of all fuel that Brazil's drivers use.

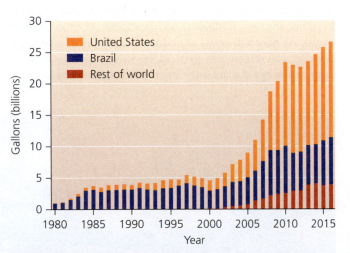

FIGURE 16.22 Ethanol production in the United States, Brazil, and elsewhere has grown rapidly in recent years. *Data from Renewable Fuels Association.*

DATA • Roughly what percentage of the world's ethanol is produced by the United States? • Based purely on the data shown in the graph, give one explanation for why we might predict that U.S. and world ethanol production will be much higher in the future. • Now explain one reason why we might predict that U.S. and world ethanol production will be about the same in the future as it is today.

Go to **Interpreting Graphs & Data** on **Mastering** Environmental Science

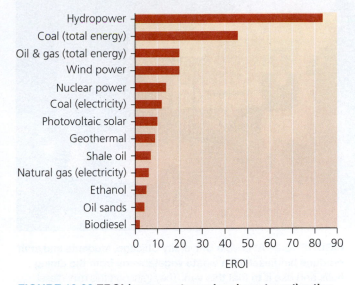

FIGURE 16.23 EROI (energy returned on investment) ratios vary from more than 80:1 for hydropower down to 2:1 for biodiesel. EROI values for ethanol vary by crop and location; estimates for U.S. corn ethanol are lower than the overall ethanol average shown here. *Adapted from Hall, C., et al., 2014. EROI of different fuels and the implications for society. Energy Policy 64: 141–152.*

The enthusiasm for ethanol shown by U.S policymakers is not widely shared by scientists. Growing corn to produce ethanol takes up millions of acres of land and intensifies the use of pesticides, fertilizers, and fresh water. It also requires substantial inputs of fossil fuel energy (for operating farm equipment, making petroleum-based pesticides and fertilizers, transporting corn to processing plants, and heating water in refineries to distill ethanol). In the end, corn ethanol yields only a modest amount of energy relative to the energy that needs to be input. Scientific estimates of the EROI (energy returned on investment) ratio (p. 345) for corn-based ethanol vary, but recent calculations place it around 1.3:1 (**FIGURE 16.23**). This means that we need to expend 1 unit of energy just to gain 1.3 units of energy from ethanol.

Biodiesel Drivers of diesel-fueled vehicles can use **biodiesel,** a fuel produced from vegetable oil, used cooking grease, or animal fat. The oil or fat is mixed with small amounts of ethanol or methanol in the presence of a chemical catalyst. In Europe, rapeseed oil is the oil of choice, whereas U.S. biodiesel producers use mostly soybean oil. Vehicles with diesel engines can run on 100% biodiesel, or biodiesel can be mixed with conventional diesel; a 20% biodiesel mix (called B20) is common. Many buses, recycling trucks, and state and federal fleet vehicles now run on biodiesel or biodiesel blends.

Replacing diesel with biodiesel cuts down on emissions. Biodiesel's fuel economy is nearly as good, it costs just slightly more, and it is nontoxic and biodegradable. Most biodiesel today, like most ethanol, comes from crops grown specifically for the purpose—and this has impacts. For example, growing soybeans in Brazil and oil palms in Southeast Asia hastens the loss of tropical rainforest (pp. 192, 197). A more sustainable option is to fuel vehicles with biodiesel

FIGURE 16.24 At Loyola University Chicago, students and staff produce biodiesel from waste vegetable oil from the dining halls and use it to fuel this van. They transport this mini-diesel reactor to local high schools to teach students about alternative fuels.

made from waste oils. Some college students are creating biodiesel from food waste from dining halls and restaurants (**FIGURE 16.24**).

Novel biofuels Because the major crops grown for biodiesel and ethanol exert heavy impacts on the land, farmers and scientists are experimenting with other crops—from wheat, sorghum, cassava, and sugar beets to less known plants such as hemp, jatropha, and the grass miscanthus. One promising next-generation biofuel crop is algae (**FIGURE 16.25a**). Several species of these photosynthetic organisms produce large amounts of lipids that can be converted to biodiesel. Alternatively, carbohydrates in algae can be fermented to create ethanol. In fact, we can use algae to produce a variety of fuels, even jet fuel (**FIGURE 16.25b**). Algae can be grown outdoors in open circulating ponds or in

labs in closed tanks or transparent tubes. Algae grow faster and produce more oil than terrestrial biofuel crops, and they can grow in seawater, saline water, or nutrient-rich wastewater from sewage treatment plants.

Relying on any monocultural crop for energy may not be sustainable. With this in mind, researchers are refining techniques to produce **cellulosic ethanol** by using enzymes to produce ethanol from cellulose, the substance that gives structure to all plant material. This would be a substantial advance because ethanol made from corn or sugarcane uses starch, which is nutritionally valuable to us. Cellulose, in contrast, is of no food value to people yet is abundant in all plants. Converting plant material to fuel would still deprive the soil of organic matter from decomposition, but if we can produce cellulosic ethanol in a commercially feasible way, then ethanol could be made from low-value crop waste (such as corn stalks and husks), rather than from high-value crops.

Is bioenergy carbon-neutral?

In principle, energy from biomass is carbon-neutral, releasing no net carbon dioxide into the atmosphere. This is because burning biomass releases carbon dioxide that plants recently pulled from the air by photosynthesis. Thus, in theory, when we replace fossil fuels with bioenergy, we reduce net carbon flux to the atmosphere, helping to fight climate change.

In practice, burning biomass for energy is not carbon-neutral if forests are destroyed to plant bioenergy crops. Forests sequester more carbon (in vegetation and soil) than croplands do, so cutting forests to grow crops increases carbon flux to the atmosphere. Bioenergy also is not carbon-neutral if we consume fossil fuel energy to produce the biomass (for instance, by using tractors, fertilizers, and pesticides to grow biofuel crops). We have a diversity of options in the realm of bioenergy, however, and researchers are working hard to develop ways of using bioenergy that are truly renewable and sustainable.

(a) Algae in the lab

(b) Fueling jets with biofuels

FIGURE 16.25 Novel biofuels are being developed for many uses. Researchers are studying algae **(a)** as a promising source of next-generation biofuels. Several commercial airlines **(b)** are already powering flights with biofuels sourced from algae and various waste products.

Hydrogen and Fuel Cells

Each renewable energy source we have discussed can be used to generate electricity more cleanly than can fossil fuels. However, electricity cannot be stored easily in large quantities for use when and where it is needed. This is why most vehicles rely on gasoline, instead of electricity, for power. The development of fuel cells and of fuel consisting of hydrogen—the simplest and most abundant element in the universe—holds promise as a way to store sizeable quantities of energy conveniently, cleanly, and efficiently. Like electricity and like batteries, hydrogen is an energy carrier, not a primary energy source. It holds energy that can be converted for use at later times and in different places.

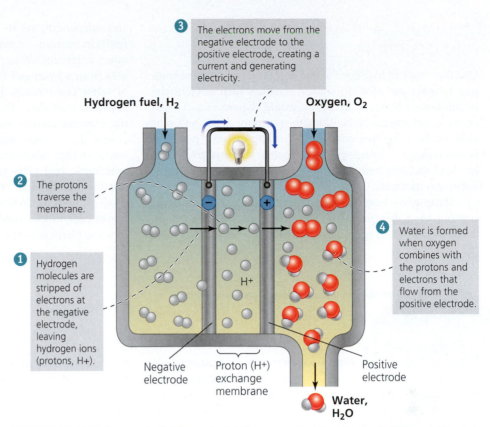

3 The electrons move from the negative electrode to the positive electrode, creating a current and generating electricity.

2 The protons traverse the membrane.

1 Hydrogen molecules are stripped of electrons at the negative electrode, leaving hydrogen ions (protons, H+).

4 Water is formed when oxygen combines with the protons and electrons that flow from the positive electrode.

Hydrogen fuel, H₂   Oxygen, O₂

H+

Negative electrode   Proton (H+) exchange membrane   Positive electrode

Water, H₂O

FIGURE 16.26 Hydrogen fuel drives electricity generation in a fuel cell. Water and heat are the only waste products that result.

Some yearn for a "hydrogen economy"

Some energy experts envision that hydrogen fuel, along with electricity, could serve as the basis for a clean, safe, and efficient energy system. In such a system, electricity generated from intermittent renewable sources, such as wind or solar energy, could be used to produce hydrogen. **Fuel cells**—essentially, hydrogen batteries (**FIGURE 16.26**)—could then use hydrogen to produce electricity to power vehicles, computers, cell phones, home heating, and more. In fact, NASA's space programs have used fuel-cell technology since the 1960s.

Basing an energy system on hydrogen could alleviate dependence on foreign fuels and help fight climate change. For these reasons, governments have funded research into hydrogen fuel and fuel-cell technology, and auto companies have developed vehicles that run on hydrogen. Today, Germany is one of several nations with hydrogen-fueled city buses, and it is planning a network of hydrogen filling stations for hydrogen cars that are being designed.

Hydrogen fuel may be produced from water or from other matter

Hydrogen gas (H_2) tends not to exist freely on Earth. Instead, hydrogen atoms bind to other molecules, becoming incorporated in everything from water to organic molecules. To obtain hydrogen gas for fuel, we must force these substances to release their hydrogen atoms, and this requires an input of energy. In **electrolysis,** electricity is used to split hydrogen atoms from the oxygen atoms of water molecules:

$$\text{electricity} + 2H_2O \longrightarrow 2H_2 + O_2$$

Electrolysis produces pure hydrogen, and it does so without emitting the carbon- or nitrogen-based pollutants of fossil fuel combustion. However, whether this strategy for producing hydrogen will cause pollution over its life-cycle depends on the source of electricity used for electrolysis. If coal is burned to generate the electricity, then the process will not reduce emissions. The "cleanliness" of a hydrogen economy would, therefore, depend largely on the source of electricity used in electrolysis.

The environmental impact of hydrogen production also depends on the source material for the hydrogen. Besides water, hydrogen can be obtained from biomass or from fossil fuels. This generally requires less energy input but results in pollution. For instance, extracting hydrogen from methane (CH_4) in natural gas produces one molecule of the greenhouse gas carbon dioxide for every four molecules of hydrogen gas.

Once isolated, hydrogen gas can be used as fuel to produce electricity with a fuel cell. The chemical reaction involved in a fuel cell is simply the reverse of that for electrolysis:

$$2H_2 + O_2 \longrightarrow 2H_2O + \text{electricity}$$

Figure 16.26 shows how this occurs within one common type of fuel cell.

Hydrogen and fuel cells have costs and benefits

One drawback of hydrogen at this point is a lack of infrastructure to make use of it. To convert a nation such as Germany or the United States to hydrogen would require massive and costly development of facilities to transport, store, and provide the fuel. Another concern, according to some research, is that leakage of hydrogen might deplete stratospheric ozone (p. 301) and lengthen the atmospheric lifetime of the greenhouse gas methane.

Hydrogen's benefits include the fact that we will never run out of it, because it is the most abundant element in the universe. Hydrogen can be clean and nontoxic to use, and—depending on its source and the source of electricity for its extraction—it may produce few greenhouse gases and other pollutants. Water and heat are the only waste products from a hydrogen fuel cell, along with negligible traces of other compounds. Fuel cells are also energy-efficient; depending on the type, 35–70% of the energy released in the reaction can be used—or up to 90% if the system is designed to capture heat as well as electricity. Unlike batteries (which also produce electricity through chemical reactions), fuel cells generate electricity whenever hydrogen fuel is supplied, without ever needing recharging. For all these reasons, hydrogen fuel cells could soon be used to power cars, much as they already power buses on the streets of some German cities.

closing THE LOOP

Rising concern over air pollution, climate change, health impacts, and security risks resulting from our reliance on fossil fuels has driven a desire to shift to renewable energy sources that pollute far less and that will not run out. Solar energy, wind power, geothermal energy, ocean energy, hydropower, and bioenergy all hold promise to sustain our civilization far into the future while exerting less impact on our environment. By using electricity from renewable sources to produce hydrogen fuel, we may be able to use fuel cells to produce electricity when and where it is needed, enabling us to manufacture nonpolluting vehicles.

Renewable energy has long been held back by limited funding for research and development and by competition with established nonrenewable fuels whose market prices have not covered external costs. Despite these obstacles, recent progress offers hope that we can replace fossil fuels with renewable energy. Germany's actions provide a prime example of how government policy can accelerate such a transition. Its feed-in tariff programs ignited the widespread adoption of PV solar, wind, and other renewable energy technologies. In fact, Germans began producing so much electricity from renewable sources that concerns rose over questions of pricing, timing, and grid capacity. In the 2016 revision of its national energy policy, Germany introduced a new strategy of competitive auctions—and once again, nations around the world are watching to see what unfolds.

Other nations are showing leadership in renewable energy as well. The United States has produced a great deal of research and technology; Denmark is a leader in wind power; and many other European nations obtain high proportions of their energy from renewable sources. Developing nations are generating power off the grid, and China has thrown its economic might behind renewable energy, producing and deploying staggering amounts of technology. With these steps by many nations around the world, our global civilization is moving faster and faster toward a future fueled by renewable energy.

TESTING Your Comprehension

1. What proportion of U.S. energy today comes from renewable sources? What is the most prevalent form of renewable energy used in the United States? What form of renewable energy is most used to generate electricity?

2. What factors and concerns are causing renewable energy use to expand? Which two renewable sources are experiencing the most rapid growth?

3. Describe several passive solar approaches. Now explain how photovoltaic (PV) cells function and are used.

4. List several advantages of solar power. What are some disadvantages?

5. Describe how wind turbines generate electricity. List several environmental and economic benefits of wind power. What are some drawbacks?

6. Define *geothermal energy*, and explain three main ways in which it is obtained and used. Describe one sense in which it is renewable and one sense in which it is not.

7. List and describe four approaches for obtaining energy from ocean water.

8. Compare and contrast the three major approaches to generating hydroelectric power. List one benefit and one negative impact of hydropower.

9. List five sources of bioenergy. What is the world's most used source of bioenergy? Describe two potential benefits and two potential drawbacks of bioenergy.

10. How is hydrogen fuel produced? What factors determine the amount of pollutants hydrogen production will emit?

SEEKING Solutions

1. For each source of renewable energy discussed in this chapter, what factors stand in the way of an expedient transition to it from fossil fuel use? In each case, what could be done to ease a shift toward these renewable sources? Would market forces alone suffice to bring about this transition, or would we also need government? Do you think such a shift would be good for our economy? Why or why not?

2. Do some research online to find out which energy sources produce the most (a) energy and (b) electricity in your own state. Create diagrams like those in Figure 16.2 showing a quantitative breakdown of the energy your state's residents use. Which renewable energy sources does your state use more than the United States as a whole, and which does it use less? What might be the reasons for these patterns?

3. There are many different sources of biomass and many ways of harnessing energy from biomass. Discuss one that seems particularly beneficial to you, and one with which you see problems. What bioenergy sources and strategies do you think our society should focus on investing in?

4. **CASE STUDY CONNECTION** Explain how Germany accelerated its development of PV solar power and other renewable energy sources by establishing a system of feed-in tariffs. What steps did it take, and what have been the results so far? What future challenges does Germany face? Do you think the United States should adopt a similar system of feed-in tariffs to promote renewable energy across the nation? Explain in detail why, or why not.

5. **THINK IT THROUGH** You are an investor seeking to invest in renewable energy. You're considering buying stock in companies that (1) build corn ethanol refineries, (2) are developing algae farms for biofuels, (3) construct turbines for hydroelectric dams, (4) produce PV solar panels, (5) install wind turbines, and (6) plan to build a wave energy facility. For each company, what questions would you research before deciding how to invest your money? How do you expect you might apportion your investments, and why?

CALCULATING Ecological Footprints

Energy sources vary tremendously in their energy returned on investment (EROI) ratios. Examine the EROI data for each of the energy sources as provided in Figure 16.23 on p. 393, and enter the data in the table below.

ENERGY SOURCE	ENERGY RETURNED ON INVESTMENT (EROI)
Coal (electricity)	12
Natural gas (electricity)	
Nuclear power	
Hydropower	
Photovoltaic solar	
Wind power	
Ethanol	
Biodiesel	

1. How many units of energy would you generate by investing 1 unit of energy into producing hydropower? To generate that same amount of energy, about how many units of energy would you need to invest into producing nuclear power? Roughly how many units of energy would you need to invest into producing electricity from coal if you wanted to generate that same amount of energy?

2. Based on EROI values, is it more efficient to obtain energy from electricity from natural gas or from wind power? Which source would you guess has a larger ecological footprint, based on EROI values?

3. Let's say you wanted to generate 100 units of energy from biodiesel. About how many units of energy would you need to invest? Explain your calculations.

4. Based on EROI ratios alone, which energy sources would you advocate that we further develop? Which would you urge that we avoid? What other issues, besides EROI, are worth considering when comparing energy sources?

Mastering Environmental Science

Students Go to Mastering Environmental Science for assignments, the etext, and the Study Area with practice tests, videos, current events, and activities.

Instructors Go to Mastering Environmental Science for automatically graded activities, current events, videos, and reading questions that you can assign to your students, plus Instructor Resources.

CHAPTER 17

Managing Our Waste

A Mania for Recycling on Campus

> **An extraterrestrial observer might conclude that conversion of raw materials to wastes is the real purpose of human economic activity.**
> —Gary Gardner and Payal Sampat, Worldwatch Institute

> **Recycling is one of the best environmental success stories of the late 20th century.**
> —U.S. Environmental Protection Agency

OHIO

Miami University • • Ohio University

At that time of year when NCAA basketball fever sweeps America's campuses, there's another kind of March Madness now taking hold: a mania for recycling.

It began in 2001, when waste managers at two Ohio campuses got the idea to use their schools' long-standing athletics rivalry to jump-start their recycling programs. Ed Newman of Ohio University, in Athens, and Stacy Edmonds Wheeler of Miami University, across the state in Oxford, challenged one another to see whose campus could recycle more in a 10-week competition. Come April, Miami University had taken the prize, recycling 41.2 pounds per student. Recyclemania was born.

Students at other colleges and universities heard about the event and wanted to get in on the action, and year by year more schools joined. Today Recyclemania pits several hundred institutions against one another, involving several million students and staff across North America. The event has grown to have a board of directors and major corporate sponsors.

Student leaders rouse their campuses to compete in two divisions and 11 categories over eight weeks each spring. Each week, recycling bins are weighed and campuses report their data, which are compiled online at the Recyclemania website. The all-around winner gets a funky trophy made of recycled materials (a figure nicknamed "Recycle Dude," whose body is a rusty propane tank)—and, more important, fame and bragging rights for a year.

In spring 2017, more than 300 colleges and universities slugged it out. In the end, the battlefield was littered with stories of the victors and the vanquished (**FIGURE 17.1**). Loyola Marymount University took top honors, recycling an impressive 84 percent of its waste, topping runners-up Walters State Community College, University of Missouri–Kansas City, and Berkshire Community College. North Lake College minimized its waste the best; students here limited their waste to just 4.14 pounds per person. The Rhode Island School of Design won the competition for most recyclables per capita, with 76.1 pounds per student. And in total weight of items recycled, Rutgers University took home the prize, having recycled a staggering 2,333,670 pounds.

Campuses also compete to see which can collect the most of certain types of items per person. In 2017, Loyola Marymount University collected the most paper, cardboard, bottles, and cans, whereas Union College saved the most food waste. Southwestern College recycled the most electronic waste per capita, and Agnes Scott College won the new category of least waste per area of building space. Lastly, Drexel University and Ohio University were champions in the competition to see who can best reduce waste at a home basketball game.

Upon completing this chapter, you will be able to:

- Summarize major approaches to managing waste, and compare and contrast the types of waste we generate

- Discuss the nature and scale of the waste dilemma

- Evaluate source reduction, reuse, composting, and recycling as approaches for reducing waste

- Describe landfills and incineration as conventional waste disposal methods

- Discuss industrial solid waste and principles of industrial ecology

- Assess issues in managing hazardous waste

Students at Pacific Lutheran University compete in the Recyclemania tournament.

RECYCLE MANIA TOURNAMENT

The world's biggest collegiate waste reduction event ▲

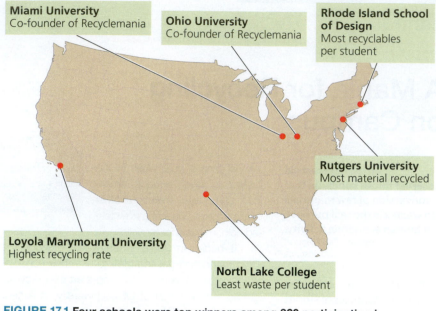

Miami University
Co-founder of Recyclemania

Ohio University
Co-founder of Recyclemania

Rhode Island School of Design
Most recyclables per student

Rutgers University
Most material recycled

Loyola Marymount University
Highest recycling rate

North Lake College
Least waste per student

FIGURE 17.1 Four schools were top winners among 320 participating in Recyclemania 2017. The event began over 15 years ago at Ohio University and Miami University.

By encouraging all this recycling, Recyclemania cuts down on pollution from the mining of new resources and the manufacture of new goods. Since the event's debut in 2001, millions of students at nearly 800 institutions have recycled and composted more than 890 million pounds of waste. This has helped to prevent the release of nearly 2.37 million metric tons of carbon dioxide—equal to removing more than 500,000 cars from the roads for a year. By focusing the attention of administrators on waste issues, Recyclemania facilitates the expansion of campus waste reduction programs. Most important, it gets a new generation of young people revved up about the benefits of recycling.

Recyclemania is the biggest of a growing number of campus competitions in the name of sustainability. Recyclemania has led this trend because recycling is the most widespread activity among campus sustainability efforts (pp. 19, 435). These efforts include water conservation, energy efficiency, green buildings, transportation options, campus gardens, and sustainable food in dining halls. Students are restoring native plants and habitats, promoting renewable energy, and advocating for carbon neutrality on campus. Campus sustainability is thriving because for students, faculty, staff, and administrators, it's satisfying to do the right thing and pitch in to help make your campus more sustainable.... And it's even more fun when you can compete and show that you can do it better than your rival school across the state!

Approaches to Waste Management

As the world's population rises, and as we produce and consume more material goods, we generate more waste. **Waste** refers to any unwanted material or substance that results from a human activity or process. Waste can degrade water quality, soil quality, air quality, and human health. Waste also indicates inefficiency—so reducing waste can save money and resources. For these reasons, waste management has become a vital pursuit.

For management purposes, we divide waste into several categories. **Municipal solid waste** is nonliquid waste that comes from homes, institutions, and small businesses. **Industrial solid waste** includes waste from production of consumer goods, mining, agriculture, and petroleum extraction and refining. **Hazardous waste** refers to solid or liquid waste that is toxic, chemically reactive, flammable, or corrosive. Another type of waste is wastewater, water we use in our households, businesses, industries, or public facilities and drain or flush down our pipes, as well as the polluted runoff from our streets and storm drains (pp. 277, 279).

There are three main components of **waste management**:

1. Minimizing the amount of waste we generate
2. Recovering discarded materials and recycling them
3. Disposing of waste effectively and safely

We have several ways to reduce the amount of material in the **waste stream**, the flow of waste as it moves from its sources toward disposal destinations (**FIGURE 17.2**). Minimizing waste at its source—called **source reduction**—is the preferred approach. The next-best strategy is **recovery**, which consists of recovering, or removing, waste from the waste stream. Recovery includes recycling and composting. **Recycling** is the process of collecting used goods and sending them to facilities that extract and reprocess raw materials that can then be used to manufacture new goods. **Composting** is the practice of recovering organic waste (such as food and yard waste) by converting it to mulch or humus (p. 144) through natural biological processes of decomposition.

Even after we decrease the waste stream through source reduction and recovery, there will still be waste left to dispose of. Disposal methods include burying waste in landfills and burning waste in incinerators. The linear movement of products from their manufacture to their disposal is described as "cradle-to-grave." As much as possible, however, the modern waste manager attempts to follow a **cradle-to-cradle** approach instead—one in which the materials from products are recovered and reused to create new products. We will first examine how waste managers use source reduction, recovery, and disposal to manage municipal solid waste, and then we will turn to industrial solid waste and hazardous waste.

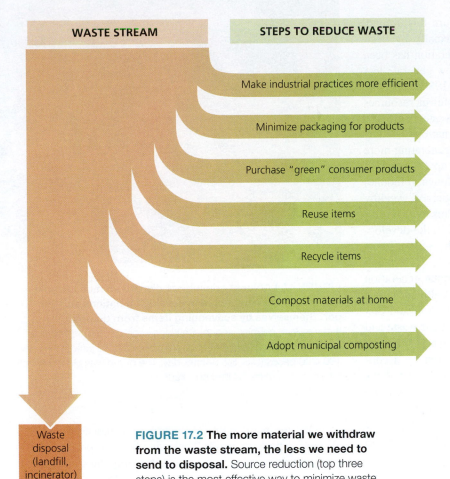

WASTE STREAM | STEPS TO REDUCE WASTE

- Make industrial practices more efficient
- Minimize packaging for products
- Purchase "green" consumer products
- Reuse items
- Recycle items
- Compost materials at home
- Adopt municipal composting

Waste disposal (landfill, incinerator)

FIGURE 17.2 The more material we withdraw from the waste stream, the less we need to send to disposal. Source reduction (top three steps) is the most effective way to minimize waste.

Municipal Solid Waste

Municipal solid waste is what we commonly refer to as "trash" or "garbage." In the United States, paper, food scraps, yard trimmings, and plastics are the principal components of municipal solid waste (**FIGURE 17.3a**). Paper is recycled at a high rate and yard trimmings are composted at a high rate, so after recycling and composting reduce the waste stream, food scraps and plastics are left as the largest components of U.S. municipal solid waste (**FIGURE 17.3b**).

Most municipal solid waste comes from packaging and nondurable goods (products meant to be discarded after a short period of use). In addition, consumers throw away old durable goods and outdated equipment as they purchase new products. Plastics, which came into wide consumer use only after 1970, have accounted for the greatest relative increase in the waste stream during the past several decades.

Consumption leads to waste

As we acquire more goods, we generate more waste. In the United States since 1960, waste generation (before recovery) has nearly tripled, and per-person waste generation has risen by 66%. Today Americans produce more than 250 million tons of municipal solid waste (before recovery) each year—close to 1 ton per person. The average U.S. resident generates 2.0 kg (4.4 lb) of trash per day—considerably more than people in most other industrialized nations. The relative wastefulness of the American lifestyle, with its excess packaging and reliance on nondurable goods, has caused critics to label the United States "the throwaway society."

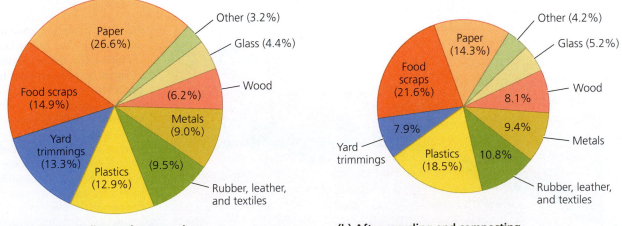

(a) Before recycling and composting

(b) After recycling and composting

FIGURE 17.3 Components of the municipal solid waste stream in the United States. Paper products make up the greatest portion by weight **(a)**, but after recycling and composting remove many items **(b)**, the waste stream becomes one-third smaller. Food scraps are now the largest contributor, because so much paper is recycled and yard waste is composted. *Data from U.S. Environmental Protection Agency, 2016.* Advancing sustainable materials management: 2014 fact sheet. *Washington, D.C.: EPA.*

However, Americans are beginning to turn this around. Thanks to source reduction and reuse (especially by businesses looking to cut costs), total waste generation has been roughly flat since about 2005. Americans now generate less waste per capita than they have since the late 1980s.

In developing nations, people consume fewer resources and goods, and as a result, generate less waste. However, consumption is intensifying in developing nations as they become more affluent, and these nations are generating more and more waste. As a result, trash is piling up and littering the landscapes of countries from Mexico to Kenya to Indonesia. Like U.S. consumers in the "throwaway society," wealthy people in developing nations often discard items that can still be used. In fact, at many dumps and landfills in the developing world, poor people support themselves by selling items that they scavenge (**FIGURE 17.4**).

In many industrialized nations, per capita generation rates have begun to decline in recent years. Wealthier nations also can afford to invest more in waste collection and disposal, so they are often better able to manage their waste and minimize impacts on health and the environment. Moreover, enhanced recycling and composting efforts—fed by a conservation ethic growing among a new generation on today's campuses—have been removing more material from the waste stream (**FIGURE 17.5**). As of 2014, U.S. waste managers were recovering 34.6% of the waste stream for composting and recycling, incinerating 12.8%, and sending the remaining 52.6% to landfills.

Reducing waste is our best option

Reducing the amount of material entering the waste stream is the preferred strategy for managing waste. Recall that preventing waste generation in this way is known as *source reduction*. This preventative approach avoids costs of disposal and recycling, helps conserve resources, minimizes pollution, and can save consumers and businesses money.

FIGURE 17.4 Affluent consumers discard so much usable material that some people in developing nations support themselves by scavenging items from dumps. Tens of thousands of people used to scavenge each day from this dump outside Manila in the Philippines, selling material to junk dealers for 100–200 pesos (U.S. $2–$4) per day. The dump was closed after an avalanche of trash killed hundreds of people.

One means of source reduction is to reduce the materials used to package goods. Packaging helps to preserve freshness, prevent breakage, protect against tampering, and provide information—yet much packaging is extraneous. Consumers can give manufacturers incentive to reduce packaging by choosing minimally packaged goods, buying unwrapped fruit and vegetables, and buying food in bulk. Manufacturers can reduce the size or weight of goods and materials, as they already have with aluminum cans, plastic soft drink bottles, personal computers, and much else.

Some policymakers have taken aim at a major source of waste—plastic grocery bags. These lightweight polyethylene

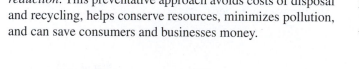

FIGURE 17.5 As recycling and composting have grown in the United States, the proportion of waste going to landfills has declined. As of 2014, 52.6% of U.S. municipal solid waste went to landfills and 12.8% to incinerators, whereas 34.6% was recovered for composting and recycling. *Data from U.S. Environmental Protection Agency, 2016.* Advancing sustainable materials management: 2014 fact sheet. *Washington, D.C.: EPA.*

 Has the amount of solid waste that is combusted (incinerated) increased or decreased since 1960?

Go to **Interpreting Graphs & Data** on **Mastering** Environmental Science

bags can persist for centuries in the environment, choking and entangling wildlife and littering the landscape—yet Americans discard 100 billion of them each year. A number of cities and more than 20 nations have now banned or restricted their use. Financial incentives are also effective. When Ireland began taxing these bags, their use dropped 90%. IKEA stores began charging for them and saw similar drops in usage. Many businesses now give discounts if you bring your own reusable canvas bags.

Reusing items helps to reduce waste

To reduce waste, you can save items to use again or substitute disposable goods with durable ones. TABLE 17.1 presents a sampling of actions we all can take to reduce waste. Habits as simple as bringing your own coffee cup to coffee shops or bringing sturdy reusable cloth bags to the grocery store can, over time, have an impact. You can also donate unwanted items and shop for used items at yard sales and at resale centers run by organizations such as Goodwill Industries or the Salvation Army. Besides reducing waste, reusing items saves money. Used items are often as functional as new ones, and they are cheaper.

TABLE 17.1 **Some Everyday Things You Can Do to Reduce and Reuse**

- Donate used items to charity
- Reuse boxes, paper, plastic wrap, plastic containers, aluminum foil, bags, wrapping paper, fabric, packing material, etc.
- Rent or borrow items instead of buying them, when possible—and lend your items to friends
- Buy groceries in bulk
- Bring reusable cloth bags shopping
- Make double-sided photocopies
- Keep electronic documents rather than printing items out
- Bring your own coffee cup to coffee shops
- Pay a bit extra for durable, long-lasting, reusable goods rather than disposable ones
- Buy rechargeable batteries
- Select goods with less packaging
- Compost kitchen and yard wastes
- Buy clothing and other items at resale stores and garage sales
- Use cloth napkins and rags rather than paper napkins and towels
- Tell businesses what you think about their packaging and products
- When solid waste policy is being debated, let your government representatives know your thoughts
- Support organizations that promote waste reduction

Adapted from U.S. Environmental Protection Agency.

On some campuses, students collect unwanted items and resell them or donate them to charity. Students at the University of Texas at Austin run a "Trash to Treasure" program. Each May, they collect 40–50 tons of items that students discard as they leave and then resell them at low prices in August to arriving students. This keeps waste out of the landfill, provides arriving students with items they need at low cost, and raises $10,000–20,000 per year that goes back into campus sustainability efforts. Hamilton College in New York runs a similar program, called "Cram & Scram." It reduces Hamilton's landfill waste by 28% (about 90 tons) each May.

Composting recovers organic waste

Composting is the conversion of organic waste into mulch or humus (p. 144) through natural decomposition. We can place waste in compost piles, underground pits, or specially constructed containers. As waste is added, heat from microbial action builds in the interior, and decomposition proceeds. Banana peels, coffee grounds, grass clippings, autumn leaves, and other organic items can be converted into rich, high-quality compost through the actions of earthworms, bacteria, soil mites, sow bugs, and other detritivores and decomposers (p. 74). The compost is then used to enrich soil.

On campus, composting is becoming popular. Ball State University in Indiana shreds surplus furniture and wood pallets and makes them into mulch to nourish campus plantings. Ithaca College in New York composts 44% of its food waste, saving $11,500 each year in landfill disposal fees. The compost is used on campus plantings, and experiments showed that the plantings grew better with the compost mix than with chemical soil amendments.

Municipal composting programs divert yard debris (and, increasingly, food waste as well) out of the waste stream and into composting facilities, where it decomposes into mulch that community residents can use for gardens and landscaping. About one-fifth of the U.S. waste stream is made up of materials that can easily be composted. Composting reduces landfill waste, enriches soil, enhances soil biodiversity, helps soil to resist erosion, makes for healthier plants and more pleasing gardens, and reduces the need for chemical fertilizers.

Recycling consists of three steps

Recycling, too, offers many benefits. It involves collecting used items and breaking them down so that their materials can be reprocessed to manufacture new items.

The recycling loop consists of three basic steps. The first step is to collect and process used goods and materials. Some towns and cities designate locations where residents can drop off recyclables or receive money for them. Others offer curbside recycling, in which trucks pick up recyclable items in front of homes, usually along with municipal trash collection. Items are taken to **materials recovery facilities (MRFs),** where workers and machines sort items using automated processes including magnetic pulleys, optical sensors, water currents, and air classifiers that separate items by weight and size. The facilities clean the materials, shred them, and prepare them for reprocessing.

Once readied, these materials are used to manufacture new goods—the second step in the recycling loop. Newspapers and many other paper products use recycled paper, many glass and metal containers are now made from recycled materials, and some plastic containers are of recycled origin. Benches, bridges, and walkways in city parks may now be made from recycled plastics, and glass can be mixed with asphalt (creating "glassphalt") to pave roads and paths.

If the recycling loop is to function, consumers and businesses must complete the third step in the cycle by purchasing ecolabeled products (p. 114) made from recycled materials. Buying recycled goods provides economic incentive for industries to recycle materials and for recycling facilities to open or expand.

Recycling has grown rapidly

Today nearly 10,000 curbside recycling programs across all 50 U.S. states serve 70% of Americans. These programs, and the 800 MRFs operating today, have sprung up only in the past few decades. Recycling in the United States rose from 6.4% of the waste stream in 1960 to 25.7% in 2014 (and 34.6% if composting is included; **FIGURE 17.6**).

Recycling rates vary greatly from one product or material type to another, ranging from nearly zero to almost 100% (**TABLE 17.2**). Recycling rates among U.S. states also vary greatly, from 1% to 48%. This variation makes clear that opportunities remain for further growth in recycling.

Many college and university campuses run active recycling programs, although attaining high recovery rates can be challenging in the campus environment. The most recent survey of campus sustainability efforts suggested that the average recycling rate was only 29%. Thus there appears to be much room for growth. Fortunately, waste management initiatives are relatively easy to conduct because they offer many opportunities for small-scale improvements and because people generally enjoy recycling and reducing waste.

weighing the ISSUES

Managing Waste on Your Campus

Does your campus have a recycling program? Does it have composting initiatives? Does it run programs to reduce or reuse materials? Think about the types and amounts of waste generated on your campus. Describe several examples of this waste that you feel could be prevented or recycled, and describe how this might be done in each case. If you could do one thing on campus to improve your school's waste management practices, what would it be?

Besides participation in Recyclemania, there are many ways to promote recycling on campus. Louisiana State University students initiated recycling efforts at home football games, and over three seasons they recycled 68 tons of refuse that otherwise would have gone to the landfill. "Trash audits" or "landfill on the lawn" events involve emptying dumpsters and sorting out recyclable items (**FIGURE 17.7**). When students at Ashland University in Ohio audited their waste, they found that 70% was recyclable, and they used this finding to press their administration to support recycling programs. On some campuses, students have even helped to conduct scientific research to find better ways of

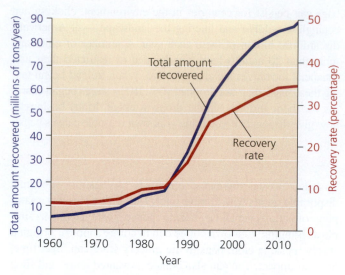

FIGURE 17.6 Recovery has risen sharply in the United States. Today more than 89 million tons of material are recovered (66 million tons by recycling and 23 million tons by municipal composting), making up one-third of the waste stream. *Data from U.S. Environmental Protection Agency.*

DATA • From the data in this graph alone, what would you infer has happened to the total amount of municipal solid waste generated (before recovery) since 1960? • Explain how you can determine this.

Go to **Interpreting Graphs & Data** on **Mastering** Environmental Science

encouraging recycling and reducing waste (see **THE SCIENCE BEHIND THE STORY**, pp. 408–409).

The growth of recycling has been propelled in part by economic forces as businesses see prospects to save money and as entrepreneurs see opportunities to start new businesses. It has also been driven by the desire of community and campus leaders to reduce waste and by the satisfaction people take in recycling. These latter two forces have driven the rise of recycling even when it has not been financially profitable. In fact, many of our popular municipal recycling programs are

TABLE 17.2 Recovery Rates for Various Materials in the United States

MATERIAL	PERCENTAGE THAT IS RECYCLED OR COMPOSTED
Lead-acid batteries	99
Newspapers	68
Paper and paperboard	65
Yard trimmings	61
Aluminum cans	55
Glass containers	33
Plastics	10

Data are for 2014, from U.S. Environmental Protection Agency.

Financial incentives help address waste

To encourage recycling, composting, and source reduction, waste managers often offer consumers economic incentives to reduce the waste stream. In "pay-as-you-throw" garbage collection programs, municipalities charge residents for home trash pickup according to the amount of trash they put out. The less waste one generates, the less one has to pay.

Bottle bills are another approach hinging on financial incentives. In the 10 U.S. states and 23 nations that have adopted these laws, consumers pay a deposit on bottles or cans upon purchase—often 5 or 10 cents per container—and then receive a refund when they return them to stores after use. U.S. states with bottle bills report that their beverage container litter has decreased by 69–84%, their total litter has decreased by 30–65%, and their per capita container recycling rates have risen 2.6-fold. Beverage container recycling rates for states with bottle bills are 3.5 times higher than for states without them.

Sanitary landfills are our main method of disposal

Material that remains in the waste stream following source reduction and recovery needs to be disposed of, and landfills provide our primary method of disposal. In modern **sanitary landfills,** waste is buried in the ground or piled up in huge mounds engineered to prevent waste from contaminating the environment and threatening public health (**FIGURE 17.8**). Most municipal landfills in the United States are regulated locally or by the states, but they must meet national standards set by the U.S. Environmental Protection Agency (EPA) under the **Resource Conservation and Recovery Act** (p. 107), a major federal law enacted in 1976 and amended in 1984.

In a sanitary landfill, waste is partially decomposed by bacteria and compresses under its own weight to take up less space. Soil is layered along with the waste to speed decomposition, reduce odor, and lessen infestation by pests. Some infiltration of rainwater into the landfill is good, because it encourages biodegradation by bacteria—yet too much is not good, because contaminants can escape if water carries them out.

To protect against environmental contamination, U.S. regulations require that landfills be located away from wetlands and

FIGURE 17.7 In a trash audit, students sort through rubbish and separate out recyclables. Events like this "Mt. Trashmore" exercise at Central New Mexico Community College show passersby just how many recyclable items are needlessly thrown away.

run at an economic loss. The expense required to collect, sort, process, and transport recycled goods is often more than recyclables are worth in the marketplace. In addition, the more people recycle, the more glass, paper, and plastic is available to manufacturers for purchase, which drives down prices. When recycling is no longer profitable for those in the recycling industry, MRFs may shut down, municipalities may cancel contracts, and recycling companies may go out of business.

Recycling advocates, however, point out that market prices do not take into account external costs (pp. 96, 104)—in particular, the environmental and health impacts of *not* recycling. Each year in the United States, recycling and composting together save energy equal to that of 230 million barrels of oil, and prevent carbon dioxide emissions equal to those of 39 million cars. Recycling aluminum cans saves 95% of the energy required to make the same amount of aluminum from mined virgin bauxite, its source material.

As more manufacturers use recycled products and as more technologies are developed to use recycled materials in new ways, markets should continue to expand, and new business opportunities may arise. We are just beginning to shift from an economy that moves linearly, from raw materials to products to waste, to a more sustainable economy that moves circularly, taking a cradle-to-cradle approach and using waste products as raw materials for new manufacturing.

weighing the ISSUES

Costs of Recycling and of Not Recycling

Should governments subsidize recycling programs even if they are run at an economic loss? What types of external costs—costs not reflected in market prices—do you think would be involved in not recycling, say, aluminum cans? Do you feel these costs justify sponsoring recycling programs even when they are not financially self-supporting? Why or why not?

FAQ

How much does garbage decompose in a landfill?

You might assume that a banana peel you throw in the trash will soon decay away to nothing in a landfill. However, it just might survive longer than you do! This is because surprisingly little decomposition occurs in landfills. Researcher William Rathje, a retired archaeologist known as "the Indiana Jones of Solid Waste," made a career out of burrowing into landfills and examining their contents to learn about what we consume and what we throw away. His research teams would routinely come across whole hot dogs, intact pastries that were decades old, and grass clippings that were still green. Newspapers 40 years old were often still legible, and the researchers used them to date layers of trash.

FIGURE 17.8 Sanitary landfills are engineered to prevent waste from contaminating soil and groundwater. Waste is laid in a large depression lined with plastic and impervious clay designed to prevent liquids from leaching out. Pipes draw out these liquids from the bottom of the landfill. Waste is layered with soil, filling the depression, and then is built into a mound until the landfill is capped. Landfill gas produced by anaerobic bacteria may be recovered, and waste managers monitor groundwater for contamination.

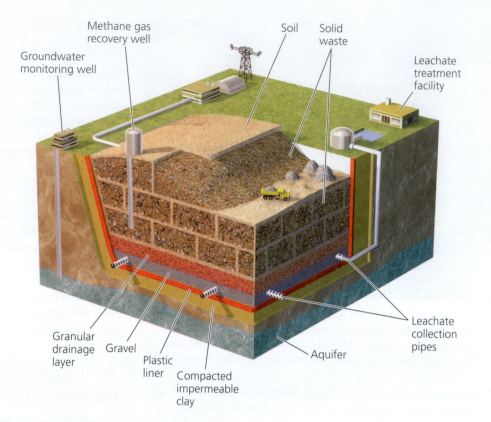

Groundwater monitoring well

Methane gas recovery well

Soil

Solid waste

Leachate treatment facility

Granular drainage layer

Gravel

Plastic liner

Compacted impermeable clay

Aquifer

Leachate collection pipes

earthquake-prone faults and be at least 6 m (20 ft) above the water table. The bottoms and sides of sanitary landfills must be lined with heavy-duty plastic and 60–120 cm (2–4 ft) of impermeable clay to help prevent contaminants from seeping into aquifers. Sanitary landfills also have systems of pipes, ponds, and treatment facilities to collect and treat **leachate,** liquid that results when substances from the trash dissolve in water as rainwater percolates downward. Once a landfill is closed, it is capped with an engineered cover consisting of layers of plastic, gravel, and soil, and managers are required to maintain leachate collection systems for 30 years thereafter.

Despite improvements in liner technology and landfill siting, however, liners can be punctured and leachate collection systems eventually cease to be maintained. Moreover, landfills are kept dry to reduce leachate, but dryness slows waste decomposition. In fact, the low-oxygen conditions of most landfills turn trash into a sort of time capsule. Researchers examining landfills find many of their contents perfectly preserved, even after years or decades.

In 1988, the United States had nearly 8000 landfills, yet today it has fewer than 2000. Waste managers have consolidated the waste stream into fewer landfills of larger size. Some landfills that were closed are now being converted into public parks or other uses (**FIGURE 17.9**). The world's

THEN: Fresh Kills Landfill in operation

NOW: Fresh Kills Landfill site today

FIGURE 17.9 Old landfills, once capped, can serve other purposes. Visitors to Freshkills Park in New York City enjoy this panoramic view of the Manhattan skyline from atop what used to be an immense mound of trash.

largest landfill conversion project is at New York City's former Fresh Kills Landfill. This site, on Staten Island, was the primary repository of New York City's garbage for half a century, and its mounds rose higher than the nearby Statue of Liberty! Today New York is transforming the site into a world-class public park—a verdant landscape of ball fields, playgrounds, jogging trails, rolling hills, and wetlands teeming with wildlife.

Incinerating trash reduces pressure on landfills

Just as sanitary landfills are an improvement over open dumping, incineration in specially constructed facilities is better than open-air burning of trash. **Incineration,** or combustion, is a controlled process in which garbage is burned at very high temperatures (**FIGURE 17.10**). At incineration facilities, waste is generally sorted and metals are removed. Metal-free waste is chopped into small pieces and then is burned in a furnace. Incinerating waste reduces its weight by up to 75% and its volume by up to 90%.

The ash remaining after trash is incinerated contains toxic components and must be disposed of in hazardous waste landfills (p. 414). Moreover, when trash is burned, hazardous chemicals—including dioxins, heavy metals, and polychlorinated biphenyls (PCBs) (Chapter 10)—may be created and released into the atmosphere. As a result, most developed nations regulate incinerator emissions, and some have banned incineration outright. Engineers have also developed technologies to mitigate emissions. Scrubbers (see Figure 13.8, p. 293) chemically treat the gases produced in combustion to neutralize acids and remove hazardous components. Particulate matter, called *fly ash*, contains some of the worst dioxin and heavy metal pollutants in incinerator emissions. To physically remove these tiny particles, facilities may use a huge system of filters known as a *baghouse*. In addition, burning garbage at especially high temperatures can destroy certain pollutants, such as PCBs. Even all these measures, however, do not fully eliminate toxic emissions.

weighing the ISSUES

Environmental Justice?

Do you know where your trash goes? Where is your landfill or incinerator located? Are the people who live closest to the facility wealthy, poor, or middle class? What race or ethnicity are they? Did the people of this neighborhood protest against the introduction of the landfill or incinerator?

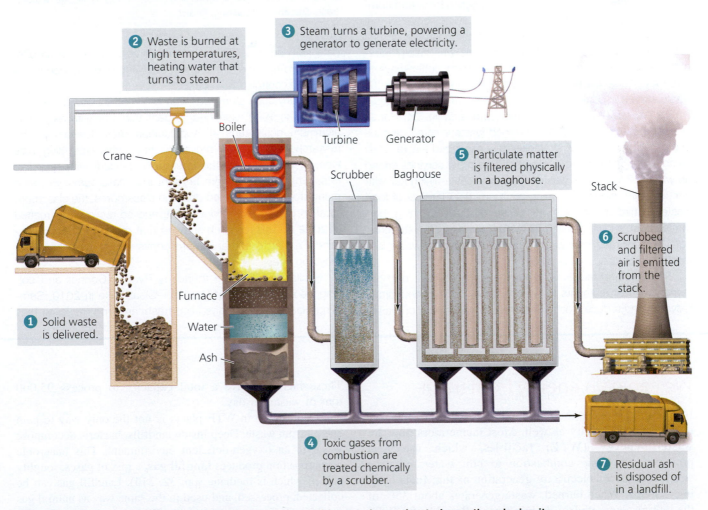

2 Waste is burned at high temperatures, heating water that turns to steam.

3 Steam turns a turbine, powering a generator to generate electricity.

Boiler

Turbine   Generator

Crane

Scrubber   Baghouse

5 Particulate matter is filtered physically in a baghouse.

Stack

Furnace

Water

Ash

6 Scrubbed and filtered air is emitted from the stack.

1 Solid waste is delivered.

4 Toxic gases from combustion are treated chemically by a scrubber.

7 Residual ash is disposed of in a landfill.

FIGURE 17.10 In a waste-to-energy (WTE) incinerator, solid waste is combusted, greatly reducing its volume and generating electricity at the same time.

Can Campus Research Help Reduce Waste?

Thousands of students on college and university campuses are engaged in efforts to reduce waste. The campus environment also provides opportunities to conduct scientific research on how to better manage waste.

The descriptive research involved in a trash audit (p. 404) is straightforward to conduct yet can yield valuable data with practical relevance. At the University of Washington, students and faculty in the UW Garbology Project studied waste on their campus for several years. Working with instructor Dr. Jack Johnson and university waste managers, students sorted through the contents of trashcans, recycling bins, and compost receptacles, and discovered that more than 80% of the material thrown in the trash was in fact recyclable or compostable (**FIGURE 1**). This was helpful information, because if the university could devise better ways to divert recyclable items and compostable food matter from the waste stream, it could save $225,000 per year on landfill fees.

At Western Michigan University, students in Dr. Harold Glasser's course in 2012 audited food waste in three campus dining halls to test what strategies best minimized waste. They found that a dining hall providing made-to-order servings ended up with 0.23 lb/meal of wasted food, whereas a dining hall with a traditional buffet-style serving produced 0.27 lb/meal of food waste. A third dining hall, which featured trayless dining, performed best, showing only 0.18 lb/meal of waste. The students' data thus seemed to support the idea that people waste less food when trays are not provided.

Students and faculty on multiple campuses have also run manipulative experiments to determine how best to encourage recycling and reduce waste. Such research involves comparing

A student does her part to recycle

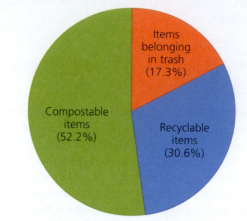

FIGURE 1 At the University of Washington, students found that most material in trashcans could be recycled or composted. As an example, the pie chart shows the average contents of trash from one floor of a building in 2014; only 17% was actually trash. *Data from UW Garbology Project.*

data from an experimental treatment condition (such as after new recycling bins are introduced) with baseline conditions used as a control.

An early such study was that of Timothy Ludwig and others, who in 1998 examined student behavior with recyclable aluminum drink cans at Appalachian State University. The researchers compared recycling rates when recycling bins were in the hallways (the baseline condition) with recycling rates when the bins were brought into classrooms. Because many students consumed drinks in classrooms, the classroom location proved more convenient, and so recycling increased (**FIGURE 2**). This research showed that making recycling containers easier to find and more convenient to use can boost recycling rates.

Similar results were found by Ryan O'Connor and colleagues at University of Houston–Clear Lake in 2010. Sampling plastic drink bottles from trashcans and recycling bins

We can gain energy from trash

Incineration reduces the volume of waste, but it can serve to generate electricity as well. Most incinerators now are **waste-to-energy (WTE) facilities,** which use heat produced by waste combustion to boil water, creating steam that drives electricity generation or that fuels heating systems. When burned, waste generates about 35% of the energy generated by burning coal. Roughly 80 WTE facilities are operating across the United States today.

These facilities have a total capacity to process 95,000 tons of waste per day.

Combustion in WTE plants is not the only way to gain energy from waste. Deep inside landfills, bacteria decompose waste in an oxygen-deficient environment. This anaerobic decomposition produces **landfill gas,** a mix of gases, roughly half of which is methane (pp. 32, 314). Landfill gas can be collected, processed, and used in the same way as natural gas (p. 346). Today hundreds of landfills are collecting landfill gas and selling it for energy.

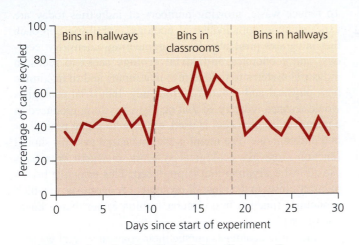

FIGURE 2 At Appalachian State University, recycling rates rose when recycling bins were moved from hallways to classrooms. *Data from Ludwig, T., et al., 1998. Increasing recycling in academic buildings: A systematic replication.* J. Appl. Behavior Analysis *31: 683–686.*

in three academic buildings, they found that making recycling bins more colorful or adding more bins had no effect on recycling rates, but that moving them from hallways and common areas into classrooms increased recycling rates greatly.

In 2017, researchers at Western Michigan University challenged the idea that moving bins to places where people use recyclable items was necessarily the best solution. Instead, Katherine Binder and others tested the hypothesis that removing trashcans from classrooms and placing them only in common areas side by side with recycling bins would lead to better recycling rates. These researchers compared recycling rates on two floors of a campus building where this was done with recycling rates on two floors where classrooms retained trashcans. Their data supported their hypothesis: Recycling rates rose when trashcans were removed from classrooms, forcing people who needed to dispose of items to walk to centrally located areas, where they encountered clearly marked recycling and trash receptacles to choose between. The research team concluded that this strategy also reduced contamination

from incorrectly sorted items and saved money because fewer receptacles were needed.

Researchers have also experimentally tested the influence of educational efforts on recycling behavior. At University of Wisconsin–Stout, student Jessica Van Der Werff ran an experiment in 2008 comparing recycling rates of freshmen who took a recycling workshop during freshman orientation with those who did not. Van Der Werff monitored the trash and recycling from two residence halls throughout the fall semester. She found that students from the residence hall who had attended the workshop showed nearly 40% higher recycling rates (**FIGURE 3**).

Taken together, campus research into waste management has revealed that we can increase recycling rates through education and strategic location of bins. Students engaged in this work have generated many practical suggestions for reducing waste; they urge that campuses provide enough receptacles, clarify how to sort items correctly, and make it convenient and easy to recycle and compost. By taking such lessons to heart, every campus should be able to significantly reduce its waste stream.

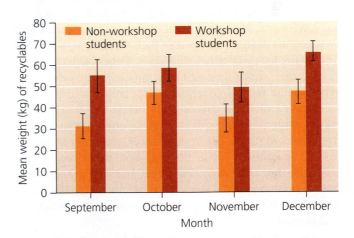

FIGURE 3 At University of Wisconsin–Stout, students who took a workshop on recycling recycled items at a higher rate than those who did not. *Data from Van Der Werff, J., 2008. Teaching recycling: The relationship between education and behavior among college freshmen and its effect on campus recycling rates.* Undergraduate project report, University of Wisconsin–Stout.

We can recycle material from landfills

Landfills offer us useful by-products beyond landfill gas. With improved technology for sorting rubbish and recyclables, businesses and entrepreneurs are weighing the economic benefits and costs of rummaging through landfills to salvage materials of value that can be recycled. Steel, aluminum, copper, and other metals are abundant enough in some landfills to make salvage operations

profitable when market prices for the metals are high enough. For instance, Americans throw out so many aluminum cans that at 2017 prices for aluminum, the nation buries $5.5 billion of this metal in landfills each year. If we could retrieve all the aluminum from U.S. landfills, it would exceed the amount the world produces from a year's worth of mining ore.

Landfills also offer soil mixed with organic waste that can be mined and sold as premium compost. In addition, old landfill waste can be incinerated in newer, cleaner-burning

WTE facilities to produce energy. Some companies are even looking into gaining carbon offset credits (p. 337) by harvesting methane (a greenhouse gas that contributes to climate change) leaking from open dumps in developing nations.

Industrial Solid Waste

Industrial solid waste includes waste from factories, mining activities, agriculture, petroleum extraction, and more. Each year, U.S. industrial facilities generate more than 7 billion tons of waste, according to the EPA, about 97% of which is wastewater. Thus, very roughly, 230 million or so tons of solid waste are generated by 60,000 facilities each year—an amount approaching that of municipal solid waste.

Regulation and economics each influence industrial waste generation

Most methods and strategies of waste disposal, reduction, and recycling by industry are similar to those for municipal solid waste. Businesses that dispose of their own waste on site must design and manage their landfills in ways that meet state, local, or tribal guidelines. Other businesses pay to have their waste disposed of at municipal disposal sites. Whereas the federal government regulates municipal solid waste, state or local governments regulate industrial solid waste (with federal guidance). Regulation varies greatly from place to place, but in most cases, state and local regulation of industrial solid waste is less strict than federal regulation of municipal solid waste. In many areas, industries are not required to have permits, install landfill liners or leachate collection systems, or monitor groundwater for contamination.

The amount of waste generated by a manufacturing process is a good measure of its efficiency; the less waste produced per unit or volume of product, the more efficient that process is, from a physical standpoint. However, physical efficiency is not always reflected in economic efficiency. Often it is cheaper for industry to manufacture its products or perform its services quickly but messily. That is, it can be cheaper to generate waste than to avoid generating waste. In such cases, economic efficiency is maximized, but physical efficiency is not. Because our market system rewards only economic efficiency, all too often industry has no financial incentive to achieve physical efficiency. The frequent mismatch between these two types of efficiency is a major reason why the output of industrial waste is so great.

Rising costs of waste disposal enhance the financial incentive to decrease waste. Once either government or the market makes the physically efficient use of raw materials economically efficient as well, businesses gain financial incentives to reduce their waste.

Industrial ecology seeks to make industry more sustainable

To reduce waste, growing numbers of industries today are experimenting with industrial ecology. A holistic approach that integrates principles from engineering, chemistry, ecology, and economics, **industrial ecology** seeks to redesign industrial systems to reduce resource inputs and to maximize both physical and economic efficiency. Industrial ecologists would reshape industry so that nearly everything produced in a manufacturing process is used, either within that process or in a different one. The intent is that industrial systems should function more like ecological systems, in which organisms use almost everything that is produced. This principle brings industry closer to the ideal of ecological economists, in which economies function in a circular fashion rather than a linear one (p. 98).

Industrial ecologists pursue their goals in several ways:

- They examine the entire life-cycle of a product—from its origins in raw materials, through its manufacturing, to its use, and finally its disposal—and look for ways to make the process more efficient. This strategy is called **life-cycle analysis.**

- They take a cradle-to-cradle approach and try to identify how waste products from one manufacturing process might be used as raw materials for another. For instance, used plastic beverage containers can be shredded and reprocessed to make items such as benches, tables, and decks.

- They seek to eliminate environmentally harmful products and materials from industrial processes.

- They study the flow of materials through industrial systems to look for ways to create products that are more durable, recyclable, or reusable.

Businesses are adopting industrial ecology

Attentive businesses are taking advantage of the insights of industrial ecology to save money while reducing waste. For example, the Swiss Zero Emissions Research and Initiatives (ZERI) Foundation sponsors dozens of innovative projects worldwide that attempt to create goods and services without generating waste. One example involves breweries in Canada, Sweden, Japan, and Namibia (**FIGURE 17.11**).

Few businesses have taken industrial ecology to heart as much as the carpet tile company Interface, which founder Ray Anderson set on the road to sustainability years ago. Interface asks customers to return used tiles for recycling and for reuse as backing for new carpet. It modified its tile design and production methods to reduce waste. It adapted its boilers to use landfill gas for energy. Through such steps, Anderson's company cut waste generation by 80%, fossil fuel use by 45%, and water use by 70%—all while saving $30 million per year, holding prices steady for customers, and raising profits by 49%.

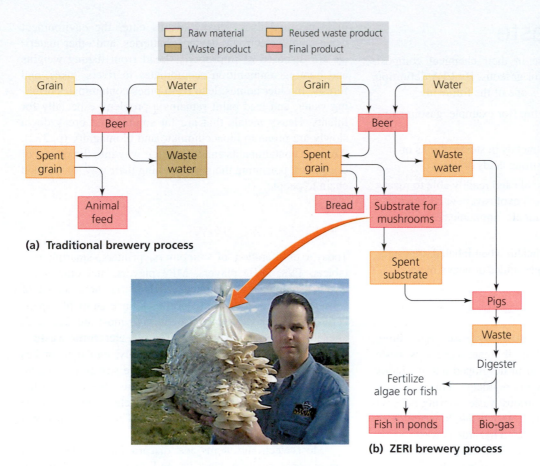

Raw material   **Reused waste product**
Waste product   **Final product**

(a) **Traditional brewery process**

(b) **ZERI brewery process**

FIGURE 17.11 Creative use of waste products can help us approach zero-waste systems. Traditional breweries **(a)** produce only beer while generating much waste, some of which goes toward animal feed. Breweries sponsored by the Zero Emissions Research and Initiatives (ZERI) Foundation **(b)** use their waste grain to make bread and to farm mushrooms **(photo)**. Waste from the mushroom farming, along with brewery wastewater, goes to feed pigs. The pigs' waste is digested in containers that capture natural gas and collect nutrients used to nourish algae for growing fish in fish farms. The brewer derives income from bread, mushrooms, pigs, gas, and fish, as well as beer.

SUCCESS STORY

Creating an Industrial Ecosystem

One place the ideals of industrial ecology have come to life is the city of Kalundborg, Denmark. Here, starting in 1972, dozens of private and public enterprises gradually formed a network of business relationships that are conserving resources while saving money. Anchoring the Kalundborg Eco-Industrial Park is a coal-fired power plant, the Asnaes Power Station. It sends its excess steam to a nearby Statoil petroleum refinery and a Novo-Nordisk pharmaceutical factory, which use the steam to run their operations. The Statoil refinery sends Asnaes its wastewater, cooling water, and waste gas, which the power plant uses to generate electricity, and also sells sulfur to a local acid manufacturer. The power plant sends its waste fly ash to a cement company and sells gypsum removed from its waste gas by a scrubber to a Gyproc factory that makes drywall. Power plants also routinely create large amounts of waste heat, and in Kalundborg, this heat is piped to more than 3000 homes as district heating and to a regional fish farm. Treated sludge from both the fish farm and the pharmaceutical plant is sent to area farms as fertilizer. By efficiently using one another's waste products, the Kalundborg Eco-Industrial Park has conserved resources like water, coal, and oil; reduced pollution; and cut greenhouse gas emissions, all while saving hundreds of millions of dollars for the enterprises involved.

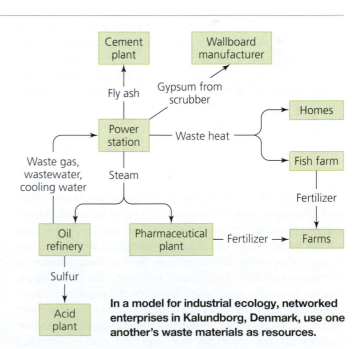

In a model for industrial ecology, networked enterprises in Kalundborg, Denmark, use one another's waste materials as resources.

EXPLORE THE DATA at **Mastering** Environmental Science

Hazardous Waste

Hazardous wastes are diverse in their chemical composition and may be liquid, solid, or gaseous. By EPA definition, hazardous waste is waste that is one of the following:

- *Ignitable.* Likely to catch fire (for example, gasoline or alcohol).
- *Corrosive.* Apt to corrode metals in storage tanks or equipment (for example, strong acids or bases).
- *Reactive.* Chemically unstable and readily able to react with other compounds, often explosively or by producing noxious fumes (for example, ammonia reacting with chlorine bleach).
- *Toxic.* Harmful to human health when inhaled, ingested, or touched (for example, pesticides or heavy metals).

Hazardous wastes are diverse

Industry, mining, households, small businesses, agriculture, utilities, and building demolition all create hazardous waste. Industry produces the most, but in developed nations industrial waste disposal is often highly regulated. This regulation has reduced the amount of hazardous waste entering the environment from industrial activities. As a result, households are now the largest source of unregulated hazardous waste.

Household hazardous waste includes a wide range of items, including paints, batteries, oils, solvents, cleaning agents, lubricants, and pesticides. Americans generate 1.6 million tons of household hazardous waste annually, and the average home contains close to 45 kg (100 lb) of it in sheds, basements, closets, and garages.

Many hazardous substances become less hazardous over time as they degrade, but some show especially persistent effects. Radioactive substances are an example, and the disposal of radioactive waste poses a serious dilemma (p. 370). Other types of persistent hazardous substances include organic compounds and heavy metals.

Organic compounds and heavy metals pose hazards

In our daily lives, we rely on synthetic organic compounds and petroleum-derived compounds to resist bacterial, fungal, and insect activity. Plastic containers, rubber tires, pesticides, solvents, and wood preservatives are useful to us precisely because they resist decomposition. We use these substances to protect buildings from decay, kill pests that attack crops, and keep stored goods intact. However, the capacity of these compounds to resist decay is a double-edged sword, for it also makes them persistent pollutants. Many synthetic organic compounds are toxic because they are readily absorbed through the skin and can act as mutagens, carcinogens, teratogens, and endocrine disruptors (p. 218).

Heavy metals such as lead, chromium, mercury, arsenic, cadmium, tin, and copper are used widely in industry for wiring, electronics, metal plating, metal fabrication, pigments, and dyes. Heavy metals enter the environment when paints, electronic devices, batteries, and other materials are disposed of improperly. Lead from fishing weights and hunting ammunition accumulates in rivers, lakes, and forests. In older homes, lead from pipes contaminates drinking water, and lead paint remains a problem, especially for infants. Heavy metals that are fat soluble and break down slowly are prone to bioaccumulate and biomagnify (p. 222). All these contaminants can make their way into the tissues of organisms, poisoning them and making their way up the food chain to people.

E-waste has grown

Today's proliferation of computers, printers, smartphones, tablets, TVs, DVD players, MP3 players, and other electronic technology has created a substantial new source of waste (**FIGURE 17.12**). These products have short life spans before people judge them obsolete, and most are discarded after just a few years. The amount of this **electronic waste**—often called **e-waste**—has grown rapidly, and now makes up more than 1% of the U.S. solid waste stream by weight. More than 7 billion electronic devices have been sold in the United States since 1980, and U.S. households discard more than 300 million per year—two-thirds of them still in working order.

Most electronic items we discard have ended up in conventional sanitary landfills and incinerators. However, electronic products contain heavy metals and toxic flame-retardants, and research suggests that e-waste should instead be treated as hazardous waste, so the EPA and a number of states are now taking steps to do so.

Fortunately, the downsizing of many electronic items and the shift toward mobile devices and tablets mean that

FIGURE 17.12 Each day, Americans throw away about 350,000 cell phones. Phones that enter the waste stream can leach toxic heavy metals into the environment. Alternatively, we can recycle phones for reuse and to recover valuable metals.

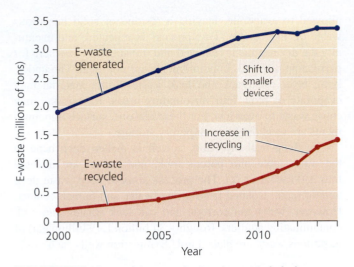

FIGURE 17.13 **More and more electronic waste is being recycled.** The total number of electronic devices sent to the waste stream each year in the United States continues to rise, but the shift to mobile devices and tablets has enabled us to hold the overall tonnage of e-waste **(blue line)** steady for several years. Meanwhile, the amount as well as proportion of e-waste recycled each year **(red line)** is growing. *Data from U.S. Environmental Protection Agency.*

fewer raw materials by weight are now going into electronics being manufactured—and, as a result, U.S. e-waste generation appears to have recently leveled off. In addition, more and more electronic waste today is being recycled (**FIGURE 17.13**; and see **Success Story** in Chapter 11, p. 250). Campus e-waste recycling drives are proving especially effective (see Figure 1.17c, p. 20). Americans now recycle 42% of their e-waste, by weight. Devices collected are shipped to facilities and taken apart, and the parts and materials are refurbished and reused in new products. There are serious concerns, however, about health risks that recycling may pose to workers doing the disassembly. Wealthy nations ship much of their e-waste to developing countries, where low-income workers disassemble the devices and handle toxic materials with minimal safety regulations.

Although these environmental justice concerns need to be resolved if electronics recycling is to be conducted safely and responsibly, e-waste recycling does help to keep toxic substances out of our waste stream. It also can help us recover trace metals used in electronics that are rare and lucrative. A typical cell phone contains up to a dollar's worth of precious metals (p. 250). By one estimate, 1 ton of computer scrap contains more gold than 16 tons of mined ore from a gold mine, while 1 ton of iPhones contains over 300 times more. Every ounce of metal we can recycle from a manufactured item is an ounce of metal we don't need to mine from the ground. Thus, "mining" e-waste for metals helps reduce the environmental impacts of mining the earth. In one example, the 2010 Winter Olympic Games in Vancouver produced its stylish gold, silver, and bronze medals (**FIGURE 17.14**) partly from metals recovered from recycled and processed e-waste!

FIGURE 17.14 **Medals awarded to athletes at the 2010 Winter Olympic Games in Vancouver were made partly from precious metals recycled from discarded e-waste.**

Several steps precede the disposal of hazardous waste

Many communities designate sites or special collection days to gather household hazardous waste (**FIGURE 17.15**). Once consolidated, the waste is transported for treatment and disposal. Under the Resource Conservation and Recovery Act, the EPA sets standards by which states manage hazardous waste. The Act also requires large generators of hazardous waste to obtain permits. Finally, it mandates that hazardous materials be tracked "from cradle to grave." As hazardous waste is generated, transported, and disposed of, the producer, carrier, and disposal facility must each report to the EPA the type and amount of material generated; its location, origin, and destination; and the way it is handled.

Because current U.S. law makes disposing of hazardous waste quite costly, irresponsible companies sometimes illegally dump waste, creating health risks for residents and financial headaches for local governments forced to deal

FIGURE 17.15 **Many communities designate collection sites or collection days for household hazardous waste.** Here, workers handle waste from a collection event in Brooklyn, New York.

FIGURE 17.16 Unscrupulous parties sometimes dump hazardous waste illegally to avoid disposal costs.

with the mess (**FIGURE 17.16**). Companies from industrialized nations sometimes dump hazardous waste illegally in developing nations—a major environmental justice issue. The Basel Convention (p. 110) was crafted to limit such practices.

High costs of disposal, however, have also encouraged conscientious businesses to invest in reducing their hazardous waste. Some biologically hazardous materials can be broken down by incineration at high temperatures. Others can be treated with bacteria that break down harmful components and synthesize them into new compounds. In addition, various plants have been bred or engineered to take up specific contaminants from soil and break down organic contaminants into safer compounds or concentrate heavy metals in their tissues. The plants are harvested and disposed of.

We use three disposal methods for hazardous waste

We have developed three primary means of hazardous waste disposal: landfills, surface impoundments, and injection wells. These do nothing to diminish the hazards of the substances, but they help keep the waste isolated from people, wildlife, and ecosystems. Design and construction standards for hazardous waste landfills are stricter than those for ordinary sanitary landfills. Hazardous waste landfills must have several impervious liners and leachate removal systems and must be located far from aquifers.

Liquid hazardous waste, or waste in dissolved form, may be stored in **surface impoundments,** shallow depressions lined with plastic and an impervious material, such as clay. The liquid or slurry is placed in the pond, and water is

allowed to evaporate, leaving a residue of solid hazardous waste on the bottom. This process is repeated, and eventually the dry residue is removed and transported elsewhere for permanent disposal. Impoundments are not ideal. The underlying layer can crack and leak waste. Some material may evaporate or blow into surrounding areas. Rainstorms may cause waste to overflow. For these reasons, surface impoundments are used only for temporary storage.

In **deep-well injection,** a well is drilled deep beneath the water table into porous rock, and wastes are injected into it (**FIGURE 17.17**). The process aims to keep waste deep underground, isolated from groundwater and human contact. However, wells can corrode and can leak wastes into soil, contaminating aquifers. Roughly 34 billion L (9 billion gal) of hazardous waste are placed in U.S. injection wells each year.

Contaminated sites are being cleaned up, slowly

Many thousands of former military and industrial sites remain contaminated with hazardous waste in the United States and many other nations. For most nations, dealing with these messes is simply too difficult, time-consuming, and expensive. In 1980, however, the U.S. Congress passed the Comprehensive Environmental Response Compensation and Liability Act (CERCLA; p. 107). This law established a federal program to clean up U.S. sites polluted with hazardous waste. The EPA administers this cleanup program, called the **Superfund.** Under EPA auspices, experts identify polluted sites, take action to protect groundwater, and clean up the pollution. Later laws also charged the EPA with cleaning up

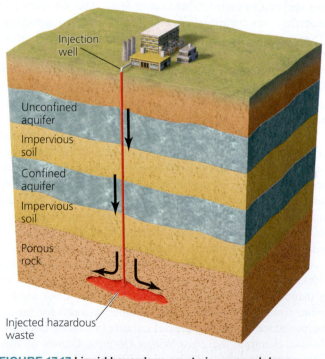

FIGURE 17.17 Liquid hazardous waste is pumped deep underground by deep-well injection. The well must be drilled below any aquifers, into porous rock isolated by impervious clay.

brownfields, lands whose reuse or development is complicated by the presence of hazardous materials.

Two well-publicized events spurred creation of the Superfund legislation. In *Love Canal*, a residential neighborhood in Niagara Falls, New York, more than 800 families were evacuated in 1978–1980 after toxic chemicals buried by a company and the city in past decades rose to the surface, contaminating homes and an elementary school and apparently leading to birth defects, miscarriages, and other health impacts (**FIGURE 17.18**). Outside of Louisville, Kentucky, at a site called *Valley of the Drums*, hazardous waste began leaking from 100,000 metal drums, contaminating waterways with 140 types of chemicals.

Once a Superfund site is identified, EPA scientists evaluate how near the site is to homes, whether wastes are confined or likely to spread, and whether the pollution threatens drinking water supplies. Sites judged to be harmful are placed on the National Priorities List, ranked by the level of risk to human health that they pose. Cleanup proceeds as funds are available. Throughout the process, the EPA is required to hold public hearings to inform area residents of its findings and to receive feedback.

The objective of CERCLA was to charge the polluting parties for the cleanup of their sites, according to the *polluter-pays principle* (p. 113). For many sites, however, the responsible parties cannot be found or held liable, and in such cases—roughly 30% so far—cleanups have been covered by taxpayers and from a trust fund established by a federal tax on industries producing petroleum and chemical raw materials. However, Congress let the tax expire and the trust fund went bankrupt in 2004, so taxpayers are now shouldering the entire burden. As the remaining cleanup jobs become more expensive, fewer are being completed.

As of 2017, 1336 Superfund sites remained on the National Priorities List, and only 393 had been cleaned up

FIGURE 17.18 This boy was one of hundreds of people evacuated from Love Canal. Outrage over the contamination of this neighborhood in Niagara Falls helped lead to the Superfund program.

or otherwise deleted from the list. The average cleanup has cost over $25 million and has taken nearly 15 years. Many sites are contaminated with hazardous chemicals we have no effective way to deal with. In such cases, cleanups aim simply to isolate waste from human contact, either by building trenches and barriers or by excavating contaminated material and shipping it to a hazardous waste disposal facility. For all these reasons, the current emphasis is on preventing hazardous waste contamination in the first place.

closing THE LOOP

We have made great strides in addressing our waste problems. Modern methods of waste management are far safer for people and gentler on the environment than past practices of open dumping and open burning. Recycling and composting efforts are advancing steadily, and Americans now divert one-third of all solid waste away from disposal. The continuing growth of recycling, driven by market forces, public policy, and consumer behavior, shows potential to further alleviate our waste problems.

Students on college and university campuses are making great contributions in accelerating these trends. The enthusiasm of students for recycling is apparent each year in the success of Recyclemania—and this competition is just the tip

of the iceberg. Most campuses have their own recycling and waste reduction programs, and these continue to grow and evolve as students and staff find new and innovative ways of inspiring people to reduce waste. Across the larger society, continued growth of recycling and composting—driven by market forces, public policy, and consumer behavior—shows potential for further advances.

Still, our prodigious consumption habits are creating as much waste as ever. Our waste management efforts are marked by a number of challenges, including the cleanup of Superfund sites and the safe disposal of hazardous waste. These dilemmas make clear that the best solution is to reduce our generation of waste and to pursue a cradle-to-cradle approach. Finding ways to reduce, reuse, and efficiently recycle the materials and goods that we use stands as a key ongoing challenge for our society.

TESTING Your Comprehension

1. Describe the three major components of managing waste. Why do we practice waste management?

2. Why have some people labeled the United States "the throwaway society"? How much solid waste do Americans generate, and how does this amount compare to that of people from other countries?

3. What is composting, and how does it help reduce the waste stream?

4. What are the three elements of the recycling process?

5. Name several guidelines by which sanitary landfills are regulated. Describe three problems with landfills.

6. Describe the process of incineration, or combustion. What is one drawback of incineration?

7. In your own words, describe the goals of industrial ecology.

8. What four criteria are used to define hazardous waste? What makes heavy metals and synthetic organic compounds particularly hazardous?

9. What are the largest sources of hazardous waste? Describe three ways to dispose of hazardous waste.

10. What is the Superfund program? How does it work?

SEEKING Solutions

1. How much waste do you generate? Look into your waste bin at the end of the day, and categorize and measure the waste there. List all other waste you generated elsewhere during the day. How much of this waste could you have avoided generating? How much could have been reused or recycled?

2. Some people have criticized current waste management practices as merely moving waste from one medium to another. How might this criticism apply to the methods now in practice? What are some potential solutions?

3. Of the various waste management approaches covered in this chapter, which ones are your community or campus pursuing? Would you suggest pursuing any new approaches? If so, which ones, and why?

4. **CASE STUDY CONNECTION** Does your college or university participate in Recyclemania? If so, describe how it has done so, how successful its efforts were, and how this success might be improved. If not, describe what events, programs, or strategies might be effective on your campus to give it a shot at winning one of the categories in Recyclemania? For information, consult the Recyclemania Web page, http://recyclemaniacs.org.

5. **THINK IT THROUGH** You are the president of your college or university and want to make the school a leader in waste reduction. Consider the industries and businesses in your community and the ways they interact with facilities on your campus. Bearing in mind the principles of industrial ecology, what novel ways might your school and local businesses mutually use and benefit from one another's services, products, or waste materials? What steps would you propose to take as president?

CALCULATING Ecological Footprints

The biennial "State of Garbage in America" survey documents the ability of U.S. residents to generate prodigious amounts of municipal solid waste (MSW). According to the most recent survey, on a per capita basis, Missouri residents generate the least MSW (4.5 lb/day), whereas Hawai'i residents (and its many tourists) generate the most (15.5 lb/day). The average for the United States as a whole is 6.8 lb MSW per person per day. Calculate the total amount of MSW generated in 1 day and in 1 year by each group listed, if they were to generate MSW at each of the rates shown in the table.

GROUPS GENERATING MUNICIPAL SOLID WASTE	AMOUNT OF MSW GENERATED, AT THREE PER CAPITA GENERATION RATES					
	U.S. AVERAGE (6.8 LB/DAY)		MISSOURI (4.5 LB/DAY)		HAWAI'I (15.5 LB/DAY)	
	DAY	YEAR	DAY	YEAR	DAY	YEAR
You	6.8	2482				
Your class						
Your state						
United States						
World						

Data from Shin, D., 2014. Generation and disposal of municipal solid waste (MSW) in the United States—A national survey. Columbia University Earth Engineering Center.

1. Suppose your town of 50,000 people has just approved construction of a landfill nearby. Estimates are that it will accommodate 1 million tons of MSW. Assuming the landfill is serving only your town, and assuming that your town's residents generate waste at the U.S. average rate, for how many years will it accept waste before filling up? How much longer would a landfill of the same capacity serve a town of the same size in Missouri?

2. One study has estimated that the average world citizen generates 1.47 pounds of trash per day. How many times more than this does the average U.S. citizen generate?

3. The same study showed that the average resident of a low-income nation generates 1.17 pounds of waste per day and that the average resident of a high-income nation generates 2.64 pounds per day. Why do you think U.S. residents generate so much more MSW than people in other "high-income" countries, when standards of living in those countries are comparable?

Mastering Environmental Science

Students Go to **Mastering** Environmental Science for assignments, the etext, and the Study Area with practice tests, videos, current events, and activities.

Instructors Go to **Mastering** Environmental Science for automatically graded activities, current events, videos, and reading questions that you can assign to your students, plus Instructor Resources.

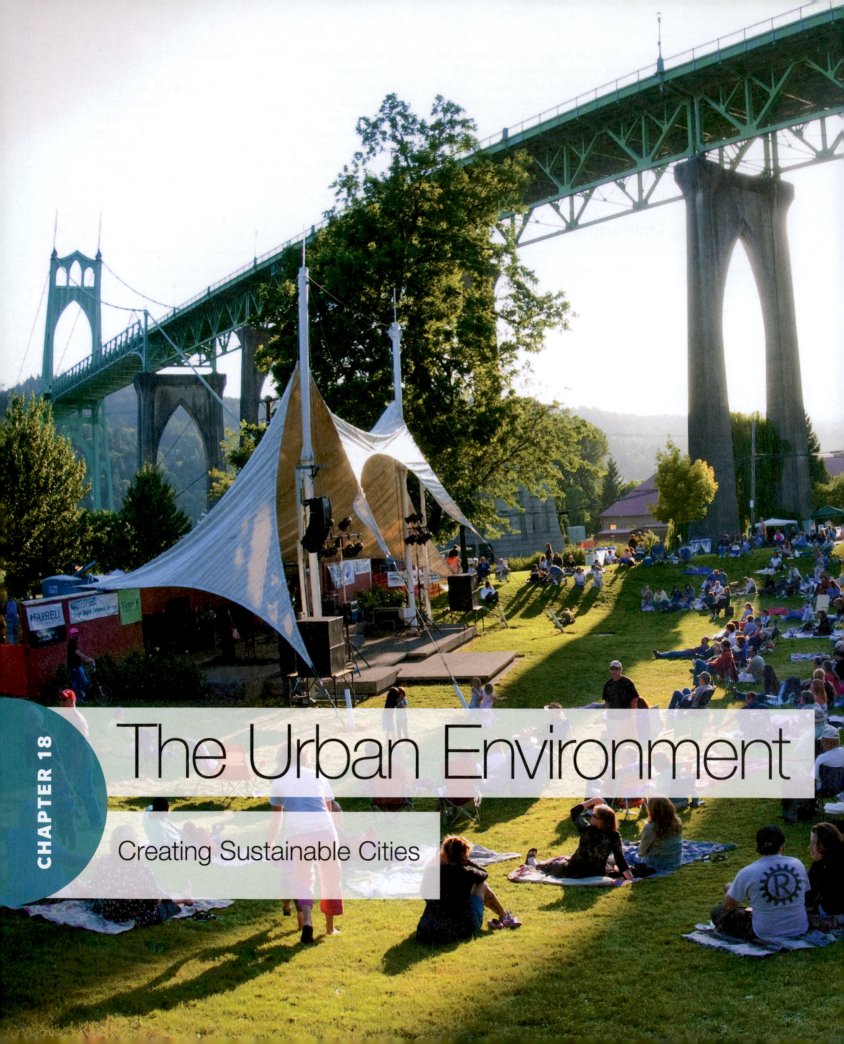

CHAPTER 18

The Urban Environment

Creating Sustainable Cities

Managing Growth in Portland, Oregon

> **Sagebrush subdivisions, coastal condomania, and the ravenous rampage of suburbia . . . all threaten to mock Oregon's status as the environmental model for the nation.**
> —Oregon Governor Tom McCall, 1973

> **We have planning boards. We have zoning regulations. We have urban growth boundaries and 'smart growth' and sprawl conferences. And we still have sprawl.**
> —Environmental scientist Donella Meadows, 1999

With fighting words, Oregon Governor Tom McCall challenged his state's legislature in 1973 to take action against runaway sprawling development, which many Oregonians feared would ruin the communities and landscapes they loved. McCall was echoing the growing concerns of state residents that farms, forests, and open space were being gobbled up and paved over.

Foreseeing a future of subdivisions, strip malls, and traffic jams engulfing the pastoral Willamette Valley, Oregon acted. The state legislature passed Senate Bill 100, a sweeping land use law that would become the focus of acclaim, criticism, and careful study for years afterward by other states and communities trying to manage their own urban and suburban growth.

Oregon's law required every city and county to draw up a comprehensive land use plan in line with statewide guidelines that had gained popular support from the state's electorate. As part of each land use plan, each metropolitan area had to establish an **urban growth boundary** (UGB), a line on a map separating areas desired to be urban from areas desired to remain rural. Development for housing, commerce, and industry would be encouraged within these urban growth boundaries but restricted beyond them. The intent was to revitalize city centers; prevent suburban sprawl; and protect farmland, forests, and open landscapes around the edges of urbanized areas.

Residents of the area around Portland, the state's largest city, established a new regional planning entity to apportion land in their region. The Metropolitan Service District, or Metro, represents 25 municipalities and three counties. Metro adopted the Portland-area urban growth boundary in 1979 and has tried to focus growth on existing urban centers and to build communities where people can walk, bike, or take mass transit between home, work, and shopping. These policies have largely worked as intended. Portland's downtown and older neighborhoods have thrived, regional urban centers are becoming denser and more community oriented, mass transit has expanded, and farms and forests have been preserved on land beyond the UGB. Portland began attracting international attention for its "livability."

To many Portlanders today, the UGB remains the key to maintaining quality of life in city and countryside alike. In the view of its critics, however, the "Great Wall of Portland" is an elitist and intrusive government

Mount Hood overlooking downtown Portland ▲

Upon completing this chapter, you will be able to:

- Describe the scale of urbanization
- Define sprawl and discuss its causes and consequences
- Outline city and regional planning and land use strategies
- Evaluate transportation options, urban parks, and green buildings
- Analyze environmental impacts and advantages of urban centers
- Assess urban ecology and the pursuit of sustainable cities

Concert below the St. John's Bridge in Portland, Oregon

regulatory tool. In 2004, Oregon voters approved a ballot measure that threatened to eviscerate their state's land use rules. Ballot Measure 37 required the state to compensate certain landowners if government regulation had decreased the value of their land. For example, regulations prevent landowners outside UGBs from subdividing their lots and selling them for housing development. Under Measure 37, the state had to pay these landowners to make up for theoretically lost income or else allow them to ignore the regulations. Because state and local governments did not have enough money to pay such claims, the measure was on track to gut Oregon's zoning, planning, and land use rules.

Landowners filed more than 7500 claims for payments or waivers affecting 295,000 ha (730,000 acres). Although the measure had been promoted to voters as a way to protect the rights of small family landowners, most claims were filed by large developers. Neighbors suddenly found themselves confronting the prospect of massive housing subdivisions, gravel mines, strip malls, or industrial facilities being developed next to their homes—and many who had voted for Measure 37 began to have misgivings.

The state legislature, under pressure from opponents and supporters alike, settled on a compromise: to introduce a new ballot measure. Oregon's voters passed Ballot Measure 49 in 2007. It restricts development outside the UGB that is on a large scale or that harms sensitive natural areas, but it protects the rights of small landowners to gain income from their property by developing small numbers of homes.

In 2010, Metro finalized a historic agreement with its region's three counties to determine where urban growth will be allowed over the next 50 years. Metro and the counties apportioned more than 121,000 ha (300,000 acres) of undeveloped land into "urban reserves" open for development and "rural reserves" where farmland and forests would be preserved. Boundaries were precisely mapped to give clarity and direction for landowners and governments alike.

People are confronting similar issues in communities throughout North America, and debates and negotiations like those in Oregon will determine how our cities and landscapes will change in the future.

Our Urbanizing World

In 2009, we passed a turning point in human history. For the first time ever, more people were living in urban areas (cities and suburbs) than in rural areas. As we undergo this historic shift from the countryside into towns and cities—a process called **urbanization**—two pursuits become ever more important. One is to make our urban areas more livable by meeting residents' needs for a safe, clean, healthy urban environment. The other is to make urban areas sustainable by creating cities that can prosper in the long term while minimizing our ecological impacts.

Industrialization has driven urbanization

Since 1950, the world's urban population has multiplied by more than five times, whereas the rural population has not even doubled. Urban populations are growing because the human population overall is growing (Chapter 6), and because more people are moving from farms to cities than are moving from cities to farms. Industrialization (p. 5) has reduced the need for farm labor while enhancing commerce and jobs in cities. Urbanization, in turn, has bred technological advances that boost production efficiencies and spur further industrialization.

United Nations demographers project that the urban population will rise by 63% between now and 2050, whereas the rural population will decline by 4%. Trends differ between developed and developing nations, however (**FIGURE 18.1**). In developed nations, urbanization has slowed because three of every four people already live in cities, towns, and **suburbs**, the smaller communities that ring cities. In contrast, today's developing nations, where most people still reside on farms, are urbanizing rapidly. In China, India, Pakistan, Nigeria,

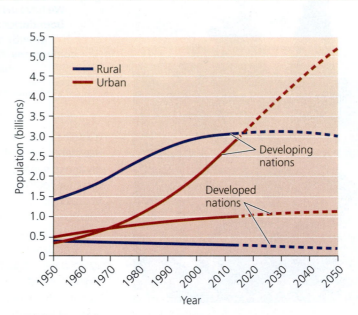

FIGURE 18.1 Population trends differ between poor and wealthy nations. In developing nations, urban populations are growing quickly, and rural populations will soon begin declining. Developed nations are already largely urbanized, so their urban populations are growing slowly, whereas rural populations are falling. Solid lines in the graph indicate past data, and dashed lines indicate future projections. *Data from UN Population Division, 2015.* World urbanization prospects: The 2014 revision. *By permission.*

 Beginning in what decade will the majority of people in developing nations be living in urban areas?

Go to **Interpreting Graphs & Data** on **Mastering** Environmental Science

and most other nations, rural people are streaming to cities in search of jobs and urban lifestyles, or to escape ecological degradation in the countryside. As a result, demographers estimate that urban areas of developing nations will absorb nearly all of the world's population growth from now on.

(a) St. Louis, Missouri

(b) Fort Worth, Texas

FIGURE 18.2 Cities develop along trade corridors. St. Louis **(a)** is situated on the Mississippi River near its confluence with the Missouri River, where river trade drove its growth in the 19th and early 20th centuries. Fort Worth, Texas **(b)**, grew in the late 20th century as a result of the interstate highway system and a major international airport.

Environmental factors influence the location of urban areas

Real estate agents use the saying, "Location, location, location," to stress how a home's setting determines its value. Location is vital for urban centers as well. Think of any major city, and chances are it's situated along a major river, seacoast, railroad, or highway—some corridor for trade that has driven economic growth (**FIGURE 18.2**).

Well-located cities often serve as linchpins in trading networks, funneling in resources from agricultural regions, processing them, manufacturing products, and shipping those products to other markets. Portland is situated where the Willamette River joins the Columbia River, just upriver from where the Columbia flows into the Pacific Ocean. The city grew as it received, processed, and shipped overseas the produce from farms of the river valleys, and as it imported products shipped in from other ports.

Today, powerful technologies and cheap transportation enabled by fossil fuels have allowed cities to thrive even in resource-poor regions. The Dallas–Fort Worth area prospers from—and relies on—oil-fueled transportation by interstate highways and a major airport. Southwestern cities such as Los Angeles, Las Vegas, and Phoenix flourish in desert regions by appropriating water from distant sources. Whether such cities

can sustain themselves as oil and water become increasingly scarce in the future is an important question.

In recent years, many cities in the southern and western United States have grown as people have moved there in search of warmer weather, more space, new economic opportunities, or places to retire. Between 1990 and 2016, the population of the Dallas–Fort Worth and Houston metropolitan areas each grew by about 80%; that of the Atlanta area grew by 96%; that of the Phoenix region grew by 108%; and that of the Las Vegas metropolitan area grew by a whopping 153%.

People have moved to suburbs

American cities grew rapidly in the 19th and early 20th centuries as a result of immigration from abroad and increased trade as the nation expanded west. The bustling economic activity of downtown districts held people in cities despite crowding, poverty, and crime. However, by the mid-20th century, many affluent city dwellers were choosing to move outward to cleaner, less crowded suburban communities. These people were pursuing more space, better economic opportunities, cheaper real estate, less crime, and better schools for their children.

As affluent people moved to the suburbs, jobs followed. This hastened the economic decline of downtown districts, and American cities stagnated. Chicago's population declined to 80% of its peak as residents moved to its suburbs. Philadelphia's population fell to 76% of its peak, Washington, D.C.'s to 71%, and Detroit's to just 55%.

Portland followed this trajectory: Its population growth stalled in the 1950s to 1970s as crowding and deteriorating economic conditions drove city dwellers to the suburbs. But the city bounced back. Policies to revitalize the city center helped reboot Portland's growth (**FIGURE 18.3**).

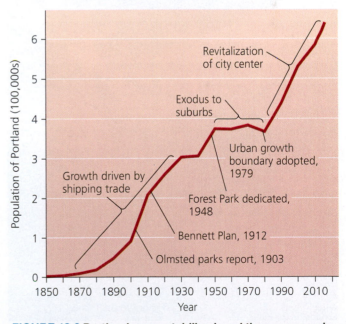

FIGURE 18.3 Portland grew, stabilized, and then grew again. Jobs in the shipping trade boosted Portland's economy and population in the 1890s–1920s. City residents began leaving for the suburbs in the 1950s–1970s, but policies to enhance the city center revitalized Portland's growth. *Data from U.S. Census Bureau.*

The exodus to the suburbs in 20th-century America was aided by the rise of the automobile, an expanding road network, and inexpensive and abundant oil. Millions of middle-class people could now commute by car to downtown workplaces from new homes in suburban "bedroom communities." By facilitating long-distance transport, fossil fuels and highway networks also made it easier for businesses to import and export resources, goods, and waste. The federal government's development of the interstate highway system was pivotal in promoting these trends.

In most ways, suburbs have delivered the qualities people sought in them. The wide spacing of homes, with each one on its own plot of land, has given families room and privacy. However, by allotting more space to each person, suburban growth has spread human impact across the landscape. Natural areas have disappeared as housing developments are constructed. Our extensive road networks have eased travel, but we find ourselves needing to climb into a car to get anywhere. People commute longer distances to get to work, shops, and amenities, and they spend more time stuck in traffic. The expanding rings of suburbs surrounding cities have grown larger than the cities themselves, and towns are merging into one another. These aspects of growth inspired a new term: *sprawl*.

Sprawl

The term *sprawl* has become laden with meanings and suggests different things to different people, but we can begin our discussion by giving **sprawl** a simple, nonjudgmental definition: the spread of low-density urban or suburban development outward from an urban center.

Urban areas spread outward

The spatial growth of urban and suburban areas is clear from maps and satellite images of rapidly spreading cities such as Las Vegas (**FIGURE 18.4**). Another example is Chicago, whose metropolitan area spreads over a region 40 times the size of the city. All in all, houses and roads supplant more than 2700 ha (6700 acres) of U.S. land every day.

Sprawl results from development approaches that place homes on spacious lots in residential tracts that spread over large areas but are far from urban centers and commercial amenities (**FIGURE 18.5**). Such approaches allot each person more space than in cities. For example, the average resident of Chicago's suburbs takes up 11 times more space than a resident of the city. As a result, the outward spatial growth of suburbs generally outpaces growth in numbers of people.

In fact, many researchers define *sprawl* as the physical spread of development at a rate that exceeds the rate of population growth. For instance, the population of Phoenix grew 12 times larger between 1950 and 2000, yet its land area grew 27 times larger. Between 1950 and 1990, the population of 58 major U.S. metropolitan areas rose by 80%, but the land area they covered rose by 305%. Even in 11 metro areas where population declined between 1970 and 1990 (as with Rust Belt cities such as Detroit, Cleveland, and Pittsburgh), the amount of land covered increased.

Sprawl has several causes

There are two main components of sprawl. One is human population growth—quite simply, more of us are alive each year.

(a) Las Vegas, Nevada, 1986   (b) Las Vegas, Nevada, 2013

FIGURE 18.4 Satellite images show the rapid urban and suburban expansion commonly called sprawl. Las Vegas, Nevada, is one of the fastest-growing cities in North America. Between 1986 **(a)** and 2013 **(b)**, its population and its developed area each tripled.

FIGURE 18.5 Sprawl is characterized by the spread of development across large areas of land. This requires people to drive cars to reach commercial amenities or community centers.

The other is per capita land consumption—each person is taking up more land than in the past, because most people desire space and privacy and dislike congestion. Better highways, inexpensive gasoline, telecommunications, and the Internet have all fostered movement away from city centers by giving workers more flexibility to live where they desire and by freeing businesses from dependence on the centralized infrastructure a city provides.

Economists and politicians have encouraged the unbridled spatial expansion of cities and suburbs. The conventional assumption has been that growth is always good and that attracting business, industry, and residents will enhance a community's economic well-being, political power, and cultural influence. Today, this assumption is being challenged as growing numbers of people feel negative effects of sprawl on their lifestyles.

What is wrong with sprawl?

To some people, the word *sprawl* evokes strip malls, traffic jams, homogenous commercial development, and tracts of cookie-cutter houses encroaching on farmland, ranchland, or forests. For other people, sprawl is simply the collective result of choices made by millions of well-meaning individuals trying to make a better life for their families. What does scientific research tell us about the impacts of sprawl?

Transportation Most studies show that sprawl constrains transportation options, essentially forcing people to own a vehicle, drive it most places, drive greater distances, and spend more time in it. Sprawling communities suffer more traffic accidents and have few or no mass transit options. Across the United States in the 1980s and 1990s, the average length of work trips rose by 36%, and total vehicle miles driven rose three times faster than population growth. A car-oriented culture encourages congestion and increases dependence on oil.

Pollution By promoting automobile use, sprawl increases pollution. Carbon dioxide emissions from vehicles contribute to climate change (Chapter 14), and nitrogen- and sulfur-containing air pollutants lead to smog and acid deposition (Chapter 13). Motor oil and road salt from roads and parking lots run off readily and pollute waterways.

Health Beyond the health impacts of pollution, some research suggests that sprawl promotes physical inactivity and obesity because driving cars largely takes the place of walking during daily errands. A 2003 study found that people from the most-sprawling U.S. counties show higher blood pressure and weigh 2.7 kg (6 lb) more for their height than people from the least-sprawling U.S. counties.

Land use As more land is developed, less is left as forests, fields, farmland, or ranchland. Natural lands and agricultural lands provide vital resources, recreation, aesthetic beauty, wildlife habitat, and air and water purification. Many children now grow up without the ability to roam through woods and fields, which used to be a normal part of childhood.

Economics Sprawl drains tax dollars from communities and funnels money into infrastructure for new development on the fringes of those communities. Funds that could be spent maintaining downtown centers are instead spent on extending the road system, water and sewer system, electricity grid, telephone lines, police and fire service, schools, and libraries. Although taxes on new development can in theory pay back the investment, studies find that in most cases taxpayers end up subsidizing new development.

weighing the ISSUES

Sprawl Near You

Is there sprawl in the area where you live? Does it bother you, or not? Has development in your area had any of the impacts described here? Do you think your city or town should encourage outward growth? Why or why not?

Creating Livable Cities

To respond to the challenges presented by sprawl, architects, planners, developers, and policymakers are trying to revitalize city centers and to plan and manage how urbanizing areas develop. They aim to make cities safer, cleaner, healthier, and more pleasant for their residents.

Planning helps to create livable urban areas

How can we design cities to maximize their efficiency, functionality, and beauty? These are the questions central to **city planning** (also known as **urban planning**). City planners advise policymakers on development options, transportation needs, public parks, and other matters.

Washington, D.C., is the earliest example of city planning in the United States. President George Washington hired French architect Pierre Charles L'Enfant in 1791 to design a capital city for the new nation on undeveloped land along the Potomac River. L'Enfant laid out a Baroque-style plan of diagonal avenues cutting across a grid of streets, with plenty of space allotted for majestic public monuments (**FIGURE 18.6**). A century later, a new generation of planners imposed a height restriction on new buildings to keep the monuments from being crowded and dwarfed by modern skyscrapers. This preserved the spacious, stately feel of the city.

City planning in North America came into its own at the turn of the 20th century, as urban leaders sought to beautify and impose order on fast-growing, unruly cities. In Portland in 1912, planner Edward Bennett's *Greater Portland Plan* proposed to rebuild the harbor; dredge the river channel; construct new docks, bridges, tunnels, and a waterfront railroad; superimpose wide radial boulevards on the old city street grid; establish civic centers downtown; and greatly expand the number of parks.

In today's world of sprawling metropolitan areas, **regional planning** has become more and more important.

(a) The L'Enfant plan, 1791

(b) Washington, D.C., today

FIGURE 18.6 Washington, D.C., is an example of early city planning. The 1791 plan **(a)** for the new U.S. capital laid out splendid diagonal avenues cutting across gridded streets, allowing space for the magnificent public monuments **(b)** that grace the city today.

Regional planners deal with the same issues as city planners, but they work on broader geographic scales and coordinate their work with multiple municipal governments. In some places, regional planning has been institutionalized in formal governing bodies; the Portland area's Metro is the epitome of such a regional planning entity. When Metro and its region's three counties in 2010 announced their collaborative plan apportioning 121,000 ha (300,000 acres) of undeveloped land into "urban reserves" and "rural reserves," it marked a historic accomplishment in regional planning. The agreement enables homeowners, farmers, developers, and policymakers to feel informed and secure knowing what kinds of land uses lie in store on and near their land over the next half-century.

weighing the
ISSUES

Your Urban Area

Think of your favorite parts of the city you know best. What do you like about them? What do you dislike about your least favorite parts of the city? What could this city do to improve quality of life for its residents?

Zoning is a key tool for planning

One tool that planners use is **zoning,** the practice of classifying areas for different types of development and land use. For instance, to preserve the cleanliness and tranquility of residential neighborhoods, industrial facilities may be kept out of districts zoned for residential use. By specifying zones for different types of development, planners can guide what gets built where. Zoning also gives home buyers and business owners security, because they know in advance what types of development can and cannot be located nearby.

Zoning involves government restriction on the use of private land and represents a top-down constraint on personal property rights. Yet most people feel that government has a proper and useful role in setting limits on property rights for the good of the community. When Oregon voters passed Ballot Measure 37 in 2004, this shackled government's ability to enforce zoning regulations with landowners who bought their land before the regulations were enacted. However, many Oregonians soon began witnessing new development they did not condone, so in 2007 they passed Ballot Measure 49 to restore public oversight over development. In general, people have supported zoning over the years because the common good it produces for communities is widely felt to outweigh the restrictions on private use.

Urban growth boundaries are now widely used

Planners intended Oregon's urban growth boundaries (UGBs) to limit sprawl by containing growth within existing urbanized areas. The UGBs aimed to revitalize downtowns; preserve working farms, orchards, ranches, and forests; and ensure urban dwellers access to open space (**FIGURE 18.7**). UGBs also save taxpayers money by reducing the amounts that municipalities need to pay for infrastructure. Since Oregon instituted

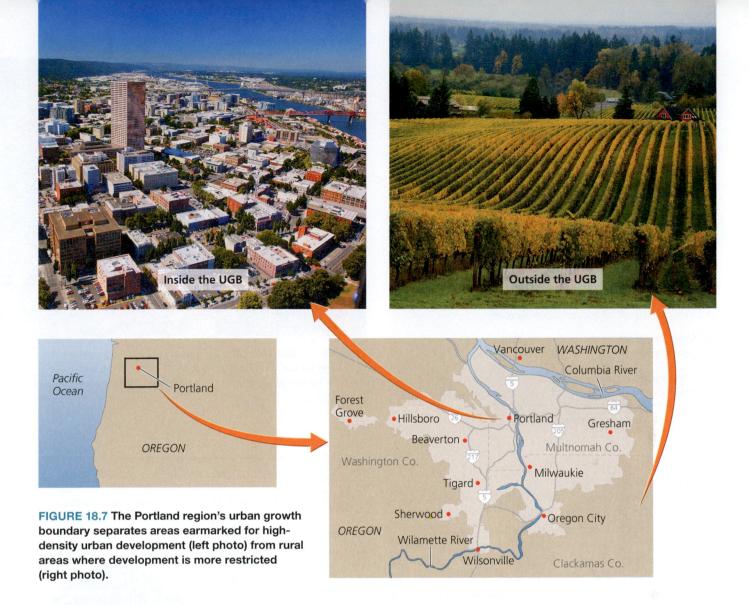

FIGURE 18.7 The Portland region's urban growth boundary separates areas earmarked for high-density urban development (left photo) from rural areas where development is more restricted (right photo).

its policies, many other states, regions, and cities have adopted UGBs—from Boulder, Colorado, to Lancaster, Pennsylvania, to many California communities.

In most ways, the Portland region's urban growth boundary has worked as intended. It has preserved farms and forests outside the UGB while increasing the density of new housing inside it. Within the UGB, homes are built on smaller lots and multistory apartments replace low-rise structures. Downtown employment has grown as businesses and residents invest in the central city.

Nonetheless, the Portland region's urbanized area grew by 101 km² (39 mi²) in the decade after its UGB was established, because 146,000 people were added to the population. Rising population pressure has led Metro to enlarge the UGB three dozen times since its establishment. In addition, UGBs tend to increase housing prices within their boundaries, and in Portland, housing has become far less affordable. Today in the city, demand for housing exceeds supply, rents are soaring, and low- and middle-income people are being forced out of neighborhoods they have lived in for years as these neighborhoods experience **gentrification,** a transformation to conditions that cater to wealthier people. These trends suggest that relentless population growth may thwart even the best anti-sprawl efforts and that livable cities can fall victim to their own success if they are in high demand as places to live.

"Smart growth" and "new urbanism" aim to counter sprawl

As more people feel impacts of sprawl on their everyday lives, efforts to manage growth are springing up throughout North America. Proponents of **smart growth** seek to rejuvenate the older existing communities that so often are drained and impoverished by sprawl. Smart growth means "building up, not out"—focusing development and economic investment in existing urban centers and favoring multistory shophouses and high-rises (**TABLE 18.1**).

A related approach among architects, planners, and developers is **new urbanism,** which seeks to design walkable neighborhoods with homes, businesses, schools, and other amenities all nearby for convenience. The aim is to create functional neighborhoods in which families can meet most of their needs close to home without using a car. These neighborhoods are often served by public transit systems, enabling people to travel most places they need to go by train and foot.

TABLE 18.1 Ten Principles of "Smart Growth"

- Mix land uses
- Take advantage of compact building design
- Create a range of housing opportunities and choices
- Create walkable neighborhoods
- Foster distinctive, attractive communities with a strong sense of place
- Preserve open space, farmland, natural beauty, and critical environmental areas
- Strengthen existing communities, and direct development toward them
- Provide a variety of transportation choices
- Make development decisions predictable, fair, and cost-effective
- Encourage community and stakeholder collaboration in development decisions

Source: U.S. Environmental Protection Agency.

FIGURE 18.8 Bicycles provide a healthy alternative to transportation by car. This Portland bicycle lot accommodates riders who commute downtown by bike, and is conveniently located at a streetcar stop.

Transit options help cities

Traffic jams on roadways cause air pollution, stress, and countless hours of lost time. They cost Americans an estimated $74 billion yearly in fuel and lost productivity. To encourage more efficient transportation, policymakers can raise fuel taxes, charge trucks for road damage, tax inefficient modes of transport, and reward carpoolers with carpool lanes. But a key component of improving the quality of urban life is to give residents alternative transportation options.

Bicycle transportation is one key option (**FIGURE 18.8**). Portland has embraced bicycles like few other American cities, and today 6% of its commuters ride to work by bike (the national average is 0.5%). The city has developed nearly 400 miles of bike lanes and paths, 5000 public bike racks, and special markings at intersections to protect bicyclists. Amazingly, all this infrastructure was created for the typical cost of just 1 mile of urban freeway. Portland also has a bike-sharing program similar to programs in cities such as Montreal, Toronto, Denver, Minneapolis, Miami, San Antonio, Boston, and Washington, D.C.

Other transportation options include **mass transit** systems: public systems of buses, trains, subways, or *light rail* (smaller rail systems powered by electricity). Mass transit systems move large numbers of passengers at once while easing traffic congestion, taking up less space than road networks, and emitting less pollution than cars. Studies show that as long as an urban center is large enough to support the infrastructure necessary, both train and bus systems are cheaper, more energy-efficient, and less polluting than roadways choked with cars (**FIGURE 18.9**).

In Portland, buses, light rail, and streetcars together carry 100 million riders per year. America's most-used train

SUCCESS STORY   Creating A World-Class Transit System

Establishing a mass transit system is not easy. Strong and visionary political leadership may be required. Such was the case in Curitiba, a metropolis of 2.5 million people in southern Brazil. Faced with an influx of immigrants from outlying farms in the 1970s, city leaders led by Mayor Jaime Lerner undertook an aggressive planning process so that they could direct their city's growth rather than being overwhelmed by it. They established a large fleet of public buses, drew up measures to encourage bicycles and pedestrians, and reconfigured Curitiba's road system to maximize its efficiency. Today Curitiba

Commuters boarding a bus in Curitiba, Brazil.

has an internationally renowned bus system with 340 routes, 250 terminals, and 1900 buses. Each day the system is used by three-quarters of its population. All of this has resulted in a steep drop in car use, despite a rapidly growing population. Besides its outstanding bus network, Curitiba provides recycling, environmental education, job training for the poor, and free health care. Surveys show that its residents are unusually happy and better off economically than people in other Brazilian cities. Curitiba's lesson for the rest of the world is that investing thoughtfully in well-planned transportation infrastructure can pay big dividends.

EXPLORE THE DATA at Mastering Environmental Science

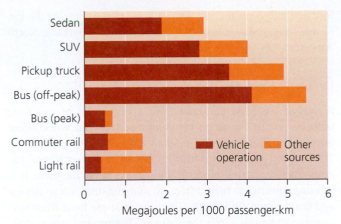

(a) Energy consumption for different modes of transit

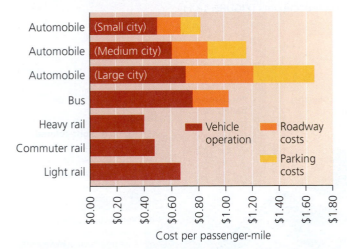

(b) Operating costs for different modes of transit

FIGURE 18.9 Rail transit consumes less energy (a) and costs less (b) than automobile transit. Bus transit is highly efficient in places and at times of high use ("peak" in figure). Data on greenhouse gas emissions (not shown) are very similar in their patterns to data in the part (a) graph. *Data from (a) Chester, M., and A. Horvath, 2009. Environmental assessment of passenger transportation should include infrastructure and supply chains.* Environmental Research Letters 4: 024008 (8 pp); and (b) Litman, T., 2005. Rail transit in America: A comprehensive evaluation of benefits. © 2005 T. Litman.

DATA Q • In part **(a)**, describe how the energy consumed by driving an SUV compares with the energy consumed by taking commuter rail. • How would you predict the greenhouse gas emissions for each mode of transport would compare? • In part **(b)**, automobile traffic creates two types of operating costs that are *not* created by rail traffic. What are they?

Go to **Interpreting Graphs & Data** on **Mastering** Environmental Science

systems are those of its largest cities, such as New York City's subways, Washington, D.C.'s Metro, the T in Boston, and the San Francisco Bay Area's BART. Each of these carries more than one-fourth of its city's daily commuters.

In general, however, the United States lags behind most nations in mass transit. Many countries, rich and poor alike, have extensive bus systems that are easily accessible to people. Japan, China, and many European nations have entire systems of modern high-speed "bullet trains." The United

States chose instead to invest in road networks for cars and trucks largely because (relative to most other nations) its population density was low and gasoline was cheap. As energy costs and population rise, however, mass transit becomes increasingly appealing, and people begin to desire train and bus systems in their communities.

Urban residents need parklands

City dwellers often desire some escape from the noise, commotion, and stress of urban life. Natural lands, public parks, and open space provide greenery, scenic beauty, freedom of movement, and places for recreation. These lands also keep ecological processes functioning by helping to regulate climate, purify air and water, and provide wildlife habitat.

One newly developed and instantly popular city park is The High Line Park in Manhattan in New York City (**FIGURE 18.10**). An elevated freight line running above the streets was going to be demolished, but a group of citizens saw its potential for a park, and they pushed the idea until city leaders came to share their vision. Today thousands of people use the 23-block-long High Line for recreation or on their commute to work.

Large city parks are vital to a healthy urban environment, but even small spaces make a difference. Playgrounds give children places to be active outdoors and interact with their peers. Community gardens allow people to grow vegetables and flowers in a neighborhood setting. "Greenways" along

FIGURE 18.10 The High Line Park was created thanks to a visionary group of Manhattan residents. They pushed to make a park out of an abandoned elevated rail line.

rivers, canals, or old railway lines can provide walking trails, protect water quality, boost property values, and serve as corridors for the movement of wildlife.

America's city parks arose in the late 19th century as urban leaders established public spaces using aesthetic ideals from European parks and gardens. The lawns, shaded groves, and curved pathways of many American city parks grew from these ideals, as interpreted by America's leading landscape architect, Frederick Law Olmsted. Olmsted designed Central Park in New York City and many other parks.

Portland's quest for parks began in 1900, when city leaders created a parks commission and hired Olmsted's son, John Olmsted, to design a park system. His 1904 plan proposed acquiring land to ring the city generously with parks, but no action was taken. A full 44 years later, residents pressured city leaders to create Forest Park along a

forested ridge on the northwest side of the city. At 11 km (7 mi) long, today it is one of the largest city parks in North America.

Green buildings bring benefits

Although we need parklands, we spend most of our time indoors, so the buildings in which we live and work affect our health. Buildings also consume 40% of our energy and 70% of our electricity, contributing to the greenhouse gas emissions that drive climate change. As a result, there is a thriving movement in architecture and construction to design and build **green buildings,** structures that are built from sustainable materials, limit their use of energy and water, control pollution, recycle waste, and minimize health impacts on their occupants (**FIGURE 18.11**).

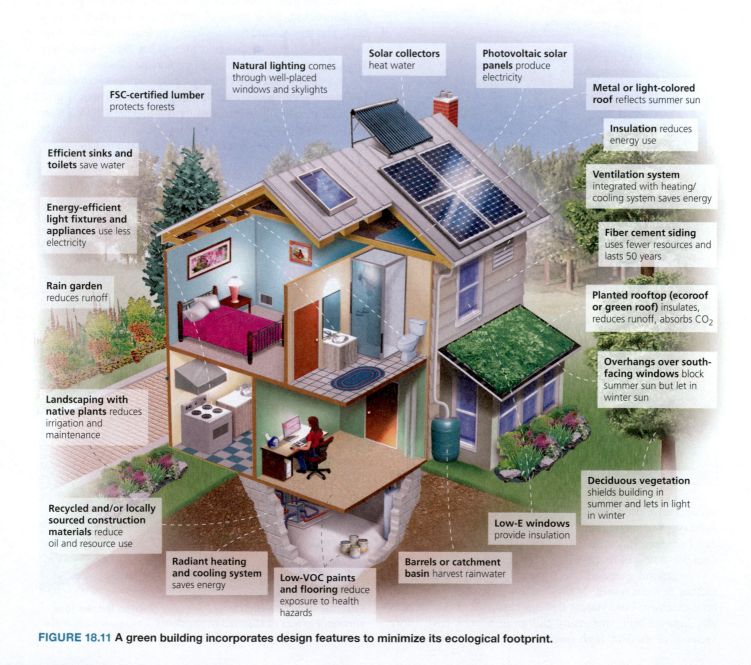

Natural lighting comes through well-placed windows and skylights

Solar collectors heat water

Photovoltaic solar panels produce electricity

FSC-certified lumber protects forests

Metal or light-colored roof reflects summer sun

Insulation reduces energy use

Efficient sinks and toilets save water

Ventilation system integrated with heating/cooling system saves energy

Energy-efficient light fixtures and appliances use less electricity

Fiber cement siding uses fewer resources and lasts 50 years

Rain garden reduces runoff

Planted rooftop (ecoroof or green roof) insulates, reduces runoff, absorbs CO_2

Landscaping with native plants reduces irrigation and maintenance

Overhangs over south-facing windows block summer sun but let in winter sun

Recycled and/or locally sourced construction materials reduce oil and resource use

Deciduous vegetation shields building in summer and lets in light in winter

Radiant heating and cooling system saves energy

Low-VOC paints and flooring reduce exposure to health hazards

Barrels or catchment basin harvest rainwater

Low-E windows provide insulation

FIGURE 18.11 A green building incorporates design features to minimize its ecological footprint.

The U.S. Green Building Council promotes sustainable building efforts through the **Leadership in Energy and Environmental Design (LEED)** certification program. Builders apply for certification (for new buildings or renovation projects) and, depending on their performance, may be granted silver, gold, or platinum status.

Green building techniques add expense to construction, yet LEED certification is booming. Portland features several dozen LEED-certified buildings, including the nation's first LEED-gold sports arena (the Moda Center, where the Trailblazers basketball team plays). The savings on energy, water, and waste at the refurbished Moda Center paid for the cost of its LEED upgrade after just one year.

Schools, colleges, and universities are leaders in sustainable building. In Portland, the Rosa Parks Elementary School was built with locally sourced and nontoxic materials, uses 24% less energy and water than comparable buildings, and diverted nearly all of its construction waste from the landfill. Schoolchildren learn about renewable energy in their own building by watching a display of the electricity produced by its photovoltaic solar system. Portland State University, the University of Portland, Reed College, and Lewis and Clark College are just a few of the many colleges and universities nationwide that are constructing green buildings as a part of their campus sustainability efforts (pp. 19, 435).

Urban Sustainability

Most of our efforts to make cities safer, cleaner, healthier, and more beautiful also help to make them more sustainable. A sustainable city is one that can function and prosper over the long term, providing generations of residents a good quality of life far into the future. In part, this entails minimizing the city's impacts on the natural systems and resources that nourish it. It also entails viewing the city itself as an ecological system (see **THE SCIENCE BEHIND THE STORY**, pp. 430–431).

Urban centers bring a mix of environmental effects

You might guess that urban living has a greater environmental impact than rural living. However, the picture is not so simple; instead, urbanization brings a complex mix of consequences.

Pollution Any location with large numbers of people is going to generate some amount of waste and pollution. Indeed, urban dwellers are exposed to smog, toxic industrial compounds, fossil fuel emissions, noise pollution, and light pollution. City residents even suffer thermal pollution, in the form of the **urban heat island effect** (**FIGURE 18.12**). Pollution and the health risks it poses are not evenly shared among residents. As environmental justice advocates point out (p. 16), those who receive the brunt of pollution are often those who are too poor to live in cleaner areas.

Cities export some of their wastes. In so doing, they transfer the costs of their activities to other regions—and mask the costs from their own residents. Citizens of Indianapolis, Columbus, or Buffalo may not realize that pollution from nearby coal-fired power plants worsens acid rain hundreds of miles to the east. New York City residents may not recognize how much garbage their city produces if it is shipped elsewhere for disposal.

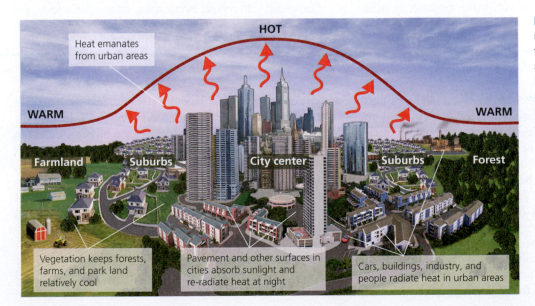

FIGURE 18.12 Cities produce urban heat islands, creating temperatures warmer than surrounding areas. People, buildings, and vehicles generate heat, and buildings and dark paved surfaces absorb daytime heat and release it at night.

HOT

Heat emanates from urban areas

WARM

WARM

Farmland   Suburbs   City center   Suburbs   Forest

Vegetation keeps forests, farms, and park land relatively cool

Pavement and other surfaces in cities absorb sunlight and re-radiate heat at night

Cars, buildings, industry, and people radiate heat in urban areas

Do Baltimore and Phoenix Act as Ecosystems?

Researchers in urban ecology examine how ecosystems function in cities and suburbs, how natural systems respond to urbanization, and how people interact with the urban environment. Today, Baltimore and Phoenix are centers for urban ecology.

These two cities are very different: Baltimore is a port city on the Chesapeake Bay with a long history, whereas Phoenix is a young and fast-growing southwestern metropolis sprawling across the desert. Each was selected by the U.S. National Science Foundation as a research site in its Long Term Ecological Research program, which funds multidecade ecological research. Since 1997, hundreds of researchers have studied Baltimore and Phoenix explicitly as ecosystems, examining nutrient cycling, biodiversity, air and water quality, environmental health threats, and more.

Sampling water beneath an overpass in Baltimore.

Research teams in both cities are combining old maps, aerial photos, and new remote sensing satellite data to reconstruct the history of landscape change. In Phoenix, one group showed how urban development spread across the desert in a "wave of advance," affecting soils, vegetation, and microclimate as it went. In Baltimore, mapping showed that development fragmented the forest into smaller patches over the past 100 years, even while the overall amount of forest remained the same.

For each city, study regions encompass both heavily urbanized central city areas and outlying rural and natural areas. To measure the impacts of urbanization, many research projects compare conditions in these two types of areas.

Baltimore scientists can see ecological effects of urbanization by comparing the urban lower end of their site's watershed with its less developed upper end. In the lower end, pavement, rooftops, and compacted soil prevent rain from infiltrating the soil, so water runs off quickly. The rapid flow cuts streambeds deeply into the earth while leaving surrounding soil drier. As a result, wetland-adapted trees and shrubs are vanishing, replaced by dry-adapted upland trees and shrubs.

The fast flow of water also worsens pollution. In natural areas, streams and wetlands filter pollution by breaking down nitrogen compounds. But in urban areas, where wetlands dry up and runoff from pavement creates flash floods, streams lose their filtering ability. In Baltimore, the resulting pollution ends up in the Chesapeake Bay, which suffers eutrophication and a large hypoxic dead zone (p. 28). Baltimore scientists studying nutrient cycling (p. 39) found that urban and suburban watersheds have far more nitrate pollution than natural forests (**FIGURE 1**).

Baltimore research also reveals impacts of applying salt to icy roads in winter. Road salt makes its way into streams, which become up to 100 times saltier. High salinity kills organisms (**FIGURE 2**), degrades habitat and water quality, and impairs streams' ability to remove nitrate.

To study contamination of groundwater and drinking water, researchers are using isotopes (p. 31) to trace where salts in the most polluted streams are coming from. Baltimore is now improving water quality substantially with a $900-million upgrade of its sewer system.

Urbanization also affects species and ecological communities. Cities and suburbs facilitate the spread of non-native species because people introduce exotic ornamental plants and because urban impacts on the soil, climate, and landscape favor weedy generalist species over more specialized native ones.

Compared with natural landscapes, cities offer steady and reliable food resources—think of people's bird feeders, or food

Resource use and efficiency Cities are sinks (p. 40) for resources, importing nearly everything they need to feed, clothe, and house their inhabitants and power their commerce. People in cities such as New York, Boston, San Francisco, and Los Angeles depend on water pumped in from faraway watersheds. Urban communities rely on large expanses of land elsewhere for resources and ecosystem services, and they burn fossil fuels to import resources and goods.

However, imagine that the world's 4 billion urban residents were instead spread evenly across the landscape. What would the transportation requirements be, then, to move resources and goods around to all those people? A world

without cities would likely require *more* transportation to provide people the same degree of access to resources and goods. Moreover, once resources arrive, cities are highly efficient in distributing goods and services. The density of cities facilitates the provision of electricity, medical care, education, water and sewer systems, waste disposal, and public transportation. Thus, although a city has a large ecological footprint (p. 5), its residents may have moderate or small footprints in per capita terms.

Land preservation Because people pack densely together in cities, more land outside cities is left undeveloped. Indeed, this is the very idea behind urban growth

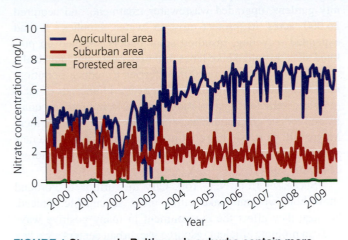

FIGURE 1 Streams in Baltimore's suburbs contain more nitrates than streams in nearby forests, but fewer than those in agricultural areas, where fertilizers are applied liberally. *Data from Baltimore Ecosystem Study, www.lternet.edu/research/keyfindings/urban-watersheds.*

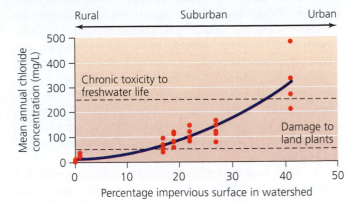

FIGURE 2 Salt concentrations in Baltimore-area streams are high enough to damage plants in the suburbs and to kill aquatic animals in urban areas. *Adapted from Kaushal, S. S., et al., 2005. Increased salinization of fresh water in the northeastern United States. Proc. Natl. Acad. Sci. USA 102: 13517–13520, Fig 2. ©2005 National Academy of Sciences, U.S.A. By permission.*

DATA • At what percentage of impervious (paved) surface would you expect to begin seeing damage to land plants as a result of salty road runoff entering streams? • At what percentage would you expect to begin seeing chronic toxicity to freshwater life?

Go to **Interpreting Graphs & Data** on **Mastering** Environmental Science

scraps from dumpsters. Growing seasons are extended and seasonal variation is buffered in cities, as well. The urban heat island effect (p. 429) raises nighttime temperatures and equalizes temperatures year-round. In a desert city like Phoenix, watering gardens boosts primary productivity and lowers daytime temperatures. Together, all these changes lead to higher population densities of animals but lower species diversity as generalists thrive and displace specialists.

Urban ecologists in Phoenix and Baltimore are studying social and demographic questions as well. Some research measures how natural amenities affect property values. One study found that proximity to a park increases a home's property values—unless crime is pervasive. If the robbery rate surpasses 6.5 times the national average, then proximity to a park begins to depress property values.

Other studies focus on environmental justice concerns (pp. 16–17). These studies have repeatedly found that sources of industrial pollution tend to be located in neighborhoods that are less affluent and that are home to people of racial and ethnic minorities. Phoenix researchers mapped patterns of air pollution and toxic chemical releases and found that minorities and the poor are exposed to a greater share of these hazards. As a result, they suffer from higher rates of childhood asthma.

Whether addressing the people, natural communities, or changing ecosystems of the urban environment, studies on urban ecology like those in Phoenix and Baltimore will be vitally informative in our ever more urban world.

boundaries. If all 7 billion of us were evenly spread across the planet's land area, no large blocks of land would be left uninhabited, and we would have far less room for agriculture, wilderness, biodiversity, or privacy. The fact that half the human population is concentrated in discrete locations helps allow space for natural ecosystems to continue functioning and provide the ecosystem services on which all of us, urban and rural, depend.

Innovation Cities promote a flourishing cultural life and, by mixing diverse people and influences, spark innovation and creativity. The urban environment can promote education and scientific research, and cities have long been viewed as engines of technological and artistic inventiveness. This inventiveness can lead to solutions to societal problems, including ways to reduce environmental impacts.

Urban ecology helps cities toward sustainability

Cities that import all their resources and export all their wastes have a linear, one-way metabolism. Linear models of production and consumption tend to destabilize environmental systems. Proponents of sustainability for cities stress the need to develop circular systems, akin to systems found in nature, which recycle materials and use renewable sources of energy. Researchers in the field of **urban ecology** hold that the fundamentals of ecosystem ecology and systems science (Chapter 2) apply to the urban environment. Urban ecologists suggest that cities follow an ecosystem-centered model by striving to maximize the efficient use of resources, recycle waste and wastewater, and develop green technologies. Urban agriculture that recycles organic waste, restores soil fertility, and produces locally consumed food is thriving in many places, from Portland to Cuba to Japan. Urban ecology research projects are ongoing in Baltimore and Phoenix, where scientists are studying these cities as ecological systems (see **The Science behind the Story.**)

In 2007, New York City unveiled an ambitious plan that then-Mayor Michael Bloomberg hoped would make it "the first environmentally sustainable 21st-century city." PlaNYC was a 132-item program to make New York City a better place to live as it accommodates 1 million more people by 2030. By 2014, the city had improved energy efficiency in 174 city buildings, planted 950,000 trees, opened or renovated 234 school playgrounds, established 129 community gardens, upgraded wastewater treatment, and acquired 36,000 acres to protect upstate water supplies. It had expanded recycling, installed solar panels, cleaned up polluted sites, installed bike lanes and bike racks and launched a bike-sharing program, retrofitted ferries to reduce pollution, introduced electric vehicles, and converted hundreds of taxis to hybrid vehicles. These actions helped to improve air quality and reduce greenhouse gas emissions by 19%. Since then, Mayor Bill de Blasio has continued the program under a new name, OneNYC, while adding new dimensions to address economic equity.

Successes from New York City to Curitiba to Portland show how we can make cities more sustainable. Indeed, because they affect the environment in many positive ways and can promote efficient resource use, urban centers are a key element in achieving progress toward global sustainability.

closing THE LOOP

As the human population shifts from rural to urban lifestyles, our environmental impacts become less direct but more far-reaching. Fortunately, we are developing sustainable solutions while making urban and suburban areas better places to live.

Portland, Oregon, is one city that has been enhancing the quality of life for its residents. However, as people stream to Portland in droves, the city risks becoming a victim of its own success. Growth forecasts estimate that the number of households in the Portland region will jump 56–74% by 2035. As density increases inside the urban growth boundary, new challenges—such as rising rents, increased traffic, and parking congestion in residential neighborhoods—are beginning to strain the smart-growth vision that has worked thus far. Portland's leaders engaged citizens in a planning process to design solutions to keep their city "prosperous, healthy, equitable, and resilient," and they recently completed a Comprehensive Plan to help guide decision making through 2035.

Portland is just one of many urban centers seeking to expand economic opportunity and enhance quality of life while protecting environmental quality. Planning and zoning, smart growth and new urbanism, mass transit, parks, and green buildings all are ingredients in sustainable cities, and we should be encouraged about our progress in these endeavors. Ongoing experimentation will help us determine how to continue creating better and more sustainable communities in which to live.

TESTING Your Comprehension

1. What factors lie behind the shift of population from rural areas to urban areas? What types of nations are experiencing the fastest urban growth today, and why?

2. Why have so many city dwellers in the United States and other developed nations moved into suburbs?

3. Give two definitions of *sprawl*. Describe five negative impacts that have been suggested to result from sprawl.

4. What is city planning? What is regional planning? Contrast planning with zoning. Give examples of some of the suggestions made by early planners such as Edward Bennett in Portland.

5. How are some people trying to prevent or slow sprawl? Describe some key elements of "smart growth." What effects, positive and negative, do urban growth boundaries tend to have?

6. Describe several apparent benefits of rail transit systems. What is a potential drawback?

7. How are parks thought to make urban areas more livable? Give three examples of types of parks or public spaces.

8. What is a green building? Describe several features a LEED-certified building may have.

9. Describe the connection between urban ecology and sustainable cities. List three actions a city can take to enhance its sustainability.

10. Name two positive effects of urban centers on the natural environment.

SEEKING Solutions

1. Describe the causes of the spread of suburbs, and outline the environmental, social, and economic impacts of sprawl. Overall, do you think the spread of urban and suburban development that is commonly labeled *sprawl* is predominantly a good thing or a bad thing? Do you think it is inevitable? Give reasons for your answers.

2. Would you personally want to live in a neighborhood developed in the new-urbanist style? Why or why not? Would you like to live in a city or region with an urban growth boundary? Why or why not?

3. All things considered, do you feel that cities are a positive thing or a negative thing for environmental quality? How much do you think we may be able to improve the sustainability of our urban areas?

4. **CASE STUDY CONNECTION** After you earn your college degree, you decide to settle in the Portland, Oregon, region, where you are being offered three equally desirable jobs, in three very different locations. If you accept the first, you will live in downtown Portland, amid commercial and cultural amenities but where population density is high and growing. If you take the second, you will live in one of Portland's suburbs, where you have more space but where commute times are long and sprawl may soon surround you for miles. If you select the third, you will live in a rural area outside the urban growth boundary with plenty of space and scenic beauty but few cultural amenities. You are a person who aims to live in an ecologically sustainable way. Where would you choose to live? Why? What considerations will you factor into your decision?

5. **THINK IT THROUGH** You are the facilities manager on your campus, and your school's administration has committed funds to retrofit one existing building with sustainable green construction techniques so that it earns LEED certification. Consider the various buildings on your campus, and select one that you feel is unhealthy or that wastes resources in some way and that you would like to see retrofitted. Describe for an architect three specific ways in which green building techniques might be used to improve this particular building.

CALCULATING Ecological Footprints

One way to reduce your ecological footprint is with alternative transportation. Each gallon of gasoline is converted during combustion to approximately 20 pounds of carbon dioxide (CO_2), which is released into the atmosphere. The table lists typical amounts of CO_2 released per person per mile using various modes of transportation, assuming typical fuel efficiencies.

For an average North American who travels 12,000 miles per year, calculate and record in the table the CO_2 emitted yearly for each transportation option. Then calculate and record the reduction in CO_2 emissions that one could achieve by relying solely on each option.

MODE OF TRANSPORT	CO₂ PER PERSON PER MILE	CO₂ PER PERSON PER YEAR	CO₂ EMISSION REDUCTION	YOUR ESTIMATED MILEAGE PER YEAR	YOUR CO₂ EMISSIONS PER YEAR
Automobile (driver only)	0.825 lb	9900 lb	0		
Automobile (2 persons)	0.413 lb				
Bus	0.261 lb				
Walking	0.082 lb				
Bicycle	0.049 lb				
				Total = 12,000	

1. Which transportation option provides the most miles traveled per unit of carbon dioxide emitted?

2. In the last two columns, estimate what proportion of the 12,000 annual miles you think that you actually travel by each method, and then calculate the CO_2 emissions that you are responsible for generating over the course of a year. Which transportation option accounts for the most emissions for you?

3. How could you reduce your CO_2 emissions? How many pounds of emissions do you think you could realistically eliminate over the course of the next year by making changes in your transportation decisions?

Mastering Environmental Science

Sustainable Solutions

The notion of sustainability has run throughout this book. In one case after another, we have seen how people are devising creative solutions to the dilemmas that arise when resources for future generations are being depleted. Our challenge as a society is to continue developing innovative and workable solutions that enhance our quality of life while protecting and restoring the natural environment that supports us.

Students are leading sustainability efforts on campus

In today's quest for sustainable solutions, students at colleges and universities are playing a crucial role. They are creating models for the wider world by leading sustainability initiatives on their campuses (TABLE E.1).

Students are pioneering ways to enhance efficiency in energy use and water use on their campuses. They are running programs to recycle and cut down on waste. They are growing organic gardens and restoring natural areas on campus. They are advocating with administrators for renewable energy, sustainable green buildings, and the reduction of greenhouse gas emissions. Most efforts are home-grown and local, like the sustainable food and dining initiatives at Kennesaw State University (pp. 139–140). Others are part of national or international programs, such as Recyclemania (pp. 399–400).

These diverse efforts by students, faculty, staff, and administrators are reducing the ecological footprints of college and university campuses. You can find links to resources for campus sustainability efforts in the *Selected Sources and References* section online at **MasteringEnvironmentalScience**.

We can develop sustainably

Whether on campus or around the world, sustainability means living in a way that can be lived far into the future. It involves conserving resources to prevent their depletion, reducing waste and pollution, and safeguarding ecological processes and ecosystem services, to ensure that our society's practices can continue and our civilization can endure. We depend on a healthy and functioning natural environment for our basic needs and our quality of life, so protecting Earth's natural capital is vital. Yet sustainability also means promoting economic well-being and social justice. Meeting this triple bottom line (p. 115) is the goal of sustainable development. It is our primary challenge for this century and likely for the rest of our species' time on Earth.

Environmental protection can enhance economic opportunity

Our society has long labored under the misconception that economic well-being and environmental protection are in conflict. In reality, our well-being depends on a healthy environment, and protecting environmental quality can improve our economic bottom line.

For individuals, businesses, and institutions, reducing resource consumption and waste often saves money. For society, promoting environmental quality can enhance economic opportunity by creating new types of employment. As we transition to a more sustainable economy, some resource-extraction industries may decline, but a variety of recycling-oriented and high-technology industries are springing up to take their place. As we reduce our dependence on fossil fuels, green-collar jobs (p. 378) and investment opportunities are opening up in renewable energy.

Moreover, people desire to live in areas with clean air, clean water, intact forests, public parks, and open space. Environmental protection enhances a region's appeal, drawing residents, increasing property values, and boosting the tax revenues that fund social services. As a result, regions that safeguard their environments tend to retain and enhance their wealth, health, and quality of life. In all these ways, environmental protection enhances economic opportunity. Indeed, a recent U.S. government review concluded that the economic benefits of environmental regulations greatly exceed their economic costs (p. 117). Both the U.S. economy and the global economy have expanded rapidly in the past 50 years, the very period during which environmental protection measures have proliferated.

TABLE E.1 Major Approaches in Campus Sustainability

Waste reduction

Reusing, recycling, and compost-ing offer abundant opportunities for tangible improvements, and people understand and enjoy these activities. Waste reduction events such as trash audits and recycling competitions can be fun and productive. Compost can be applied to plantings to beautify the campus. Some schools aim to become "zero-waste."

Wayne State University

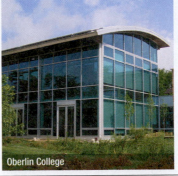

Green buildings

Constructed from sustainable materi-als, "green buildings" feature designs and technologies to reduce pollution, use renewable energy, and encour-age efficiency in water and energy use. These sustainable buildings are certified according to LEED standards (pp. 428–429).

Oberlin College

Water conservation

Indoors, schools are installing water-saving faucets, toilets, urinals, and showers in dorms and classroom buildings to reduce water waste. Out-doors, students are helping to land-scape gardens that harvest rainwater and to build facilities to treat and reuse wastewater.

Loyola University Chicago

Energy efficiency

Campuses today are installing energy-efficient lighting, motion detectors to shut off lights when rooms are empty, and sensors to record and display a building's energy consumption. Students are mounting campaigns to reduce thermostat settings, distribute efficient bulbs, and publicize energy-saving tips to their peers.

Xavier University of Louisiana

Renewable energy

Many schools are switching from fossil fuels to renewable heating and elec-tricity. Some are installing solar panels; others use biomass in power plants; a few have built wind turbines on cam-pus. Students have persuaded admin-istrators—and have voted for student fees—to buy "green tags" or carbon offsets to fund renewable energy.

The Catholic University of America

Food and dining

More and more schools grow their own food on campus farms and in gardens, supplying students with local, healthy, organic food. In dining halls, trayless dining cuts down on waste (on average, 25% of food taken is wasted otherwise) because people take only what they really want. Food scraps are composted on many campuses.

Yale University

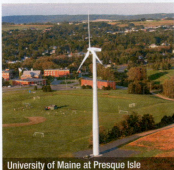

Transportation

Half the greenhouse gas emissions of the average college or university come from commuting to and from cam-puses in motor vehicles. To combat pollution, traffic congestion, delays, and parking shortages, schools are investing in bus and shuttle systems; hybrid and alternative-fuel fleet vehi-cles; and programs to promote car-pooling, walking, and bicycling.

University of Virginia

Plants and landscaping

Students are helping to restore native plants and communities, remove invasive species, improve habitat for wildlife, and enhance soil and water quality. Some schools have green roofs, greenhouses, and botanical gardens. Enhancing a campus's natu-ral environment creates healthier, more attractive surroundings, and makes for educational opportunities in ecology and natural resources.

University of Arizona

Carbon-neutrality

To combat climate change, many campuses aim to become carbon-neutral, emitting no net greenhouse gases. These schools seek to power themselves with clean renewable energy, although for most, buying car-bon offsets is also necessary. Nearly 700 presidents have signed the Ameri-can College and University Presidents' Climate Commitment, with carbon-neutrality as a goal.

University of Maine at Presque Isle

Fossil fuel divestment

Endowment money in U.S. higher education is estimated at more than $400 billion, and some of this is invested in stocks of coal, oil, and gas corporations. Since 2012, students have lobbied their administrators, boards of trustees, and fund manag-ers to divest from, or sell off, stocks in fossil fuel companies.

Tufts University

We can follow ten strategies toward sustainable solutions

Truly lasting win-win solutions for humanity and our environment are numerous, and we have seen specific examples throughout this book. Let's now summarize 10 broad strategies that can help generate sustainable solutions (**TABLE E.2**).

Political engagement Sustainable solutions often require policymakers to usher them through, and policymakers respond to whoever exerts influence. You can exercise your power at the ballot box, by attending public hearings, by volunteering or donating to advocacy groups, and by writing letters and making phone calls to officeholders. The environmental and consumer protection laws we all benefit from today came about because citizens pressured their representatives to act. As we enjoy today's cleaner air, cleaner water, and greater prosperity, we owe a debt to the people who fought for the policies that enabled these advances (pp. 106–108). We owe it to future generations to engage ourselves so that they, too, can enjoy a better world.

Consumer power Each of us also wields influence through the choices we make as consumers. When products such as recycled paper, organic produce, or sustainable seafood are ecolabeled (p. 114), we can "vote with our wallets" by purchasing these products.

Quality of life Economic growth has largely been driven by rising consumption of material goods and services (and thus the use of resources for their manufacture). Advertisers seek to sell us more goods more quickly, but accumulating possessions does not always bring contentment. Affluent people often fail to find happiness in their material wealth.

We can enhance our quality of life by prioritizing friends, family, leisure time, and memorable experiences over material consumption. Economists and policymakers can help shift the current focus on economic growth toward a focus on people's quality of life by incorporating external costs into market prices, introducing green taxes, eliminating harmful subsidies, and adopting full cost accounting practices

TABLE E.2 Major Strategies for Sustainability

- Be politically active
- Vote with our wallets
- Pursue quality of life, not just economic growth
- Limit population growth
- Encourage green technologies
- Mimic natural systems by promoting closed-loop processes
- Enhance local self-sufficiency, yet embrace some aspects of globalization
- Think in the long term
- Pursue systemic solutions
- Promote research and education

(Chapter 5). The market might then become a truly free market and a powerful tool for improving the quality of our lives.

Population stability Just as continued growth in consumption is unsustainable, so is growth in the human population. Sooner or later our population, like all populations, will stop growing. The question is how: through war, plagues, and famine, or through voluntary means? Thanks to urbanization, wealth, education, and the empowerment of women, the demographic transition (p. 128) is already well advanced in many developed nations. If today's developing nations also pass through the demographic transition, then humanity may be able to rein in population growth while creating a more prosperous and equitable society.

Green technologies Technology has facilitated our population growth and has magnified our impacts on Earth's systems—yet it also gives us ways to reduce our impacts. Scrubbers and catalytic converters have lowered emissions (pp. 293–294). Recycling technology and wastewater treatment are reducing waste. Solar, wind, and geothermal energy technologies are producing cleaner, renewable energy. Technological advances such as these help explain why people of the United States and western Europe today enjoy cleaner environments—although they consume far more—than people of eastern Europe or rapidly industrializing nations such as China.

Mimicking nature As industries develop green technologies and sustainable practices, they have an excellent model: nature itself. Environmental systems tend to operate in cycles featuring feedback loops and the circular flow of materials. Forward-thinking industrialists are transforming linear pathways into circular ones, in which waste is recycled and reused (pp. 410–411). Their ultimate vision is to create truly closed-loop processes, generating no waste.

Local and global approaches Encouraging local self-sufficiency is an important element of building sustainable societies. When people feel closely tied to the area in which they live, they value it and seek to sustain its environment and its human community. Relying on locally made products also cuts down on fossil fuel use from long-distance transportation.

At the same time, globalization enables people of the world's diverse cultures to communicate, making us more aware of one another's cultures and more likely to respect and celebrate, rather than fear, cultural differences. Moreover, globalization may foster sustainability because Western democracy, as imperfect as it is, serves as a model and a beacon for people living under repressive governments. Open societies allow for entrepreneurship and the flowering of creativity in business, art, science, and education. When millions of minds can think freely, we are more likely to come up with sustainable solutions to our challenges.

Long-term perspective To be sustainable, a solution must work in the long term (**FIGURE E.1**). Policymakers often act for short-term good, seeking to produce quick results that will help them be reelected. Yet many environmental dilemmas

FIGURE E.1 Sustainable solutions require long-term thinking. These students in India are caring for tree seedlings at their school's nursery and, by planting them on eroded hillsides, are helping to reforest the valley in which they live. In doing so, they are investing in their own future.

are cumulative, worsen gradually, and can be resolved only over long periods. Often the costs of addressing an environmental problem are short term but the benefits are long term. For this reason, keeping pressure on policymakers is vital.

Systemic solutions There are two general ways to respond to a problem. One is to address the symptoms (a symptomatic approach), and the other is to address the root cause (a systemic approach). Addressing symptoms is easier, but generally it is not effective in resolving a problem. For instance, as we deplete easily accessible fossil fuel deposits, we are choosing to reach further for new fossil fuel sources, even in the face of mounting environmental and health repercussions. A systemic solution to our energy demands would involve developing clean renewable sources instead. For many issues we face, it will prove best to pursue systemic solutions.

Research and education Finally, we each can magnify our influence by educating others and by serving as role models through our actions. Environmental science provides information we all can use to make wise decisions on a diversity of issues. By promoting scientific research and by educating others about environmental science, we all can assist in the pursuit of sustainable solutions.

Time is precious

We can bring sustainable solutions within reach, but time is getting short, and human impacts continue to intensify. Even if we can visualize sustainable solutions, how can we find the time to implement them before we do irreparable damage to our environment and our future?

In 1961, U.S. President John F. Kennedy announced that within a decade the United States would be "landing a man on the moon and returning him safely to the Earth." It was a bold and astonishing statement; the technology to achieve this feat did not yet exist. Yet just eight years later, astronauts walked on the moon. America accomplished this historic milestone by harnessing public enthusiasm for a goal and by supporting its scientists and engineers.

Today humanity faces a challenge more important than any other. Attaining sustainability is a larger and more complex process than traveling to the moon. However, it is one to which every person can contribute, and toward which all nations can work together. If America was able to reach the moon in a mere eight years, then certainly humanity can pursue sustainability with comparable speed.

Fortunately, in our society today we have many thousands of scientists who study Earth's processes and resources. Thanks to their efforts, we are amassing a detailed knowledge and a growing understanding of our dynamic planet, what it offers us, and what impacts it can bear. Environmental science, this study of Earth and of ourselves, offers us hope for our future.

Earth is an island

We began this book with the vision of Earth as an island, and indeed it is (**FIGURE E.2**). Islands can be paradise, as Easter Island (pp. 8–9) likely was when Polynesians first reached it. Yet when Europeans arrived centuries later, they witnessed the aftermath of a civilization that had collapsed once its island's resources were depleted and its environment degraded.

It would be tragic folly to let such a fate befall our planet as a whole. By recognizing this fact, by encouraging sustainable practices, and by employing science to help us achieve these ends, we may yet be able to live happily and sustainably on our wondrous island, Earth.

FIGURE E.2 This photo of Earth, taken by astronauts orbiting the moon, shows our planet as it truly is—an island in space. Everything we know, need, love, and value comes from and resides on this small sphere—so we had best treat it well.

APPENDIX A

Answers to Data Analysis Questions

NOTE: *The calculations shown in the answers below use actual data values that are sometimes more precise than those you will be able to approximate from your visual inspection of the graphs involved. As a result, your answers may differ slightly from those given here. The most important thing is that the steps or reasoning you use to arrive at the answer match the steps or reasoning described.*

Chapter 1

Fig. 1.3 The graph shows that nearly 1 billion people were alive in 1800, whereas over 7 billion are alive today. Thus, for every person alive in 1800, there are slightly more than 7 alive today.

Fig. 1.5 The global ecological footprint today is roughly 1.68 planet Earths. The global ecological footprint half a century ago (1961) was roughly 0.73 planet Earth. This makes for a difference of 0.95 planet Earth, and it means that today's footprint is 2.3 times larger than the footprint half a century ago.

Fig. 1.9 In part **(a)**, time in weeks is shown on the *x*-axis, the horizontal axis. In part **(b)**, the dependent variable is the percentage of pond surface area covered by algae, because this is on the *y*-axis and depends on whether ponds were fertilized. In part **(c)**, the data show a positive correlation, with values of pond cover increasing along with increases in fertilizer application. In part **(d)**, Species #1 is most numerous because it has the largest percentage and the largest pie slice, whereas Species #5 is least numerous. The thin black lines in part **(b)** are called *error bars*. They indicate the amount of variation the data show around the mean, or average, value (which is indicated by the height of each colored bar).

Fig. 1.16 Of the nations shown in the figure, Canada has the largest per capita footprint, and Haiti has the smallest per capita footprint. The Canadian footprint (8.8 ha) is 14.7 times larger than Haiti's footprint (0.6 ha), because 8.8 divided by 0.6 equals 14.7.

Chapter 2

Fig. 2.16 Tropical rainforest has a net primary productivity of about 2200 g $C/m^2/yr$, whereas cultivated land has around 600 g $C/m^2/yr$. Dividing 2200 g $C/m^2/yr$ by 600 g $C/m^2/yr$ shows that the hectare of land would have been around 3.7 times more productive as rainforest than as cultivated land. As productive ecosystems absorb more carbon dioxide than less productive ecosystems, one approach for combating climate change is to slow or eliminate the conversion of highly productive rainforest to less productive farmland in tropical regions, such as the Amazon.

Fig. 2.24 Reducing nitrogen inputs into the Chesapeake Bay through enhanced nutrient management programs costs $21.90 per pound, whereas a 1-pound reduction from forest buffers costs only $3.10. Dividing $21.90 per pound by $3.10 per pound shows us that for the same price, we could keep about 7 pounds of nitrogen out of waterways by using forested buffers versus 1 pound of nitrogen by using nutrient management programs.

Chapter 3

Fig. 3.6 All vertebrate groups shown, except for lampreys, are shown branching off "after" (to the right of) the hash mark for jaws. This indicates that lampreys diverged before jaws originated (and thus lack them), whereas all other vertebrate groups possess jaws. Birds are more closely related to crocodiles than to amphibians. We can tell this because birds and crocodiles share a recent common ancestor, having diverged from this ancestor much more recently than the lineages that lead to birds and to amphibians diverged in the tree.

Fig. 3.15 Because exponential growth cannot last forever, we would expect that growth of the western U.S. population of Eurasian collared doves will eventually slow down and that the population will reach carrying capacity. Thus the population growth graph for the western United States would come to have a shape more like the current graph for Florida, showing logistic growth.

Chapter 4

Fig. 4.7 In the generalized example shown, there are 100 units of energy among primary consumers (such as grasshoppers) for every 10 units among secondary consumers (such as rodents). Therefore, there is 1/10 as much energy among secondary consumers as among primary consumers. Thus, if a system had 3000 kcal/m^2/yr of energy among primary consumers, we would expect 300 kcal/m^2/yr of energy among secondary consumers (300/3000, a 1:10 ratio). Likewise, in the example shown there is 1 unit of energy among tertiary consumers (such as hawks) for every 10 units among secondary consumers—again, a 1:10 ratio. Thus if there were 300 kcal/m^2/yr of energy among secondary consumers, there would be 30 kcal/m^2/yr of energy among tertiary consumers.

Fig. 4.17 Average monthly temperatures for temperate grassland are similar to those for temperate deciduous forest in the summertime, but in the wintertime, grassland temperatures tend to get colder. As for precipitation, temperate grassland receives substantially less precipitation all year than temperate deciduous forest. This is the driving factor causing it to be grassland, because trees require more water than grass. There is also slightly more variability (difference between low and high extremes) in precipitation in the grassland biome.

Chapter 5

Fig. 5.2 The arrows show the directions in which items are moving. When you work at a job, you give labor and you receive wages. When you buy a product, you pay money and you receive the product in terms of goods or services. The environment provides to the economy both ecosystem goods (natural resources) and ecosystem services (such as waste acceptance). Some ecosystem services (such as climate regulation, nutrient cycling, and air and water purification) act as natural recycling processes, helping to repurpose waste materials into the creation of new resources.

Fig. 5.4 The ecosystem services that provide the most benefits in dollar terms are: treating waste

($22.6 trillion/yr), enabling recreation ($20.6 trillion/yr), and controlling erosion ($16.2 trillion/yr).

Fig. 5.5 In 1950, GDP was just over $3000 per capita and GPI was just over $2000 per capita, making for a ratio of about 1.5 to 1. In 2004, GDP was nearly $10,000 per capita and GPI was almost $4000 per capita, making for a ratio of about 2.5 to 1. For most people, the ratio for the year they were born will be between these two values. Together these changes indicate that GDP has been growing faster than GPI.

Fig. 5.12 Oil, coal, and natural gas have together received about $594 billion, while all other energy sources received about $244 billion in the United States during the period covered by the figure. If we divide 594 by 244, we get 2.43. Thus, about $2.43 has been spent on fossil fuel subsidies for every $1.00 that has gone to all other energy sources combined. Renewable energy sources consisting of wind, solar, geothermal, and biofuels together received about $81 billion in the United States during the period covered by the figure. If we divide 594 by 81, we get 7.33. Thus, about $7.33 has been spent on fossil fuel subsidies for every $1.00 that has gone to renewable energy.

Chapter 6

Fig. 6.4 By examining the key that links colors on the map to growth rates, we see that Africa has the highest overall growth rate of any region. Europe has the lowest growth rate of any region, as evidenced by its many nations with very low or negative growth rates.

Fig. 6.12 The transitional stage has the greatest growth of any stage in the demographic transition, because that is the period when birth and death rates are far apart and population increase is substantial. The growth is the greatest toward the end of the transitional stage, about three-fourths of the way through, when the difference between birth and death rates is the greatest.

Fig. 6.14 The best approach to answer a question such as this one is to draw a "best-fit" line (see Appendix B) through the points on the figure that minimizes the distance between each point and the line you draw. Doing this produces a line that slopes downward from left to right, suggesting a negative relationship between total fertility rate and the rate of enrollment of girls in secondary school. This relationship makes sense—one would expect that as more girls pursue education, they delay childbirth and reduce the nation's TFR.

Fig. 6.16 Africa will add about 1.3 billion people to the global population by 2050, more than the roughly 900 million people added by Asia. Africa will also increase by the largest percentage, some 108%. This value is calculated by dividing the number of people added to Africa's population (1.3 billion) by its 2016 population (1.2 billion) and then multiplying by 100 to convert the resulting proportion to a percentage: (1.3/1.2) × 100 = 108%. Possible explanations for Africa growing the fastest of any world region would include relatively lower levels of women's rights than other regions, significantly less contraceptive use than other regions, and the lowest per capita income of any world region. Because all of these factors are correlated with high fertility, it follows that Africa's population growth will likely surpass that of other regions in coming decades.

Chapter 7

Fig. 7.2 Overall growth of the human population provides the answer. Between the two periods specified, our global population increased by several hundred million people, with the vast majority of this growth occurring in developing nations. Thus, although the absolute number of undernourished people in the developing world rose slightly, a great many people were added to the total population, so the percentage of people who were undernourished still fell.

Fig. 7.15 Beef requires 17.5 times more land to produce than chicken (245 m² / 14 m² = 17.5). Beef requires 15 times more water to produce than chicken (750 kg / 50 kg = 15). Beef releases 8.6 times more greenhouse gases than chicken (342 kg / 40 kg = 8.6).

Fig. 7.18 In 2007, around 50 million hectares of GM crops were growing in developing nations, and by 2011 it had risen to roughly 80 million hectares—an increase of 30 million hectares over those 5 years. In 2012, about 85 million hectares of GM crops were being cultivated in developing nations, and by 2016 it had grown to nearly 100 million hectares—an increase of around 15 million hectares. Comparing the two values, we see that the global increase in GM crop acreage was therefore about half as fast in recent years as it was in the previous five year period. While GM crops have been readily embraced by farmers in Latin American nations, they have been slow to gain adoption in Africa and Asia. In some cases, African and Asian farmers choose not to plant GM crops due to local concerns about their safety and potential ecological impacts. In other cases, farmers do not grow GM plants because the nations to which they export their harvest—Japan and European nations—refuse to import GM foods.

Chapter 8

Fig. 8.3 The rightmost pie chart in the figure shows that there are 5900 species of mammals. The middle pie chart shows that there are about 66,000 species of vertebrates. Because 5900 is 0.089 of 66,000, this means that mammals make up about 8.9% of all vertebrates. The leftmost pie chart shows that there are about 1,552,000 known and described species of animals. Because 5900 is 0.0038 of 1,552,000, this means that mammals make up just 0.38% of all animals. One can add up the numbers in the leftmost pie chart to find that there are about 2,118,000 likely species of organisms in total. Because 5900 is 0.28% of 2,118,000, this means that mammals make up just 0.28% of all organisms (or about 1 out of 350). In reality, the percentage is actually much lower, because virtually all mammal species have already been discovered, yet most species of other types of organisms have not yet been discovered. Finally, the center pie chart shows 1,014,000 insect species; therefore, there are 1,014,000/5900 = 172 insect species for every mammal species.

Fig. 8.12 In the winter of 2016–2017, monarchs occupied just 2.91 hectares (ha), whereas in 1994–1995 (the first year of data), they occupied 7.81 ha. Thus in 2016–2017 they occupied just 37.3% (2.91/7.81) of their original area. Compared with the year of greatest area occupied (18.19 ha in 1996–1997), monarchs in 2016–2017 occupied just 16.0% of that area (2.91/18.19).

Fig. 8.13 The bar for pollution stretches to a value of nearly 1200 species, second only to habitat loss; this indicates that pollution is the second-greatest cause of amphibian declines overall. For threatened species alone, we need to look at the red portions of the bars. Comparing red portions of the bars, we can see that habitat loss is the primary cause of declines for threatened species of amphibians. For non-threatened species, we need to look at the yellow portions of the bars. Comparing these, we see that the portion for fires is shorter than that for pollution; thus, pollution is a greater cause of declines.

Fig. 8.17 In 2016, there were 276 condors in the wild—more than the 170 that were in captivity. In the 1980s and 1990s, there were far more birds in captivity than in the wild. The proportion and number of wild birds have generally increased since then, and wild birds have outnumbered captive birds since 2011. Today there are about 446 condors alive, which is over 22 times more than in the 1980s, when the number was down to about 20. The wild condor population in 1890 was about 500 birds, so today's wild population is about 55% of that (276/500), and today's total (wild plus captive) population is about 89% of the 1890 total (446/500).

Chapter 9

Fig. 9.13 The ratio of growth to removal is greatest for the land type for which the relative height of the two bars is most different. This is the case for the national forests, where annual growth exceeds 4 billion ft³ while annual removal totals less than 0.4 billion ft³.

Fig. 2 (SBS) According to the data in the graph, the forest plot held about 55 bird species in the 4 years researchers monitored it before its fragmentation in 1984. After the plot became a fragment, the average number of bird species dropped to about 20 species.

Fig. 3 (SBS) According to the data in the graph, a tree 275 meters in from the edge of a forest fragment would be susceptible to elevated tree mortality (an edge effect that extends 300 m in) and increased wind disturbance (which extends 400 m in).

Chapter 10

Fig. 10.3 In 2015, respiratory infections claimed approximately 3.2 million lives and diarrheal diseases about 1.8 million, for a total of 5 million lives. There were about 1.7 million deaths from AIDS. Dividing 5 million by 1.7 million reveals that about three times more lives were lost to respiratory infections and diarrheal diseases than were lost to AIDS. As this example shows, the diseases that garner the most attention are not always the ones that cause the greatest impacts on human health.

Fig. 10.11 The odds of perishing in a motor vehicle accident is 1 in 113, whereas the chance of dying in an air and space transport incident is 1 in 9737. Dividing 9737 by 113 shows that the odds of dying in a car accident are about 86 times those of dying in a plane crash—even though our instinctive risk assessment often makes us feel safer "behind the wheel."

Chapter 11

Fig. 11.5 Comparison of the two figures reveals that this belt of intense earthquake and volcanic activity corresponds closely to the subduction zones at the boundaries of the tectonic plates that surround the Pacific Ocean. As shown in Figure 11.2, convergent plate boundaries dominate the length of the ring of fire. Note that other locations that experience earthquakes and volcanic activity—such as Indonesia, Iran, Turkey, and southern Italy—are similarly located along convergent plate boundaries.

Fig. 11.22 At present rates of consumption, molybdenum has technically recoverable reserves that would last around 85 years, of which about 66 years of reserves are economically recoverable. Dividing 66 years of economically recoverable reserves by 85 years of technically recoverable reserves yields the value 0.78, indicating that 78% (multiply 0.78 by 100 to present the value as a percentage) of molybdenum reserves are economically recoverable. Nickel (60%) comes in second when similar calculations are performed. The metal with the lowest percentage of economically recoverable reserves is lead. Lead's 18 years of economically recoverable reserves divided by its 415 years of technically recoverable reserves show that only 4% of the world's known lead reserves are currently economically recoverable.

Fig. 2 (SBS) Answering this question requires some estimation to determine the values shown by the height of the bars in the figure. Make your estimate more accurate by using a ruler to find the spot on the y-axis that corresponds with the top of the bar you're measuring. If you do this, you should see that the greatest change in the number of earthquakes was from 2013 (approximately 125 earthquakes) to 2014 (around 675 earthquakes). Dividing 675 by 125 yields 5.4, showing that there were more than five times more earthquakes of magnitude 3 or greater in the central and Eastern United States in 2014 than in 2013. As shown in the figure, both of these values are significantly higher than the historic average of 10 to 20 large quakes per year that occurred from the early 1970s to the late 2000s.

Chapter 12

Fig. 12.2 Consulting the figure, note that 2.5% of the water on Earth is fresh water and that 1% of all fresh water is surface water. Within this surface water, 52% is found in lakes. To determine the percentage of water found in freshwater lakes, multiply 2.5% (0.025) by 1% (0.01) and by 52% (0.52). Multiplying these three values reveals that although freshwater lakes (such as the Great Lakes) seem massive, all of the world's freshwater lakes *combined* contain only 0.013% of Earth's water.

Fig. 12.7 Released off the southeastern coast of Japan, the buoy would be carried northeast by the Kuroshio Current and then eastward across the ocean on the North Pacific Current. Upon reaching North America, it could turn southward on the California Current and float by the western coast of the United States, passing Washington, Oregon, and California. Alternatively, the buoy might follow the Alaska Current northward upon reaching North America and pass Alaska, then return to Japan. So although Japan is closer to Australia, ocean currents would carry the buoy to the United States first.

Fig. 12.17 Among the three regions, Latin America and the Caribbean have the greatest amount of water per capita. With its abundant river systems (including the mighty Amazon) and relatively small population compared to that of Africa and Asia, the per-person water quantities in the nations of this region are consistently large. Although Africa and Asia do contain abundant river systems, they also have larger populations, which means that less water is available per person. Africa and Asia are also home to large regions with arid climates, another factor that reduces the quantities of available water.

Fig. 12.29 In 2014, the total global fisheries production was about 170 million metric tons, with around 90 million metric tons coming from capture fisheries and about 80 million metric tons from aquaculture. To determine the percentage of the total derived from aquaculture, you divide 80 million metric tons by 170 million metric tons (and then multiply the answer by 100 to convert it to a percentage) and find that about 47% of the world's seafood originated from aquaculture operations in that year. Similar calculations for 1980 (65 million metric tons from capture, 5 million metric tons from aquaculture) reveals that only about 7% of total fisheries production came from aquaculture. One explanation for this increase is that as ocean stocks of wild fish dwindled due to overharvesting, it became more costly to locate schools of large fish, opening the door for aquaculture as an economically viable alternative to wild capture.

Chapter 13

Fig. 13.7 Population has increased by 57% since 1970. Emissions have decreased by 71%. Thus, emissions per person have decreased by over five times. (Imagine a population rising from 100 to 157, and emission dropping from 100 to 29. 29/157 = 0.185, or less than one-fifth of the original 1-unit-per-person rate.)

Fig. 13.10 Answers will vary. For example, a person living in Los Angeles would be breathing dirtier air than people in most other cities, yet would find that L.A.'s air had improved over time, with only about 40% as many unhealthy days in 2015 as in 1985. All 29 cities on the map have improved their air quality in recent years. Factors influencing air quality could include topography and climate, pollution sources such as power plants and the type of fuel they use, intensity of vehicle traffic, and more. Steps for reduction could include various policy measures and adoption of better pollution control technology.

Fig. 13.14 According to the data in the bars of the graph, in the late 1970s the L.A. basin suffered about 210–215 days per year of unhealthy air, and in recent years it has suffered about 125 such days—this is roughly a 41% reduction. According to the data in the line of the graph, in the late 1970s peak daily ozone levels averaged about 0.30 ppm, and in recent years they have averaged about 0.12 ppm, representing about a 60% reduction. Thus, both data sets show similar declines, with the decline of peak daily ozone levels being somewhat greater. One can tell this at a glance because the slopes of the two data sets each head downward through time, but the slope of the red line is a bit steeper than the slope of the bars.

Fig. 13.22 Answers will vary, but in virtually all locations, precipitation has become less acidic. For example, in many parts of the northeastern United States, pH increased from about 4.3 to about 5.0.

Chapter 14

Fig. 14.3 Since 1750 the atmospheric carbon dioxide concentration has increased from about 280 ppm to more than 400 ppm—a 43% increase.

Fig. 14.4 Changing land use accounts for 6 metric tons of carbon dioxide emissions per year, and industry emits 26 metric tons of carbon dioxide annually. Because 26/6 = 4.33, this means that for every 1 metric ton released by changing land use, 4.33 tons are released by industry.

Fig. 14.9 Answers will vary. In most regions temperature rose. In some areas of the Southeast it was stable or fell slightly.

Fig. 14.29 Many of the nations that reduced emissions are European. Three of the nations where emissions increased (Australia, Canada, and the United States) are large and less densely populated. Because they are geographically more spread out, long-distance transportation consumes more petroleum, giving rise to more emissions. In addition, these nations are more politically conservative than most European nations, and many conservatives tend to fear that emissions reductions will suppress economic activity.

Fig. 3 (SBS) The red-shaded area reflects modeling results of temperature change with both natural factors and human factors considered. Human impacts (in particular, greenhouse gas emissions) increased greatly during the course of the 20th century due to steep growth in our population and resource consumption. Because the right-hand side of the graph shows data for the latter portion of the century, it is here that the influence of human impacts moves the red-shaded area upward,

diverging from the blue-shaded area. The observed data track with the red-shaded area because human-caused emissions have raised global temperatures, so it is to be expected that the data actually observed on Earth would match what is simulated by a model that takes human, as well as natural, impacts into account.

Chapter 15

Fig. 15.3 Answers will vary. One should take the value at the far right end of the data line for oil (4.42 billion tons in 2016) and divide it by the value of that line in the year one was born. For example, for a person born in 1999, when oil consumption was about 3.5 billion tons per year, the percentage change by 2016 would be roughly 4.42/3.5 = 1.26, or about a 26% increase. For most people, coal has risen fastest since they were born; the same type of calculation can be performed, and note how the data line for coal rises more steeply than those for oil or gas.

Fig. 15.18 Answers will vary. One should take the value at the far right end of the black ("Total") data line and divide it by the value of that line in the year one's mother or father was born. For example, if one's parent were born in 1970, when emissions were about 4.0 billion tons per year, then the percentage of change by 2014 would be about 9.8/4.0 = 2.45, or roughly a 145% increase.

Fig. 15.21 The United States extracts 15.1 million barrels of oil per day and consumes 19.5 million barrels per day. Thus, for every barrel extracted, 19.5/15.1 = 1.29 barrels are consumed.

Chapter 16

Fig 16.2 Note that in both part (a) and part (b), the bar on the right gives a breakdown of data from the pie slice for renewable energy. In part (a), the pie chart for energy consumption tells us that renewable energy as a whole provides 10.5% of U.S. energy consumption. The bar tells us that solar energy provides 5.8% of renewable energy. Therefore, solar contributes 5.8% of 10.5%—or 0.6%—of total U.S. energy consumption. Similarly, in part (b), the pie chart for electricity generation tells us that renewable energy as a whole provides 15.4% of U.S. electricity generation. The bar tells us that wind power provides 36.1% of renewable energy. Therefore, wind contributes 36.1% of 15.4%—or 5.6%—of total U.S. electricity generation.

Fig 16.3 Using median values indicated by the thin black vertical lines within the colored bars, 1 kilowatt-hour of electricity from PV solar results in roughly 35 g CO_2-equivalent emissions, whereas 1 kilowatt-hour of electricity from coal results in roughly 1000 g CO_2-equivalent emissions. Thus, for every unit of emissions from PV solar, we would expect roughly 1000/35 = 29 units from coal.

Fig 16.11 On average, southern Arizona receives roughly 2400–2600 kilowatt-hours per square meter per year, and most of Germany receives fewer than 1200 kWh/m^2/yr. Thus, southern Arizona receives more than twice as much sunlight as does Germany. For one's own state, answers will vary, but the comparison would be made in the same way.

Fig 16.16 Answers will vary, but regions that appear underutilized for wind power include (on land) North and South Dakota, Nebraska, Montana, Wyoming, and New Mexico; and (offshore) the Great Lakes and the entire Atlantic, Pacific, and Gulf coasts.

Fig. 16.22 In 2016, 26.6 billion gallons of ethanol were produced across the world, of which the United States

produced 15.3 billion gallons. Thus U.S. production is about 15.3/26.6 = 0.58. Therefore, U.S. production makes up about 58% of the world total. From the data alone, we might reasonably predict ethanol production to rise sharply in the future because that has been the overall trend since 1980 and especially since 2000. However, we might just as reasonably predict that ethanol production will level off in the future, because production has remained fairly constant for the most recent seven years. This illustrates the uncertainties of extrapolating data into the future; your predictions of trends can depend on how far back in time you reach to assess past data. Ideally, you would want to go beyond the data and learn as much as you can about the many factors driving ethanol production before making predictions.

Chapter 17

Fig. 17.5 The amount of solid waste that is combusted (incinerated) is shown in green. To determine increase or decrease, note whether the green band widens or narrows over time. (This is separate from the overall height of the graph data, which reflects cumulative, summed, totals.) Examining the green band, we see that the amount of solid waste that is combusted (incinerated) decreased from 1960 until about 1985, and then it increased until about 2002. As of 2014, the amount was roughly equal to the amount back in 1960.

Fig. 17.6 Between 1960 and 2014, the total amount of waste that was recovered increased by almost 15 times (from about 6 to about 89 million tons). However, in that same time period the recovery rate (percentage of waste generated that is recovered) increased by just under 6 times (from about 6% to nearly 35%). From this we can infer that the total amount of waste generated must also have risen. This is because had the total amount generated stayed the same, the amount recovered and percentage recovered would have changed by the same amount. Had the total amount generated fallen, the percentage recovered would have increased by more than the amount recovered.

Chapter 18

Fig. 18.1 The dashed red line (which projects the urban population) for developing nations surpasses the dashed blue line (which projects the rural population) for developing nations between the year 2010 and the year 2020.

Fig. 18.9 In part (a), driving an SUV consumes about 4 MJ/passenger-km of energy, whereas riding commuter rail consumes only about 1.4 MJ/passenger-km—a difference of 2.6, and a ratio of 2.9 to 1. Greenhouse gas emissions from vehicles are generally the result of fossil fuel combustion, so all else being equal, we can predict that the difference in greenhouse gas emissions per passenger-mile from driving an SUV versus taking commuter rail would be roughly the same as the difference in energy consumption. Indeed, as the figure caption notes, data on greenhouse gas emissions look very similar to the energy data in part (a). Roadway costs and parking costs are created by automobile traffic, but not by rail traffic. Note the yellow and orange portions of the bars in the figure for part (b). These costs make automobile traffic more costly overall than rail traffic.

Fig. 2 (SBS) The blue curve represents the expected distribution of values based on the data obtained. The blue curve crosses the threshold for damage to land plants at about 14% of impervious surface, so that is the percentage at which one would expect to begin seeing this effect. The blue curve crosses the threshold for chronic toxicity to freshwater life at about 36% of impervious surface, so that is the percentage at which one would expect to begin seeing this effect.

How to Interpret Graphs

Presenting data in ways that help make trends and patterns visually apparent is a vital element of science. For scientists, businesspeople, journalists, policymakers, and others, the primary tool for expressing patterns in data is the graph. The ability to interpret graphs is a vital skill that you will want to cultivate. This appendix guides you in how to read graphs, introduces a few key conceptual points, and surveys the most common types of graphs, giving rationales for their use.

Navigating a Graph

A graph is a type of diagram that shows relationships among *variables,* which are factors that can change in value. The most common types of graphs relate values of a *dependent variable* to those of an *independent variable*. As explained in Chapter 1 (p. 10), a dependent variable is so named because its values "depend on" the values of an independent variable. In other words, as the values of an independent variable change, the values of the dependent variable change in response. In a manipulative experiment (p. 11), changes that a researcher specifies in the value of the independent variable *cause* changes in the value of the dependent variable. In observational studies, there may be no causal relationship, and scientists may plot a correlation (p. 11). In a positive correlation, values of one variable go up or down along with values of another. In a negative correlation, values of one variable go up when values of the other go down. Whether we are graphing a correlation or a causal relationship, the values of the independent variable are known or specified by the researcher, whereas the values of the dependent variable are unknown until the research has taken place. The values of the dependent variable are what we are interested in observing or measuring.

By convention, independent variables are generally represented on the horizontal axis, or *x*-axis, of a graph, while dependent variables are represented on the vertical axis, or *y*-axis. Numerical values of variables generally become larger as one proceeds rightward on the *x*-axis or upward on the *y*-axis. Note that the tick marks along the axes must be uniformly spaced so that when the data are plotted, the graph gives an accurate visual representation of the scale of quantitative change in the data.

As a simple example, **FIGURE B.1** shows data from the Breeding Bird Survey that reflect population growth of the Eurasian collared dove following its introduction to North America. The *x*-axis shows values of the independent variable, which in this case is time, expressed in units of years. The dependent variable, presented on the *y*-axis, is the average number of doves detected on each route. For each year, a data point is plotted on the graph to show the average number of doves detected. In this particular graph, a line (dark red curve) was then drawn through the actual data points (orange dots), showing how closely the empirical data match an exponential growth curve (p. 63), a theoretical phenomenon of importance in ecology.

Now that you're familiar with the basic building blocks of a graph, let's survey the most common types of graphs you'll see, and examine a few vital concepts in graphing.

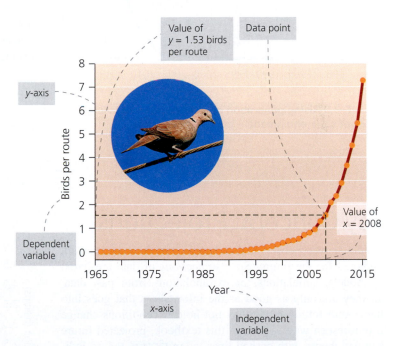

FIGURE B.1 Exponential population growth, demonstrated by the Eurasian collared dove in North America in recent years. (Figure 3.13, p. 63)

Mastering Environmental Science

Once you've explored this appendix, take advantage of the graphing resources at **www.masteringenvironmentalscience.com**. The **GRAPHit!** tutorials allow you to plot your own data. The **Interpreting Graphs and Data** exercises guide you through critical-thinking questions on graphed data from recent research in environmental science. The **Data Analysis Questions** help you hone your skills in reading graphed data. All these features will help you expand your comprehension and use of graphs.

Graph Type: Line Graph

A line graph is used when a data set involves a sequence of some kind, such as a series of values that occur one by one and change through time or across distance. In a line graph, a line runs from one data point to the next. Line graphs are most appropriate when the *y*-axis expresses a continuous numerical variable, and the *x*-axis expresses either continuous numerical data or discrete sequential categories (such as years). **FIGURE B.2** shows values for the size of the ozone hole over Antarctica in recent years. Note how the data show that the size of the hole increases until 1987, when the Montreal Protocol (p. 303) came into force, and then begins to stabilize afterwards.

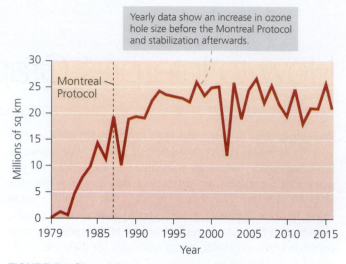

Yearly data show an increase in ozone hole size before the Montreal Protocol and stabilization afterwards.

FIGURE B.2 Size of the Antarctic ozone hole before and after a treaty that was designed to address it. (Figure 13.17, p. 303)

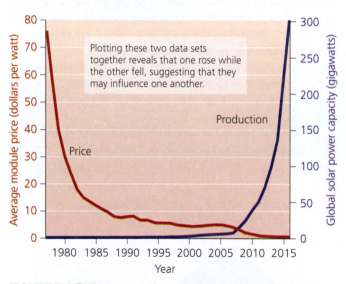

Plotting these two data sets together reveals that one rose while the other fell, suggesting that they may influence one another.

FIGURE B.3 Rising generation of solar power and falling prices of solar equipment. (Figure 16.12, p. 385)

One useful technique is to plot two or more data sets together on the same graph. This allows us to compare trends in the data sets to see whether they may be related and, if so, the nature of that relationship. For example, **FIGURE B.3** shows how the generation of electrical power from solar energy has increased as prices for solar equipment have decreased, suggesting a possible connection.

Key Concept: Projections

Besides showing observed data, graphs can show data that are predicted for the future. Such *projections* of data are based on models, simulations, or extrapolations from past data, but they are only as good as the information that goes into them—and future trends may not hold if conditions change in unforeseen ways. Thus, in this textbook, projected future data are shown with dashed lines, as in **FIGURE B.4**, to indicate that they are less certain than data that have already been observed. Be careful when interpreting graphs in the popular media and on the Internet, however; often newspapers, magazines, websites, and advertisements will show projected future data in the same way as known past data!

FIGURE B.4 Past population change and projected future population change for rural and urban areas in more developed and less developed nations. (Figure 18.1, p. 420)

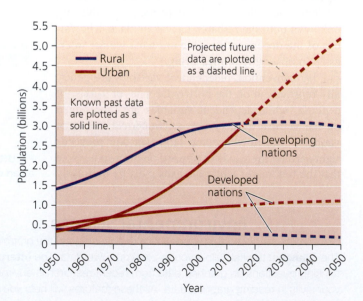

Projected future data are plotted as a dashed line.

Known past data are plotted as a solid line.

Developing nations

Developed nations

Graph Type: Bar Chart

A bar chart is most often used when one variable is a category and the other is a number. In such a chart, the height (or length) of each bar represents the numerical value of a given category. Higher or longer bars signify larger values. In **FIGURE B.5**, the bar for the category "Respiratory infections" is higher than that for "Malaria," indicating that respiratory infections cause more deaths each year (the numerical variable on the *y*-axis) than does malaria.

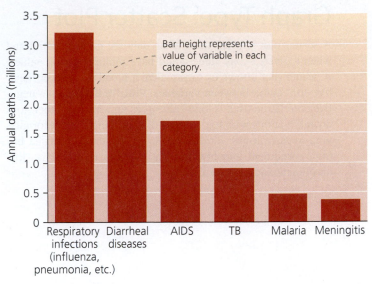

FIGURE B.5 **Leading causes of death from infectious disease.** (Figure 10.3b, p. 214)

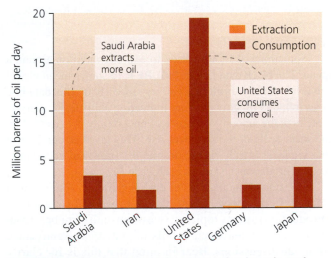

FIGURE B.6 **Oil extraction and consumption by selected nations.** (Figure 15.21, p. 362)

As we saw with line graphs, it is often instructive to graph two or more data sets together to reveal patterns and relationships. A bar chart such as **FIGURE B.6** lets us compare two data sets (oil extraction and oil consumption) both within and among nations. A graph that does double duty in this way allows for higher-level analysis (in this case, suggesting which nations depend on others for petroleum imports). Most bar charts in this book illustrate multiple types of information at once in this manner.

Graph Type: Pie Chart

A pie chart is used when we wish to compare the numerical proportions of some whole that are taken up by each of several categories. Each category is represented visually like a slice from a pie, with the size of the slice reflecting the percentage of the whole that is taken up by that category. For example, **FIGURE B.7** compares the percentages of genetically modified crops worldwide that are soybeans, corn, cotton, and canola.

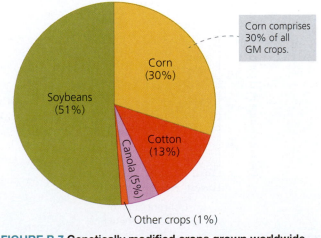

FIGURE B.7 **Genetically modified crops grown worldwide, by type.** (Figure 7.19a, p. 161)

Graph Type: Scatter Plot

A scatter plot is often used when data are not sequential and when a given *x*-axis value could have multiple *y*-axis values. A scatter plot allows us to visualize a broad positive or negative correlation between variables. **FIGURE B.8** shows a negative correlation (that is, one value goes up while the other goes down): Nations with higher rates of school enrollment for girls tend to have lower fertility rates. For example, Jamaica has high enrollment and low fertility, whereas Ethiopia has low enrollment and high fertility.

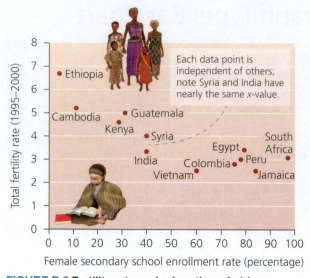

Each data point is independent of others; note Syria and India have nearly the same *x*-value.

FIGURE B.8 Fertility rate and education of girls, by nation. (Figure 6.14, p. 133)

Key Concept: Statistical Uncertainty

Most data sets involve some degree of uncertainty. When a graphed value represents the *mean* (average) of many measurements, the researcher may want to use mathematical techniques to show the degree to which the raw data vary around this mean. Results from such statistical analyses may be expressed in a number of ways, and the two graphs in this section show methods used in this book.

In a bar chart or a scatter plot (**FIGURE B.9**), thin black lines called *error bars* may be shown extending above and/or below each mean data value. In this example of likelihood of death from air pollution, error bars show the most variation at the highest measured concentration of pollutants and no variation at the lowest measured concentration.

Sometimes shading is used to express variation around a mean. The black data line in **FIGURE B.10** shows mean global sea level readings from tide gauges since 1880. This data line is surrounded by gray shading indicating statistical variation. Note how the amount of statistical uncertainty is exceeded by the sheer scale of the sea level rise. This gives us confidence that sea level is truly rising, despite the statistical uncertainty we find around mean values each year.

The statistical analysis of data is critically important in science. In this book, we provide a broad and streamlined introduction to many topics, so we often omit error bars from our graphs and leave out details of statistical significance from our discussions. Bear in mind that this is for clarity of presentation only; the research we discuss analyzes its data in far more depth than any textbook could possibly cover.

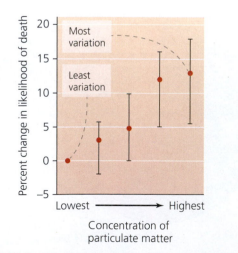

FIGURE B.9 Likelihood of death due to air pollution. (Figure 3, SBS, p. 299)

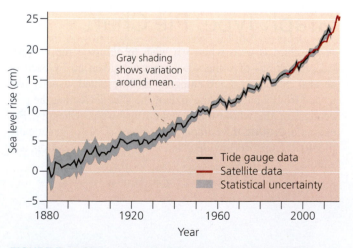

FIGURE B.10 Change in global sea level, measured since 1880. (Figure 14.15, p. 324)

Metric System

MEASUREMENT	UNIT AND ABBREVIATION	METRIC EQUIVALENT	METRIC TO ENGLISH CONVERSION FACTOR	ENGLISH TO METRIC CONVERSION FACTOR
Length	1 kilometer (km)	= 1000 (10^3) meters	1 km = 0.62 mile	1 mile = 1.61 km
	1 meter (m)	= 100 (10^2) centimeters	1 m = 1.09 yards	1 yard = 0.914 m
			= 3.28 feet	1 foot = 0.305 m
			= 39.37 inches	= 30.5 cm
	1 centimeter (cm)	= 10 millimeters	1 cm = 0.394 inch	1 inch = 2.54 cm
		= 0.01 (10^{-2}) meter		
	1 millimeter (mm)	= 0.01 (10^{-2}) centimeter	1 mm = 0.039 inch	
Area	1 square meter (m^2)	= 10,000 square centimeters	1 m^2 = 1.196 square yards	1 square yard = 0.8361 m^2
			= 10.764 square feet	1 square foot = 0.0929 m^2
	1 hectare (ha)	= 10,000 square meters	1 ha = 2.47 acres	1 acre = 0.405 hectare
	1 square kilometer (km^2)	= 1,000,000 square meters	1 km^2 = 0.386 square mile	1 square mile = 2.59 km^2
Mass	1 metric ton (t)	= 1000 kilograms	1 t = 1.103 tons	1 ton = 0.907 t
	1 kilogram (kg)	= 1000 grams	1 kg = 2.205 pounds	1 pound (lb) = 0.4536 kg
	1 gram (g)	= 1000 milligrams	1 g = 0.0353 ounce	1 ounce = 28.35 g
	1 milligram (mg)	= 0.001 gram		
Volume (solids)	1 cubic meter (m^3)	= 1,000,000 cubic centimeters	1 m^3 = 1.3080 cubic yards	1 cubic yard = 0.7646 m^3
	1 cubic centimeter (cm^3 or cc)	= 0.000001 cubic meter	1 cm^3 = 0.0610 cubic inch	1 cubic inch = 16.387 cm^3
	1 cubic millimeter (mm^3)	= 0.001 cubic centimeter		
Volume (liquids and gases)	1 kiloliter (kl or kL)	= 1000 liters	1 kL = 264.17 gallons	1 gallon = 3.785 L
	1 liter (l or L)	= 1000 milliliters	1 L = 0.264 gallon	1 quart = 0.946 L
			= 1.057 quarts	= 946 ml
	1 milliliter (ml or mL)	= 0.001 liter	1 ml = 0.034 fluid ounce	1 fluid ounce = 29.57 ml
		= 1 cubic centimeter	= approx. 1/5 teaspoon	1 teaspoon = approx. 5 ml
Temperature	Degrees Celsius (°C)		$°C = \dfrac{5}{9}\,(°F - 32)$	$°F = \dfrac{9}{5}\,(°C) + 32$
Energy and Power	1 gigawatt (GW)	= 1,000,000,000 (10^9) watts		
	1 megawatt (MW)	= 1,000,000 (10^6) watts		
	1 kilowatt (kW)	= 1000 (10^3) watts		
	1 watt (W)	= 0.001 kilowatt		
		= 1 joule/second		
	1 kilowatt-hour (kWh)	= 3,600,000 joules		
		= 3412 BTU		
		= 860,400 calories		
	1 calorie (cal)	= The amount of energy needed to raise the temperature of 1 gram (1 cm^3) of water by 1 degree Celsius		
	1 joule	= 0.239 cal		
		= 2.778×10^{-7} kilowatt-hours		
Pressure	1 atmosphere (atm)	= 1013.25 millibars (mbar)		
		= 14.696 pounds per square inch (psi)		
		= 760 millimeters of mercury (mmHg)		

Periodic Table of the Elements

Legend

6	— Atomic number
C	— Chemical symbol
12.011	— Atomic weight
Carbon	— Name

Representative (main group) elements · **Transition metals** · **Rare earth elements**

IA	IIA	IIIB	IVB	VB	VIB	VIIB	VIIIB	VIIIB	VIIIB	IB	IIB	IIIA	IVA	VA	VIA	VIIA	VIIIA
1 **H** 1.0079 Hydrogen																	2 **He** 4.003 Helium
3 **Li** 6.941 Lithium	4 **Be** 9.012 Beryllium											5 **B** 10.811 Boron	6 **C** 12.011 Carbon	7 **N** 14.007 Nitrogen	8 **O** 15.999 Oxygen	9 **F** 18.998 Fluorine	10 **Ne** 20.180 Neon
11 **Na** 22.990 Sodium	12 **Mg** 24.305 Magnesium											13 **Al** 26.982 Aluminum	14 **Si** 28.086 Silicon	15 **P** 30.974 Phosphorus	16 **S** 32.066 Sulfur	17 **Cl** 35.453 Chlorine	18 **Ar** 39.948 Argon
19 **K** 39.098 Potassium	20 **Ca** 40.078 Calcium	21 **Sc** 44.956 Scandium	22 **Ti** 47.88 Titanium	23 **V** 50.942 Vanadium	24 **Cr** 51.996 Chromium	25 **Mn** 54.938 Manganese	26 **Fe** 55.845 Iron	27 **Co** 58.933 Cobalt	28 **Ni** 58.69 Nickel	29 **Cu** 63.546 Copper	30 **Zn** 65.39 Zinc	31 **Ga** 69.723 Gallium	32 **Ge** 72.61 Germanium	33 **As** 74.922 Arsenic	34 **Se** 78.96 Selenium	35 **Br** 79.904 Bromine	36 **Kr** 83.8 Krypton
37 **Rb** 85.468 Rubidium	38 **Sr** 87.62 Strontium	39 **Y** 88.906 Yttrium	40 **Zr** 91.224 Zirconium	41 **Nb** 92.906 Niobium	42 **Mo** 95.94 Molybdenum	43 **Tc** 98 Technetium	44 **Ru** 101.07 Ruthenium	45 **Rh** 102.906 Rhodium	46 **Pd** 106.42 Palladium	47 **Ag** 107.868 Silver	48 **Cd** 112.411 Cadmium	49 **In** 114.82 Indium	50 **Sn** 118.71 Tin	51 **Sb** 121.76 Antimony	52 **Te** 127.60 Tellurium	53 **I** 126.905 Iodine	54 **Xe** 131.29 Xenon
55 **Cs** 132.905 Cesium	56 **Ba** 137.327 Barium	57* **La** 138.906 Lanthanum	72 **Hf** 178.49 Hafnium	73 **Ta** 180.948 Tantalum	74 **W** 183.84 Tungsten	75 **Re** 186.207 Rhenium	76 **Os** 190.23 Osmium	77 **Ir** 192.22 Iridium	78 **Pt** 195.08 Platinum	79 **Au** 196.967 Gold	80 **Hg** 200.59 Mercury	81 **Tl** 204.383 Thallium	82 **Pb** 207.2 Lead	83 **Bi** 208.980 Bismuth	84 **Po** 209 Polonium	85 **At** 210 Astatine	86 **Rn** 222 Radon
87 **Fr** 223 Francium	88 **Ra** 226.025 Radium	89** **Ac** 227.028 Actinium	104 **Rf** 267 Rutherfordium	105 **Db** 268 Dubnium	106 **Sg** 269 Seaborgium	107 **Bh** 270 Bohrium	108 **Hs** 269 Hassium	109 **Mt** 278 Meitnerium	110 **Ds** 281 Darmstadtium	111 **Rg** 281 Roentgenium	112 **Cn** 285 Copernicium	113 **Nh** 286 Nihonium	114 **Fl** 289 Flerovium	115 **Mc** 289 Moscovium	116 **Lv** 293 Livermorium	117 **Ts** 293 Tennessine	118 **Og** 294 Oganesson

*Lanthanides

58 **Ce** 140.115 Cerium	59 **Pr** 140.908 Praseodymium	60 **Nd** 144.24 Neodymium	61 **Pm** 145 Promethium	62 **Sm** 150.36 Samarium	63 **Eu** 151.964 Europium	64 **Gd** 157.25 Gadolinium	65 **Tb** 158.925 Terbium	66 **Dy** 162.5 Dysprosium	67 **Ho** 164.93 Holmium	68 **Er** 167.26 Erbium	69 **Tm** 168.934 Thulium	70 **Yb** 173.04 Ytterbium	71 **Lu** 174.967 Lutetium

**Actinides

90 **Th** 232.038 Thorium	91 **Pa** 231.036 Protactinium	92 **U** 238.029 Uranium	93 **Np** 237.048 Neptunium	94 **Pu** 244 Plutonium	95 **Am** 243 Americium	96 **Cm** 247 Curium	97 **Bk** 247 Berkelium	98 **Cf** 251 Californium	99 **Es** 252 Einsteinium	100 **Fm** 257 Fermium	101 **Md** 258 Mendelevium	102 **No** 259 Nobelium	103 **Lr** 262 Lawrencium

The periodic table arranges elements by atomic number and atomic weight into horizontal rows called *periods* and vertical columns called *groups*.

Elements of each group in Class A have similar chemical and physical properties. This reflects the fact that members of a particular group have the same number of valence shell electrons, which is indicated by the group's number. For example, group IA elements have one valence shell electron, group IIA elements have two, and group VA elements have five. In contrast, as you progress across a period from left to right, properties of the elements change, varying from the very metallic properties of groups IA and IIA to the nonmetallic properties of group VIIA to the inert elements (noble gases) in group VIIIA. This reflects changes in the number of valence shell electrons.

Class B elements, or transition elements, are metals and generally have one or two valence shell electrons. In these elements, some electrons occupy more distant electron shells before the deeper shells are filled.

In this periodic table, elements with symbols printed in black exist as solids under standard conditions (25°C and 1 atmosphere of pressure); elements in red exist as gases; and those in dark blue exist as liquids. Elements with symbols in green do not exist in nature and must be created by some type of nuclear reaction.

Geologic Time Scale

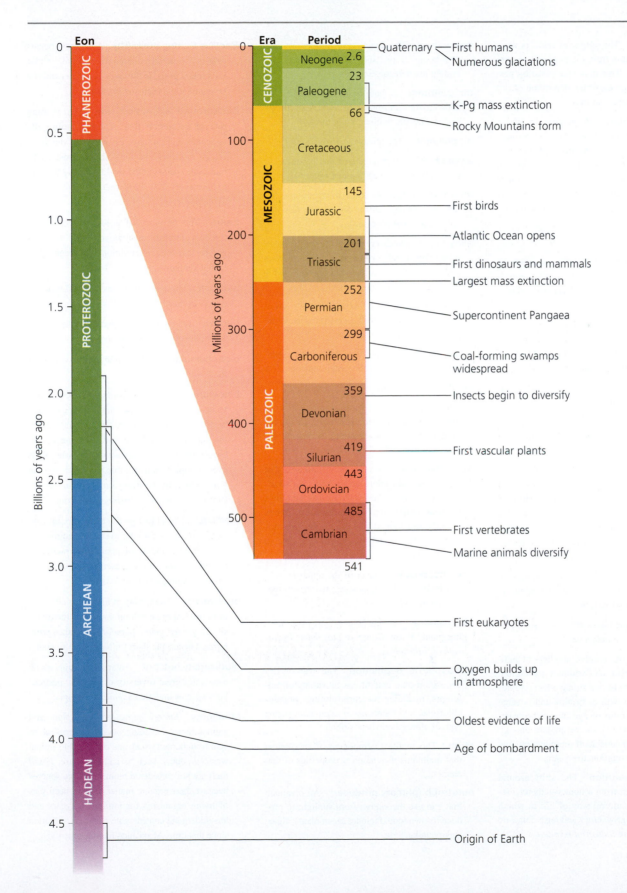

Glossary

acid deposition   The settling of acidic or acid-forming pollutants from the *atmosphere* onto Earth's surface. This may take place by precipitation, fog, gases, or the deposition of dry particles. Compare *acid rain*.

acid mine drainage   A process in which sulfide *minerals* in newly exposed *rock* surfaces react with *oxygen* and rainwater to produce sulfuric acid, which causes chemical *runoff* as it *leaches* metals from the *rocks*. Acid drainage is a natural phenomenon, but mining greatly accelerates it by exposing many new surfaces.

acid rain   *Acid deposition* that takes place through rain.

acidic   The property of a *solution* in which the concentration of *hydrogen* (H⁺) *ions* is greater than the concentration of hydroxide (OH⁻) ions. Compare *basic*.

active solar energy collection   An approach in which technological devices are used to focus, move, or store *solar energy*. Compare *passive solar energy collection*.

acute exposure   Exposure to a *toxicant* occurring in high amounts for short periods of time. Compare *chronic exposure*.

adaptation (re: *climate change*)   The pursuit of strategies to protect ourselves from the impacts of climate change. Compare *mitigation*.

adaptation (re: *evolution*)   (1) The process by which traits that lead to increased reproductive success in a given environment evolve in a *population* through *natural selection*. (2) A trait that confers greater likelihood that an individual will reproduce.

adaptive management   The systematic testing of different management approaches to improve methods over time.

aerosols   Very fine liquid droplets or solid particles aloft in the atmosphere.

age structure   The relative numbers of individuals of different ages within a *population*. Age structure can have a strong effect on rates of population growth or decline and is often expressed as a ratio of age classes, consisting of organisms (1) not yet mature enough to reproduce, (2) capable of reproduction, and (3) beyond their reproductive years.

agricultural revolution   The shift around 10,000 years ago from a hunter-gatherer lifestyle to an agricultural way of life in which people began to grow crops and raise domestic animals. Compare *industrial revolution*.

agriculture   The practice of cultivating *soil*, producing crops, and raising livestock for human use and consumption.

air pollutant   A gas or particulate material added to the atmosphere that can affect climate or harm people or other living things.

air pollution   The release of *air pollutants*.

airshed   The geographic area that produces air pollutants likely to end up in a waterway.

allergen   A *toxicant* that overactivates the immune system, causing an immune response when one is not necessary.

alloy   A substance created by fusing a *metal* with other metals or nonmetals. Bronze is an alloy of the metals copper and tin, and steel is an alloy of iron and the nonmetal *carbon*.

anthropocentrism   A human-centered view of our relationship with the *environment*. Compare *biocentrism; ecocentrism*.

aquaculture   The cultivation of aquatic organisms for food in controlled environments.

aquifer   An underground water reservoir.

artificial selection   *Natural selection* conducted under human direction. Examples include the selective breeding of crop plants, pets, and livestock.

asbestos   Any of several types of *mineral* that form long, thin microscopic fibers—a structure that allows asbestos to insulate buildings for heat, muffle sound, and resist fire. When inhaled and lodged in lung tissue, asbestos scars the tissue and may eventually lead to lung cancer.

asthenosphere   A layer of the upper *mantle*, just below the *lithosphere*, consisting of especially soft *rock*.

atmosphere   The thin layer of gases surrounding planet Earth. Compare *biosphere; hydrosphere; lithosphere*.

atmospheric deposition   The wet or dry deposition onto land of a wide variety of pollutants, including mercury, nitrates, organochlorines, and others. *Acid deposition* is one type of atmospheric deposition.

atom   The smallest component of an *element* that maintains the chemical properties of that element.

autotroph (primary producer)   An organism that can use the *energy* from sunlight to produce its own food. Includes green plants, algae, and cyanobacteria.

Bacillus thuringiensis (Bt)   A naturally occurring *soil* bacterium that produces a protein that kills many pests, including caterpillars and the larvae of some flies and beetles.

background extinction rate   The average rate of *extinction* that occurred before the appearance of humans. For example, the *fossil record* indicates that for both birds and mammals, one *species* in the world typically became extinct every 500–1000 years. Compare *mass extinction event*.

basic   The property of a *solution* in which the concentration of *hydroxide* (OH⁻) *ions* is greater than the concentration of hydrogen (H⁺) ions. Compare *acidic*.

bedrock   The continuous mass of solid *rock* that makes up Earth's *crust*.

benthic   Of, relating to, or living on the bottom of a water body. Compare *pelagic*.

bioaccumulation   The buildup of *toxicants* in the tissues of an animal.

biocentrism   A philosophy that ascribes relative values to actions, entities, or properties on the basis of their effects on all living things or on the integrity of the *biotic* realm in general. The biocentrist evaluates an action in terms of its overall impact on living things, including—but not exclusively focusing on—human beings. Compare *anthropocentrism; ecocentrism*.

biodiesel   Diesel fuel produced by mixing vegetable oil, used cooking grease, or animal fat with small amounts of *ethanol* or methanol (wood alcohol) in the presence of a chemical catalyst. A major type of *biofuel*.

biodiversity   The variety of life across all levels of biological organization, including the diversity of *species, genes, populations,* and *communities*. The term is short for *biological diversity*.

biodiversity hotspot   An area that supports an especially great diversity of *species*, particularly species that are *endemic* to the area.

bioenergy   *Energy* harnessed from plant and animal matter, including wood from trees, charcoal from burned wood, and combustible animal waste products, such as cattle manure. *Fossil fuels* are not considered biomass energy sources because their organic matter has not been part of living organisms for millions of years and has undergone considerable chemical alteration since that time. Also called biomass energy.

biofuel Liquid fuel produced from *biomass* sources and used primarily to power automobiles. Examples include *ethanol* and *biodiesel*.

biogeochemical cycle See *nutrient cycle*.

biological control Control of pests and weeds with organisms that prey on or parasitize them, rather than with chemical *pesticides*. Commonly called biocontrol.

biological diversity See *biodiversity*.

biomagnification The magnification of the concentration of *toxicants* in an organism caused by its consumption of other organisms in which toxicants have *bioaccumulated*.

biomass (1) In *ecology,* organic material that makes up living organisms; the collective mass of living matter in a given place and time. (2) In *energy,* organic material derived from living or recently living organisms, containing chemical *energy* that originated with *photosynthesis*.

biome A major regional complex of similar plant *communities;* a large *ecological* unit defined by its dominant plant type and vegetation structure.

biophilia An inherent love for and fascination with nature and an instinctive desire people have to affiliate with other living things. Defined by biologist E.O. Wilson as "the connections that human beings subconsciously seek with the rest of life."

biopower Power attained by combusting *biomass* sources to generate *electricity*.

biosphere The sum total of all the planet's living organisms and the nonliving portions of the *environment* with which they interact.

biosphere reserve A tract of land with exceptional *biodiversity* that couples preservation with *sustainable development* to benefit local people. Biosphere reserves are designated by UNESCO (the *United Nations* Educational, Scientific, and Cultural Organization) following application by local stakeholders.

biotechnology The material application of biological *science* to create products derived from organisms. The creation of *transgenic* organisms is one type of biotechnology.

birth control The effort to control the number of children one bears, particularly by reducing the frequency of pregnancy. Compare *contraception; family planning*.

bisphenol A An endocrine-disrupting chemical found in plastics.

boreal forest A *biome* of northern coniferous forest that stretches in a broad band across much of Canada, Alaska, Russia, and Scandinavia. Also known as *taiga,* boreal forest consists of a limited number of *species* of evergreen trees, such as black spruce, that dominate large regions of forests interspersed with occasional bogs and lakes.

bottle bill A law establishing a program whereby consumers pay a deposit on bottles or cans upon purchase—often 5 or 10 cents per container—and then receive a refund when they return them to stores after use. Bottle bills reduce litter, raise *recycling* rates, and decrease the *waste stream*.

bottleneck In environmental science, a step in a process that limits the progress of the overall process.

breakdown product A compound that results from the degradation of a toxicant.

brownfield An area of land whose redevelopment or reuse is complicated by the presence or potential presence of hazardous material.

bycatch (1) The accidental capture of nontarget organisms while fishing for target species. (2) That portion of a commercial fishing catch consisting of animals caught unintentionally. Bycatch kills many thousands of fish, sharks, marine mammals, and birds each year.

campus sustainability A term describing a wide array of efforts taking place on college and university campuses by which students, faculty, staff, and administrators are trying to reduce the environmental impacts of their institutions.

cap-and-trade A type of *emissions trading* system in which government determines an acceptable level of *pollution* and then issues polluting parties permits to pollute. A company receives credit for amounts it does not emit and can then sell this credit to other companies.

captive breeding The practice of keeping members of *threatened* and *endangered species* in captivity so that their young can be bred and raised in controlled *environments* and subsequently reintroduced into the wild.

carbohydrate An *organic compound* consisting of *atoms* of *carbon, hydrogen,* and *oxygen*.

carbon The chemical *element* with six protons and six neutrons. A key element in *organic compounds*.

carbon capture and storage Technologies or approaches to remove *carbon dioxide* from emissions of power plants or other facilities, and sequester, or store, it (generally in liquid form) underground under pressure in locations where it will not seep out, in an effort to mitigate *global climate change*. We are still a long way from developing adequate technology and secure storage space to accomplish this reliably.

carbon cycle A major *nutrient cycle* consisting of the routes that *carbon atoms* take through the nested networks of environmental *systems*.

carbon dioxide (CO_2) A colorless gas used by plants for *photosynthesis,* given off by *respiration,* and released by burning *fossil fuels*. A primary *greenhouse gas* whose buildup contributes to *global climate change*.

carbon footprint The cumulative amount of carbon, or *carbon dioxide,* that a person or institution emits (directly or indirectly) into the *atmosphere,* contributing to *global climate change*. Compare *ecological footprint*.

carbon monoxide (CO) A colorless, odorless gas produced primarily by the incomplete combustion of fuel. An EPA *criteria pollutant*.

carbon-neutrality The state in which an individual, business, or institution emits no net carbon to the atmosphere. This may be achieved by reducing carbon emissions and/or employing *carbon offsets* to offset emissions.

carbon offset A voluntary payment to another entity intended to enable that entity to reduce the *greenhouse gas* emissions that one is unable or unwilling to reduce oneself. The payment thus offsets one's own emissions.

carbon pricing The practice of putting a price on the emission of *carbon dioxide,* through either *carbon trading* or a *carbon tax,* as a means to address *global climate change*. Carbon pricing compensates the public for the external costs of fossil fuel use by shifting costs to emitters, and creates financial incentives to reduce emissions.

carbon tax A type of *green tax* charged to entities that pollute by emitting *carbon dioxide*. Carbon taxation is one approach to *carbon pricing,* and gives polluters a financial incentive to reduce emissions in order to address *global climate change*. Compare *carbon trading; fee-and-dividend; revenue-neutral carbon tax*.

carbon trading A form of *emissions trading* that focuses on the emission of *carbon dioxide*. In a carbon trading market, emitters buy and sell permits to emit CO_2. Carbon trading is one approach to *carbon pricing,* and gives polluters a financial incentive to reduce emissions in order to address *global climate change*. Compare *carbon tax*.

carcinogen A chemical or type of radiation that causes cancer.

carrying capacity The maximum *population size* of a given organism that a given *environment* can sustain.

case history Medical approach involving the observation and analysis of individual patients.

catalytic converter Automotive technology that chemically treats engine exhaust to reduce *air pollution*. Reacts exhaust with metals that convert *hydrocarbons,* CO, and NO_X into *carbon dioxide, water* vapor, and *nitrogen* gas.

cellular respiration The process by which a *cell* uses the chemical reactivity of *oxygen* to split glucose into its constituent parts, water and *carbon dioxide,* and thereby release chemical energy that can be used to form chemical bonds or to perform other tasks within the cell. Compare *photosynthesis*.

cellulosic ethanol *Ethanol* produced from the cellulose in plant tissues by treating it with enzymes. Techniques for producing cellulosic ethanol are being developed because of the desire to make ethanol from low-value crop waste (residues such as corn stalks and husks), rather than from the sugars of high-value crops.

chaparral A *biome* consisting mostly of densely thicketed evergreen shrubs occurring in limited small patches. Its "Mediterranean" *climate* of mild, wet winters and warm, dry summers is induced by oceanic influences. In addition to ringing the Mediterranean Sea, chaparral occurs along the coasts of California, Chile, and southern Australia.

chemical energy *Potential energy* held in the bonds between *atoms*.

chemistry The study of the different types of *matter* and how they interact.

chemosynthesis The process by which bacteria in *hydrothermal vents* use the chemical energy of hydrogen sulfide (H_2S) to transform inorganic *carbon* into *organic compounds*. Compare *photosynthesis*.

Chernobyl Site of a nuclear power plant in Ukraine (then part of the Soviet Union), where in 1986 an explosion caused the most severe *nuclear reactor* accident the world has yet seen. The term is also often used to denote the accident itself. Compare *Fukushima Daiichi; Three Mile Island.*

chlorofluorocarbon (CFC) A type of *halocarbon* consisting of only chlorine, fluorine, carbon, and hydrogen. CFCs were used as refrigerants, as fire extinguishers, as propellants for aerosol spray cans, as cleaners for electronics, and for making polystyrene foam. They were phased out under the *Montreal Protocol* because they are *ozone-depleting substances* that destroy stratospheric *ozone*.

chronic exposure Exposure to a *toxicant* occurring in low amounts for long periods of time. Compare *acute exposure*.

city planning The professional pursuit that attempts to design cities in such a way as to maximize their efficiency, functionality, and beauty. Also known as *urban planning*.

classical economics Founded by Adam Smith, the study of the behavior of buyers and sellers in a capitalist market economy. Holds that individuals acting in their own self-interest may benefit society, provided their behavior is constrained by the rule of law and by private property rights and operates within competitive markets. Compare *neoclassical economics*.

Clean Air Act U.S. *legislation* to control *air pollution*, first enacted in 1963 and amended multiple times since, most significantly in 1970 and 1990. Funds research into pollution control, sets standards for air quality, encourages emissions standards for automobiles and for stationary point sources such as industrial plants, imposes limits on emissions from new sources, funds a nationwide air quality monitoring system, enables citizens to sue parties violating the standards, and introduced an *emissions trading* program for *sulfur dioxide*.

clean coal technologies An array of techniques, equipment, and approaches to remove chemical contaminants (such as sulfur) during the process of generating *electricity* from *coal*.

clear-cutting The harvesting of timber by cutting all the trees in an area. Although it is the most cost-efficient method, clear-cutting is also the most ecologically damaging.

climate The pattern of atmospheric conditions that typifies a geographic region over long periods of time (typically years, decades, centuries, or millennia). Compare *weather*.

climate change See *global climate change*.

climate diagram A visual representation of a region's average monthly temperature and *precipitation*. Also known as a *climatograph*.

climate model A computer program that combines what is known about weather patterns, atmospheric circulation, atmosphere–ocean interactions, and feedback mechanisms, to simulate *climate* processes.

coal A solid blackish *fossil fuel* formed from organic matter (generally, woody plant material) that was compressed under very high pressure and with little decomposition, creating dense, solid carbon structures.

coevolution The process by which two or more *species* evolve in response to one another. Parasites and hosts may coevolve, as may flowering plants and their pollinators.

cogeneration A practice in which the extra heat generated in the production of *electricity* is captured and put to use heating workplaces and homes, as well as producing other kinds of power.

colony collapse disorder A mysterious malady afflicting honeybees, which has destroyed roughly one-third of all honeybees in the United States annually over the past decade. Likely caused by chemical insecticides, pathogens and parasites, habitat and resource loss, or combinations of these factors.

command-and-control A top-down approach to policy, in which a legislative body or a regulating agency sets rules, standards, or limits and threatens punishment for violations of those limits.

community In *ecology*, an assemblage of *populations* of interacting organisms that live in the same area at the same time.

community-based conservation The practice of engaging local people to protect land and wildlife in their own region.

community ecology The scientific study of patterns of species diversity and interactions among *species*, ranging from one-to-one interactions to complex interrelationships involving entire *communities*.

community-supported agriculture (CSA) A system in which consumers pay farmers in advance for a share of their yield, usually in the form of weekly deliveries of produce.

competition A relationship in which multiple organisms seek the same limited resource. Competition can take place among members of the same *species* or among members of different species.

compost A mixture produced when *decomposers* break down organic matter, such as food and crop waste, in a controlled environment.

composting The conversion of organic *waste* into mulch or humus by encouraging, in a controlled manner, the natural biological processes of decomposition.

compound A *molecule* whose *atoms* are composed of two or more *elements*.

concentrated solar power (CSP) A means of generating *electricity* at a large scale by focusing sunlight from a large area onto a smaller area. Several approaches are used.

concession The right to extract a resource, granted by a government to a corporation. Compare *conservation concession*.

confined (artesian) aquifer A water-bearing, porous layer of *rock, sand,* or gravel that is trapped between an upper and lower layer of less permeable substrate, such as *clay*. The water in a confined aquifer is under pressure because it is trapped between two impermeable layers. Compare *unconfined aquifer*.

conservation biologist A scientist who studies the factors, forces, and processes that influence the loss, protection, and restoration of *biodiversity* within and among *ecosystems*.

conservation biology A scientific discipline devoted to understanding the factors, forces, and processes that influence the loss, protection, and restoration of *biodiversity* within and among *ecosystems*.

conservation ethic An *ethic* holding that people should put *natural resources* to use but also have a responsibility to manage them wisely. Compare *preservation ethic*.

Conservation Reserve Program U.S. policy in farm bills since 1985 that pays farmers to stop cultivating highly erodible cropland and instead place it in conservation reserves planted with grasses and trees.

conservation tillage *Agriculture* that limits the amount of tilling (plowing, disking, harrowing, or chiseling) of *soil*. Compare *no-till*.

consumptive use Use of *fresh water* in which water is removed from a particular *aquifer* or surface water body and is not returned to it.

Irrigation for *agriculture* is an example of consumptive use. Compare *nonconsumptive use.*

continental collision The meeting of two tectonic plates of continental *lithosphere* at a *convergent plate boundary,* wherein the continental *crust* on both sides resists *subduction* and instead crushes together, bending, buckling, and deforming layers of *rock* and forcing portions of the buckled crust upward, often creating mountain ranges.

contour farming The practice of plowing furrows sideways across a hillside, perpendicular to its slope, to help prevent the formation of rills and gullies. The technique is so named because the furrows follow the natural contours of the land.

contraception The deliberate attempt to prevent pregnancy despite sexual intercourse. Compare *birth control.*

control The portion of an *experiment* in which a *variable* has been left unmanipulated, to serve as a point of comparison with the *treatment.*

controlled experiment An *experiment* in which a *treatment* is compared against a *control* in order to test the effect of a *variable.*

convective circulation A circular *current* (of air, water, magma, etc.) driven by temperature differences. In the atmosphere, warm air rises into regions of lower *atmospheric pressure,* where it expands and cools and then descends and becomes denser, replacing warm air that is rising. The air picks up heat and moisture near ground level and prepares to rise again, continuing the process.

Convention on Biological Diversity A 1992 treaty that aims to conserve *biodiversity,* use biodiversity in a *sustainable* manner, and ensure the fair distribution of biodiversity's benefits.

Convention on International Trade in Endangered Species of Wild Fauna and Flora (CITES) A 1973 treaty facilitated by the *United Nations* that protects *endangered species* by banning the international transport of their body parts.

conventional law International law that arises from *conventions,* or treaties, that nations agree to enter into. Compare *customary law.*

convergent evolution The evolutionary process by which very unrelated species acquire similar traits as they adapt to similar selective pressures from similar environments.

convergent plate boundary The area where tectonic plates converge or come together. Can result in *subduction* or *continental collision.* Compare *divergent plate boundary; transform plate boundary.*

coral Tiny marine animals that build *coral reefs.* Corals attach to *rock* or existing reef and capture passing food with stinging tentacles. They also derive nourishment from photosynthetic symbiotic algae known as *zooxanthellae.*

coral reef A mass of calcium carbonate composed of the skeletons of tiny colonial marine organisms called *corals.*

core The innermost part of Earth, made up mostly of iron, that lies beneath the *crust* and *mantle.*

correlation Statistical association (positive or negative) among variables. The association may be causal or may occur by chance.

corridor A passageway of protected land established to allow animals to travel between islands of protected *habitat.*

cost-benefit analysis A method commonly used in *neoclassical economics,* in which estimated costs for a proposed action are totaled and then compared to the sum of benefits estimated to result from the action.

covalent bond A type of chemical bonding in which atoms share electrons in chemical bonds. An example is a water molecule, which forms when an oxygen atom shares electrons with two hydrogen atoms.

cradle-to-cradle An approach to *waste management* and industrial design in which the materials from products are recovered and reused to create new products.

criteria pollutant One of six *air pollutants*—carbon monoxide, sulfur dioxide, nitrogen dioxide, tropospheric ozone, particulate matter, and lead—for which the *Environmental Protection Agency* has established maximum allowable concentrations in ambient outdoor air because of the threats they pose to human health.

crop rotation The practice of alternating the kind of crop grown in a particular field from one season or year to the next.

cropland Land that people use to raise plants for food and fiber.

crude oil *Oil* in its natural state, as it occurs once extracted from the ground but before processing and refining.

crust The lightweight outer layer of Earth, consisting of *rock* that floats atop the malleable *mantle,* which in turn surrounds a mostly iron *core.*

current The flow of a liquid or gas in a certain direction.

customary law International law that arises from long-standing practices, or customs, held in common by most cultures. Compare *conventional law.*

dam Any obstruction placed in a river or stream to block the flow of water so that water can be stored in a *reservoir.* Dams are built to prevent floods, provide drinking water, facilitate *irrigation,* and generate *electricity.*

Darwin, Charles (1809–1882) English naturalist who proposed the concept of *natural selection* as a mechanism for *evolution* and as a way to explain the great variety of living things. Compare *Wallace, Alfred Russel.*

data Information, generally quantitative information.

deep-well injection A *hazardous waste* disposal method in which a well is drilled deep beneath an area's *water table* into porous *rock* below an impervious *soil* layer. Wastes are then injected into the well, so that they will be absorbed into the porous *rock* and remain deep underground, isolated from *groundwater* and human contact. Compare *surface impoundment.*

deforestation The clearing and loss of *forests.*

demographer A social scientist who studies the population size; density; distribution; age structure; sex ratio; and rates of birth, death, immigration, and emigration of human populations. See *demography.*

demographic fatigue An inability on the part of governments to address overwhelming challenges related to population growth.

demographic transition A theoretical *model* of economic and cultural change that explains the declining death rates and birth rates that occurred in Western nations as they became industrialized. The model holds that industrialization caused these rates to fall naturally by decreasing mortality and by lessening the need for large families. Parents would thereafter choose to invest in quality of life rather than quantity of children.

demography A *social science* that applies the principles of *population ecology* to the study of statistical change in human *populations.*

denitrifying bacteria Bacteria that convert the nitrates in *soil* or water to gaseous *nitrogen* and release it back into the *atmosphere.*

density-dependent The condition of a *limiting factor* whose effects on a *population* increase or decrease depending on the *population density.* Compare *density-independent.*

density-independent The condition of a *limiting factor* whose effects on a *population* are constant regardless of *population density.* Compare *density-dependent.*

deoxyribonucleic acid (DNA) A double-stranded *nucleic acid* composed of four nucleotides, each of which contains a sugar (deoxyribose), a phosphate group, and a nitrogenous base. DNA carries the hereditary information for living organisms and is responsible for passing traits from parents to offspring. Compare *RNA.*

dependent variable The *variable* that is affected by manipulation of the *independent variable* in an *experiment.*

deposition The arrival of eroded *soil* at a new location. Compare *erosion.*

desalination (desalinization) The removal of salt from seawater to generate *fresh water* for human use.

descriptive science Research in which scientists gather basic information about organisms, materials, systems, or processes that are not yet well known. Compare *hypothesis-driven science*.

desert The driest *biome* on Earth, with annual *precipitation* of less than 25 cm. Because deserts have relatively little vegetation to insulate them from temperature extremes, sunlight readily heats them in the daytime, but daytime heat is quickly lost at night, so temperatures vary widely from day to night and in different seasons.

desertification A form of *land degradation* in which more than 10% of a land's productivity is lost due to *erosion, soil* compaction, forest removal, *overgrazing*, drought, *salinization, climate* change, water depletion, or other factors. Severe desertification can result in the expansion of desert areas or creation of new ones. Compare *land degradation; soil degradation*.

development The use of natural resources for economic advancement (as opposed to simple subsistence, or survival).

directional drilling A drilling technique (e.g., for oil or natural gas) in which a drill bores down vertically and then bends horizontally to follow layered deposits for long distances from the drilling site. This enables extracting more *fossil fuels* with less environmental impact on the surface.

discounting A practice in *neoclassical economics* by which short-term costs and benefits are granted more importance than long-term costs and benefits. Future effects are thereby "discounted," because the idea is that an impact far in the future should count much less than one in the present.

disturbance An event that affects environmental conditions rapidly and drastically, resulting in changes to the *community* and *ecosystem*. Disturbance can be natural or can be caused by people.

divergent plate boundary The area where tectonic plates push apart from one another as *magma* rises upward to the surface, creating new *lithosphere* as it cools and spreads. A prime example is the Mid-Atlantic Ridge. Compare *convergent plate boundary; transform plate boundary*.

dose The amount of *toxicant* a test animal receives in a dose-response test. Compare *response*.

dose-response analysis A set of experiments that measure the *response* of test animals to different *doses* of a *toxicant*. The response is generally quantified by measuring the proportion of animals exhibiting negative effects.

dose-response curve A curve that plots the *response* of test animals to different *doses* of a *toxicant*, as a result of *dose-response analysis*.

downwelling In the ocean, the flow of warm *surface water* toward the ocean floor. Downwelling occurs where surface *currents* converge. Compare *upwelling*.

drainage basin The entire area of land from which water drains into a given body of water.

Dust Bowl An area that loses huge amounts of *topsoil* to wind *erosion* as a result of drought and/or human impact. First used to name the region in the North American Great Plains severely affected by drought and topsoil loss in the 1930s. The term is now also used to describe that historical event and others like it.

dynamic equilibrium The state reached when processes within a *system* are moving in opposing directions at equivalent rates so that their effects balance out.

e-waste See *electronic waste*.

earthquake A release of energy that occurs as Earth relieves accumulated pressure between masses of *lithosphere* and that results in shaking at the surface.

ecocentrism A philosophy that considers actions in terms of their damage or benefit to the integrity of whole ecological systems, including both living and nonliving elements. For an ecocentrist, the well-being of an individual is less important than the long-term well-being of a larger integrated ecological system. Compare *anthropocentrism; biocentrism*.

ecolabeling The practice of designating on a product's label how the product was grown, harvested, or manufactured, so that consumers can judge which brands use more sustainable processes.

ecological economics A school of *economics* that applies the principles of *ecology* and *systems* thinking to the description and analysis of *economies*. Compare *environmental economics; neoclassical economics*.

ecological footprint A metric that measures the cumulative area of biologically productive land and water required to provide the resources a person or population consumes and to dispose of or recycle the waste the person or population produces. The total area of Earth's biologically productive surface that a given person or population "uses" once all direct and indirect impacts are summed together.

ecological modeling The practice of constructing and testing *models* that aim to explain and predict how ecological *systems* function.

ecological restoration Efforts to reverse the effects of human disruption of ecological systems and to restore *communities* to their condition before the disruption. The practice that applies principles of *restoration ecology*.

ecologist A scientist who studies *ecology*.

ecology The *science* that deals with the distribution and abundance of organisms, the interactions among them, and the interactions between organisms and their nonliving *environments*.

economic growth An increase in an economy's activity—that is, an increase in the production and consumption of goods and services.

economics The study of how we decide to use potentially scarce resources to satisfy demand for goods and services.

economy A social *system* that converts resources into *goods* and *services*.

ecosystem In *ecology,* an assemblage of all organisms and nonliving entities that occur and interact in a particular area at the same time.

ecosystem-based management The attempt to manage the harvesting of resources in ways that minimize impact on the *ecosystems* and ecological processes that provide the resources.

ecosystem diversity The number and variety of ecosystems in a particular area. One way to express *biodiversity*. Related concepts consider the geographic arrangement of *habitats, communities,* or *ecosystems* at the landscape level, including the sizes, shapes, and interconnectedness of patches of these entities.

ecosystem ecology The scientific study of how the living and nonliving components of *ecosystems* interact.

ecosystem services Processes or the outcomes of processes that naturally result from the normal functioning of ecological systems and from which human beings draw benefits. Examples include nutrient cycling, air and water purification, climate regulation, *pollination,* waste recycling, and more.

ecotone A transitional zone where *ecosystems* meet.

ecotourism Visitation of natural areas for tourism and recreation. Most often involves tourism by more-affluent people, which may generate *economic* benefits for less-affluent communities near natural areas and thereby provide economic incentives for conservation of natural areas.

ED_{50} (effective dose–50%) The amount of a *toxicant* it takes to affect 50% of a *population* of test animals. Compare LD_{50}; *threshold dose*.

edge effect An impact on organisms, populations, or communities that results because conditions along the edge of a habitat fragment differ from conditions in the interior.

El Niño–Southern Oscillation (ENSO) A systematic shift in atmospheric pressure, sea surface temperature, and ocean circulation in the tropical Pacific Ocean. ENSO cycles give rise to *El Niño* and *La Niña* conditions.

electricity A secondary form of energy that can be transferred over long distances and applied for a variety of uses.

electrolysis A process in which electrical current is passed through a *compound* to release *ions*. Electrolysis offers one way to produce *hydrogen* for use as fuel: Electrical current is passed through water, splitting the water *molecules* into hydrogen and *oxygen atoms*.

electron A negatively charged particle that moves about the nucleus of an *atom*.

electronic waste Discarded electronic products such as computers, monitors, printers, televisions, DVD players, cell phones, and other devices. *Heavy metals* in these products mean that this *waste* may be judged hazardous. Also known as *e-waste*.

element A fundamental type of *matter;* a chemical substance with a given set of properties, which cannot be broken down into substances with other properties. Chemists currently recognize 92 elements that occur in nature, as well as more than 20 others that have been artificially created.

emissions trading The practice of buying and selling government-issued marketable permits to emit pollutants. Under a *cap-and-trade* emissions trading system, the government determines an acceptable level of *pollution* and then issues permits to pollute. A company receives credit for amounts it does not emit and can then sell this credit to other companies.

endangered In danger of becoming extinct in the near future.

Endangered Species Act The primary *legislation,* enacted in 1973, for protecting *biodiversity* in the United States. It forbids the government and private citizens from taking actions (such as developing land) that would destroy *threatened* and *endangered species* or their *habitats,* and it prohibits trade in products made from *threatened* and *endangered* species.

endemic Restricted to a particular geographic region. An endemic *species* occurs in one area and nowhere else on Earth.

endocrine disruptor A *toxicant* that interferes with the *endocrine (hormone) system.*

energy The capacity to change the position, physical composition, or temperature of *matter;* a force that can accomplish work.

energy conservation The practice of reducing *energy* use as a way of extending the lifetime of our *fossil fuel* supplies, of being less wasteful, and of reducing our impact on the *environment.* Conservation can result from behavioral decisions or from technologies that demonstrate *energy efficiency.*

energy efficiency The ability to obtain a given result or amount of output while using less energy input. Technologies permitting greater

energy efficiency are one main route to *energy conservation.*

energy intensity A measure of energy use per dollar of Gross Domestic Product (GDP). Lower energy intensity indicates greater efficiency.

energy returned on investment See *EROI.*

enhanced geothermal systems A new approach whereby engineers drill deeply into *rock,* fracture it, pump in water, and then pump it out once it is heated belowground. This approach would enable us to obtain *geothermal energy* in many locations.

environment The sum total of our surroundings, including all of the living things and nonliving things with which we interact.

environmental economics A school of *economics* that modifies the principles of *neoclassical economics* to address environmental challenges. Most environmental economists believe that we can attain *sustainability* within our current economic systems. Whereas ecological economists call for revolution, environmental economists call for reform. Compare *ecological economics; neoclassical economics.*

environmental ethics The application of *ethical standards* to environmental questions.

environmental health The study of environmental factors that influence human health and quality of life and the health of *ecological* systems essential to environmental quality and long-term human well-being.

environmental impact statement (EIS) A report of results from detailed scientific studies that assess the potential effects on the *environment* that would likely result from development projects or other actions undertaken by the government.

environmental justice The fair and equitable treatment of all people with respect to environmental policy and practice, regardless of their income, race, or ethnicity. This principle is a response to the perception that minorities and the poor suffer more pollution than the majority and the more affluent.

environmental literacy A basic understanding of Earth's physical and living systems and how we interact with them. Some people take the term further and use it to refer to a deeper understanding of society and the environment and/or a commitment to advocate for *sustainability.*

environmental policy *Public policy* that pertains to human interactions with the *environment.* It generally aims to regulate resource use or reduce *pollution* to promote human welfare and/or protect natural systems.

Environmental Protection Agency (EPA) An administrative agency of the U.S. federal government charged with conducting and evaluating research, monitoring environmental

quality, setting standards, enforcing those standards, assisting the states in meeting standards and goals for environmental protection, and educating the public.

environmental science The scientific study of how the natural world functions, how our *environment* affects us, and how we affect our environment.

environmental studies An academic *environmental science* program that emphasizes the social sciences as well as the natural sciences.

environmental toxicology The study of *toxicants* that come from or are discharged into the *environment,* including the study of health effects on humans, other animals, and *ecosystems.*

environmentalism A social movement dedicated to protecting the natural world and, by extension, people.

epidemiological study A study that involves large-scale comparisons among groups of people, usually contrasting a group known to have been exposed to some *toxicant* and a group that has not.

EROI (energy returned on investment) The ratio determined by dividing the quantity of *energy* returned from a process by the quantity of energy invested in the process. Higher EROI ratios mean that more energy is produced from each unit of energy invested. Compare *net energy.*

erosion The removal of material from one place and its transport to another by the action of wind or water. Compare *deposition.*

estuary An area where a river flows into the ocean, mixing *fresh water* with saltwater.

ethanol The alcohol in beer, wine, and liquor, produced as a *biofuel* by fermenting biomass, generally from *carbohydrate*-rich crops such as corn or sugarcane.

ethical standard A criterion that helps differentiate right from wrong.

ethics The academic study of good and bad, right and wrong. The term can also refer to a person's or group's set of moral principles or values.

eutrophic Term describing a water body that has high-nutrient and low-oxygen conditions. Compare *oligotrophic.*

eutrophication The process of *nutrient* enrichment, increased production of organic matter, and subsequent *ecosystem* degradation in a water body.

evaporation The conversion of a substance from a liquid to a gaseous form.

evolution Genetically based change in *populations* of organisms across generations. Changes in *genes* may lead to changes in the appearance, physiology, and/or behavior of organisms across generations, often by the process of *natural selection*.

experiment An activity designed to test the validity of a *hypothesis* by manipulating *variables*. See *controlled experiment*.

exponential growth The increase of a *population* (or of anything) by a fixed percentage each year. Results in a J-shaped curve on a graph. Compare *logistic growth*.

external cost A cost borne by someone not involved in an economic transaction. Examples include harm to citizens from *water pollution* or *air pollution* discharged by nearby factories.

extinction The disappearance of an entire *species* from Earth. Compare *extirpation*.

extirpation The disappearance of a particular *population* from a given area, but not the entire *species* globally. Compare *extinction*.

family planning The effort to plan the number and spacing of one's children to offer children and parents the best quality of life possible.

farmers' market A market at which local farmers and food producers sell fresh, locally grown items.

fee-and-dividend A *carbon tax* program in which proceeds from the tax are paid to consumers as a tax refund or "dividend." This strategy seeks to prevent consumers from losing money if polluters pass their costs along to them.

feedback loop A circular process in which a *system*'s output serves as input to that same system. See *negative feedback loop; positive feedback loop*.

feed-in tariff A program of public policy intended to promote renewable energy investment, whereby utilities are mandated to purchase electricity from homeowners or businesses that generate power from renewable energy sources and feed it into the electrical grid. Under such a system, utilities must pay guaranteed premium prices for this power under long-term contract. Compare *net metering*.

feedlot A huge indoor or outdoor pen designed to deliver *energy*-rich food to animals living at extremely high densities. Also called a factory farm or concentrated animal feeding operation.

fertilizer A substance that promotes plant growth by supplying essential *nutrients* such as *nitrogen* or *phosphorus*.

first law of thermodynamics The physical law stating that *energy* can change from one form to another, but cannot be created or lost. The total energy in the universe remains constant and is said to be conserved.

flooding The spillage of water over a river's banks due to heavy rain or snowmelt.

floodplain The region of land over which a river has historically wandered and periodically floods.

flux The movement of nutrients among *pools* or *reservoirs* in a *nutrient cycle*.

food chain A linear series of feeding relationships. As organisms feed on one another, energy and matter are transferred from lower to higher *trophic levels*. Compare *food web*.

food security The guaranteed availability of an adequate, safe, nutritious, and reliable food supply to all people at all times.

food web A visual representation of feeding interactions within an *ecological community* that shows an array of relationships between organisms at different *trophic levels*. Compare *food chain*.

forensic science The scientific analysis of evidence to make an identification or answer a question relating to a crime or an accident. Often called forensics.

forest Any ecosystem characterized by a high density of trees.

forest type A category of *forest* defined by its predominant tree *species*.

forestry The professional management of *forests*.

fossil The remains, impression, or trace of an animal or plant of past geologic ages that has been preserved in *rock* or *sediments*.

fossil fuel A *nonrenewable natural resource*, such as *crude oil, natural gas,* or *coal,* produced by the decomposition and compression of organic matter from ancient life. Fossil fuels have provided most of society's *energy* since the *industrial revolution*.

fossil record The cumulative body of *fossils* worldwide, which paleontologists study to infer the history of past life on Earth.

fracking See *hydraulic fracturing*.

Frank R. Lautenberg Chemical Safety for the 21st Century Act U.S. legislation, enacted in 2016, that updates the *Toxic Substances Control Act* and directs the EPA to monitor and regulate industrial chemicals.

free rider A party that fails to invest in conserving resources, controlling *pollution,* or carrying out other responsible activities and instead relies on the efforts of other parties to do so. For example, a factory that fails to control its emissions gets a "free ride" on the efforts of other factories that do.

fresh water Water that is relatively pure, holding very few dissolved salts.

fuel cell A device that can store and transport energy to produce electricity, much as a battery can. A hydrogen fuel cell generates electricity by the input of hydrogen fuel and oxygen, producing only water and heat as waste products.

Fukushima Daiichi Japanese nuclear power plant severely damaged by the tsunami associated with the March 2011 Tohoku earthquake that rocked Japan. Most radiation drifted over the ocean away from population centers, but the event was history's second most serious nuclear accident. Compare *Chernobyl; Three Mile Island*.

full cost accounting An accounting approach that attempts to summarize all costs and benefits by assigning monetary values to entities without market prices and then generally subtracting costs from benefits. Examples include the *Genuine Progress Indicator,* the Happy Planet Index, and others. Also called true cost accounting.

gene A stretch of *DNA* that represents a unit of hereditary information.

General Mining Act of 1872 U.S. law that legalized and promoted *mining* by private individuals on public lands for just $5 per acre, subject to local customs, with no government oversight.

generalist A *species* that can survive across a wide array of *habitats* or can use a wide array of resources. Compare *specialist*.

genetic diversity A measurement of the differences in *DNA* composition among individuals within a given *species*.

genetic engineering Any process scientists use to manipulate an organism's genetic material in the lab by adding, deleting, or changing segments of its *DNA*.

genetically modified organism (GMO) An organism that has been *genetically engineered* using recombinant DNA technology.

gentrification The transformation of a neighborhood to conditions (such as expensive housing and high-end shops and restaurants) that cater to wealthier people. Often results in longtime lower-income residents being "priced out" of their homes or apartments.

Genuine Progress Indicator (GPI) A *full cost accounting* indicator that attempts to differentiate between desirable and undesirable economic activity. The GPI accounts for benefits such as volunteerism and for costs such as environmental degradation and social upheaval. Compare *Gross Domestic Product (GDP)*.

geoengineering Any of a suite of proposed efforts to cool Earth's *climate* by removing *carbon dioxide* from the atmosphere or reflecting sunlight away from Earth's surface. Such ideas are controversial and are not nearly ready to implement.

geographic information system (GIS) Computer software that takes multiple types of data (for instance, on geology, hydrology, vegetation, animal species, and human development) and overlays them on a common set of geographic coordinates. GIS is used to create a complete picture of a landscape and to analyze how elements of the different datasets are arrayed spatially and how they may be correlated. A common tool of geographers, landscape ecologists, resource managers, and conservation biologists.

geology The scientific study of Earth's physical features, processes, and history.

geothermal energy Thermal energy that arises from beneath Earth's surface, ultimately from the radioactive decay of elements amid high pressures deep underground. Can be used to generate electrical power in power plants, for direct heating via piped water, or in *ground-source heat pumps.*

global climate change Systematic change in aspects of Earth's *climate,* such as temperature, *precipitation,* and storm intensity. Generally refers today to the current warming trend in global temperatures and the many associated climatic changes. Compare *global warming.*

global warming An increase in Earth's average surface temperature. The term is most frequently used in reference to the pronounced warming trend of recent decades. Global warming is one aspect of *global climate change* and in turn drives other components of climate change.

globalization The ongoing process by which the world's societies have become more interconnected, linked in many ways by diplomacy, commercial trade, and communication technologies.

Great Pacific Garbage Patch A portion of the North Pacific *gyre* where *currents* concentrate plastics and other floating debris that pose danger to marine organisms.

green building (1) A structure that minimizes the ecological footprint of its construction and operation by using sustainable materials, using minimal energy and water, reducing health impacts, limiting pollution, and recycling waste. (2) The pursuit of constructing or renovating such buildings.

green-collar job A job resulting from an employment opportunity in a more sustainably oriented *economy,* such as a job in *renewable energy.*

Green Revolution An intensification of the industrialization of *agriculture* in the developing world in the latter half of the 20th century that dramatically increased crop yields produced per unit area of farmland. Practices include devoting large areas to *monocultures* of crops specially bred for high yields and rapid growth; heavy use of *fertilizers, pesticides,* and *irrigation* water; and sowing and harvesting on the same parcel of land more than once per year or per season.

green tax A levy on environmentally harmful activities and products aimed at providing a market-based incentive to correct for *market failure.* Compare *subsidy.*

greenhouse effect The warming of Earth's surface and *atmosphere* (especially the *troposphere*) caused by the *energy* emitted by *greenhouse gases.*

greenhouse gas A gas that absorbs infrared radiation released by Earth's surface and then warms the surface and *troposphere* by emitting *energy,* thus giving rise to the *greenhouse effect.* Greenhouse gases include *carbon dioxide* (CO_2), water vapor, *ozone* (O_3), nitrous oxide (N_2O), halocarbon gases, and *methane* (CH_4).

greenwashing A public relations effort by a corporation or institution to mislead customers or the public into thinking it is acting more sustainably than it actually is.

Gross Domestic Product (GDP) The total monetary value of final goods and services produced in a country each year. GDP sums all *economic* activity, whether good or bad, and does not account for benefits such as volunteerism or for *external costs* such as environmental degradation and social upheaval. Compare *Genuine Progress Indicator (GPI).*

gross primary production The *energy* that results when *autotrophs* convert solar energy (sunlight) to energy of chemical bonds in sugars through *photosynthesis.* Autotrophs use a portion of this production to power their own metabolism, which entails oxidizing *organic compounds* by *cellular respiration.* Compare *net primary production.*

ground-source heat pump A pump that harnesses *geothermal energy* from near-surface sources of earth and water to heat and cool buildings. Operates on the principle that temperatures belowground are less variable than temperatures aboveground.

groundwater Water held in *aquifers* underground. Compare *surface water.*

gyre An area of the ocean where *currents* converge and floating debris accumulates.

habitat The specific *environment* in which an organism lives, including both biotic (living) and abiotic (nonliving) elements.

habitat fragmentation The process by which an expanse of natural *habitat* becomes broken up into discontinuous fragments, often as a result of farming, logging, road building, and other types of human development and land use.

habitat selection The process by which organisms select *habitats* from among the range of options they encounter.

habitat use The process by which organisms use *habitats* from among the range of options they encounter.

half-life The amount of time it takes for one-half the atoms of a *radioisotope* to emit radiation and decay. Different radioisotopes have different half-lives, ranging from fractions of a second to billions of years.

halocarbon A class of human-made chemical *compounds* derived from simple *hydrocarbons* in which *hydrogen* atoms are replaced by halogen atoms such as bromine, fluorine, or chlorine. Many halocarbons are *ozone-depleting substances* and/or *greenhouse gases.*

harmful algal bloom A *population* explosion of toxic algae caused by excessive *nutrient* concentrations.

hazardous waste Liquid or solid *waste* that is toxic, chemically reactive, flammable, or corrosive. Compare *industrial solid waste; municipal solid waste.*

herbivory The consumption of plants by animals.

heterotroph (consumer) An organism that consumes other organisms. Includes most animals, as well as fungi and microbes that decompose organic matter.

homeostasis The tendency of a *system* to maintain constant or stable internal conditions.

humus A dark, spongy, crumbly mass of material made up of complex organic compounds, resulting from the partial decomposition of organic matter.

hydraulic fracturing A process to extract tight oil or *shale gas,* in which a drill is sent deep underground and angled horizontally into a shale formation; water, sand, and chemicals are pumped in under great pressure, fracturing the *rock;* and gas migrates up through the drilling pipe as sand holds the fractures open. Also called hydrofracking or simply *fracking.*

hydrocarbon An *organic compound* consisting solely of *hydrogen* and *carbon* atoms.

hydroelectric power The generation of *electricity* using the *kinetic energy* of moving water. Also called *hydropower.*

hydrogen The chemical *element* with one proton. The most abundant element in the universe. Also a possible fuel for our future economy.

hydrologic cycle The flow of water—in liquid, gaseous, and solid forms—through our *biotic* and *abiotic environment.*

hydropower See *hydroelectric power.*

hydrosphere All water—salt or fresh, liquid, ice, or vapor—in surface bodies, underground, and in the *atmosphere.* Compare *biosphere; lithosphere.*

hypothesis A statement that attempts to explain a phenomenon or answer a *scientific* question. Compare *theory.*

hypothesis-driven science Research in which scientists pose questions that seek to explain how and why things are the way they are. Generally proceeds in a somewhat structured manner, using *experiments* to test *hypotheses*. Compare *descriptive science*.

hypoxia The condition of extremely low dissolved *oxygen* concentrations in a body of water.

igneous rock One of the three main categories of *rock*. Formed from cooling *magma*. Granite and basalt are examples of igneous *rock*. Compare *metamorphic rock; sedimentary rock*.

incineration A controlled process of burning solid *waste* for disposal in which mixed garbage is combusted at very high temperatures. Compare *sanitary landfill*.

independent variable The *variable* that a scientist manipulates in an *experiment*.

indoor air pollution *Air pollution* that occurs indoors.

industrial agriculture *Agriculture* that uses large-scale mechanization and *fossil fuel* combustion, enabling farmers to replace horses and oxen with faster and more powerful means of cultivating, harvesting, transporting, and processing crops. Other aspects include large-scale *irrigation* and the use of *inorganic fertilizers*. Use of chemical herbicides and *pesticides* reduces *competition* from weeds and *herbivory* by insects. Compare *traditional agriculture*.

industrial ecology A holistic approach to industry that integrates principles from engineering, chemistry, *ecology, economics*, and other disciplines and seeks to redesign industrial *systems* in order to reduce resource inputs and minimize inefficiency.

industrial revolution The shift beginning in the mid-1700s from rural life, animal-powered agriculture, and manufacturing by craftsmen to an urban society powered by *fossil fuels*. Compare *agricultural revolution*.

industrial smog "Gray-air" *smog* caused by the incomplete combustion of *coal* or oil when burned. Compare *photochemical smog*.

industrial solid waste Nonliquid *waste* that is not especially hazardous and that comes from production of consumer goods, mining, *petroleum* extraction and *refining*, and *agriculture*. Compare *hazardous waste; municipal solid waste*.

industrial stage The third stage of the *demographic transition* model, characterized by falling birth rates that close the gap with falling death rates and reduce the rate of *population* growth. Compare *post-industrial stage; pre-industrial stage; transitional stage*.

infant mortality rate The number of deaths of infants under 1 year of age per 1000 live births in a population.

infectious disease A disease in which a pathogen attacks a host. Compare *noninfectious disease*.

inorganic fertilizer A *fertilizer* that consists of mined or synthetically manufactured mineral supplements. Inorganic fertilizers are generally more susceptible than *organic fertilizers* to *leaching* and *runoff* and may be more likely to cause unintended off-site impacts.

integrated pest management (IPM) The use of multiple techniques in combination to achieve long-term suppression of pests, including *biological control*, use of *pesticides*, close monitoring of *populations, habitat* alteration, *crop rotation, transgenic* crops, alternative tillage methods, and mechanical pest removal.

intercropping Planting different types of crops in alternating bands or other spatially mixed arrangements.

interdisciplinary Involving or borrowing techniques from multiple traditional fields of study and bringing together research results from these fields into a broad synthesis.

intertidal Of, relating to, or living along shorelines between the highest reach of the highest *tide* and the lowest reach of the lowest tide.

introduced species A *species* introduced by human beings from one place to another (whether intentionally or by accident). Some introduced species may become *invasive species*.

invasive species A *species* that spreads widely and rapidly becomes dominant in a *community*, interfering with the community's normal functioning.

inversion layer A band of air in which temperature rises with altitude (that is, in which the normal direction of temperature change is inverted). Cool air at the bottom of the inversion layer is denser than the warm air above, so it resists vertical mixing and remains stable. A key feature of a *temperature inversion*.

ion An electrically charged *atom* or combination of atoms.

ionic bond A type of chemical bonding in which electrons are transferred between atoms, creating oppositely charged ions that bond due to their differing electrical charges. Table salt, sodium chloride, is formed by the bonding of positively charged sodium ions with negatively charged chloride ions.

IPAT model A formula that represents how humans' total impact (I) on the *environment* results from the interaction among three factors: *population* (P), affluence (A), and technology (T).

irrigation The artificial provision of water to support *agriculture*.

isotope One of several forms of an *element* having differing numbers of *neutrons* in the nucleus of its *atoms*. Chemically, isotopes of an element behave almost identically, but they have different physical properties because they differ in mass.

kelp Large brown algae, or seaweed, that can form underwater "forests," providing habitat for marine organisms.

keystone species A *species* that has an especially far-reaching effect on a *community*.

kinetic energy *Energy* of motion. Compare *potential energy*.

Kyoto Protocol An international agreement drafted in 1997 that called for reducing, by 2012, emissions of six *greenhouse gases* to levels lower than those in 1990. It was extended to 2020 as nations worked toward the Paris Accord. An outgrowth of the *U.N. Framework Convention on Climate Change*.

La Niña An exceptionally strong cooling of *surface water* in the equatorial Pacific Ocean that occurs every 2–8 years and has widespread climatic consequences. Compare *El Niño*.

land trust A local or regional organization that preserves lands valued by its members. In most cases, land trusts purchase land outright with the aim of preserving it in its natural condition.

landfill gas A mix of gases that consists of roughly half *methane* and that is produced by anaerobic decomposition deep inside *landfills*.

landscape ecology The study of how landscape structure affects the abundance, distribution, and interaction of organisms. This approach to the study of organisms and their *environments* at the landscape scale focuses on broad geographic areas that include multiple *ecosystems*.

landslide The collapse and downhill flow of large amounts of *rock* or *soil*. A severe and sudden form of *mass wasting*.

lava *Magma* that is released from the *lithosphere* and flows or spatters across Earth's surface.

law of conservation of matter The physical law stating that *matter* may be transformed from one type of substance into others, but that it cannot be created or destroyed.

LD$_{50}$ (lethal dose–50%) The amount of a *toxicant* it takes to kill 50% of a *population* of test animals. Compare *ED$_{50}$; threshold dose*.

leachate Liquid that results when substances from waste dissolve in water as rainwater percolates downward. Leachate may sometimes seep through liners of a *sanitary landfill* and leach into the *soil* underneath.

leaching The process by which *minerals* dissolved in a liquid (usually water) are transported to another location (generally downward through *soil horizons*).

lead (Pb) A heavy metal that may be ingested through water or paint, or that may enter the *atmosphere* as a particulate pollutant through combustion of leaded gasoline or other processes. Atmospheric lead deposited on land and water can enter the *food chain*, accumulate within body tissues, and cause *lead poisoning* in animals and people. An EPA *criteria pollutant*.

lead poisoning Poisoning by ingestion or inhalation of the heavy metal lead, causing an array of maladies including damage to the brain, liver, kidney, and stomach; learning problems and behavioral abnormalities; anemia; hearing loss; and even death. Lead poisoning can result from drinking water that passes through old lead pipes or ingesting dust or chips of old lead-based paint.

Leadership in Energy and Environmental Design (LEED) The leading set of standards for certification of a *green building*.

legislation Statutory law passed by a legislative body.

Leopold, Aldo (1887–1949) American scientist, scholar, philosopher, and author. His book *The Land Ethic* argued that humans should view themselves and the land itself as members of the same *community* and that humans are obligated to treat the land *ethically*.

levee A long raised mound of earth erected along a river bank to protect against floods by holding rising water in the main channel. Synonymous with dike.

life-cycle analysis A quantitative analysis of inputs and outputs across the entire life-cycle of a product—from its origins, through its production, transport, sale, and use, and finally its disposal—in an attempt to judge the sustainability of the process and make it more ecologically efficient.

life expectancy The average number of years that individuals in particular age groups are likely to continue to live.

limiting factor A physical, chemical, or biological characteristic of the *environment* that restrains *population* growth.

lipids A class of chemical compounds that do not dissolve in water and are used in organisms for energy storage, for structural support, and as key components of cellular membranes.

lithosphere The outer layer of Earth, consisting of *crust* and uppermost *mantle* and located just above the *asthenosphere*. More generally, the solid part of Earth, including the *rocks, sediment,* and *soil* at the surface and extending down many miles underground. Compare *atmosphere; biosphere; hydrosphere*.

logistic growth The pattern of population growth that results as a *population* at first grows exponentially and then is slowed and finally brought to a standstill at *carrying capacity* by *limiting factors*. Results in an S-shaped curve on a graph. Compare *exponential growth*.

macromolecule A very large *molecule*, such as a *protein, nucleic acid, carbohydrate,* or *lipid*.

magma Molten, liquid *rock*.

malnutrition The condition of lacking *nutrients* the body needs, including a complete complement of vitamins and minerals.

mangrove A tree with a unique type of roots that curve upward to obtain *oxygen,* which is lacking in the mud in which they grow, or that curve downward to serve as stilts to support the tree in changing water levels. Mangrove forests grow on the coastlines of the tropics and subtropics.

mantle The malleable layer of *rock* that lies beneath Earth's *crust* and surrounds a mostly iron *core*.

marine protected area (MPA) An area of the ocean set aside to protect marine life from fishing pressures. An MPA may be protected from some human activities but be open to others. Compare *marine reserve*.

marine reserve An area of the ocean designated as a "no-fishing" zone, allowing no extractive activities. Compare *marine protected area (MPA)*.

market failure The failure of markets to take into account the *environment*'s positive effects on *economies* (for example, *ecosystem services*) or to reflect the negative effects of economic activity on the environment and thereby on people (*external costs*).

mass extinction event The *extinction* of a large proportion of the world's *species* in a very short time period due to some extreme and rapid change or catastrophic event. Earth has seen five mass extinction events in the past half-billion years.

mass transit A public transportation system for a metropolitan area that moves large numbers of people at once. Buses, trains, subways, streetcars, trolleys, and light rail are types of mass transit.

mass wasting The downslope movement of *soil* and *rock* due to gravity. Compare *landslide*.

materials recovery facility A *recycling* facility where items are sorted, cleaned, shredded, and prepared for reprocessing into new items. Often abbreviated as MRF.

matter All material in the universe that has mass and occupies space. See *law of conservation of matter*.

maximum sustainable yield The maximal harvest of a particular *renewable natural resource* that can be accomplished while still keeping the resource available for the future.

meltdown The accidental melting of the uranium fuel rods inside the core of a *nuclear reactor,* causing the release of radiation.

metal A type of chemical *element,* or a mass of such an element, that typically is lustrous, opaque, and malleable and that can conduct heat and electricity.

metamorphic rock One of the three main categories of *rock*. Formed by great heat and/or pressure that reshapes crystals within the *rock* and changes its appearance and physical properties. Common metamorphic *rocks* include marble and slate. Compare *igneous rock; sedimentary rock*.

methane hydrate An ice-like solid consisting of molecules of *methane* embedded in a crystal lattice of water molecules. Most is found in sediments on the continental shelves and in the Arctic. Methane hydrate is an unconventional *fossil fuel*.

Milankovitch cycle One of three types of variations in Earth's rotation and orbit around the sun that result in slight changes in the relative amount of solar radiation reaching Earth's surface at different latitudes. As the cycles proceed, they change the way solar radiation is distributed over Earth's surface and contribute to changes in atmospheric heating and circulation that have triggered *glaciations* and other *climate* changes.

mineral A naturally occurring solid *element* or inorganic *compound* with a crystal structure, a specific chemical composition, and distinct physical properties. Compare *ore; rock*.

mining (1) In the broad sense, the extraction of any resource that is nonrenewable on the timescale of our society (such as *fossil fuels* or *groundwater*). (2) In relation to *mineral* resources, the systematic removal of *rock, soil,* or other material for the purpose of extracting minerals of economic interest.

mitigation (re: *climate change*) The pursuit of strategies to lessen the severity of climate change, notably by reducing emissions of *greenhouse gases*. Compare *adaptation*.

model A simplified representation of a complex natural process, designed by scientists to help understand how the process occurs and to make predictions.

molecule A combination of two or more *atoms*.

monoculture The uniform planting of a single crop over a large area. Characterizes *industrial agriculture*. Compare *polyculture*.

Montreal Protocol International treaty ratified in 1987 in which 180 (now 196) signatory nations agreed to restrict production of *chlorofluorocarbons (CFCs)* in order to halt stratospheric *ozone* depletion. This was a protocol of the Vienna Convention for the Protection of the Ozone Layer. The Montreal Protocol is widely considered the most successful effort to date in addressing a global *environmental* problem.

mosaic In *landscape ecology,* a spatial configuration of *patches* arrayed across a landscape.

mountaintop removal mining A large-scale form of *coal* mining in which entire mountaintops are blasted away in order to extract the resource. While this process is economically efficient, large volumes of *rock* and *soil* slide downhill, causing extensive impacts on surrounding *ecosystems* and human residents.

Muir, John (1838–1914) Scottish immigrant to the United States who eventually settled in California and made the Yosemite Valley his wilderness home. Today, he is most strongly associated with the *preservation ethic.* He argued that nature deserved protection for its own *intrinsic value* (an *ecocentrist* argument) but also claimed that nature facilitated human happiness and fulfillment (an *anthropocentrist* argument).

municipal solid waste Nonliquid *waste* that is not especially hazardous and that comes from homes, institutions, and small businesses. Compare *hazardous waste; industrial solid waste.*

mutagen A *toxicant* that causes *mutations* in the *DNA* of organisms.

mutation An accidental change in *DNA* that may range in magnitude from the deletion, substitution, or addition of a single nucleotide to a change affecting entire sets of chromosomes. Mutations provide the raw material for evolutionary change.

mutualism A relationship in which all participating organisms benefit from their interaction. Compare *parasitism.*

National Environmental Policy Act (NEPA) A U.S. law enacted on January 1, 1970, that created an agency called the Council on Environmental Quality and requires that an *environmental impact statement* be prepared for any major federal action.

national forest An area of forested public land managed by the U.S. Forest Service. The system consists of 191 million acres (more than 8% of the nation's land area) in many tracts spread across all but a few states.

national park A scenic area set aside for recreation and enjoyment by the public and managed by the National Park Service. The U.S. national park system today numbers 397 sites totaling 84 million acres and includes national historic sites, national recreation areas, national wild and scenic rivers, and other areas.

national wildlife refuge An area of public land set aside to serve as a haven for wildlife and also sometimes to encourage hunting, fishing, wildlife observation, photography, environmental education, and other uses. The system of more than 560 sites is managed by the U.S. Fish and Wildlife Service.

natural capital Earth's accumulated wealth of *natural resources* and *ecosystem services.*

natural gas A *fossil fuel* consisting primarily of *methane* (CH_4) and including varying amounts of other volatile hydrocarbons.

natural resource Any of the various substances and *energy* sources that we take from our environment and that we need in order to survive.

natural sciences Academic disciplines that study the natural world. Compare *social sciences.*

natural selection The process by which traits that enhance survival and reproduction are passed on more frequently to future generations of organisms than traits that do not, thereby altering the genetic makeup of populations through time. Natural selection acts on genetic variation and is a primary driver of *evolution.*

negative feedback loop A *feedback loop* in which output of one type acts as input that moves the *system* in the opposite direction. The input and output essentially neutralize each other's effects, stabilizing the system. Compare *positive feedback loop.*

neoclassical economics A mainstream economic school of thought that explains market prices in terms of consumer preferences for units of particular commodities and that uses *cost-benefit analysis.* Compare *classical economics; ecological economics; environmental economics.*

net energy The quantitative difference between *energy* returned from a process and *energy* invested in the process. Positive net energy values mean that a process produces more energy than is invested. Compare *EROI.*

net metering Process by which homeowners or businesses with photovoltaic systems or *wind turbines* can sell their excess *solar energy* or *wind power* to their local utility. Whereas *feed-in tariffs* award producers with prices above market rates, net metering offers market-rate prices.

net primary production The *energy* or biomass that remains in an ecosystem after *autotrophs* have metabolized enough for their own maintenance through *cellular respiration.* Net primary production is the energy or biomass available for consumption by *heterotrophs.* Compare *gross primary production; secondary production.*

net primary productivity The rate at which *net primary production* is produced. See *gross primary production; net primary production; productivity; secondary production.*

neurotoxin A *toxicant* that assaults the nervous system. Neurotoxins include heavy metals, *pesticides,* and some chemical weapons developed for use in war.

neutron An electrically neutral (uncharged) particle in the nucleus of an *atom.*

new urbanism An approach among architects, planners, and developers that seeks to design neighborhoods in which homes, businesses, schools, and other amenities are within walking distance of one another, so that families can meet most of their needs close to home without the use of a car.

niche The functional role of a *species* in a community.

nitrification The conversion by bacteria of ammonium ions (NH_4^+) first into nitrite ions (NO_2^-) and then into nitrate ions (NO_3^-).

nitrogen The chemical *element* with seven *protons* and seven *neutrons.* The most abundant element in the *atmosphere,* a key element in *macromolecules,* and a crucial plant *nutrient.*

nitrogen cycle A major *nutrient cycle* consisting of the routes that *nitrogen atoms* take through the nested networks of environmental *systems.*

nitrogen dioxide (NO_2) A foul-smelling reddish brown gas that contributes to *smog* and *acid deposition.* It results when atmospheric *nitrogen* and *oxygen* react at the high temperatures created by combustion engines. An EPA *criteria pollutant.*

nitrogen fixation The process by which inert *nitrogen* gas combines with *hydrogen* to form ammonium ions (NH_4^+), which are chemically and biologically active and can be taken up by plants.

nitrogen-fixing bacteria Bacteria that live independently in the *soil* or water and can fix nitrogen, or those that form *mutualistic* relationships with many types of plants and provide *nutrients* to the plants by converting gaseous *nitrogen* to a usable form.

nitrogen oxide (NO_X) One of a family of compounds that include nitric oxide (NO) and *nitrogen dioxide (NO_2).*

nonconsumptive use Use of *fresh water* in which the water from a particular *aquifer* or surface water body either is not removed or is removed only temporarily and then returned. The use of water to generate electricity in hydroelectric *dams* is an example. Compare *consumptive use.*

nongovernmental organization (NGO) An organization not affiliated with any national government, and frequently international in scope, that pursues a particular mission or advocates for a particular cause.

noninfectious disease A disease that develops as a result of the interaction of an individual organism's genes, lifestyle, and environmental exposures, rather than by pathogenic infection. Compare *infectious disease.*

nonmarket value A value that is not usually included in the price of a good or service.

non-point source A diffuse source of *pollutants,* often consisting of many small sources. Compare *point source.*

nonrenewable natural resource A *natural resource* that is in limited supply and is formed much more slowly than we use it. Compare *renewable natural resource.*

North Atlantic Deep Water (NADW) The deep portion of the *thermohaline circulation* in the northern Atlantic Ocean.

no-till *Agriculture* that does not involve tilling (plowing, disking, harrowing, or chiseling) the *soil*. The most intensive form of *conservation tillage*.

novel community An ecological *community* composed of a novel mixture of organisms, with no current analog or historical precedent.

nuclear energy The *energy* that holds together *protons* and *neutrons* within the nucleus of an *atom*. Several processes, each of which involves transforming *isotopes* of one *element* into isotopes of other elements, can convert nuclear energy into thermal energy, which is then used to generate *electricity*. See also *nuclear fission; nuclear power; nuclear reactor*.

nuclear fission The conversion of the *energy* within an *atom*'s nucleus to usable thermal energy by splitting apart atomic nuclei.

nuclear power The use of *nuclear energy* to generate *electricity*. This is accomplished using *nuclear fission* within *nuclear reactors* in power plants.

nuclear reactor A facility within a nuclear power plant that initiates and controls the process of *nuclear fission* to generate electricity.

nucleic acid A *macromolecule* that directs the production of *proteins*. Includes *DNA* and *RNA*.

nutrient An *element* or *compound* that organisms consume and require for survival.

nutrient cycle The comprehensive set of cyclical pathways by which a given *nutrient* moves through the *environment*.

ocean acidification The process by which today's oceans are becoming more *acidic* (attaining lower *pH*) as a result of increased *carbon dioxide* concentrations in the atmosphere. Ocean acidification occurs as ocean water absorbs CO_2 from the air and forms carbonic acid. This impairs the ability of corals and other organisms to build exoskeletons of calcium carbonate, imperiling coral reefs and the many organisms that depend on them.

ocean thermal energy conversion An *energy* source (not yet commercially used) that involves harnessing the solar radiation absorbed by tropical ocean water by strategically manipulating the movement of warm surface water and cold deep water.

oil A *fossil fuel* produced by the slow underground conversion of *organic compounds* by heat and pressure. Oil is a mixture of hundreds of different types of *hydrocarbon* molecules characterized by *carbon* chains of different lengths. Compare *crude oil; petroleum*.

oil sands *Fossil fuel* deposits that can be mined from the ground, consisting of moist sand and clay containing 1–20% bitumen. Oil sands represent *crude oil* deposits that have been degraded and chemically altered by water *erosion* and bacterial decomposition. Also called *tar sands*.

oil shale *Sedimentary rock* filled with kerogen that can be processed to produce liquid *petroleum*. Oil shale is formed by the same processes that form *crude oil* but occurs when kerogen was not buried deeply enough or subjected to enough heat and pressure to form oil.

oligotrophic Term describing a water body that has low-nutrient and high-oxygen conditions. Compare *eutrophic*.

open pit mining A *mining* technique that involves digging a gigantic hole and removing the desired *ore*, along with waste *rock* that surrounds the ore.

ore A *mineral* or grouping of minerals from which we extract *metals*.

organic agriculture *Agriculture* that uses no synthetic *fertilizers* or *pesticides* but instead relies on biological approaches such as *composting* and *biological control*.

organic compound A *compound* made up of *carbon atoms* (and, generally, *hydrogen* atoms) joined by covalent bonds and sometimes including other *elements*, such as *nitrogen, oxygen,* sulfur, or *phosphorus*. The unusual ability of carbon to build elaborate *molecules* has resulted in millions of different organic compounds showing various degrees of complexity.

organic fertilizer A *fertilizer* made up of natural materials (largely the remains or wastes of organisms), such as animal manure, crop residues, charcoal, fresh vegetation, and *compost*. Compare *inorganic fertilizer*.

outdoor air pollution *Air pollution* that occurs outdoors. Also called ambient air pollution.

overgrazing The consumption by too many animals of plant cover, impeding plant regrowth and the replacement of *biomass*. Overgrazing can worsen damage to *soils*, natural *communities*, and the land's productivity for further grazing.

overnutrition A condition of excessive food intake in which people receive more than their daily caloric needs.

overshoot The amount by which humanity's resource use, as measured by its *ecological footprint*, has surpassed Earth's long-term capacity to support us.

oxygen The chemical *element* with eight *protons* and eight *neutrons*. A key element in the atmosphere that is produced by *photosynthesis*.

ozone-depleting substance One of a number of airborne chemicals, such as *halocarbons,* that destroy *ozone* molecules and thin the *ozone layer* in the *stratosphere*.

ozone hole Term popularly used to describe the thinning of the stratospheric *ozone layer* that occurs over Antarctica each year, as a result of *chlorofluorocarbons (CFCs)* and other *ozone-depleting substances*.

ozone layer A portion of the *stratosphere*, roughly 17–30 km (10–19 mi) above sea level, that contains most of the *ozone* in the *atmosphere*.

paradigm A dominant philosophical and theoretical framework within a scientific discipline.

parasitism A relationship in which one organism, the *parasite*, depends on another, the *host*, for nourishment or some other benefit while simultaneously doing the host harm. Compare *mutualism; predation*.

parent material The base geologic material in a particular location.

particulate matter Solid or liquid particles small enough to be suspended in the *atmosphere* and able to damage respiratory tissues when inhaled. Includes *primary pollutants,* such as dust and soot, as well as *secondary pollutants,* such as sulfates and nitrates. An *EPA criteria pollutant*.

passive solar energy collection An approach in which buildings are designed and building materials are chosen to maximize direct absorption of sunlight in winter and to keep the interior cool in the summer. Compare *active solar energy collection*.

patch In *landscape ecology,* spatial areas within a landscape. Depending on a researcher's perspective, patches may consist of habitat for a particular organism, or communities, or ecosystems. An array of patches forms a *mosaic*.

pathway inhibitor A *toxicant* that interrupts vital biochemical processes in organisms by blocking one or more steps in important biochemical pathways. Compounds in the herbicide atrazine kill plants by blocking key steps in the process of *photosynthesis*.

peak oil Term used to describe the point of maximum production of *petroleum* in the world (or for a given nation), after which oil production declines. This is expected to be roughly the midway point of extraction of the world's oil supplies.

peer review The process by which a scientific manuscript submitted for publication in an academic journal is examined by specialists in the field, who provide comments and criticism (generally anonymously) and judge whether the work merits publication in the journal.

pelagic Of, relating to, or living between the surface and floor of the ocean. Compare *benthic*.

pest A pejorative term for any organism that damages crops that we value. The term is subjective and defined by our own economic interests, and is not biologically meaningful. Compare *weed*.

pesticide An artificial chemical used to kill insects (called an insecticide), plants (called an herbicide), or fungi (called a fungicide).

petroleum See *oil*. However, the term is also used to refer to both *oil* and *natural gas* together.

pH A measure of the concentration of *hydrogen ions* in a *solution*. The pH scale ranges from 0 to 14: A solution with a pH of 7 is neutral; solutions with a pH below 7 are *acidic*, and those with a pH higher than 7 are *basic*. Because the pH scale is logarithmic, each step on the scale represents a 10-fold difference in hydrogen ion concentration.

phosphorus The chemical *element* with 15 *protons* and 15 *neutrons*. An abundant element in the *lithosphere*, a key element in *macromolecules*, and a crucial plant *nutrient*.

phosphorus cycle A major *nutrient cycle* consisting of the routes that *phosphorus atoms* take through the nested networks of environmental *systems*.

photochemical smog "Brown-air" *smog* formed by light-driven reactions of *primary pollutants* with normal atmospheric *compounds* that produce a mix of over 100 different chemicals, *tropospheric ozone* often being the most abundant among them. Compare *industrial smog*.

photosynthesis The process by which *autotrophs* produce their own food. Sunlight powers a series of chemical reactions that convert *carbon dioxide* and water into sugar (glucose), thus transforming low-quality *energy* from the sun into high-quality energy the organism can use. Compare *cellular respiration*.

photovoltaic (PV) cell A technology designed to collect sunlight and directly convert it to electrical *energy*. When light strikes one of a pair of metal plates in the cell, this causes the release of *electrons*, which are attracted by electrostatic forces to the opposing plate. The flow of electrons from one plate to the other creates an electrical current. This is the basis of PV solar power technology.

phthalates A class of endocrine-disrupting chemicals found in fire retardants and plasticizers.

phylogenetic tree A treelike diagram that represents the history of divergence of *species* or other taxonomic groups of organisms.

Pinchot, Gifford (1865–1946) The first professionally trained American *forester*, Pinchot helped establish the U.S. Forest Service. Today, he is the person most closely associated with the *conservation ethic*.

pioneer species A *species* that arrives earliest, beginning the ecological process of *succession* in a terrestrial or aquatic *community*.

placer mining A *mining* technique that involves sifting through material in modern or ancient riverbed deposits, generally using running water to separate lightweight mud and gravel from heavier *minerals* of value.

plate tectonics The process by which Earth's surface is shaped by the extremely slow movement of tectonic plates, or sections of *crust*. Earth's surface includes about 15 major tectonic plates. Their interaction gives rise to processes that build mountains, cause *earthquakes*, and otherwise influence the landscape.

poaching The illegal killing of wildlife, usually for meat or body parts.

point source A specific spot—such as a factory—where large quantities of *air pollutants* or *water pollutants* are discharged. Compare *non-point source*.

policy A rule or guideline that directs individual, organizational, or societal behavior.

pollination A plant-animal interaction in which one organism (for example, a bee or a hummingbird) transfers pollen (containing male sex cells) from flower to flower, fertilizing ovaries (containing female sex cells) that grow into fruits with seeds.

polluter-pays principle Principle specifying that the party responsible for producing *pollution* should pay the costs of cleaning up the pollution or mitigating its impacts.

polybrominated diphenyl ethers (PBDEs) A class of synthetic compounds that provide fire-retardant properties and are used in a diverse array of consumer products, including computers, televisions, plastics, and furniture. Released during production, disposal, and use of products, these chemicals persist and accumulate in living tissue and appear to be *endocrine disruptors*.

polyculture The planting of multiple crops in a mixed arrangement or in close proximity. An example is some traditional Native American farming that mixed maize, beans, squash, and peppers. Compare *monoculture*.

polymer A chemical *compound* or mixture of compounds consisting of long chains of repeated *molecules*. Important biological molecules, such as *DNA* and *proteins*, are examples of polymers.

population A group of organisms of the same *species* that live in the same area. Species are often composed of multiple populations.

population density The number of individuals within a *population* per unit area. Compare *population size*.

population distribution The spatial distribution of organisms in an area. Three common patterns are random, uniform, and clumped.

population ecology The scientific study of the quantitative dynamics of population change and the factors that affect the distribution and abundance of members of a population.

population growth rate The rate of change in a *population*'s size per unit time (generally expressed in percent per year), taking into accounts births, deaths, immigration, and emigration. Compare *rate of natural increase*.

population size The number of individual organisms present at a given time in a *population*.

positive feedback loop A *feedback loop* in which output of one type acts as input that moves the *system* in the same direction. The input and output drive the system further toward one extreme or another. Compare *negative feedback loop*.

post-industrial stage The fourth and final stage of the *demographic transition* model, in which both birth and death rates have fallen to a low level and remain stable there, and *populations* may even decline slightly. Compare *industrial stage; pre-industrial stage; transition stage*.

potential energy *Energy* of position. Compare *kinetic energy*.

precautionary principle The idea that one should not undertake a new action until the ramifications of that action are well understood.

precipitation Water that condenses out of the *atmosphere* and falls to Earth in droplets or crystals.

precision agriculture The use of technology to precisely monitor crop conditions, crop needs, and resource use to maximize production while minimizing waste of resources.

predation The process by which one *species* (the *predator*) searches for, captures, and ultimately kills and consumes its *prey*. Compare *parasitism*.

prediction A specific statement, generally arising from a *hypothesis*, that can be tested directly and unequivocally.

pre-industrial stage The first stage of the *demographic transition* model, characterized by conditions that defined most of human history. In pre-industrial societies, both death rates and birth rates are high. Compare *industrial stage; post-industrial stage; transitional stage*.

prescribed burn A low-intensity fire set by managers in *forest* or grassland under carefully controlled conditions to improve the health of *ecosystems*, return them to a more natural state, reduce fuel loads, and help prevent uncontrolled catastrophic fires.

preservation ethic An ethic holding that we should protect the natural *environment* in a pristine, unaltered state. Compare *conservation ethic*.

primary forest Natural *forest* uncut by people. Compare *secondary forest*.

primary pollutant A hazardous substance, such as soot or *carbon monoxide,* that is emitted into the *troposphere* directly from a source. Compare *secondary pollutant.*

primary production The conversion of solar energy to the energy of chemical bonds in sugars during *photosynthesis,* performed by *autotrophs.* Compare *secondary production.*

primary succession A stereotypical series of changes as an ecological *community* develops over time, beginning with a lifeless substrate. In terrestrial *systems,* primary succession begins when a bare expanse of *rock, sand,* or *sediment* becomes newly exposed to the atmosphere and *pioneer species* arrive. Compare *secondary succession.*

primary treatment A stage of *wastewater* treatment in which contaminants are physically removed. Wastewater flows into tanks in which sewage solids, grit, and particulate matter settle to the bottom. Greases and oils float to the surface and can be skimmed off. Compare *secondary treatment.*

productivity The rate at which plants convert solar *energy* (sunlight) to *biomass. Ecosystems* whose plants convert solar energy to biomass rapidly are said to have high productivity. See *gross primary production; net primary production; net primary productivity.*

protein A *macromolecule* made up of long chains of amino acids.

proton A positively charged particle in the nucleus of an *atom.*

proven recoverable reserve The amount of a given *fossil fuel* in a deposit that is technologically and economically feasible to remove under current conditions.

proxy indicator A source of indirect evidence that serves as a proxy, or substitute, for direct measurement and that sheds light on past *climate.* Examples include data from ice cores, sediment cores, tree rings, packrat middens, and coral reefs.

public policy *Policy* made by governments, including those at the local, state, federal, and international levels; it may consist of *legislation, regulations,* orders, incentives, and practices intended to advance societal welfare. See also *environmental policy.*

public trust doctrine A legal philosophy holding that natural resources such as air, water, *soil,* and wildlife should be held in trust for the public and that government should protect them from exploitation by private interests. This doctrine has its roots in ancient Roman law and in the Magna Carta.

pumped storage A technique used to generate *hydroelectric power,* in which water is pumped from a lower reservoir to a higher reservoir when power demand is weak and prices are low. When demand is strong and prices are high, water is allowed to flow downhill through a turbine, generating *electricity.* Compare *run-of-river; storage.*

radiative forcing The amount of change in thermal energy that a factor (such as a *greenhouse gas* or an *aerosol*) causes in influencing Earth's temperature. Positive forcing warms Earth's surface, whereas negative forcing cools it.

radioactive The quality by which some *isotopes* "decay," changing their chemical identity as they shed atomic particles and emit high-energy radiation.

radioisotope A radioactive *isotope* that emits subatomic particles and high-*energy* radiation as it "decays" into progressively lighter isotopes until becoming a stable isotope.

radon A highly *toxic,* radioactive, colorless gas that seeps up from the ground in areas with certain types of bedrock and that can build up inside basements and homes with poor air circulation.

rangeland Land used for grazing livestock.

rate of natural increase The rate of change in a *population*'s size resulting from birth and death rates alone, excluding migration. Compare *population growth rate.*

REACH Program of the European Union that shifts the burden of proof for testing chemical safety from national governments to industry and requires that chemical substances produced or imported in amounts of over 1 metric ton per year be registered with a new European Chemicals Agency. *REACH,* which stands for *R*egistration, *E*valuation, *A*uthorization, and restriction of *CH*emicals, went into effect in 2007.

rebound effect The phenomenon by which gains in efficiency from better technology are partly offset when people engage in more energy-consuming behavior as a result. This common psychological effect can sometimes reduce conservation and efficiency efforts substantially.

recharge zone An area where water infiltrates Earth's surface and reaches an *aquifer* below.

reclamation The act of restoring a *mining* site to an approximation of its pre-mining condition. To reclaim a site, companies are required to remove all mining structures, replace *overburden,* fill in mine shafts, and replant the area with vegetation.

recovery Waste management strategy composed of *recycling* and *composting.*

recycling The process by which materials are collected and then broken down and reprocessed to manufacture new items.

red tide A *harmful algal bloom* consisting of algae that produce reddish pigments that discolor surface waters.

refining The process of separating *molecules* of the various *hydrocarbons* in *crude oil* into different-sized classes and transforming them into various fuels and other petrochemical products.

regime shift A fundamental shift in the overall character of an ecological community, generally occurring after some extreme *disturbance,* and after which the community may not return to its original state. Also known as a *phase shift.*

regional planning Planning similar to *city planning* but conducted across broader geographic scales, generally involving multiple municipal governments.

regulation A specific rule issued by an administrative agency, based on the more broadly written statutory law passed by Congress and enacted by the president.

relativist An ethicist who maintains that *ethics* do and should vary with social context. Compare *universalist.*

renewable natural resource A *natural resource* that is virtually unlimited or that is replenished by the *environment* over relatively short periods (hours to weeks to years). Compare *nonrenewable natural resource.*

replacement fertility The *total fertility rate (TFR)* that maintains a stable *population* size.

reproductive window The portion of a woman's life between sexual maturity and menopause during which she may become pregnant.

reserves-to-production ratio The total remaining reserves of a *fossil fuel* divided by the annual rate of production (extraction and processing). Abbreviated as R/P ratio.

reservoir (1) An artificial water body behind a dam that stores water for human use. (2) A location in which nutrients in a *biogeochemical cycle* remain for a period of time before moving to another reservoir. Can be living or nonliving entities. Compare *flux; residence time.*

residence time (1) In a *biogeochemical cycle,* the amount of time a nutrient typically remains in a given *pool* or *reservoir* before moving to another. Compare *flux.* (2) In the *atmosphere,* the amount of time a gas molecule or a *pollutant* typically remains aloft.

resilience The ability of an ecological *community* to change in response to disturbance but later return to its original state. Compare *resistance.*

resistance The ability of an ecological *community* to remain stable in the presence of a disturbance. Compare *resilience.*

Resource Conservation and Recovery Act U.S. law (enacted in 1976 and amended in 1984) that specifies, among other things, how to manage *sanitary landfills* to protect against environmental contamination. Often abbreviated as RCRA.

resource management Strategic decision making about how to extract resources, so that resources are used wisely and conserved for the future.

resource partitioning The process by which *species* adapt to *competition* by evolving to use slightly different resources, or to use their shared resources in different ways, thus minimizing interference with one another.

response The type or magnitude of negative effects an animal exhibits in response to a *dose* of *toxicant* in a *dose-response analysis*. Compare *dose*.

restoration ecology The study of the historical conditions of ecological *communities* as they existed before humans altered them. Principles of restoration ecology are applied in the practice of *ecological restoration*.

revenue-neutral carbon tax A type of *fee-and-dividend* program in which funds from the *carbon tax* that a government collects are disbursed to citizens in the form of payments or tax refunds. It is "revenue-neutral" because the government neither gains nor loses revenue in the end.

ribonucleic acid (RNA) A usually single-stranded *nucleic acid* composed of four nucleotides, each of which contains a sugar (ribose), a phosphate group, and a nitrogenous base. RNA carries the hereditary information for living organisms and is responsible for passing traits from parents to offspring. Compare *DNA*.

risk The mathematical probability that some harmful outcome (for instance, injury, death, *environmental* damage, or *economic* loss) will result from a given action, event, or substance.

risk assessment The quantitative measurement of *risk*, together with the comparison of risks involved in different activities or substances.

risk management The process of considering information from scientific *risk assessment* in light of economic, social, and political needs and values, to make decisions and design strategies to minimize *risk*.

rock A solid aggregation of *minerals*.

rock cycle The very slow process in which *rocks* and the *minerals* that make them up are heated, melted, cooled, broken, and reassembled, forming *igneous, sedimentary,* and *metamorphic* rocks.

runoff The water from *precipitation* that flows into streams, rivers, lakes, and ponds, and (in many cases) eventually to the ocean.

run-of-river Any of several methods used to generate *hydroelectric power* without greatly disrupting the flow of river water. Run-of-river approaches eliminate much of the *environmental* impact of large *dams*. Compare *pumped storage; storage*.

salinization The buildup of salts in surface *soil* layers.

salt marsh Flat land that is intermittently flooded by the ocean where the *tide* reaches inland. Salt marshes occur along temperate coastlines and are thickly vegetated with grasses, rushes, shrubs, and other herbaceous plants.

salvage logging The removal of dead trees following a natural disturbance. Although it may be economically beneficial, salvage logging can be ecologically destructive, because *snags* provide food and shelter for wildlife and because removing timber from recently burned land can cause *erosion* and damage to *soil*.

sanitary landfill A site at which solid waste is buried in the ground or piled up in large mounds for disposal, designed to prevent the waste from contaminating the *environment*. Compare *incineration*.

savanna A *biome* characterized by grassland interspersed with clusters of acacias and other trees. Savanna is found across parts of Africa (where it was the ancestral home of our *species*), South America, Australia, India, and other dry tropical regions.

science (1) A systematic process for learning about the world and testing our understanding of it. (2) The accumulated body of knowledge that arises from this dynamic process.

scientific method A formalized method for testing ideas with observations that involves a more-or-less consistent series of interrelated steps.

scrubber Technology to chemically treat gases produced in combustion in order to reduce smokestack emissions. These devices typically remove hazardous components and neutralize acidic gases, such as *sulfur dioxide* and hydrochloric acid, turning them into water and salt.

second law of thermodynamics The physical law stating that the nature of *energy* tends to change from a more-ordered state to a less-ordered state; that is, *entropy* increases.

secondary forest *Forest* that has grown back after *primary forest* has been cut. Consists of second-growth trees.

secondary pollutant A hazardous substance produced through the reaction of primary pollutants with one another or with other constituents of the *atmosphere*. Compare *primary pollutant*.

secondary production The total *biomass* that *heterotrophs* generate by consuming *autotrophs*. Compare *primary production*.

secondary succession A stereotypical series of changes experienced by an ecological *community* as it develops over time, beginning when some *disturbance* disrupts or dramatically alters an existing community. Compare *primary succession*.

secondary treatment A stage of *wastewater* treatment in which biological means are used to remove contaminants remaining after *primary treatment*. Wastewater is stirred up in the presence of *aerobic* bacteria, which degrade organic pollutants in the water. The wastewater then passes to another settling tank, where remaining solids drift to the bottom. Compare *primary treatment*.

sediment The eroded remains of *rocks*.

sedimentary rock One of the three main categories of *rock*. Formed when dissolved *minerals* seep through *sediment* layers and act as a kind of glue, crystallizing and binding sediment particles together. Sandstone and shale are examples of sedimentary *rock*. Compare *igneous rock; metamorphic rock*.

seed bank A storehouse for samples of seeds representing the world's crop diversity.

septic system A *wastewater* disposal method, common in rural areas, consisting of an underground tank and series of drainpipes. Wastewater runs from the house to the tank, where solids precipitate out. The water proceeds downhill to a drain field of perforated pipes laid horizontally in gravel-filled trenches, where microbes decompose the remaining waste.

sex ratio The proportion of males to females in a *population*.

shale gas *Natural gas* trapped deep underground in tiny bubbles dispersed throughout formations of shale, a type of *sedimentary rock*. Shale gas is often extracted by *hydraulic fracturing*.

shale oil A liquid form of petroleum extracted from deposits of *oil shale*.

shelterbelt A row of trees or other tall perennial plants that are planted along the edges of farm fields to break the wind and thereby minimize wind *erosion*.

sick building syndrome A building-related illness produced by indoor *pollution* in which the specific cause is not identifiable.

silicon The chemical *element* with 14 *protons* and 14 *neutrons*. An abundant element in *rocks* in Earth's crust.

sink In a *nutrient cycle*, a *reservoir* that accepts more *nutrients* than it releases.

sinkhole An area where the ground has given way with little warning as a result of subsidence caused by depletion of water from an *aquifer*.

slash-and-burn A mode of agriculture frequently used in the tropics in which natural vegetation is cut and then burned, adding nutrition to the *soil*, before farming begins. Generally, farmers move on to another plot once the *soil* fertility is depleted.

smart growth A *city planning* concept in which a community's growth is managed in ways intended to limit *sprawl* and maintain or improve residents' quality of life.

smelting A process in which *ore* is heated beyond its melting point and combined with other *metals* or chemicals to form metal with desired characteristics. Steel is created by smelting iron ore with carbon, for example.

smog Term popularly used to describe unhealthy mixtures of air *pollutants* that often form over urban and industrial areas as a result of fossil fuel combustion. See *industrial smog; photochemical smog.*

social cost of carbon An estimate of the total economic cost of damages resulting from the emission of *carbon dioxide* (from *fossil fuel* burning, *deforestation*, etc.) and resulting *global climate change*, on a per-ton basis. Estimates vary widely, but the U.S. government has recently used an estimate of roughly $40/ton CO_2.

social sciences Academic disciplines that study human interactions and institutions. Compare *natural sciences.*

soil A complex plant-supporting *system* consisting of disintegrated *rock*, organic matter, air, water, *nutrients*, and microorganisms.

soil degradation A deterioration of *soil* quality and decline in *soil* productivity, resulting primarily from forest removal, cropland agriculture, and *overgrazing* of livestock. Compare *desertification; land degradation.*

soil horizon A distinct layer of *soil.*

soil profile The cross-section of a *soil* as a whole, including all *soil horizons* from the surface to the *bedrock.*

solar energy Energy from the sun. Solar energy is perpetually renewable and may be harnessed in several ways.

solution mining A *mining* technique in which a narrow borehole is drilled deep into the ground to reach a *mineral* deposit, and water, acid, or another liquid is injected down the borehole to *leach* the resource from the surrounding *rock* and dissolve it in the liquid. The resulting solution is then sucked out, and the desired resource is isolated.

source In a *nutrient cycle*, a *reservoir* that releases more *nutrients* than it accepts.

source reduction The reduction of the amount of material that enters the *waste stream* to avoid the costs of disposal and *recycling*, help conserve resources, minimize *pollution*, and save consumers and businesses money.

specialist A *species* that can survive only in a narrow range of *habitats* or that depends on very specific resources. Compare *generalist.*

speciation The process by which new *species* are generated. In one common mechanism, allopatric speciation, species form in the

aftermath of the physical separation of populations over some geographic distance.

species A *population* or group of populations of a particular type of organism whose members share certain characteristics and can breed freely with one another and produce fertile offspring. Biologists may differ in their approaches to diagnosing species boundaries.

species diversity The number and variety of *species* in the world or in a particular region.

sprawl The unrestrained spread of urban or suburban development outward from a city center and across the landscape. Often specified as growth in which the area of development outpaces *population* growth.

steady-state economy An *economy* that does not grow or shrink but remains stable.

storage Technique used to generate *hydroelectric power*, in which large amounts of water are impounded in a reservoir behind a concrete *dam* and then passed through the dam to turn *turbines* that generate *electricity*. Compare *pumped storage; run-of-river.*

stratosphere The layer of the *atmosphere* above the *troposphere*; it contains the *ozone layer* and extends 11–50 km (7–31 mi) above sea level.

strip mining The use of heavy machinery to remove huge amounts of earth to expose *coal* or *minerals*, which are mined out directly. Compare *subsurface mining.*

subduction The *plate tectonic* process by which denser *crust* slides beneath lighter crust at a *convergent plate boundary*. Often results in *volcanism.*

subsidy A government grant of money or resources to a private entity, intended to support and promote an industry or activity. A tax break is one type of subsidy. Compare *green tax.*

subsurface mining Method of *mining* underground deposits of *coal, minerals*, or fuels, in which shafts are dug deeply into the ground and networks of tunnels are dug or blasted out to follow coal seams. Compare *strip mining.*

suburb A smaller community located at the outskirts of a city.

succession A stereotypical series of changes in the composition and structure of an ecological *community* through time. See *primary succession; secondary succession.*

sulfur dioxide (SO₂) A colorless gas that can result from the combustion of *coal*. In the *atmosphere*, it may react to form sulfur trioxide and sulfuric acid, which may return to Earth in *acid deposition*. An EPA *criteria pollutant.*

Superfund A program administered by the *Environmental Protection Agency* in which experts identify sites polluted with hazardous chemicals, protect *groundwater* near these

sites, and clean up the *pollution*. Established by the Comprehensive Environmental Response Compensation and Liability Act (CERCLA) in 1980.

surface impoundment (1) A disposal method for *hazardous waste* or mining waste in which waste in liquid or slurry form is placed into a shallow depression lined with impervious material such as clay and allowed to evaporate, leaving a solid residue on the bottom. (2) The site of such disposal. Compare *deep-well injection.*

surface water Water located atop Earth's surface. Compare *groundwater.*

sustainability A guiding principle of *environmental science*, entailing conserving resources, maintaining functional ecological systems, and developing long-term solutions, such that Earth can sustain our civilization and all life for the future, allowing our descendants to live at least as well as we have lived.

sustainable agriculture *Agriculture* that can be practiced in the same way and in the same place far into the future. Sustainable agriculture does not deplete *soils* faster than they form, nor reduce the clean water, genetic diversity, pollinators, and other resources essential to long-term crop and livestock production.

sustainable development Development that satisfies our current needs without compromising the future availability of *natural capital* or our future quality of life.

Sustainable Development Goals A program of targets for *sustainable development* set by the international community in 2015 through the *United Nations.*

sustainable forest certification A form of *ecolabeling* that identifies timber products that have been produced using *sustainable* methods. The Forest Stewardship Council (FSC) and several other organizations issue such certification.

symbiosis A relationship between different *species* of organisms that live in close physical proximity. People most often use the term "symbiosis" when referring to a *mutualism*, but symbiotic relationships can be either parasitic or mutualistic.

synergistic effect An interactive effect (as of *toxicants*) that is more than or different from the simple sum of their constituent effects.

system A network of relationships among a group of parts, elements, or components that interact with and influence one another through the exchange of *energy*, matter, and/or information.

tailings Portions of *ore* left over after *metals* have been extracted in *mining.*

tar sands See *oil sands.*

temperate deciduous forest A *biome* consisting of midlatitude *forests* characterized by broad-leafed trees that lose their leaves each fall and remain dormant during winter. These forests occur in areas where *precipitation* is spread relatively evenly throughout the year: much of Europe, eastern China, and eastern North America.

temperate grassland A *biome* whose vegetation is dominated by grasses and features more extreme temperature differences between winter and summer and less *precipitation* than *temperate deciduous forests*. Also known as *steppe* or *prairie*.

temperate rainforest A *biome* consisting of tall coniferous trees, cooler and less *species*-rich than *tropical rainforest* and milder and wetter than *temperate deciduous forest*.

temperature inversion A departure from the normal temperature distribution in the *atmosphere*, in which a pocket of relatively cold air occurs near the ground, with warmer air above it. The cold air, denser than the air above it, traps *pollutants* near the ground and can thereby cause a buildup of *smog*. Also called a thermal inversion.

teratogen A *toxicant* that causes harm to the unborn, resulting in birth defects.

terracing The cutting of level platforms, sometimes with raised edges, into steep hillsides to contain water from *irrigation* and *precipitation*. Terracing transforms slopes into series of steps like a staircase, enabling farmers to cultivate hilly land while minimizing their loss of *soil* to water *erosion*.

theory A widely accepted, well-tested explanation of one or more cause-and-effect relationships that have been extensively validated by a great amount of research. Compare *hypothesis*.

thermohaline circulation A worldwide system of ocean *currents* in which warmer, fresher water moves along the surface and colder, saltier water (which is denser) moves deep beneath the surface.

thin-film solar cell A photovoltaic material compressed into an ultra-thin lightweight sheet that may be incorporated into various surfaces to produce photovoltaic solar power.

threatened Likely to become *endangered* soon.

Three Mile Island Nuclear power plant in Pennsylvania that in 1979 experienced a partial *meltdown*. The term is often used to denote the accident itself, the most serious *nuclear reactor* malfunction that the United States has thus far experienced. Compare *Chernobyl*; *Fukushima Daiichi*.

threshold dose The amount of a *toxicant* at which it begins to affect a *population* of test animals. Compare ED_{50}; LD_{50}.

tidal energy *Energy* harnessed by erecting a *dam* across the outlet of a tidal basin. Water flowing with the incoming or outgoing *tide* through sluices in the dam turns turbines to generate *electricity*.

tide The periodic rise and fall of the ocean's height at a given location, caused by the gravitational pull of the moon and sun.

topsoil That portion of the *soil* that is most nutritive for plants and is thus of the most direct importance to *ecosystems* and to *agriculture*. A *soil horizon* also known as the A horizon.

total fertility rate (TFR) The average number of children born per female member of a *population* during her lifetime.

Toxic Substances Control Act (TSCA) A 1976 U.S. law that directs the *Environmental Protection Agency* to monitor thousands of industrial chemicals and gives the EPA authority to regulate and ban substances found to pose excessive risk.

toxicant A substance that acts as a poison to humans or wildlife.

toxicity The degree of harm a chemical substance can inflict.

toxicology The scientific field that examines the effects of poisonous chemicals and other agents on humans and other organisms.

toxin A *toxic* chemical stored or manufactured in the tissues of living organisms. For example, a chemical that plants use to ward off *herbivores* or that insects use to deter predators.

traditional agriculture *Agriculture* in which human and animal muscle power, along with hand tools and simple machines, performs the work of cultivating, harvesting, storing, and distributing crops. Compare *industrial agriculture*.

tragedy of the commons The process by which publicly accessible resources open to unregulated use tend to become damaged and depleted through overuse. Term was coined by Garrett Hardin and is widely applicable to resource issues.

transform plate boundary The area where two tectonic plates meet and slip and grind alongside one another, creating *earthquakes*. For example, the Pacific Plate and the North American Plate rub against each other along California's San Andreas Fault. Compare *convergent plate boundary*; *divergent plate boundary*.

transgene A *gene* that has been extracted from the *DNA* of one organism and transferred into the DNA of an organism of another *species*.

transgenic Term describing an organism that contains *DNA* from another *species*.

transitional stage The second stage of the *demographic transition* model, which occurs during the transition from the *pre-industrial stage* to the *industrial stage*. It is characterized by declining death rates but continued high birth rates. Compare *industrial stage*; *post-industrial stage*; *pre-industrial stage*.

transpiration The release of water vapor by plants through their leaves.

treatment The portion of an *experiment* in which a *variable* has been manipulated in order to test its effect. Compare *control*.

triple bottom line An approach to sustainability that attempts to meet environmental, economic, and social goals simultaneously.

trophic cascade A series of changes in the *population* sizes of organisms at different *trophic levels* in a *food chain*, occurring when *predators* at high trophic levels indirectly promote populations of organisms at low trophic levels by keeping *species* at intermediate trophic levels in check. Trophic cascades may become apparent when a top predator is eliminated from a system.

trophic level Rank in the feeding hierarchy of a *food chain*. Organisms at higher trophic levels consume those at lower trophic levels.

tropical dry forest A *biome* that consists of deciduous trees and occurs at tropical and subtropical latitudes where wet and dry seasons each span about half the year. Widespread in India, Africa, South America, and northern Australia. Also known as *tropical deciduous forest*.

tropical rainforest A *biome* characterized by year-round rain and uniformly warm temperatures. Found in Central America, South America, Southeast Asia, West Africa, and other tropical regions. Tropical rainforests have dark, damp interiors; lush vegetation; and highly diverse biotic communities.

troposphere The bottommost layer of the *atmosphere*; it extends to 11 km (7 mi) above sea level.

tropospheric ozone *Ozone* that occurs in the *troposphere*, where it is a *secondary pollutant* created by the interaction of sunlight, heat, *nitrogen oxides*, and volatile *carbon*-containing chemicals. A major component of *photochemical smog*, it can injure living tissues and cause respiratory problems. An EPA *criteria pollutant*. Also called ground-level ozone.

tsunami An immense swell, or wave, of ocean water triggered by an *earthquake*, *volcano*, or *landslide* that can travel long distances across oceans and inundate coasts.

tundra A *biome* that is nearly as dry as *desert* but is located at very high latitudes along the northern edges of Russia, Canada, and Scandinavia. Extremely cold winters with little daylight and moderately cool summers with lengthy days characterize this landscape of lichens and low, scrubby vegetation.

U.N. Framework Convention on Climate Change (FCCC) An international treaty sponsored by the United Nations and signed in

1992 that outlined a plan to reduce emissions of *greenhouse gases*. Gave rise to the *Kyoto Protocol*.

unconfined aquifer A water-bearing, porous layer of *rock, sand,* or gravel that lies atop a less permeable substrate. The water in an unconfined aquifer is not under pressure because there is no impermeable upper layer to confine it. Compare *confined aquifer*.

undernutrition A condition of insufficient nutrition in which people receive fewer calories than are needed on a daily basis for a healthy diet.

United Nations (UN) Organization founded in 1945 to promote international peace and to cooperate in solving international economic, social, cultural, and humanitarian problems.

universalist An ethicist who maintains that there exist objective notions of right and wrong that hold across cultures and situations. Compare *relativist*.

upwelling In the ocean, the flow of cold, deep water toward the surface. Upwelling occurs in areas where surface *currents* diverge. Compare *downwelling*.

urban ecology A scientific field of study that views cities explicitly as *ecosystems*. Researchers in this field apply the fundamentals of *ecosystem ecology* and *systems* science to urban areas.

urban growth boundary (UGB) A line on a map established to separate areas zoned to be high-density and urban from areas intended to remain low-density and rural. The aim is to control *sprawl*, revitalize cities, and preserve the rural character of outlying areas.

urban heat island effect The phenomenon whereby a city becomes warmer than outlying areas because of the concentration of heat-generating buildings, vehicles, and people, and because buildings and dark paved surfaces absorb heat and release it at night.

urban planning See *city planning*.

urbanization A population's shift from rural living to city and suburban living.

variable In an *experiment*, a condition that can change. See *dependent variable; independent variable*.

vector An organism that transfers a *pathogen* to its host. An example is a mosquito that transfers the malaria pathogen to humans.

volatile organic compound (VOC) One of a large group of potentially harmful organic chemicals used in industrial processes. One of six major pollutants whose emissions are monitored by the *EPA* and state agencies.

volcano A site where molten *rock*, hot gas, or ash erupts through Earth's surface, often creating a mountain over time as cooled *lava* accumulates.

Wallace, Alfred Russel (1823–1913) English naturalist who proposed, independently of *Charles Darwin*, the concept of *natural selection* as a mechanism for *evolution* and as a way to explain the great variety of living things.

waste Any unwanted material or substance that results from a human activity or process.

waste management Strategic decision making to minimize the amount of *waste* generated and to dispose of waste safely and effectively.

waste stream The flow of *waste* as it moves from its sources toward disposal destinations.

waste-to-energy (WTE) facility An incinerator that uses heat from its furnace to boil water to create steam that drives *electricity* generation or that fuels heating systems.

wastewater Any water that is used in households, businesses, industries, or public facilities and is drained or flushed down pipes, as well as the polluted *runoff* from streets and storm drains.

water A *compound* composed of two hydrogen atoms bonded to one oxygen atom, denoted by the chemical formula H_2O.

water pollution The release of matter or *energy* into waters that causes undesirable impacts on the health and well-being of humans or other organisms. Water pollution can be physical, chemical, or biological.

water table The upper limit of *groundwater* held in an *aquifer*.

waterlogging The saturation of *soil* by water, in which the *water table* is raised to the point that water bathes plant roots. Waterlogging deprives roots of access to gases, essentially suffocating them and damaging or killing the plants.

watershed See *drainage basin*.

wave energy *Energy* harnessed from the motion of ocean waves. Many designs for machinery to harness wave energy have been invented, but few are commercially operational.

weather The local physical properties of the *troposphere*, such as temperature, pressure, humidity, cloudiness, and wind, over relatively short time periods (typically minutes, hours, days, or weeks). Compare *climate*.

weathering The process by which *rocks* and *minerals* are broken down, turning large particles into smaller particles. Weathering may proceed by physical, chemical, or biological means.

weed A pejorative term for any plant that competes with our crops. The term is subjective and defined by our own economic interests, and is not biologically meaningful. Compare *pest*.

wetland A system in which the *soil* is saturated with water and which generally features shallow standing water with ample vegetation. These biologically productive systems include freshwater marshes, swamps, bogs, and seasonal wetlands such as vernal pools.

wilderness area Federal land that is designated off-limits to development of any kind but is open to public recreation, such as hiking, nature study, and other activities that have minimal impact on the land.

wind farm A development involving a group of *wind turbines*.

wind power A form of *renewable energy*, in which *kinetic energy* from the passage of wind through *wind turbines* is used to generate *electricity*.

wind turbine A mechanical assembly that converts the wind's *kinetic energy*, or energy of motion, into electrical energy for the generation of *wind power*.

World Bank Institution founded in 1944 that serves as one of the globe's largest sources of funding for economic development, including such major projects as *dams, irrigation* infrastructure, and other undertakings.

world heritage site A location internationally designated by the *United Nations* for its cultural or natural value. There are more than 1000 such sites worldwide.

World Trade Organization (WTO) Organization based in Geneva, Switzerland, that represents multinational corporations and promotes free trade by reducing obstacles to international commerce and enforcing fairness among nations in trading practices.

xeriscaping Landscaping using plants that are adapted to arid conditions.

zoning The practice of classifying areas for different types of development and land use.

zooxanthellae Symbiotic algae that inhabit the bodies of *corals* and produce food through *photosynthesis*.

Photo Credits

Index